中国科学院大学本科生教材系列

理性力学教程
A Course in Rational Mechanics

赵亚溥 著
ZHAO Ya-Pu

科学出版社
北京

内 容 简 介

本书是为中国科学院大学工程科学学院与数学科学学院二年级本科生所撰写的教材，在体系结构上，本书较为平衡地讨论了固体力学和流体力学问题，增加了现代数学对张量的定义、拓扑、群论、微分流形上的张量分析、狄拉克符号，流体动力学客观性，脑科学的连续介质力学张量图，人工智能中的张量流 (TensorFlow)、张量网络 (TensorNetwork) 以及连续介质力学在思维动力学、金融动力学、社会动力学和管理动力学中的应用等新内容，在使理性连续介质力学教材的现代化方面做出了深入探索.

本书既注重对相关科学史的深入挖掘，又十分关注本学科领域的最新发展，很多文献已经更新到 2019 年. 本书的创作过程充分地遵循了培根的名言"历史使人聪明，诗歌使人机智，数学使人精细，哲学使人深邃，道德使人严肃，逻辑与修辞使人善辩."努力使本书成为既在专业上艰深、前沿，又活泼有趣，使本科生能够萌发对该学科的兴趣和进行深入探索的冲动!

本书可用作工程科学、应用数学、应用物理、管理科学和相关专业的本科生或研究生教材及参考书.

图书在版编目(CIP)数据

理性力学教程/赵亚溥著. —北京：科学出版社，2020.3
中国科学院大学本科生教材系列
ISBN 978-7-03-064633-0

Ⅰ. ①理⋯ Ⅱ. ①赵⋯ Ⅲ. ①理性力学–高等学校–教材 Ⅳ. ①O331

中国版本图书馆 CIP 数据核字 (2020) 第 037709 号

责任编辑：刘信力 / 责任校对：邹慧卿
责任印制：肖　兴 / 封面设计：陈　敬

科学出版社 出版
北京东黄城根北街 16 号
邮政编码：100717
http://www.sciencep.com

天津市新科印刷有限公司 印刷
科学出版社发行　各地新华书店经销

*

2020 年 3 月第 一 版　开本：850×1168 1/16
2022 年 2 月第二次印刷　印张：34 1/2
字数：840 000
定价：158.00 元
(如有印装质量问题，我社负责调换)

前　言

这部《理性力学教程》是为中国科学院大学 (国科大) 工程科学学院与数学科学学院本科生所撰写的教材. 著者于 2016 年出版《近代连续介质力学》后, 虽然已经连续印刷了十余次, 在多所双一流大学被用作教材或主要参考书, 但仍有很多意犹未尽之处, 想借本次到国科大玉泉路校区为本科二年级开设《连续介质力学》课程来弥补.

本书和《近代连续介质力学》的最大不同之处是在体系结构上有所突破.《理性力学教程》较为平衡地讨论了固体和流体问题, 增加了现代数学对张量的定义、拓扑、群论、微分流形上的张量分析、狄拉克符号, 流体动力学客观性, 脑科学的连续介质力学张量图, 人工智能中的张量流 (TensorFlow)、张量网络 (TensorNetwork) 以及连续介质力学在思维动力学、社会动力学、金融动力学和管理动力学中的应用等内容, 在使理性连续介质力学教材的现代化方面做出了深入探索.

纵观泰勒、普朗特、冯·卡门、钱学森这些近代力学大师们, 他们的一个共同特点是固体和流体力学均融会贯通, 在不同的领域均贡献卓著, 属于典型的"教科书式"的人物. 杰弗里·泰勒 (Geoffrey Ingram Taylor, 1886—1975) 共计 11 年次被 20 位科学家提名过诺贝尔物理学奖, 是一位典型的固体和流体力学的通才, 在位错理论、湍流和冲击动力学等方向上均作出过开创性的历史性贡献. 在他于 1954 年出版的《著作集》(*Collected Works*) 中, 第一卷为包含了 41 篇论文的"固体力学", 第二卷为包含了 45 篇文章的"气象学、海洋学和湍流", 第三卷为包含了 58 篇文章的"气体力学、弹丸和爆炸力学", 第四卷为包含了 49 篇论文的"其他有关流体力学的论文". 泰勒从不申请科研经费 (never applied for a research grant), 也不跟踪文献 (did not keep up with the literature), 他所需要的设备就是一间实验室和一个助手. 泰勒的学生巴切勒教授 (见图 62.10) 曾说过, 泰勒是个极其可爱的人, 不像现今研究机构的科学家因为压力遭受着对环境适应失调和自我关注过高的问题, 这就让泰勒创造性的能量发挥到了极致.

路德维希·普朗特 (Ludwig Prandtl, 1875—1953) 在弗普尔教授 (见图 27.4) 指导下攻读弹塑性力学方向的博士学位, 其 1899 年完成的博士论文 *Tilting Phenomena, a Case of Unstable Elastic Balance* 探讨了狭长矩形截面梁的侧向稳定性, 属于典型的固体力学方向; 1903 年提出的柱体扭转问题的薄膜比拟法, 早已进入了

69 岁仍坚守实验台的
力学大师泰勒

实验中的普朗特 (29 岁)

黑板的力量 —— 授课中的冯·卡门 (75 岁)

黑板的力量 —— 授课中的钱学森

弹塑性力学的教材；他继承并推广了圣维南所开创的塑性流动的研究；普朗特还于 1921 年解决了半无限体受狭条均匀压力时的塑性流动分析. 其后, 普朗特在边界层理论、机翼理论和湍流等流体力学领域作出过举世公认的开创性贡献, 被誉为"近代流体力学之父". 普朗特共两年次被四位科学家提名过诺贝尔物理学奖.

冯·卡门 (Theodore von Kármán, 1881—1963) 在普朗特的指导下, 于 1908 年完成了题为《屈曲强度研究》(*Investigations on buckling strength*) 的固体力学领域的博士论文, 研究的是关于柱体塑性区的屈曲问题. 其后, 在卡门涡街、超声速和高超声速方面作出了奠基性的贡献. 1963 年被授予美国首枚国家科学奖章.

钱学森 (Hsue-Shen Tsien, 1911—2009) 在冯·卡门指导下于 1938 年完成的博士论文是流体力学方向《可压缩流体运动和反力推进中的问题》(*Problems in motion of compressible fluids and reaction propulsion*), 但一年后的 1939 年, 他和冯·卡门一起发现了薄壳的非线性失稳的载荷只有线性失稳的载荷的三分之一, 这当然是在固体力学领域的重要贡献, 钱学森更是在空气动力学等方面作出过更为杰出的贡献. 钱学森是"国家杰出贡献科学家"荣誉称号的唯一获得者.

上述案例给我们的启示是, 力学的大厦是一个整体 (从理性力学出发, 柯西弹性固体和牛顿流体作为简单物质特例见图 0.1), 学科分得太细, 往往会限制人的思维, 难以作出重大的科学发现和贡献.

根据本书著者多年的研究和教学心得, 连续介质力学中 (同其他学科之间的联系见前言图 0.2) 有三部分内容在后续的研究和应用中特别行之有效, 这三部分内容分别是: 对称性、叠加原理、量纲分析和数量级估计. 其中, "对称性"和"叠加原理"属于性质, 而"量纲分析和数量级估计"则属于方法.

一、关于对称性, 正如《力学讲义》中诺特定理所深刻阐明的, 对称性导致守恒律. 所以说, 连续介质力学中的核心内容之一"守恒律"的本源就是相对应的对称性. 这在本书 §32.1、§33.1、§34.1 中给予了透彻的说明. "优美的"空间平移对称性、时间平移对称性和空间旋转对称性将导致"壮美的"动量守恒、能量守恒和角动量守恒. 另外, 应力、应变 (率)、刚度张量、柔度张量的对称性为理论分析和工程实际应用提供了极大方便.

二、叠加原理的本质是线弹性, 也就是函数的一阶泰勒展开. 这在本书的多个章节中得以淋漓尽致的展现.

三、量纲分析和粗略数量级估计. 在本书 §54、§56.2~§56.4 和 §61.2 中得以展现, 和在《力学讲义》中所提出的"量纲分析的快速匹配法"一样, 本书中的很多例子是本书著者的研究心得. 连续介质力学中量纲分析应用的"华彩乐章"就是本

书 §59.5 的湍流的柯尔莫哥洛夫 K41 标度律和 §59.6 的奥布霍夫 –5/3 标度律.

爱因斯坦在 1916 年发表其广义相对论的当年, 还发表了一篇题为《水波与飞行的基本理论》连续介质力学方面的文章. 1917 年, 爱因斯坦进而还设计了一种类似于猫的驼背的翼型, 被称为猫背翼. 爱因斯坦把这个设计交给了飞机制造商 LVG 并建造了新型的飞机. 一名试飞员报告说, 这架飞机在空中像 "一只怀孕的鸭子 (a pregnant duck)" 一样摇摆不定. 在 1954 年也就是他去世的前一年, 爱因斯坦称他自己年轻时有关航空问题的研究是 "年轻人的愚蠢行为 (youthful folly)". 被公认为是继伽利略、牛顿以来最伟大物理学家的爱因斯坦, 却没有对升力的理解做出积极的贡献, 更没有能设计出一个实用的翼型.

对于像国科大这样的高校, 著者部分赞赏数学大师雅可比 (Carl Gustav Jacob, 1804—1851) 对教学的观点: "把年轻学生们扔进冰水里, 由他们自己去学会游泳或者淹死 (Pitching young men into the icy water to learn to swim or drown by themselves)". 事实上, 那些 "自己学会游泳" 存活下来的学生自然就会成为人才.

在欧美的一些著名高校, 连续介质力学的本科课程常常在数学系和工程系同时开设, 这也表明了连续介质力学的两重性: 一种是作为应用数学, 另一种是工程科学的 "大统一理论". 正如冯·卡门所指出的 "科学家研究世界的本来面目, 工程师创造出从未有过的世界 (The scientist describes what is; the engineer creates what never was)". 冯元桢 (1919—2019) 也深刻地指出: "工程学和科学有很大的不同. 科学家试图理解自然; 工程师试图创造出自然中不存在的东西, 工程师强调的是发明". 作为本科教育, 学生们首先要深刻地理解该学科的学术经典, 通过批判性和创造性思维, 能够灵活地运用该学科的主要内容, 从而为工程科学其他学科的学习奠定坚实的基础.

什么是好的本科教学? 曾担任过英国皇家学会会长的约瑟夫·约翰·汤姆逊 (Joseph John Thomson, 1856—1940, 1906 年诺贝尔物理学奖获得者) 在回忆他在曼彻斯特的欧文学院读本科学习工程学的老师——流体力学大师雷诺——时, 汤姆逊写道: "我打交道最多的教授就是奥斯鲍恩·雷诺 (Osborne Reynolds). 他的行为处事以及上课表达与其他教授是大相径庭的. 结果就是我们根本没法在他的课上记笔记, 我们只能依赖兰金 (Rankine) 的教科书. 有时候他会忘了来上高年级的课, 我们就会等上十分钟然后让保洁员告诉他学生们还在等着上课呢. 他这时就会冲到教室, 进门的时候还不忘整理一下衣袍, 然后将桌上兰金的一卷书拿起, 随手翻到一页, 见到公式什么的就说它们错了. 然后他就上黑板证明, 背朝我们, 自言自语, 时不时又把黑板全擦了, 还一边说这是错的呀. 他这时会在黑板上新起一行,又重新开始证明了. 通常在快要下课时, 他会留下一条没擦掉的证明, 说 (我前面这些推导是白做功, 但) 这毕竟还是证明兰金是对的啊."

从汤姆逊的上述回忆来看, 雷诺的授课很有批判性思维, 他经常性地质疑当时的学术经典——兰金的工程学教材上很多公式的正确性. 从他的回忆我们还可以引申出, 雷诺的授课并不是按照课程大纲来"一板一眼"地、循序渐进地进行, 但为什么这样既不系统而且跳跃性很大的授课也能培养出学术大师呢? 这很值得我们反思.

1884 年 12 月 22 日, 时年仅 28 岁的 J. J. 汤姆逊被遴选为卡文迪许教授, 亦即卡文迪许实验室主任. J. J. 汤姆逊是继麦克斯韦 (James Clerk Maxwell, 1831—1879) 和瑞利勋爵 (Lord Rayleigh, 1842—1919) 之后, 卡文迪许实验室的第三位主任. 根据当时的规定, 卡文迪许教授 (也即剑桥大学物理系主任) 的首要职责是每两学期讲授至少一门关于热、电和磁相关的课程. 麦克斯韦每学年讲授三门, 瑞利则讲授两门. 而 J. J. 汤姆逊并不满足于仅仅是完成最低的教学要求, 他每学年讲授四到七门课程. 在 J. J. 汤姆逊做卡文迪许实验室主任的头十年间, 他教课的数量大约是麦克斯韦和瑞利十四年间教课数量之和的两倍. J. J. 汤姆逊对教学的热忱, 与很多对教学是勉为其难的剑桥教授形成了鲜明对比. 在汤姆逊看来, "讲课对于一个研究人员极为有益, 可促使自己重新考查基本概念." 同时, "他讲课时既热情, 又严格, 思维敏捷并富有启发性." 不仅如此, "他常以改进大学和中学的物理教学而自娱".

通过多年的教学实践, J. J. 汤姆逊于 1895 年出版了第一部教材《电磁学数学理论基础》(*Elements of the Mathematical Theory of Electricity and Magnetism*). 随后, 他又和他的老朋友坡印亭 (John Henry Poynting, 1852—1914) 合作, 出版过三部教材.

人们自然关心的是, J. J. 汤姆逊所承担的如此繁重的教学是否"耽误了"他的科研工作? 还是让事实说话吧! J. J. 汤姆逊于 1906 年独享了诺贝尔物理学奖, 他的学生中有九位获得了诺贝尔奖, 其子 G. P. 汤姆逊 (George Paget Thomson, 1892—1975) 也于 1937 年与他人分享了诺贝尔物理学奖.

J. J. 和 G. P. 汤姆逊父子 (1909 年)

在著者于 2018 年出版的《力学讲义》中, 多次谈到朗道《理论物理教程》十卷和费曼的《物理学讲义》. 朗道在 1938 年出版十卷中的《统计物理学 I》时仅仅 30 岁, 在 1940 年出版《力学》时也仅仅 32 岁. 一套或一部好教程的重要性是显而易见的. 正像数学大师阿诺尔德所强调的: "一个数学教师, 如果至今还没有掌握至少几卷朗道和栗弗席兹著的物理学教程, 他必将成为一个数学界的稀罕的残存者, 就好似如今一个仍不知道开集与闭集差别的人 (A teacher of mathematics, who has not got to grips with at least some of the volumes of the course by Landau and Lifshitz, will then become a relict like the one nowadays who does not know the difference between an open and a closed set)."

  费曼自己相信，他对物理学最重要的贡献不是量子电动力学，也不会是超流体氦的理论，或极子或成子. 他这辈子最大的贡献就是《费曼物理学讲义》的三本红皮书 (Feynman himself believed that his most important contribution to physics would not be QED, or the theory of superfluid helium, or polarons, or partons. His foremost contribution would be the three red books of *The Feynman Lectures on Physics*).

  在本书交稿和校对过程中，有一幅照片经常性地映入到著者的脑海，那就是在 1967 年，时年 47 岁的巴切勒在完成《流体动力学引论》(*Introduction to Fluid Dynamics*) 的创作后，他提着手稿，骑着自行车，从家里前往剑桥大学出版社付梓的动人情景. 这当然是一个庄严的历史性时刻，迄今这部教材仍然发挥着不可替代的作用，已经成为学术经典.

1967 年，巴切勒提着书稿，准备交给剑桥大学出版社出版

  本书在国科大工程科学学院的 2019 年春季的《连续介质力学》的本科生教学中完整地试用过，取得了很好的效果. 本书的出版得到了中国科学院大学教材出版中心的资助，也感谢中国科学院战略性先导科技专项 (B 类，XDB22040401) 和中国科学院前沿科学重点研究计划 (QYZDJ-SSW-JSC019) 的支持.

  最后，让著者以近代分析学之父巴拿赫 (Stefan Banach, 1892—1945) 在富有传奇色彩的利沃夫咖啡馆所说过的一段发人深省的话来结束本前言："数学家能找到定理之间的相似之处，优秀的数学家能看到证明之间的相似之处，卓越的数学家能察觉到数学分支之间的相似之处. 可以想象，顶级的数学家能俯瞰这些相似之处之间的相似之处 (A mathematician is a person who can find analogies between theorems; a better mathematician is one who can see analogies between proofs; the best mathematician can notice analogies between theories; one can imagine that the ultimate mathematician is one who can see analogies between analogies)."

  爱因斯坦则能洞察出顶级物理学家之间的相似性，1949 年，爱因斯坦在他的自述中谈道："法拉第–麦克斯韦这一对同伽利略–牛顿这一对之间有着非常惊人的内在相似性 —— 每一对中的第一位都直观地把握了事物之间的关系，而第二位则严格地将这些关系公式化，并且定量地应用它们 (The pair Faraday-Maxwell has a most remarkable inner similarity with the pair Galileo-Newton — the former of each pair grasping the relations intuitively, and the second one formulating those relations exactly and applying them quantitatively)."

赵亚溥
2020 年 2 月于中关村

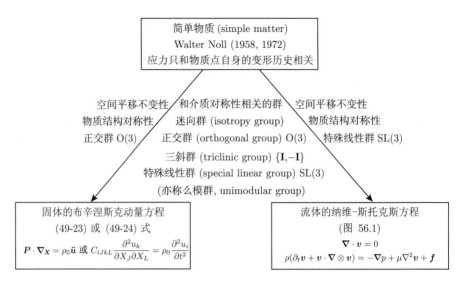

图 0.1 力学大厦是一个整体的例子：线弹性体和牛顿流体均是简单物质的特例

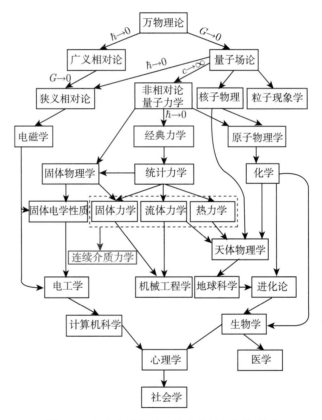

图 0.2 连续介质力学同其他学科之间的联系

图中符号意义见 (2-1) 式

目 录

前言

第1章 基础知识 1

§1. 理性力学、相关数学术语和空间 1
    1.1 什么是理性力学? 1
    1.2 数学术语 5
    1.3 理性力学常用的空间 11

§2. 连续介质假设 20
    2.1 物质世界的特征尺度 20
    2.2 努森数与四种流动区间 22
    2.3 连续介质假定 24

§3. 运动的两种描述方法——欧拉和拉格朗日观点 26

§4. 从质点和刚体动力学到连续介质力学 30
    4.1 质点和刚体力学概述 30
    4.2 从离散到连续的微元体方法 32
    4.3 连续介质力学研究的几位先驱性的工作 36

思考题和补充材料 38

参考文献 60

第2章 矢量分析 62

§5. 矢量分析中的狄拉克符号 63

§6. 刚体平衡的两个条件 64

§7. 质点动力学中速度和加速度的合成 65

§8. 克罗内克 δ 和置换符号 68
    8.1 克罗内克 δ 符号 68
    8.2 置换符号的基本定义 70
    8.3 混合积和置换符号 71

思考题和补充材料 75

参考文献 83

第 3 章 张量代数和微积分 · · · · · · 84

§9. 张量的引入，二阶投影张量 · · · · · · 84
9.1 二阶张量作为并矢的引入 · · · · · · 84
9.2 现代数学对张量的引入——多重线性映射 · · · · · · 86
9.3 二阶投影张量的引入 · · · · · · 88

§10. 爱丁顿张量和张量方程的广义量纲原理 · · · · · · 91

§11. 张量的缩并 · · · · · · 94
11.1 简单缩并 · · · · · · 94
11.2 张量的幂 · · · · · · 95
11.3 双缩并 · · · · · · 95
11.4 三缩并 · · · · · · 97

§12. 张量的转置、逆、对称化、反对称化，科恩不等式 · · · · · · 97

§13. 谱定理 · · · · · · 104
13.1 谱分解定理的基本内容 · · · · · · 104
13.2 谱定理在应力分解中的应用 · · · · · · 107

§14. 数学算子——拉普拉斯和黑森算子 · · · · · · 108
14.1 拉普拉斯算子 $\nabla^2 = \nabla \cdot \nabla = \Delta$ · · · · · · 109
14.2 黑森算子 $\nabla\nabla = \nabla \otimes \nabla$ · · · · · · 109

§15. 黎曼度量张量 · · · · · · 112

§16. 克里斯托费尔符号和联络 · · · · · · 124
16.1 曲纹坐标系的基矢量、黎曼度规张量 · · · · · · 124
16.2 第二类克里斯托费尔符号 · · · · · · 125
16.3 克里斯托费尔符号的应用 · · · · · · 126

§17. 张量的标量函数的泰勒展开 · · · · · · 129
17.1 多元矢量函数的泰勒展开 · · · · · · 129
17.2 张量的标量函数的一阶泰勒展开 · · · · · · 130

§18. 张量的标量函数的微分 · · · · · · 131
18.1 直角坐标系下的时间导数 · · · · · · 132
18.2 张量的标量函数的微分 · · · · · · 132
18.3 物质时间导数和空间时间导数之间的关系 · · · · · · 137

§19. 四阶单位张量、对称的四阶单位张量、四阶投影张量 · · · · · · 138
19.1 四阶单位张量、对称的四阶单位张量 · · · · · · 138

  19.2 四阶投影张量 ·················· 140

思考题和补充材料 ·················· 141

参考文献 ·················· 151

第 4 章　旋转群，拓扑，微分流形上的张量分析 ·················· 153

§20.　理性力学中常用的群论 ·················· 153
  20.1 对称与群 ·················· 153
  20.2 外尔的小册子——《对称》 ·················· 155
  20.3 旋转李群——SO(n) 和 SU(n) ·················· 157
  20.4 理性力学中常用的正交群 ·················· 161
  20.5 阿贝尔、伽罗瓦在建立群论过程中曲折经历 ·················· 165

§21.　微分流形 ·················· 167
  21.1 高斯与内蕴微分几何 ·················· 168
  21.2 黎曼与流形 ·················· 169
  21.3 流形的定义 ·················· 170

§22.　拓扑学与拓扑相变 ·················· 171
  22.1 拓扑学与亏格 ·················· 171
  22.2 拓扑相变 ·················· 173

§23.　微分同胚，坐标图册，切空间，余切空间 ·················· 176
  23.1 微分同胚 ·················· 176
  23.2 坐标图册 ·················· 178
  23.3 切空间，余切空间 ·················· 180

§24.　微分形式，外微分运算 ·················· 181
  24.1 微分形式的引入 ·················· 181
  24.2 外微分运算，庞加莱引理 ·················· 182
  24.3 斯托克斯定理 ·················· 186
  24.4 霍奇星算子和对偶 ·················· 187
  24.5 霍奇星号在麦克斯韦方程组中的应用 ·················· 190

§25.　流形上的矢量和张量分析 ·················· 192
  25.1 推前和拉回映射 ·················· 192
  25.2 李导数 ·················· 193
  25.3 黎曼度量与黎曼流形 ·················· 194

思考题和补充材料 ·················· 195

参考文献 · 197

第 5 章　变形运动学、功共轭 · 198

　§26.　变形梯度 F 及其极分解 · 198

　　　26.1　变形梯度张量 F 及其转置、逆、逆的转置的详细推导 · · · · · · · · · · · · · · 198

　　　26.2　变形梯度的极分解 · 203

　§27.　拉格朗日描述下的格林应变与欧拉描述下的阿尔曼西应变 · · · · · · · · · · · 204

　　　27.1　拉格朗日描述下有限变形的格林应变 · 204

　　　27.2　欧拉描述下有限变形的阿尔曼西应变 · 213

　§28.　赛斯–希尔应变度量 · 214

　　　28.1　希尔应变度量 · 214

　　　28.2　赛斯应变度量 · 216

　§29.　功共轭 · 218

　　　29.1　面元变换的南森公式 · 218

　　　29.2　基尔霍夫应力、第一类和第二类皮奥拉–基尔霍夫应力

　　　　　　(PK1 和 PK2) · 220

　　　29.3　功共轭 · 221

　　思考题和补充材料 · 227

　　参考文献 · 237

第 6 章　守恒律与场方程 · 238

　§30.　雷诺输运定理 · 238

　§31.　质量守恒 · 240

　　　31.1　欧拉描述下的质量守恒方程 · 240

　　　31.2　拉格朗日描述下的质量守恒方程 · 242

　　　31.3　笛卡儿坐标系、柱坐标系和球坐标系下的质量守恒方程 · · · · · · · · · · 243

　§32.　动量守恒 · 243

　　　32.1　动量守恒与空间平移不变性 · 243

　　　32.2　欧拉描述下流体的动量守恒方程 · 244

　　　32.3　欧拉描述下固体的动量守恒方程 · 244

　　　32.4　拉格朗日描述下固体的动量守恒方程 · 245

　§33.　动量矩守恒 · 246

　　　33.1　角动量守恒与空间旋转不变性 · 247

　　　33.2　柯西应力的对称性 · 247

  33.3 用 PK1 表示应力的对称性条件 ································ 247

§34. 能量守恒 ·· 248
  34.1 能量守恒与时间平移不变性 ································· 248
  34.2 欧拉描述下流体力学的能量守恒 ····························· 248
  34.3 固体力学中的动能定理 ·· 250
  34.4 固体力学中的能量守恒律 ····································· 250

§35. 熵守恒和热力学不等式 ··· 251
  35.1 熵平衡方程和熵不等式 ·· 251
  35.2 热力学第二定律在固体力学中的应用 ······················· 252

思考题和补充材料 ··· 255

参考文献 ·· 257

第 7 章 连续介质力学中的客观性 ··· 258

§36. 标量，位移、速度、加速度矢量的欧几里得客观性 ················ 258
  36.1 欧几里得变换 ·· 258
  36.2 标量和位移矢量的欧几里得客观性 ·························· 259
  36.3 速度矢量的欧几里得客观性 ·································· 259
  36.4 加速度矢量的欧几里得客观性 ······························· 259

§37. 张量的欧几里得客观性和客观率 ··· 261
  37.1 变形梯度张量的欧几里得客观性 ····························· 261
  37.2 柯西应力的欧几里得客观性 ·································· 262
  37.3 PK1 和 PK2 应力张量的欧几里得客观性 ··················· 262
  37.4 速度梯度、应变率、旋率张量的欧几里得客观性 ········ 263
  37.5 客观矢量率的定义 ··· 263
  37.6 客观张量率的定义 ··· 264

§38. 流体动力学的客观性 ··· 265
  38.1 基本方程组 ·· 265
  38.2 基本方程组的无量纲化 ······································· 267
  38.3 雷诺数相似性 ··· 268
  38.4 时空不变性 ·· 268
  38.5 时间反演不变性 ·· 269
  38.6 旋转和反射不变性 ··· 269
  38.7 伽利略不变性 ··· 270

- 38.8 扩展伽利略不变性 ··· 270
- 38.9 标架旋转 ··· 271
- 38.10 关于虚拟力的进一步讨论 ································ 272

思考题和补充材料 ·· 273

参考文献 ··· 274

第 8 章 本构关系 ··· 275

§39. 理性力学中的公理 ··· 275
- 39.1 本构公理的提出与建立 ······································ 275
- 39.2 里夫林等学者对连续介质力学公理的批评 ············· 277

§40. 线弹性本构关系——广义胡克定律 ······················ 278
- 40.1 材料力学和弹性力学中的广义胡克定律 (应变–应力关系式) ··········· 279
- 40.2 应力–应变关系式 ··· 282
- 40.3 对弹性常数的限制 ·· 283
- 40.4 固体力学材料常数常用关系的简单证明 ················ 284

§41. 流体力学本构关系 ··· 285
- 41.1 帕斯卡定律和帕斯卡水桶实验 ····························· 286
- 41.2 流体本构关系的一般形式 ··································· 286
- 41.3 牛顿流体本构关系的一般形式 ····························· 287

§42. 不可压缩超弹性材料的新胡克本构模型 ················ 288

§43. 超弹性材料的本构方程和应力 ···························· 292
- 43.1 用 PK1 表示的不可压缩 ($J=1$) 和可压缩 ($J \neq 1$) 的超弹性本构关系 ··········· 292
- 43.2 用 PK2 和 PK1 表示的可压缩和不可压缩的超弹性本构关系 ············ 293
- 43.3 用柯西应力表示的可压缩和不可压缩的超弹性本构关系 ··············· 294

§44. 可压缩超弹性体材料的穆尼–里夫林本构模型 ········ 299

思考题和补充材料 ·· 301

参考文献 ··· 303

第 9 章 虚功原理在连续介质力学中的应用 ··················· 305

§45. 微元长度、面积、体积和雅可比的变分 ················ 305
- 45.1 知道虚位移后如何确定虚体积? ··························· 305
- 45.2 知道虚位移后如何确定雅可比的变分? ·················· 305
- 45.3 知道虚位移后如何确定微线段矢量的变分? ············ 306

    45.4 知道虚位移后如何确定微面积矢量的变分？ ················307

§46. 虚功原理在连续介质力学中的应用 ························308
    46.1 当前构形中的虚位移 ·································308
    46.2 和变形梯度张量相关的变分 ···························309
    46.3 格林和柯西应变张量的变分 ···························309
    46.4 虚功原理 ···310

§47. 贝蒂定理与材料弹性模量对称性之间的关系 ··················311
    47.1 积分法 ···311
    47.2 微分法 ···312

思考题和补充材料 ···313

参考文献 ··314

第 10 章　固体力学要义 ···································315

§48. 材料力学之提纲挈领 ···································315

§49. 弹性力学提法和方程 ···································324
    49.1 弹性力学平衡方程 ··································324
    49.2 弹性模量独立分量的个数 ····························325
    49.3 勒让德-阿达玛不等式 ·······························331

思考题和补充材料 ···334

参考文献 ··338

第 11 章　流体动力学 ·····································340

§50. 从哈维的血液循环学说到血压计的发明 ······················340

§51. 伯努利方程的建立 ·····································342
    51.1 星光灿烂的伯努利家族 ······························342
    51.2 伯努利定律 ······································344

§52. 流体力学势流问题 ·····································350
    52.1 势流的特点 ······································350
    52.2 不可压缩流体的特性与势流方程 ······················351
    52.3 势流的分析 ······································351
    52.4 基本流 ··352

§53. 流变体与牛顿流体 ·····································357
    53.1 流变体的定义 ····································357
    53.2 牛顿流体 ··358

  53.3 牛顿流体的本构关系 · 359

§54. 哈根–泊肃叶流动定律 · 361

§55. 达朗贝尔佯谬 · 364

  55.1 马略特有关流体阻力的研究 · 364

  55.2 达朗贝尔佯谬 · 365

§56. 纳维–斯托克斯方程 · 367

  56.1 用拉格朗日方程推导纳维–斯托克斯方程 · 367

  56.2 纳维–斯托克斯方程的无量纲化及无量纲数 · 371

  56.3 用快速匹配法获得纳维–斯托克斯方程的无量纲数 · · · · · · · · · · · · · · · · 373

  56.4 相似律 · 374

§57. 马赫数、马赫锥、马赫角 · 378

§58. 斯托克斯阻力 · 379

  58.1 斯托克斯流动 · 379

  58.2 斯托克斯阻力公式 · 380

§59. 从层流到湍流的转捩 · 386

  59.1 雷诺 1883 年的经典论文 · 386

  59.2 卡门涡街——科学与艺术结合的典范 · 388

  59.3 费曼等对湍流的论述 · 394

  59.4 理查森的串级 · 394

  59.5 柯尔莫哥洛夫的 K41 理论的 2/3 标度律 · 395

  59.6 奥布霍夫的 $-5/3$ 标度律, 柯尔莫哥洛夫–奥布霍夫标度 · · · · · · · · · · · · · 398

§60. 杨–拉普拉斯方程 · 400

  60.1 谁最先提出了表面张力的概念? · 400

  60.2 应用能量法推导杨–拉普拉斯方程 · 401

  60.3 应用力平衡法推导杨–拉普拉斯方程 · 402

  60.4 应用矢量法推导杨–拉普拉斯方程 · 402

§61. 润滑近似和液滴铺展的动力学方程 · 404

  61.1 润滑近似下膜厚方程的推导 · 404

  61.2 薄膜铺展的标度律 · 406

§62. 流体动力学中的不稳定性理论 · 407

  62.1 里克特迈耶–梅什科夫 (RM) 不稳定性 · 407

  62.2 瑞利–泰勒 (RT) 不稳定性 · 408

	62.3 开尔文-亥姆霍兹 (KH) 不稳定性	409
	62.4 普拉托-瑞利 (PR) 不稳定性	410
	62.5 萨夫曼-泰勒 (ST) 不稳定性	416
思考题和补充材料 420
参考文献 430

第 12 章 连续介质力学新发展——思维动力学、金融动力学、社会动力学、管理动力学 436

§63. 脑科学中的首张连续介质力学张量图 436
　　63.1 左脑的批判性思维与右脑的创造性思维的对比 436
　　63.2 大脑发育过程的首张连续介质力学张量图 437
　　63.3 思维动力学 441

§64. 连续介质力学在人工智能中的应用——张量流和张量网络 443
　　64.1 麦卡锡是如何受冯·诺依曼启发创始人工智能这一学科的? 443
　　64.2 机器学习在材料设计中的应用 448
　　64.3 深度学习 448
　　64.4 张量流 451
　　64.5 维度诅咒与张量网络 451

§65. 连续介质力学在社会动力学、金融动力学和管理学动力学中的应用 453
　　65.1 社会心理学中的场论和生活空间 453
　　65.2 连续介质力学在社会动力学中的应用 455
　　65.3 金融动力学 460
　　65.4 弹性、刚性、柔性、韧性等概念在社会及管理动力学上的推广和延深 467
　　65.5 黏性概念在社会及管理动力学上的推广和延深 469
　　65.6 金融动力学中的艾略特波浪理论 470
　　65.7 金融数学 472

思考题和补充材料 474
参考文献 478

附录 A　和本书内容相关的科学大事年表 481

附录 B　连续介质力学中的相关物理量 485

附录 C　连续介质力学中的无量纲数 490

附录 D　弗雷歇导数和加托导数 493

- D.1 可微、可偏导、连续的关系 ··· 493
- D.2 弗雷歇导数、加托导数 ··· 495

附录 E 玻尔兹曼动理学方程、BBGKY 级联、利用玻尔兹曼方程对连续介质力学守恒律的证明 ·································· 499
- E.1 玻尔兹曼动理学方程 (1872) ·· 499
- E.2 BBGKY 级联 ·· 500
- E.3 应用玻尔兹曼方程对连续介质力学守恒律的证明 ······················· 502
- E.4 结束语 ··· 506
- 参考文献 ·· 508

索引 ·· 509

人像索引 ·· 529

第1章 基础知识

§1. 理性力学、相关数学术语和空间

1.1 什么是理性力学?

为了使读者对理性力学有一个全局性的把握,首先让我们提纲挈领地来回顾具有里程碑性质的牛顿力学、电动力学、相对论力学和量子力学的创立以及所伴随产生的新的数学分支,见表 1.1.

表 1.1 牛顿力学等新学科的创立和伴随产生的新的数学分支

创立学科	年代	学者	所实现的统一	新数学
牛顿力学	1687	牛顿	牛顿力学和万有引力得到统一	微积分
电动力学	1865	麦克斯韦	电、磁、光得到统一	纤维丛、规范场论
相对论力学	1905	爱因斯坦	狭义相对论使时间和空间得到统一	黎曼几何
	1916		广义相对论使引力和时空曲率得到统一	
量子力学	1920s	薛定谔等	粒子和波得到统一	线性代数

"理性力学 (rational mechanics)" 一词来源于牛顿 1687 年出版的《自然哲学的数学原理》(*Principia*)[1.1]. 牛顿在 1686 年 5 月 8 日为 *Principia* 所写的序言中明确地指出:"理性力学是一门精确地提出问题并加以演示的科学,旨在研究某种力所产生的运动,以及某种运动所需要的力."

理性力学亦称为力学公理化的非线性场论,其基本宗旨、纲领或者范式是用精确的数学描述来揭示力学理论的本质. 特别是当代,在多学科交叉、融合和多场耦合的大背景下,统一性的而不是碎片化的、框架型的而不是枝枝节节的理论描述将在更本质的层面上揭示现有理论的本质. 数学的一个重要发展趋势是不断地进行抽象,一层一层地进行抽象,抽象得越好、适用面越宽就越深刻.

为什么对于经典力学而言,对应于欧氏几何的牛顿力学已经足够了,我们还需要对应于黎曼几何的拉格朗日力学和对应于辛几何的哈密顿力学这样的分析力学? 其原因是分析力学能将问题的实质进一步抽象出来,找拉格朗日量和哈密顿量比受力分析观点更高、更深刻,也更简洁优雅些. 还可以进一步将相关方程和物理、力学规律抽象成最小作用量原理,所以说抽象本身就是在做 "统一" 这件事情.

理性力学就是针对整个连续介质力学学科做抽象.

在 20 世纪中叶, 克利福德·特鲁斯德尔 (Clifford Ambrose Truesdell III, 1919—2000) 倡导并致力于理性力学的复兴 [1.2]. 正像他所指出的那样, 理性力学作为自然哲学的一个分支, 要用 "最适度的数学概念 (by the most *fit* mathematical concept)" 去描述力学现象. 最适度的或最合适的数学概念不见得是最 "现代" 的, 但也可能是 (The most fit need not be the most modern, but they may be); 确实, 我们既不刻意地寻求, 也不刻意地避开最抽象的数学概念 (We neither seek nor avoid the most abstract mathematics).

特鲁斯德尔对理性力学所用到的 "最适度的数学概念" 范围做了如下限定 [1.2,1.3]: 流形 (manifold)、光滑映射 (smooth mapping)、欧几里得空间 (Euclidean space)、向量 (vector)、张量 (tensor)、泛函和群论 (functional and group). 在半个多世纪已经过去的今天, 拓扑学、微分几何和非欧几何等亦应该进入到理性力学的最适度的数学概念之列.

理性力学的研究内容包括: (1) 建立像几何学般的公理体系, 强调力学理论的严格数学演绎与数学结论的严格数学证明; (2) 寻求同族问题的统一解, 并研究解的存在性和唯一性; (3) 经典连续介质力学理论的发展与扩充.

事实上, 理性力学的一些重要进展可视为是近代物理学和数学相关理论的针对力学问题的进一步延拓和延深, 如: (1) 理性力学和连续介质力学的公理化则可追溯到 1900 年希尔伯特 (David Hilbert, 1862—1943) 所提出的著名的二十三个问题之六——"物理公理化的数学处理"; (2) 由沃尔特·诺尔 (Walter Noll, 1925—2017) 和罗纳德·塞缪尔·里夫林 (Ronald Samuel Rivlin, 1915—2005) 等人所做的以群论为不同的物质做分类, 则与固体物理中以群论做晶体的分类是一脉相承的; (3) 理性连续介质力学中十分关键的 "客观性" 的概念可上溯至爱因斯坦及其更早期的物理学家的相对论思想中; (4) 位错和裂纹的极限速度的确定也类比于相对论中的光速不变性原理; (5) 微分几何、非欧几何、微分流形、拓扑学等在理性力学中的广泛应用; (6) 统计力学和量子力学理论在流体和固体力学中的深入应用, 等等.

理性力学的生命力在于在经典连续介质力学和现代物理学、数学间的桥梁作用. 客观地说, 是理性力学使力学的理论逐步走向了现代化.

需要对两个方面进行强调: 其一, 理性并不等于决定论, 甚至有理性的理论专门用来处理不确定现象, 例如概率论. 其二, 理性并不排斥实验, 实验可以为理性提供大量素材. 自伽利略 (Galileo Galilei, 1564—1642) 以来, 理论的发展一直是实验的发展相辅相成, 没有实验, 理性就成了无源之水, 空中楼阁; 而没有理性, 实验的结果则永远不能升华为指导行动的可靠理论. 爱因斯坦于 1953 年曾高度概括道: "西方科学的发展基于两个伟大的成就: 希腊哲学家 (在欧几里得几何学中) 发明的形式逻辑体系, 和 (在文艺复兴时期) 通过系统的实验找到因果关系的可能性的发现 (Development of Western science is based on two great achievements: the invention of the formal logical system (in Euclidean geometry) by the Greek philosophers, and the discovery of the possibility to find out causal relationships by systematic experiment (during the Renaissance))."

钱学森 (1911—2009) 于 1978 年在全国力学规划会议上所作的题为 "现代力学"[1.4] 的发言中, 对理性力学做出如下定位: "研究具有复杂物性物质的运动, 必然联系到比以前我们习用的弹性力学方程式、纳维-斯托克斯方程式以及流变学的一些方程式更复杂得多的基本方程. 我们建立了这些宏观的方程式后, 还该仔细地看一看跟热力学、跟力学的基本定理有没有不符合的地方. 如果跟热力学、跟力学的基本定理有不符合的, 这个方程式当然是不对的, 不能用. 我们需要这样一个把关的工作, 这就是理性力学的任务. 它是有十分重要的实际意义的. 理性力学就是连续介质力学的基础理论.

我认为, 从事理性力学这样一类能概括地提高我们认识的科学研究, 不但重要, 也是一种精神享受 …… 我们的享受来源于感到自己站得更高了, 能洞察事物的本质了, 不单是知其所以然, 而且是透彻地知其所以然了. 这样的科学工作是很有用处的, 它使我们提高认识, 不是在那些枝枝节节的问题上钻进去拔不出来. 已故的物理大师沃尔夫冈·泡利 (Wolfgang Pauli) 受到推崇, 也是这个缘故."

从其对理性力学 "把关的工作""精神享受" 和 "连续介质力学的基础" 的三个要点的评价反映出, 钱学森的确是一位站得高、看得远的战略科学家.

必须坦言的是, 理性力学的发展也遇到了诸多困难和困境.

一方面是学科的定位问题. 正像特鲁斯德尔所描述的那样, 一些著名数学家鲜明地指出, "任何一篇论文, 只要出现 '应力' 或 '旋涡' 这样的字眼, 就显然属于力学或物理领域 (Any paper in which the words 'stress' or 'vorticity' appeared was clearly engineering or physics)". 言外之意是, 国际数学界对理性力学是漠视的.

另一方面是, 在欧美的一些名校, 除了工程系外, 很多数学系也开设理性力学或连续介质力学课程. 数学系所开设的这些课程高度抽象化, 使得理科学生很难理解其内在的工程科学的内涵; 而另一方面, 在工程系开设的这些课程由于流形、拓扑、微分几何等数学内容远超出工科学生的知识范围和接受能力. 从而势必形成这样一种十分尴尬的局面：在数学上能读懂的不一定能深入理解理性力学和连续介质力学的工程科学内涵，而实际中应用理性力学或连续介质力学的工程师则难以读懂其抽象的数学，因而应用乏力．因此，理性力学的发展处于两难的境地．

事实上，早在 1900 年，时年 38 岁的希尔伯特在第二届世界数学家大会的开幕式报告中 [1.5]，除了提出了具有指路明灯意义的二十三个问题外，还深入阐述了一个学科是否具有生命力的标志，那就是 "只要一门科学分支能提出大量的问题，它就充满生命力，而问题缺乏则预示着独立发展的衰亡或中止 (As long as a branch of science offers an abundance of problems, so long is it alive; a lack of problems foreshadows extinction or the cessation of independent development)."

希尔伯特继而在其开幕式报告中豪迈地指出："正是通过这些问题的解决，研究者锻炼其钢铁意志，发现新观点，达到更为广阔的自由的境界 (It is by the solution of problems that the investigator tests the temper of his steel; he finds new methods and new outlooks, and gains a wider and freer horizon)."

那什么是好的问题？希尔伯特给出了两个标准 [1.5]．

一是问题本身清晰易懂，正像拉格朗日所说的："应该能向在大街上遇到的第一个人解释清楚 (You can explain it to the first man whom you meet on the street)"．爱因斯坦也曾经令人深思地指出："除非你能向你的祖母解释清楚，否则你不会真正理解某件事 (You do not really understand something unless you can explain it to your grandmother)"．费曼也曾经不无夸张地指出："如果你不能向一个 6 岁的孩子解释清楚的话，你就不可能真正理解它 (If you can't explain it to a six year old, you don't really understand it)."

第二个标准是问题本身应该是困难的，具有很大的挑战性，能作为试金石，检验研究者的价值，衡量其能力．

希尔伯特当年的话仍然让我们激动不已："在通向那隐藏的真理的曲折道路上，它 (具有挑战性的问题) 应该是指引我们前进的一盏明灯，最终并以成功的喜悦作为对我们的报偿 (It should be to us a guide post on the mazy paths to hidden truths, and ultimately a reminder of our pleasure in the successful solution)".

我们不禁要问，理性力学是否能持续地提出学术界公认的学术难题以吸引青年才俊献身于这些难题的解决，从而证明其价值？这些问题的解决是否能真正推动工程科学的进步？

1.2 数学术语

在本书的开篇有必要首先复习和明确相关的数学术语. 表 1.2 给出了相关数学术语的定义、若干例子和优先级. 给出优先级的目的是, 提醒同学们在打牢基础的前提下, 关注并在高优先级的问题上进行选题并开展研究, 避免沉浸于细枝末节的工作而不能自拔.

表 1.2 数学术语: 定义、例子和优先级

序号	数学名词	定义	例子	优先级
1	理论、学说 (theory)	又称学说或学说理论, 指人类对自然、社会现象, 按照已有的实证知识、经验、事实、法则、认知以及经过验证的假说, 经由一般化与演绎推理等等的方法, 进行合乎逻辑的推论性总结. A theory is a contemplative and rational type of abstract or generalizing thinking, or the results of such thinking. Depending on the context, the results might, for example, include generalized explanations of how nature works.	(1) 相对论 (theory of relativity) (2) 进化论 (theory of evolution) (3) 精神分析理论 (psychoanalytic theory) (4) 血液循环学说 (blood circulation theory)	1
2	数学理论 (mathematical theory)	数学理论是数学的一个分支, 是数学研究的一个领域. 一个理论可以是一个知识体, 因此在这个意义上, "数学理论" 指的是数学研究领域. 这与数学模型的思想不同. A mathematical theory is a subfield of mathematics that is an area of mathematical research. A theory can be a body of knowledge, and so in this sense a "mathematical theory" refers to an area of mathematical research. This is distinct from the idea of mathematical models.	(1) 群论 (group theory) (2) 数论 (number theory)	1

续表

序号	数学名词	定义	例子	优先级
3	纲领 (program)	数学中一系列影响深远的构想，它联系或沟通了多门学科.	(1) 埃尔朗根纲领 (Erlangen program) (2) 朗兰兹纲领 (Langlands program)	1
4	公理系统 (axiomatic system)	一个公理系统是一个公理的集合，从中一些或全部公理可以一并用来逻辑地导出定理. An axiomatic system is any set of axioms from which some or all axioms can be used in conjunction to logically derive theorems. 欧几里得的五条公理奠定了欧氏几何学的基础. 1889 年, 意大利数学家皮亚诺 (Giuseppe Peano, 1858—1932) 提出了一个算术公理系统: (1) 0 是自然数; (2) 每一个确定的自然数 a, 都有一个确定的后继数 a', a' 也是自然数; (3) 0 不是任何自然数的后继数; (4) 不同的自然数有不同的后继数, 如果两个自然数的后继数相等, 那么它们是同一个数; (5) 任意关于自然数的命题, 如果证明: 它对自然数 0 是真的, 且假定它对自然数 a 为真时, 可以证明对 a' 也真. 那么, 命题对所有自然数都真. 公理化运动的最大成就则是希尔伯特在 1899 年对于初等几何的公理化.	(1) 欧氏几何公理 (axioms of Euclidean geometry) (2) 皮亚诺公理 (Peano axioms) (3) 理性力学公理 (Axioms of rational mechanics)	1
5	公理 (axiom)	公理是一个不证自明的表述, 并且被当做演绎及推论其他 (理论相关) 事实的起点. An axiom is a statement that is accepted without proof and regarded as fundamental to a subject.	(1) 等同于相同事物的事物会相互等同; (2) 若等同物加上等同物, 则整体会相等; (3) 若等同物减去等同物, 则其差会相等; (4) 相互重合的事物会相互等同; (5) 整体大于部分.	2

续表

序号	数学名词	定义	例子	优先级
6	公设 (postulate)	在各种科学领域的基础中，或许会有某些未经证明而被接受的附加假定，其有效性必须建立在现实世界的经验上； 公理是许多科学分支所共有的，而各个科学分支中的公设则是不同的. At the foundation of the various sciences lay certain additional hypotheses which were accepted without proof. Such a hypothesis was termed a postulate. Their validity had to be established by means of real-world experience. While the axioms were common to many sciences, the postulates of each particular science were different.	平面几何的五大公设： (1) 能从任一点画一条直线到另外任一点上去； (2) 能在一条直线上造出一条连续的有限长线段； (3) 能以圆心和半径来描述一个圆. (4) 每个直角都相互等值； (5) 平行公设：若一条直线与两条直线相交，在某一侧的内角和小于两个直角，那么这两条直线在各自不断地延伸后，会在内角和小于两直角的一侧相交.	2
7	定律 (law)	定律，或称科学定律、科学法则，为研究宇宙间不变的事实规律所归纳出的结论，不同于理论、假设、定义、定理，是对客观事实的一种表达形式，通过大量具体的客观事实经验累积归纳而成的结论. The laws of science, scientific laws, or scientific principles are statements that describe or predict a range of phenomena as they appear in nature. Scientific laws summarize and explain a large collection of facts determined by experiment, and are tested based on their ability to predict the results of future experiments. They are developed either from facts or through mathematics, and are strongly supported by empirical evidence.	(1) 守恒律 (conservation laws); (2) 热力学定律 (laws of thermodynamics); (3) 牛顿定律 (Newton's laws) (4) 摩尔定律 (Moore's law) (5) 贝尔定律 (Bell's law) (6) 吉尔德定理 (Gilder's law)	3

续表

序号	数学名词	定义	例子	优先级
8	原理 (principle)	原理是基于实验和观察的结论，是适用于多种情况的定理，是基本规律. principles are typically conclusions based on repeated experiments and observations. A principle is a theorem that applies in a wide range of circumstances, is a fundamental (basic) law.	(1) 阿基米德原理 (Arichimedes' principle) (2) 达朗伯原理 (D'Alembert's principle) (3) 最小作用量原理 (principle of least action) (4) 圣维南原理 (principle of Saint Venant's law)	3
9	法则 (rule)	法则是一个建立有用公式的定理. A rule is a theorem that establishes a useful formula.	(1) 右手法则 (right-hand rule) (2) 链式法则 (chain rule) (3) 洛比塔法则 (L'Hôpital's rule)	3
10	准则 (criterion, rule)	一个陈述或定理成立所必须的要求. A requirement necessary for a given statement or theorem to hold. Also called a condition.	柯西-玻恩准则 (Cauchy-Born rule)	3
11	定理 (theorem)	定理是指在既有命题的基础上证明出来的命题，这些既有命题可以是其他定理，也可以是被广为接受的陈述. 定理是公理的逻辑推论. A theorem is a statement that has been proved on the basis of previously established statements, such as other theorems, and generally accepted statements, such as axioms. A theorem is a logical consequence of the axioms.	(1) 勾股定理 (Pythagorean theorem) (2) 夹挤定理 (Squeeze theorem) (3) 费马大定理 (Fermat's last theorem) (4) 费马小定理 (Fermat's little theorem)	3

续表

序号	数学名词	定义	例子	优先级
12	命题 (proposition)	命题是一个次要的定理,意味着一个有简单证明的表述,这个术语有时意味着一个简单的证明,而定理通常是有复杂证明或者保留重要结果的. 经典几何中,欧几里得《几何原本》中的所有定理和几何构造,无论其重要性,都被称为"命题". A proposition is a theorem of lesser importance. This term sometimes connotes a statement with a simple proof, while the term theorem is usually reserved for the most important results or those with long or difficult proofs. In classical geometry, this term was used differently: In Euclid's *Elements* (c. 300 BCE), all theorems and geometric constructions were called "propositions" regardless of their importance.	欧氏几何命题: (1) 若直线 a 平行于 b,且分别平行于圆的 A、B 两点,则 AB 为直径. (2) 经典几何命题,蝴蝶定理: 过圆的定弦 XY 的中点 M,任作其他二弦 AB, CD,连接 AD, CB 交 XY 于点 P, Q,则 $PM = MQ$.	4
13	引理 (lemma)	引理是一个"帮助定理",为证明其他定理而使用的已被证明的定理,是它构成了一个大定理证明的一部分. 在某些情况下,当定理间的相对重要性变得更加清晰时,曾经被认为的引理现在被当做一个定理. A lemma is a "helping theorem", a proposition with little applicability except that it forms part of the proof of a larger theorem. In some cases, as the relative importance of different theorems becomes more clear, what was once considered a lemma is now considered a theorem.	(1) 高斯引理 (Gauss's lemma) (2) 费马引理 (Fermat's lemma) (3) 庞加莱引理 (Poincaré's lemma) 见 §24.2	4

续表

序号	数学名词	定义	例子	优先级
14	推论 (corollary)	推论是从另一个定理或定义简单推导得到的命题,更严格的特殊情况下,推论也可以是一个定理. A corollary is a proposition that follows with little proof from another theorem or definition. Also a corollary can be a theorem restated for a more restricted special case.	(1) 垂径定理推论:平分弧的直径垂直平分这条弧所对的弦. (2) 阿伏伽德罗定律的推论 (corollary of Avogadro's law): 同温同压下,气体的体积比等于它们的物质的量之比.	4
15	猜想 (conjecture, hypothesis)	未被证实但被认为正确的表述. An unproved statement that is believed true.	(1) 庞加莱猜想 (Poincaré's conjecture); (2) 哥德巴赫猜想 (Goldbach's conjecture); (3) 黎曼猜想(假设) (Riemann's hypothesis); (4) 开普勒猜想 (Kepler's conjecture).	5
16	假设 (hypothesis, premise)	定理结论的证明用到一些条件,这些条件称为假设或前提. The proof of a theorem deduces the conclusion from conditions called hypotheses or premises.	(1) 定理 "如果 A 则 B" 中, A 是假设. Theorem: if A, then B. A is called the hypothesis of the theorem. (2) 连续介质假设 (continuum hypothesis) 见 §2.3.	5

续表

序号	数学名词	定义	例子	优先级

17	数学模型 (mathematical model)	数学模型是运用数学概念和语言对一个系统的描述.
		A mathematical model is a description of a system using mathematical concepts and languages.
		在物理科学中，一个经典的数学模型大多包含下列元素：
		(1) 控制方程；
		(2) 辅助亚模型，包括：(i) 定义方程；(ii) 本构方程；
		(3) 假设和约束，包括：(i) 初始和边界条件；(ii) 约束和运动学方程.
		In the physical sciences, a traditional mathematical model contains most of the following elements:
		(1) Governing equations
		(2) Supplementary sub-models
		   √ Defining equations
		   √ Constitutive equations
		(3) Assumptions and constraints
		   √ Initial and boundary conditions
		   √ Classical constraints and kinematic equations
			数学模型可进行如下分类：
			(1) 线性与非线性 (linear vs. nonlinear)，见例 1.7；
			(2) 离散与连续 (discrete vs. continuous)；
			(3) 静态与动态 (static vs. dynamic)；
			(4) 确定性与概率性或随机性 (deterministic vs. probabilistic or stochastic)，见附录 E；
			(5) 显式与隐式 (explicit vs. implicit)；
			(6) 演绎，归纳与漂移 (deductive, inductive, or floating).
				6

1.3 理性力学常用的空间

空间是带有结构的集合，比如线性空间就是定义了两种代数运算结构的集合. 举例说明，没有内积就不能定义角度以及正交，空间就不能正交分解 —— 这正是 n 维欧氏空间的结构，也没有投影定理，所以，就有了内积空间. 把有结构的集合定义为空间，既是归纳以方便使用的需要，也是布尔巴基学派结构化数学的影响.

至于同学们经常问到的，为什么学习力学要用到那么多的空间 (见表 1.3)？一是抽象、提炼、概括的需要；二是公理化完备的需要；三是便于分类；四是便于研究；五是循于历史惯例.

在实际工程问题的数值求解中，工程师们关心解的存在性、数值方法的收敛性和收敛速度. 有限元法是对偏微分方程 (PDE) 中常用的离散方法. 为了衡量数值解的质量，需要将其与精确解比较，即求 "误差"——以连续介质力学的主要分支之一的弹性力学为例，位移是空间的函数，我们需要度量数值解和精确解这两个函数之间的距离. 这就是数学中度量的概念，要放在特定的度量空间或者赋范线性空间下去谈. 那么，我们想要求什么空间里的解呢？对于弱形式下的变分原理，回答就是索伯列夫空间 (Sobolev space). 确定了空间之后，我们才能在此空间下对函数进行逼近，并分析收敛性；否则收敛性无从谈起.

表 1.3 经典力学和连续介质力学中常用的空间 (如图 1.1 所示)

序号	空间名称	定义和说明
1	拓扑空间 topological space	拓扑空间是最基本的，由集合 + 开集构成，该空间中没有距离. 换言之，拓扑 "弱化了" 距离. 因此，拓扑空间和度量空间相比在概念上更加抽象和宽泛. 拓扑空间是一个集合 X 和其上定义的拓扑结构 τ 组成的二元组 (X,τ). X 的元素 x 通常称为拓扑空间 (X,τ) 的点. 而拓扑结构一词 τ 涵盖了开集，闭集，邻域，开核，闭包，导集，滤子等若干概念. 豪斯多夫于 1914 年基于四个豪斯多夫公理定义了拓扑空间的概念.
2	度量空间 metric space	度量空间是个具有距离函数的集合，该距离函数定义集合内所有元素间之距离. 此一距离函数被称为集合上的度量. 该空间是个有序对 (M,d), M 是集合而 d 是在 M 上的度量，即为函数： $d: M \times M \to \mathbb{R}$ 使得对于任何在 M 内的 x、y、z, 下列条件均成立： $d(x,y) \geqslant 0$ (非负性或半正定性) $d(x,y) = 0$ 当且仅当 $x = y$ (非退化性) 同时满足非负性和非退化性则称为正定性. $d(x,y) = d(y,x)$ (对称性) $d(x,z) \leqslant d(x,y) + d(y,z)$ (三角不等式). 度量空间是由弗雷歇 (见 D.2 小节) 于 1906 年在其博士论文中引入的.

续表

序号	空间名称	定义和说明
3	向量空间 vector space 线性空间 linear space	给定域 $\mathcal{F}$ (在连续介质力学中，一般为实数域 $\mathbb{R}$)，$\mathcal{F}$ 上的向量空间 $\mathcal{V}$ 是一个集合，其上定义了两种二元运算： "向量加法 +"：$\mathcal{V}\times\mathcal{V}\to\mathcal{V}$，把 $\mathcal{V}$ 中的两个元素 $\boldsymbol{u}$ 和 $\boldsymbol{v}$ 映射到 $\mathcal{V}$ 中另一个元素，记作 $\boldsymbol{u}+\boldsymbol{v}$； "标量乘法 ·"：$\mathcal{F}\times\mathcal{V}\to\mathcal{V}$，把 $\mathcal{F}$ 中的一个元素 λ 和 $\mathcal{V}$ 中的一个元素 $\boldsymbol{u}$ 变为 $\mathcal{V}$ 中的另一个元素，记作 $\lambda\cdot\boldsymbol{u}$ 或 $\lambda\boldsymbol{u}$； $\mathcal{V}$ 中的元素称为向量，相对地，$\mathcal{F}$ 中的元素称为标量. 对 $\mathcal{F}$ 中的任意元素 a、b 以及 $\mathcal{V}$ 中的任意元素 $\boldsymbol{u}$、$\boldsymbol{v}$、$\boldsymbol{w}$，向量空间满足如下运算规则： (1) 向量加法的结合律：$(\boldsymbol{u}+\boldsymbol{v})+\boldsymbol{w}=\boldsymbol{u}+(\boldsymbol{v}+\boldsymbol{w})$； (2) 向量加法的交换律：$\boldsymbol{u}+\boldsymbol{v}=\boldsymbol{v}+\boldsymbol{u}$； (3) 向量加法的单位元：存在一个零向量元素 $\boldsymbol{0}\in\mathcal{V}$，使得对任意的向量 $\boldsymbol{u}\in\mathcal{V}$ 满足：$\boldsymbol{u}+\boldsymbol{0}=\boldsymbol{u}$； (4) 向量加法的逆元素：$\forall\boldsymbol{v}\in\mathcal{V}$，都存在其逆元素 $-\boldsymbol{v}\in\mathcal{V}$，使得 $\boldsymbol{v}+(-\boldsymbol{v})=\boldsymbol{0}$； (5) 标量乘法与标量的域乘法相容：$a(b\boldsymbol{v})=(ab)\boldsymbol{v}$； (6) 标量乘法的单位元：域 $\mathcal{F}$ 存在单位元 1 使得：$1\boldsymbol{v}=\boldsymbol{v}$； (7) 标量乘法对向量加法的分配律：$a(\boldsymbol{u}+\boldsymbol{v})=a\boldsymbol{u}+a\boldsymbol{v}$； (8) 标量乘法对域加法的分配律：$(a+b)\boldsymbol{v}=a\boldsymbol{v}+b\boldsymbol{v}$ 作为向量空间的具体常用的例子，在例 1.7 后面，将给出位置空间、动量空间和波矢空间的相关讨论.
4	仿射空间 affine space	仿射空间是没有原点的向量空间 (线性空间). 宇宙是一个四维仿射空间 $\mathbb{A}^4$，其中的点称为世界点或事件. 伽利略空间是经典力学中典型的 $\mathbb{A}^4$.
5	赋范向量空间 normed vector space	赋范向量空间是具有"长度"概念的向量空间，是通常的欧几里得空间 $\mathcal{E}^n$ 的推广.
6	巴拿赫空间 Banach space	巴拿赫空间是一个完备赋范向量空间. 更精确地说，巴拿赫空间是一个具有范数并对此范数完备的向量空间.
7	索伯列夫空间 Sobolev space	索伯列夫空间是一个由函数组成的赋范向量空间. 该空间主要用来研究偏微分方程理论，因此，在计算力学中常常遇到.
8	内积空间 inner product space	内积空间是增添了一个额外的结构的向量空间. 这个额外的结构叫做内积，或标量积，或点积. 这个增添的结构允许我们谈论向量的角度和长度.

续表

序号	空间名称	定义和说明
9	三维欧氏空间 three-dimensional Euclidean space	欧几里得空间就是具有有限维的内积空间. 欧几里得空间 $\mathcal{E}^n$, 简称欧氏空间, 亦称平直空间, 在数学中是对欧几里得所研究的二维和三维空间的一般化. 该一般化把欧几里得对于距离、以及相关的概念长度和角度, 转换成任意数维的坐标系. 本课程将问题的讨论限制在三维欧氏空间. 设 $\mathcal{V}$ 是实数域 $\mathbb{R}$ 上的线性空间 (向量空间), 若 $\mathcal{V}$ 上定义着正对称双线性型 g (称为内积), 则 $\mathcal{V}$ 称为对于 g 的内积空间或欧几里得空间. $g(\boldsymbol{x}, \boldsymbol{y})$ 是 $\mathcal{V}$ 上的二元实值函数, 满足如下关系: (1) $g(\boldsymbol{x}, \boldsymbol{y}) = g(\boldsymbol{y}, \boldsymbol{x})$; (2) $g(\boldsymbol{x}+\boldsymbol{y}, \boldsymbol{z}) = g(\boldsymbol{x}, \boldsymbol{z}) + g(\boldsymbol{y}, \boldsymbol{z})$; (3) $g(k\boldsymbol{x}, \boldsymbol{y}) = kg(\boldsymbol{x}, \boldsymbol{y})$; (4) $g(\boldsymbol{x}, \boldsymbol{x}) \geqslant 0$, 而且 $g(\boldsymbol{x}, \boldsymbol{x}) = 0$ 当且仅当 $\boldsymbol{x} = \boldsymbol{0}$ 时成立. 这里 $\boldsymbol{x}, \boldsymbol{y}, \boldsymbol{z}$ 是 $\mathcal{V}$ 中任意向量, k 是任意实数. 欧几里得空间 $\mathcal{E}^n$ 在 n 维实向量空间 $\mathbb{R}^n$ 中定义了内积 $(\boldsymbol{x}, \boldsymbol{y}) = x_1 y_1 + \cdots + x_n y_n$, 则 $\mathbb{R}^n$ 为欧几里得空间.
10	希尔伯特空间 Hilbert space	希尔伯特空间就是完备的内积空间. 其上所有的柯西列等价于收敛列, 从而微积分中的大部分概念都可以无障碍地推广到希尔伯特空间中. 量子力学用希尔伯特空间作为数学基础的原因如下: (1) 量子力学实验基础就是各种粒子的波粒二象性, 能自洽地描述波粒二象性的说法就是几率解释. (2) 态应该具有可加性, 态用矢量描述最方便. (3) 态是矢量, 几率是数, 由矢量到数的映射, 数学上就是内积了, 但内积有正有负, 所以取内积模方为几率, 数学基础目前为内积空间. (4) 独立的物理态有无穷多个, 所以内积空间维数无穷大. 无穷大涉及收敛的问题, 某些参数取无穷大时, 相应的物理态不能跑出空间去, 所以数学上需要任何一个序列的极限仍在空间内, 即空间要满足完备性.
11	豪斯多夫空间 Hausdorff space	假设 X 是拓扑空间. 设 x 和 y 是 X 中的点. 称 x 和 y 可以由邻域分离, 如果存在 x 的邻域 U 和 y 的邻域 V 使得 U 和 V 是不相交的 $(U \cap V = \varnothing)$, 且 X 中的任意两个不同的点都可以由这样的邻域分离, 那么称 X 是豪斯多夫空间. 在数学分析所遇到的几乎所有空间都是豪斯多夫空间; 最重要的实数是豪斯多夫空间. 更一般地说, 所有度量空间都是豪斯多夫空间. 事实上, 在分析中用到的很多空间, 比如拓扑群和拓扑流形在其定义中明确地声明了豪斯多夫条件.
12	位形空间 configuration space	由广义坐标组成的空间. 例如, 拉格朗日力学就是用位形空间来描述力学系统的运动.

续表

序号	空间名称	定义和说明
13	相空间 phase space	由广义坐标和广义动量组成的空间. 哈密顿力学就是相空间的几何学.
14	对偶空间 dual space	对偶空间构造是行向量 ($1\times n$) 与列向量 ($n\times 1$) 的关系的抽象化. 例如, 位移向量和力向量组成一对对偶向量空间 (dual vector space), 简称对偶空间.
15	共轭空间 conjugate space	复数域的对偶空间, 或统称为共轭空间.

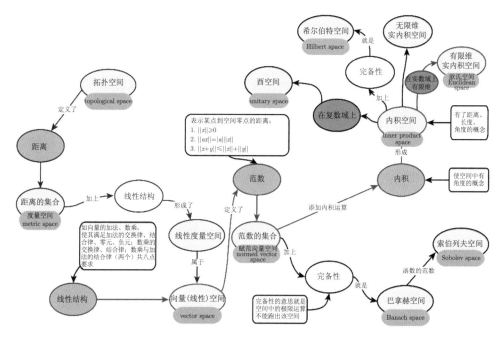

图 1.1 连续介质力学常用各种空间之间的关系图

例 1.1 用映射的语言表示一个向量和一个标量的乘积.

解：一个标量和一个向量的乘积可通过映射语言表示如下：

$$\begin{aligned} \mathbb{R} \times \mathcal{V} &\to \mathcal{V} \\ (\lambda, \boldsymbol{a}) &\mapsto \lambda \boldsymbol{a} = \boldsymbol{a}\lambda, \ \forall \lambda \in \mathbb{R} \end{aligned} \tag{1-1}$$

式 (1-1) 中, 箭头符号 "$\mapsto$" 被称为 "maplet 或 maplet arrow", 意为 "映射到 (maps to)". 例如: $x \mapsto f(x)$ 表示将 x 映射到 $f(x)$; 具体地, $x \mapsto x^2$ 表示将所有实数 x 映射到其平方. $\mathbb{R}$ 表示实数集合 (set of real numbers).

式 (1-1) 中, 箭头符号 "$\to$" 表示 "线性映射 (linear map)".

叉号 "×" 表示 "双线性映射 (bilinear map)"，比如：$\mathcal{V} \times \mathcal{V} \to \mathbb{R}$ 表示来自于两个向量空间中元素进行内积 (inner product) 的双映射.

例 1.2　用双映射的语言表示两个向量空间中向量的内积.

解：可表示如下：

$$\langle \cdot, \cdot \rangle : \mathcal{V} \times \mathcal{V} \to \mathbb{R}$$
$$(\boldsymbol{a}, \boldsymbol{b}) \mapsto \langle \boldsymbol{a}, \boldsymbol{b} \rangle = \langle \boldsymbol{b}, \boldsymbol{a} \rangle = \boldsymbol{a} \cdot \boldsymbol{b} = |a||b|\cos\theta \tag{1-2}$$

上式第一排，冒号前面表示的是即将进行的内积运算 $\langle \cdot, \cdot \rangle$，冒号后面表示的是双线性映射；上式第二排，$\mapsto$ 前面是即将参与运算的两个参量，后面表示的是具体的运算过程.

例 1.3　给出线性变换、线性映射或同形的定义.

答：令 $\mathcal{V}$ 和 $\mathcal{W}$ 为向量空间，一个满足具有如下性质的映射 $\boldsymbol{A} : \mathcal{V} \to \mathcal{W}$

$$(1) \ \boldsymbol{A}(\boldsymbol{v}_1 + \boldsymbol{v}_2) = \boldsymbol{A}\boldsymbol{v}_1 + \boldsymbol{A}\boldsymbol{v}_2 \in \mathcal{W}, \quad \forall \boldsymbol{v}_1, \boldsymbol{v}_2 \in \mathcal{V}$$
$$(2) \ \boldsymbol{A}(\lambda \boldsymbol{v}) = \lambda(\boldsymbol{A}\boldsymbol{v}) \in \mathcal{W}, \quad \forall \lambda \in \mathbb{R} \tag{1-3}$$

就称为线性变换 (linear transformation)、线性映射 (linear map) 或者共形 (homomorphism).

如果 $\boldsymbol{A} : \mathcal{V} \to \mathcal{W}$ 是双射的 (bijective)，$\boldsymbol{A}^{-1}$ 则是 $\boldsymbol{A}$ 的逆，$\mathcal{V}$ 和 $\mathcal{W}$ 拥有相同的维数的话，则 $\boldsymbol{A}$ 被称为 "同构 (isomorphism)". 当同构满足 $\mathcal{V} = \mathcal{W}$ 时，被称为 "自同构 (automorphism)".

线性变换 $\boldsymbol{A} : \mathcal{V} \to \mathcal{V}$ 被称为 "自同态 (endomorphism)".

例 1.4　恒等映射 (identity map) 的引入和定义.

答：如果 $\forall \boldsymbol{v} \in \mathcal{V}$ 满足：

$$\boldsymbol{I}\boldsymbol{v} = \boldsymbol{v} \tag{1-4}$$

则线性变换 $\boldsymbol{I} : \mathcal{V} \to \mathcal{V}$ 被称为 "恒等映射". 如果 $\boldsymbol{A}$ 在向量空间 $\mathcal{V}$ 上是自同构的话，则有下列基本常用关系式：

$$\boldsymbol{A}^{-1}\boldsymbol{A} = \boldsymbol{A}\boldsymbol{A}^{-1} = \boldsymbol{I} \tag{1-5}$$

值得指出的是，$\boldsymbol{I}\boldsymbol{v}$ 表示的是二阶单位张量 (identity tensor, 亦称为：等同张量) $\boldsymbol{I}$ 和向量 $\boldsymbol{v}$ 之间的点积；而 $\boldsymbol{A}^{-1}\boldsymbol{A}$ 表示的是二阶张量 $\boldsymbol{A}^{-1}$ 和 $\boldsymbol{A}$ 之间的点积.

如图 1.2 所示，如果 $\mathcal{Q}$ 是一个流形 (manifold)，(a) 图中底部的映射 $i : \mathcal{Q} \to \mathcal{Q}$ 就是一个流形间的恒等映射 (identity map)，而 (b) 图中的两个流形 $\mathcal{B}$ 和 $\mathcal{S}$ 间的映射 $\phi : \mathcal{B} \to \mathcal{S}$ 则为流形间的普通映射. 向量场 $\boldsymbol{v}$ 和 $\boldsymbol{V}$ 的相应映射 $v : \mathcal{Q} \to T\mathcal{Q}$

和 $\boldsymbol{V}: \mathcal{B} \to T\mathcal{S}$ 是从流形到相应切丛间的映射，而图中的 π 则是丛投影 (bundle projection).

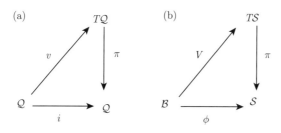

图 1.2 流形间的映射

也就是说，在图 1.2(a) 中，恒等映射就是一种复合映射 (composite mapping)：$i = \pi \circ v$，使得 $\mathcal{Q} \xrightarrow{v} T\mathcal{Q} \xrightarrow{\pi} \mathcal{Q}$.

例 1.5 用等距映射 (isometry) 定义正交映射 (orthogonal map).

答：若 $\mathcal{V}$ 和 $\mathcal{W}$ 为两个欧几里得向量空间 (Euclidean vector spaces)，对于任意的 $\boldsymbol{a}, \boldsymbol{b} \in \mathcal{V}$，一个满足下列条件的映射 $\boldsymbol{Q}: \mathcal{V} \to \mathcal{W}$

$$\langle \boldsymbol{Qa}, \boldsymbol{Qb} \rangle_{\mathcal{W}} = \langle \boldsymbol{a}, \boldsymbol{b} \rangle_{\mathcal{V}} \tag{1-6}$$

等距映射. 所谓等距映射是黎曼流形间保持弧长的映射. 一个等距映射 $\boldsymbol{Q}: \mathcal{V} \to \mathcal{V}$ 是正交映射的条件是满足下列两个条件：

$$\det \boldsymbol{Q} = \pm 1 \quad \text{和} \quad \boldsymbol{Q}^{-1} = \boldsymbol{Q}^{\mathrm{T}} \tag{1-7}$$

(1-6) 式中，$\boldsymbol{Qa}$ 和 $\boldsymbol{Qb}$ 分别表示矩阵 $\boldsymbol{Q}$ 与矢量 $\boldsymbol{a}$ 和 $\boldsymbol{b}$ 之间的点积. (1-7) 式中，$\det \boldsymbol{Q}$ 表示的是矩阵 $\boldsymbol{Q}$ 的矩阵 (determinant)，$\boldsymbol{Q}^{\mathrm{T}}$ 表示的是 $\boldsymbol{Q}$ 的转置 (transpose). 正交映射在理性力学中有着十分重要的应用，详见 §20.4 的讨论.

例 1.6 给出描述函数光滑性 (smoothness) 的可微性分类 (differentiability class).

答：光滑函数 (smooth function) 在数学中特指无穷可导的函数，也就是说，存在任意有限阶导数. 分类如下：(1) 若一函数是连续的，则称其为 C^0 类函数；(2) 若函数存在导函数，且其导函数连续，则称为连续可导，记为 C^1 类函数；在 20 世纪，人们发现 C^1 函数空间不是研究微分方程的解的恰当的空间. 而索伯列夫空间正是 C^1 空间的替代品，用于研究偏微分方程的解；(3) 若一函数 n 阶可导，并且其 n 阶导函数连续，则为 C^n 类函数 ($n \geqslant 1$)；(4) 而光滑函数是对任意阶都存在连续偏导数，则称其为 C^∞ 类函数；(5) 如果函数在其开集中每一个点的邻域均能表示成收敛的幂级数，即该函数是解析的 (analytic)，则称其是 C^ω 类的函数.

光滑流形之间的光滑映射可以用坐标图的方式来定义. 因为函数的光滑性的概念和特定的坐标图的选取无关. 这样的映射有一个一阶导数, 定义在切向量上; 它给出了在切丛的级别上的对应纤维间的线性映射.

例 1.7 在连续介质力学中, 经常说一个方程是线性的, 而另外一个方程是非线性的. 通过证明薛定谔方程是一个线性方程, 从而给出方程是否是线性的定义.

答: 由于含时的薛定谔方程比不含时的复杂些, 这里只给出针对含时情况的证明. 含时的薛定谔方程为

$$-\frac{\hbar^2}{2m}\nabla^2\psi + V\psi = i\hbar\frac{\partial \psi}{\partial t} \tag{1-8}$$

式中, $\hbar$ 为约化普朗克常数, m 粒子为质量, $\psi(\boldsymbol{r},t)$ 为波函数, V 为势能. 注意到 (1-8) 式中, 拉普拉斯算子 ∇^2 和偏导 $\dfrac{\partial}{\partial t}$ 均为线性算子.

称薛定谔方程 (1-8) 是一个线性方程, 是指满足薛定谔方程的波函数拥有线性关系. 设波函数 ψ_A 和 ψ_B 都是薛定谔方程的解, 则其任意的线性组合:

$$\psi_C = a\psi_A + b\psi_B \tag{1-9}$$

也是薛定谔方程 (1-8) 式的解. 上式中, a 和 b 为常量. 简要证明如下:

由于 ψ_A 和 ψ_B 都是薛定谔方程 (1-8) 的解, 则有

$$\begin{cases} -\dfrac{\hbar^2}{2m}\nabla^2(a\psi_A) + V(a\psi_A) = i\hbar\dfrac{\partial(a\psi_A)}{\partial t} \\ -\dfrac{\hbar^2}{2m}\nabla^2(b\psi_B) + V(b\psi_B) = i\hbar\dfrac{\partial(b\psi_B)}{\partial t} \end{cases} \tag{1-10}$$

上述两个方程相加, 自然得到 $-\dfrac{\hbar^2}{2m}\nabla^2\psi_C + V\psi_C = i\hbar\dfrac{\partial \psi_C}{\partial t}$, 亦即薛定谔方程为一个线性方程. 上述线性组合自然可以延伸至任意多个波函数的叠加. 因此, 波函数的叠加也同样是薛定谔方程的解. 这种叠加性质是量子力学最为奥妙的性质之一. 量子系统可以同时处于两个以上的经典状态; 一个粒子可以同时出现在几个不同位置, 可以同时拥有不同的能量.

容易证明, 下列含时格罗斯-皮塔耶夫斯基方程 (time-dependent Gross-Pitaevskii equation) 则为非线性方程:

$$\left(-\frac{\hbar^2}{2m}\nabla^2 + V(\boldsymbol{r}) + g|\psi(\boldsymbol{r},t)|^2\right)\psi(\boldsymbol{r},t) = i\hbar\frac{\partial \psi(\boldsymbol{r},t)}{\partial t} \tag{1-11}$$

位置空间 (position space) 与动量空间 (momentum space) 是物理学中一对联系紧密的向量空间.

通常是三维的位置空间，亦称实空间 (real space)、坐标空间 (coordinate space)，是空间中所有物体的位置矢量 r 的集合. 位置矢量定义了空间中的一个点. 如果位置矢量随时间发生变化，那么它就可以描绘出一个路径或一个面，如粒子的运动轨迹.

类比地，通常也是三维的动量空间是空间中所有物体的动量矢量 p 的集合. 一个物体的动量 p 可以反映它的运动情况. 依据量子力学的德布罗意关系：$p = \hbar k$，一个自由粒子的动量 p 正比于波矢 k. 系统的所有波矢的集合构成波矢空间 (wave-vector space). 在不严格区分动量与波矢时，这两个概念可以混用；因此，动量空间和波矢空间也是等价的. 有关动量空间在经典力学中应用的例子见本章思考题 1.7.

欧氏空间是配有内积的 n 维线性空间，在有限维空间中，内积与欧氏标准范数可以相互导出，而范数就是向量的模——长度，内积可以给出正交的概念，进一步可以把欧氏空间的一组基化为正交归一的基.

为什么要研究欧氏空间？因为维数相同的线性空间 (有限维) 是同构的，就是它们几乎一模一样，这样我们只要研究 n 维欧氏空间就可以得到一般的 n 维线性空间的性质了.

范数 (norm) 是用来度量某个向量空间 (或矩阵) 中的每个向量的长度或大小. 常用范数的分类如表 1.4 所示.

表 1.4　范数的分类

向量范数	1-范数	$\|x\|_1 = \sum\limits_{i=1}^{N} \|x_i\|$，即向量元素绝对值之和
	2-范数	$\|x\|_2 = \sqrt{\sum\limits_{i=1}^{N} x_i^2}$，即向量元素绝对值平方和再开方
欧几里得范数	p-范数	$\|x\|_p = \left(\sum\limits_{i=1}^{N} \|x_i\|^p\right)^{\frac{1}{p}}$，即向量元素绝对值的 p 次方和再 $1/p$ 次幂
	∞-范数	$\|x\|_\infty = \max\limits_i \|x_i\|$，即所有向量元素绝对值中的最大值
	$-\infty$-范数	$\|x\|_{-\infty} = \min\limits_i \|x_i\|$，即所有向量元素绝对值中的最小值

§2. 连续介质假设

2.1 物质世界的特征尺度

特征尺度在各学科中具有举足轻重的作用, 往往起到 "纲举目张" 和 "牛鼻子" 的作用.

图 2.1 给出了物质世界的特征尺度和各学科领域所研究的大致尺度. 其中, 作为最小尺度的普朗克长度 (Planck length) 为

$$l_{\mathrm{P}} = \sqrt{\frac{\hbar G}{c^3}} \approx 1.616 \times 10^{-35} \text{ m} \tag{2-1}$$

式中, $\hbar = 1.05457266(63) \times 10^{-34}$ J·s 为约化普朗克常数 (reduced Planck constant), $G = 6.67408 \pm 0.0031 \times 10^{-11}$ m³kg⁻¹s⁻² 为万有引力常数, $c = 299792458$ m/s 为真空中的光速.

理论上的最小可测时间间隔称为普朗克时间 (Planck time), 其数量级为

$$t_{\mathrm{P}} = \frac{l_{\mathrm{P}}}{c} = \sqrt{\frac{\hbar G}{c^5}} \approx 5.39 \times 10^{-44} \text{ s} \tag{2-2}$$

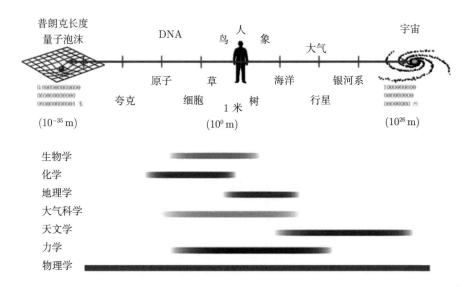

图 2.1 物质世界的特征尺度和各学科领域的大致研究尺度

下面用粗略数量级估计来估算宇宙半径 (10^{26} m). 为了摆脱一个引力质量为 M, 半径为 r 的星球, 所需要的第二宇宙速度为

$$v_2 = \sqrt{\frac{2GM}{r}} \tag{2-3}$$

式中, G 为引力常数, (2-3) 式也被称为 "逃逸速度 (escape velocity)". 如果星球的质量如此之大, 以至于其第二宇宙速度达到了光速 $v_2 = c$, 这时就连光子也不能克服其引力的作用而发射出来. 因此, 外界看不到这个星体, 这类星体被称为黑洞. 将 $v_2 = c$ 代入 (2-3) 式, 便得到引力半径 (gravitational radius) 为

$$r_s = \frac{2GM}{c^2} \tag{2-4}$$

上式亦称为史瓦西半径 (Schwarzschild radius), 其原因是天文学家史瓦西 (Karl Schwarzschild, 1873—1916, 终年 42 岁) 于 1916 年由广义相对论而严格推得, 亦可见本书 (15-20) 式. 但这里所给出的方法是完全基于经典的牛顿力学, 该方法是由英国物理学家约翰·米歇尔 (John Michell, 1724—1793) 于 1784 年和法国天文学家拉普拉斯 (Pierre-Simon Laplace, 1749—1827) 彼此独立创立的. 凑巧的是, 由经典力学和广义相对论所给出的引力半径的表达式 (2-4) 完全一致.

当代天文学观测业已证明, 宇宙在大尺度上物质分布是相当均匀的. 我们来考虑一个半径为 r、密度为 ρ 的均匀的球体, 则其质量为 $M = 4\pi r^3 \rho/3$, 如果这个球体的半径恰好达到了自己的引力半径 $r = r_s$, 则该球体的引力半径为

$$r_s = \sqrt{\frac{3c^2}{8\pi G\rho}} \tag{2-5}$$

则生活在此球内部的人不可能将光子发射到数量级比 (2-5) 式中 r_s 更大的范围之外. 将宇宙的平均密度

$$\rho \sim 10^{-29} \text{ g/cm}^3 \tag{2-6}$$

代入 (2-5) 式, 便得到宇宙的引力半径 (简称宇宙半径) 的数量级为

$$r_s \sim 10^{26} \text{ m} \tag{2-7}$$

列奥纳多·达·芬奇 (Leonardo da Vinci, 1452—1519) 于 1487 年前后, 在意大利的威尼斯创作的世界著名素描 ——《维特鲁威人》(*Uomo vitruviano*). 如图 2.2 所示, 它是钢笔和墨水绘制的手稿. 描绘了一个男人在同一位置上的 "十" 字型和 "火" 字型的姿态, 并同时被分别嵌入到一个矩形和一个圆形中.

图 2.2 达·芬奇的作品《维特鲁威人》

"人体中自然的中心点是肚脐. 因为如果人把手脚张开, 作仰卧姿势, 然后以他的肚脐为中心用圆规画出一个圆, 那么他的手指和脚趾就会与圆周接触. 不仅可以在人体中这样地画出圆形, 而且可以在人体中画出方形. 即若由脚底量到头顶, 并把这一量度移到张开的两手, 那么就会发现高和宽相等, 恰似平面上用直尺确定方形一样."《维特鲁威人》也是达·芬奇以比例最精准的男性为蓝本, 这种 "完美比例" 也即是数学上所谓的 "黄金分割".

如图 2.3 所示, 宇宙星系的螺旋、地球生物的韵律, 都以黄金比例为基础树叶的生长模式、鹦鹉螺壳、公羊角、人的上半身与下半身的比例、人的手骨、人的面部、手掌、DNA 的螺旋、星球的大气云层旋窝、沙滩上的贝颌等都有黄金比例. 人类的每只手有五个手指, 每个手指的前一指节与后一指节的长度之比仍为黄金比值.

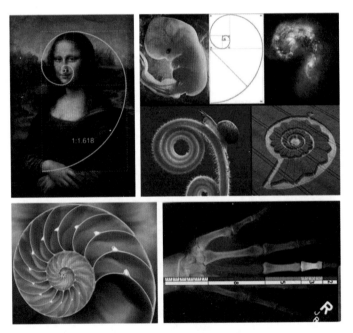

图 2.3 黄金分割无处不在

2.2 努森数与四种流动区间

丹麦物理学家努森 (Martin Knudsen, 1871—1949, 如图 2.4 所示) 于 1911 年建议了无量纲数——被后人称为 "努森数 (Knudsen number)":

$$Kn = \frac{\lambda}{L} \tag{2-8}$$

Kn 数是气体的平均自由程 (mean free path) λ 和物体特征尺度 L 之比. 平均自由程的概念是克劳修斯 (Rudolf Clausius, 1822—1888) 于 1858 年引入的. 对于大气中的例子动力学而言, 标准状况下 (standard temperature and pressure, STP, 25°C, 1 atm), 气体的平均自由程大约为 $\lambda \approx 8\times 10^{-8}$ m (80 nm).

图 2.4　左图: 马丁·努森; 中图: 1911 年索尔维会议; 右图: 1911 年索尔维会议中的努森

1946 年, 钱学森按照 Kn 数, 将流动划分为四种情形, 如表 2.1 所示.

表 2.1　按照 Kn 数, 对流动区间进行的划分

连续流区 (continuous flow zone) 非稀薄 (non rarefied)	$Kn < 10^{-2}$
滑移流区 (slip flow zone) 轻微稀薄 (slightly rarefied)	$10^{-2} < Kn < 10^{-1}$
过渡流区 (transition flow zone) 中等稀薄 (moderately rarefied)	$10^{-1} < Kn < 10$
自由分子流区 (free macular flow zone) 高度稀薄 (highly rarefied)	$Kn > 10$

按照努森数 Kn 的划分, 如图 2.5 所示, 只有当飞行器的特征尺度 L 要比气体的平均自由程 λ 大两个数量级时, 气体的流动才属于连续介质, 此时气体和飞行物体是无滑移边界, 该门学科被称为流体力学, 是连续介质力学的一个重要组成部分.

图 2.5　按照努森数对流动性质划分的示意图

在滑流区, 气体流动与连续介质的差别主要表现在气体和飞行物体的边界附近, 即所谓的速度滑移和温度跳跃现象; 在自由分子流区, 分子之间的碰撞机会很

少,从物面反射的分子流几乎不受来流影响,近似服从物面条件下的麦克斯韦分布,这类问题容易处理,只要了解分子在物面是如何反射的就得到了问题的解;在过渡流区,分子之间的碰撞与分子和物面的碰撞同等重要,这是稀薄大气动力学的核心的,也是最困难的问题之一.

2.3 连续介质假定

下面我们按照不同的尺度对固体的行为进行相应地讨论.

如图 2.6 所示,当我们所讨论的介质的尺度小于某一临界体积 V_* 时,也就是仅仅包含数十个原子或分子时,此时介质为非均质 (inhomogeneous) 且介质的行为由于涨落过大,尚不能进行平均;当介质的特征尺度大到一定尺度时,如介于 V_* 和 $V_{\min}$ 之间时,此时称为含有微结构的非均质材料 (inhomogeneous materials with microstructures),亦可称为含有微形态的多相材料 (micromorphic multiphase materials),此时的介质行为可能呈现出明显的尺寸效应 (size effect). 而经典的连续介质力学研究的是宏观介质的力学行为,一般为均质的 (homogeneous)、各向同性的 (isotropic) 的力学行为.

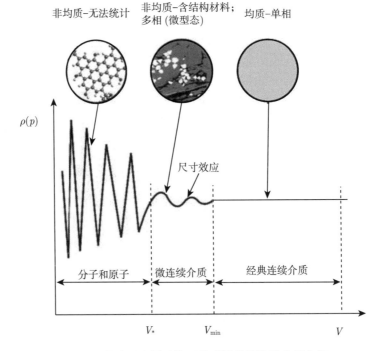

图 2.6 从分子和原子物理学到经典的连续介质力学

连续介质的概念和流形是紧密相连的，可参阅 §21.3 的定义和讨论.

1946 年，博戈柳博夫 (Nikolay Nikolayevich Bogoliubov, 1909—1992) 针对气体，提出了关于空间、时间上大致有三种不同尺度的描述方法，称为 "博戈柳博夫级联 (Bogoliubov hierarchy)"，又称为三种标度: (1) 微观描述或动力学标度; (2) 动理学描述或标度; (3) 流体力学描述或标度.

需要区分三个特征尺度: (1) 粒子间作用力程; (2) 粒子的平均自由程; (3) 密度等宏观量非均匀性的量程.

在动力学标度 (微观标度) 上: 分布函数随时间有急剧的变化, 系统需要有多粒子的分布函数来描述;

在动理学标度上, 系统的分布函数迅速地开始 "同步" 化, 这时多粒子分布函数可表示为单粒子分布函数的泛函, 只用单粒子分布函数就能描述系统的行为;

在流体力学标度 (事实上, 就是连续介质力学标度) 上, 则只需要分布函数的若干个矩即可描述.

以常温、常压的氢气为例, 说明博戈柳博夫的三个标度的划分.

一、微观 (动力学) 层次, 特征尺度为粒子间的作用力程, 可取为化学键键长的特征尺度 10^{-10} m (Å). 若气体分子热运动的特征速度为 10^3 m/s, 则该标度的特征时间大致为 10^{-13} s.

二、动理学层次, 气体分子发生一到两次碰撞后进入到动理学描述阶段, 特征尺度为粒子的平均自由程 (mean free path), 在 10^{-7} m 量级. 特征时间 (时标) 为两次碰撞之间自由飞行的时间, 约为 10^{-10} s.

三、流体力学层次, 当时标 $\gg 10^{-10}$ s, 每个气体分子 (原子) 都已经过多次碰撞, 它们之间已建立了新的局部平衡, 进入到可进行宏观平均的流体力学阶段. 特征尺度为密度等非均匀性的量程, 约为 10^{-2} m 量级.

本教程只讨论连续介质作为可变性的宏观力学行为, 也就是介质在三维欧几里得空间和均匀流逝时间下, 受牛顿力学支配的物质行为.

连续介质力学这门学科有着两重性. 一方面, 是这门学科的基础性, 在 §1 中已经提及过, 在欧美的很多名校中, 该课程的开设一般是在数学系. 理论物理学家对这门学科也是相当地重视, 著者已经在《近代连续介质力学》[1.6] 中着重述及过爱因斯坦对连续介质力学的评价, 此不赘述. 这里著者想补充的是, 从 1912 年到 1914 年的三年里, 爱因斯坦在苏黎世联邦理工学院 (ETH Zurich) 担任理论物理教授, 他主要教授分析力学和热力学的课程 (He taught analytical mechanics and thermodynamics), 期间, 他还研究了连续介质力学、热的分子理论和引力理论 (He also studied continuum mechanics, the molecular theory of heat, and the problem of gravitation).

这门学科的基础性还在于，很多理论物理的著名教程均包含连续介质力学的内容. 最著名的是，朗道和栗弗席兹的十卷理论物理的教程中，有《流体力学》[1.7] 和《弹性理论》[1.8] 两卷的连续介质力学内容；在《费曼物理学讲义》[1.9] 的第 2 卷中，第 38—41 共四章均为连续介质力学的内容，分别讨论了弹性理论和流体力学的相关内容；索末菲的 6 卷本的《理论物理教程》中，第 2 卷即为《可变形体的力学》[1.10].

连续介质力学更重要的是其作为工程科学的 "大统一理论 (Grand Unified Theory, GUT)" 的应用性，此点已经在《近代连续介质力学》[1.6] 中做了深入阐述.

§3. 运动的两种描述方法——欧拉和拉格朗日观点

着眼于场的欧拉描述和着眼于粒子的拉格朗日描述 (Eulerian and Lagrangian descriptions) 又被广泛地称为欧拉和拉格朗日观点 (Eulerian and Lagrangian viewpoints). 如图 3.1 所示，欧拉是拉格朗日的学术导师.

图 3.1 欧拉 (左) 和拉格朗日 (右)

以测量温度为例来说明. 如图 3.2 所示. 站在桥上测量温度的欧拉和在河中运动的船上测量温度的拉格朗日为何所测量到的温度值不同呢？事实上，即使拉格朗日所乘坐的船穿过欧拉所站的桥下时，他们所测量到的结果也很可能不同. 原因何在？

图 3.2 欧拉和拉格朗日描述的区别

按照一般常识，温度场是时间 t 和位置 $\boldsymbol{r}$ 的函数：$T = T(t, \boldsymbol{r})$，当然，位置 $\boldsymbol{r}$ 也是时间的函数：$T = T(t, \boldsymbol{r}(t))$，为了便于理解，温度场还可表示为如下坐标分量的形式：

$$T = T(t, x(t), y(t), z(t)) \tag{3-1}$$

温度的变化率是指其随时间的变化率，也就是对时间的导数 (time derivative). 即使是对同一位置 $\boldsymbol{r}$ 也有两种不同的导数. 拉格朗日导数 (Lagrangian derivative) 又称物质时间导数 (material time derivative) dT/dt, 很多文献和书中采用 DT/Dt；欧拉导数 (Eulerian derivative) 又称为局部时间导数，是空间某点处函数对时间的偏导数.

如果 $\partial T/\partial t = 0$, 则温度场不依赖于时间，仅仅是位置的函数，$T = T(\boldsymbol{r})$, 此时的一类流场被称为是"定常的 (steady)".

对 (3-1) 式应用莱布尼兹链式法则，有

$$\frac{dT}{dt} = \frac{\partial T}{\partial t} + \frac{\partial T}{\partial x}\frac{\partial x}{\partial t} + \frac{\partial T}{\partial y}\frac{\partial y}{\partial t} + \frac{\partial T}{\partial z}\frac{\partial z}{\partial t} = \frac{\partial T}{\partial t} + v_x\frac{\partial T}{\partial x} + v_y\frac{\partial T}{\partial y} + v_z\frac{\partial T}{\partial z}$$
$$= \frac{\partial T}{\partial t} + \boldsymbol{v} \cdot \boldsymbol{\nabla} T \tag{3-2}$$

式中，$\boldsymbol{v} \cdot \boldsymbol{\nabla}$ 被称为对流导数 (convective derivative)，该项来自于如下内积：

$$(\boldsymbol{v} \cdot \boldsymbol{\nabla})(\cdot) = \left(v_x\hat{\boldsymbol{i}} + v_y\hat{\boldsymbol{j}} + v_z\hat{\boldsymbol{k}}\right) \cdot \left(\frac{\partial}{\partial x}\hat{\boldsymbol{i}} + \frac{\partial}{\partial y}\hat{\boldsymbol{j}} + \frac{\partial}{\partial z}\hat{\boldsymbol{k}}\right)(\cdot)$$
$$= v_x\frac{\partial(\cdot)}{\partial x} + v_y\frac{\partial(\cdot)}{\partial y} + v_z\frac{\partial(\cdot)}{\partial z} \tag{3-3}$$

(3-2) 式表明，连续介质给定质点的温度随时间的变化 (dT/dt) 等于质点所在位置温度的变化 ($\partial T/\partial t$) 加上由于质点运动所导致的温度变化 ($\boldsymbol{v} \cdot \boldsymbol{\nabla} T$).

正确地理解对流导数项 (3-3) 式对于理性力学十分关键. 一般而言, 对流导数不为零. 只有在以下三种情形时对流导数为零:

(1) 质点不存在运动, 也就是 $v = \mathbf{0}$;

(2) 两矢量间的点积 $v \cdot \nabla T$ 为零, 意味着两个矢量相互垂直, 此时要求质点运动方向沿着等值面运动, 由于梯度方向垂直于过该点的等值面, 故满足要求. 等值面为场内数值相同的点的集合所构成的空间曲面. 若场为连续分布, 则等值面是不相交的;

(3) 标量场为均匀场, 梯度为零: $\nabla(\cdot) = \mathbf{0}$.

特别地, 当将 (3-2) 和 (3-3) 两式应用于速度流场时, 得到加速度的表达式为

$$a = \frac{\mathrm{d}v}{\mathrm{d}t} = \underbrace{\frac{\partial v}{\partial t}}_{\text{非定常加速度}} + \underbrace{v \cdot \nabla v}_{\substack{\text{对流加速度} \\ \text{非线性项}}} \tag{3-4}$$

上式说明, 质点的加速度可分解为非定常加速度 (unsteady acceleration) 和对流加速度 (convective acceleration). 其中, 对流加速度来自于流体流动随空间的变化所产生的速度改变. ∇v 是速度梯度的左梯度, 它是一个二阶张量, 在理性连续介质力学中, 一般写为 $\nabla \otimes v$, 这里的 $\otimes$ 为并矢或张量矢. 容易理解, 由于对流加速度项 $v \cdot \nabla v$ 的非线性, 因此, 将 (3-4) 式应用于牛顿第二定律所得到的流体动力学的基本方程, 纳维–斯托克斯方程, 从本质上就是一个非线性的方程. 当然地, 在某些特殊的情况下, 问题可以得到适当的简化.

当牛顿第二定律应用于流体微元时, 应用 (3-4) 式, 得到

$$\int \rho a \mathrm{d}v = \int \rho \left(\frac{\partial v}{\partial t} + v \cdot \nabla \otimes v \right) \mathrm{d}v = \int \rho f \mathrm{d}v \tag{3-5}$$

式中, f 为流体微元单位质量所受到的合力矢量, 包括静水压强、黏性力、重力等矢量, ρ 为微元体的密度. 特别地, 当流体所受到的外力合力为零时, 也就是牛顿第一定理成立的情形, 此时, 在 (3-4) 和 (3-5) 两式中, 由于加速度为零, 此时有

$$\frac{\partial v}{\partial t} + v \cdot \nabla \otimes v = \mathbf{0} \Rightarrow \frac{\partial v}{\partial t} = -v \cdot \nabla \otimes v \tag{3-6}$$

此时, 流体的例子将于匀速运动, 也就是拉格朗日加速度或物质加速度为零: $\mathrm{d}v/\mathrm{d}t = \mathbf{0}$; 但 (3-6) 式说明, 局部加速度或欧拉加速度 $\partial v/\partial t$ 未必为零. 这说明, 一个空间固定点速度的改变是由于被流场输运的拥有不同速度的流体质点流过该点.

例 3.1 在笛卡儿坐标系中求速度矢量 v 的对流加速度 $v \cdot \nabla v$ 或 $v \cdot \nabla \otimes v$.

第 1 章 基 础 知 识

解: 由于速度矢量: $\boldsymbol{v} = \boldsymbol{v}(v_x, v_y, v_z)$,其梯度为一个二阶张量,可用矩阵表示为

$$[\boldsymbol{\nabla} \otimes \boldsymbol{v}] = \begin{bmatrix} \dfrac{\partial v_x}{\partial x} & \dfrac{\partial v_y}{\partial x} & \dfrac{\partial v_z}{\partial x} \\ \dfrac{\partial v_x}{\partial y} & \dfrac{\partial v_y}{\partial y} & \dfrac{\partial v_z}{\partial y} \\ \dfrac{\partial v_x}{\partial z} & \dfrac{\partial v_y}{\partial z} & \dfrac{\partial v_z}{\partial z} \end{bmatrix} \tag{3-7}$$

则,速度矢量的对流导数为

$$[\boldsymbol{v} \cdot \boldsymbol{\nabla} \otimes \boldsymbol{v}] = \begin{bmatrix} v_x & v_y & v_z \end{bmatrix} \begin{bmatrix} \dfrac{\partial v_x}{\partial x} & \dfrac{\partial v_y}{\partial x} & \dfrac{\partial v_z}{\partial x} \\ \dfrac{\partial v_x}{\partial y} & \dfrac{\partial v_y}{\partial y} & \dfrac{\partial v_z}{\partial y} \\ \dfrac{\partial v_x}{\partial z} & \dfrac{\partial v_y}{\partial z} & \dfrac{\partial v_z}{\partial z} \end{bmatrix} = \begin{bmatrix} v_x\dfrac{\partial v_x}{\partial x} + v_y\dfrac{\partial v_x}{\partial y} + v_z\dfrac{\partial v_x}{\partial z} \\ v_x\dfrac{\partial v_y}{\partial x} + v_y\dfrac{\partial v_y}{\partial y} + v_z\dfrac{\partial v_y}{\partial z} \\ v_x\dfrac{\partial v_z}{\partial x} + v_y\dfrac{\partial v_z}{\partial y} + v_z\dfrac{\partial v_z}{\partial z} \end{bmatrix}^{\mathrm{T}} \tag{3-8}$$

(3-8) 式还将在例 56.2 中用张量表示出, 见 (56-32) 式.

为了进一步便于理解欧拉和拉格朗日描述, 表 3.1 给出两种描述的多种典型的例子.

表 3.1 欧拉和拉格朗日描述的一些例子

情景	对问题的解释
经典力学	固体力学一般为拉格朗日描述,流体力学一般为欧拉描述.
气象观测	在气象观测中广泛使用的是欧拉法. 在世界各地 (空间点) 设立星罗棋布的气象站. 根据统一时间各气象站把同一时间观测到的气象要素迅速报到规定的通讯中心, 然后发至世界各地, 绘制成同一时刻的气象图, 据此做出天气预报.
球类比赛	"人盯人" 防守类比于拉格朗日观点, "区域防守" 或 "联防" 类比于欧拉观点.
历史研究	跟踪特定家谱类比于拉格朗日观点, 查地方志则类比于欧拉观点.
警察跟踪	尾随其人或跟踪身份证、移动电话、信用卡等类比于拉格朗日观点, 而用监控录像则类比于欧拉观点.
大学等机构	"铁打的营盘"(大学、军营、监狱等) 为欧拉描述, "流水的兵"(学生、军人、犯人等) 则属于拉格朗日描述.
环境和流行病检测	着眼于雾霾颗粒的运动轨迹属于拉格朗日描述, 着眼于某一特定区域的数据变化则属于欧拉描述; 着眼于某个传染病人研究类比于拉格朗日观点, 而着眼于某个区域的流行病的演化规律则类比于欧拉观点.
特定区域人员流动情况	以银行为例, 从每个人一踏进门开始盯防其轨迹类比于拉格朗日描述, 用监控录像研究某个时间段人员的演化和流动情况类比于欧拉描述.

§4. 从质点和刚体动力学到连续介质力学

4.1 质点和刚体力学概述

质点和刚体动力学可分为两大类：(1) 牛顿力学；(2) 分析力学. 后者又可分为拉格朗日力学和哈密顿力学. 质点力学的特点如表 4.1 所示, 三位代表性科学家如图 4.1 所示. 表中, $\boldsymbol{F}$ 为力矢量, $\boldsymbol{p}$ 为动量矢量, $L=T-V$ 为拉格朗日量, T 和 V 分别为动能和势能, q 和 p 分别为广义坐标和广义动量, t 为时间. $\alpha=1,2,\cdots,s$, s 为自由度数.

表 4.1　质点动力学的特点一览表

类型	创立年代	方程	空间	几何	方程类型
牛顿力学	1687 年 牛顿时年 45 岁	$\boldsymbol{F}=\dfrac{\mathrm{d}\boldsymbol{p}}{\mathrm{d}t}$ (4-1)	三维欧氏空间	欧氏几何	二阶微分方程
拉格朗日力学	1788 年 拉格朗日时年 52 岁	$\dfrac{\mathrm{d}}{\mathrm{d}t}\dfrac{\partial L}{\partial \dot{q}_\alpha}-\dfrac{\partial L}{\partial q_\alpha}=0$ (4-2)	位形空间	黎曼几何	s 个二阶微分方程组
哈密顿力学	1834—1835 年 哈密顿时年 29 和 30 岁	$\begin{cases}\dot{q}_\alpha=\dfrac{\partial H}{\partial p_\alpha}\\ \dot{p}_\alpha=-\dfrac{\partial H}{\partial q_\alpha}\end{cases}$ (4-3)	相空间	辛几何	$2s$ 个一阶微分方程组

图 4.1　左起：牛顿 (1642—1727)、拉格朗日 (1736—1813)、哈密顿 (1805—1865)

本书著者已经在《力学讲义》[1.11] 中对牛顿力学为何对应于欧氏空间和欧氏几何, 以及哈密顿力学为何对应于相空间和辛几何做了深入、细致的阐述.《力学讲义》[1.11] 中对拉格朗日力学所对应的位形空间也已经阐述过, 本书将在 §15 的黎曼度规张量一节中作为例题，对拉格朗日力学为何对应于黎曼几何做适当补充.

拉格朗日是在 19 岁上在意大利的都灵构思了他的《分析力学》[1.12] 和变分法. 拉格朗日在 19 岁上成为都灵皇家炮兵学院的数学教授，尚有一些孩子气的数学教授在都灵给年纪都比他大的学生上课. 在欧拉的大力举荐下，拉格朗日于 1759 年 10 月 2 日，在 23 岁上当选为柏林科学院的外籍院士，拉格朗日的学术地位得以确认.

由于长期的过度用功和工作而导致神经衰弱，拉格朗日作为他那个年代最杰出的数学家之一，在 40 岁上就对数学研究的热情开始锐减. 拉格朗日曾在柏林从事研究工作 20 年，《分析力学》[1.12] 就成书于柏林. 拉格朗日于 1786 年辞别柏林而到巴黎从事研究工作后，《分析力学》[1.12] 的出版提到了日程. 数学家勒让德 (Adrien-Marie Legendre, 1752—1833) 承担了这部著作的前期编辑，拉格朗日的老朋友马里神父在向一位巴黎的出版商承诺在某一设定的日期后买下所有未售出的存书后，这位出版商才于 1788 年勉强出版了拉格朗日的巨著《分析力学》[1.12]. 但当书送到时年 52 岁的拉格朗日手中时，他已经对科学和数学都十分冷漠了 [1.13]，令人吃惊的是，他竟然在足足两年的时间里都没有劳神去打开此书!

拉格朗日压抑的心情和他妻子 Vittoria 经过多年的病痛于 1783 年在柏林的去世有关，Vittoria 是拉格朗日的表妹. 1792 年，当拉格朗日 56 岁时，时年 24 岁芳龄的 Renée-Françoise-Adélaïde Le Monnier 小姐由于仰慕拉格朗日，坚持要嫁给他 (insisted on marrying him). 这位小姐是拉格朗日的一位天文学家朋友 Pierre Charles Le Monnier 的女儿. 这段幸福的婚姻使拉格朗日又重新振作起来，以至于拉格朗日又重新开始了对他的巨著《分析力学》[1.12] 的校对再版工作.

拉格朗日在《分析力学》[1.12] 的前言中指出："读者在本书中将找不到任何插图. 我所阐明的方法既不需要构图，也不需要几何或力学的推理，只需要遵循统一规定的步骤进行代数运算 (The reader will find no figures in this work. The methods which I set forth do not require either constructions or geometrical or mechanical reasonings: but only algebraic operations, subject to a regular and uniform rule of procedure)."

拉格朗日量 L 对时间的积分被称为作用量 (action):

$$S = \int_{t_1}^{t_2} L\mathrm{d}t = \int_{t_1}^{t_2} (T - V)\,\mathrm{d}t \tag{4-4}$$

哈密顿原理 (Hamilton's principle) 表明，一个真实的运动使得系统的作用量取驻值 (最小值): $\delta S = 0$，由此可得到拉格朗日方程 (4-2) 式. 哈密顿原理亦被称为"哈密顿最小作用量原理". 事实上, 这里的 "最小" 只是局部最小, 而不是通常的全局意义上的.

普朗克 (Max Planck, 1858—1947, 1918 年诺贝尔物理学奖获得者) 对最小作用量原理恰如其分地评价道: "科学最崇高且最为人们梦寐以求的目标, 是把业已观察到并行将观察到的一切自然现象缩并成单独一个原理……在那些标志着过去几百年物理科学成就的, 多少带有一般性的定律中, 最小作用量原理, 就其内容和形式而论, 可能最接近于理论研究上这一理想的最终目标 (It [science] has as its highest principle and most coveted aim the solution of the problem to condense all natural phenomena which have been observed and are still to be observed into one simple principle, that allows the computation of past and more especially of future processes from present ones. ⋯ Amid the more or less general laws which mark the achievements of physical science during the course of the last centuries, the principle of least action is perhaps that which, as regards form and content, may claim to come nearest to that ideal final aim of theoretical research)."

值得特别指出的是, 英国数学家、以其命名的 "泰勒展开" 的布鲁克·泰勒 (Brook Taylor, 1685—1731) 曾于 1713 年将牛顿定律应用于弦线的一个微元, 他当时并不明白把牛顿第二定律应用于物体无穷小的微元就可以推出运动方程, 正像理性力学大师克利福德·特鲁斯德尔 (Clifford Ambrose Truesdell III, 1919—2000) 所指出的: "他并不把这个结果认作运动的微分方程, 因此他进一步的工作便建立在混乱和部分错误的假设之上."

1747 年, 达朗贝尔 (Jean-Baptiste le Rond d'Alembert, 1717—1783) 在研究弦的振动的基础上发表了题为《张紧的弦振动时形成的曲线研究》的论文, 这是现代偏微分方程的经典文献. 迄今大部分偏微分方程的教程都是由弦振动方程开篇的. 所有弦乐器的理论基础是达朗贝尔提出的弦振动理论.

4.2 从离散到连续的微元体方法

1750 年, 欧拉明确指出: "连续介质力学的真正基础在于牛顿第二定律作用于物体的微元体 (The true basis of continuum mechanics was Newton's second law applied to the infinitesimal elements of bodies)". "微元体" 又被称为 "单元体". 可见, 从质点动力学到连续介质力学, 是一个从离散到连续的自下而上的过程.

欧拉是著者的几本书中始终都绕不过去的 "教科书式" 的大科学家, 他被人形容为 "计算起来毫不费劲, 就像人呼吸或鹰在风中翱翔一样 (calculated without apparent effort, as men breathe, or as eagles sustain themselves in the wind)".

下面我们通过弦振动的例子来说明如何将牛顿第二定律作用于连续介质的微元体,从而建立连续体得动力学方程.

例 4.1 通过微元体法,用牛顿第二定律建立弦的振动方程.

解法一： 如图 4.2,设弦中的张力为 T,弦的线密度 (单位长度的质量) 为 μ,弦振动的位移为 $u(x,t)$,应用小角度近似：$\sin\theta \approx \tan\theta$,弦中沿纵轴的合力为 $\sum F_u = T(\sin\theta_B - \sin\theta_A) \approx T(\tan\theta_B - \tan\theta_A)$,对微元应用牛顿第二定律,有如下关系式：

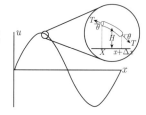

图 4.2 弦的振动图

$$T(\tan\theta_B - \tan\theta_A) = \mu \cdot \Delta x \frac{\partial^2 u}{\partial t^2} \quad \Rightarrow \quad \mu \frac{\partial^2 u}{\partial t^2} = T\frac{(\partial u/\partial x)_B - (\partial u/\partial x)_A}{\Delta x}$$

对上面的第二式取极限 $\Delta x \to 0$,应用偏导数的定义,得到弦振动的偏微分方程 (PDE) 为

$$\frac{1}{c^2}\frac{\partial^2 u}{\partial t^2} = \frac{\partial^2 u}{\partial x^2} \tag{4-5}$$

式中

$$c = \sqrt{\frac{T}{\mu}} \tag{4-6}$$

为弦振动的横波波速. (4-5) 式是一个典型的双曲型波动方程.

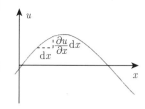

图 4.3 弦的振动示意图

解法二、分析力学方法：

弦振动的动能为

$$E_k = \int_a^b \frac{1}{2}\mu\left(\frac{\partial u}{\partial t}\right)^2 dx \tag{4-7}$$

式中,a 和 b 为弦的两个端点.

如图 4.3 所示,弦在振动中的伸长为

$$dl = \sqrt{(dx)^2 + \left(\frac{\partial u}{\partial x}dx\right)^2} - dx = dx\left[\sqrt{1 + \left(\frac{\partial u}{\partial x}\right)^2} - 1\right] \approx \frac{1}{2}\left(\frac{\partial u}{\partial x}\right)^2 dx \tag{4-8}$$

弦在张力 T 的作用下发生伸长所储存的势能为

$$E_p = \int_a^b T dl = \int_a^b \frac{T}{2}\left(\frac{\partial u}{\partial x}\right)^2 dx \tag{4-9}$$

在获得了弦振动的动能和势能后,其拉格朗日量可由 (4-7) 和 (4-9) 两式获得

$$L = E_k - E_p = \frac{1}{2}\int_a^b \left[\mu\left(\frac{\partial u}{\partial t}\right)^2 - T\left(\frac{\partial u}{\partial x}\right)^2\right] dx \tag{4-10}$$

则弦一维振动的作用量可表示为

$$S(u) = \frac{1}{2}\iint_D \left[\mu\left(\frac{\partial u}{\partial t}\right)^2 - T\left(\frac{\partial u}{\partial x}\right)^2\right]\mathrm{d}x\mathrm{d}t = \frac{1}{2}\int \mathrm{d}t \int \left[\mu(u_t)^2 - T(u_x)^2\right]\mathrm{d}x \tag{4-11}$$

式中，$u_t = \dfrac{\partial u}{\partial t}$, $u_x = \dfrac{\partial u}{\partial x}$，则 (4-11) 式可改写为

$$S = \int \mathrm{d}t \int \mathcal{L}(u, u_t, u_x; x, t)\mathrm{d}x = \frac{1}{2}\int \mathrm{d}t \int (\mu u_t^2 - T u_x^2)\mathrm{d}x, \tag{4-12}$$

式中，$\mathcal{L}(u, u_t, u_x; x, t)$ 是拉格朗日量的线密度：$L = \int \mathcal{L}\mathrm{d}x$. $\mathcal{L}(u, u_t, u_x; x, t)$ 是一个单函数 (u) 双变量 (x,t) 问题，由于两个变量 x 和 t 的地位是平等的，对应于最小作用量原理 $\delta S = 0$, 此时的欧拉–拉格朗日方程为

$$\frac{\partial}{\partial t}\frac{\partial \mathcal{L}}{\partial u_t} + \frac{\partial}{\partial x}\frac{\partial \mathcal{L}}{\partial u_x} - \frac{\partial \mathcal{L}}{\partial u} = 0 \tag{4-13}$$

从而对于拉格朗日量密度：$\mathcal{L} = \mu u_t^2 - T u_x^2$，则有如下关系式：

$$\frac{\partial \mathcal{L}}{\partial u_t} = \mu u_t, \quad \frac{\partial}{\partial t}\frac{\partial \mathcal{L}}{\partial u_t} = \mu u_{tt}, \quad \frac{\partial \mathcal{L}}{\partial u_x} = -T u_x, \quad \frac{\partial}{\partial x}\frac{\partial \mathcal{L}}{\partial u_x} = -T u_{xx}, \quad \frac{\partial \mathcal{L}}{\partial u} = 0$$

将上式代入 (4-13) 式有

$$\frac{1}{c^2}u_{tt} = u_{xx} \tag{4-14}$$

(4-14) 和 (4-5) 两式完全相同.

对比下质点或刚体的作用量 (4-4) 式和一维弦振动的作用量 (4-12) 式可知，对于一维的连续介质力学问题，需要将质点的拉格朗日量表达成为沿一维线元的积分，换句话说，对应于最小作用量的欧拉–拉格朗日方程是针对拉格朗日量的线密度 $\mathcal{L}$ 而言的.

推而广之，对于二维和三维的连续介质力学问题，欧拉–拉格朗日方程将分别针对拉格朗日量的面密度 $L = \iint \mathcal{L}\mathrm{d}x\mathrm{d}y$ 和体密度 $L = \iiint \mathcal{L}\mathrm{d}x\mathrm{d}y\mathrm{d}z$. 表 4.2 给出了从质点动力学到连续介质力学的拉格朗日量和欧拉 – 拉格朗日方程的变化和对比.

表 4.2　从质点动力学到连续介质力学的拉格朗日量和欧拉–拉格朗日方程的对比

质点力学	由 (4-4) 式 $S = \int_{t_1}^{t_2} L \mathrm{d}t = \int_{t_1}^{t_2}(T-V)\mathrm{d}t$, 取驻值 $\delta S = 0$, 由变分法得到欧拉–拉格朗日方程 (4-2) 式: $\dfrac{\mathrm{d}}{\mathrm{d}t}\dfrac{\partial L}{\partial \dot{q}_\alpha} - \dfrac{\partial L}{\partial q_\alpha} = 0$.
连续介质力学	一维, $S = \int \mathrm{d}t \int \mathcal{L}\mathrm{d}x$, 取驻值 $\delta S = 0$, 欧拉–拉格朗日方程为 (4-13) 式: $\dfrac{\partial}{\partial t}\dfrac{\partial \mathcal{L}}{\partial u_t} + \dfrac{\partial}{\partial x}\dfrac{\partial \mathcal{L}}{\partial u_x} - \dfrac{\partial \mathcal{L}}{\partial u} = 0$.
	二维, $S = \int \mathrm{d}t \iint \mathcal{L}\mathrm{d}x\mathrm{d}y$, 相应的欧拉–拉格朗日方程为 $\dfrac{\partial}{\partial t}\dfrac{\partial \mathcal{L}}{\partial u_t} + \dfrac{\partial}{\partial x}\dfrac{\partial \mathcal{L}}{\partial u_x} + \dfrac{\partial}{\partial y}\dfrac{\partial \mathcal{L}}{\partial u_y} - \dfrac{\partial \mathcal{L}}{\partial u} = 0$　(4-15)
	三维, $S = \int \mathrm{d}t \iiint \mathcal{L}\mathrm{d}x\mathrm{d}y\mathrm{d}z$, 相应的欧拉–拉格朗日方程为 $\dfrac{\partial}{\partial t}\dfrac{\partial \mathcal{L}}{\partial u_t} + \dfrac{\partial}{\partial x}\dfrac{\partial \mathcal{L}}{\partial u_x} + \dfrac{\partial}{\partial y}\dfrac{\partial \mathcal{L}}{\partial u_y} + \dfrac{\partial}{\partial z}\dfrac{\partial \mathcal{L}}{\partial u_z} - \dfrac{\partial \mathcal{L}}{\partial u} = 0$　(4-16)

见表 3.1, 固体力学一般用拉格朗日描述, 微元体要宏观无限小, 微观无限大, 以包含有足够多的分子以足够进行统计平均. 流体力学一般用欧拉描述, 控制体积 (control volume, CV) 是流体力学及热力学中, 为一物理现象建立数学模型时会用到的一个专有名词. 在惯性参考系中, 控制体积可能是一固定的区域, 或者是随着流体运动. 如图 4.4 所示, 控制体积的表面也称为控制表面 (control surface, CS).

稳态时, 控制体积可以视为一个其中流体体积为定值的任意空间. 流体可能会流进或流出控制体积, 但流入控制体积的流体质量等于流出控制体积的流体质量. 在稳态且没有功或能量的交换, 控制体积内的能量也是一个定值. 控制体积的概念类似经典力学的自由体图 (free body diagram).

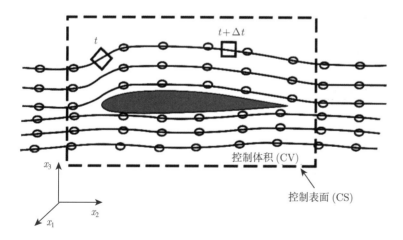

图 4.4　流体力学的控制体积和控制表面示意图

4.3 连续介质力学研究的几位先驱性的工作

列奥纳多·达·芬奇 (Leonardo da Vinci, 1452—1519) 是意大利文艺复兴时期的一个天才：除了是画家，他还是雕刻家、建筑师、音乐家、数学家、工程师、发明家、解剖学家、地质学家、制图师，植物学家和作家. 他是文艺复兴时期人文主义的代表人物，也使得他成为文艺复兴时期典型的艺术家，也是历史上最著名的画家之一，与米开朗基罗和拉斐尔并称文艺复兴后三杰. 本书在 §2 中已经对其高超绘画艺术中的科学原理做了介绍.

本节要强调的是，达·芬奇还是一位力学家，在流体力学和固体力学两个方面均有很高的成就. 达·芬奇作为一个私生子，还终身未婚 (见题表 3.1)，是一位名副其实的百科全书式的人物! 达·芬奇在 500 多年前就十分深刻地指出："力学是数学科学的天堂，因为我们在这里获得数学的成果 (Mechanics is the paradise of the mathematical sciences because by means of it one comes to the fruits of mathematics)." 这句话直到今天仍然意义深远.

和大多数人一样，儿时的达·芬奇，每当思绪飞往云外，总幻想着自己也有一双可以飞翔的翅膀. 1483—1486 年间，达·芬奇绘制了一幅飞行器工作草图，如图 4.5 所示. 飞机的外形由木头、帆布等当时的材料制成，在飞行器两侧是一双结构和形状酷似蝙蝠或翼龙的膜状翅膀，这双翅膀展翼最大可以达到 11 米. 这些工作说明了他在 30 岁以前就开始研究飞行器了. 到 1519 年去世前，达·芬奇都还在试验改进自己发明的扑翼机. 他对飞行的探索，可以说是贯穿一生. 可以毫不夸张地说，达·芬奇是人类航空的奠基者.

图 4.5 达·芬奇对飞机的概念设计

如图 4.6 所示，有三幅画被认为和湍流有着非常密切的关系. 图 4.6 左图便是达·芬奇的涡流；中间的图是梵高 (Vincent van Gogh, 1853—1890) 于 1889 年在阿尔勒圣雷米的一家精神病院创作的《星夜》(*Starry Night*)，一年后的 1890 年梵高在抑郁中自尽，年仅 37 岁. 研究发现梵高的后期作品，包括《星夜》在内，包含有一种"湍流"的神韵 (见思考题 1.17)，并推测此神韵来源于梵高由于长期处于癫狂状态中而得到超于常人的感悟能力和绘画表述能力. 右图是创作于 1832 年的日本名画《神奈川冲浪里》，梵高非常赞赏这幅画.

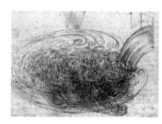 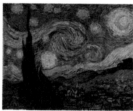 

图 4.6　三幅画作和湍流有着惊人的相似性

在流体力学方面，达·芬奇总结出在同一河深的前提下，河水的流速同河道宽度成反比 (A river of uniform depth will have a more rapid flow at the narrower section than at the wider)，并用这一结论说明血液在血管中的流动. 达·芬奇这个发现就是我们后来所说的"连续性定律 (law of continuity)"或"质量守恒定律 (law of mass conservation)".

在固体力学方面，达·芬奇也成就斐然. 如图 4.7 所示，他就梁的变形进行了深入研究，以至于有学者提出将材料力学中的"欧拉–伯努利梁"更名为"达·芬奇–欧拉–伯努利梁"[1.14].

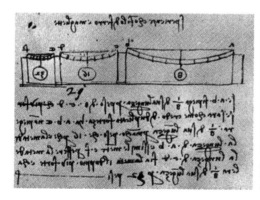 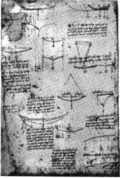

图 4.7　达·芬奇对梁的变形进行了深入研究

达·芬奇还是实验力学的先驱. 如图 4.8 所示, 他进行过不同长度铁丝的强度的实验研究. 实验样品是一定长度的铁丝; 载荷施加方式是, 将装有沙子的容器下面开一个小孔, 使沙子能够漏到铁丝下面所悬挂的篮子中. 在铁丝断裂时刻立即将漏沙孔堵上, 测得铁丝断裂时的长度和所承受的荷载. 实验重复若干次以校验其结果. 后续的实验分别用原来铁丝长度的 1/2、1/4 进行实验, 并详细记录其极限强度和断裂的位置.

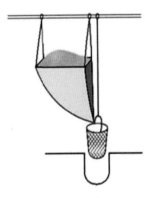

图 4.8 达·芬奇的铁丝强度的实验研究

伯努利家族在连续介质力学方面的研究成果亦已彪炳史册. 限于篇幅, 本节只简要地提及他们的两个贡献:

(1) 在固体力学方面, 伯努利梁 (Bernoulli beam) 是以雅克布·伯努利 (Jacob Bernoulli, 1654—1705) 命名的, 他在 1694 年的里程碑式的论文中[1.15], 最早用微积分研究了梁的变形, 他假定梁在变形时梁的横截面保持平面, 这就是平截面假定的最早提法. 丹尼尔·伯努利 (Daniel Bernoulli, 1700—1782) 于 1725—1733 年的八年间在俄罗斯的圣彼得堡科学院工作, 1727 年欧拉在丹尼尔·伯努利的推荐下来到圣彼得堡科学院工作, 1728 年丹尼尔·伯努利和欧拉合作, 进一步发展了梁的模型, 进而被称为 "欧拉-伯努利梁".

(2) 在流体力学方面, 丹尼尔·伯努利于 1726 年在圣彼得堡科学院提出并于 1738 年出版了 "伯努利原理", 这是在流体力学的连续介质理论方程建立之前, 水力学所采用的基本原理. 详见本书 §51 的详细讨论.

思考题和补充材料

1.1 挪威科学与文学院 2018 年 3 月 20 日宣布, 将 2018 年度阿贝尔奖 (Abel prize) 授予 81 岁的数学家罗伯特·朗兰兹 (Robert Phelan Langlands, 1936—), 以表彰以他名字命名的 "朗兰兹纲领"(见表 1.2) 将数学中的表示论和数论联系了起来, 成为 "数学的大统一理论 (grand unified theory of mathematics)", 能够统一数学中三个核心学科: 算术、几何和数学分析.

目前, 朗兰兹作为荣休教授 (emeritus professor) 仍在普林斯顿高等研究院工作, 他的办公室曾被爱因斯坦使用过. 挪威科学与文学院在一份声明中说, 朗兰兹早在 1967 年就提出了一项全新的数学理论, 认为数学中一些表面上看起来毫无关系的领域之间可能存在深刻的联系. 他提出的见解大胆而内涵丰富, 在数学中的很多分支领域之间架起了 "桥梁", 数学界将这一理论命名为 "朗兰兹纲领". "朗兰兹纲领" 在几十年间深刻影响了世界各地的数学家, 启发产生了一系列新的数学研究成果. 罗伯特·朗兰兹的名言是: "当然, 最美妙的时光是我与数学独处时, 没有野心，无需伪装，忘怀天地 (Certainly the best times were when I was alone with mathematics, free of ambition and pretense, and indifferent to the world.)"

1.2 为了呼应 1900 年德国数学家大卫·希尔伯特在巴黎提出的二十三个数学问题，美国克雷数学研究所 (Clay Mathematics Institute, CMI) 于 2000 年 5 月 24 日公布了七个数学猜想 —— 千禧年大奖难题 (millennium prize problems), 又称世界七大数学难题. 拟定这七个问题的数学家之一是怀尔斯, 费马大定理这个有 300 多年历史的难题没被选入的唯一理由就是已经被他解决了. 其他的专家, 除了克雷促进会会长贾菲 (Arthur Jaffe, 1937—), 还有阿蒂亚 (Michael Francis Atiyah, 1929—2019) 和泰特 (John Tate, 1925—2019), 以及孔涅 (Alain Connes, 1947—) 和威滕 (Edward Witten, 1951—). 根据克雷数学研究所制定的规则, 任何一个猜想的解答, 只要发表在数学期刊上, 并经过两年的验证期, 解决者就会被颁发一百万美元奖金.

这七个难题分别是：(1) 贝赫和斯维讷通–戴尔猜想 (Birch and Swinnerton-Dyer conjecture); (2) 霍奇猜想 (Hodge conjecture); (3) 纳维–斯托克斯 (方程解的) 存在性与光滑性 (Navier–Stokes existence and smoothness); (4) P = NP? (P versus NP problem); (5) 庞加莱猜想 (Poincaré conjecture); (6) 黎曼猜想 (Riemann hypothesis); (7) 杨–米尔斯 (规范场) 存在性和质量间隔 (Yang–mills existence and mass gap).

题图 1.2　黑板的力量
——数学家佩雷尔曼

迄今，七个千禧年大奖难题中, 只有庞加莱猜想被最终解决, 俄罗斯圣彼得堡数学家佩雷尔曼 (Grigori Yakovlevich Perelman, 1966— , 题图 1.2) 于 2003 年将庞加莱猜想盖棺定论. 在 2006 年, 佩雷尔曼拒绝了世界数学家大会颁发的 "菲尔兹奖"(Fields medal), 还于 2010 年 7 月拒绝领取 100 万美元奖金. 佩雷尔曼说："大家应该理解如果证明是对的, 那么其他的认可都是不需要的."

"纳维–斯托克斯方程解的存在性与光滑性" 是七个难题中唯一和连续介质力学有直接关系的问题, 有关其进一步的讨论参见思考题 11.22.

1.3 被誉为 "业余数学家王子 (The Prince of Amateurs)"[1.13] 的皮埃尔·费马 (Pierre de Fermat, 1607—1665) 的职业是律师和法国图卢兹议会的议员 (councillor at the Parlement de Toulouse), 据说是一个买来的职位. 费马于 1631 年 5 月 14 日就任议员职务, 从而将其名字从 "Pierre Fermat" 改为 "Pierre de Fermat", 中间的 de 是贵族姓氏的标志.

费马研究数学纯粹是他的消遣 (his recreation) 或业余爱好 (more of a hobby than a profession). 费马是一个一流的数学家, 一个无可指摘的诚实的人, 一个历史上无以伦比的算术学家 (Fermat was a mathematician of the first rank, a man of unimpeachable honesty, and an arithmetician without a superior in history). 费马的名言是："我确信已找到了 (这个一般定理的) 一个极佳的证明, 但书的空白太

窄，写不下 (I have discovered a truly marvellous demonstration (of this general theorem) which this margin is too narrow to contain)." 这是 1637 年费马针对"费马大定理 (Fermat's last theorem)"写下的著名文字. 有人戏言：费马这位训练有素的律师、业余数学家之王在其后的 358 年间，将数学家们玩弄得团团转！

之所以说是 358 年，是由于数学家安德鲁·怀尔斯 (Andrew Wiles, 1953—) 于 1994 年宣布成功证明了"费马大定理"，最终于 1995 年发表 (如题图 1.3 右图). 怀尔斯也因此获得下列重要奖励：沃尔夫奖 (Wolf prize, 1995/6)、邵逸夫奖 (Shaw prize, 2005)、阿贝尔奖 (Abel prize, 2016)，等.

题图 1.3 和费马大定理相关的三张邮票

1.4 法国 18 世纪后期到 19 世纪初数学、力学界著名的三位著名科学家：拉格朗日 (Joseph Louis Lagrange)、拉普拉斯 (Pierre-Simon Laplace) 和勒让德 (Adrien-Marie Legendre) 三个人的姓氏的第一个字母为"L"，又生活在同一时代，所以人们并称他们为"3L"，如题图 1.4 所示. 他们为 18 世纪末 19 世纪初法国数学力学的复兴作出重要贡献，是典型的"教科书"式的人物，他们还均高寿，并曾担任众多的官方职务. 他们三位的比较由题表 1.4 给出.

题表 1.4 法国的 3L 比较

姓名	生卒年	主要学术贡献	经典著作	成名年代	官方职务
拉格朗日	1736—1813 享年 77 岁	1788 年创立分析力学——拉格朗日力学，被哈密顿誉为"科学诗"，变分法的"欧拉-拉格朗日方程"	分析力学	1759 年 23 岁当选柏林科学院外籍院士. 被拿破仑誉为"数学科学高耸的金字塔". 1813 年葬入先贤祠.	参议员、伯爵、度量衡委员会主席
拉普拉斯	1749—1827 享年 77 岁	拉普拉斯变换、拉普拉斯方程、拉普拉斯妖	天体力学	1785 年 36 岁当选科学院院士. 被誉为"法国的牛顿".	内政大臣、伯爵、侯爵、贵族院
勒让德	1752—1833 享年 80 岁	勒让德变换、勒让德多项式	几何学基础	1785 年 33 岁当选科学院院士.	经度局主席

题图 1.4 法国 3L 的学术贡献汇总

1.5 英年早逝的数学、物理学和力学家一览表. 每个人对 "英年早逝" 年龄的理解可能不同, 因为大物理学家麦克斯韦和大数学家闵可夫斯基的原因, 本书将 "英年早逝" 限定在 50 岁以下. 但是, 任何以年龄来进行划分的事情, 往往都有一定的遗憾. 比如,近代力学的奠基人之一,克劳德–路易·纳维 (Claude-Louis Navier, 1785—1836, 终年 51 岁),不但于 1822 年用分子理论建立了流体动力学的基本方程"纳维–斯托克斯方程"的雏形,而且弹性力学的基本方程"拉梅–纳维方程 (Lamé–Navier equation)"也在很大程度上基于他的贡献,像这样一位伟大的力学家的寿命仅仅 51 岁,也是令人唏嘘不已的! 然而令人欣慰的是,纳维在生前获得了他应得的荣誉,他于 1824 年 39 岁上当选为法兰西科学院院士,并培养出像圣维南 (Saint Venant) 那样杰出力学家的学生.

题表 1.5 英年早逝的代表性科学家

姓名	生卒年	主要学术贡献	人生传奇或坎坷遭遇	早逝原因
托里拆利 Evangelista Torricelli	1608—1647 终年 39 岁	水银气压计	托里拆利是伽利略的学生和晚年的助手,1642 年继承伽利略任佛罗伦萨学院数学教授	伤寒
帕斯卡 Blaise Pascal	1623—1662 终年 39 岁	帕斯卡定律	压强或应力的单位用"帕斯卡 (Pa)"命名	从 17 岁直到 39 岁去世,几乎每天都在病痛中度过
格雷果里 James Gregory	1638—1675 终年 36 岁	反射望远镜 级数展开	在其黄金时期,他就任仍完全处于中世纪传统的苏格兰圣安得烈大学的数学教授. 他是这所大学中唯一知晓他那个年代数学新进展的人,尽管他充满着新的数学思想,但周围无人可讨论问题,也看不到新的文献,更无需在课堂上讲授新内容,事实上格雷果里在学术上被孤立了起来.	过度的太阳观测损及目力,终至失明,不久去世
罗杰·柯特斯 Roger Cotes	1682—1716 终年 33 岁	牛顿–柯特斯公式	科特斯去世后,牛顿惋惜地说"如果他还活着,我们还能知道更多些事!"	高烧
布鲁克·泰勒 Brook Taylor	1685—1731 终年 46 岁	泰勒级数展开、有限差分法	两任妻子均死于生产	体质脆弱
华伦海特 Daniel Gabriel Fahrenheit	1686—1736 终年 50 岁	于 1724 年创立华氏温标	1701 年,其父母因食用毒蘑菇双亡	待考

续表

姓名	生卒年	主要学术贡献	人生传奇或坎坷遭遇	早逝原因
摄尔修斯 Anders Celsius	1701—1744 终年 42 岁	于 1742 年创立摄氏温标	他父亲和他本人均为瑞典乌普萨拉大学的教授	肺结核
菲涅尔 Augustin-Jean Fresnel	1788—1827 终年 39 岁	物理光学的缔造者反射定律和折射定律的菲涅耳公式	生于乱世,悲惨的童年	肺结核
格林 George Green	1793—1841 终年 47 岁	格林应变、格林定理、格林函数	曾以业余人士身份发表了一度无人问津的《论数学分析在电磁理论上的应用》. 此文引入了几个重要概念,其中有一条定理与现在的格林定理相似,还有物理中势函数和格林函数的概念. 格林出生贫苦,小时候只读了 1 年的书,几乎全靠自学成才,而且格林在世时,其工作在数学界并不知名.	剑桥大学的传闻是,格林沉溺于酒精而死 (succumbed to alcohol)
卡诺 Sadi Carnot	1796—1832 终年 36 岁	卡诺循环	在 1824—1878 年间,卡诺的热机理论一直没有得到广泛传播. 卡诺生前的好友罗贝林(Robelin)在法国《百科评论》杂志上曾经这样写道:卡诺孤独地生活、凄凉地死去,他的著作无人阅读,无人承认. 卡诺理论的蒙难有如下几个原因:社会的政治压抑;卡诺的早逝;卡诺生前未能进入某一学派.	流行性霍乱
阿贝尔 Niels Henrik Abel	1802—1829 终年 26 岁	群论的先驱	阿贝尔 1826 年 10 月的论文呈交给巴黎科学院,数学家柯西负责审阅,将阿贝尔的手稿遗失. 在挪威驻巴黎的领事就该遗失的手稿提出外交抗议后,柯西于 1830 年阿贝尔去世之后,才将其手稿翻出,1841 年才得以最终出版. 另外, 高斯、勒让德等的疏忽也害了这位天才数学家.	长期贫困、肺结核、出血

续表

姓名	生卒年	主要学术贡献	人生传奇或坎坷遭遇	早逝原因
雅可比 Carl Gustav Jacob Jacobi	1804—1851 终年 46 岁	雅可比行列式、哈密顿–雅可比方程	1843 年时因工作过度而导致健康极度恶化，后前往意大利休养. 回来时，雅可比搬到了柏林，开始领取养老金. 在 1848 年革命期间，雅可比卷入了政治斗争并代表自由党派选举议会候选人失败. 导致雅可比的皇家补助金被取消，但因为他的名气和声誉，很快补助金被恢复.	天花
伽罗瓦 Évariste Galois	1811—1832 终年 20 岁	群论的创立者	1829 年，伽罗瓦将群论的结果呈交给法国科学院，负责审阅的柯西却将文章弄丢了，造成群论晚问世约半个世纪. 详见本书 §20.5.	决斗
傅科 Léon Foucault	1819—1868 终年 48 岁	傅科摆	1862 年他测量到光速为每秒 298,000 千米，与精确值差仅 0.6%.	发展迅速的多发性硬化症
艾森斯坦 Gotthold Eisenstein	1823—1852 终年 29 岁	艾森斯坦判别法	曾因民主聚会而被捕	肺结核
黎曼 Georg Friedrich Bernhard Riemann	1826—1866 终年 39 岁	黎曼几何、黎曼流形、黎曼假设（猜想）	有人将数学家分为五类，其中第一类是指那些拥有永恒创造力的绝世天才，他们能够不断攻克基本的难题，变革现有领域，开创新方向. 而一位第一类数学家对于年轻数学家的影响超过所有其他类别数学家的总和，黎曼无疑属于第一类.	长期贫困、胸膜炎和肺结核
麦克斯韦 James Clerk Maxwell	1831—1879 终年 48 岁	麦克斯韦方程组	麦克斯韦死、爱因斯坦生 (1879 年)，爱因斯坦说他站在了麦克斯韦的肩膀上.	肺结核

续表

姓名	生卒年	主要学术贡献	人生传奇或坎坷遭遇	早逝原因
克莱布什 Alfred Clebsch	1833—1872 终年39岁	19世纪代数几何德国学派的领导者之一	1854年以流体动力学方面的论文获柯尼斯堡大学博士学位	白喉 (diphteria)
克利福德 William Kingdon Clifford	1845—1879 终年33岁	克利福德代数	1874年成为英国皇家学会会员,时年29岁. 克利福德去世后的第11天, 爱因斯坦出生.	肺结核
卡斯提利亚诺 Carlo Alberto Castigliano	1847—1884 终年37岁	卡斯提利亚诺第一定理和第二定理	出生于贫困家庭.	肺炎
霍普金森 Bertram Hopkinson	1874—1918 终年44岁	霍普金森压杆, 材料的动态力学行为	其父John Hopkinson和一弟和两妹在1898年8月27日登山时全部罹难. 1918年8月26日, 他驾驶的Bristol战斗机在飞往伦敦的途中因气候恶劣而失事. 此时他年仅44岁, 殉难的日子同他父亲和弟妹在登山事故中去世的日子几乎正好相距20年! 令人唏嘘不止.	飞机失事
科瓦列夫斯卡娅 Sofia Kovalevskaya	1850—1891 终年41岁	科瓦列夫斯卡娅陀螺	斯德哥尔摩大学讲师, 1889年晋升为教授, 是世界上第一位女数学教授; 1889年, 她还成为了俄国圣彼得堡科学院第一位女院士.	肺炎
赫兹 Heinrich Hertz	1857—1894 终年36岁	证实电磁波的存在、接触力学	频率单位以他的姓氏命名. 他的墓志铭: 才华横溢, 性格坚毅, 用其短暂的一声, 解决了很长时间以来许多物理学家想解决又没有解决的许多重大问题.	败血症
闵可夫斯基 Hermann Minkowski	1864—1909 终年44岁	闵可夫斯基时空	爱因斯坦的数学教授, 因爱因斯坦经常逃课, 称其为 "lazy dog".	急性阑尾炎

续表

姓名	生卒年	主要学术贡献	人生传奇或坎坷遭遇	早逝原因
威尔伯·莱特 Wilbur Wright	1867—1912 终年45岁	现代航空之父 (fathers of modern aviation)	弥尔顿·莱特后来在日记中是这么描述他儿子 (Wilbur Wright) "短暂而富有成果的一生，聪明绝顶而又性情冷静，自力更生而又十分谦虚，明辨是非而又坚持真理，（总而言之）他生的伟大死的光荣." Milton Wright wrote later about his son in his diary: "A short life, full of consequences. An unfailing intellect, imperturbable temper, great self-reliance and as great modesty, seeing the right clearly, pursuing it steadfastly, he lived and died."	伤寒症 (typhoid fever)
加托 René Gâteaux	1889—1914 终年25岁	加托导数	加托去世后，他的部分研究成果由数学家 Paul Lévy 帮助其出版.	1914年10月3日凌晨1点被德军杀害
约尔旦 Philip Jourdain	1879—1919 终年39岁	1908年创立约尔旦原理（虚速度原理、虚功率原理）	将马赫的德文版的《力学史评》翻译为英文版.	
拉马努扬 Srinivasa Ramanujan	1887—1920 终年32岁	数学家 沉迷数论，尤爱牵涉 π、质数等数学常数的求和公式，以及整数分拆.	未受过正规教育，自学成才. 大数学家哈代曾提到他本人对数学最大的贡献，就是"发现了拉马努扬". 他认为拉马努扬的天才可以和欧拉等巨匠相比. 拉马努扬最后在剑桥大学三一学院获得教席，亦成为英国皇家学会会员 (FRS).	自幼体弱多病、肺结核
邦德 Wilfrid Noel Bond	1897—1937 终年39岁	无量纲数邦德数 Bo 以其命名		急性肠道疾病

续表

姓名	生卒年	主要学术贡献	人生传奇或坎坷遭遇	早逝原因
马约拉纳 Ettore Majorana	1906—1938? 1938年失踪时不满32岁	马约拉纳费米子	1938年3月25日他给家人和他任职的那不勒斯大学物理研究所所长卡瑞利 (Antonio Carrelli) 各留了一封短信后,就登上了一艘开往西西里首府巴勒莫的邮船.一般人和警方都把这两封信解读为绝命书.	
达姆科勒 Gerhard Damköhler	1908—1944 终年36岁	达姆科勒数	被形容为"每天工作25小时".	工作狂
图灵 Alan Turing	1912—1954 终年41岁	计算机科学之父 人工智能之父	以他名字命名的"图灵奖"堪称是计算机科学的诺贝尔奖.	同性恋
富兰克林 Rosalind Elsie Franklin	1920—1958 终年37岁	她所拍摄的DNA晶体衍射图片"照片51号",以及关于此物质的相关数据,是沃森与克里克建构DNA结构模型的关键	克里克在一篇纪念DNA结构发现40周年的文章中说道:"富兰克林的贡献没有受到足够的肯定,她清楚的阐明了两种型态的DNA,并且确定出A型DNA的密度、大小与对称性.";46年后沃森演讲里吐露:"我看到了罗莎琳的X射线照片……我并没有跑进去翻抽屉,把它偷走,是别人拿来给我看的,而且也是别人告诉我螺距为34埃……一个月后我们便做出这个结构……罗莎琳的照片确实是关键……"	卵巢癌
别列津斯基 Vadim L'vovich Berezinskii	1935—1980 终年44岁	别列津斯基-科斯特利茨-索利斯相变		长期困难疾病 (long difficult illness)
米尔扎哈尼 Maryam Mirzakhani	1977—2017 终年40岁	对黎曼曲面和及其模空间的动力学和几何学研究被授予菲尔兹奖	2014年获得菲尔兹奖,是获得该奖的唯一的女数学家.	乳腺癌

时年 44 岁的科里奥利 (Gustave Gaspard de Coriolis, 1792—1843, 终年 51 岁) 在 1836 年纳维去世后, 接替了纳维在巴黎路桥学校 (École Nationale des Ponts et Chaussées) 和法国科学院的位置. 十分巧合的是, 科里奥利的寿命也仅仅 51 岁.

然而活得长就一定好吗? 1943 年的情人节那一天 (2 月 14 日), 一位以前所未有的方式深刻影响了 20 世纪数学的 81 岁老人希尔伯特在德国哥廷根的家中孤独地告别了他毕生的 "情人"——数学——与世长辞. 希尔伯特堪称是数学史上最伟大的导师, 被广泛地称为 "数学界的亚历山大" 和 "数学的无冕之王", 用他学生和学术继承人赫尔曼·外尔的话说: "我仿佛听到, 希尔伯特作为花衣魔笛手, 所吹奏的来自远方的甜蜜笛声, 引诱着如此众多的像鼠一般的追随者和他一起跳进数学的深河徜徉."

虽然希尔伯特的学生以及学生的学生遍布世界各地, 但他的葬礼却只有十来个人出席, 而且只有两人来自学术界. 他最早的学生之一, 理论物理学家索末菲 (Arnold Sommerfeld, 1868—1951), 从慕尼黑赶来, 站在棺柩边讲述了他的工作. 战争阻碍了世界对这位数学领袖的及时悼念, 很多地方 (比如遭轰炸的英国) 是半年多以后才得到了不准确的消息. 因而, 希尔伯特的去世也被数学界称为 "最凄凉的死".

他一生中最知己的挚友闵可夫斯基早在 1909 年就已先他而去了. 希尔伯特和闵可夫斯基早期的学生玻恩 (Max Born, 1882—1970, 1954 年诺贝尔物理学奖获得者) 评价道: "希尔伯特比他的朋友多活了三十九年. 这使他又做出了重大的成果. 但是, 谁会说他在黑暗的纳粹时代孤独地死去不是比闵可夫斯基死于全盛的壮年时期更具有悲剧性呢?"

1.6 有关黎曼猜想的一个生动的典故. 英国大数学家哈代 (Godfrey Hardy, 1877—1947) 当时在丹麦有一位很要好的数学家朋友叫做哈罗德·玻尔 (Harald Bohr, 1887—1951), 他是大物理学家尼尔斯·玻尔 (Niels Bohr, 1885—1962, 1922 年诺贝尔物理学奖获得者) 的弟弟. 哈罗德·玻尔对黎曼猜想也有浓厚的兴趣, 曾与德国数学家兰道 (Edmund Landau, 1877—1938) 一起研究黎曼猜想. 后来哈代很喜欢与哈罗德·玻尔共度暑假, 一起讨论黎曼猜想. 他们对讨论都很投入, 哈代常常要待到假期将尽才匆匆赶回英国. 结果有一次当他赶到码头时, 很不幸地发现只剩下一条小船可以乘坐了. 从丹麦到英国要跨越宽达几百公里的北海 (North Sea), 在那样的汪洋大海中乘坐小船可不是闹着玩的事情, 弄得好算是浪漫刺激, 弄不好就得葬身鱼腹. 为了旅途的平安, 信奉上帝的乘客们大都忙着祈求上帝的保佑. 哈代却是一个坚决不信上帝的人, 不仅不信, 有一年他还把向大众证明上帝不存在列入自己的年度六大心愿之中, 且排名第三 (排名第一的是证明黎曼猜想). 不过在面临生死攸关的旅程之时哈代也没闲着, 他给哈罗德·玻尔发去了一张简短的明信片, 上面只有一句话: "我已经证明了黎曼猜想." 哈代果真已经证明了黎曼猜想吗? 当然不是. 那他为什么要发那样一张明信片呢? 回到英国后他向哈罗德·玻尔解释了原

第 1 章 基础知识

因，他说如果那次他乘坐的小船真的沉没了，那人们就只好相信他真的证明了黎曼猜想. 但他知道上帝是肯定不会把这么巨大的荣誉送给他这样一个坚决不信上帝的人的，因此上帝是一定不会让他的小船沉没的 (典故的英文版本: Hardy stayed in Denmark with Bohr until the very end of the summer vacation, and when he was obliged to return to England to start his lectures there was only a very small boat available · · · . The North Sea can be pretty rough, and the probability that such a small boat would sink was not exactly zero. Still, Hardy took the boat, but sent a postcard to Bohr: "I proved the Riemann Hypothesis. G.H. Hardy." If the boat sinks and Hardy drowns, everybody must believe that he has proved the Riemann Hypothesis. Yet God would not let Hardy have such a great honor and so He will not let the boat sink).

1.7 动量空间在拉格朗日方程中的应用.

在拉格朗日力学中 [1.11]，由表 4.1，拉格朗日量是在位形空间中给出: $\mathcal{L}(\boldsymbol{q}, \dot{\boldsymbol{q}}, t)$，所谓位形空间是指由 $\boldsymbol{q} = \boldsymbol{q}(q_1, q_2, \cdots, q_n)$ n 元广义坐标组成的空间. 拉格朗日方程由 (4-2) 式给出. 本思考题的出发点是: 拉格朗日方程 (4-2) 式是否可在动量空间中给出? 问题的关键是将 $\mathcal{L}(\boldsymbol{q}, \dot{\boldsymbol{q}}, t)$ 变换为 $\mathcal{L}'(\boldsymbol{p}, \dot{\boldsymbol{p}}, t)$，这里 $\boldsymbol{p} = \dfrac{\partial \mathcal{L}}{\partial \dot{\boldsymbol{q}}}$ 为广义动量. 所谓动量空间是指由 $\boldsymbol{p} = \boldsymbol{p}(p_1, p_2, \cdots, p_n)$ n 元广义动量组成的空间. 拉格朗日量 $\mathcal{L}(\boldsymbol{q}, \dot{\boldsymbol{q}}, t)$ 的全微分为

$$\mathrm{d}\mathcal{L} = \sum_{i=1}^{n} \left(\frac{\partial \mathcal{L}}{\partial q_i} \mathrm{d}q_i + \frac{\partial \mathcal{L}}{\partial \dot{q}_i} \mathrm{d}\dot{q}_i \right) + \frac{\partial \mathcal{L}}{\partial t} \mathrm{d}t \tag{s1-1}$$

由于 $\dfrac{\partial \mathcal{L}}{\partial \dot{q}_i} = p_i$, $\dfrac{\partial \mathcal{L}}{\partial q_i} = \dfrac{\partial (T-V)}{\partial q_i} = -\dfrac{\partial V}{\partial q_i} = \dot{p}_i$, 代入 (s1-1) 式，得到

$$\begin{aligned}\mathrm{d}\mathcal{L} &= \sum_{i=1}^{n} (\dot{p}_i \mathrm{d}q_i + p_i \mathrm{d}\dot{q}_i) + \frac{\partial \mathcal{L}}{\partial t} \mathrm{d}t \\ &= \sum_{i=1}^{n} \mathrm{d}(q_i \dot{p}_i + \dot{q}_i p_i) - \sum_{i=1}^{n} (q_i \mathrm{d}\dot{p}_i + \dot{q}_i \mathrm{d}p_i) + \frac{\partial \mathcal{L}}{\partial t} \mathrm{d}t\end{aligned} \tag{s1-2}$$

将 (s1-2) 式移项，有

$$\mathrm{d}\left(\mathcal{L} - \sum_{i=1}^{n}(q_i \dot{p}_i + \dot{q}_i p_i)\right) = -\sum_{i=1}^{n}(q_i \mathrm{d}\dot{p}_i + \dot{q}_i \mathrm{d}p_i) + \frac{\partial \mathcal{L}}{\partial t}\mathrm{d}t \tag{s1-3}$$

引入勒让德变换 [1.11]:

$$\mathcal{L}' = \mathcal{L} - \sum_{i=1}^{n}(q_i \dot{p}_i + \dot{q}_i p_i) \tag{s1-4}$$

由 (s1-4) 式得到 $q_i = -\dfrac{\partial \mathcal{L}'}{\partial \dot{p}_i}$，$\dot{q}_i = -\dfrac{\partial \mathcal{L}'}{\partial p_i}$，观察此两个关系式，得到在动量空间中表示的拉格朗日方程：

$$\frac{\mathrm{d}}{\mathrm{d}t}\frac{\partial \mathcal{L}'}{\partial \dot{p}_i} - \frac{\partial \mathcal{L}'}{\partial p_i} = 0 \qquad (\text{s1-5})$$

1.8 薛定谔方程准确地描述了我们今天所知原子的每一种行为，这是整个化学和绝大部分物理学的基础. 虚数 $\mathrm{i} = \sqrt{-1}$ 意味着大自然是以复数而不是实数的方式运行. 这一发现让薛定谔和其他所有人耳目一新. 薛定谔记得，当时，他 14 岁大的 "女朋友" 伊萨·荣格尔 (Itha Junger) 曾对他说："嗨, 开始时, 你从来没想过会出现这么多有意义的结果吧？(Hey, you never even thought when you began that so much sensible stuff would come out of it.)"[1.16]

1.9 拉格朗日力学和哈密尔顿力学，从几何方面考虑，本质上也是流形理论. 流形 (manifold) 是局部具有欧氏空间性质的空间. 而实际上欧氏空间就是流形最简单的实例. 流形在数学中用于描述几何形体，它们提供了研究可微性的最自然的舞台. 物理上, 经典力学的相空间和构造广义相对论的时空模型的四维伪黎曼流形都是流形的实例. 他们也用于位形空间 (configuration space). 环面 (torus) 就是双摆的位形空间.

1.10 题表 1.10 中几个有关信息和网络的定律.

题表 1.10

定律名称	定理内容	提出人和提出年代	提出者肖像
摩尔定律 Moore's Law	当价格不变时，集成电路上可容纳的元器件的数目，约每隔 18—24 个月便会增加一倍，性能也将提升一倍.	戈登·摩尔 (Gordon Moore, 1929—), 1965 年提出, 1975 年修正	
贝尔定律 Gordon Bell's Law	计算机每 10 年产生新一代，其设备或用户数增加 10 倍.	戈登·贝尔 (Gordon Bell, 1934—), 1972 年	
吉尔德定律 Gilder's Law	在未来 25 年，主干网的带宽每 6 个月增长一倍，其增长速度是莫尔定律预测的 CPU 增长速度的 3 倍并预言将来上网会免费. Total bandwidth of communications systems triples every 12 months.	乔治·吉尔德 (George Gilder, 1939—), 数字时代三大思想家之一	

续表

定律名称	定理内容	提出人和提出年代	提出者肖像
迈特卡夫定律 Metcalfe's Law	网络的价值与网络使用者数量的平方成正比 (物以多为贵).	罗伯特 · 迈特卡夫 (Robert Metcalfe, 1947—)	

1.11 有关科学与诗歌之间的类比.

题表 1.11

科学家	名人名言	头像
培根 Francis Bacon 1561—1626 终年 65 岁	Histories make men wise; poets, witty; the mathematics, subtle; natural philosophy, deep; moral, grave; logic and rhetoric, able to contend. 历史使人聪明, 诗歌使人机智, 数学使人精细, 哲学使人深邃, 道德使人严肃, 逻辑与修辞使人善辩.	
哈密顿 William Rowan Hamilton 1805—1865 终年 60 岁	拉格朗日展现出以一个惊世骇俗的公式描述了系统运动万变的结果; 拉格朗日方法的美在于它完全容纳了其结果的尊严, 以至于他的伟大工作仿佛像一种科学诗篇. (the beauty of the method so suiting the dignity of the results, as to make of his great work a kind of scientific poem).	
魏尔斯特拉斯 Karl Weierstrass 1815—1897 终年 81 岁	It is true that a mathematician who is not also something of a poet will never be a perfect mathematician. 确实, 一个没有几分诗人气质的数学家, 永远不会成为一个完美的数学家.	
麦克斯韦 James Clerk Maxwell 1831—1879 终年 48 岁	James Clerk Maxwell describes Fourier's work as a great mathematical poem. 詹姆斯 · 克拉克 · 麦克斯韦将傅里叶的工作形容为一部伟大的数学史诗.	

续表

科学家	名人名言	头像
柯瓦列夫斯卡娅 Sofia Kovalevskaya 1850—1891 终年 41 岁	It is impossible to be a mathematician without being a poet in soul. 数学家的灵魂深处就是诗人.	
普朗克 Max Planck 1858—1947 终年 89 岁	Experiments are the only means of knowledge at our disposal. The rest is poetry, imagination. 实验是我们掌握知识的唯一手段，其余的都是诗歌和想象力.	
达西·汤姆森 D'Arcy Wentworth Thompson 1860—1948 终年 88 岁	The harmony of the world is made manifest in Form and Number, and the heart and soul and all the poetry of Natural Philosophy are embodied in the concept of mathematical beauty. 世界的和谐表现在形式和数量上，而自然哲学的心灵和一切诗意都体现在数学美的概念中.	
哈代 Godfrey Harold Hardy 1877—1947 终年 70 岁	A mathematician, like a painter or a poet, is a maker of patterns. If his patterns are more permanent than theirs, it is because they are made with ideas. A painter makes patterns with shapes and colors, a poet with words. Beauty is the first test: there is no permanent place in the world for ugly mathematics. 数学家，就像画家、诗人一样，都是模式的创制者. 要说数学家的模式比画家、诗人的模式更长久，那是因为数学家的模式由思想组成，而画家以形状和色彩创制模式，诗人则以言语和文字造型. 数学的优美至关重要，丑陋参差的数学，在世界毫无立足之地.	
爱因斯坦 Albert Einstein 1879—1955 终年 76 岁	Pure mathematics is, in its way, the poetry of logical ideas. 纯粹数学，就其本质而言，是逻辑思想的诗篇.	

科学家	名人名言	头像
狄拉克 Paul Adrien Maurice Dirac 1902—1984 终年 82 岁	The aim of science is to make difficult things understandable in a simpler way; the aim of poetry is to state simple things in an incomprehensible way. The two are incompatible. 科学的目的是用更简单的方法使困难的事情变得容易理解;诗歌的目的是用一种难以理解的方式来表达简单的事情. 这两者是不相容的. Anecdotally, when Oppenheimer was working at Göttingen, Dirac supposedly came to him one day and said: "Oppenheimer, they tell me you are writing poetry. I do not see how a man can work on the frontiers of physics and write poetry at the same time. They are in opposition. In science you want to say something that nobody knew before, in words which everyone can understand. In poetry you are bound to say··· something that everybody knows already in words that nobody can understand." 有一则逸事说的是奥本海默在哥廷根大学工作期间, 有一天狄拉克来造访并对他说: "奥本海默, 我听人说你在写诗. 我不明白一个人怎么能同时从事物理前沿的研究和写诗. 它们是完全对立的. 科学研究是你要用大家都能理解的语言来表述大家都不知道的东西. 而诗歌创作是你要用谁也不理解的词汇来表达大家都知道的东西."	

1.12 有关 "简单性 (simplicity)" 的论述.

题表 1.12

科学家	名人名言	头像
达·芬奇 Leonardo da Vinci 1452—1519 终年 67 岁	Simplicity is the ultimate of sophistication. 简单 (简约) 是复杂的最高境界. 大道至简; 至繁归于至简; 极致的简约.	
牛顿 Isaac Newton 1642—1726 终年 84 岁	Nature is simple, and affects not the pomp of superfluous causes. 自然是简单的, 其运行机理/原因并不会纷繁复杂.	

续表

科学家	名人名言	头像
希尔伯特 David Hilbert 1862—1943 终年 81 岁	Mathematics is a game played according to certain simple rules with meaningless marks on pape. 数学是根据一些简单规则把玩纯粹符号的纸上游戏.	
爱因斯坦 Albert Einstein 1879—1955 终年 76 岁	Everything should be made as simple as possible, but not simpler. 凡事应力求简单，但不能过分简单.	

1.13 语惊四座的名人名言.

题表 1.13

科学家	名人名言	头像
达·芬奇 Leonardo da Vinci 1452—1519 终年 67 岁	In the stufy of the sciences which depend upon mathematics, those who do not consult nature but authors, are not the children of nature but only her grandchildren. 在以数学为依据的科学的研究中，如果有些人不直接向自然界请教而是向书本的作者请教，那么，他就不是自然界的儿子而只是孙子了.	
克洛德·贝尔纳 Claude Bernard 1813—1878 终年 67 岁	Art is I, science is we. 艺术是我，科学是我们.	
切比雪夫 Pafnuty Chebyshev 1821—1894 终年 73 岁	To isolate mathematics from the practical demands of the sciences is to invite the sterility of a cow shut away from the bulls. 使数学脱离实际需要，就好比把母牛关起来不让她接触公牛.	

续表

科学家	名人名言	头像
贝多芬 Ludwig van Beethoven 1770—1827 终年 56 岁	I love a tree more than a man. 比起人，我更喜欢树.	
韦伊 André Weil 1906—1998 终年 92 岁	Every mathematician worthy of the name has experienced, if only rarely, the state of lucid exaltation in which one thought succeeds another as if miraculously, and in which the unconscious (however one interprets this word) seems to play a role. In a famous passage, Poincaré describes how he discovered Fuchsian functions in such a moment. About such states, Gauss is said to have remarked as follows: "Procreare jucundum (to conceive is a pleasure)"; he added, however, "sed parturire molestum (but to give birth is painful)." Unlike sexual pleasure, this feeling may last for hours at a time, even for days. Once you have experienced it, you are eager to repeat it but unable to do so at will, unless perhaps by dogged work which it seems to reward with its appearance. 每一位名副其实的数学家都曾经历过一种极少出现的亢奋状态，在这种状态下，一个思想奇迹般地接替着另一个思想，潜意识似乎在其中扮演了一定角色. 在一篇著名的文章中，庞加莱描述了他是如何在这样的时刻发现富克斯函数的. 关于这种状态，高斯说过这样的话："怀孕是一种乐趣"；然而，他补充道："但分娩是痛苦的". 与性快感不同，这种感觉可能会持续数小时，乃至数天. 一旦你经历了它，你渴望重复它，但却不能随心所欲地重复，除非你坚持不懈地工作，它似乎会以它的出现作为回报.	

1.14 达·芬奇的局限性 [1.17].

在克莱因《古今数学思想》第 11 章 "文艺复兴" 第 5 节 "要求科学改革的呼声" 中，提到了达·芬奇虽然作为一个文艺复兴时期的天才人物，但同样具有局限性，克莱因指出："达·芬奇并没有掌握真正的科学方法. 他没有方法论，也没有以任何哲学作为基础. 他不像伽利略那样自觉追求定量规律.""当你审阅达·芬奇的笔记本时,你会发现他的数学知识是多么少,而他处理问题的方法完全是经验的、直观的."

1.15 著名科学家个性鲜明的称谓.

题表 1.15

科学家	个性化称谓	历史背景
赫尔曼·外尔 Hermann Weyl 1885—1955 终年 70 岁	A lone wolf in Zürich 苏黎世一只孤独的狼	外尔曾经用"苏黎世一只孤独的狼"来描述被自己崇拜的偶像爱因斯坦批评时感觉失望和迷茫的心态. 外尔在黎曼几何二次型度规的基础上, 加上了一个一次形式来包容电磁场, 这在数学上看起来是非常美妙的一招, 并且还有它的"纯粹无穷小几何"之解释, 也闪耀着新思想的火花. 因此, 外尔兴致勃勃地将他的文章寄给了爱因斯坦. 爱因斯坦一方面赞赏外尔几何是"天才之作、神来之笔", 一方面又从物理的角度, 强烈批评这篇文章脱离了物理的真实性. 面对爱因斯坦的反对意见, 外尔失望和落寞, 他以朋友海塞的小说《荒原狼》的主人翁自比, 他在给爱因斯坦的回信中说:"我们之间已经战斧高悬, 但仍然阻挡不了我对您真诚的敬意."
阿基米德 Archimedes c.287— c.212 BC 终年约 75 岁	Archimedes was a lonely sort of eagle [1.13]. 阿基米德是一只孤独的鹰.	他与牛顿和高斯被西方世界评价为有史以来最伟大的三位数学家.

赫尔曼·外尔的肖像

1.16 杰出科学家和工程师中著名的"三冯".

题图 1.16　左起:冯·卡门、冯·诺依曼、冯·布劳恩

题表 1.16

科学家	学术荣誉	学术贡献和名言
冯·卡门 Theodore von Kármán 1881—1963 享年 81 岁	1962 年美国首枚国家科学奖章.	匈牙利裔. The scientist describes what is; the engineer creates what never was. 科学家研究已有的世界, 工程师创造未来的世界.
冯·诺依曼 John von Neumann 1903—1957 终年 53 岁	20 世纪最重要的数学家之一, 在现代计算机、博弈论、核武器和生化武器等领域内的科学全才之一, 被称为"计算机之父"(冯·诺依曼结构) 和"博弈论之父". 戴森将冯·诺依曼划归到"青蛙"[1.16]而不是"鸟", 这令很多人感到意外.	匈牙利裔. 冯·诺依曼尽快名声十分显赫, 但生前确实没有获得过与其贡献相称的奖励. 在他离世后的第六年, 匈牙利裔 1963 年度诺贝尔物理学奖获得者尤金·维格纳在获奖演讲时, 被问及匈牙利如何在同时代培养出那么多的天才时, 维格纳回答说: 冯·诺依曼是唯一的天才 (When asked why the Hungary of his generation had produced so many geniuses, Wigner, who won the Nobel Prize in Physics in 1963, replied that von Neumann was the only genius).
冯·布劳恩 Wernher von Braun 1912—1977 终年 65 岁	1975 年获美国国家科学奖章. 他的墓碑上刻有圣经诗篇 19:1 "诸天述说神的荣耀, 穹苍传扬他的手段."(Von Braun's gravestone mentions Psalm 19:1: "The heavens declare the glory of God; and the firmament sheweth his handywork.")	德裔. 20 世纪航天事业的先驱之一. 曾是 V2 火箭的总设计师. 战败后, 美国将他和他的设计小组带到美国. 移居美国后任美国国家航空航天局 (NASA) 的空间研究开发项目的主设计师, 主持设计了阿波罗 4 号的运载火箭土星 5 号. NASA 用以下的话来形容冯·布劳恩:"毋庸置疑, 他是史上最伟大的火箭科学家. 他最大成就是在担任 NASA 马歇尔太空飞行中心总指挥时, 主持土星 5 号的研发, 成功地在 1969 年 7 月首次达成人类登陆月球的壮举."

1.17 研究表明[1.18], 一些激情洋溢的梵高画作如《星夜》《麦田群鸦》《夕阳下两位农妇开掘积雪覆盖的田地》等, 显示出类似于湍流 (turbulence) 中观察到的标度特性, 这表明这些画作可以以这样一种现实主义的方式来反映湍流的指纹, 这种方法甚至与表征这一现象的数学模型一致. 具体地说, 该文的作者们发现以距离 R 分开的点 (像素) 的亮度波动的概率分布函数 (PDF) 与湍流中的柯尔莫哥洛夫标度律 (Kolmogorov scaling, 见本书 §59.5 和 §59.6) 一致. 此外, 梵高最紊乱的画作恰逢创作于这位艺术家长期的精神错乱 (turbulent mind) 时期.

1.18 一些著名学者关于"孤独"的论述, 见题表 1.18.

题表 1.18

学者	肖像	名言
亚里士多德 Aristotle 384—322 BC 终年约 62 岁		Whosoever is delighted in solitude is either a wild beast or a god. 喜爱孤独者，非兽即神.
阿道司·赫胥黎 Aldous Huxley 1894—1963 终年 69 岁		The more powerful and original a mind, the more it will incline towards the religion of solitude. 越伟大、越有独创精神的人，就越喜欢孤独.
爱迪生 Thomas A. Edison 1847—1931 终年 84 岁		The best thinking has been done in solitude. The worst has been done in turmoil. 最好的思想在孤独中诞生；最糟的思想在混乱中产生.
爱因斯坦 Albert Einstein 1879—1955 终年 76 岁		Be a loner. That gives you time to wonder, to search for the truth. Have holy curiosity. Make you life worth living. 做一个孤独者. 这给了你时间去思考，去寻找真相. 有神圣的好奇心. 让你的生命有价值. 爱因斯坦对牛顿的评价： He stands before us strong, certain and alone. 他以坚强、自信和孤独的姿态屹立在我们面前.

1.19 有关"上帝–魔鬼"的著名论述，见题表 1.19.

题表 1.19

学者	肖像	名言
克罗内克 Leopold Kronecker 1823—1891 终年 68 岁		"God made the integers; all else is the work of man". (But surely the devil let them conceive partial differential equations!) 上帝创造了整数；其余一切都是人造的. (但魔鬼肯定让他们构思出偏微分方程！) 注：引号部分为克罗内克所言.

第 1 章 基础知识

续表

学者	肖像	名言
劳伦斯·布拉格 William Lawrence Bragg 1890—1971 终年 81 岁		God runs electromagnetics on Monday, Wednesday, and Friday by the wave theory, and the devil runs it by quantum theory on Tuesday, Thursday, and Saturday. 上帝在周一、周三、周五用波理论来操控电磁学；而魔鬼则在周二、周四和周六用量子理论来操控电磁学.
沃尔冈·泡利 Wolfgang Pauli 1900—1958 终年 58 岁		God made the bulk; the surface was invented by the devil. 上帝创造了固体，魔鬼发明了表面.
韦伊 André Weil 1906–1998 终年 92 岁		God exists since mathematics is consistent, and the Devil exists since we cannot prove it. 上帝存在是因为数学具有一致性，魔鬼存在是因为我们无法证明它.
里夫林 Ronald Samuel Rivlin 1915—2005 终年 90 岁		God created the professor, and the devil created the colleague[1.19]. 上帝创造了教授，魔鬼创造了同事.

1.20 有关科学家类型的论述，见题表 1.20.

题表 1.20

论述者	肖像	有关科学家类型的论述
约瑟夫·汤姆逊爵士 Sir Joseph John Thomson 1856—1940 终年 83 岁		有一些科学巨匠的魅力在于他们对一个课题说了第一个词，引入了后来被证明是卓有成效的新思想；还有一些科学巨匠的魅力也许在于他们在那个课题上说了最后一个词，使其变得具有逻辑一致性和明确性. 我认为从气质上讲，瑞利勋爵确实属于第二类. There are some great men of science whose charm consists in having said the first word on a subject, in having introduced some new idea which has proved fruitful; there are others whose charm consists perhaps in having said the last word on a subject, and who reduced the subject to logical consistency and clearness. I think, by temperament, Lord Rayleigh really belonged to the second group.

续表

论述者	肖像	有关科学家类型的论述
弗里曼·戴森 Freeman John Dyson 1923—2020		有些数学家是鸟，其他的则是青蛙. 鸟翱翔在高高的天空，俯瞰延伸至遥远地平线的广袤的数学远景. 他们喜欢那些统一我们思想、并将不同领域的诸多问题整合起来的概念. 青蛙生活在天空下的泥地里，只看到周围生长的花儿. 他们乐于探索特定问题的细节，一次只解决一个问题. 我碰巧是一只青蛙，但我的许多最好朋友都是鸟[1.16]. Some mathematicians are birds, others are frogs. Birds fly high in the air and survey broad vistas of mathematics out to the far horizon. They delight in concepts that unify our thinking and bring together diverse problems from different parts of the landscape. Frogs live in the mud below and see only the flowers that grow nearby. They delight in the details of particular objects, and they solve problems one at a time. I happen to be a frog, but many of my best friends are birds.

1.21 经济学家张五常认为，在社会科学中只有经济学是公理性的实证科学. 正因为是公理性，经济学可以做事前推断，也就是常说的预测. 经济科学只有三个公理：一是需求定律，二是成本概念，三是竞争含义. 张五常特别强调说，五十年来他写下的无数的推断或解释的经济学文章，来来去去不过是用上述的三个简单公理作为基础，只是他要花几十年的心血才能一般性地做出上述的理论归纳.

参 考 文 献

[1.1] Newton I. Philosophiae Naturalis Principia Mathematica. London, 1687.

[1.2] Truesdell C. Six Lectures on Modern Natural Philosophy. Berlin: Springer, 1966: 83-108.

[1.3] Noll W. On the past and future of natural philosophy. Journal of Elasticity, 2006, 84: 1-11.

[1.4] 钱学森. 现代力学 —— 在 1978 年全国力学规划会上的发言. 力学与实践, 1979, 1: 4-9.

[1.5] Hilbert D. Mathematische probleme. Göttinger Nachrichten, 1900, 3: 253-297.

[1.6] 赵亚溥. 近代连续介质力学. 北京: 科学出版社, 2016.

[1.7] Landau L D, Lifshitz E M. Fluid Mechanics (2nd English Edition). Oxford: Pergamon Press, 1987.

[1.8] Landau L D, Lifshitz E M. Theory of Elasticity (3rd English Edition). Oxford: Pergamon Press, 1986.

[1.9] Feynman R P, Leighton R B, Sands M. The Feynman Lectures on Physics, Vol. 2: Mainly Electromagnetism and Matter. Reading, MA: Addison-Wesley, 1979.

[1.10] Sommerfeld A. Mechanics of Deformable Bodies – Lectures on Theoretical Physics Volume II. New York: Academic Press, 1964.

[1.11] 赵亚溥. 力学讲义. 北京: 科学出版社, 2018.

[1.12] Lagrange J L. Méchanique Analitique. Paris: La Veuve Desaint, 1788.

[1.13] Bell E T. Men of Mathematics: The Lives and Achievements of the Great Mathematicians from Zero to Poincaré. New York: Simon and Schuster, 1986. (中译本: 埃里克·坦普尔·贝尔. 数学大师: 从芝诺到庞加莱 (徐源译). 上海: 上海科技教育出版社, 2012.)

[1.14] Ballarini R. The da Vinci-Euler-Bernoulli beam theory? Mechanical Engineering Magazine Online, 2003.

[1.15] Bernoulli J. Curvatura Laminae Elasticae. Acta Eruditorum, 1694, 262-276.

[1.16] Dyson F. Birds and frogs. Notices of the AMS, 2009, 56: 212-223.

[1.17] Kline M. Mathematical Thought from Ancient to Modern Times: Volumes 1-3. New York: Oxford University Press, 1972.

[1.18] Aragón J L, Naumis G G, Bai M, Torres M, Maini P K. Turbulent luminance in impassioned van Gogh paintings. Journal of Mathematical Imaging and Vision, 2008, 30: 275-283.

[1.19] Rivlin R S, Barenblatt G I, Joseph D D. Collected Papers of RS Rivlin. New York: Springer Science & Business Media, 1997.

第 2 章 矢量分析

一方面，牛顿力学又被称为矢量力学 (vector mechanics)，连续介质力学又是牛顿第二定律作用于微元体后，再对时空进行积分；第二方面，矢量是一阶张量. 所以，能够熟练掌握矢量运算对于质点力学和连续介质力学均十分重要.

"Vector" 中国数学界称作 "向量"，而在中国物理界和工程界则被广泛地称作 "矢量". 在中国力学界 "矢量" 和 "向量" 二者混用，究其原因，中国力学的发展有的起源与数学力学，有的则是偏重于工程应用. 本书和本课程主要采用 "矢量" 一词，但有时必须尊重数学界的传统，用 "向量" 一词，比如：向量空间.

国家名词委员会为了能够统一 "vector" 的汉译，曾在 20 世纪 90 年代初，专门召开会议，由名词委员会主任钱三强 (1913—1992) 亲自主持，结果没有能够统一，在此次会议上，钱三强作为物理学家，并未发表任何有倾向性的意见. 由此可见科技名词的统一并不是一件轻松的事情.

首先让我们从作为二阶张量的应力的定义说起：

$$\sigma \triangleq \lim_{\Delta A \to 0} \frac{f}{\Delta A} \tag{A2-1}$$

式中，f 为矢量，$\Delta A = (\Delta A) n$ 为方向沿其外法线 n 的面积矢量. (A2-1) 式明确地告诉我们，一个矢量被另外一个矢量除，便得到一个二阶张量，像柯西应力张量、第一类皮奥拉–基尔霍夫应力 (Piola–Kirchhoff stress, PK1) 张量等均是这样得到的.

推而广之，一个标量对一个二阶张量的微分将得到一个二阶张量，可参阅 (18-25) 式；一个标量对一个二阶张量的二次微分将得到一个四阶张量，可参阅 (18-26) 式.

一个二阶张量被另外一个二阶张量除或者微分的话，将得到一个四阶张量，如四阶单位张量、对称的四阶单位张量等便是这样获得的，请参阅 (19-7) 式和 (19-17) 式. 另外的典型例子是四阶弹性模量张量和四阶弹性柔度张量的获得：四阶弹性张量是通过二阶应力张量对二阶应变张量的微分获得的，线弹性情况是二者直接相除.

那二阶应变张量是如何获得的呢？同样. 应变定义为位移矢量对原长矢量的微分：

$$\varepsilon \triangleq \frac{\partial u}{\partial X} = \frac{\partial (x - X)}{\partial X} \tag{A2-2}$$

式中，$u = x - X$ 为位移矢量，x 和 X 分别为某点在当前和参考构形中的矢径. (A2-2) 式再一次明确地告诉我们：一个矢量被另一个矢量除或微分，将得到一个二阶张量 ε.

上述的经验告诉我们，一个经相除或微分所获得的张量的阶数由分子中物理量的阶数加上分母中物理量的阶数.

§5. 矢量分析中的狄拉克符号

狄拉克符号或狄拉克标记 (Dirac notation) $\langle \varphi | \phi \rangle$ 是量子力学中广泛应用于描述量子态的一套标准符号系统. 在这套系统中，每一个量子态都被描述为希尔伯特空间中的态矢量，定义为右矢 (ket) $|\phi\rangle$，通常表示一个列向量 (column vector)；每一个右矢的共轭转置定义为其左矢 (bra) $\langle \varphi |$，通常表示一个行向量 (row vector). 这是 1939 年狄拉克 (1902—1984, 1933 年诺贝尔物理学奖获得者) 将 "bracket"(括号) 这个词拆开后所创造的 [2.1].

例 5.1 三维欧氏空间中一个矢量 $\boldsymbol{A}$ 可用分量形式表示为 $\boldsymbol{A} = (A^x, A^y, A^z)$，这种表达形式在英文文献中被称为 "ordered triplet (有序三元组)". 该矢量还可用左矢表示为 $\langle A | = \begin{bmatrix} A^x & A^y & A^z \end{bmatrix}$，给出其右矢.

解：矢量 $\boldsymbol{A}$ 的右矢为

$$|A\rangle = \begin{bmatrix} A^x \\ A^y \\ A^z \end{bmatrix} \tag{5-1}$$

例 5.2 将牛顿第二定律用狄拉克符号来表示.

解：作为矢量方程的牛顿第二定律 $\boldsymbol{F} = m\boldsymbol{a}$ 中的力矢量 $\boldsymbol{F}$ 和加速度矢量 $\boldsymbol{a}$ 其实均为列矢量. 故而，牛顿第二定律可以用狄拉克符号表示为

$$|F\rangle = m |a\rangle \tag{5-2}$$

值得注意的是，在狄拉克符号中已经不用再对符号是否用黑体进行区分 [2.2].

例 5.3 已知：左矢 $\langle A | = \begin{bmatrix} A^x & A^y & A^z \end{bmatrix}$ 和右矢 $|B\rangle = \begin{bmatrix} B^x \\ B^y \\ B^z \end{bmatrix}$，给出：$\langle A | B \rangle$.

解：这两个矢量用三元组可分别表示为 $\boldsymbol{A} = (A^x, A^y, A^z)$ 和 $\boldsymbol{B} = (B^x, B^y, B^z)$，事实上 $\langle A | B \rangle$ 表示的是这两个矢量的内积 (inner product) 或缩并 (contraction)：

$A \cdot B$，用狄拉克记号则表示为

$$\langle A \mid B \rangle = \begin{bmatrix} A^x & A^y & A^z \end{bmatrix} \begin{bmatrix} B^x \\ B^y \\ B^z \end{bmatrix} = A^x B^x + A^y B^y + A^z B^z \tag{5-3}$$

而两个矢量的"外积 (outer product)" $|A\rangle\langle B|$ 给出的则是一个 3×3 的矩阵：

$$|A\rangle\langle B| = \begin{bmatrix} A^x \\ A^y \\ A^z \end{bmatrix} \begin{bmatrix} B^x & B^y & B^z \end{bmatrix} = \begin{bmatrix} A^x B^x & A^x B^y & A^x B^z \\ A^y B^x & A^y B^y & A^y B^z \\ A^z B^x & A^z B^y & A^z B^z \end{bmatrix} \tag{5-4}$$

如果用 M 来表示外积：$M = |A\rangle\langle B|$，则矩阵 M 的第 i 行第 j 列元素 M^{ij} 可表示为

$$M^{ij} = \langle i | M | j \rangle \tag{5-5}$$

例如，对于 (5-4) 式而言，$M^{23} = \langle 2 | A \rangle \langle B | 3 \rangle = A^y B^z$.

§6. 刚体平衡的两个条件

设在欧氏空间的 n 个质点约束在同一刚体上，其虚位移由平动和转动两部分组成：

$$\delta \boldsymbol{u}_i = \delta \boldsymbol{u}_0 + \delta \boldsymbol{\Theta} \times \boldsymbol{r}_i \tag{6-1}$$

式中，$\delta \boldsymbol{u}_0$ 为平动虚位移矢量；$\delta \boldsymbol{\Theta}$ 为刚体的虚转动角度矢量，其满足右手法则，四指指向刚体的转动转向，大拇指指向 $\delta \boldsymbol{\Theta}$ 矢量的方向；$\boldsymbol{r}_i$ 为第 i 个质点的位矢 (position vector).

应用表 3.1 的对偶矢量空间的概念，也就是每个质点上所作用的力矢量 $\boldsymbol{F}_i$ 和上述虚位移矢量 $\delta \boldsymbol{u}_i$ 组成对偶矢量空间，因此，刚体的平衡条件可由总虚功为零的等价条件给出：

$$\sum_{i=1}^{n} \boldsymbol{F}_i \cdot \delta \boldsymbol{u}_i = 0 \tag{6-2}$$

由于功为标量，故上式右端的 0 为标量. 将 (6-1) 式代入 (6-2) 式，得到

$$\sum_{i=1}^{n} \boldsymbol{F}_i \cdot \delta \boldsymbol{u}_i = \sum_{i=1}^{n} \boldsymbol{F}_i \cdot (\delta \boldsymbol{u}_0 + \delta \boldsymbol{\Theta} \times \boldsymbol{r}_i) = \sum_{i=1}^{n} \boldsymbol{F}_i \cdot \delta \boldsymbol{u}_0 + \sum_{i=1}^{n} \boldsymbol{F}_i \cdot (\delta \boldsymbol{\Theta} \times \boldsymbol{r}_i) = 0 \tag{6-3}$$

针对 (6-3) 式中的项：$\boldsymbol{F}_i \cdot (\delta\boldsymbol{\Theta} \times \boldsymbol{r}_i)$，可利用标量的三重积公式 (scalar triple product formula)：

$$\begin{bmatrix} \boldsymbol{a} & \boldsymbol{b} & \boldsymbol{c} \end{bmatrix} = \boldsymbol{a} \cdot (\boldsymbol{b} \times \boldsymbol{c}) = \boldsymbol{b} \cdot (\boldsymbol{c} \times \boldsymbol{a}) = \boldsymbol{c} \cdot (\boldsymbol{a} \times \boldsymbol{b}) = \begin{vmatrix} a_x & a_y & a_z \\ b_x & b_y & b_z \\ c_x & c_y & c_z \end{vmatrix} \quad (6\text{-}4)$$

上述三重积得到的是以三个矢量 $\boldsymbol{a}$、$\boldsymbol{b}$ 和 $\boldsymbol{c}$ 为棱的平行六面体的体积，如图 6.1 所示. (6-4) 式中的运算规则如图 6.2 所示.

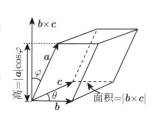

图 6.1 六面体的体积示意图

利用标量的三重积公式 (6-4) 式，(6-3) 式可等价地表示为

$$\sum_{i=1}^{n} \boldsymbol{F}_i \cdot \delta\boldsymbol{u}_i = \sum_{i=1}^{n} \boldsymbol{F}_i \cdot \delta\boldsymbol{u}_0 + \sum_{i=1}^{n} \boldsymbol{F}_i \cdot (\delta\boldsymbol{\Theta} \times \boldsymbol{r}_i) = \sum_{i=1}^{n} \boldsymbol{F}_i \cdot \delta\boldsymbol{u}_0 + \sum_{i=1}^{n} (\boldsymbol{r}_i \times \boldsymbol{F}_i) \cdot \delta\boldsymbol{\Theta} = 0 \quad (6\text{-}5)$$

由于平动虚位移矢量 $\delta\boldsymbol{u}_0$ 和虚转动角度矢量 $\delta\boldsymbol{\Theta}$ 的任意性，则有

$$\boldsymbol{F} = \sum_{i=1}^{n} \boldsymbol{F}_i = \boldsymbol{0}, \quad \boldsymbol{\tau} = \sum_{i=1}^{n} \boldsymbol{r}_i \times \boldsymbol{F}_i = \boldsymbol{0} \quad (6\text{-}6)$$

上式表明，合力和合力矩矢量均为零是刚体平衡的条件. 上式中，$\boldsymbol{0}$ 为零矢量. 在连续介质力学中，应特别注意 0 标量、$\boldsymbol{0}$ 矢量和 $\boldsymbol{0}$ 张量的区别，后两者均用黑体表示.

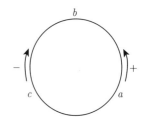

图 6.2 三重积的运算规则

反之，设刚体上的质点系在任意力矢量 $\boldsymbol{F}_i$ 的作用下满足平衡条件 (6-6) 式，求刚体的虚位移 $\delta\boldsymbol{u}_i$ 的表达式. 由于

$$\sum_{i=1}^{n} \boldsymbol{F}_i \cdot \delta\boldsymbol{u}_i - \sum_{i=1}^{n} \boldsymbol{F}_i \cdot \delta\boldsymbol{u}_0 - \sum_{i=1}^{n} (\boldsymbol{r}_i \times \boldsymbol{F}_i) \cdot \delta\boldsymbol{\Theta} = 0 \quad (6\text{-}7)$$

再由标积的三重积公式 (6-4) 式，(6-7) 式可进一步表示为

$$\sum_{i=1}^{n} \boldsymbol{F}_i \cdot \delta\boldsymbol{u}_i - \sum_{i=1}^{n} \boldsymbol{F}_i \cdot \delta\boldsymbol{u}_0 - \sum_{i=1}^{n} \boldsymbol{F}_i \cdot (\delta\boldsymbol{\Theta} \times \boldsymbol{r}_i) = \sum_{i=1}^{n} \boldsymbol{F}_i \cdot (\delta\boldsymbol{u}_i - \delta\boldsymbol{u}_0 - \delta\boldsymbol{\Theta} \times \boldsymbol{r}_i) = 0 \quad (6\text{-}8)$$

再由于力矢量 $\boldsymbol{F}_i$ 的任意性，则得到刚体的虚位移为平动和转动合成的 (6-1) 式.

§7. 质点动力学中速度和加速度的合成

质点的绝对运动等于相对运动加牵连运动. 如图 7.1 所示，我们来研究质点 P 的运动. 首先建立两个笛卡儿坐标系：定坐标系 $O\text{-}x_ny_nz_n$ 和动坐标系 $O'\text{-}x_my_mz_m$. 质点 P 在定和动坐标系中的位矢 (position vector) 分别为 $\boldsymbol{R}_n$ 和 $\boldsymbol{R}_m$.

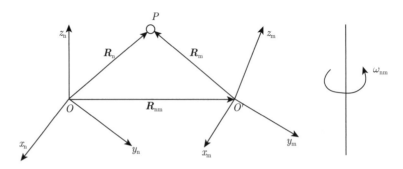

图 7.1 质点运动坐标系

比如我们要研究行进中的航空母舰夹板上一架战斗机起飞的运动,此时地面相当于定坐标系,航空母舰相当于动坐标系,我们可以通过研究战斗机和航空母舰的相对运动,航空母舰作为动坐标系的牵连运动,然后进行复合,从而获得战斗机起飞过程相对于地面这个定坐标系的绝对运动.

因此,质点的复合运动分为如下三种:(1) 绝对运动:质点相对于定坐标系的运动,其位矢用 $\boldsymbol{R}_n$ 表示;(2) 相对运动:质点相对于动坐标系的运动,其位矢用 $\boldsymbol{R}_m$ 表示;(3) 牵连运动:动坐标系相对于定坐标系的运动,其位矢用 $\boldsymbol{R}_{mn}$ 表示.

首先需要明确的是,固定坐标系 $O\text{-}x_n y_n z_n$ 的单位基矢 $\hat{\boldsymbol{i}}$、$\hat{\boldsymbol{j}}$ 和 $\hat{\boldsymbol{k}}$ 是不随时间变化的,亦即,它们对时间的导数为零;但是,运动坐标系 $O'\text{-}x_m y_m z_m$ 的单位基矢 $\hat{\boldsymbol{i}}'$、$\hat{\boldsymbol{j}}'$ 和 $\hat{\boldsymbol{k}}'$ 是随时间变化的,它们对时间的变化率满足如下关系式:

$$\frac{\mathrm{d}\hat{\boldsymbol{i}}'}{\mathrm{d}t} = \boldsymbol{\omega} \times \hat{\boldsymbol{i}}', \quad \frac{\mathrm{d}\hat{\boldsymbol{j}}'}{\mathrm{d}t} = \boldsymbol{\omega} \times \hat{\boldsymbol{j}}', \quad \frac{\mathrm{d}\hat{\boldsymbol{k}}'}{\mathrm{d}t} = \boldsymbol{\omega} \times \hat{\boldsymbol{k}}' \tag{7-1}$$

三个位矢之间满足:

$$\boldsymbol{R}_n = \boldsymbol{R}_{nm} + \boldsymbol{R}_m \tag{7-2}$$

上式对时间求一次导数,即

$$\frac{\mathrm{d}\boldsymbol{R}_n}{\mathrm{d}t} = \frac{\mathrm{d}\boldsymbol{R}_{mn}}{\mathrm{d}t} + \frac{\mathrm{d}\boldsymbol{R}_m}{\mathrm{d}t} \tag{7-3}$$

由于 $\boldsymbol{R}_m = R_{mx}\hat{\boldsymbol{i}}' + R_{my}\hat{\boldsymbol{j}}' + R_{mz}\hat{\boldsymbol{k}}'$,应用 (7-1) 式,有

$$\begin{aligned}\frac{\mathrm{d}\boldsymbol{R}_m}{\mathrm{d}t} &= \frac{\mathrm{d}R_{mx}}{\mathrm{d}t}\hat{\boldsymbol{i}}' + \frac{\mathrm{d}R_{my}}{\mathrm{d}t}\hat{\boldsymbol{j}}' + \frac{\mathrm{d}R_{mz}}{\mathrm{d}t}\hat{\boldsymbol{k}}' + R_{mx}\frac{\mathrm{d}\hat{\boldsymbol{i}}'}{\mathrm{d}t} + R_{my}\frac{\mathrm{d}\hat{\boldsymbol{j}}'}{\mathrm{d}t} + R_{mz}\frac{\mathrm{d}\hat{\boldsymbol{k}}'}{\mathrm{d}t} \\ &= \frac{\mathrm{d}R_{mx}}{\mathrm{d}t}\hat{\boldsymbol{i}}' + \frac{\mathrm{d}R_{my}}{\mathrm{d}t}\hat{\boldsymbol{j}}' + \frac{\mathrm{d}R_{mz}}{\mathrm{d}t}\hat{\boldsymbol{k}}' + \boldsymbol{\omega} \times \boldsymbol{R}_m\end{aligned} \tag{7-4}$$

将 (7-4) 式代入 (7-3) 式,可知质点的绝对速度可分解为相对速度和牵连速度之和:

第 2 章 矢量分析

$$v_a = \frac{d\boldsymbol{R}_n}{dt} = \underbrace{\frac{dR_{mx}}{dt}\hat{\boldsymbol{i}}' + \frac{dR_{my}}{dt}\hat{\boldsymbol{j}}' + \frac{dR_{mz}}{dt}\hat{\boldsymbol{k}}'}_{\text{相对速度 } v_r} + \underbrace{\frac{d\boldsymbol{R}_{mn}}{dt} + \boldsymbol{\omega} \times \boldsymbol{R}_m}_{\text{牵连速度 } v_e} \tag{7-5}$$

其中，相对速度为

$$v_r = \frac{dR_{mx}}{dt}\hat{\boldsymbol{i}}' + \frac{dR_{my}}{dt}\hat{\boldsymbol{j}}' + \frac{dR_{mz}}{dt}\hat{\boldsymbol{k}}' \tag{7-6}$$

牵连速度由如下两部分组成：

$$v_e = \underbrace{\frac{d\boldsymbol{R}_{mn}}{dt}}_{\substack{\text{牵连速度}\\\text{平动部分}}} + \underbrace{\boldsymbol{\omega} \times \boldsymbol{R}_m}_{\substack{\text{牵连速度}\\\text{转动部分}}} \tag{7-7}$$

为了得到加速度的合成，需要对 (7-4) 式对时间进行求导，首先看其中的第一部分 $\frac{dR_{mx}}{dt}\hat{\boldsymbol{i}}' + \frac{dR_{my}}{dt}\hat{\boldsymbol{j}}' + \frac{dR_{mz}}{dt}\hat{\boldsymbol{k}}'$ 对时间的导数，再次应用 (7-1) 式，有

$$\begin{aligned}
&\frac{d}{dt}\left(\frac{dR_{mx}}{dt}\hat{\boldsymbol{i}}' + \frac{dR_{my}}{dt}\hat{\boldsymbol{j}}' + \frac{dR_{mz}}{dt}\hat{\boldsymbol{k}}'\right) \\
&= \frac{d^2R_{mx}}{dt^2}\hat{\boldsymbol{i}}' + \frac{d^2R_{my}}{dt^2}\hat{\boldsymbol{j}}' + \frac{d^2R_{mz}}{dt^2}\hat{\boldsymbol{k}}' + \frac{dR_{mx}}{dt}\frac{d\hat{\boldsymbol{i}}'}{dt} + \frac{dR_{my}}{dt}\frac{d\hat{\boldsymbol{j}}'}{dt} + \frac{dR_{mz}}{dt}\frac{d\hat{\boldsymbol{k}}'}{dt} \\
&= \frac{d^2R_{mx}}{dt^2}\hat{\boldsymbol{i}}' + \frac{d^2R_{my}}{dt^2}\hat{\boldsymbol{j}}' + \frac{d^2R_{mz}}{dt^2}\hat{\boldsymbol{k}}' + \boldsymbol{\omega} \times v_r
\end{aligned} \tag{7-8}$$

而 (7-4) 式中的第二部分 $\boldsymbol{\omega} \times \boldsymbol{R}_m$ 对时间的导数为

$$\begin{aligned}
\frac{d(\boldsymbol{\omega} \times \boldsymbol{R}_m)}{dt} &= \frac{d\boldsymbol{\omega}}{dt} \times \boldsymbol{R}_m + \boldsymbol{\omega} \times \frac{d\boldsymbol{R}_m}{dt} \\
&= \frac{d\boldsymbol{\omega}}{dt} \times \boldsymbol{R}_m + \boldsymbol{\omega} \times \left(\underbrace{\frac{dR_{mx}}{dt}\hat{\boldsymbol{i}}' + \frac{dR_{my}}{dt}\hat{\boldsymbol{j}}' + \frac{dR_{mz}}{dt}\hat{\boldsymbol{k}}'}_{\text{相对速度 } v_r} + \boldsymbol{\omega} \times \boldsymbol{R}_m\right) \\
&= \frac{d\boldsymbol{\omega}}{dt} \times \boldsymbol{R}_m + \boldsymbol{\omega} \times v_r + \boldsymbol{\omega} \times (\boldsymbol{\omega} \times \boldsymbol{R}_m)
\end{aligned} \tag{7-9}$$

故绝对加速度为

$$\begin{aligned}
\boldsymbol{a}_a = \frac{d^2\boldsymbol{R}_n}{dt^2} &= \underbrace{\frac{d^2R_{mx}}{dt^2}\hat{\boldsymbol{i}}' + \frac{d^2R_{my}}{dt^2}\hat{\boldsymbol{j}}' + \frac{d^2R_{mz}}{dt^2}\hat{\boldsymbol{k}}'}_{\text{相对运动加速度 } \boldsymbol{a}_r} + \underbrace{2\boldsymbol{\omega} \times v_r}_{\text{柯氏加速度 } \boldsymbol{a}_k} \\
&+ \underbrace{\frac{d^2\boldsymbol{R}_{mn}}{dt^2} + \frac{d\boldsymbol{\omega}}{dt} \times \boldsymbol{R}_m + \boldsymbol{\omega} \times (\boldsymbol{\omega} \times \boldsymbol{R}_m)}_{\text{牵连运动加速度 } \boldsymbol{a}_e}
\end{aligned} \tag{7-10}$$

其中，牵连加速度可进一步分解为如下三个部分：

$$\bm{a}_\mathrm{e} = \underbrace{\frac{\mathrm{d}^2 \bm{R}_\mathrm{mn}}{\mathrm{d}t^2}}_{\text{牵连运动}\atop\text{平动加速度}} + \underbrace{\frac{\mathrm{d}\bm{\omega}}{\mathrm{d}t} \times \bm{R}_\mathrm{m}}_{\text{牵连运动转动}\atop\text{的切向加速度}} + \underbrace{\bm{\omega} \times (\bm{\omega} \times \bm{R}_\mathrm{m})}_{\text{牵连运动转动}\atop\text{的法向加速度}} \tag{7-11}$$

§8. 克罗内克 δ 和置换符号

8.1 克罗内克 δ 符号

1866 年，德国数学家利奥波德·克罗内克 (Leopold Kronecker, 1823—1891) 最先应用了下列符号：

$$\delta_{ij} = \begin{cases} 1, & \text{当 } i = j \\ 0, & \text{当 } i \neq j \end{cases} \tag{8-1}$$

δ_{ij} 被广泛地称为 "克罗内克 δ 符号 (Kronecker delta)"。它事实上就是一个单位矩阵：

$$[\bm{\delta}] = \begin{bmatrix} 1 & 0 & 0 \\ 0 & 1 & 0 \\ 0 & 0 & 1 \end{bmatrix} \tag{8-2}$$

根据爱因斯坦求和约定 (Einstein's summation convention)，也就是指标重复就要求和，便有

$$\delta_{ii} = \delta_{11} + \delta_{22} + \delta_{33} = 1 + 1 + 1 = 3 \tag{8-3}$$

重复出现的指标被称为 "哑标 (dummy index, silent index)"。

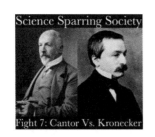

图 8.1 科学辩论社所展现的克罗内克和康托尔斗争的场景

克罗内克是一个极其罕见的专业数学家中的实业家 (businessmen)，他在 30 岁以前就在经济上实现了完全独立，换句话说，他可以不再为经济收入等担忧，从而可以无忧无虑地进行数学研究。克罗内克身材不高 (约 1.52 米) 但结实，晚年还喜欢登山运动。

克罗内克曾对康托尔 (Georg Cantor, 1845—1918) 的集合论进行了 "猛烈地、恶毒地攻击"[1.8]，从而使康托尔进入了疯人院。康托尔在 40 岁时第一次经历了精神崩溃，在其以后的漫长岁月中，以不同的强度和周期发作。克罗内克于 1891 年去世，享年 69 岁。康托尔则于 1918 年初在精神病院去世，时年 73 岁，他最后获得了自己应该享有的荣誉。图 8.1 给出了科学辩论社针对两个数学家之间斗争的场景。

第 2 章 矢量分析

数学力学和物理学大师庞加莱称康托尔的思想为感染着数学界的一种"烈性传染病"(Poincaré referred to his ideas as a "grave disease" infecting the discipline of mathematics), 克罗内克则形容康托尔为一位"科学骗子 (scientific charlatan)"、"叛徒 (renegade)"和"年轻人的腐蚀剂 (corrupter of youth)".

让我们记住马克思的名言: "在科学的入口处, 正像在地狱的入口处一样, 必须提出这样的要求: '这里必须根绝一切犹豫; 这里任何怯懦都无济于事.'" 对于第一个去征服科学某一领域征途上的险隘雄关的科学家来说, 更需要有超乎常人的勇气和毅力, 有时甚至需要为其献身!

例 8.1 应用 §5 中的狄拉克符号来表示三维欧氏空间的三个单位基矢.

解: 三维欧氏空间的三个单位基矢 $\hat{i}$、$\hat{j}$ 和 $\hat{k}$ 可用狄拉克记号分别表示为

$$\hat{i} = |1\rangle = \begin{bmatrix} 1 \\ 0 \\ 0 \end{bmatrix}, \quad \hat{j} = |2\rangle = \begin{bmatrix} 0 \\ 1 \\ 0 \end{bmatrix}, \quad \hat{k} = |3\rangle = \begin{bmatrix} 0 \\ 0 \\ 1 \end{bmatrix} \tag{8-4}$$

因此, 两个任意单位基矢的内积可表示为

$$\langle i | j \rangle = \delta^{ij} \tag{8-5}$$

例 8.2 给出右矢 $|1\rangle$, $|2\rangle$ 和 $|3\rangle$ 的厄米共轭 (Hermitian conjugate) 左矢.

解: 右矢 $|1\rangle$, $|2\rangle$ 和 $|3\rangle$ 的厄米共轭左矢分别为

$$\langle 1| = \begin{bmatrix} 1 & 0 & 0 \end{bmatrix}, \quad \langle 2| = \begin{bmatrix} 0 & 1 & 0 \end{bmatrix}, \quad \langle 3| = \begin{bmatrix} 0 & 0 & 1 \end{bmatrix} \tag{8-6}$$

这样, 对于例 5.3 中的左矢 $\langle A| = \begin{bmatrix} A_x & A_y & A_z \end{bmatrix}$ 和右矢 $|B\rangle = \begin{bmatrix} B_x \\ B_y \\ B_z \end{bmatrix}$, 还可通过 (8-4) 式和 (8-6) 式将其厄米共轭量分别表示为

$$\begin{cases} |A\rangle = A_x |1\rangle + A_y |2\rangle + A_z |3\rangle \\ \langle B| = B_x \langle 1| + B_y \langle 2| + B_z \langle 3| \end{cases} \tag{8-7}$$

8.2 置换符号的基本定义

以笛卡儿直角坐标系为例说明. 三个坐标轴的正交基单位矢量 (e_x, e_y, e_z) 满足:

$$\left.\begin{cases} e_x \cdot e_x = e_y \cdot e_y = e_z \cdot e_z = 1 \\ e_x \cdot e_y = e_y \cdot e_z = e_z \cdot e_x = 0 \end{cases}\right\} \text{Kronecker delta 的定义 } \delta_{ij}$$

$$\left.\begin{cases} e_x \times e_y = e_z, \ e_y \times e_z = e_x, \ e_z \times e_x = e_y \\ e_y \times e_x = -e_z, \ e_z \times e_y = -e_x, \ e_x \times e_z = -e_y \\ e_x \times e_x = e_y \times e_y = e_z \times e_z = 0 \end{cases}\right\} \text{Levi-Civita 置换符号的定义 } \varepsilon_{ijk} \tag{8-8}$$

置换符号 (permutation symbol, alternating symbol) ε_{ijk} 一般地常被称为 "列维–奇维塔符号 (Levi-Civita symbol)" 或 "Epsilon 符号" 或 "反对称符号 (antisymmetric symbol)", 其定义为

$$e_i \times e_j = \varepsilon_{ijk} e_k \tag{8-9}$$

如图 8.2 所示, (8-9) 式中,

$$\varepsilon_{ijk} = \begin{cases} 1, & \text{当 } i,j,k \text{ 按照偶置换取值, 即 } 123, 231, 312 \\ -1, & \text{当 } i,j,k \text{ 按照奇置换取值, 即 } 132, 213, 321 \\ 0, & \text{当两个或两个以上指标取值相同时} \end{cases} \tag{8-10}$$

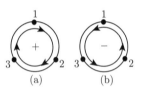

图 8.2 (a) "+" 为偶置换; (b) "−" 为奇置换

列维–奇维塔符号是一个 "赝张量 (pseudo-tensor)". 有需要深究此概念的读者请参阅《近代连续介质力学》[1.6] 中的 §3.10.

爱因斯坦对列维–奇维塔 (Tullio Levi-Civita, 1873—1941) 的学术贡献倍加赞赏. 在一封信中, 在谈到列维–奇维塔的新工作时, 爱因斯坦谦虚地写道: "我倾慕你计算方法的优雅; 骑一匹真正的 '数学' 之马驰骋这些领域的感觉肯定很好; 而我们这班人只能徒步艰难跋涉 (I admire the elegance of your method of computation; it must be nice to ride through these fields upon the horse of true mathematics while the like of us have to make our way laboriously on foot)."

张量分析之前被称为 "绝对微分学", 源出黎曼几何, 这一学科的创立应归功于里奇 (Gregorio Ricci-Curbastro, 1853—1925). 他在黎曼 (Bernhard Riemann, 1826—1866)、贝尔特拉米 (Eugenio Beltrami, 1835—1900)、克里斯托费尔 (Elwin Bruno Christoffel, 1829—1900)、利普希茨 (Rudolf Lipschitz, 1832—1903) 等人开创微分不变量研究的基础上, 在 1887—1896 年的 10 年间创建了绝对微分学理论. 1900 年, 里奇和他的学生列维–奇维塔 (如图 8.3) 合写了《绝对微分法及其应用》[2.3], 成为

张量分析的经典著作, 为张量分析和拓扑学的发展开辟了道路, 成为爱因斯坦广义相对论的有效的数学工具. 1916 年爱因斯坦发表了《广义相对论的基础》[2.4] 一文, 成功地运用这一理论表述他的广义相对论, 论文几乎用一半的篇幅解说这种绝对微分学.

  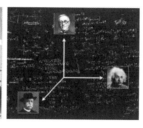

图 8.3　左: 意大利帕多瓦大学的里奇和列维–奇维塔师徒;
中: 里奇和列维–奇维塔师徒是对爱因斯坦产生过重大影响的意大利数学家
右: 由里奇、列维–奇维塔和爱因斯坦组成的坐标系

列维–奇维塔被认为是 20 世纪主要数学家之一. 他与阿马尔迪 (Amaldi) 合著的《理论力学讲义》[2.5] 被公认为是经典著作.

1914 年 4 月, 时年 41 岁的列维–奇维塔与 24 岁的 Libera Trevisani (1890—1973) 结婚. 列维–奇维塔是 Libera Trevisani 的博士导师, Libera Trevisani 的论文题目是 "About the average motion within the three body problem (三体问题中的平均运动)". 从图 8.4 可以看出, 列维–奇维塔的夫人身材高挑、有魅力并十分优雅. Libera Trevisani 于 1944 年当选为意大利高等教育妇女联合会 (Federation of Italian Graduated Women in Higher Education) 的主席, 任职到 1953 年, 她在维护妇女权利和地位方面作出了贡献.

图 8.4　右一和右二: 列维–奇维塔和夫人; 左一至左三: 沃尔泰拉和其家属

图 8.4 中的维多·沃尔泰拉 (Vito Volterra, 1860—1940) 是意大利著名数学家与物理学家, 他在位错理论方面作了奠基性的贡献, 请参阅《近代连续介质力学》[1.6] 中第 310—311 页中对其的介绍.

8.3　混合积和置换符号

三个坐标轴的正交基单位矢量 (e_x, e_y, e_z) 所组成的平行六面体的体积 (混合积) 为

$$\begin{cases} \begin{bmatrix} e_x & e_y & e_z \end{bmatrix} = e_x \cdot (e_y \times e_z) = e_x \cdot e_x = 1 \\ \begin{bmatrix} e_x & e_y & e_z \end{bmatrix} = e_y \cdot (e_z \times e_x) = e_y \cdot e_y = 1 \\ \begin{bmatrix} e_x & e_y & e_z \end{bmatrix} = e_z \cdot (e_x \times e_y) = e_z \cdot e_z = 1 \end{cases} \quad (8\text{-}11)$$

所以，可通过混合积来定义置换符号：

$$\begin{bmatrix} \boldsymbol{e}_i & \boldsymbol{e}_j & \boldsymbol{e}_k \end{bmatrix} = \boldsymbol{e}_i \cdot (\boldsymbol{e}_j \times \boldsymbol{e}_k) = \varepsilon_{jkl}\boldsymbol{e}_i \cdot \boldsymbol{e}_l = \varepsilon_{jkl}\delta_{il} = \varepsilon_{jki} = \varepsilon_{ijk} \tag{8-12}$$

在上式的最后的运算中，应用了置换符号偶置换的性质．

例 8.3 求证连续介质力学常用关系式：

$$\mathbf{curl}\boldsymbol{v} = \boldsymbol{\nabla} \times (\boldsymbol{\omega} \times \boldsymbol{r}) = 2\boldsymbol{\omega} \tag{8-13}$$

证明方法一：矩阵求解．距离矢量为 $\boldsymbol{r}=(x,y,z)$，角速度矢量 $\boldsymbol{\omega}=(\omega_x,\omega_y,\omega_z)$，速度与角速度的关系为

$$\boldsymbol{v} = \boldsymbol{\omega} \times \boldsymbol{r} = \begin{vmatrix} \hat{\boldsymbol{i}} & \hat{\boldsymbol{j}} & \hat{\boldsymbol{k}} \\ \omega_x & \omega_y & \omega_z \\ x & y & z \end{vmatrix}$$

$$= (\omega_y z - \omega_z y)\hat{\boldsymbol{i}} + (\omega_z x - \omega_x z)\hat{\boldsymbol{j}} + (\omega_x y - \omega_y x)\hat{\boldsymbol{k}} \tag{8-14}$$

对速度求旋度，将 (8-14) 式代入：

$$\mathbf{curl}\boldsymbol{v} = \boldsymbol{\nabla} \times (\boldsymbol{\omega} \times \boldsymbol{r}) = \begin{vmatrix} \hat{\boldsymbol{i}} & \hat{\boldsymbol{j}} & \hat{\boldsymbol{k}} \\ \dfrac{\partial}{\partial x} & \dfrac{\partial}{\partial y} & \dfrac{\partial}{\partial z} \\ \omega_y z - \omega_z y & \omega_z x - \omega_x z & \omega_x y - \omega_y x \end{vmatrix}$$

$$= 2\omega_x\hat{\boldsymbol{i}} + 2\omega_y\hat{\boldsymbol{j}} + 2\omega_z\hat{\boldsymbol{k}}$$

$$= 2\boldsymbol{\omega} \tag{8-15}$$

上式表明，速度的旋度是旋转角速度的 2 倍，从而 (8-13) 式得证．

证明方法二：张量指标．由 (8-9) 式可知，两个向量的叉乘可表示为

$$\boldsymbol{v} \times \boldsymbol{w} = v_i w_j \boldsymbol{e}_i \times \boldsymbol{e}_j = \varepsilon_{ijk} v_i w_j \boldsymbol{e}_k \tag{8-16}$$

从而向量三重积 (vector triple product) 为

$$\boldsymbol{u} \times (\boldsymbol{v} \times \boldsymbol{w}) = u_m \boldsymbol{e}_m \times \varepsilon_{ijk} v_i w_j \boldsymbol{e}_k = \varepsilon_{mkn}\varepsilon_{ijk} u_m v_i w_j \boldsymbol{e}_n \tag{8-17}$$

则速度的旋度为

$$\mathbf{curl}\boldsymbol{v} = \boldsymbol{\nabla} \times (\boldsymbol{\omega} \times \boldsymbol{r}) = \boldsymbol{\nabla} \times (\omega_i x_j \boldsymbol{e}_i \times \boldsymbol{e}_j)$$

$$= \boldsymbol{\nabla} \times (\varepsilon_{ijk}\omega_i x_j \boldsymbol{e}_k) = \varepsilon_{mkn}\frac{\partial(\varepsilon_{ijk}\omega_i x_j)}{\partial x_m}\boldsymbol{e}_n$$

$$= \varepsilon_{mkn}\varepsilon_{ijk}\omega_i \boldsymbol{e}_n \frac{\partial x_j}{\partial x_m} \tag{8-18}$$

只有下标 j 和 m 相同时，$\frac{\partial x_j}{\partial x_m} = \delta_{jm}$ 才不为零，则

$$\mathbf{curl}\boldsymbol{v} = \varepsilon_{mkn}\varepsilon_{ijk}\omega_i \boldsymbol{e}_n \frac{\partial x_j}{\partial x_m} = \varepsilon_{jkn}\varepsilon_{ijk}\omega_i \boldsymbol{e}_n \tag{8-19}$$

考虑到 $\varepsilon_{jkn}\varepsilon_{ijk} = \varepsilon_{njk}\varepsilon_{ijk} = 2\delta_{in}$，则

$$\mathbf{curl}\boldsymbol{v} = 2\delta_{in}\omega_i \boldsymbol{e}_n = 2\omega_i \boldsymbol{e}_i = 2\boldsymbol{\omega} \tag{8-20}$$

例 8.4 由如下真空中的麦克斯韦方程组：

$$\begin{cases} \boldsymbol{\nabla} \cdot \boldsymbol{E} = 0 \\ \boldsymbol{\nabla} \cdot \boldsymbol{B} = 0 \\ \boldsymbol{\nabla} \times \boldsymbol{E} = -\dfrac{\partial \boldsymbol{B}}{\partial t} \\ \boldsymbol{\nabla} \times \boldsymbol{B} = \varepsilon_0\mu_0 \dfrac{\partial \boldsymbol{E}}{\partial t} \end{cases} \tag{8-21}$$

推导电磁波方程.

解：对 (8-21) 式中第三式两端取旋度，由矢量三重积的关系式 (s2-10)，有

$$\boldsymbol{\nabla} \times (\boldsymbol{\nabla} \times \boldsymbol{E}) = \boldsymbol{\nabla}(\boldsymbol{\nabla} \cdot \boldsymbol{E}) - \nabla^2 \boldsymbol{E} = -\frac{\partial}{\partial t}(\boldsymbol{\nabla} \times \boldsymbol{B})$$

将 (8-21) 式中第一式和 (8-21) 式中第四式代入上式，并考虑到光速 c 和介电常数之间满足 $c = 1/\sqrt{\varepsilon_0\mu_0}$，则电场所满足的波动方程为

$$\frac{1}{c^2}\frac{\partial^2 \boldsymbol{E}}{\partial t^2} - \nabla^2 \boldsymbol{E} = \boldsymbol{0} \tag{8-22}$$

同理，对 (8-21) 式中第四式两端取旋度，再将 (8-21) 式中第二式和 (8-21) 式中第三式代入，则可得到磁场的波动方程：

$$\frac{1}{c^2}\frac{\partial^2 \boldsymbol{B}}{\partial t^2} - \nabla^2 \boldsymbol{B} = \boldsymbol{0} \tag{8-23}$$

用达朗贝尔算子或盒子算子表示的波动方程见 (15-31) 式. 和麦克斯韦方程组相关的电磁张量见思考题 2.5.

例 8.5 用克罗内克符号表示量子力学交换子 (commutator).

答：著者已经在《力学讲义》[1.11] 中应用泊松括号给出过经典力学的如下关系：

$$[x, p_x] = [y, p_y] = [z, p_z] = 1 \tag{8-24}$$

式中,(x,y,z) 为坐标分量,(p_x,p_y,p_z) 为动量分量. 一般地, (8-24) 式可更广泛和简洁地表示为

$$[r_i, p_j] = \sum_k \left(\frac{\partial r_i}{\partial r_k}\frac{\partial p_j}{\partial p_k} - \frac{\partial r_i}{\partial p_k}\frac{\partial p_j}{\partial r_k}\right) = \sum_k (\delta_{ik}\delta_{jk} - 0) = \delta_{ij} \tag{8-25}$$

下面我们用带帽子的符号表示相应的量子力学参量. 对于量子力学而言, (8-24) 式将改写为如下常用关系式:

$$[\hat{x}, \hat{p}_x] = [\hat{y}, \hat{p}_y] = [\hat{z}, \hat{p}_z] = \mathrm{i}\hbar \tag{8-26}$$

式中,i 为虚数,$\hbar$ 为约化普朗克常数. 对于量子力学场论, (8-25) 式改写为

$$[\hat{r}_i, \hat{p}_j] = \mathrm{i}\hbar\delta_{ij} \tag{8-27}$$

下面给出由玻恩 (Max Born, 1882—1970) 于 1925 年建立的量子力学正则对易关系 (canonical commutation relation) (8-27) 式的证明. 将 $[\hat{r}_i, \hat{p}_j]$ 作用于 ψ,考虑到量子力学动量算子 (20-1) 式中第一式和 $\dfrac{\partial r_i}{\partial r_j} = \delta_{ij}$,有

$$\begin{aligned}
[\hat{r}_i, \hat{p}_j]\psi &= (\hat{r}_i\hat{p}_j - \hat{p}_j\hat{r}_i)\psi = -\mathrm{i}\hbar r_i\frac{\partial \psi}{\partial r_j} - (-\mathrm{i}\hbar)\frac{\partial (r_i\psi)}{\partial r_j}\\
&= -\mathrm{i}\hbar r_i\frac{\partial \psi}{\partial r_j} + \mathrm{i}\hbar\frac{\partial r_i}{\partial r_j}\psi + \mathrm{i}\hbar r_i\frac{\partial \psi}{\partial r_j}\\
&= \mathrm{i}\hbar\delta_{ij}\psi
\end{aligned}$$

因此 (8-27) 式得证.

例 8.6 用列维–奇维塔置换符号表示量子力学的角动量的对易关系.

答: 著者在《力学讲义》[1.11] 的 (44-43) 式中, 给出了经典场论中角动量所满足的关系:

$$[L_x, L_y] = L_z \tag{8-28}$$

应用列维–奇维塔置换符号, 上式可推而广之为

$$[L_i, L_j] = \varepsilon_{ijk}L_k \tag{8-29}$$

将上式推广到量子力学情形:

$$\left[\hat{L}_i, \hat{L}_j\right] = \mathrm{i}\hbar\varepsilon_{ijk}\hat{L}_k \tag{8-30}$$

例如, 在偶置换时, (8-30) 式可更为具体地写成

$$\left[\hat{L}_x, \hat{L}_y\right] = \mathrm{i}\hbar\hat{L}_z, \quad \left[\hat{L}_y, \hat{L}_z\right] = \mathrm{i}\hbar\hat{L}_x, \quad \left[\hat{L}_z, \hat{L}_x\right] = \mathrm{i}\hbar\hat{L}_y \tag{8-31}$$

第 2 章 矢量分析

下面作为例子来证明 (8-31) 式中第一式. 由下列经典角动量的表达式：

$$\begin{cases} L_x = yp_z - zp_y \\ L_y = zp_x - xp_z \\ L_z = xp_y - yp_x \end{cases} \tag{8-32}$$

对上式应用量子力学动量算子 (20-1) 式中第一式，便可得到量子力学角动量算子 $\hat{\boldsymbol{L}} = \hat{\boldsymbol{r}} \times \hat{\boldsymbol{p}} = -\mathrm{i}\hbar\,(\hat{\boldsymbol{r}} \times \boldsymbol{\nabla})$ 的三个分量为

$$\begin{cases} \hat{L}_x = \hat{y}\hat{p}_z - \hat{z}\hat{p}_y = -\mathrm{i}\hbar\left(y\dfrac{\partial}{\partial z} - z\dfrac{\partial}{\partial y}\right) \\ \hat{L}_y = \hat{z}\hat{p}_x - \hat{x}\hat{p}_z = -\mathrm{i}\hbar\left(z\dfrac{\partial}{\partial x} - x\dfrac{\partial}{\partial z}\right) \\ \hat{L}_z = \hat{x}\hat{p}_y - \hat{y}\hat{p}_x = -\mathrm{i}\hbar\left(x\dfrac{\partial}{\partial y} - y\dfrac{\partial}{\partial x}\right) \end{cases} \tag{8-33}$$

由于

$$\begin{aligned}
\hat{L}_x\hat{L}_y &= (-\mathrm{i}\hbar)^2 \left(y\dfrac{\partial}{\partial z} - z\dfrac{\partial}{\partial y}\right)\left(z\dfrac{\partial}{\partial x} - x\dfrac{\partial}{\partial z}\right) \\
&= (-\mathrm{i}\hbar)^2 \left[y\dfrac{\partial}{\partial z}\left(z\dfrac{\partial}{\partial x} - x\dfrac{\partial}{\partial z}\right) - z\dfrac{\partial}{\partial y}\left(z\dfrac{\partial}{\partial x} - x\dfrac{\partial}{\partial z}\right)\right] \\
&= (-\mathrm{i}\hbar)^2 \left(y\dfrac{\partial}{\partial x} + yz\dfrac{\partial^2}{\partial z\partial x} - yx\dfrac{\partial^2}{\partial z^2} - z^2\dfrac{\partial^2}{\partial y\partial x} + zx\dfrac{\partial^2}{\partial y\partial z}\right)
\end{aligned} \tag{8-34}$$

同理，有

$$\begin{aligned}
\hat{L}_y\hat{L}_x &= (-\mathrm{i}\hbar)^2 \left(z\dfrac{\partial}{\partial x} - x\dfrac{\partial}{\partial z}\right)\left(y\dfrac{\partial}{\partial z} - z\dfrac{\partial}{\partial y}\right) \\
&= (-\mathrm{i}\hbar)^2 \left(zy\dfrac{\partial^2}{\partial x\partial z} - z^2\dfrac{\partial^2}{\partial x\partial y} - xy\dfrac{\partial^2}{\partial z^2} + x\dfrac{\partial}{\partial y} + xz\dfrac{\partial^2}{\partial z\partial y}\right)
\end{aligned} \tag{8-35}$$

考虑到求偏导顺序的可交换性，(8-34) 和 (8-35) 两式相减，有

$$\begin{aligned}
\left[\hat{L}_x, \hat{L}_y\right] &= \hat{L}_x\hat{L}_y - \hat{L}_y\hat{L}_x = \mathrm{i}\hbar\,(-\mathrm{i}\hbar)\left(x\dfrac{\partial}{\partial y} - y\dfrac{\partial}{\partial x}\right) \\
&= \mathrm{i}\hbar\hat{L}_z
\end{aligned} \tag{8-36}$$

亦即 (8-31) 式中第一式得证.

思考题和补充材料

2.1 进一步对梯度、散度和旋度进行对比，见题图 2.1.

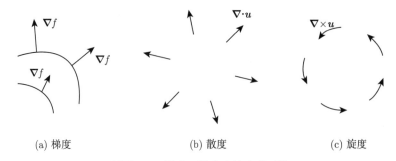

(a) 梯度　　　　　　(b) 散度　　　　　　(c) 旋度

题图 2.1　梯度、散度和旋度的对比

2.2 证明介质不可压缩 (incompressible) 的条件为 div$v = 0$.

2.3 一个无散度的流动 (divergence-free flow) 被称为 "等容的 (isochoric)". 横波的传播往往被称为 "等容波" 是由于在波的传播过程中, 其体积不变, 见表 14.1.

2.4 一个满足 **curl**$v = 0$ 的矢量场被称为 "无旋的 (irrotational)". 纵波的传播往往被称为 "无旋波" 是由于其位移矢量 u 满足 **curl**$u = 0$.

2.5 电磁张量 (electromagnetic tensor) 或电磁场张量 (electromagnetic field tensor) 是描述一物理系统中电磁场的数学客体, 所根据的是麦克斯韦的电磁学理论. 该张量是在赫尔曼·闵可夫斯基 (Hermann Minkowski, 1864—1909) 提出狭义相对论的四维张量形式之后被首次使用. 电磁张量可用 4×4 矩阵形式表示为

$$[\boldsymbol{F}] = \begin{bmatrix} 0 & E_x & E_y & E_z \\ -E_x & 0 & -B_z & B_y \\ -E_y & B_z & 0 & -B_x \\ -E_z & -B_y & B_x & 0 \end{bmatrix} \tag{s2-1}$$

请证明 (s2-1) 式中的电磁张量满足如下性质: (1) 反对称性: $F_{ij} = -F_{ji}$; (2) 无迹 (traceless), 亦即其迹 (对角和) 为零; (3) 六个独立分量.

2.6 表面张力或界面张力定义为每增加单位新的表面所需要的功, 也就是说, 表面张力是作为标量的功被作为矢量的面积除, 按照本书提出的法则, 表面张力为矢量. 所以, 界面张力有三个分量: 液–气、固–气、固-液界面张力. 有很多书中错误地说表面张力是一个标量, 而标量是没有分量的.

2.7 证明角动量 (动量矩) 关系式: $\boldsymbol{L}^2 = (\boldsymbol{r} \times \boldsymbol{p})^2 = r^2 p^2 - (\boldsymbol{r} \cdot \boldsymbol{p})^2$. 进一步讨论: 对于量子系统, 相应角动量为 $\hat{\boldsymbol{L}}^2 = (\boldsymbol{r} \times \hat{\boldsymbol{p}})^2 = r^2 \hat{p}^2 - (\boldsymbol{r} \cdot \hat{\boldsymbol{p}})^2 + i\hbar \boldsymbol{r} \cdot \hat{\boldsymbol{p}}$.

2.8 爱因斯坦被认为是科学与艺术结合的典范, 他从六岁开始学习小提琴, 长大成人后很少不携带小提琴旅行. 爱因斯坦演奏小提琴的情景如题图 2.8(a)—(c) 所示. 爱因斯坦的名言是 "如果我不是物理学家, 我很可能会成为一名音乐家. 我经常用音乐思考. 我活在音乐的白日梦中. 我用音乐的角度来看待我的生活 …… 我从音乐中得到了生活中最大的乐趣 (If I were not a physicist, I would probably be a musician. I often think in music. I live my daydreams in music. I see my life in term of music ⋯ I got most joy in life out of music)."

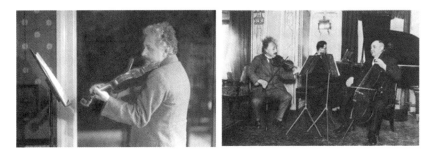

题图 2.8(a)　爱因斯坦是莫扎特奏鸣曲的忠实粉丝 (particular fan of Mozart sonatas, 左)；爱因斯坦在室内乐演奏中 (右)

题图 2.8(b)　爱因斯坦在演奏中 (左、中)；爱因斯坦赠给清洁工儿子的小提琴 2018 年 3 月 11 日纽约拍卖出 516500 美元

题图 2.8(c)　爱因斯坦在苏黎世建立广义相对论的初期，经常到同事阿道夫·赫维兹 (Adolf Hurwitz, 1859—1919) 家和其女儿 Lisi 进行小提琴二重奏 (violin duet) 来放松，赫维兹假装在进行指挥 (pretending to conduct)

2.9 对于任意矢量 $\boldsymbol{A} = (A_x, A_y, A_z)$,经典场论中最常用的矢量点积和叉积运算的另外一种方式:

$$\boldsymbol{\nabla} \cdot \boldsymbol{A} = \begin{pmatrix} \dfrac{\partial}{\partial x} & \dfrac{\partial}{\partial y} & \dfrac{\partial}{\partial z} \end{pmatrix} \begin{bmatrix} A_x \\ A_y \\ A_z \end{bmatrix} \tag{s2-2}$$

$$[\boldsymbol{\nabla} \times \boldsymbol{A}] = \begin{bmatrix} 0 & -\dfrac{\partial}{\partial z} & \dfrac{\partial}{\partial y} \\ \dfrac{\partial}{\partial z} & 0 & -\dfrac{\partial}{\partial x} \\ -\dfrac{\partial}{\partial y} & \dfrac{\partial}{\partial x} & 0 \end{bmatrix} \begin{bmatrix} A_x \\ A_y \\ A_z \end{bmatrix} \tag{s2-3}$$

问题: 总结 (8-14) 式和 (s2-3) 式中两种叉积运算的各自特点.

2.10 在麦克斯韦去世的 1879 年, 当时的麦克斯韦方程组还不是我们现如今所看到的四个简洁无比的方程. 奥利弗·亥维赛 (Oliver Heaviside, 1850—1925) 于 1885 年发表了一篇论文, 把麦克斯韦方程组里的 20 个方程, 简化到了 4 个, 就是我们现在所看到的优美形式. 亥维赛是一个自学成才的英国人, 被称为是 "第一等古怪" (a first-rate oddity)[2.6] 的传奇人物. 他从小家境贫寒, 并且耳聋. 亥维赛虽未受过正规的教育, 但通过自学数学特别是麦克斯韦的《电磁通论》, 独自创立了专门的矢量微积分学 (vector calculus), 极大地简化了场论的符号表述.

题图 2.10　自学成才的 "第一等古怪" 和 "被遗忘的天才"——亥维赛

2.11 亥维赛曾长期处于学术界的远边缘 (on the far fringes of the scientific community), 但因其在创立矢量微积分学, 特别是在麦克斯韦方程组形式现代化方面的杰出贡献, 逐渐被学术界所承认, 他于 1891 年当选为英国皇家学会会员 (FRS). 亥维赛的巨大学术影响和其贡献及时地进入教科书有密切的关系. 特别值得指出的是, 奥古斯特·弗普尔 (August Föppl, 1854—1924, 见本书例题 27.6) 在于 1894 年出版的教材《麦克斯韦电学理论的介绍》(*Einführung in die Maxwellsche Theorie der Electricität*) 中介绍了亥维赛使麦克斯韦方程组现代化的贡献, 使得爱因斯坦等德国青年得以学

第 2 章 矢量分析

习现代化的电磁理论. 请见《力学讲义》[1.11] 中第 176 页思考题 3.11 的深入论述.

2.12 亥维赛由于未受到过正规的教育, 不太重视严格的数学论证, 善于直觉进行论述和演算, 以致于在数学和工程上所做出了众多原创性成就被同行们轻视和泼冷水. 亥维赛的名言是: "数学分两种, 严格的和物理直观化的. 前者的范围窄; 后者粗犷而广泛. 拘泥于公式的严格证明只会令绝大多数的数学物理探究止步不前. 难道我要在完全理解消化反应的所有机理以前, 拒绝进食吗? (Mathematics is of two kinds, Rigorous and Physical. The former is Narrow: the latter Bold and Broad. To have to stop to formulate rigorous demonstrations would put a stop to most physico-mathematical inquiries. Am I to refuse to eat because I do not fully understand the mechanism of digestion?)".

2.13 在亥维赛曾经写过的一篇关于麦克斯韦的文章中的一些有关灵魂不朽的观点同样也适用于他自己[2.6]. 亥维赛从 "更为高尚意义上的"(far nobler sense) 层次上反思了他所谓的灵魂不朽的真正教义 (true doctrine of the immortality of the soul), 即每个人对世界产生的持久影响, 并在他或她死后继续存在. 他说, 从这个意义上说, 有些灵魂是非常伟大的: "莎士比亚或牛顿的作品是令人叹为观止的. 这样的人在摆脱尘世的盘绕, 步入坟墓之后, 就会过上他们生命中最美好的时光. 麦克斯韦就是其中之一. 他的灵魂将会长久地存在和增长, 几千年后, 它将会像过去的一颗耀眼的恒星一样闪耀, 它的光芒需要漫长的很长时间才能照耀到我们 (这些能够理解麦克斯韦方程组的人), 而对于其他 (不大理解麦克斯韦方程组的) 大众而言, 这星光甚至还有些暗淡 (That of a Shakespeare or a Newton is stupendous. Such men live the best parts of their lives after they shuffle off the mortal coil and fall into the grave. Maxwell was one of those men. His soul will live and grow for long to come, and, thousands of years hence, it will shine as one of the bright stars of the past, whose light takes ages to reach us, amongst the crowd of others, not the least bright)."

2.14 和例 8.3 中所定义的旋转角速度有密切关系的是描述旋涡运动的 "涡量 (vorticity)", 其定义为

$$\boldsymbol{\Omega} = \nabla \times \boldsymbol{v} = \mathbf{curl}\boldsymbol{v} \tag{s2-4}$$

显然, 对比 (8-13) 和 (s2-4) 两式, 易知: 涡量是角速度的 2 倍, 涡量和角速度均为赝矢量. 验证涡量场是无源管式场, 也就是满足如下关系式:

$$\nabla \cdot \boldsymbol{\Omega} = \nabla \cdot (\nabla \times \boldsymbol{v}) = 0 \tag{s2-5}$$

2.15 流体力学中常用的兰姆矢量 (Lamb vector)[2.7] 定义为 (s2-4) 式中的涡量矢量和速度矢量之间的叉积:

$$\boldsymbol{l} = \boldsymbol{\Omega} \times \boldsymbol{v} = (\nabla \times \boldsymbol{v}) \times \boldsymbol{v} \tag{s2-6}$$

(1) 验证 (3-4) 式中的对流加速度可通过兰姆矢量表示为

$$\boldsymbol{v} \cdot \boldsymbol{\nabla} \otimes \boldsymbol{v} = \boldsymbol{l} + \boldsymbol{\nabla}\left(\frac{1}{2}v^2\right) \tag{s2-7}$$

(s2-7) 式在纳维–斯托克斯方程中的应用见思考题 11.3.

(2) 验证兰姆矢量的散度关系式：

$$\boldsymbol{\nabla} \cdot \boldsymbol{l} = \boldsymbol{v} \cdot (\boldsymbol{\nabla} \times \boldsymbol{\Omega}) - \boldsymbol{\Omega} \cdot \boldsymbol{\Omega} \tag{s2-8}$$

(3) 兰姆矢量为零的流动被称为 "贝尔特拉米流 (Beltrami flows)"，按照 (s2-6) 式，贝尔特拉米流的涡量矢量 $\boldsymbol{\Omega}$ 和速度矢量 $\boldsymbol{v}$ 相平行，亦即涡量可表示为一个标量函数 $\alpha(\boldsymbol{r},t)$ 和速度矢量的乘积: $\boldsymbol{\Omega} = \alpha(\boldsymbol{r},t)\boldsymbol{v}$. 问题：继续深入分析贝尔特拉米流的其他性质. 贝尔特拉米流以意大利数学力学家贝尔特拉米 (Eugenio Beltrami, 1835—1899, 有少数文献说其逝世于 1900 年 2 月) 命名. 贝特拉米长期担任理性力学教授，图 48.2 中有他的名字.

2.16 矢量运算中的 "费曼下标符号 (Feynman subscript notation)" 是《费曼物理学讲义》第二卷第 27.3 节中所介绍的内容. 如果 $\boldsymbol{a}$ 和 $\boldsymbol{b}$ 为两个矢量，则通过费曼下标符号有下列重要运算规则：

$$\boldsymbol{\nabla}(\boldsymbol{a} \cdot \boldsymbol{b}) = \boldsymbol{\nabla}_a(\boldsymbol{b} \cdot \boldsymbol{a}) + \boldsymbol{\nabla}_b(\boldsymbol{a} \cdot \boldsymbol{b}) \tag{s2-9}$$

式中，$\boldsymbol{\nabla}_a(\boldsymbol{b} \cdot \boldsymbol{a})$ 是将 $\boldsymbol{b}$ 作为了矢量 $\boldsymbol{a}$ 的系数，梯度算子 $\boldsymbol{\nabla}_a$ 表示只对矢量 $\boldsymbol{a}$ 求梯度，余此类推. 由矢量三叉积的运算公式：

$$\boldsymbol{p} \times (\boldsymbol{q} \times \boldsymbol{r}) = \boldsymbol{q}(\boldsymbol{p} \cdot \boldsymbol{r}) - \boldsymbol{r}(\boldsymbol{p} \cdot \boldsymbol{q}) \tag{s2-10}$$

则有

$$\boldsymbol{\nabla}_a(\boldsymbol{b} \cdot \boldsymbol{a}) = \boldsymbol{b} \times (\boldsymbol{\nabla}_a \times \boldsymbol{a}) + \boldsymbol{a}(\boldsymbol{b} \cdot \boldsymbol{\nabla}_a) = \boldsymbol{b} \times (\boldsymbol{\nabla}_a \times \boldsymbol{a}) + \boldsymbol{b} \cdot \boldsymbol{\nabla}_a \boldsymbol{a} \tag{s2-11}$$

同理有

$$\boldsymbol{\nabla}_b(\boldsymbol{a} \cdot \boldsymbol{b}) = \boldsymbol{a} \times (\boldsymbol{\nabla}_b \times \boldsymbol{b}) + \boldsymbol{a} \cdot \boldsymbol{\nabla}_b \boldsymbol{b} \tag{s2-12}$$

将 (s2-11) 和 (s2-12) 两式代回 (s2-9) 式中，整理得到两矢量点积后再求梯度的公式：

$$\begin{aligned}\boldsymbol{\nabla}(\boldsymbol{a} \cdot \boldsymbol{b}) &= \boldsymbol{\nabla}_b(\boldsymbol{a} \cdot \boldsymbol{b}) + \boldsymbol{\nabla}_a(\boldsymbol{b} \cdot \boldsymbol{a}) \\ &= \boldsymbol{a} \times (\boldsymbol{\nabla} \times \boldsymbol{b}) + \boldsymbol{a} \cdot \boldsymbol{\nabla} \boldsymbol{b} + \boldsymbol{b} \times (\boldsymbol{\nabla} \times \boldsymbol{a}) + \boldsymbol{b} \cdot \boldsymbol{\nabla} \boldsymbol{a}\end{aligned} \tag{s2-13}$$

式中，将梯度算子的下标去掉的原因是梯度算子已经作用到相关的矢量上.

验证：在 (s2-13) 式中，令 $\boldsymbol{a} = \boldsymbol{b} = \boldsymbol{v}$，即可得到关系式 (s2-7).

2.17 将上题中的费曼下标符号用于两个矢量叉积的散度运算，有

$$\nabla \cdot (a \times b) = \nabla_a \cdot (a \times b) + \nabla_b \cdot (a \times b) \tag{s2-14}$$

式中，∇_a 是要作用到矢量 a 上的算子，∇_b 是要作用到矢量 b 上的算子. 由矢量的标量三重积 (scalar triple product) 公式：

$$p \cdot (q \times r) = q \cdot (r \times p) = r \cdot (p \times q) \tag{s2-15}$$

则有

$$\begin{cases} \nabla_a \cdot (a \times b) = b \cdot (\nabla_a \times a) = b \cdot (\nabla \times a) \\ \nabla_b \cdot (a \times b) = a \cdot (b \times \nabla_b) = -a \cdot (\nabla \times b) \end{cases} \tag{s2-16}$$

将 (s2-16) 式代回 (s2-14) 式中，得到

$$\nabla \cdot (a \times b) = b \cdot (\nabla \times a) - a \cdot (\nabla \times b) \tag{s2-17}$$

2.18 证明高斯引力定律 (Gauss's law for gravity) 或称高斯引力通量定理 (Gauss's flux theorem for gravity)：

$$\begin{cases} \oiint_{\partial V} g \cdot n \, \mathrm{d}A = -4\pi GM \text{ (积分形式)} \\ \nabla \cdot g = -4\pi G\rho \text{ (微分形式)} \end{cases} \tag{s2-18}$$

式中，g 为引力加速度矢量，n 为外法线矢量，G 为万有引力常数，M 为包含在表面 ∂V 内的总质量，ρ 为任意点的质量密度.

证明：由牛顿万有引力定律，引力加速度矢量表示为

$$g = -GM\frac{\hat{r}}{r^2} \tag{s2-19}$$

式中，$\hat{r}$ 为单位矢量. 对于一个球体而言，对其应用 (s2-18) 式中第一式的左端，则积分形式的高斯引力定律证明如下：

$$\oiint_{\partial V} g \cdot n \, \mathrm{d}A = 4\pi r^2 g \cdot n = -4\pi GM \tag{s2-20}$$

在上式的运算中，由于 $\hat{r}$ 和 n 均为单位矢量且方向相同，则有 $\hat{r} \cdot n = 1$.

由散度定理 (详见本书 (24-25) 式)，亦即

$$\oiint_{\partial V} g \cdot n \, \mathrm{d}A = \iiint_V \nabla \cdot g \, \mathrm{d}V = -4\pi GM = -4\pi G \iiint_V \rho \, \mathrm{d}V \tag{s2-21}$$

则 (s2-18) 式中第二式得证.

由于引力场为保守场，则引力加速度矢量满足无旋性 (irrotationality) 条件：

$$\nabla \times g = 0 \tag{s2-22}$$

引入引力势：
$$g = -\nabla\phi \tag{s2-23}$$

则高斯引力定律的微分形式 (s2-18) 式中第二式满足如下泊松方程 (Poisson's equation)：
$$\nabla^2\phi = 4\pi G\rho \tag{s2-24}$$

2.19 接上题，讨论牛顿超距作用 (第二定律和万有引力定律)、高斯引力定律 (泊松引力势理论) 和爱因斯坦场方程之间的联系，如题图 2.19 所示.

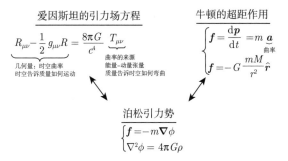

题图 2.19 　三种引力理论之间的联系

2.20 有关兰姆矢量的如下关系式，也就是 (s2-7) 式的等价关系式：
$$l = (\nabla \times v) \times v = v \cdot \nabla \otimes v - \nabla\left(\frac{1}{2}v^2\right) \tag{s2-25}$$

已经在思考题 2.16 中得到了验证.

在电动力学中，针对电场矢量 E 也存在形式上和 (s2-25) 式完全相同的关系式：
$$(\nabla \times E) \times E = E \cdot \nabla \otimes E - \nabla\left(\frac{1}{2}E^2\right) \tag{s2-26}$$

问题：运用列维–奇维塔符号 ε_{ijk} 来证明 (s2-25) 或 (s2-26) 两式.

证明：两个列维–奇维塔符号的乘积满足如下关系式：
$$\varepsilon_{ijk}\varepsilon_{lmn} = \begin{vmatrix} \delta_{il} & \delta_{im} & \delta_{in} \\ \delta_{jl} & \delta_{jm} & \delta_{jn} \\ \delta_{kl} & \delta_{km} & \delta_{kn} \end{vmatrix} \tag{s2-27}$$

如果两个列维–奇维塔符号的乘积中前两个指标相等，则有
$$\varepsilon_{lij}\varepsilon_{lkm} = \delta_{ik}\delta_{jm} - \delta_{im}\delta_{jk} \tag{s2-28}$$

利用 $\varepsilon_{ijl} = \varepsilon_{lij}$ 和 (s2-28) 式，则 (s2-25) 式详细证明如下：
$$(\nabla \times v) \times v = \left(\frac{\partial}{\partial x_i}e_i \times v_j e_j\right) \times v_k e_k$$

$$= \frac{\partial v_j}{\partial x_i} v_k \varepsilon_{ijl} \boldsymbol{e}_l \times \boldsymbol{e}_k = \frac{\partial v_j}{\partial x_i} v_k \varepsilon_{ijl} \varepsilon_{lkm} \boldsymbol{e}_m = \frac{\partial v_j}{\partial x_i} v_k \varepsilon_{lij} \varepsilon_{lkm} \boldsymbol{e}_m$$

$$= \frac{\partial v_j}{\partial x_i} v_k \left(\delta_{ik} \delta_{jm} - \delta_{im} \delta_{jk} \right) \boldsymbol{e}_m = v_i \frac{\partial v_j}{\partial x_i} \boldsymbol{e}_j - v_j \frac{\partial v_j}{\partial x_i} \boldsymbol{e}_i$$

$$= \boldsymbol{v} \cdot \boldsymbol{\nabla} \otimes \boldsymbol{v} - \boldsymbol{\nabla} \left(\frac{1}{2} \boldsymbol{v}^2 \right) \tag{s2-29}$$

2.21 凯斯·莫法特 (Keith Moffatt, 1935—　) 于 1969 年定义了三维流场中流体流动的螺旋度 (helicity)[2.8]：

$$\mathcal{H} = \int_V \boldsymbol{v} \cdot (\boldsymbol{\nabla} \times \boldsymbol{v}) \, \mathrm{d}V = \int_V \boldsymbol{v} \cdot \boldsymbol{\Omega} \mathrm{d}V \tag{s2-30}$$

式中，$\boldsymbol{v}$ 为流速，$\boldsymbol{\Omega} = \boldsymbol{\nabla} \times \boldsymbol{v}$ 为由 (s2-4) 式定义的涡量.

问题：(1) 欧拉方程有四个已知的积分不变量：动量、角动量、能量、螺旋度. 证明螺旋度的守恒性 $\frac{\mathrm{d}\mathcal{H}}{\mathrm{d}t} = 0$ 并给出该守恒性所需要满足的条件；(2) 根据诺特定理，守恒律一定对应着某种对称性，螺旋度守恒对应着何种对称性？(3) 莫法特于 2017 年在《科学》期刊著文称《螺旋度 —— 即使在黏性流体中也具有不变性》[2.9]，请阅读该文并解释原因.

参 考 文 献

[2.1] Dirac P A M. A new notation for quantum mechanics. Mathematical Proceedings of the Cambridge Philosophical Society, 1939, 35: 416-418.

[2.2] Neuenschwander D E. Tensor Calculus for Physics: A Concise Guide. Baltimore: Johns Hopkins University Press, 2015.

[2.3] Ricci M M G, Levi-Civita T. Méthodes de calcul différentiel absolu et leurs applications. Mathematische Annalen, 1900, 54: 125-201.

[2.4] Einstein A. Die grundlage der allgemeinen relativitätstheorie. Annalen der Physik, 1916, 354: 769-822.

[2.5] Levi-Civita T, Amaldi U. Lezioni di meccanica razionale. N. Zanichelli, 1923.

[2.6] Hunt B J. Oliver Heaviside. Physics Today, 2012, 65: 48-54.

[2.7] Lamb H. Hydrodynamics. Cambridge: Cambridge University Press, 1993.

[2.8] Moffatt H K. The degree of knottedness of tangled vortex lines. Journal of Fluid Mechanics, 1969, 35: 117-129.

[2.9] Moffatt H K. Helicity-invariant even in a viscous fluid. Science, 2017, 357: 448-449.

第 3 章 张量代数和微积分

§9. 张量的引入，二阶投影张量

9.1 二阶张量作为并矢的引入

为了方便一二年级本科生理解，本小节将用一种最为通俗的语言来引入二阶张量. 考虑三维欧几里得空间的两个矢量：

$$\boldsymbol{a} = 3\hat{\boldsymbol{i}} + 2\hat{\boldsymbol{j}} + 5\hat{\boldsymbol{k}}, \quad \boldsymbol{b} = 7\hat{\boldsymbol{i}} + 0\hat{\boldsymbol{j}} + 11\hat{\boldsymbol{k}} \tag{9-1}$$

定义并矢 (dyad, $\otimes$)，记为 $\boldsymbol{a} \otimes \boldsymbol{b}$ 或 $\boldsymbol{ab}$，是指如下运算：

$$\begin{aligned}
\boldsymbol{a} \otimes \boldsymbol{b} &= \left(3\hat{\boldsymbol{i}} + 2\hat{\boldsymbol{j}} + 5\hat{\boldsymbol{k}}\right) \otimes \left(7\hat{\boldsymbol{i}} + 0\hat{\boldsymbol{j}} + 11\hat{\boldsymbol{k}}\right) \\
&= 21\hat{\boldsymbol{i}} \otimes \hat{\boldsymbol{i}} + 0\hat{\boldsymbol{i}} \otimes \hat{\boldsymbol{j}} + 33\hat{\boldsymbol{i}} \otimes \hat{\boldsymbol{k}} \\
&\quad + 14\hat{\boldsymbol{j}} \otimes \hat{\boldsymbol{i}} + 0\hat{\boldsymbol{j}} \otimes \hat{\boldsymbol{j}} + 22\hat{\boldsymbol{j}} \otimes \hat{\boldsymbol{k}} \\
&\quad + 35\hat{\boldsymbol{k}} \otimes \hat{\boldsymbol{i}} + 0\hat{\boldsymbol{k}} \otimes \hat{\boldsymbol{j}} + 55\hat{\boldsymbol{k}} \otimes \hat{\boldsymbol{k}}
\end{aligned} \tag{9-2}$$

式中的并矢即为一个二阶张量，可以表示为 $\boldsymbol{T} = T_{ij}\boldsymbol{e}_i \otimes \boldsymbol{e}_j$，其中 $\boldsymbol{e}_1 = \hat{\boldsymbol{i}}, \boldsymbol{e}_2 = \hat{\boldsymbol{j}}, \boldsymbol{e}_3 = \hat{\boldsymbol{k}}$，且 $i, j = 1, 2, 3$ 为哑标 (dummy index，亦有用 silent index 的)，满足爱因斯坦求和约定 (Einstein summation convention)，即同侧指标重复时，便循环求和. 即

$$\begin{aligned}
\boldsymbol{T} = &T_{11}\boldsymbol{e}_1 \otimes \boldsymbol{e}_1 + T_{12}\boldsymbol{e}_1 \otimes \boldsymbol{e}_2 + T_{13}\boldsymbol{e}_1 \otimes \boldsymbol{e}_3 \\
&+ T_{21}\boldsymbol{e}_2 \otimes \boldsymbol{e}_1 + T_{22}\boldsymbol{e}_2 \otimes \boldsymbol{e}_2 + T_{23}\boldsymbol{e}_2 \otimes \boldsymbol{e}_3 \\
&+ T_{31}\boldsymbol{e}_3 \otimes \boldsymbol{e}_1 + T_{32}\boldsymbol{e}_3 \otimes \boldsymbol{e}_2 + T_{33}\boldsymbol{e}_3 \otimes \boldsymbol{e}_3
\end{aligned} \tag{9-3}$$

与 (9-3) 式对应，该张量的系数可用矩阵 (matrix) 表示为

$$[\boldsymbol{a} \otimes \boldsymbol{b}] = [\boldsymbol{T}] = \begin{bmatrix} 21 & 0 & 33 \\ 14 & 0 & 22 \\ 35 & 0 & 55 \end{bmatrix} \tag{9-4}$$

第 3 章 张量代数和微积分

并矢运算满足结合律 (associative law) 与分配律 (distributive law), 对于任意的实数 m 和 n, 在三维欧氏空间用基矢量表示:

$$\text{结合律}: \begin{cases} m\left(\hat{i} \otimes \hat{j}\right) = \left(m\hat{i}\right) \otimes \hat{j} = \hat{i} \otimes \left(m\hat{j}\right) \\ \left(\hat{i} \otimes \hat{j}\right) \otimes \hat{k} = \hat{i} \otimes \left(\hat{j} \otimes \hat{k}\right) = \hat{i} \otimes \hat{j} \otimes \hat{k} \\ \left(m\hat{i}\right) \otimes \left(n\hat{j}\right) = mn\left(\hat{i} \otimes \hat{j}\right) \end{cases} \tag{9-5}$$

$$\text{分配律}: \begin{cases} \left(\hat{i}+\hat{j}\right) \otimes \hat{k} = \hat{i} \otimes \hat{k} + \hat{j} \otimes \hat{k} \\ m\left(\hat{i} \otimes \hat{j} + \hat{k} \otimes \hat{l}\right) = m\left(\hat{i} \otimes \hat{j}\right) + m\left(\hat{k} \otimes \hat{l}\right) \\ \left(\hat{i}+\hat{j}\right) \otimes \left(\hat{k}+\hat{l}\right) = \hat{i} \otimes \hat{k} + \hat{i} \otimes \hat{l} + \hat{j} \otimes \hat{k} + \hat{j} \otimes \hat{l} \end{cases} \tag{9-6}$$

事实上, $b \otimes a$ 或 ba 也是一个并矢, 其运算如下:

$$\begin{aligned} b \otimes a &= \left(7\hat{i} + 0\hat{j} + 11\hat{k}\right) \otimes \left(3\hat{i} + 2\hat{j} + 5\hat{k}\right) \\ &= 21\hat{i} \otimes \hat{i} + 14\hat{i} \otimes \hat{j} + 35\hat{i} \otimes \hat{k} \\ &\quad + 0\hat{j} \otimes \hat{i} + 0\hat{j} \otimes \hat{j} + 0\hat{j} \otimes \hat{k} \\ &\quad + 33\hat{k} \otimes \hat{i} + 22\hat{k} \otimes \hat{j} + 55\hat{k} \otimes \hat{k} \end{aligned} \tag{9-7}$$

则 (9-7) 式中的二阶张量 $b \otimes a$ 的系数用矩阵表示为

$$[b \otimes a] = \begin{bmatrix} 21 & 14 & 35 \\ 0 & 0 & 0 \\ 33 & 22 & 55 \end{bmatrix} \tag{9-8}$$

对比 (9-4) 式, 可见两个并矢 $a \otimes b$ 和之间的关系互为转置 (transpose):

$$b \otimes a = (a \otimes b)^{\mathrm{T}} \tag{9-9}$$

张量 $b \otimes a$ 可表示为 $b \otimes a = (a \otimes b)^{\mathrm{T}} = T^{\mathrm{T}} = T_{ji} e_i \otimes e_j = T_{ij} e_j \otimes e_i$, 上述验证说明并矢运算不满足交换律 (commutation law), 用基矢量表示即 $\hat{i} \otimes \hat{j} \neq \hat{j} \otimes \hat{i}$, $\hat{i} \otimes \hat{j} = \left(\hat{j} \otimes \hat{i}\right)^{\mathrm{T}}$.

为讨论张量与矢量的点积运算, 再引入三维欧几里得空间的一个矢量:

$$c = 2\hat{i} + 3\hat{j} + 4\hat{k} \tag{9-10}$$

首先注意到基矢量点积满足 (8-5) 式的克罗内克符号的运算规则:

$$e_i \cdot e_j = \delta_{ij} = \begin{cases} 1, & i = j \\ 0, & i \neq j \end{cases}, \quad (i,j = 1,2,3) \tag{9-11}$$

我们来看如下运算：

$$(c \cdot a) b = \left(2\hat{i} + 3\hat{j} + 4\hat{k}\right) \cdot \left(3\hat{i} + 2\hat{j} + 5\hat{k}\right) b$$
$$= (6 + 6 + 20) \left(7\hat{i} + 0\hat{j} + 11\hat{k}\right)$$
$$= 224\hat{i} + 352\hat{k} \tag{9-12}$$

再来进行如下运算：$c \cdot (a \otimes b)$，也就是 (9-10) 式和二阶张量 (9-2) 式进行点积，则有

$$c \cdot (a \otimes b) = \left(2\hat{i} + 3\hat{j} + 4\hat{k}\right) \cdot \left(21\hat{i} \otimes \hat{i} + 33\hat{i} \otimes \hat{k} + 14\hat{j} \otimes \hat{i} + 22\hat{j} \otimes \hat{k}\right.$$
$$\left. + 35\hat{k} \otimes \hat{i} + 55\hat{k} \otimes \hat{k}\right)$$
$$= (42 + 3 \times 14 + 4 \times 35)\hat{i} + (66 + 3 \times 22 + 4 \times 55)\hat{k}$$
$$= 224\hat{i} + 352\hat{k} \tag{9-13}$$

对比 (9-12) 和 (9-13) 两式，表明：

$$(c \cdot a) b = c \cdot (a \otimes b) \tag{9-14}$$

矢量与张量进行点积运算，即二阶张量 $(a \otimes b)$ 将矢量 c 投影到矢量 b 的方向，投影的大小为 $(c \cdot a)$。

在学习《普通物理·力学》或《理论力学》时，刚体转动的角动量我们通常地表示为 $L = I\omega$，这里 ω 为角速度，I 为欧拉于 1765 年首次提出的转动惯量[3.1]。但角动量 L 和角速度 ω 均为矢量，所以，转动惯量 I 则必然是一个二阶张量，因为一个二阶张量 I 和一个矢量 ω 点积缩并后才能得到一个矢量（一阶张量）：$L = I \cdot \omega$，按照现代张量分析的一般约定，二阶张量和矢量的点积中的点可略去，则可表示为 $L = I\omega$，其指标形式为 $L_i = I_{ik}\omega_k$，这里，指标 i 为在等号两端均出现一次的自由指标 (free index)，而 k 为在等号右端重复出现，从而按照爱因斯坦求和约定进行求和的哑标。其实，在本科一年级学习《普通物理·力学》时，就不自觉地用到了转动惯量是一个二阶张量的概念。

9.2 现代数学对张量的引入——多重线性映射

现代数学对张量做出如下定义：张量是多重线性映射或函数，即

$$f: \mathcal{V}^r \times (\mathcal{V}^*)^s = \underbrace{\mathcal{V} \times \mathcal{V} \times \cdots \times \mathcal{V}}_{r \text{ 重协变}} \times \underbrace{\mathcal{V}^* \times \mathcal{V}^* \times \cdots \times \mathcal{V}^*}_{s \text{ 重逆变}} \to \mathbb{R}$$

其中，$\mathbb{R}$ 是实数域，$\mathcal{V}$ 是 $\mathbb{R}$ 上的向量空间 (vector space)，$\mathcal{V}^*$ 是 $\mathcal{V}$ 的对偶空间 (dual space) 或共轭空间 (conjugate space). r 和 s 都是大于等于零的整数. $\mathcal{V}^r \times (\mathcal{V}^*)^s$ 是 r 重空间 $\mathcal{V}$ 和 s 重空间 $\mathcal{V}^*$ 的笛卡儿积 (Cartesian product) 或直积 (direct product). 所有的 $r+s$ 重线性映射：$f: \mathcal{V}^r \times (\mathcal{V}^*)^s \to \mathbb{R}$ 都称为 (r,s) 型，$r+s$ 阶的张量. 亦称 f 是一个 r 次协变 (covariant, 共变) 且 s 次逆变 (contravariant, 反变) 的混合张量. 例如，一个 $r+s$ 阶的张量可表示为

$$\boldsymbol{T} = T_{i_1 \cdots i_r}^{k_1 \cdots k_s} \boldsymbol{e}^{i_1} \otimes \boldsymbol{e}^{i_2} \otimes \cdots \otimes \boldsymbol{e}^{i_r} \otimes \boldsymbol{e}_{k_1} \otimes \boldsymbol{e}_{k_2} \cdots \otimes \boldsymbol{e}_{k_s}$$

特别地，当 $s=0$ 时就称张量是协变的；反之，当 $r=0$ 时则称张量是逆变的. 可对 (r,s) 型张量举例如下：

(1) (0,0) 型零阶张量为标量；

(2) (1,0) 和 (0,1) 型一阶张量分别为向量 $f_i \boldsymbol{e}^i$ 和 $f^i \boldsymbol{e}_i$；

(3) (1,1) 型二阶张量为 $f_i^k \boldsymbol{e}^i \otimes \boldsymbol{e}_k$ 或 $f_k^i \boldsymbol{e}^k \otimes \boldsymbol{e}_i$；

(4) (2,0) 和 (0,2) 型二阶张量分别为 $f_{ik} \boldsymbol{e}^i \otimes \boldsymbol{e}^k$ 和 $f^{ik} \boldsymbol{e}_i \otimes \boldsymbol{e}_k$.

当然对于工程力学中常用的笛卡儿直角坐标系而言，向量空间的本身就是其对偶空间，此时没有必要区分协变和逆变.

多重线性 (multilinearity) 是张量的核心性质，多重线性是指张量对于每个参数都是线性的. 张量的旧定义 (按坐标变换方式定义)、张量的分量形式 (如二阶张量的矩阵表示) 等诸多性质，都可以从多重线性这一条推出.

多重线性用数学语言表述为：令 $\mathcal{V}$ 为向量空间 (vector space)，有 n 个向量 $\boldsymbol{u}_1, \boldsymbol{u}_2, \cdots, \boldsymbol{u}_n \in \mathcal{V}$，在 $\mathcal{K}$ 域中存在 n 个标量 $c_1, c_2, \cdots, c_n \in \mathcal{K}$，存在如下一阶齐次函数的线性映射：

$$f(c_1 \boldsymbol{u}_1 + c_2 \boldsymbol{u}_2 + \cdots + c_n \boldsymbol{u}_n) = c_1 f(\boldsymbol{u}_1) + c_2 f(\boldsymbol{u}_2) + \cdots + c_n f(\boldsymbol{u}_n)$$

作为多重线性映射定义的一种退化形式，就是人工智能中的深度学习开源软件 TensorFlow 中对张量的定义：张量是多维数组. TensorFlow 是这么定义张量的：A tensor is a generalization of vectors and matrices to potentially higher dimensions (张量是向量和矩阵向更高维度的推广). 也就是说，张量是多维数组，目的是把向量、矩阵推向更高的维度. 事实上，在深度学习中，为了方便，也经常会把 n 维数组都统称为张量，如图 9.1 所示. 另外，TensorFlow 中还有张量处理器 (tensor processing unit, TPU).

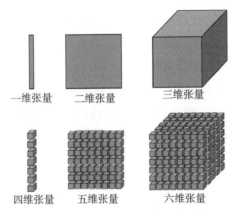

一维张量　二维张量　三维张量

四维张量　五维张量　六维张量

图 9.1　多维数组与多维张量

9.3　二阶投影张量的引入

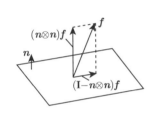

图 9.2　投影张量示意图

如图 9.2 所示，某截面上有一个一般不垂直于该截面的力矢量 $\bm{f}$. 力矢量 $\bm{f}$ 沿截面外法线 $\bm{n}$ 的投影的大小为 $\bm{f}\cdot\bm{n}$，由于标积 $\bm{f}\cdot\bm{n}$ 的方向是沿外法线 $\bm{n}$，故其可表示为 $(\bm{f}\cdot\bm{n})\bm{n}$，由 (9-14) 式，则该沿着外法线 $\bm{n}$ 的投影分量可通过引入所谓并矢表示为

$$\bm{f}_n = (\bm{f}\cdot\bm{n})\bm{n} = \bm{f}\cdot(\bm{n}\otimes\bm{n}) \tag{9-15}$$

由于二阶张量和矢量的点积中的点可略去，(9-15) 式则可表示为 $(\bm{f}\cdot\bm{n})\bm{n} = \bm{f}(\bm{n}\otimes\bm{n})$. 注意到省去点的操作，(9-15) 式还可以等价地表示为

$$\bm{f}_n = \bm{n}(\bm{n}\cdot\bm{f}) = (\bm{n}\otimes\bm{n})\cdot\bm{f} = (\bm{n}\otimes\bm{n})\bm{f} \tag{9-16}$$

截面上合力为 $\bm{f}$，外法线分量为 (9-15) 或 (9-16) 式，按照平行四边形法则，切线分量为

$$\bm{f}_\tau = \bm{f} - (\bm{n}\otimes\bm{n})\bm{f} = (\mathbf{I} - \bm{n}\otimes\bm{n})\bm{f} \tag{9-17}$$

二阶投影张量 (projection tensor)

$$\mathcal{P} = \mathbf{I} - \bm{n}\otimes\bm{n} \tag{9-18}$$

就是通过上式引入的. 提醒：$\bm{f}(\mathbf{I} - \bm{n}\otimes\bm{n})$ 表示的是矢量 $\bm{f}$ 和二阶投影张量的点积. 上式中，$\mathbf{I}$ 为二阶单位张量 (2nd order identity tensor)，可通过 Kronecker delta δ_{ij} 来定义：

$$\mathbf{I} = \delta_{ij}\bm{e}_i\otimes\bm{e}_j \tag{9-19}$$

用矩阵表示，二阶单位张量为元素为 1 的单位对角阵：

$$[\mathbf{I}] = \begin{bmatrix} 1 & 0 & 0 \\ 0 & 1 & 0 \\ 0 & 0 & 1 \end{bmatrix} \tag{9-20}$$

例 9.1 结合 (5-4) 式和例 8.1 和例 8.2，用单位基矢的外积表示二阶单位张量 (9-20) 式.

解：显然有

$$|1\rangle\langle 1| + |2\rangle\langle 2| + |3\rangle\langle 3| = \begin{bmatrix} 1 & 0 & 0 \\ 0 & 1 & 0 \\ 0 & 0 & 1 \end{bmatrix} = [\mathbf{I}]$$

例 9.2 推导刚体的转动惯量张量.

解：刚体转动的动量矩矢量为

$$\begin{aligned} \boldsymbol{L} &= \int \boldsymbol{r} \times (\rho \boldsymbol{v}) \mathrm{d}V \\ &= \rho \int \boldsymbol{r} \times (\boldsymbol{\omega} \times \boldsymbol{r}) \mathrm{d}V \end{aligned} \tag{9-21}$$

式中，$\boldsymbol{\omega}$ 为刚体的旋转角速度矢量. 由矢量三叉积公式 (s2-10)，有

$$\begin{aligned} \boldsymbol{L} &= \rho \int \boldsymbol{r} \times (\boldsymbol{\omega} \times \boldsymbol{r}) \mathrm{d}V = \rho \int [\boldsymbol{\omega}(\boldsymbol{r} \cdot \boldsymbol{r}) - \boldsymbol{r}(\boldsymbol{r} \cdot \boldsymbol{\omega})] \mathrm{d}V \\ &= \rho \int [r^2 \boldsymbol{\omega} - (\boldsymbol{r} \otimes \boldsymbol{r}) \cdot \boldsymbol{\omega}] \mathrm{d}V \\ &= \left[\rho \int (r^2 \mathbf{I} - \boldsymbol{r} \otimes \boldsymbol{r}) \mathrm{d}V \right] \cdot \boldsymbol{\omega} = \boldsymbol{I} \boldsymbol{\omega} \end{aligned} \tag{9-22}$$

式中，$\boldsymbol{I}\boldsymbol{\omega}$ 表示转动惯量张量 $\boldsymbol{I}$ 和旋转角速度 $\boldsymbol{\omega}$ 之间的点积，二阶转动惯量张量为

$$\boldsymbol{I} = \rho \int (r^2 \mathbf{I} - \boldsymbol{r} \otimes \boldsymbol{r}) \mathrm{d}V \tag{9-23}$$

其分量形式为

$$I_{ik} = \rho \int (r^2 \delta_{ik} - r_i r_k) \mathrm{d}V \tag{9-24}$$

即

$$\begin{cases} I_{xx} = \rho \int (x^2 + y^2 + z^2 - x^2) \mathrm{d}V = \rho \int (y^2 + z^2) \mathrm{d}V \\ I_{yy} = \rho \int (x^2 + y^2 + z^2 - y^2) \mathrm{d}V = \rho \int (x^2 + z^2) \mathrm{d}V \\ I_{zz} = \rho \int (x^2 + y^2 + z^2 - z^2) \mathrm{d}V = \rho \int (x^2 + y^2) \mathrm{d}V \\ I_{xy} = I_{yx} = -\rho \int (0 - xy) \mathrm{d}V = \rho \int xy \mathrm{d}V \\ I_{yz} = I_{zy} = -\rho \int (0 - yz) \mathrm{d}V = \rho \int yz \mathrm{d}V \\ I_{zx} = I_{xz} = -\rho \int (0 - zx) \mathrm{d}V = \rho \int zx \mathrm{d}V \end{cases} \tag{9-25}$$

上式表明，交叉项一般取其负值. 从而，二阶转动惯量可用矩阵形式表示为

$$[I_{ik}] = \begin{bmatrix} I_{xx} & I_{xy} & I_{xz} \\ I_{yx} & I_{yy} & I_{yz} \\ I_{zx} & I_{zy} & I_{zz} \end{bmatrix} \tag{9-26}$$

例 9.3 用狄拉克符号表示如下转动惯量的协变形式：

$$I^{ij} = \int \rho \left(r^2 \delta^{ij} - x^i x^j \right) \mathrm{d}V \tag{9-27}$$

解：由 (8-5) 式：

$$\langle i \mid j \rangle = \langle j \mid i \rangle = \delta^{ij} \tag{9-28}$$

再由 (5-5) 等式，有

$$I^{ij} = \langle i \mid \boldsymbol{I} \mid j \rangle, \quad r^2 = \boldsymbol{r} \cdot \boldsymbol{r} = \langle r \mid r \rangle, \quad x^i = \langle i \mid r \rangle = \langle r \mid i \rangle \tag{9-29}$$

则转动惯量张量的 (9-27) 式可用狄拉克符号表示为

$$\langle i \mid \boldsymbol{I} \mid j \rangle = \int \rho \left(\underbrace{\langle r \mid r \rangle}_{r^2} \underbrace{\langle i \mid j \rangle}_{\delta^{ij}} - \underbrace{\langle i \mid r \rangle}_{x^i} \underbrace{\langle r \mid j \rangle}_{x^j} \right) \mathrm{d}V \tag{9-30}$$

将积分括号中的 $\langle i |$ 和 $| j \rangle$ 提到积分号外，则转动惯量张量可进一步表示为

$$\langle i \mid \boldsymbol{I} \mid j \rangle = \left\langle i \left| \int \rho \left(\langle r \mid r \rangle \mathbf{I} - |r\rangle \langle r| \right) \mathrm{d}V \right| j \right\rangle \tag{9-31}$$

式中，$\mathbf{I}$ 为二阶单位张量. 上式中利用了下列狄拉克符号的重要分解：

$$\langle i \mid j \rangle = \langle j \mid i \rangle = \langle i | \cdot \mathbf{I} \cdot | j \rangle = \delta^{ij} \tag{9-32}$$

观察 (9-31) 式，转动惯量张量的可表示为如下紧凑形式：

$$\boldsymbol{I} = \int \rho \left(\langle r \mid r \rangle \mathbf{I} - |r\rangle \langle r| \right) \mathrm{d}V \tag{9-33}$$

例 9.4 二阶单位张量 $\mathbf{I}$ 在二阶张量求迹中的应用.

定义：对于笛卡儿坐标系下的任意二阶张量 $\boldsymbol{A} = A_{ij} \boldsymbol{e}_i \otimes \boldsymbol{e}_j$，其迹 (trace) 定义为

$$\mathrm{tr} \boldsymbol{A} = \mathrm{tr} \left(A_{ij} \boldsymbol{e}_i \otimes \boldsymbol{e}_j \right) = A_{ij} \boldsymbol{e}_i \cdot \boldsymbol{e}_j = A_{ij} \delta_{ij} = A_{ii} \tag{9-34}$$

事实上，二阶张量的迹 (9-34) 式还可以方便地表示为

$$\mathrm{tr} \boldsymbol{A} = \mathbf{I} : \boldsymbol{A} = \boldsymbol{A} : \mathbf{I} \tag{9-35}$$

可进行如下简要证明：

$$\boldsymbol{A}:\boldsymbol{I} = (A_{ij}\boldsymbol{e}_i \otimes \boldsymbol{e}_j):(\boldsymbol{e}_k \otimes \boldsymbol{e}_k) = A_{ij}\delta_{ik}\delta_{jk} = A_{kk} \tag{9-36}$$

比较 (9-36) 和 (9-34) 两式，故得证.

例 9.5 二阶投影张量在流体无滑移边界条件中的应用.

解：对于外单位法线为 $\boldsymbol{n}$ 的静止的固体壁面，常用的流–固界面边界条件有两个，一是无渗条件 (impermeability condition)：

$$\boldsymbol{n} \cdot \boldsymbol{u} = 0 \tag{9-37}$$

式中，$\boldsymbol{u}$ 为流体在固体边界处的流速；二是无滑移边界条件 (non-slip boundary condition)：

$$\begin{aligned}\boldsymbol{u} - \boldsymbol{n}\,(\boldsymbol{n} \cdot \boldsymbol{u}) &= \boldsymbol{u}\,(\boldsymbol{I} - \boldsymbol{n} \otimes \boldsymbol{n}) \\ &= \boldsymbol{u}\mathcal{P} = \boldsymbol{0}\end{aligned} \tag{9-38}$$

上式中的 $\mathcal{P}$ 就是由 (9-18) 式定义的二阶投影张量. 由 (9-37) 和 (9-38) 两式得出流–固界面边界条件为 $\boldsymbol{u} = \boldsymbol{0}$.

§10. 爱丁顿张量和张量方程的广义量纲原理

在 §8.3，我们通过混合积定义了列维–奇维塔置换符号，也称为 Epsilon 符号：

$$\begin{bmatrix}\boldsymbol{e}_i & \boldsymbol{e}_j & \boldsymbol{e}_k\end{bmatrix} = \boldsymbol{e}_i \cdot (\boldsymbol{e}_j \times \boldsymbol{e}_k) = \varepsilon_{jkl}\boldsymbol{e}_i \cdot \boldsymbol{e}_l = \varepsilon_{jkl}\delta_{il} = \varepsilon_{jki} = \varepsilon_{ijk} \tag{10-1}$$

则直角坐标系下爱丁顿张量 (Eddington tensor) 定义为

$$\boldsymbol{\varepsilon} = \varepsilon_{ijk}\boldsymbol{e}_i \otimes \boldsymbol{e}_j \otimes \boldsymbol{e}_k \tag{10-2}$$

式中，

$$\varepsilon_{ijk} = \begin{cases} 1, & \text{当 } i,j,k \text{ 按照偶置换取值，即 } 123, 231, 312 \\ -1, & \text{当 } i,j,k \text{ 按照偶置换取值，即 } 132, 213, 321 \\ 0, & \text{当两个或两个以上指标取值相同时} \end{cases} \tag{10-3}$$

由于 i, j, k 可分别取为 1, 2, 3，爱丁顿张量就有 3×3×3 个分量，引用数学排列的概念，三个指标皆不相同共有 A_3^3 即 6 种排列方式，其余 21 项为零，则三阶爱丁顿张量的显式包含有如下 6 项：

$$\boldsymbol{\varepsilon} = \boldsymbol{e}_1 \otimes \boldsymbol{e}_2 \otimes \boldsymbol{e}_3 + \boldsymbol{e}_2 \otimes \boldsymbol{e}_3 \otimes \boldsymbol{e}_1 + \boldsymbol{e}_3 \otimes \boldsymbol{e}_1 \otimes \boldsymbol{e}_2$$

$$-e_1 \otimes e_3 \otimes e_2 - e_2 \otimes e_1 \otimes e_3 - e_3 \otimes e_2 \otimes e_1 \tag{10-4}$$

由爱丁顿张量的显式 (10-4)，我们可以容易地证明：

$$\begin{aligned}
\varepsilon_{ijk}\varepsilon_{ijk} = \boldsymbol{\varepsilon} &\vdots \boldsymbol{\varepsilon} = (e_1 \otimes e_2 \otimes e_3 + e_2 \otimes e_3 \otimes e_1 + e_3 \otimes e_1 \otimes e_2) \\
&\vdots (e_1 \otimes e_2 \otimes e_3 + e_2 \otimes e_3 \otimes e_1 + e_3 \otimes e_1 \otimes e_2) \\
&+ (-e_1 \otimes e_3 \otimes e_2 - e_2 \otimes e_1 \otimes e_3 - e_3 \otimes e_2 \otimes e_1) \\
&\vdots (-e_1 \otimes e_3 \otimes e_2 - e_2 \otimes e_1 \otimes e_3 - e_3 \otimes e_2 \otimes e_1) \\
&= 6
\end{aligned} \tag{10-5}$$

顺便指出的是，可对物理量进行如下分类 (专有名词)：

在同构的意义下，零阶张量 (其秩或阶 $r=0$) 为标量 (scalars)；一阶张量 ($r=1$) 为向量 (vectors)；二阶张量 (dyadics, $r=2$) 则成为矩阵 (matrix)；三阶张量 (triadics, $r=3$)，如爱丁顿张量；四阶张量 (tetradics, $r=4$)，如四阶单位张量、四阶对称单位张量、四阶弹性张量、四阶柔度张量等.

由于指标方式的不同，张量分成协变张量 (covariant tensor，指标在下者)、逆变张量 (contravariant tensor，指标在上者)、混合张量 (指标在上和指标在下两者都有) 三类.

爱丁顿 (Arthur Eddington, 1882—1944, 如图 10.1 和爱因斯坦的合影) 是最早尝试应用纯理论方法计算精细结构常数的科学家. 他用纯逻辑证明，精细结构常数应该等于：$1/\alpha = (162-16)/2 + 16 = 136$. 这与当时的实验结果相符合. 但随着实验精确性的提高，发现精细结构常数更接近于 1/137，于是爱丁顿宣称自己在计算中犯了个错误，他重新计算后断定一定等于整数的 137. 爱丁顿在这个事情上的摇摆为他带来了名誉上的损害，也就是从 136 到 137，为他 "赢得" 了绰号："爱丁旺"(Adding-one). 图 10.1 右图中的威廉·德西特 (Willem de Sitter, 1872—1934) 和爱因斯坦于 1932 年共同发表论文，声称宇宙可能有大量不发光的物质，就是今日所说的暗物质. 他还提出了爱因斯坦广义相对论的解，即德西特宇宙 (de Sitter universe) 和德西特空间 (de Sitter space). 有关精细结构常数的进一步讨论见本章思考题 3.7—3.11.

作为天文学家的爱丁顿利用 1919 年 5 月 29 日持续时间长达 6 分 51 秒的超长日全食，对广义相对论中光线的引力偏折效应进行了成功检验. 他后来把获得观测结果的那一刻称为自己一生最伟大的时刻. 那一刻不仅是他个人的伟大时刻，而且也使爱因斯坦几乎在一夜之间获得了世界性的公众影响.

图 10.1　左图：爱因斯坦 (左) 和爱丁顿 (右)；右图：1923 年 9 月荷兰莱顿, 前排左起：爱丁顿、洛伦兹；后排左起：爱因斯坦、埃伦费斯特、德西特

提到爱丁顿就不能不提他首版于 1923 年的名著《相对论的数学理论》[3.2]. 爱丁顿在这部巨著中, 将我们所熟知的对于代数方程适用的量纲一致性原理推广到张量方程情形.

代数方程的量纲一致性原理表明, 方程中的每一项必须要有相同的量纲, 如后面我们要着重提到的伯努利方程等. 但是对于有协变和逆变分量的张量方程又如何呢? 例如, 对于测地线方程 (geodesic equation, 见例 16.4 的详细证明):

$$\frac{\mathrm{d}^2 x^i}{\mathrm{d} t^2} + \Gamma^i_{jk} \frac{\mathrm{d} x^j}{\mathrm{d} t} \frac{\mathrm{d} x^k}{\mathrm{d} t} = 0 \tag{10-6}$$

以及杨-米尔斯规范场 (非阿贝尔规范场):

$$F_{\mu\nu} = \underbrace{\left(\frac{\partial B_\nu}{\partial x_\mu} - \frac{\partial B_\mu}{\partial x_\nu}\right)}_{B \text{ 是 } 2\times 2 \text{ 的矩阵}} + \underbrace{(B_\mu B_\nu - B_\nu B_\mu)}_{\text{对易子}} \tag{10-7}$$

如何理解上述方程中每一项的量纲一致性？

爱丁顿在其名著《相对论的数学理论》[3.2] 的第 21 节中开宗明义地指出：

▶ 张量微积分将量纲原理推广至度量法则 (measure-code) 的改变, 而这比仅仅改变单位要更普遍的多 (The tensor calculus extends this *principle of dimensions* to changes of measure-code much more general than mere changes of units).

▶ 协变和逆变是一种广义的量纲, 其展示了世界一种情况的度量如何向另外一种情况的度量改变 (Covariance and contravariance are a kind of generalized dimension, showing how the measure of one condition of the world is changed when the measure of another condition is changed).

> 一般原则是，方程的两边必须有协变和逆变的相同成分 (The general rule is that both sides of the equation must have the same elements of covariance and contravariance).

> 作为只是改变单位的经典的量纲原理只是广义量纲的一个特例 (The ordinary theory of change of units is merely an elementary case of this).

阅读名著心中经常有一种不可名状的豁然开朗的感觉——或称 "顿悟 (the intoxicating feeling of sudden understanding)"，爱丁顿首版于 1923 年的上述著者之所以近百年之后仍然受人追捧，自然有其独到之见解. 著者读爱丁顿出版于 90 多年前的这本 "老书" 仍然能产生共鸣！

对于张量方程，需要满足两个一致性：不但要满足每一项的物理量纲的一致性，还必须要满足协变和逆变的一致性.

§11. 张量的缩并

11.1 简单缩并

需要提醒的是，现代一般文献中，将一任意二阶张量 $\boldsymbol{T} = T_{ij}\boldsymbol{e}_i \otimes \boldsymbol{e}_j$ 和一任意矢量 $\boldsymbol{a} = a_k\boldsymbol{e}_k$ 之间的点积之间的 "·" 省略掉：

$$\boldsymbol{T}\boldsymbol{a} = (T_{ij}\boldsymbol{e}_i \otimes \boldsymbol{e}_j) \cdot (a_k\boldsymbol{e}_k) = T_{ij}a_k\boldsymbol{e}_i(\boldsymbol{e}_j \cdot \boldsymbol{e}_k) = T_{ij}a_k\delta_{jk}\boldsymbol{e}_i = T_{ij}a_j\boldsymbol{e}_i \tag{11-1}$$

一任意二阶张量 $\boldsymbol{T} = T_{ij}\boldsymbol{e}_i \otimes \boldsymbol{e}_j$ 和另一任意二阶张量 $\boldsymbol{S} = S_{kl}\boldsymbol{e}_k \otimes \boldsymbol{e}_l$ 之间的简单缩并 (simple contraction)：

$$\begin{aligned}\boldsymbol{T}\boldsymbol{S} &= (T_{ij}\boldsymbol{e}_i \otimes \boldsymbol{e}_j) \cdot (S_{kl}\boldsymbol{e}_k \otimes \boldsymbol{e}_l) = T_{ij}S_{kl}(\boldsymbol{e}_j \cdot \boldsymbol{e}_k)(\boldsymbol{e}_i \otimes \boldsymbol{e}_l) \\ &= T_{ij}S_{kl}\delta_{jk}\boldsymbol{e}_i \otimes \boldsymbol{e}_l = T_{ij}S_{jl}\boldsymbol{e}_i \otimes \boldsymbol{e}_l\end{aligned} \tag{11-2}$$

同理，上面的运算可以推广到两个任意阶张量间的简单缩并，两个张量之间的简单缩并是两个最相邻的基矢量间的点积，点积后成为克罗内克 δ：

$$\begin{aligned}(\boldsymbol{a} \otimes \boldsymbol{b})\boldsymbol{c} &= (\boldsymbol{b} \cdot \boldsymbol{c})\boldsymbol{a} \\ (\boldsymbol{a} \otimes \boldsymbol{b})(\boldsymbol{c} \otimes \boldsymbol{d}) &= (\boldsymbol{b} \cdot \boldsymbol{c})(\boldsymbol{a} \otimes \boldsymbol{d}) \\ (\boldsymbol{a} \otimes \boldsymbol{b})(\boldsymbol{c} \otimes \boldsymbol{d} \otimes \boldsymbol{e}) &= (\boldsymbol{b} \cdot \boldsymbol{c})(\boldsymbol{a} \otimes \boldsymbol{d} \otimes \boldsymbol{e}) \\ (\boldsymbol{a} \otimes \boldsymbol{b} \otimes \boldsymbol{c})(\boldsymbol{d} \otimes \boldsymbol{e} \otimes \boldsymbol{f}) &= (\boldsymbol{c} \cdot \boldsymbol{d})(\boldsymbol{a} \otimes \boldsymbol{b} \otimes \boldsymbol{e} \otimes \boldsymbol{f})\end{aligned} \tag{11-3}$$

一般地，一个任意 m 阶张量和一任意 n 阶张量之间的简单缩并，所得到新的张量的阶数为 $m + n - 2$.

11.2 张量的幂

一任意二阶张量 $\boldsymbol{T} = T_{ij}\boldsymbol{e}_i \otimes \boldsymbol{e}_j$ 的零次幂为二阶单位张量：

$$\boldsymbol{T}^0 = \mathbf{I} \tag{11-4}$$

通过简单缩并，可定义二阶张量的任意次幂：

$$\begin{aligned} \boldsymbol{T}^2 &= \boldsymbol{TT} \\ \boldsymbol{T}^3 &= \boldsymbol{TTT} \\ \boldsymbol{T}^n &= \boldsymbol{T}^{n-1}\boldsymbol{T} \end{aligned} \tag{11-5}$$

11.3 双缩并

双缩并 (double contraction) 一般用 ":" 表示，如：$\boldsymbol{T} : \boldsymbol{S}$，英文是 "colon product"。

两对并矢 (dyadic) 之间的双缩并为

$$(\boldsymbol{a} \otimes \boldsymbol{b}) : (\boldsymbol{c} \otimes \boldsymbol{d}) = (\boldsymbol{a} \cdot \boldsymbol{c})(\boldsymbol{b} \cdot \boldsymbol{d}) \tag{11-6}$$

两对三阶张量之间的双缩并为

$$(\boldsymbol{a} \otimes \boldsymbol{b} \otimes \boldsymbol{c}) : (\boldsymbol{d} \otimes \boldsymbol{e} \otimes \boldsymbol{f}) = (\boldsymbol{b} \cdot \boldsymbol{d})(\boldsymbol{c} \cdot \boldsymbol{e})(\boldsymbol{a} \otimes \boldsymbol{f}) \tag{11-7}$$

两个二阶张量 $\boldsymbol{T} = T_{ij}\boldsymbol{e}_i \otimes \boldsymbol{e}_j$ 和 $\boldsymbol{S} = S_{kl}\boldsymbol{e}_k \otimes \boldsymbol{e}_l$ 之间的双缩并为标量：

$$\boldsymbol{T} : \boldsymbol{S} = (T_{ij}\boldsymbol{e}_i \otimes \boldsymbol{e}_j) : (S_{kl}\boldsymbol{e}_k \otimes \boldsymbol{e}_l) = T_{ij}S_{kl}(\boldsymbol{e}_i \cdot \boldsymbol{e}_k)(\boldsymbol{e}_j \cdot \boldsymbol{e}_l) = T_{ij}S_{kl}\delta_{ik}\delta_{jl} = T_{ij}S_{ij} \tag{11-8}$$

两个二阶张量 $\boldsymbol{A}$ 和 $\boldsymbol{B}$ 之间的双缩并和迹的计算存在如下重要关系：

$$\begin{aligned} \boldsymbol{A} : \boldsymbol{B} &= \operatorname{tr}\left(\boldsymbol{A}^{\mathrm{T}}\boldsymbol{B}\right) = \operatorname{tr}\left(\boldsymbol{B}^{\mathrm{T}}\boldsymbol{A}\right) \\ &= \operatorname{tr}\left(\boldsymbol{A}\boldsymbol{B}^{\mathrm{T}}\right) = \operatorname{tr}\left(\boldsymbol{B}\boldsymbol{A}^{\mathrm{T}}\right) \\ &= \boldsymbol{B} : \boldsymbol{A} \end{aligned} \tag{11-9}$$

例 11.1 证明：$\boldsymbol{A}^{-\mathrm{T}} : \boldsymbol{A} = 3$.

证明：应用 (11-9) 式，$\boldsymbol{A}^{-\mathrm{T}} : \boldsymbol{A} = \operatorname{tr}\left\{\left(\boldsymbol{A}^{-\mathrm{T}}\right)^{\mathrm{T}}\boldsymbol{A}\right\} = \operatorname{tr}\left(\boldsymbol{A}^{-1}\boldsymbol{A}\right) = \operatorname{tr}\mathbf{I} = 3$. 证毕.

作为 (11-9) 式后续的一个直接应用，是功的共轭对之间的双缩并：

$$\begin{cases} \int_{\Omega} \boldsymbol{\sigma} : \boldsymbol{d}\,\mathrm{d}v = \int_{\Omega} \operatorname{tr}\left(\boldsymbol{\sigma}^{\mathrm{T}}\boldsymbol{d}\right)\mathrm{d}v \\ \int_{\Omega_0} \boldsymbol{P} : \dot{\boldsymbol{F}}\,\mathrm{d}V = \int_{\Omega_0} \operatorname{tr}\left(\boldsymbol{P}^{\mathrm{T}}\dot{\boldsymbol{F}}\right)\mathrm{d}V \end{cases} \tag{11-10}$$

双缩并一些常用的关系式如下：

$$\mathbf{I}:\mathbf{A} = \operatorname{tr}\mathbf{A} = \mathbf{A}:\mathbf{I} \tag{11-11}$$

$$\mathbf{A}:(\mathbf{BC}) = \left(\mathbf{B}^{\mathrm{T}}\mathbf{A}\right):\mathbf{C} = \left(\mathbf{A}\mathbf{C}^{\mathrm{T}}\right):\mathbf{B} \tag{11-12}$$

$$\mathbf{A}:(\mathbf{u}\otimes\mathbf{v}) = \mathbf{u}\cdot\mathbf{A}\mathbf{v} = (\mathbf{u}\otimes\mathbf{v}):\mathbf{A} \tag{11-13}$$

$$(\mathbf{u}\otimes\mathbf{v}):(\mathbf{w}\otimes\mathbf{x}) = (\mathbf{u}\cdot\mathbf{w})(\mathbf{v}\cdot\mathbf{x}) \tag{11-14}$$

$$(\mathbf{e}_i\otimes\mathbf{e}_j):(\mathbf{e}_k\otimes\mathbf{e}_l) = (\mathbf{e}_i\cdot\mathbf{e}_k)(\mathbf{e}_j\cdot\mathbf{e}_l) = \delta_{ik}\delta_{jl} \tag{11-15}$$

例 11.2 给出连续介质力学中最为常用的关系式 (11-12) 的两种证明.

方法一：

根据两个二阶张量的内积定义 (11-9) 式：

$$\begin{aligned}\mathbf{A}:\mathbf{B} &= (A_{ij}\mathbf{e}_i\otimes\mathbf{e}_j):(B_{mn}\mathbf{e}_m\otimes\mathbf{e}_n) = A_{ij}B_{mn}(\mathbf{e}_i\cdot\mathbf{e}_m)(\mathbf{e}_j\cdot\mathbf{e}_n)\\ &= A_{ij}B_{mn}\delta_{im}\delta_{jn} = A_{ij}B_{ij} = \operatorname{tr}\left(\mathbf{A}^{\mathrm{T}}\mathbf{B}\right)\end{aligned}$$

且张量的迹满足性质 $\operatorname{tr}\mathbf{A}^{\mathrm{T}} = \operatorname{tr}\mathbf{A}$ 和 $\operatorname{tr}(\mathbf{AB}) = \operatorname{tr}(\mathbf{BA})$，则有

$$\mathbf{A}:\mathbf{B} = \operatorname{tr}\left(\mathbf{A}^{\mathrm{T}}\mathbf{B}\right) = \operatorname{tr}\left(\mathbf{B}^{\mathrm{T}}\mathbf{A}\right) = \operatorname{tr}\left(\mathbf{A}\mathbf{B}^{\mathrm{T}}\right) = \operatorname{tr}\left(\mathbf{B}\mathbf{A}^{\mathrm{T}}\right) = \mathbf{B}:\mathbf{A} \tag{11-16}$$

即为 $A_{ij}B_{ij} = B_{ij}A_{ij}$. 根据 (11-16) 式，则常用关系式满足

$$\mathbf{A}:(\mathbf{BC}) = \operatorname{tr}\left[\mathbf{A}^{\mathrm{T}}(\mathbf{BC})\right] = \operatorname{tr}\left[\left(\mathbf{A}^{\mathrm{T}}\mathbf{B}\right)\mathbf{C}\right] = \left(\mathbf{B}^{\mathrm{T}}\mathbf{A}\right):\mathbf{C} = \mathbf{C}:\left(\mathbf{B}^{\mathrm{T}}\mathbf{A}\right)$$

$$\mathbf{A}:(\mathbf{BC}) = \operatorname{tr}\left[\mathbf{A}\left(\mathbf{C}^{\mathrm{T}}\mathbf{B}^{\mathrm{T}}\right)\right] = \operatorname{tr}\left[\left(\mathbf{A}\mathbf{C}^{\mathrm{T}}\right)\mathbf{B}^{\mathrm{T}}\right] = \left(\mathbf{A}\mathbf{C}^{\mathrm{T}}\right):\mathbf{B} = \mathbf{B}:\left(\mathbf{A}\mathbf{C}^{\mathrm{T}}\right)$$

得证.

方法二：

采用下标形式：

$$\begin{aligned}\mathbf{A}:(\mathbf{BC}) &= (A_{ij}\mathbf{e}_i\otimes\mathbf{e}_j):[(B_{mn}\mathbf{e}_m\otimes\mathbf{e}_n)(C_{kl}\mathbf{e}_k\otimes\mathbf{e}_l)]\\ &= (A_{ij}\mathbf{e}_i\otimes\mathbf{e}_j):(B_{mn}C_{kl}\delta_{nk}\mathbf{e}_m\otimes\mathbf{e}_l)\\ &= A_{ij}B_{mn}C_{nl}(\mathbf{e}_i\otimes\mathbf{e}_j):(\mathbf{e}_m\otimes\mathbf{e}_l)\\ &= A_{ij}B_{mn}C_{nl}(\mathbf{e}_i\cdot\mathbf{e}_m)(\mathbf{e}_j\cdot\mathbf{e}_l) = A_{ij}B_{mn}C_{nl}\delta_{im}\delta_{jl}\\ &= A_{ij}B_{in}C_{nj}\end{aligned} \tag{11-17}$$

同理,有

$$\begin{aligned}
\left(\boldsymbol{B}^{\mathrm{T}}\boldsymbol{A}\right):\boldsymbol{C} &= [(B_{nm}\boldsymbol{e}_m\otimes\boldsymbol{e}_n)(A_{ij}\boldsymbol{e}_i\otimes\boldsymbol{e}_j)]:(C_{kl}\boldsymbol{e}_k\otimes\boldsymbol{e}_l) \\
&= (A_{ij}B_{nm}\delta_{ni}\boldsymbol{e}_m\otimes\boldsymbol{e}_j):(C_{kl}\boldsymbol{e}_k\otimes\boldsymbol{e}_l) \\
&= A_{ij}B_{im}C_{kl}(\boldsymbol{e}_m\cdot\boldsymbol{e}_k)(\boldsymbol{e}_j\cdot\boldsymbol{e}_l) = A_{ij}B_{im}C_{kl}\delta_{mk}\delta_{jl} \\
&= A_{ij}B_{im}C_{mj}
\end{aligned} \quad (11\text{-}18)$$

$$\begin{aligned}
\left(\boldsymbol{A}\boldsymbol{C}^{\mathrm{T}}\right):\boldsymbol{B} &= [(A_{ij}\boldsymbol{e}_i\otimes\boldsymbol{e}_j)(C_{lk}\boldsymbol{e}_k\otimes\boldsymbol{e}_l)]:(B_{mn}\boldsymbol{e}_m\boldsymbol{e}_n) \\
&= (A_{ij}C_{lk}\delta_{jk}\boldsymbol{e}_i\otimes\boldsymbol{e}_l):(B_{mn}\boldsymbol{e}_m\otimes\boldsymbol{e}_n) \\
&= A_{ij}B_{mn}C_{lj}(\boldsymbol{e}_i\cdot\boldsymbol{e}_m)(\boldsymbol{e}_l\cdot\boldsymbol{e}_n) = A_{ij}B_{mn}C_{lj}\delta_{im}\delta_{nl} \\
&= A_{ij}B_{in}C_{nj}
\end{aligned} \quad (11\text{-}19)$$

由 (11-17)—(11-19) 三式,显然,$\boldsymbol{A}:(\boldsymbol{B}\boldsymbol{C}) = \left(\boldsymbol{B}^{\mathrm{T}}\boldsymbol{A}\right):\boldsymbol{C} = \left(\boldsymbol{A}\boldsymbol{C}^{\mathrm{T}}\right):\boldsymbol{B}$ 成立.

11.4 三缩并

作为三阶张量的爱丁顿张量通过三缩并 (triple contraction) 在混合积中的应用,如下两种运算是等价的:

$$\begin{aligned}
\begin{bmatrix} \boldsymbol{a} & \boldsymbol{b} & \boldsymbol{c} \end{bmatrix} &= \boldsymbol{\varepsilon}\vdots(\boldsymbol{a}\otimes\boldsymbol{b}\otimes\boldsymbol{c}) = (\varepsilon_{ijk}\boldsymbol{e}_i\otimes\boldsymbol{e}_j\otimes\boldsymbol{e}_k)\vdots(a_l b_m c_n\boldsymbol{e}_l\otimes\boldsymbol{e}_m\otimes\boldsymbol{e}_n) \\
&= \varepsilon_{ijk}a_l b_m c_n\delta_{il}\delta_{jm}\delta_{kn} = \varepsilon_{ijk}a_i b_j c_k
\end{aligned} \quad (11\text{-}20)$$

$$\begin{aligned}
\begin{bmatrix} \boldsymbol{a} & \boldsymbol{b} & \boldsymbol{c} \end{bmatrix} &= (\boldsymbol{a}\otimes\boldsymbol{b}\otimes\boldsymbol{c})\vdots\boldsymbol{\varepsilon} = (a_i b_j c_k\boldsymbol{e}_i\otimes\boldsymbol{e}_j\otimes\boldsymbol{e}_k)\vdots(\varepsilon_{lmn}\boldsymbol{e}_l\otimes\boldsymbol{e}_m\otimes\boldsymbol{e}_n) \\
&= a_i b_j c_k\varepsilon_{lmn}\delta_{il}\delta_{jm}\delta_{kn} = a_i b_j c_k\varepsilon_{ijk}
\end{aligned} \quad (11\text{-}21)$$

§12. 张量的转置、逆、对称化、反对称化,科恩不等式

二阶张量 $\boldsymbol{F}$ 的转置 (transpose) 记为 $\boldsymbol{F}^{\mathrm{T}}$,也就是其矩阵的行和列调换后的张量. 在后续章节中,将给出很多转置张量的具体例子. 有三个重要概念和张量的转置相关:

1. 对称张量 (symmetrical tensor):当张量 $\boldsymbol{A}$ 满足 $\boldsymbol{A} = \boldsymbol{A}^{\mathrm{T}}$ 时.

2. 反对称张量 (skew symmetrical, antisymmetrical tensor):当 $\boldsymbol{A}$ 满足 $\boldsymbol{A} = -\boldsymbol{A}^{\mathrm{T}}$ 时.

3. 张量的对称化和反对称化:对于任意张量 $\boldsymbol{S}$,可通过下列简单操作将其对称化和反对称化:

$$\boldsymbol{S}_{\mathrm{sym}} = \frac{\boldsymbol{S}+\boldsymbol{S}^{\mathrm{T}}}{2}, \quad \boldsymbol{S}_{\mathrm{anti}} = \frac{\boldsymbol{S}-\boldsymbol{S}^{\mathrm{T}}}{2} \quad (12\text{-}1)$$

对于两个任意二阶张量 $\boldsymbol{A}$ 和 $\boldsymbol{B}$，其转置的运算规则为

$$(\alpha\boldsymbol{A}+\beta\boldsymbol{B})^{\mathrm{T}} = \alpha\boldsymbol{A}^{\mathrm{T}}+\beta\boldsymbol{B}^{\mathrm{T}} \tag{12-2}$$

$$(\boldsymbol{AB})^{\mathrm{T}} = \boldsymbol{B}^{\mathrm{T}}\boldsymbol{A}^{\mathrm{T}} \tag{12-3}$$

$$\left(\boldsymbol{A}^{\mathrm{T}}\right)^{\mathrm{T}} = \boldsymbol{A} \tag{12-4}$$

(12-2) 式中，α 和 β 为标量.

张量 $\boldsymbol{A}$ 的行列式 (determinant) 是一个标量：

$$\det\boldsymbol{A} = \begin{vmatrix} A_{11} & A_{12} & A_{13} \\ A_{21} & A_{22} & A_{23} \\ A_{31} & A_{32} & A_{33} \end{vmatrix} = \frac{1}{6}\varepsilon_{ijk}\varepsilon_{pqr}A_{ip}A_{jq}A_{kr} \tag{12-5}$$

当且仅当 (if and only if, iff) $\det\boldsymbol{A}=0$ 时，张量 $\boldsymbol{A}$ 是奇异的 (singular).

二阶张量的行列式有如下常用的性质：

$$\det(\boldsymbol{AB}) = (\det\boldsymbol{A})(\det\boldsymbol{B}) \tag{12-6}$$

$$\det\boldsymbol{A}^{\mathrm{T}} = \det\boldsymbol{A} \tag{12-7}$$

请参阅例 1.4 有关恒等映射的定义，并对下面的运算做对比. 当张量 $\boldsymbol{A}$ 非奇异时，也就是当 $\det\boldsymbol{A}\neq 0$ 时，张量 $\boldsymbol{A}$ 存在逆 (inverse) $\boldsymbol{A}^{-1}$，满足下列关系式：

$$\boldsymbol{AA}^{-1} = \boldsymbol{A}^{-1}\boldsymbol{A} = \mathbf{I} \tag{12-8}$$

$$(\boldsymbol{AB})^{-1} = \boldsymbol{B}^{-1}\boldsymbol{A}^{-1} \tag{12-9}$$

$$\left(\boldsymbol{A}^{-1}\right)^{-1} = \boldsymbol{A} \tag{12-10}$$

$$(\alpha\boldsymbol{A})^{-1} = \frac{1}{\alpha}\boldsymbol{A}^{-1} \tag{12-11}$$

$$\left(\boldsymbol{A}^{-1}\right)^{\mathrm{T}} = \left(\boldsymbol{A}^{\mathrm{T}}\right)^{-1} = \boldsymbol{A}^{-\mathrm{T}} \tag{12-12}$$

$$(\boldsymbol{AB})^{-\mathrm{T}} = \left[(\boldsymbol{AB})^{\mathrm{T}}\right]^{-1} = \left[(\boldsymbol{AB})^{-1}\right]^{\mathrm{T}} = \boldsymbol{A}^{-\mathrm{T}}\boldsymbol{B}^{-\mathrm{T}} \tag{12-13}$$

$$\det\left(\boldsymbol{A}^{-1}\right) = (\det\boldsymbol{A})^{-1} \tag{12-14}$$

下面让我们利用二阶张量逆的定义式 (12-8) 式来证明 (12-9) 式. 由于：$\left(\boldsymbol{B}^{-1}\boldsymbol{A}^{-1}\right)(\boldsymbol{AB}) = \boldsymbol{B}^{-1}\left(\boldsymbol{A}^{-1}\boldsymbol{A}\right)\boldsymbol{B} = \boldsymbol{B}^{-1}\mathbf{I}\boldsymbol{B} = \boldsymbol{B}^{-1}\boldsymbol{B} = \mathbf{I}$，故得证. 对 (24-30) 式的逆操作比喻为："出门时，先穿袜子再穿鞋；回家时，先脱鞋，再脱袜子."

例 12.1 作为理性连续介质力学常用性质之一,如果张量 $\boldsymbol{A}$ 为满足 $\boldsymbol{A} = \boldsymbol{A}^{\mathrm{T}}$ 的二阶对称张量,而 $\boldsymbol{S}$ 为任意张量,证明两个张量之间的双缩并满足

$$\boldsymbol{A} : \boldsymbol{S} = \boldsymbol{A} : \boldsymbol{S}_{\mathrm{sym}} = \boldsymbol{A} : \frac{\boldsymbol{S} + \boldsymbol{S}^{\mathrm{T}}}{2} \tag{12-15}$$

证明:容易理解,二阶对称张量 $\boldsymbol{A}$ 和一个二阶反对称张量的双缩并的结果为零,由 (12-1) 式,也就是

$$\boldsymbol{A} : \boldsymbol{S}_{\mathrm{anti}} = \boldsymbol{A} : \frac{\boldsymbol{S} - \boldsymbol{S}^{\mathrm{T}}}{2} = 0 \tag{12-16}$$

故 (12-15) 式得证.

例 12.2 求二阶张量的球量和偏量.

解:二阶张量 $\boldsymbol{A}$ 的球量部分 (spherical part),也就是相当于静水压强,可通过张量的迹用矩阵表示为

$$[\alpha \mathbf{I}] = \begin{bmatrix} \frac{1}{3}(A_{11} + A_{22} + A_{33}) & 0 & 0 \\ 0 & \frac{1}{3}(A_{11} + A_{22} + A_{33}) & 0 \\ 0 & 0 & \frac{1}{3}(A_{11} + A_{22} + A_{33}) \end{bmatrix} \tag{12-17}$$

α 为一标量,而二阶张量 $\boldsymbol{A}$ 的球型张量则表示为

$$\alpha \mathbf{I} = \frac{1}{3}(\mathrm{tr}\boldsymbol{A})\mathbf{I} = \frac{1}{3}(\mathbf{I} : \boldsymbol{A})\mathbf{I} \tag{12-18}$$

二阶张量 $\boldsymbol{A}$ 的偏量 (deviatoric part) 用矩阵表示为

$$[\mathrm{dev}\boldsymbol{A}] = \left[\boldsymbol{A} - \frac{1}{3}(\mathbf{I} : \boldsymbol{A})\mathbf{I}\right]$$

$$= \begin{bmatrix} A_{11} - \frac{A_{11} + A_{22} + A_{33}}{3} & A_{12} & A_{13} \\ A_{21} & A_{22} - \frac{A_{11} + A_{22} + A_{33}}{3} & A_{23} \\ A_{31} & A_{32} & A_{33} - \frac{A_{11} + A_{22} + A_{33}}{3} \end{bmatrix} \tag{12-19}$$

从 (12-17) 和 (12-19) 两式明显地看出,一个张量的球量部分具有各向同性的对称性,而其偏量部分的迹为零.

例 12.3 应用张量的对称化和反对称化,验证位移可分解为变形和刚体转动两部分.

解: 本例题将为 §53.3 的内容做铺垫，两部分内容可结合起来理解和领会. 考察任意点 $A(\boldsymbol{r})$ 邻域内的点 $A'(\boldsymbol{r}+\mathrm{d}\boldsymbol{r})$ 的位移 $\boldsymbol{u}$，进行一阶泰勒展开，有

$$\boldsymbol{u}(\boldsymbol{r}+\mathrm{d}\boldsymbol{r}) \approx \boldsymbol{u}(\boldsymbol{r}) + (\boldsymbol{u}\otimes\boldsymbol{\nabla})\cdot\mathrm{d}\boldsymbol{r} \tag{12-20}$$

式中，二阶张量：

$$\boldsymbol{u}\otimes\boldsymbol{\nabla} = \frac{\partial u_i}{\partial x_j}\boldsymbol{e}_i\otimes\boldsymbol{e}_j = u_{i,j}\boldsymbol{e}_i\otimes\boldsymbol{e}_j \tag{12-21}$$

为位移矢量 $\boldsymbol{u}$ 的右梯度. 利用张量的对称化和反对称化的 (12-1) 式，(12-20) 式可写为

$$\mathrm{d}\boldsymbol{u} = \boldsymbol{u}(\boldsymbol{r}+\mathrm{d}\boldsymbol{r}) - \boldsymbol{u}(\boldsymbol{r}) = \frac{1}{2}(\boldsymbol{u}\otimes\boldsymbol{\nabla}+\boldsymbol{\nabla}\otimes\boldsymbol{u})\cdot\mathrm{d}\boldsymbol{r} + \frac{1}{2}(\boldsymbol{u}\otimes\boldsymbol{\nabla}-\boldsymbol{\nabla}\otimes\boldsymbol{u})\cdot\mathrm{d}\boldsymbol{r} \tag{12-22}$$

或者，可等价地写为

$$\begin{aligned}\mathrm{d}\boldsymbol{u} &= \frac{1}{2}\left[\boldsymbol{u}\otimes\boldsymbol{\nabla}+(\boldsymbol{u}\otimes\boldsymbol{\nabla})^{\mathrm{T}}\right]\cdot\mathrm{d}\boldsymbol{r} + \frac{1}{2}\left[\boldsymbol{u}\otimes\boldsymbol{\nabla}-(\boldsymbol{u}\otimes\boldsymbol{\nabla})^{\mathrm{T}}\right]\cdot\mathrm{d}\boldsymbol{r} \\ &= \underbrace{\boldsymbol{\varepsilon}\cdot\mathrm{d}\boldsymbol{r}}_{\text{变形}} + \underbrace{\boldsymbol{\Omega}\cdot\mathrm{d}\boldsymbol{r}}_{\text{刚体转动}}\end{aligned} \tag{12-23}$$

式中，反映变形的二阶张量：

$$\boldsymbol{\varepsilon} = \frac{1}{2}\left[\boldsymbol{u}\otimes\boldsymbol{\nabla}+(\boldsymbol{u}\otimes\boldsymbol{\nabla})^{\mathrm{T}}\right] = \frac{1}{2}(\boldsymbol{u}\otimes\boldsymbol{\nabla}+\boldsymbol{\nabla}\otimes\boldsymbol{u}) \tag{12-24}$$

为满足如下对称性的柯西应变 (Cauchy strain)：

$$\boldsymbol{\varepsilon} = \boldsymbol{\varepsilon}^{\mathrm{T}} \tag{12-25}$$

而 (12-23) 式中反映刚体转动的二阶张量：

$$\boldsymbol{\Omega} = \frac{1}{2}\left[\boldsymbol{u}\otimes\boldsymbol{\nabla}-(\boldsymbol{u}\otimes\boldsymbol{\nabla})^{\mathrm{T}}\right] = \frac{1}{2}(\boldsymbol{u}\otimes\boldsymbol{\nabla}-\boldsymbol{\nabla}\otimes\boldsymbol{u}) \tag{12-26}$$

为满足如下反对称性质的转动张量：

$$\boldsymbol{\Omega} = -\boldsymbol{\Omega}^{\mathrm{T}} \tag{12-27}$$

由例 8.3 可知，存在一个赝矢量 (pseudo vector)：

$$\boldsymbol{\omega} = \frac{1}{2}\boldsymbol{\nabla}\times\boldsymbol{u} \tag{12-28}$$

则 (12-23) 式可等价地表示为

$$\mathrm{d}\boldsymbol{u} = \boldsymbol{\varepsilon}\cdot\mathrm{d}\boldsymbol{r} + \boldsymbol{\Omega}\cdot\mathrm{d}\boldsymbol{r}$$

$$= \boldsymbol{\varepsilon} \cdot \mathrm{d}\boldsymbol{r} + \boldsymbol{\omega} \times \mathrm{d}\boldsymbol{r} \tag{12-29}$$

例 12.4 将一个三阶张量 A_{ijk} 对称化.

解：先讨论 N 阶张量的普遍情况. 将张量的同类 N 个指标进行 $N!$ 次不同的置换，再取所得到的 $N!$ 个新张量的算术平均值，这就是任意阶张量的对称化，用圆括号表示. 作为特殊的例子，二阶张量 A_{ij} 和三阶张量 A_{ijk} 的对称化张量可分别表示为

$$A_{(ij)} = \frac{1}{2!}(A_{ij} + A_{ji}) \tag{12-30}$$

$$A_{(ijk)} = \frac{1}{3!}(A_{ijk} + A_{jki} + A_{kij} + A_{ikj} + A_{kji} + A_{jik}) \tag{12-31}$$

例 12.5 将一个三阶张量 A_{ijk} 反对称化.

解：将指标经奇数次置换的新张量取反符号后再取平均值，这样的运算被称为反对称化，用方括号表示. 作为已经熟知的例子，二阶张量 A_{ij} 的反对称化张量可分别表示为

$$A_{[ij]} = \frac{1}{2!}(A_{ij} - A_{ji}) \tag{12-32}$$

参考 (8-10) 式和图 8.2，易知三个指标的奇置换为 ikj, jik 和 kji，三阶张量 A_{ijk} 的反对称化张量可表示为

$$A_{[ijk]} = \frac{1}{3!}(A_{ijk} + A_{jki} + A_{kij} - A_{ikj} - A_{kji} - A_{jik}) \tag{12-33}$$

例 12.6 给出对 (12-24) 式中的柯西应变所满足如下协调方程 (compatibility equation) 的等价条件：

$$\varepsilon_{[i|[j,k]|l]} = 0 \tag{12-34}$$

解法一：首先考虑指标 j 和 k 的反对称化：

$$\begin{aligned}\varepsilon_{[i|[j,k]|l]} &= \frac{1}{2}\left[\varepsilon_{[i|j,k|l]} - \varepsilon_{[i|k,j|l]}\right] \\ &= \frac{1}{2}\left\{\frac{1}{2}\left[\varepsilon_{ij,kl} - \varepsilon_{lj,ki}\right] - \frac{1}{2}\left[\varepsilon_{ik,jl} - \varepsilon_{lk,ji}\right]\right\} \\ &= \frac{1}{4}\left[\varepsilon_{ij,kl} + \varepsilon_{lk,ji} - \varepsilon_{lj,ki} - \varepsilon_{ik,jl}\right]\end{aligned} \tag{12-35}$$

由柯西应变的对称条件 (12-25) 式，再由二阶偏导数的可交换性，有

$$\varepsilon_{lk,ji} = \varepsilon_{kl,ij}, \quad \varepsilon_{lj,ki} = \varepsilon_{jl,ik} \tag{12-36}$$

则 (12-35) 式可进一步表示为

$$\varepsilon_{[i|\,[j,k]\,|l]} = \frac{1}{4}\left[\varepsilon_{ij,kl} + \varepsilon_{kl,ij} - \varepsilon_{jl,ik} - \varepsilon_{ik,jl}\right] \tag{12-37}$$

最后，(12-34) 式可等价地表示为

$$\varepsilon_{ij,kl} + \varepsilon_{kl,ij} = \varepsilon_{ik,jl} + \varepsilon_{jl,ik} \tag{12-38}$$

解法二：首先考虑指标 i 和 l 的反对称化：

$$\begin{aligned}
\varepsilon_{[i|\,[j,k]\,|l]} &= \frac{1}{2}\left[\varepsilon_{i[j,k]l} - \varepsilon_{l[j,k]i}\right] \\
&= \frac{1}{2}\left\{\frac{1}{2}\left[\varepsilon_{ij,kl} - \varepsilon_{ik,jl}\right] - \frac{1}{2}\left[\varepsilon_{lj,ki} - \varepsilon_{lk,ji}\right]\right\} \\
&= \frac{1}{4}\left[\varepsilon_{ij,kl} + \varepsilon_{lk,ji} - \varepsilon_{ik,jl} - \varepsilon_{lj,ki}\right] \\
&= \frac{1}{4}\left[\varepsilon_{ij,kl} + \varepsilon_{kl,ij} - \varepsilon_{ik,jl} - \varepsilon_{jl,ik}\right]
\end{aligned} \tag{12-39}$$

在上式的最后一步运算中，已经应用了 (12-36) 式中的对称关系. 上述两种解法是完全等价的.

例 12.7 给出应变张量 (12-24) 式和转动张量 (12-26) 式之间所满足的一些重要关系式以及科恩不等式 (Korn's inequality).

解：从应变张量 (12-24) 和转动张量 (12-26) 两个式子，我们可以立即给出它们之间所满足的下列关系式：

$$\begin{cases} |\boldsymbol{\varepsilon}|^2 + |\boldsymbol{\Omega}|^2 = |\boldsymbol{u} \otimes \boldsymbol{\nabla}_{\boldsymbol{X}}|^2 = (\boldsymbol{u} \otimes \boldsymbol{\nabla}_{\boldsymbol{X}}) : (\boldsymbol{u} \otimes \boldsymbol{\nabla}_{\boldsymbol{X}}) = |\boldsymbol{H}|^2 \\ |\boldsymbol{\varepsilon}|^2 - |\boldsymbol{\Omega}|^2 = (\boldsymbol{u} \otimes \boldsymbol{\nabla}_{\boldsymbol{X}}) : (\boldsymbol{\nabla}_{\boldsymbol{X}} \otimes \boldsymbol{u}) = \boldsymbol{H} : \boldsymbol{H}^{\mathrm{T}} \end{cases} \tag{12-40}$$

式中，$\boldsymbol{H} = \boldsymbol{u} \otimes \boldsymbol{\nabla}_{\boldsymbol{X}}$ 为位移梯度，其转置为 $\boldsymbol{H}^{\mathrm{T}} = \boldsymbol{\nabla}_{\boldsymbol{X}} \otimes \boldsymbol{u}$.

数学界流行如下说法："所有分析数学家都要花费几乎一半的研究时间，去到处查阅那些尚未能证明但却要使用的有关不等式的文献资料." 这充分说明了不等式在数学分析中的极端重要性. 作为力学理论数学化的理性力学，不等式亦应占有重要的一席之地 [3.3]. 在本例题中将给出连续介质力学中的科恩不等式，在 §49.3 中将给出勒让德–阿达玛不等式.

阿瑟·科恩 (Arthur Korn, 1870—1945) 于 1909 年在题为《关于弹性理论与转轴弯曲的不等式》[3.4] 一文中给出了弹性力学能量正定性的不等式，后被广泛地称为 "科恩不等式". 弹性力学中的科恩不等式表明在弹性体的边界上位移为零的前提下，下列不等式成立：

科恩

$$\int_{\mathcal{B}} |\boldsymbol{u} \otimes \boldsymbol{\nabla}_{\boldsymbol{X}}|^2 \, \mathrm{d}V = \int_{\mathcal{B}} |\boldsymbol{H}|^2 \, \mathrm{d}V \leqslant 2 \int_{\mathcal{B}} |\boldsymbol{\varepsilon}|^2 \, \mathrm{d}V \tag{12-41}$$

事实上, 将 (12-40) 式中第一式代入 (12-41) 式中, 得到等价的科恩不等式为

$$\int_{\mathcal{B}} |\boldsymbol{\Omega}|^2 \, \mathrm{d}V \leqslant \int_{\mathcal{B}} |\boldsymbol{\varepsilon}|^2 \, \mathrm{d}V \tag{12-42}$$

(12-42) 式说明, 物体的整体旋转量要小于等于其整体变形量.

德国犹太数学家、物理学家阿瑟·科恩是法国数学、物理学、哲学大师庞加莱的学生, 科恩是传真机的主要发明人之一 [3.5].

赛德里克·维拉尼 (Cédric Villani, 1973— , 2010 年菲尔兹奖获得者) 在完成博士答辩三年后, 于 2002 年同合作者朗洛·德维莱特 (Laurent Desvillettes) 一起发现了弹性理论中的科恩不等式与统计力学玻尔兹曼理论中熵增的联系 [3.6].

我们首先来证明 (12-40) 式. 由于

$$\begin{aligned}
|\boldsymbol{\varepsilon}|^2 &= \boldsymbol{\varepsilon} : \boldsymbol{\varepsilon} = \frac{1}{4} \left(\boldsymbol{H} + \boldsymbol{H}^{\mathrm{T}} \right) : \left(\boldsymbol{H} + \boldsymbol{H}^{\mathrm{T}} \right) \\
&= \frac{1}{4} \left(|\boldsymbol{H}|^2 + \boldsymbol{H} : \boldsymbol{H}^{\mathrm{T}} + \boldsymbol{H}^{\mathrm{T}} : \boldsymbol{H} + \left|\boldsymbol{H}^{\mathrm{T}}\right|^2 \right)
\end{aligned} \tag{12-43}$$

$$\begin{aligned}
|\boldsymbol{\Omega}|^2 &= \boldsymbol{\Omega} : \boldsymbol{\Omega} = \frac{1}{4} \left(\boldsymbol{H} - \boldsymbol{H}^{\mathrm{T}} \right) : \left(\boldsymbol{H} - \boldsymbol{H}^{\mathrm{T}} \right) \\
&= \frac{1}{4} \left(|\boldsymbol{H}|^2 - \boldsymbol{H} : \boldsymbol{H}^{\mathrm{T}} - \boldsymbol{H}^{\mathrm{T}} : \boldsymbol{H} + \left|\boldsymbol{H}^{\mathrm{T}}\right|^2 \right)
\end{aligned} \tag{12-44}$$

位移向量 $\boldsymbol{u}$ 的右梯度 $\boldsymbol{u} \otimes \boldsymbol{\nabla}_{\boldsymbol{X}} = \boldsymbol{H}$ 和左梯度 $\boldsymbol{\nabla}_{\boldsymbol{X}} \otimes \boldsymbol{u} = \boldsymbol{H}^{\mathrm{T}}$ 满足如下关系式:

$$|\boldsymbol{H}|^2 = \left|\boldsymbol{H}^{\mathrm{T}}\right|^2, \quad \boldsymbol{H} : \boldsymbol{H}^{\mathrm{T}} = \boldsymbol{H}^{\mathrm{T}} : \boldsymbol{H} \tag{12-45}$$

将 (12-45) 式代入 (12-43) 和 (12-44) 两式, 得到

$$\begin{cases} |\boldsymbol{\varepsilon}|^2 = \dfrac{1}{2} \left(|\boldsymbol{H}|^2 + \boldsymbol{H} : \boldsymbol{H}^{\mathrm{T}} \right) \\ |\boldsymbol{\Omega}|^2 = \dfrac{1}{2} \left(|\boldsymbol{H}|^2 - \boldsymbol{H} : \boldsymbol{H}^{\mathrm{T}} \right) \end{cases} \tag{12-46}$$

则 (12-40) 式得证.

由于下列两个等式存在:

$$\boldsymbol{\nabla}_{\boldsymbol{X}} \cdot [(\boldsymbol{\nabla}_{\boldsymbol{X}} \cdot \boldsymbol{u}) \boldsymbol{u}] = \boldsymbol{u} \cdot \boldsymbol{\nabla}_{\boldsymbol{X}} (\boldsymbol{\nabla}_{\boldsymbol{X}} \cdot \boldsymbol{u}) + (\boldsymbol{\nabla}_{\boldsymbol{X}} \cdot \boldsymbol{u})^2 \tag{12-47}$$

和

$$\begin{aligned}
\boldsymbol{\nabla}_{\boldsymbol{X}} \cdot [(\boldsymbol{u} \otimes \boldsymbol{\nabla}_{\boldsymbol{X}}) \cdot \boldsymbol{u}] &= (\boldsymbol{u} \otimes \boldsymbol{\nabla}_{\boldsymbol{X}}) : (\boldsymbol{\nabla}_{\boldsymbol{X}} \otimes \boldsymbol{u}) + \boldsymbol{u} \cdot \boldsymbol{\nabla}_{\boldsymbol{X}} (\boldsymbol{\nabla}_{\boldsymbol{X}} \cdot \boldsymbol{u}) \\
&= \boldsymbol{H} : \boldsymbol{H}^{\mathrm{T}} + \boldsymbol{u} \cdot \boldsymbol{\nabla}_{\boldsymbol{X}} (\boldsymbol{\nabla}_{\boldsymbol{X}} \cdot \boldsymbol{u})
\end{aligned} \tag{12-48}$$

上两式相减, 得到 [3.7]

$$H : H^{\mathrm{T}} = \nabla_X \cdot [(u \otimes \nabla_X) \cdot u - (\nabla_X \cdot u) u] + (\nabla_X \cdot u)^2 \tag{12-49}$$

假设在物体的边界 $\partial \mathcal{B}$ 上位移场为零: $u|_{\partial \mathcal{B}} = 0$, 通过散度定理, 有

$$\int_{\mathcal{B}} \nabla_X \cdot [(u \otimes \nabla_X) \cdot u - (\nabla_X \cdot u) u] \, \mathrm{d}V = \int_{\partial \mathcal{B}} [(u \otimes \nabla_X) \cdot u - (\nabla_X \cdot u) u] \, \mathrm{d}A = 0 \tag{12-50}$$

则 (12-49) 式对物体 $\mathcal{B}$ 的积分有

$$\int_{\mathcal{B}} \left(H : H^{\mathrm{T}} \right) \mathrm{d}V = \int_{\mathcal{B}} (\nabla_X \cdot u)^2 \mathrm{d}V \geqslant 0 \tag{12-51}$$

则由 (12-40) 式中第二式易得

$$\int_{\mathcal{B}} \left(|\varepsilon|^2 - |\Omega|^2 \right) \mathrm{d}V = \int_{\mathcal{B}} \left(H : H^{\mathrm{T}} \right) \mathrm{d}V = \int_{\mathcal{B}} (\nabla_X \cdot u)^2 \mathrm{d}V \geqslant 0 \tag{12-52}$$

亦即由 (12-42) 式表示的科恩不等式得证, 同理可证由 (12-41) 式表示的科恩不等式.

§13. 谱 定 理

13.1 谱分解定理的基本内容

在数学上, 特别是线性代数和泛函分析中, 谱定理 (spectral theorem) 给出了算子或者矩阵可以对角化的条件. 在泛函分析中, 算子的谱的概念是有限维矩阵的本征值概念的推广; 在线性代数中, 谱定理提供了一种向量空间的标准分解, 即矩阵分解成正则形式的因式分解. 利用矩阵的本征值 (eigenvalue) 和本征向量 (eigenvector) 表示矩阵, 称为谱分解 (spectral decomposition), 也叫本征分解. 只有可对角化矩阵才能用这种方法分解.

如果存在单位向量 e 使得 $Se = \omega e$ 则标量 ω 是张量 S 的本征值, e 为本征向量, S 对应于 ω 的本征空间是相应的本征向量组成的线性空间 $\mathcal{V}$ 的子空间, S 的谱为 $(\omega_1, \omega_2, \omega_3, \cdots)$. 需要对初学者再一次强调的是, Se 表示的是张量 S 和单位向量 e 之间的点积.

谱定理: 若 S 是对称张量, 则存在线性空间 $\mathcal{V}$ 中的一个标准正交基, 它完全由 S 的本征向量组成. 此外, 对任一个这样的基 e_1, e_2, e_3 按次序排列的相应的本征值 $\omega_1, \omega_2, \omega_3$ 构成 S 的整个谱, 且

$$S = \sum_i \omega_i e_i \otimes e_i \tag{13-1}$$

反之，如果 S 具有形式 (13-1)，其中 $\{e_i\}$ 是标准正交的，则 $\omega_1, \omega_2, \omega_3$ 是相对应于本征向量 e_1, e_2, e_3 的 S 的本征值. 此外:

(a) 当且仅当 S 的本征空间是通过 0 的三个互相垂直的直线时，S 恰好有三个不同的本征值，且 S 表示为 $S = \omega_1 e_1 \otimes e_1 + \omega_2 e_2 \otimes e_2 + \omega_3 e_3 \otimes e_3$;

(b) 当且仅当 S 有两个不同的本征值 ω_1 和 ω_2，其对应的本征空间分别是一条通过 0 的线 l 以及通过 0 并垂直于 l 的平面 (如图 13.1 所示)，则 S 表示为

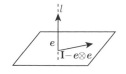

图 13.1　直线 l 和垂直于 l 的平面

$$S = \omega_1 e \otimes e + \omega_2 (\mathbf{I} - e \otimes e) \tag{13-2}$$

反之当且仅当 S 有 (13-2) 式的形式，S 确切地有两个不同的本征值;

(c) 当且仅当 S 有一个本征值 ω，其对应的本征空间为 $\mathcal{V}$，则 S 表示为

$$S = \omega \mathbf{I} \tag{13-3}$$

反之，当且仅当 S 有 (13-3) 式的形式，S 确切地有一个本征值.

所谓一个张量 S 的本征向量，即为对 $\mathbf{I}$ 做 S 变换 (伸缩、旋转、斜切) 之后，方向不发生改变的向量，相对应的本征值为该本征向量伸缩的倍数.

以伸缩为例分别对谱分解的三种情况举例:

(a) $[S] = \begin{bmatrix} 1 & 0 & 0 \\ 0 & 2 & 0 \\ 0 & 0 & 3 \end{bmatrix}$，此时根据 $\det(Se - \omega e) = 0$ 可求得 S 的本征值分别为 $\omega_1 = 1, \omega_2 = 2$ 和 $\omega_3 = 3$，相对应的本征矢量分别为 $e_1 = \begin{bmatrix} 1 & 0 & 0 \end{bmatrix}$, $e_2 = \begin{bmatrix} 0 & 1 & 0 \end{bmatrix}$ 和 $e_3 = \begin{bmatrix} 0 & 0 & 1 \end{bmatrix}$，则根据 (13-1) 式可知，$S$ 可表示为

$$\begin{aligned} S &= e_1 \otimes e_1 + 2 e_2 \otimes e_2 + 3 e_3 \otimes e_3 \\ &= \begin{bmatrix} 1 & 0 & 0 \\ 0 & 0 & 0 \\ 0 & 0 & 0 \end{bmatrix} + 2 \begin{bmatrix} 0 & 0 & 0 \\ 0 & 1 & 0 \\ 0 & 0 & 0 \end{bmatrix} + 3 \begin{bmatrix} 0 & 0 & 0 \\ 0 & 0 & 0 \\ 0 & 0 & 1 \end{bmatrix} \end{aligned} \tag{13-4}$$

此时，S 的变换即为对 $\mathbf{I}$ 三个基矢量按照 1, 2, 3 不同比例的伸长.

(b) $[S] = \begin{bmatrix} 1 & 0 & 0 \\ 0 & 2 & 0 \\ 0 & 0 & 2 \end{bmatrix}$，此时 S 的本征值分别为 $\omega_1 = 1, \omega_2 = 2$ (重度为 2)，相对应的本征空间为 $e_1 = \begin{bmatrix} 1 & 0 & 0 \end{bmatrix}$ 和 $\mathbf{I} - e_1 \otimes e_1$ (即 $x = 0$ 平面内所有矢量方向

在变换之后均不改变），则根据 (13-2) 式可知，S 可表示为

$$S = e_1 \otimes e_1 + 2(I - e_1 \otimes e_1) = \begin{bmatrix} 1 & 0 & 0 \\ 0 & 0 & 0 \\ 0 & 0 & 0 \end{bmatrix} + 2\begin{bmatrix} 1 & 0 & 0 \\ 0 & 1 & 0 \\ 0 & 0 & 1 \end{bmatrix} - 2\begin{bmatrix} 1 & 0 & 0 \\ 0 & 0 & 0 \\ 0 & 0 & 0 \end{bmatrix}$$
(13-5)

此时，S 的变换即为对 I 三个基矢量按照 $1, 2, 2$ 不同比例的伸长.

(c) $[S] = \begin{bmatrix} 2 & 0 & 0 \\ 0 & 2 & 0 \\ 0 & 0 & 2 \end{bmatrix}$，此时 S 的本征值为 $\omega = 2$（重度为 3），相对应的本征空间是 $\mathcal{V}$，即所有矢量方向在变换之后同比伸缩，方向均不改变，则根据 (13-3) 式可知，S 可表示为

$$[S] = [2I] = 2\begin{bmatrix} 1 & 0 & 0 \\ 0 & 1 & 0 \\ 0 & 0 & 1 \end{bmatrix}$$

此时，S 的变换即为对 I 三个基矢量按照 $2, 2, 2$ 同比例的伸长.

当 S 包含旋转变换，举例如 $[S] = \begin{bmatrix} 1 & 1 & 0 \\ 1 & 2 & 0 \\ 0 & 0 & 3 \end{bmatrix}$，此时未旋转的方向 z 轴基矢量 $e_1 = \begin{bmatrix} 0 & 0 & 1 \end{bmatrix}$ 依旧为本征矢量（对应于本征值 $\omega_1 = 3$），伸长比例为 3，Oxy 平面内的本征矢量不再为原坐标系的基矢量 $\begin{bmatrix} 1 & 0 & 0 \end{bmatrix}$ 与 $\begin{bmatrix} 0 & 1 & 0 \end{bmatrix}$，而变为 $-\frac{1+\sqrt{5}}{2}x + y = 0$ 和 $-\frac{1-\sqrt{5}}{2}x + y = 0$ 两条直线上的单位向量 $e_2 = \begin{bmatrix} \frac{2}{\sqrt{10+2\sqrt{5}}} & \frac{\sqrt{5}+1}{\sqrt{10+2\sqrt{5}}} & 0 \end{bmatrix}$ 和 $e_3 = \begin{bmatrix} \frac{2}{\sqrt{10-2\sqrt{5}}} & \frac{1-\sqrt{5}}{\sqrt{10-2\sqrt{5}}} & 0 \end{bmatrix}$（分别对应于本征值 $\omega_2 = \frac{3+\sqrt{5}}{2}$ 和 $\omega_3 = \frac{3-\sqrt{5}}{2}$）. 此时，$S$ 对应于谱分解的第一种情况，可以写为

$$S = \omega_1 e_1 \otimes e_1 + \omega_2 e_2 \otimes e_2 + \omega_3 e_3 \otimes e_3$$

$$= 3\begin{bmatrix} 0 & 0 & 0 \\ 0 & 0 & 0 \\ 0 & 0 & 1 \end{bmatrix} + \frac{3+\sqrt{5}}{2}\begin{bmatrix} \frac{4}{10+2\sqrt{5}} & \frac{2(1+\sqrt{5})}{10+2\sqrt{5}} & 0 \\ \frac{2(1+\sqrt{5})}{10+2\sqrt{5}} & \frac{6+2\sqrt{5}}{10+2\sqrt{5}} & 0 \\ 0 & 0 & 0 \end{bmatrix}$$

$$+\frac{3-\sqrt{5}}{2}\begin{bmatrix} \dfrac{4}{10+2\sqrt{5}} & \dfrac{2(1-\sqrt{5})}{10+2\sqrt{5}} & 0 \\ \dfrac{2(1-\sqrt{5})}{10+2\sqrt{5}} & \dfrac{6-2\sqrt{5}}{10-2\sqrt{5}} & 0 \\ 0 & 0 & 0 \end{bmatrix}$$

$$=\begin{bmatrix} 1 & 1 & 0 \\ 1 & 2 & 0 \\ 0 & 0 & 3 \end{bmatrix} \tag{13-6}$$

当 S 做斜切变换,如 $[S]=\begin{bmatrix} 1 & 1 & 0 \\ 0 & 2 & 0 \\ 0 & 0 & 3 \end{bmatrix}$,此时矩阵不对称,基矢量不再为标准正交基,在此不做讨论.

13.2 谱定理在应力分解中的应用

请先复习 §9 中的相关内容.

应力 T 为对称的二阶应力张量,如果 $Tn=\sigma n$,且 n 为单位矢量,则 σ 就是主应力,n 就是主方向,即主应力和主方向分别为 T 的本征值和本征向量. 由于 T 是对称张量,故而存在三个相互垂直的主方向和相应的主应力.

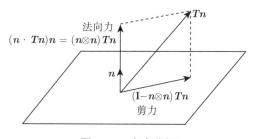

图 13.2 应力分解

若考虑一个如图 13.2 所示的任意平面,其单位法向量为 n,则面力 Tn,也就是应力张量和 n 之间的点积,可以分解为法向力:

$$(n\cdot Tn)n=(n\otimes n)Tn \tag{13-7}$$

和剪力

$$Tn-(n\otimes n)Tn=(I-n\otimes n)Tn \tag{13-8}$$

之和,当且仅当相应的剪力为零时,n 就是主方向.

对应于谱定理的三种情况, 分析三种常见的应力状态:

(a) **纯拉 (压):** 在方向 e_x 上拉应力为 σ, 此时 $T = \sigma e_x \otimes e_x$.

如图 13.3(a) 在 xy 平面应力状态, 在 $e_x = \begin{bmatrix} 1 & 0 & 0 \end{bmatrix}$ 上拉应力为 σ, 此时

$$T = \sigma e_x \otimes e_x = \begin{bmatrix} \sigma & 0 & 0 \\ 0 & 0 & 0 \\ 0 & 0 & 0 \end{bmatrix} \tag{13-9}$$

(b) **纯剪:** 对应于平面内两个正交的方向 (e_x, e_y) 的剪应力 τ, 此时

$$T = \tau (e_x \otimes e_y + e_y \otimes e_x)$$

如图 13.3(b) 在 xy 平面上的应力状态, 在 $e_x = \begin{bmatrix} 1 & 0 & 0 \end{bmatrix}$, $e_y = \begin{bmatrix} 0 & 1 & 0 \end{bmatrix}$ 上剪应力 τ, 此时

$$\begin{aligned} T &= \tau (e_x \otimes e_y + e_y \otimes e_x) \\ &= \tau \begin{bmatrix} 0 & 1 & 0 \\ 0 & 0 & 0 \\ 0 & 0 & 0 \end{bmatrix} + \tau \begin{bmatrix} 0 & 0 & 0 \\ 1 & 0 & 0 \\ 0 & 0 & 0 \end{bmatrix} = \begin{bmatrix} 0 & \tau & 0 \\ \tau & 0 & 0 \\ 0 & 0 & 0 \end{bmatrix} \end{aligned} \tag{13-10}$$

关于纯剪切和简单剪切区别的讨论, 见例 48.3 和思考题 10.4 和思考题 10.5.

(c) **静水压:** 对静态流体不存在剪力作用, 每一个单位向量均为 T 的本征向量, 如图 13.3(c) 所示, 此时

$$T = -p\mathbf{I} \tag{13-11}$$

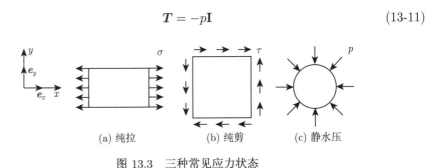

图 13.3 三种常见应力状态

§14. 数学算子 —— 拉普拉斯和黑森算子

小应变并忽略体力下, 用位移矢量 u 表示的弹性力学运动方程由著者的《近

代连续介质力学》[1.6] 中的 (22-23) 式给出：

$$(\lambda + \mu)\boldsymbol{\nabla}(\boldsymbol{\nabla} \cdot \boldsymbol{u}) + \mu\nabla^2 \boldsymbol{u} = \rho\ddot{\boldsymbol{u}} \tag{14-1}$$

上式的分量形式为《近代连续介质力学》[1.6] 中的 (25-3) 式：

$$(\lambda + \mu)u_{j,ji} + \mu u_{i,jj} = \rho\ddot{u}_i \tag{14-2}$$

为了使初学者进一步熟悉相关数学算子的运算，下面将分别介绍 (14-1) 式中两种数学算子：拉普拉斯算子 (Laplacian operator, 简称 Laplacian) 和黑森算子 (Hessian).

黑森算子是由德国数学家奥托·黑森 (Otto Hesse, 1811—1874) 提出的.

14.1 拉普拉斯算子 $\nabla^2 = \boldsymbol{\nabla} \cdot \boldsymbol{\nabla} = \Delta$

拉普拉斯算子 (如图 14.1 所示) 为两个梯度算子的点积：$\nabla^2 = \boldsymbol{\nabla} \cdot \boldsymbol{\nabla}$，有些文献和专著中亦广泛地记为 Δ. 在笛卡儿坐标系中，其详细运算过程为

$$\nabla^2(\cdot) = \boldsymbol{\nabla} \cdot \boldsymbol{\nabla}(\cdot) = \left(\frac{\partial}{\partial x_i}\boldsymbol{e}_i\right) \cdot \left(\frac{\partial(\cdot)}{\partial x_j}\boldsymbol{e}_j\right) = \frac{\partial^2(\cdot)}{\partial x_i x_j}\boldsymbol{e}_i \cdot \boldsymbol{e}_j = \frac{\partial^2(\cdot)}{\partial x_i x_j}\delta_{ij} = \frac{\partial^2(\cdot)}{\partial x_i^2} \tag{14-3}$$

如果某个标量场 Φ 满足：

$$\nabla^2 \Phi = \Psi \tag{14-4}$$

图 14.1 邮票上的拉普拉斯算子和拉普拉斯方程

如果 $\Psi = 0$，则 (14-4) 式为拉普拉斯方程 (Laplace's equation)，在弹性力学中，满足拉普拉斯方程 $\nabla^2 \Phi = 0$ 的标量场 Φ 被称为是协调的 (harmonic). 如果一个标量场 Θ 满足 $\nabla^2\nabla^2\Theta = \nabla^4\Theta = 0$，则称标量场 Θ 是双协调的 (bi-harmonic).

如果 $\Psi \neq 0$，则 (14-4) 式为泊松方程 (Poisson's equation).

14.2 黑森算子 $\boldsymbol{\nabla}\boldsymbol{\nabla} = \boldsymbol{\nabla} \otimes \boldsymbol{\nabla}$

黑森算子 (Hessian)

$$\boldsymbol{\nabla}\boldsymbol{\nabla} = \boldsymbol{\nabla} \otimes \boldsymbol{\nabla} = \frac{\partial}{\partial x_i \partial x_j}\boldsymbol{e}_i \otimes \boldsymbol{e}_j \tag{14-5}$$

显然是一个二阶张量. 方程 (14-1) 中的 $\boldsymbol{\nabla}(\boldsymbol{\nabla} \cdot \boldsymbol{u})$ 事实上表示的是作为二阶张量的黑森算子和一个向量场 $\boldsymbol{u}$ 之间的点积：

$$\boldsymbol{\nabla}(\boldsymbol{\nabla} \cdot \boldsymbol{u}) = \boldsymbol{\nabla}\boldsymbol{\nabla} \cdot \boldsymbol{u} = \boldsymbol{\nabla} \otimes \boldsymbol{\nabla} \cdot \boldsymbol{u}$$

上式的指标形式为

$$\nabla(\nabla \cdot \boldsymbol{u}) = \nabla\nabla \cdot \boldsymbol{u} = \nabla \otimes \nabla \cdot \boldsymbol{u}$$
$$= \left(\frac{\partial}{\partial x_i}\boldsymbol{e}_i\right) \otimes \left(\frac{\partial}{\partial x_j}\boldsymbol{e}_j\right) \cdot (u_k\boldsymbol{e}_k) = \left(\frac{\partial}{\partial x_i}\boldsymbol{e}_i\right)\frac{\partial u_k}{\partial x_j}\delta_{jk}$$
$$= \frac{\partial^2 u_j}{\partial x_i \partial x_j}\boldsymbol{e}_i = \frac{\partial^2 u_j}{\partial x_j \partial x_i}\boldsymbol{e}_i = u_{j,ji}\boldsymbol{e}_i \qquad (14\text{-}6)$$

一般地，上述运算还经常地表示为

$$\nabla(\nabla \cdot \boldsymbol{u}) = \nabla\nabla \cdot \boldsymbol{u} = \nabla \otimes \nabla \cdot \boldsymbol{u} = \mathbf{grad}\operatorname{div}\boldsymbol{u} \qquad (14\text{-}7)$$

如在朗道和栗弗席兹教程的第 7 卷《弹性理论》(*Theory of Elasticity*)[1.8] 中，大量地、经常性地将各向同性弹性力学的平衡方程表示为

$$\mathbf{grad}\operatorname{div}\boldsymbol{u} + (1 - 2\nu)\Delta\boldsymbol{u} = \mathbf{0} \qquad (14\text{-}8)$$

将各向同性的弹性力学运动方程表示为

$$\frac{E}{2(1+\nu)(1-2\nu)}\mathbf{grad}\operatorname{div}\boldsymbol{u} + \frac{E}{2(1+\nu)}\Delta\boldsymbol{u} = \rho\ddot{\boldsymbol{u}} \qquad (14\text{-}9)$$

例 14.1 证明如下数学算子的恒等式：

$$\mathbf{grad}\operatorname{div}\boldsymbol{u} = \mathbf{curl}\operatorname{curl}\boldsymbol{u} + \nabla^2\boldsymbol{u}$$
$$= \nabla \times \nabla \times \boldsymbol{u} + \Delta\boldsymbol{u} \qquad (14\text{-}10)$$

证明：由三个矢量的三重叉积的关系式：$\boldsymbol{a} \times (\boldsymbol{b} \times \boldsymbol{c}) = (\boldsymbol{a} \cdot \boldsymbol{c})\boldsymbol{b} - (\boldsymbol{a} \cdot \boldsymbol{b})\boldsymbol{c}$，则有

$$\mathbf{curl}\operatorname{curl}\boldsymbol{u} = \nabla \times \nabla \times \boldsymbol{u} = (\nabla \cdot \boldsymbol{u})\nabla - (\nabla \cdot \nabla)\boldsymbol{u}$$
$$= \mathbf{grad}\operatorname{div}\boldsymbol{u} - \nabla^2\boldsymbol{u} \qquad (14\text{-}11)$$

在朗道和栗弗席兹教程的第 7 卷《弹性理论》[1.8] 中，还经常性地将平衡方程写为

$$\mathbf{grad}\operatorname{div}\boldsymbol{u} - \frac{1-2\nu}{2(1-\nu)}\mathbf{curl}\operatorname{curl}\boldsymbol{u} = \mathbf{0} \qquad (14\text{-}12)$$

例 14.2 给出弹性波动方程，用声学张量和谱分解定理来确定三维无限大体中弹性波的纵波和横波的波速．

解：在弹性动力学方程 (14-1) 式中位移矢量可表示为

$$\boldsymbol{u} = \boldsymbol{a}\sin[k(\boldsymbol{r} \cdot \boldsymbol{m} - ct)] \qquad (14\text{-}13)$$

式中，$\boldsymbol{a}$ 为振幅矢量，k 为波数 (单位为长度的倒数)，$\boldsymbol{r}$ 为矢径，$\boldsymbol{m}$ 为单位方向矢量 (满足 $|\boldsymbol{m}|=1$ 或者 $\boldsymbol{m}^2=1$)，c 为波速. 如果 $\boldsymbol{a}//\boldsymbol{m}$ 则为纵波；反之如果 $\boldsymbol{a}\perp\boldsymbol{m}$，则为横波.

由 (14-13) 式，可得

$$\begin{cases} \mathbf{grad}\boldsymbol{u} = \boldsymbol{\nabla}\otimes\boldsymbol{u} = k\,(\boldsymbol{m}\otimes\boldsymbol{a})\cos[k\,(\boldsymbol{r}\cdot\boldsymbol{m}-ct)] \\ \mathrm{div}\boldsymbol{u} = \boldsymbol{\nabla}\cdot\boldsymbol{u} = k\,(\boldsymbol{m}\cdot\boldsymbol{a})\cos[k\,(\boldsymbol{r}\cdot\boldsymbol{m}-ct)] \\ \mathbf{curl}\boldsymbol{u} = \boldsymbol{\nabla}\times\boldsymbol{u} = k\,(\boldsymbol{m}\times\boldsymbol{a})\cos[k\,(\boldsymbol{r}\cdot\boldsymbol{m}-ct)] \end{cases} \quad (14\text{-}14)$$

横波的充要条件 $\boldsymbol{a}\perp\boldsymbol{m}$ 等价于 $\boldsymbol{a}\cdot\boldsymbol{m}=0$；而纵波的充要条件 $\boldsymbol{a}//\boldsymbol{m}$ 等价于 $\boldsymbol{m}\times\boldsymbol{a}=\boldsymbol{0}$. 则两种波的充要条件可等价地表示于表 14.1 中.

表 14.1 纵波和横波之间的比较

弹性波的类型	充要条件	等价条件
纵波 longitudinal wave	$\boldsymbol{a}//\boldsymbol{m}$	$\mathbf{curl}\boldsymbol{u}=\boldsymbol{0}$ (矢量零) 纵波又被称为无旋波 (irrotational wave).
横波 transverse wave	$\boldsymbol{a}\perp\boldsymbol{m}$	$\mathrm{div}\boldsymbol{u}=0$ (标量零) 由于 $\mathrm{div}\boldsymbol{u}$ 表征的是体积，故横波又称为等容波 (equivoluminal waves, isochoric wave).

由 (14-13) 式，进一步可得

$$\begin{cases} \nabla^2\boldsymbol{u} = \Delta\boldsymbol{u} = -k^2\boldsymbol{a}\sin[k\,(\boldsymbol{r}\cdot\boldsymbol{m}-ct)] \\ \mathbf{grad}\,\mathrm{div}\boldsymbol{u} = -k^2\,(\boldsymbol{a}\cdot\boldsymbol{m})\,\boldsymbol{m}\sin[k\,(\boldsymbol{r}\cdot\boldsymbol{m}-ct)] \\ \qquad\qquad\quad = -k^2\boldsymbol{a}\,(\boldsymbol{m}\otimes\boldsymbol{m})\sin[k\,(\boldsymbol{r}\cdot\boldsymbol{m}-ct)] \\ \ddot{\boldsymbol{u}} = -k^2c^2\boldsymbol{a}\sin[k\,(\boldsymbol{r}\cdot\boldsymbol{m}-ct)] \end{cases} \quad (14\text{-}15)$$

请特别注意 (14-15) 式中第二式和 (9-15) 式的类似性.

将 (14-15) 式代回 (14-1) 式，两端约掉 k^2，得到

$$(\lambda+\mu)\,\boldsymbol{a}\,(\boldsymbol{m}\otimes\boldsymbol{m}) + \mu\boldsymbol{a} = \rho c^2\boldsymbol{a} \quad (14\text{-}16)$$

亦即

$$\frac{\lambda+2\mu}{\rho}\boldsymbol{a}\,(\boldsymbol{m}\otimes\boldsymbol{m}) + \frac{\mu}{\rho}\boldsymbol{a}\,(\mathbf{I}-\boldsymbol{m}\otimes\boldsymbol{m}) = c^2\boldsymbol{a} \quad (14\text{-}17)$$

(14-17) 式可进一步简写为

$$\boldsymbol{A}\,(\boldsymbol{m})\,\boldsymbol{a} = c^2\boldsymbol{a} \quad (14\text{-}18)$$

式中，二阶张量

$$\boldsymbol{A}\,(\boldsymbol{m}) = \frac{\lambda+2\mu}{\rho}\,(\boldsymbol{m}\otimes\boldsymbol{m}) + \frac{\mu}{\rho}\,(\mathbf{I}-\boldsymbol{m}\otimes\boldsymbol{m}) \quad (14\text{-}19)$$

被称为声学张量 (acoustic tensor) [3.7]. 因此, (14-18) 式说明: (14-13) 式中的位移矢量满足弹性动力学方程 (14-1) 式的充要条件是, c^2 为声学张量 $\boldsymbol{A}(\boldsymbol{m})$ 的本征值, 而振幅矢量 $\boldsymbol{a}$ 为其本征向量. 谱分解定理告诉我们, 纵波波速和横波波速分别为

$$c_l = \sqrt{\frac{\lambda + 2\mu}{\rho}}; \quad c_t = \sqrt{\frac{\mu}{\rho}} \tag{14-20}$$

§15. 黎曼度量张量

度量张量 (metric tensor) 在连续介质力学中是一个处理曲面问题十分基础的概念, 在物理界也通常被称为度规张量. 为了增加读者对该名词的感性认识, 让我们来讲一个和度量张量相关的著名科学典故 ——"费曼靠告诉出租车司机度规张量 $g_{\mu\nu}$ 找到开会地点".

读过费曼 (Richard Feynman, 1918—1988, 1965 年诺贝尔物理学奖获得者, 如图 15.1 所示) 自传体《别闹了, 费曼先生》(*Surely You're Joking, Mr. Feynman!*) 的读者会对一则有趣的故事留有深刻印象: 费曼去北卡罗来纳大学参加一广义相对论会议, 却不幸迟到了一天, 因此既不会有人接机, 也没有同伴可问, 更糟糕的是, 出租车司机告诉费曼 "北卡罗来纳大学" 在当地有两个校区, 而费曼不知道自己该去哪个. 眼看就要出现令人无比尴尬的局面, 好在费曼是个超级聪明的家伙, 灵机一动地告诉出租车司机, 他的目的地跟前一天的某批客人相同, 那批客人的特点是相互间频繁地念叨着 "g-mu-nu, g-mu-nu". 出租车司机当然不会知道 "g-mu-nu" 是度规张量 $g_{\mu\nu}$ 的读音, 但对那批满口怪语的客人显然印象深刻, 于是立刻明白了费曼要去哪里.

这则故事发生在 1957 年, 正是广义相对论复苏期间的小插曲, 而费曼所要参加的会议被称为 "教堂山会议 (Chapel Hill Conference)", 地点在北卡罗来纳大学教堂山分校, 是那一时期的重要会议.

在创建于 1637 年 [3.8] 的笛卡儿直角坐标系 (x, y, z) 中, 一个线元可以简单地表示为

$$\begin{aligned} \mathrm{d}s^2 &= \mathrm{d}\boldsymbol{r} \cdot \mathrm{d}\boldsymbol{r} = \left(\mathrm{d}x\hat{\boldsymbol{i}} + \mathrm{d}y\hat{\boldsymbol{j}} + \mathrm{d}z\hat{\boldsymbol{k}}\right) \cdot \left(\mathrm{d}x\hat{\boldsymbol{i}} + \mathrm{d}y\hat{\boldsymbol{j}} + \mathrm{d}z\hat{\boldsymbol{k}}\right) \\ &= \mathrm{d}x^2 + \mathrm{d}y^2 + \mathrm{d}z^2 \end{aligned} \tag{15-1}$$

图 15.1　费曼课后在解答学生的问题

被誉为 "数学王子 (prince of mathematics)" 的高斯创造了 "非欧几何"(non-Euclidean geometry) 的术语，在 (u,v,w) 的曲纹坐标系 (curvilinear coordinates) 中，高斯 (Carl Friedrich Gauss, 1777—1855, 如图 15.2 左图所示) 于 1827 年将曲面上两点的距离表示为如下二次型 [3.9]：

$$ds^2 = Adu^2 + Bdv^2 + Cdw^2 + 2Edu \cdot dv + 2Fdv \cdot dw + 2Gdw \cdot du \tag{15-2}$$

式中，A, B, C, E, F, G 为坐标 u, v, w 的函数.

业已在题表 1.5 中给出的 39 岁英年早逝的黎曼 (如图 15.2 右图所示) 于 1854 年 6 月 10 日答辩的有关在哥廷根大学的无俸讲师 (privatdozent) 的论文中给出了后来被称为 "黎曼度量张量" 的表达式：

$$ds^2 = g_{ij}dx^i dx^j \tag{15-3}$$

在黎曼去世两年后，也就是在黎曼于 1854 年宣布他的上述结果的 14 年后的 1868 年才得以正式出版 [3.10].

图 15.2　高斯 (左) 和黎曼 (右) 师徒

空间两点间的一条以 t 为参数的曲线的长度为

$$\mathcal{L} = \int_0^\tau \sqrt{g_{ij}\frac{\mathrm{d}\lambda^i}{\mathrm{d}t}\frac{\mathrm{d}\lambda^j}{\mathrm{d}t}}\mathrm{d}t \tag{15-4}$$

空间中两点之间的距离是最短曲线的长度，具有局部最短距离的曲线被称为 "测地线 (geodesics)"。

度量张量的基本性质如下：

- 度量张量是对称的 (symmetrical)：$g_{ij} = g_{ji}$；
- 度量张量是可逆的 (invertible)；
- 欧几里得度量张量 (Euclidean metric) 其实就是克罗内克符号：$g_{ij} = \delta_{ij}$.

例 15.1 给出欧几里得度量张量的分量.

解：在欧氏空间中有 $\mathrm{d}s^2 = \mathrm{d}x^2 + \mathrm{d}y^2 + \mathrm{d}z^2 = g_{ij}\mathrm{d}x^i\mathrm{d}x^j$，按照欧几里得度量张量其实就是克罗内克符号的基本性质，自然有

$$\begin{cases} g_{xx} = \delta_{xx} = 1 \\ g_{yy} = \delta_{yy} = 1 \\ g_{zz} = \delta_{zz} = 1 \end{cases} \quad \begin{cases} g_{xy} = \delta_{xy} = 0 \\ g_{yz} = \delta_{yz} = 0 \\ g_{zx} = \delta_{zx} = 0 \end{cases}$$

例 15.2 给出洛伦兹变换 (Lorentz transformation) 的度量张量.

解：洛伦兹变换可表示为

$$\mathrm{d}s^2 = g_{\alpha\beta}\mathrm{d}x^\alpha\mathrm{d}x^\beta = \left(\mathrm{d}x^0\right)^2 - \left(\mathrm{d}x^1\right)^2 - \left(\mathrm{d}x^2\right)^2 - \left(\mathrm{d}x^3\right)^2 \tag{15-5}$$

通过 (15-4) 式很容易给出洛伦兹变换的度量张量的分量为

$$g_{00} = 1; \quad g_{11} = g_{22} = g_{33} = -1 \tag{15-6}$$

上述分量可用矩阵方便地表示为

$$[g_{\alpha\beta}] = \begin{bmatrix} 1 & 0 & 0 & 0 \\ 0 & -1 & 0 & 0 \\ 0 & 0 & -1 & 0 \\ 0 & 0 & 0 & -1 \end{bmatrix} \tag{15-7}$$

由此得出，洛伦兹变换的度量张量不但对称，而且为对角阵 (diagonal).

例 15.3 柱坐标 (cylindrical polar, 如图 15.3 所示) 中的度量张量

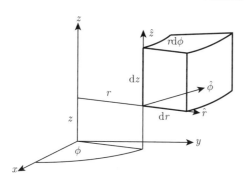

图 15.3 柱坐标系中的线元 $\mathrm{d}s$

解：由柱坐标和直角坐标系间的变换关系：

$$\begin{cases} x = r\cos\phi \\ y = r\sin\phi \\ z = z \end{cases} \tag{15-8}$$

则柱坐标中的线元可表示为

$$\mathrm{d}s^2 = g_{\mu\nu}\mathrm{d}x^\mu \mathrm{d}x^\nu = \mathrm{d}r^2 + r^2\mathrm{d}\phi^2 + \mathrm{d}z^2 \tag{15-9}$$

则度量张量的对角分量分别为

$$g_{rr} = 1, \quad g_{\phi\phi} = r^2, \quad g_{zz} = 1; \quad g_{r\phi} = g_{\phi z} = g_{zr} = 0 \tag{15-10}$$

(15-10) 式的矩阵形式 (matrix form) 为

$$[g_{\mu\nu}] = \begin{bmatrix} 1 & 0 & 0 \\ 0 & r^2 & 0 \\ 0 & 0 & 1 \end{bmatrix} \tag{15-11}$$

例 15.4 球坐标 (spherical polar, 如图 15.4 所示) 中的度量张量

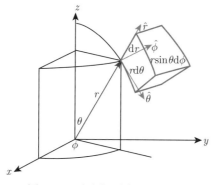

图 15.4 球坐标系中的线元 $\mathrm{d}s$

解：由球坐标和直角坐标系间的变换关系：

$$\begin{cases} x = r\sin\theta\cos\phi \\ y = r\sin\theta\sin\phi \\ z = r\cos\theta \end{cases} \tag{15-12}$$

则柱坐标中的线元可表示为

$$ds^2 = g_{\mu\nu}dx^\mu dx^\nu = dr^2 + r^2 d\theta^2 + r^2 \sin^2\theta d\phi^2 \tag{15-13}$$

则球坐标系度量张量的对角分量分别为

$$g_{rr} = 1, \quad g_{\theta\theta} = r^2, \quad g_{\phi\phi} = r^2\sin^2\theta; \quad g_{r\theta} = g_{\theta\phi} = g_{\phi r} = 0 \tag{15-14}$$

(15-14) 式的矩阵形式为

$$[g_{\alpha\beta}] = \begin{bmatrix} 1 & 0 & 0 \\ 0 & r^2 & 0 \\ 0 & 0 & r^2\sin^2\theta \end{bmatrix} \tag{15-15}$$

例 15.5 计算三维欧氏空间如下坐标变化的度量张量：

$$\begin{cases} u = x + 2y \\ v = x - y \\ w = z \end{cases} \tag{15-16}$$

解：(15-16) 式的逆变换为

$$\begin{cases} x = (u + 2v)/3 \\ y = (u - v)/3 \\ z = w \end{cases} \tag{15-17}$$

从而，线元可表示为

$$ds^2 = \left(\frac{1}{3}du + \frac{2}{3}dv\right)^2 + \left(\frac{1}{3}du - \frac{1}{3}dv\right)^2 + dw^2 = \frac{2}{9}du^2 + 2 \times \frac{1}{9}du \cdot dv + \frac{5}{9}dv^2 + dw^2$$

我们通过上式可立即写出相关度量张量的分量为

$$g_{uu} = 2/9, \quad g_{uv} = g_{vu} = 1/9, \quad g_{vv} = 5/9, \quad g_{ww} = 1 \tag{15-18}$$

图 15.5　左图：史瓦西在德国波茨坦的书房中；右图：爱因斯坦、史瓦西等为引力波作出奠基性贡献的几位著名物理学家

例 15.6　半经典史瓦西度规 (The semi-classical Schwarzschild metric)[3.11]

解：一则传说是，第一次世界大战期间应征入伍的德国天文学家卡尔·史瓦西 (Karl Schwarzschild, 1873—1916, 如图 15.5 所示. 穿军服的史瓦西如图 15.6 所示) 是 1915 年在与俄国作战前线的战壕中推导获得的该度规 (The legend that Schwarzschild found the solution while in a trench on the battlefield).

图 15.6　穿军装的史瓦西 (左一) 和其他参战军官在一起

1915 年 12 月 22 日，史瓦西从冰天雪地的作战前线把论文寄给了爱因斯坦. 在寄给爱因斯坦的信中，史瓦西这样写道："如您所见，除了干扰思路的重机枪枪声，战争已经很善待我了，它允许我摆脱周遭这一切，而在您的思维领地上进行这样一场漫步."

经爱因斯坦推荐，史瓦西的上述论文于 1916 年 1 月 13 日发表在《普鲁士科学院会刊》[3.11]. 爱因斯坦给在前线的史瓦西回信道："我抱着最大的兴趣阅读了您的论文. 我没有想到，能有人以这样简洁的形式求出精确解. 我非常喜欢您对那些对象的数学处理手法."

不幸的是,史瓦西在俄罗斯的作战前线得了一种称为 "天疱疮 (pemphigus)" 的皮肤病,由于病情严重,史瓦西于 1916 年 3 月被从前线送回家,2 个月后的 5 月 11 日就在波茨坦 (Potsdam) 去世了.

史瓦西度规方程 (Schwarzschild metric equation) 为

$$\mathrm{d}s^2 = \frac{1}{1-\frac{r_s}{r}}\mathrm{d}r^2 + r^2\mathrm{d}\theta^2 + r^2\sin^2\theta\mathrm{d}\phi^2 - \left(1-\frac{r_s}{r}\right)c^2\mathrm{d}t^2 \tag{15-19}$$

式中,史瓦西半径 (Schwarzschild radius) 为

$$r_s = \frac{2MG}{c^2} \tag{15-20}$$

式中,G 是引力常数,M 为质量,c 为光速. 对于太阳,$r_s = 2.95$ km;对于地球,则 $r_s = 8.87$ mm. 当 $\frac{r_s}{r} \to 0$ 时,(15-19) 式将退化为狭义相对论中球坐标下的闵可夫斯基度规. $r = 0$ 处为黑洞的引力奇点,$r = r_s$ 处为黑洞的事件视界 (event horizon).

从 (15-19) 式,我们很容易提取出用矩阵表示的史瓦西协变度规 (covariant metric) 为

$$[g_{ij}] = \begin{bmatrix} \dfrac{1}{1-\dfrac{r_s}{r}} & 0 & 0 & 0 \\ 0 & r^2 & 0 & 0 \\ 0 & 0 & r^2\sin^2\theta & 0 \\ 0 & 0 & 0 & -\left(1-\dfrac{r_s}{r}\right) \end{bmatrix} \tag{15-21}$$

史瓦西协变度规 (15-21) 式的行列式为其对角线诸量的乘积:

$$\det g_{ij} = -r^4\sin^2\theta \tag{15-22}$$

下面给出史瓦西逆变度规 (contravariant metric) g^{ij}. 由于协变和与之相应的逆变度规间满足下列关系:

$$g_{ij}g^{jk} = \delta_i^k \tag{15-23}$$

式中,$\mathbf{I}$ 为单位矩阵. 则,史瓦西逆变度规为

$$[g^{ij}] = \begin{bmatrix} 1-\dfrac{r_s}{r} & 0 & 0 & 0 \\ 0 & \dfrac{1}{r^2} & 0 & 0 \\ 0 & 0 & \dfrac{1}{r^2\sin^2\theta} & 0 \\ 0 & 0 & 0 & -\dfrac{1}{1-\dfrac{r_s}{r}} \end{bmatrix} \tag{15-24}$$

例 15.7 为什么拉格朗日力学对应于黎曼几何?

解: 著者在《力学讲义》[1.11] 第 13 页图 3.2 中,给出了薛定谔有关哈密顿光学-力学类比的深刻讨论. 按照薛定谔的原文,系统的动能 T 和线元 $\mathrm{d}s$ 可通过黎曼度规张量 g_{ik} 分别表示为

$$T = g_{ik}\dot{q}_i\dot{q}_k, \quad \mathrm{d}s^2 = g_{ik}\mathrm{d}q_i\mathrm{d}q_k \tag{15-25}$$

薛定谔作为大物理学家,当然深知在动能项中应有系数 $1/2$,亦即 $T = \dfrac{1}{2}g_{ik}\dfrac{\mathrm{d}q_i}{\mathrm{d}t}\dfrac{\mathrm{d}q_k}{\mathrm{d}t}$. 但在类比和数量级的估计中,该系数已不重要. 在大数学家阿诺尔德的《经典力学中的数学方法》[3.12] 中,在动能项中往往会将质量等常数省略掉.

从 (15-25) 式中得到重要关系式:

$$T = \left(\frac{\mathrm{d}s}{\mathrm{d}t}\right)^2 \tag{15-26}$$

从而有

$$\begin{aligned}T\mathrm{d}t &= \left(\frac{\mathrm{d}s}{\mathrm{d}t}\right)^2\mathrm{d}t = \frac{\mathrm{d}s}{\mathrm{d}t}\mathrm{d}s = \sqrt{T}\mathrm{d}s \\ &= \sqrt{E-U}\mathrm{d}s\end{aligned} \tag{15-27}$$

式中,E 和 U 分别为系统的总能和势能. 这样,最小作用量原理可等价地表示为 (可参阅《力学讲义》[1.11] 中的表 3.1)

$$\delta\int_{t_1}^{t_2}T\mathrm{d}t = \delta\int_{s_1}^{s_2}\sqrt{E-U}\mathrm{d}s = 0 \tag{15-28}$$

使上述作用量取最小值的真实运动,便是黎曼空间的测地线 (geodesic) 或短程线.

因此,拉格朗日力学的位形空间 $(q_1, q_2, \cdots, q_n)$ 也就是一个 n 维流形,所以说拉格朗日力学是流形上的力学,拉格朗日力学最终归纳为黎曼几何问题.

例 15.8 黎曼度量张量在引力波中的应用.

解: 由狭义相对论,平直时空间隔 (flat space-time interval) 为

$$\mathrm{d}s^2 = -c^2\mathrm{d}t^2 + \mathrm{d}x^2 + \mathrm{d}y^2 + \mathrm{d}z^2 = \eta_{\mu\nu}\mathrm{d}x^\mu\mathrm{d}x^\nu \tag{15-29}$$

由广义相对论,弯曲时空间隔 (curved space-time interval) $\mathrm{d}s^2 = g_{\mu\nu}\mathrm{d}x^\mu\mathrm{d}x^\nu$ 中的黎曼度规张量 $g_{\mu\nu}$ 可看做是平直时空间隔度规 $\eta_{\mu\nu}$ 再加上一个度规摄动 (metric perturbation) $h_{\mu\nu}$ (如图 15.7 所示):

$$g_{\mu\nu} = \eta_{\mu\nu} + h_{\mu\nu} \quad (h_{\mu\nu} \ll 1) \tag{15-30}$$

图 15.7 弯曲空间与平直空间的简化过程,黑洞事件视界附近网格的大小是普朗克长度

对于弱引力场和横向无迹规范场 (transverse traceless gauge, TT gauge)，将 (15-30) 式代入爱因斯坦场方程：$G_{ij} = 8\pi T_{ij}$，则得到用度规摄动量 $h_{\mu\nu}$ 满足的波动方程：

$$\left(\frac{1}{c^2}\frac{\partial^2}{\partial t^2} - \nabla^2\right) h_{\mu\nu} = \Box h_{\mu\nu} = g^{ij}\partial_j\partial_i h_{\mu\nu} = 0 \tag{15-31}$$

式中，$\Box = \frac{1}{c^2}\frac{\partial^2}{\partial t^2} - \nabla^2$ 被称为"达朗贝尔算子 (D'Alembertian)"或者"盒子算子 (box operator)"，黎曼度规张量 g^{ij} 和 (15-6) 式相同：$g^{00} = 1, g^{11} = g^{22} = g^{33} = -1$.

上式表明，引力波的波速确实是光速 c，这是大数学家、物理学家、力学家、哲学家庞加莱于 1905 年 6 月 5 日就已经预言的！

爱因斯坦场方程已经在《力学讲义》[1,11] 的例 B.14 中做了十分详细的说明，此不赘述.

例 15.9 在黎曼空间中，用狄拉克符号表示 (15-3) 式：$ds^2 = g_{ij}dx^i dx^j$.

解：在三维直角坐标系中，不必区分上下指标，但在弯曲空间中，则必须区分上下指标. 黎曼度量张量的一个重要作用是进行指标的升降，如

$$g_{ij}dx^j = dx_i \tag{15-32}$$

因此有

$$ds^2 = g_{ij}dx^i dx^j = dx_i dx^i = \langle dx \mid dx \rangle \tag{15-33}$$

同理，在黎曼空间中，两个矢量 **A** 和 **B** 的内积可相应地表示为

$$\langle A \mid B \rangle = g_{\mu\nu}A^\mu B^\nu = A_\nu B^\nu = A^\mu B_\mu \tag{15-34}$$

例 15.10 在上例中，已经利用黎曼度量张量进行了指标升降. 进一步展开讨论之.

解：在曲纹坐标中，与 (15-1) 式不同，位矢的增量可表示为

$$d\boldsymbol{r} = dx_\mu \boldsymbol{e}^\mu = dx^\mu \boldsymbol{e}_\mu \tag{15-35}$$

式中，$\boldsymbol{e}^\mu$ 和 $\boldsymbol{e}_\mu$ 分别为曲纹坐标系中的基矢的逆变和协变分量. 则黎曼度量张量的定义式 (15-3) 式可详细地表示为

$$ds^2 = d\boldsymbol{r} \cdot d\boldsymbol{r} = dx^\mu \boldsymbol{e}_\mu \cdot dx^\nu \boldsymbol{e}_\nu = (\boldsymbol{e}_\mu \cdot \boldsymbol{e}_\nu) dx^\mu dx^\nu$$
$$= g_{\mu\nu}dx^\mu dx^\nu \tag{15-36}$$

上式还可表示为

$$ds^2 = d\boldsymbol{r} \cdot d\boldsymbol{r} = dx_\mu \boldsymbol{e}^\mu \cdot dx_\nu \boldsymbol{e}^\nu = (\boldsymbol{e}^\mu \cdot \boldsymbol{e}^\nu) dx_\mu dx_\nu$$

$$= g^{\mu\nu}\mathrm{d}x_\mu \mathrm{d}x_\nu \tag{15-37}$$

或者

$$\mathrm{d}s^2 = \mathrm{d}\boldsymbol{r}\cdot\mathrm{d}\boldsymbol{r} = \mathrm{d}x_\mu \boldsymbol{e}^\mu \cdot \mathrm{d}x^\nu \boldsymbol{e}_\nu = (\boldsymbol{e}^\mu \cdot \boldsymbol{e}_\nu)\mathrm{d}x_\mu \mathrm{d}x^\nu$$
$$= g_\nu^\mu \mathrm{d}x_\mu \mathrm{d}x^\nu = g_\mu^\nu \mathrm{d}x^\mu \mathrm{d}x_\nu \tag{15-37'}$$

例 15.11 在曲纹坐标系中，如何定义基矢量？拉梅系数和黎曼度量张量间有何关系？

解：笛卡儿坐标系 (Cartesian coordinates) 是法国数学家、哲学家笛卡儿 (René Descartes, 1596—1650) 于 1637 年在其哲学巨著《谈谈方法》中引入的 [3.8]. 笛卡儿的该论著，对西方人的思维方式，思想观念和科学研究方法有极大的影响. 笛卡儿在该书中不但第一次提出了 "我思故我在 (Je pense, donc je suis; I think, therefore I am)" 的名言，还第一次引入了笛卡儿坐标系.

曲线坐标系则是由法国数学、力学家拉梅 (Gabriel Lamé, 1795—1870) 于 1852 年在其出版的《固体弹性的数学理论讲义》一书中引入的 [3.13]. 拉梅在该书中研究了球形弹性包络线的平衡条件，该包络线的平衡条件取决于给定的载荷在球面上的分布.

正如我们十分熟悉的，在直角坐标系 (x^1, x^2, x^3) 中，坐标线 (lines of coordinates) x^i 都是直线，坐标基 $(\boldsymbol{e}_1, \boldsymbol{e}_2, \boldsymbol{e}_3)$ 都是单位正交基，满足 $\boldsymbol{e}_i \cdot \boldsymbol{e}_k = \delta_{ik}$. 则矢径 $\boldsymbol{r}(x^1, x^2, x^3)$ 可表示为

$$\boldsymbol{r} = x^1 \boldsymbol{e}_1 + x^2 \boldsymbol{e}_2 + x^3 \boldsymbol{e}_3 = x^i \boldsymbol{e}_i \tag{15-38}$$

亦即，在笛卡儿坐标系中，单位正交基可以表示为

$$\boldsymbol{e}_i = \frac{\partial \boldsymbol{r}}{\partial x^i} \tag{15-39}$$

所谓曲线坐标系 (q^1, q^2, q^3)，就是说在坐标线中，至少有一条坐标线 q^i 是曲线. 由于存在如下一对一的映射关系：

$$x^i = x^i(q^1, q^2, q^3) \tag{15-40}$$

则在曲线坐标系中，协变基矢 $\boldsymbol{g}_i$ 为

$$\boldsymbol{g}_i = \frac{\partial \boldsymbol{r}}{\partial q^i} = \frac{\partial \boldsymbol{r}}{\partial x^j}\frac{\partial x^j}{\partial q^i} = \frac{\partial x^j}{\partial q^i}\boldsymbol{e}_j \tag{15-41}$$

上式可用直角坐标 (x^1, x^2, x^3) 和其单位正交基 (e_1, e_2, e_3) 表示为

$$\begin{cases} g_1 = \dfrac{\partial x^1}{\partial q^1} e_1 + \dfrac{\partial x^2}{\partial q^1} e_2 + \dfrac{\partial x^3}{\partial q^1} e_3 \\ g_2 = \dfrac{\partial x^1}{\partial q^2} e_1 + \dfrac{\partial x^2}{\partial q^2} e_2 + \dfrac{\partial x^3}{\partial q^2} e_3 \\ g_3 = \dfrac{\partial x^1}{\partial q^3} e_1 + \dfrac{\partial x^2}{\partial q^3} e_2 + \dfrac{\partial x^3}{\partial q^3} e_3 \end{cases} \tag{15-42}$$

注意到由 (15-41) 式定义的协变基矢 g_i 并非单位基矢, 正交曲线坐标系单位化后的基矢 (b_1, b_2, b_3) 可通过拉梅系数 H_i 表示为

$$b_1 = \frac{g_1}{H_1}, \quad b_2 = \frac{g_2}{H_2}, \quad b_3 = \frac{g_3}{H_3} \tag{15-43}$$

式中, 拉梅系数的表达式为

$$H_i = \left| \frac{\partial r}{\partial q^i} \right| = \sqrt{\left(\frac{\partial x^1}{\partial q^i} \right)^2 + \left(\frac{\partial x^2}{\partial q^i} \right)^2 + \left(\frac{\partial x^3}{\partial q^i} \right)^2} \tag{15-44}$$

由于 b_i 为单位正交基, 不再区分逆变与协变, 即 $b_i = b^i$, 根据基本关系 $g_i \cdot g^j = \delta_i^j$, 得到 $|g^i| = \dfrac{1}{|g_i|}$, 进一步可以得到单位基矢 b_i 与逆变基矢 g^i 的关系

$$b_1 = b^1 = H_1 g^1, \quad b_2 = b^2 = H_2 g^2, \quad b_3 = b^3 = H_3 g^3 \tag{15-43'}$$

下面以极坐标 (b_ρ, b_θ) 为例说明. 由于矢径可通过笛卡儿坐标系的单位基矢表示为

$$r = x^1 e_1 + x^2 e_2 = \rho \cos\theta e_1 + \rho \sin\theta e_2 \tag{15-45}$$

式中,

$$\begin{cases} \rho = \sqrt{(x^1)^2 + (x^2)^2} \\ \theta = \mathrm{actan}\dfrac{x^2}{x^1} \end{cases} \tag{15-46}$$

由 (15-44) 式可得到极坐标系的拉梅系数为

$$\begin{cases} H_\rho = \sqrt{\left(\dfrac{\partial x^1}{\partial \rho} \right)^2 + \left(\dfrac{\partial x^2}{\partial \rho} \right)^2} = \sqrt{\left(\dfrac{\partial (\rho \cos\theta)}{\partial \rho} \right)^2 + \left(\dfrac{\partial (\rho \sin\theta)}{\partial \rho} \right)^2} = 1 \\ H_\theta = \sqrt{\left(\dfrac{\partial x^1}{\partial \theta} \right)^2 + \left(\dfrac{\partial x^2}{\partial \theta} \right)^2} = \sqrt{\left(\dfrac{\partial (\rho \cos\theta)}{\partial \theta} \right)^2 + \left(\dfrac{\partial (\rho \sin\theta)}{\partial \theta} \right)^2} = \rho \end{cases} \tag{15-47}$$

则可得到极坐标系的单位基矢为

$$\begin{cases} b_\rho = \dfrac{1}{H_\rho} \dfrac{\partial r}{\partial \rho} = \cos\theta e_1 + \sin\theta e_2 \\ b_\theta = \dfrac{1}{H_\theta} \dfrac{\partial r}{\partial \theta} = -\sin\theta e_1 + \cos\theta e_2 \end{cases} \tag{15-48}$$

或者可用矩阵形式表示为

$$\begin{bmatrix} \boldsymbol{b}_\rho \\ \boldsymbol{b}_\theta \end{bmatrix} = \begin{bmatrix} \cos\theta & \sin\theta \\ -\sin\theta & \cos\theta \end{bmatrix} \begin{bmatrix} \boldsymbol{e}_1 \\ \boldsymbol{e}_2 \end{bmatrix} \tag{15-49}$$

容易验证极坐标系单位基矢之间的正交性:

$$\boldsymbol{b}_\rho \cdot \boldsymbol{b}_\theta = 0 \tag{15-50}$$

下面求曲线坐标系的速度和加速度矢量的表达式. 按照定义, 速度矢量为

$$\boldsymbol{v} = \frac{\mathrm{d}\boldsymbol{r}}{\mathrm{d}t} = \frac{\partial \boldsymbol{r}}{\partial q^i}\frac{\mathrm{d}q^i}{\mathrm{d}t} = \frac{\mathrm{d}q^i}{\mathrm{d}t}H_i\boldsymbol{b}_i = \dot{q}^i H_i \boldsymbol{b}_i \tag{15-51}$$

而加速度矢量的分量 a_i 为加速度 $\boldsymbol{a} = \dfrac{\mathrm{d}\boldsymbol{v}}{\mathrm{d}t}$ 沿单位基矢 $\boldsymbol{b}_i = \dfrac{\boldsymbol{g}_i}{H_i}$ 的投影:

$$\begin{aligned} a_i &= \frac{\mathrm{d}\boldsymbol{v}}{\mathrm{d}t} \cdot \frac{1}{H_i}\frac{\partial \boldsymbol{r}}{\partial q^i} = \frac{1}{H_i}\frac{\mathrm{d}\boldsymbol{v}}{\mathrm{d}t}\cdot\frac{\partial \boldsymbol{r}}{\partial q^i} \\ &= \frac{1}{H_i}\left[\frac{\mathrm{d}}{\mathrm{d}t}\left(\boldsymbol{v}\cdot\frac{\partial \boldsymbol{r}}{\partial q^i}\right) - \boldsymbol{v}\cdot\frac{\mathrm{d}}{\mathrm{d}t}\frac{\partial \boldsymbol{r}}{\partial q^i}\right] \\ &= \frac{1}{H_i}\left[\frac{\mathrm{d}}{\mathrm{d}t}\left(\boldsymbol{v}\cdot\frac{\partial \boldsymbol{v}}{\partial \dot{q}^i}\right) - \boldsymbol{v}\cdot\frac{\partial \boldsymbol{v}}{\partial q^i}\right] \\ &= \frac{1}{mH_i}\left(\frac{\mathrm{d}}{\mathrm{d}t}\frac{\partial T}{\partial \dot{q}^i} - \frac{\partial T}{\partial q^i}\right) \end{aligned} \tag{15-52}$$

式中, $T = \dfrac{mv^2}{2} = m\dfrac{\boldsymbol{v}\cdot\boldsymbol{v}}{2}$ 为动能. 在上式中, 我们还应用了关系式[1.11]: $\dfrac{\partial \boldsymbol{r}}{\partial q^i} \equiv \dfrac{\partial \boldsymbol{v}}{\partial \dot{q}^i}$. 根据 (15-43) 式与 (15-43′) 式, 即 $\boldsymbol{a} = a_i\boldsymbol{b}_i = \dfrac{a_i}{H_i}\boldsymbol{g}_i = a_iH_i\boldsymbol{g}^i$. 下面仍以极坐标系为例说明. 由于动能为

$$T = \frac{mv^2}{2} = \frac{m}{2}\left|\dot{\rho}\boldsymbol{b}_\rho + \rho\dot{\theta}\boldsymbol{b}_\theta\right|^2 = \frac{m}{2}\left[\dot{\rho}^2 + \left(\rho\dot{\theta}\right)^2\right] \tag{15-53}$$

将 (15-53) 式代入 (15-52) 式, 得到

$$\begin{cases} a_\rho = \dfrac{1}{2H_\rho}\left[\dfrac{\mathrm{d}}{\mathrm{d}t}\dfrac{\partial}{\partial \dot{\rho}}\left(\dot{\rho}^2 + \rho^2\dot{\theta}^2\right) - \dfrac{\partial}{\partial \rho}\left(\dot{\rho}^2 + \rho^2\dot{\theta}^2\right)\right] = \ddot{\rho} - \rho\dot{\theta}^2 \\ a_\theta = \dfrac{1}{2H_\theta}\left[\dfrac{\mathrm{d}}{\mathrm{d}t}\dfrac{\partial}{\partial \dot{\theta}}\left(\dot{\rho}^2 + \rho^2\dot{\theta}^2\right) - \dfrac{\partial}{\partial \theta}\left(\dot{\rho}^2 + \rho^2\dot{\theta}^2\right)\right] = \dfrac{1}{\rho}\dfrac{\mathrm{d}\left(\rho^2\dot{\theta}\right)}{\mathrm{d}t} = 2\dot{\rho}\dot{\theta} + \rho\ddot{\theta} \end{cases} \tag{15-54}$$

事实上, 可从 (15-52) 式来构造出质点运动方程. 对于保守力, 从 (15-52) 式构造出如下关系式:

$$mH_ia_i = \frac{\mathrm{d}}{\mathrm{d}t}\frac{\partial T}{\partial \dot{q}^i} - \frac{\partial T}{\partial q^i} = F_i = -\frac{\partial V}{\partial q^i} \tag{15-55}$$

从而得出两套等价的运动方程. 一是曲线坐标系中牛顿运动方程:

$$F_i = mH_i a_i \quad \text{或} \quad F_i \boldsymbol{g}^i = mH_i a_i \boldsymbol{g}^i = mA_i \boldsymbol{g}^i \tag{15-56}$$

式中, F_i 和 $A_i = H_i a_i$ 分别为沿基矢 $\boldsymbol{g}^i$ 的力分量和加速度分量.

二是在位形空间的欧拉–拉格朗日方程:

$$\frac{\mathrm{d}}{\mathrm{d}t}\frac{\partial L}{\partial \dot{q}^i} - \frac{\partial L}{\partial q^i} = 0 \tag{15-57}$$

式中, 拉格朗日量为 $L = T - V$.

§16. 克里斯托费尔符号和联络

1869 年, 克里斯托费尔符号 (Christoffel symbols) 由德国数学家和物理学家克里斯托费尔 (Elwin Bruno Christoffel, 1829—1900, 如图 16.1 所示) 引入 [3.14]. 克里斯托费尔是本书中所述及的终身未婚的著名科学家之一, 见题表 3.1.

图 16.1　克里斯托费尔

16.1　曲纹坐标系的基矢量、黎曼度规张量

如图 16.2 所示的曲纹坐标系 (x^1, x^2, x^3), 其协变基矢量:

$$\boldsymbol{g}_i = \frac{\partial \boldsymbol{r}}{\partial x^i} \tag{16-1}$$

是随点变化的. 式中, $\boldsymbol{r}$ 为矢径.

下面我们首先给出协变基矢量 $\boldsymbol{g}_i$ 和逆变基矢量 $\boldsymbol{g}^i$ 与直角坐标系基矢量的关系. 由于曲纹坐标与直角坐标的关系 $x^k = x^k(\tilde{x}^m)$ 或相应直角坐标与曲纹坐标的逆关系 $\tilde{x}^m = \tilde{x}^m(x^k)$, 由链式法则, 有

$$\boldsymbol{g}_k = \frac{\partial \boldsymbol{r}}{\partial x^k} = \frac{\partial \boldsymbol{r}}{\partial \tilde{x}^p}\frac{\partial \tilde{x}^p}{\partial x^k} = \frac{\partial \tilde{x}^p}{\partial x^k}\hat{\boldsymbol{i}}_p \tag{16-2}$$

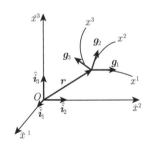

图 16.2　曲纹坐标系

第 3 章 张量代数和微积分

由协变基矢量和逆变基矢量之间的基本关系：

$$\boldsymbol{g}_k \cdot \boldsymbol{g}^l = \delta_k^l \quad \text{或} \quad \langle g_k \mid g^l \rangle = \delta_k^l \tag{16-3}$$

将 (16-2) 式代入 (16-3) 式，则有

$$\frac{\partial \tilde{x}^p}{\partial x^k} \hat{\boldsymbol{i}}_p \cdot \boldsymbol{g}^l = \delta_k^l \;\Rightarrow\; \hat{\boldsymbol{i}}_p \cdot \boldsymbol{g}^l = \delta_k^l \frac{\partial x^k}{\partial \tilde{x}^p} = \frac{\partial x^l}{\partial \tilde{x}^p} \tag{16-4}$$

对 (16-4) 式中第二式两端点积 $\hat{\boldsymbol{i}}^p$，则有

$$\boldsymbol{g}^l = \frac{\partial x^l}{\partial \tilde{x}^p} \hat{\boldsymbol{i}}^p \tag{16-5}$$

有了曲纹坐标系的协变基矢量 (16-2) 式和逆变基矢量 (16-5) 式后，很容易给出其相应的黎曼度量张量：

$$g_{kl} = \boldsymbol{g}_k \cdot \boldsymbol{g}_l = \langle g_k \mid g_l \rangle = \frac{\partial \tilde{x}^p}{\partial x^k} \frac{\partial \tilde{x}^p}{\partial x^l} \tag{16-6}$$

$$g^{kl} = \boldsymbol{g}^k \cdot \boldsymbol{g}^l = \langle g^k \mid g^l \rangle = \frac{\partial x^k}{\partial \tilde{x}^p} \frac{\partial x^l}{\partial \tilde{x}^p} \tag{16-7}$$

16.2 第二类克里斯托费尔符号

由于协变基矢量对坐标的导数 $\dfrac{\partial \boldsymbol{g}_j}{\partial x^i}$ 为矢量，故可令

$$\frac{\partial \boldsymbol{g}_j}{\partial x^i} = \Gamma_{ij}^k \boldsymbol{g}_k \tag{16-8}$$

式中，Γ_{ij}^k 为第二类克里斯托费尔符号. 由于

$$\frac{\partial \boldsymbol{g}_j}{\partial x^i} = \frac{\partial}{\partial x^i}\left(\frac{\partial \boldsymbol{r}}{\partial x^j}\right) = \frac{\partial^2 \boldsymbol{r}}{\partial x^i \partial x^j} = \frac{\partial^2 \boldsymbol{r}}{\partial x^j \partial x^i} = \frac{\partial \boldsymbol{g}_i}{\partial x^j} = \Gamma_{ji}^k \boldsymbol{g}_k \tag{16-9}$$

亦即，由于二阶偏导数的可交换性，第二类克里斯托费尔符号满足对称性：

$$\Gamma_{ij}^k = \Gamma_{ji}^k \tag{16-10}$$

例 16.1 平直空间中是否存在克里斯托费尔符号？

答：克里斯托费尔符号反映的是曲纹坐标系下基矢量随坐标变化的行为，故在平直坐标系下，克里斯托费尔符号均等于零. 因此也说明克里斯托费尔符号和坐标系有关，故不是张量.

例 16.2 第二类克里斯托费尔符号共有多少分量？

解: 首先由 (16-8) 式给出第二类克里斯托费尔符号的具体表达式. (16-8) 式两端点乘 $\boldsymbol{g}^k$, 考虑到: $\boldsymbol{g}_k \cdot \boldsymbol{g}^k = 1$, 则有

$$\Gamma_{ij}^k = \frac{\partial \boldsymbol{g}_j}{\partial x^i} \cdot \boldsymbol{g}^k \tag{16-11}$$

由于 (16-11) 式所给出的第二类克里斯托费尔符号的对称性, 故 Γ_{ij}^k 共有 $\dfrac{3^2 \times (3+1)}{2}$ = 18 个分量. 计算过程为 $N^3 - C_N^1 C_N^2 = N^3 - N\dfrac{N(N-1)}{2} = \dfrac{N^2(N+1)}{2}$.

例 16.3 求逆变基矢量对坐标的导数 $\dfrac{\partial \boldsymbol{g}^i}{\partial x^j}$.

解: 由对偶条件:

$$\boldsymbol{g}^i \cdot \boldsymbol{g}_p = \delta_p^i \tag{16-12}$$

上式对 x^j 求偏导, 有

$$\frac{\partial \boldsymbol{g}^i}{\partial x^j} \cdot \boldsymbol{g}_p + \boldsymbol{g}^i \cdot \frac{\partial \boldsymbol{g}_p}{\partial x^j} = 0 \tag{16-13}$$

按照定义 (16-11) 式, 则有

$$\frac{\partial \boldsymbol{g}^i}{\partial x^j} \cdot \boldsymbol{g}_p = -\boldsymbol{g}^i \cdot \frac{\partial \boldsymbol{g}_p}{\partial x^j} = -\Gamma_{jp}^i \tag{16-14}$$

因此, 有下列关系式:

$$\frac{\partial \boldsymbol{g}^i}{\partial x^j} = -\Gamma_{jp}^i \boldsymbol{g}^p \tag{16-15}$$

16.3 克里斯托费尔符号的应用

对于曲纹坐标系中的任意随时间变化的矢量 $\boldsymbol{u}(t) = u^i(t)\, \boldsymbol{g}_i\left(x^k(t)\right)$, 其时间变化率:

$$\begin{aligned}
\frac{\mathrm{d}\boldsymbol{u}}{\mathrm{d}t} &= \frac{\mathrm{d}u^i}{\mathrm{d}t}\boldsymbol{g}_i + u^m \frac{\mathrm{d}\boldsymbol{g}_m}{\mathrm{d}t} \\
&= \frac{\mathrm{d}u^i}{\mathrm{d}t}\boldsymbol{g}_i + u^m \frac{\partial \boldsymbol{g}_m}{\partial x^k}\frac{\mathrm{d}x^k}{\mathrm{d}t} = \frac{\mathrm{d}u^i}{\mathrm{d}t}\boldsymbol{g}_i + u^m v^k \Gamma_{mk}^i \boldsymbol{g}_i \\
&= \left(\frac{\mathrm{d}u^i}{\mathrm{d}t} + u^m v^k \Gamma_{mk}^i\right)\boldsymbol{g}_i = \frac{\mathrm{D}u^i}{\mathrm{D}t}\boldsymbol{g}_i
\end{aligned} \tag{16-16}$$

而

$$\frac{\mathrm{D}u^i}{\mathrm{D}t} = \frac{\mathrm{d}u^i}{\mathrm{d}t} + u^m v^k \Gamma_{mk}^i \tag{16-17}$$

称为矢量分量 u^i 对参数 (时间) 的全导数. 作为一个例子, 加速度的逆变分量可表示为

$$a^i = \frac{\mathrm{D}v^i}{\mathrm{D}t} = \frac{\mathrm{d}v^i}{\mathrm{d}t} + v^m v^k \Gamma_{mk}^i \tag{16-18}$$

第 3 章　张量代数和微积分

同理，利用 (16-15) 式，可得

$$\frac{\mathrm{D}u_i}{\mathrm{D}t} = \frac{\mathrm{d}u_i}{\mathrm{d}t} - u_m v^k \Gamma^m_{ki} \tag{16-19}$$

$$a_i = \frac{\mathrm{D}v_i}{\mathrm{D}t} = \frac{\mathrm{d}v_i}{\mathrm{d}t} - v_m v^k \Gamma^m_{ki} \tag{16-20}$$

例 16.4　详细推导测地线方程 (10-6) 式.

解：对于位形空间，由 (15-3) 式可知，任意两点间的一条连线 l 的长度为

$$s(l) = \int_l \mathrm{d}s = \int_l \sqrt{g_{\mu\nu}\mathrm{d}q^\mu \mathrm{d}q^\nu} \tag{16-21}$$

所谓测地线，就是使 (16-21) 式取得极小值的连线，亦即 $\delta s(l) = 0$.

类似于 (15-4) 式，取一参数 λ，广义坐标可表示为参数方程 $q^k = q^k(\lambda)$. 这里的参数 λ 一般地取为固有时 (proper time)，可参阅《力学讲义》[1.11] §39.2.

$$s(l) = \int_{\lambda_0}^{\lambda_1} \frac{\mathrm{d}s}{\mathrm{d}\tau} \mathrm{d}\tau = \int_{\lambda_0}^{\lambda_1} \underbrace{\sqrt{g_{\mu\nu}\frac{\mathrm{d}q^\mu}{\mathrm{d}\lambda}\frac{\mathrm{d}q^\nu}{\mathrm{d}\lambda}}}_{\sqrt{\mathcal{L}}} \mathrm{d}\lambda \tag{16-22}$$

为了运算上的方便，一个巧妙的技巧 (neat trick) 是令

$$\mathcal{L} = g_{\mu\nu}\frac{\mathrm{d}q^\mu}{\mathrm{d}\lambda}\frac{\mathrm{d}q^\nu}{\mathrm{d}\lambda} \tag{16-23}$$

而不是令 $\mathcal{L} = \sqrt{g_{\mu\nu}\frac{\mathrm{d}q^\mu}{\mathrm{d}\lambda}\frac{\mathrm{d}q^\nu}{\mathrm{d}\lambda}}$，这样做的好处是使微分运算得以简化. 将 (16-23) 式代入欧拉–拉格朗日方程：

$$\frac{\mathrm{d}}{\mathrm{d}\lambda}\frac{\partial \mathcal{L}}{\partial \frac{\mathrm{d}q^k}{\mathrm{d}\lambda}} - \frac{\partial \mathcal{L}}{\partial q^k} = 0 \tag{16-24}$$

式中，$\frac{\mathrm{d}q^k}{\mathrm{d}\lambda} = \dot{q}^k$ 为广义速度. 由于

$$\frac{\partial \mathcal{L}}{\partial q^k} = \frac{\partial g_{\mu\nu}}{\partial q^k}\frac{\mathrm{d}q^\mu}{\mathrm{d}\lambda}\frac{\mathrm{d}q^\nu}{\mathrm{d}\lambda} \tag{16-25}$$

和

$$\frac{\partial \mathcal{L}}{\partial \frac{\mathrm{d}q^k}{\mathrm{d}\lambda}} = g_{\mu\nu}\delta_{\mu k}\frac{\mathrm{d}q^\nu}{\mathrm{d}\lambda} + g_{\mu\nu}\delta_{\nu k}\frac{\mathrm{d}q^\mu}{\mathrm{d}\lambda} = g_{\mu k}\frac{\mathrm{d}q^\mu}{\mathrm{d}\lambda} + g_{k\nu}\frac{\mathrm{d}q^\nu}{\mathrm{d}\lambda} \tag{16-26}$$

将 (16-25) 和 (16-26) 两式代入 (16-24) 式，得到

$$\frac{\mathrm{d}}{\mathrm{d}\lambda}\left(g_{\mu k}\frac{\mathrm{d}q^\mu}{\mathrm{d}\lambda} + g_{k\nu}\frac{\mathrm{d}q^\nu}{\mathrm{d}\lambda}\right) - \frac{\partial g_{\mu\nu}}{\partial q^k}\frac{\mathrm{d}q^\mu}{\mathrm{d}\lambda}\frac{\mathrm{d}q^\nu}{\mathrm{d}\lambda} = 0 \tag{16-27}$$

可对上式中第一项进一步展开：

$$\frac{\mathrm{d}}{\mathrm{d}\lambda}\left(g_{\mu k}\frac{\mathrm{d}q^\mu}{\mathrm{d}\lambda}+g_{k\nu}\frac{\mathrm{d}q^\nu}{\mathrm{d}\lambda}\right)=g_{\mu k}\frac{\mathrm{d}^2 q^\mu}{\mathrm{d}\lambda^2}+g_{k\nu}\frac{\mathrm{d}^2 q^\nu}{\mathrm{d}\lambda^2}+\frac{\partial g_{\mu k}}{\partial q^\nu}\frac{\mathrm{d}q^\mu}{\mathrm{d}\lambda}\frac{\mathrm{d}q^\nu}{\mathrm{d}\lambda}$$
$$+\frac{\partial g_{k\nu}}{\partial q^\mu}\frac{\mathrm{d}q^\mu}{\mathrm{d}\lambda}\frac{\mathrm{d}q^\nu}{\mathrm{d}\lambda} \tag{16-28}$$

式中右端的前两项为 $g_{\mu k}\frac{\mathrm{d}^2 q^\mu}{\mathrm{d}\lambda^2}+g_{k\nu}\frac{\mathrm{d}^2 q^\nu}{\mathrm{d}\lambda^2}$，注意到这两项中的 μ 和 ν 均为哑标，因此，第二项中的哑标 ν 可换为 μ，考虑到黎曼度量张量的对称性，则有

$$g_{\mu k}\frac{\mathrm{d}^2 q^\mu}{\mathrm{d}\lambda^2}+\underbrace{g_{k\nu}\frac{\mathrm{d}^2 q^\nu}{\mathrm{d}\lambda^2}}_{\text{更换哑标}}=2g_{\mu k}\frac{\mathrm{d}^2 q^\mu}{\mathrm{d}\lambda^2} \tag{16-29}$$

结合 (16-28) 和 (16-29) 两式，(16-27) 式变为

$$2g_{\mu k}\frac{\mathrm{d}^2 q^\mu}{\mathrm{d}\lambda^2}+\left(\frac{\partial g_{\mu k}}{\partial q^\nu}+\frac{\partial g_{k\nu}}{\partial q^\mu}-\frac{\partial g_{\mu\nu}}{\partial q^k}\right)\frac{\mathrm{d}q^\mu}{\mathrm{d}\lambda}\frac{\mathrm{d}q^\nu}{\mathrm{d}\lambda}=0 \tag{16-30}$$

上式两端均乘以 $g^{k\gamma}$，考虑到

$$g^{k\gamma}\cdot g_{\mu k}=\delta^\gamma_\mu \tag{16-31}$$

和

$$\Gamma^\gamma_{\mu\nu}=\frac{1}{2}g^{k\gamma}\left(\frac{\partial g_{\mu k}}{\partial q^\nu}+\frac{\partial g_{k\nu}}{\partial q^\mu}-\frac{\partial g_{\mu\nu}}{\partial q^k}\right) \tag{16-32}$$

将 (16-31) 和 (16-32) 两式代入 (16-30) 式，稍加整理后便得到测地线方程 (equation of the geodesics)：

$$\frac{\mathrm{d}^2 q^\gamma}{\mathrm{d}\lambda^2}+\Gamma^\gamma_{\mu\nu}\frac{\mathrm{d}q^\mu}{\mathrm{d}\lambda}\frac{\mathrm{d}q^\nu}{\mathrm{d}\lambda}=0 \tag{16-33}$$

测地线 (geodesic) 又称大地线或短程线，可以定义为空间中两点的局域最短或最长路径. 例如，地球的表面是一弯曲的二维空间. 如图 16.3 所示，地球上的测地线称为大圆 (great circles)，是两点之间最近的路径. 由于测地线是两个机场之间的最短程，这正是领航员让飞行员飞行的航线.

克里斯托费尔符号在微分几何中被公认为是难懂的 (elusive). 克里斯托费尔一生共发表 35 篇期刊论文，而其更多的工作是他在斯特拉斯堡大学 (Strasbourg Univ.) 数学系的教学和演讲中公布的.

爱因斯坦曾多次强调克里斯托费尔、里奇和列维–奇维塔相关工作的重要性. 爱因斯坦在其 1916 年的广义相对论的著名原文 [3.15] 中陈述道："那些真正明了该理论的人将会对之终生痴迷. 它代表了由高斯、黎曼、克里斯托费尔、里奇和列

维-奇维塔所开创的一般微分学方法的真正胜利 (the fascination of this theory would hardly leave anybody who has really grasped it. It represents a real triumph of the method of the general differential calculus founded by Gauss, Riemann, Christoffel, Ricci and Levi-Civita)."

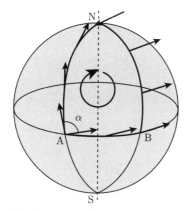

图 16.3　地球上的测地线为大圆, 其切矢量关于曲线本身是平行移动的

有学者对克里斯托费尔联络方面的工作高度评价道[3.16]: "在我看来, 克里斯托费尔在微分几何方面的巨大成就在于他发现了最为重要的微分几何结构, 也就是一个流形中的联络, 即使是他本人也很难察觉到这一发现的巨大意义 (Christoffel's great achievement for differential geometry is, in my opinion, that he discovered the most important differential-geometrical structure, namely that of connection in a manifold, even if he himself could hardly have been aware of the significance of this discovery)."

克里斯托费尔在斯特拉斯堡大学工作的 22 年中, 从未给大一的本科生授过课, 他授课的对象是 2—4 年级. 克里斯托费尔在教学上是个完美主义者, 人们基本上可以从一个人捡起粉笔的方式中来判断此人是否是克里斯托费尔的学生 (One can almost recognize Christoffel's students by the way they pick up the chalk), 他的学生们即使在这些外在的方面都要展现出这位大师对他们的影响.

§17. 张量的标量函数的泰勒展开

17.1　多元矢量函数的泰勒展开

设所研究的问题为多元函数 $(x_1, x_2, \cdots, x_n)$, 将其用矢量 $\boldsymbol{x}$ 表示为列 (column):

$$\boldsymbol{x} = \begin{bmatrix} x_1 \\ \vdots \\ x_n \end{bmatrix} \tag{17-1}$$

则多元函数的标量函数 $f(\boldsymbol{x})$ 在 $\boldsymbol{a}$ 点的一阶泰勒展开为

$$f(\boldsymbol{x}) \approx f(\boldsymbol{a}) + (\boldsymbol{x} - \boldsymbol{a})^{\mathrm{T}} \cdot \boldsymbol{\nabla} f(\boldsymbol{a}) \tag{17-2}$$

式中，请注意右端第二项中的点积关系，而转置项为

$$(\boldsymbol{x} - \boldsymbol{a})^{\mathrm{T}} = \begin{bmatrix} x_1 - a_1, & \cdots, & x_n - a_n \end{bmatrix} \tag{17-3}$$

利用黑森矩阵 (Hessian matrix)，$f(\boldsymbol{x})$ 在 $\boldsymbol{a}$ 点的二阶泰勒展开为

$$f(\boldsymbol{x}) \approx f(\boldsymbol{a}) + (\boldsymbol{x} - \boldsymbol{a})^{\mathrm{T}} \cdot \boldsymbol{\nabla} f(\boldsymbol{a}) + \frac{1}{2}(\boldsymbol{x} - \boldsymbol{a})^{\mathrm{T}} \cdot \boldsymbol{H}(\boldsymbol{a}) \cdot (\boldsymbol{x} - \boldsymbol{a}) \tag{17-4}$$

式中，$\boldsymbol{H}(\boldsymbol{a})$ 为一个二阶张量，可用矩阵表示为

$$[\boldsymbol{H}(\boldsymbol{a})] = \begin{bmatrix} \dfrac{\partial^2 f}{\partial x_1^2} & \dfrac{\partial^2 f}{\partial x_1 \partial x_2} & \cdots & \dfrac{\partial^2 f}{\partial x_1 \partial x_n} \\ \dfrac{\partial^2 f}{\partial x_2 \partial x_1} & \dfrac{\partial^2 f}{\partial x_2^2} & \cdots & \dfrac{\partial^2 f}{\partial x_2 \partial x_n} \\ & & \vdots & \\ \dfrac{\partial^2 f}{\partial x_n \partial x_1} & \dfrac{\partial^2 f}{\partial x_n \partial x_2} & \cdots & \dfrac{\partial^2 f}{\partial x_n^2} \end{bmatrix} \tag{17-5}$$

由于二阶偏导数满足 $\dfrac{\partial^2 f}{\partial x_k \partial x_j} = \dfrac{\partial^2 f}{\partial x_j \partial x_k}$，所以，黑森矩阵满足对称性。

17.2 张量的标量函数的一阶泰勒展开

设 $\boldsymbol{A}$ 为一任意二阶张量，其标量函数为 $\Psi(\boldsymbol{A})$，则其一阶泰勒展开为

$$\begin{aligned} \Psi(\boldsymbol{A} + \mathrm{d}\boldsymbol{A}) &= \Psi(\boldsymbol{A}) + \frac{\partial \Psi(\boldsymbol{A})}{\partial \boldsymbol{A}} : \mathrm{d}\boldsymbol{A} + o(\mathrm{d}\boldsymbol{A}) \\ &= \Psi(\boldsymbol{A}) + \mathrm{tr}\left[\left(\frac{\partial \Psi(\boldsymbol{A})}{\partial \boldsymbol{A}}\right)^{\mathrm{T}} \mathrm{d}\boldsymbol{A}\right] + o(\mathrm{d}\boldsymbol{A}) \end{aligned} \tag{17-6}$$

式中，$o(\mathrm{d}\boldsymbol{A})$ 被称为小 O 符号，是由德国数学家爱德蒙·兰道 (Edmund Landau, 1877—1938, 如图 17.1 所示) 引入的，也被称为 "兰道阶符号 (Landau order symbol)"，它是指比 $\mathrm{d}\boldsymbol{A}$ 更高一阶的小量，因此，有

$$\lim_{\mathrm{d}\boldsymbol{A} \to 0} \frac{o(\mathrm{d}\boldsymbol{A})}{|\mathrm{d}\boldsymbol{A}|} = \boldsymbol{0} \tag{17-7}$$

图 17.1 哥廷根大学数学家的主要代表性人物之一——爱德蒙·兰道

式中，$\boldsymbol{0}$ 为二阶零张量.

例 17.1 将 $\det(\boldsymbol{A} + \mathrm{d}\boldsymbol{A})$ 进行一阶泰勒展开.

解：按照 (17-6) 式的定义，有

$$\det(\boldsymbol{A} + \mathrm{d}\boldsymbol{A}) = \det \boldsymbol{A} + \mathrm{tr}\left[\left(\frac{\partial \det \boldsymbol{A}}{\partial \boldsymbol{A}}\right)^{\mathrm{T}} \mathrm{d}\boldsymbol{A}\right] + o(\mathrm{d}\boldsymbol{A}) \tag{17-8}$$

由 (18-15) 式，$\dfrac{\partial \det \boldsymbol{A}}{\partial \boldsymbol{A}} = (\det \boldsymbol{A})\boldsymbol{A}^{-\mathrm{T}}$，将其代入上式，得到

$$\begin{aligned}\det(\boldsymbol{A} + \mathrm{d}\boldsymbol{A}) &= \det \boldsymbol{A} + \mathrm{tr}\left[(\det \boldsymbol{A})\boldsymbol{A}^{-1}\mathrm{d}\boldsymbol{A}\right] + o(\mathrm{d}\boldsymbol{A}) \\ &= \det \boldsymbol{A} + (\det \boldsymbol{A})\boldsymbol{A}^{-\mathrm{T}} : \mathrm{d}\boldsymbol{A} + o(\mathrm{d}\boldsymbol{A})\end{aligned} \tag{17-9}$$

例 17.2 在 (17-6) 式的运算中，$\Psi(\boldsymbol{A})$ 仅仅是 $\boldsymbol{A}$ 的函数，$\Psi(\boldsymbol{A})$ 对 $\boldsymbol{A}$ 的导数为何写成偏导形式 $\dfrac{\partial \Psi(\boldsymbol{A})}{\partial \boldsymbol{A}}$ 而非 $\dfrac{\mathrm{d}\Psi(\boldsymbol{A})}{\mathrm{d}\boldsymbol{A}}$？

答：因为 $\boldsymbol{A}$ 是二阶张量，一般情况下其有九个分量，$\Psi(\boldsymbol{A})$ 对 $\boldsymbol{A}$ 的导数隐含着对其九个分量求导数

$$\left[\frac{\partial \Psi(\boldsymbol{A})}{\partial \boldsymbol{A}}\right] = \begin{bmatrix} \dfrac{\partial \Psi(\boldsymbol{A})}{\partial A_{11}} & \dfrac{\partial \Psi(\boldsymbol{A})}{\partial A_{12}} & \dfrac{\partial \Psi(\boldsymbol{A})}{\partial A_{13}} \\ \dfrac{\partial \Psi(\boldsymbol{A})}{\partial A_{21}} & \dfrac{\partial \Psi(\boldsymbol{A})}{\partial A_{22}} & \dfrac{\partial \Psi(\boldsymbol{A})}{\partial A_{23}} \\ \dfrac{\partial \Psi(\boldsymbol{A})}{\partial A_{31}} & \dfrac{\partial \Psi(\boldsymbol{A})}{\partial A_{32}} & \dfrac{\partial \Psi(\boldsymbol{A})}{\partial A_{33}} \end{bmatrix} \tag{17-10}$$

所以 $\Psi(\boldsymbol{A})$ 对 $\boldsymbol{A}$ 的导数写成偏导数形式更为合理. 事实上，对于矢量求导情况类似.

本书以后章节中会经常出现 $\dfrac{\partial \mathrm{tr}\boldsymbol{A}}{\partial \boldsymbol{A}}$，$\dfrac{\partial \det \boldsymbol{F}}{\partial \boldsymbol{F}}$，$\dfrac{\partial w(\boldsymbol{B})}{\partial \boldsymbol{B}}$，$\dfrac{\partial w(\boldsymbol{C})}{\partial \boldsymbol{C}}$ 等类似情况，不再一一进行解释.

§18. 张量的标量函数的微分

张量微积分是张量分析乃至整个连续介质力学中较难的问题，为了本节不占用太多的篇幅，在附录 D 中给出了弗雷歇导数和加托导数的详细定义和具体例子，先仔细阅读附录 D 对于本节内容的理解十分关键.

18.1 直角坐标系下的时间导数

为了直观明了，我们首先讨论单位正交基矢量为常量的直角坐标系，此时有 $\dfrac{\mathrm{d}}{\mathrm{d}t}\boldsymbol{e}_i = 0$.

例 18.1 对于二阶任意张量 $\boldsymbol{A}$，证明恒等式：$\overline{\boldsymbol{A}^{-1}} = -\boldsymbol{A}^{-1}\dot{\boldsymbol{A}}\boldsymbol{A}^{-1}$.

证明：首先说明，$\overline{\boldsymbol{A}^{-1}} = \dfrac{\mathrm{d}}{\mathrm{d}t}\left(\boldsymbol{A}^{-1}\right)$. 由于 $\boldsymbol{A}^{-1}\boldsymbol{A} = \boldsymbol{I}$，所以两端对时间求导，有 $\overline{\boldsymbol{A}^{-1}}\boldsymbol{A} + \boldsymbol{A}^{-1}\dot{\boldsymbol{A}} = \boldsymbol{0}$，该式中的 $\boldsymbol{0}$ 为二阶零张量，亦即 $\overline{\boldsymbol{A}^{-1}}\boldsymbol{A} = -\boldsymbol{A}^{-1}\dot{\boldsymbol{A}}$，对该式两端右侧点积 $\boldsymbol{A}^{-1}$，得证.

18.2 张量的标量函数的微分

18.2.1 张量迹的导数

为了直观起见，让我们先看一个例子，那就是迹 $\operatorname{tr}\boldsymbol{A} = A_{ii} = A_{11} + A_{22} + A_{33}$ 显然是一个任意二阶张量 $\boldsymbol{A}$ 的标量函数. 我们先给出一个任意二阶张量 $\boldsymbol{A}$ 的标量函数导数的定义. 设 $f(\boldsymbol{A})$ 是二阶张量 $\boldsymbol{A}$ 的标量函数，则 $f(\boldsymbol{A})$ 关于 $\boldsymbol{A}$ 的在方向 $\boldsymbol{B}$ 的导数为

$$\frac{\partial f(\boldsymbol{A})}{\partial \boldsymbol{A}} : \boldsymbol{B} = \lim_{\alpha \to 0} \frac{f(\boldsymbol{A}+\alpha\boldsymbol{B}) - f(\boldsymbol{A})}{\alpha} = \left.\frac{\mathrm{d}f(\boldsymbol{A}+\alpha\boldsymbol{B})}{\mathrm{d}\alpha}\right|_{\alpha=0} \tag{18-1}$$

可见一个标量函数对一个二阶张量的导数是一个二阶张量.

例 18.2 求：$\dfrac{\partial \operatorname{tr}\boldsymbol{A}}{\partial \boldsymbol{A}}$

解法一、直观的分量法：

$$\begin{aligned}\frac{\partial \operatorname{tr}\boldsymbol{A}}{\partial \boldsymbol{A}} &= \frac{\partial A_{11}}{\partial A_{ij}}\boldsymbol{e}_i \otimes \boldsymbol{e}_j + \frac{\partial A_{22}}{\partial A_{ij}}\boldsymbol{e}_i \otimes \boldsymbol{e}_j + \frac{\partial A_{33}}{\partial A_{ij}}\boldsymbol{e}_i \otimes \boldsymbol{e}_j \\ &= \boldsymbol{e}_1 \otimes \boldsymbol{e}_1 + \boldsymbol{e}_2 \otimes \boldsymbol{e}_2 + \boldsymbol{e}_3 \otimes \boldsymbol{e}_3 = \boldsymbol{I}\end{aligned} \tag{18-2}$$

解法二、按照导数的定义：

首先按照迹的定义，有

$$\operatorname{tr}\boldsymbol{A} = A_{ii} = \boldsymbol{I} : \boldsymbol{A} \tag{18-3}$$

按照标量函数导数的定义 (18-1) 式，有

$$\frac{\partial \operatorname{tr}\boldsymbol{A}}{\partial \boldsymbol{A}} : \boldsymbol{B} = \lim_{\alpha \to 0} \frac{\operatorname{tr}(\boldsymbol{A}+\alpha\boldsymbol{B}) - \operatorname{tr}\boldsymbol{A}}{\alpha} = \lim_{\alpha \to 0} \frac{\boldsymbol{I}:(\boldsymbol{A}+\alpha\boldsymbol{B}) - \boldsymbol{I}:\boldsymbol{A}}{\alpha} = \boldsymbol{I}:\boldsymbol{B} \tag{18-4}$$

由上式便可立即得到

$$\frac{\partial \operatorname{tr}\boldsymbol{A}}{\partial \boldsymbol{A}} = \mathbf{I} \tag{18-5}$$

由上式还可得到如下关系式：

$$\frac{\partial (\operatorname{tr}\boldsymbol{A})^2}{\partial \boldsymbol{A}} = 2(\operatorname{tr}\boldsymbol{A})\mathbf{I} \tag{18-6}$$

$$\frac{\partial (\operatorname{tr}\boldsymbol{A})^n}{\partial \boldsymbol{A}} = n(\operatorname{tr}\boldsymbol{A})^{n-1}\mathbf{I} \tag{18-7}$$

例 18.3 求：$\dfrac{\partial \operatorname{tr}\boldsymbol{A}^2}{\partial \boldsymbol{A}}$

解法一：利用迹的定义 (18-3) 式，有

$$\operatorname{tr}\boldsymbol{A}^2 = \mathbf{I} : \boldsymbol{A}^2 \tag{18-8}$$

由二阶张量双点积的恒等式：

$$\mathbf{I} : \boldsymbol{A}^2 = \mathbf{I} : (\boldsymbol{A}\boldsymbol{A}) = \left(\mathbf{I}\boldsymbol{A}^{\mathrm{T}}\right) : \boldsymbol{A} = \boldsymbol{A}^{\mathrm{T}} : \boldsymbol{A} \tag{18-9}$$

故，由导数的定义 (18-1) 式，有

$$\begin{aligned}
\frac{\partial \operatorname{tr}\boldsymbol{A}^2}{\partial \boldsymbol{A}} : \boldsymbol{B} &= \lim_{\alpha \to 0} \frac{(\boldsymbol{A}+\alpha\boldsymbol{B})^{\mathrm{T}} : (\boldsymbol{A}+\alpha\boldsymbol{B}) - \boldsymbol{A}^{\mathrm{T}} : \boldsymbol{A}}{\alpha} \\
&= \lim_{\alpha \to 0} \frac{\left(\boldsymbol{A}^{\mathrm{T}}+\alpha\boldsymbol{B}^{\mathrm{T}}\right) : (\boldsymbol{A}+\alpha\boldsymbol{B}) - \boldsymbol{A}^{\mathrm{T}} : \boldsymbol{A}}{\alpha} \\
&= \lim_{\alpha \to 0} \frac{\alpha\left(\boldsymbol{A}^{\mathrm{T}} : \boldsymbol{B} + \boldsymbol{B}^{\mathrm{T}} : \boldsymbol{A}\right) + \alpha^2 \left(\boldsymbol{B}^{\mathrm{T}} : \boldsymbol{B}\right)}{\alpha} \\
&= \boldsymbol{A}^{\mathrm{T}} : \boldsymbol{B} + \boldsymbol{B}^{\mathrm{T}} : \boldsymbol{A} = 2\boldsymbol{A}^{\mathrm{T}} : \boldsymbol{B}
\end{aligned} \tag{18-10}$$

因此，有

$$\frac{\partial \operatorname{tr}\boldsymbol{A}^2}{\partial \boldsymbol{A}} = 2\boldsymbol{A}^{\mathrm{T}} \tag{18-11}$$

同理，有

$$\frac{\partial \operatorname{tr}\boldsymbol{A}^n}{\partial \boldsymbol{A}} = n\left(\boldsymbol{A}^{n-1}\right)^{\mathrm{T}} \tag{18-12}$$

解法二：利用迹的定义和 (11-12) 式，有 $\mathbf{I} : (\boldsymbol{AB}+\boldsymbol{BA}) = \mathbf{I} : (2\boldsymbol{AB}) = 2\boldsymbol{A}^{\mathrm{T}} : \boldsymbol{B}$，故，(18-11) 式得证. 有关解法二中的具体步骤的细节，请见思考题 3.6.

18.2.2 张量不变量的导数

由《近代连续介质力学》[1.6] 中的 (4-5) 式，二阶张量 $\boldsymbol{A}$ 的三个不变量为

$$\begin{cases} \mathrm{I}_1(\boldsymbol{A}) = \mathrm{tr}\boldsymbol{A} = \boldsymbol{A} : \mathbf{I} = A_{ii} \\ \mathrm{I}_2(\boldsymbol{A}) = \dfrac{1}{2}\left[(\mathrm{tr}\boldsymbol{A})^2 - \mathrm{tr}\boldsymbol{A}^2\right] = \mathrm{tr}(\mathbf{Cof}\boldsymbol{A}) = \det\boldsymbol{A}\,\mathrm{tr}(\boldsymbol{A}^{-1}) \\ \mathrm{I}_3(\boldsymbol{A}) = \det\boldsymbol{A} = \dfrac{1}{6}\left[(\mathrm{tr}\boldsymbol{A})^3 - 3(\mathrm{tr}\boldsymbol{A})(\mathrm{tr}\boldsymbol{A}^2) + 2\mathrm{tr}\boldsymbol{A}^3\right] \end{cases} \quad (18\text{-}13)$$

由 (18-5)—(18-8)、(18-11) 和 (18-12) 诸式，得到上述三个不变量的导数为

$$\begin{cases} \dfrac{\partial \mathrm{I}_1(\boldsymbol{A})}{\partial \boldsymbol{A}} = \dfrac{\partial \mathrm{tr}\boldsymbol{A}}{\partial \boldsymbol{A}} = \mathbf{I} \\ \dfrac{\partial \mathrm{I}_2(\boldsymbol{A})}{\partial \boldsymbol{A}} = \dfrac{1}{2}\left[\dfrac{\partial (\mathrm{tr}\boldsymbol{A})^2}{\partial \boldsymbol{A}} - \dfrac{\partial \mathrm{tr}\boldsymbol{A}^2}{\partial \boldsymbol{A}}\right] = \mathrm{I}_1 \mathbf{I} - \boldsymbol{A}^{\mathrm{T}} \\ \dfrac{\partial \mathrm{I}_3(\boldsymbol{A})}{\partial \boldsymbol{A}} = \dfrac{1}{6}\left\{\dfrac{\partial (\mathrm{tr}\boldsymbol{A})^3}{\partial \boldsymbol{A}} - 3\dfrac{\partial}{\partial \boldsymbol{A}}\left[(\mathrm{tr}\boldsymbol{A})(\mathrm{tr}\boldsymbol{A}^2)\right] + 2\dfrac{\partial \mathrm{tr}\boldsymbol{A}^3}{\partial \boldsymbol{A}}\right\} \\ \qquad\quad = \left(\boldsymbol{A}^{\mathrm{T}}\right)^2 - \mathrm{I}_1 \boldsymbol{A}^{\mathrm{T}} + \mathrm{I}_2 \mathbf{I} \end{cases} \quad (18\text{-}14)$$

在 (18-13) 式中第三式中，考虑到 $\mathrm{I}_3(\boldsymbol{A}) = \det\boldsymbol{A}$，以及 $\left(\boldsymbol{A}^{\mathrm{T}}\right)^2 - \mathrm{I}_1 \boldsymbol{A}^{\mathrm{T}} + \mathrm{I}_2 \mathbf{I} = \mathrm{I}_3 \boldsymbol{A}^{-\mathrm{T}}$，亦即

$$\dfrac{\partial \det \boldsymbol{A}}{\partial \boldsymbol{A}} = (\det\boldsymbol{A})\boldsymbol{A}^{-\mathrm{T}} = \mathbf{Cof}\boldsymbol{A} \quad (18\text{-}15)$$

式中，$\mathbf{Cof}\boldsymbol{A}$ 为 $\boldsymbol{A}$ 的余子式矩阵 (cofactor matrix)。如果令 $\boldsymbol{A}$ 为变形梯度张量 $\boldsymbol{F}$，则有如下常用关系式：

$$\dfrac{\partial \det \boldsymbol{F}}{\partial \boldsymbol{F}} = \dfrac{\partial J}{\partial \boldsymbol{F}} = (\det\boldsymbol{F})\boldsymbol{F}^{-\mathrm{T}} = J\boldsymbol{F}^{-\mathrm{T}} = \mathbf{Cof}\boldsymbol{F} \quad (18\text{-}16)$$

例 18.4 下面用标量函数导数的定义 (18-1) 式，来严格证明 (18-16) 式.

证明：设 λ 为 $\boldsymbol{F}$ 的本征值，按照定义有

$$\det(\boldsymbol{F} + \lambda\mathbf{I}) = \begin{vmatrix} F_{11}+\lambda & F_{12} & F_{13} \\ F_{21} & F_{22}+\lambda & F_{23} \\ F_{31} & F_{32} & F_{33}+\lambda \end{vmatrix} = \lambda^3 + \mathrm{I}_1(\boldsymbol{F})\lambda^2 + \mathrm{I}_2(\boldsymbol{F})\lambda + \mathrm{I}_3(\boldsymbol{F}) \quad (18\text{-}17)$$

再者，如果行列式的每一个元素均乘以一个常数的话，有

$$\det(\lambda\boldsymbol{F}) = \lambda^3 \det(\boldsymbol{F}) = \lambda^3 J \quad (18\text{-}18)$$

根据张量微分的定义式，对任意一个二阶张量 $\boldsymbol{A}$ 有

第 3 章　张量代数和微积分

$$\frac{\partial J}{\partial \boldsymbol{F}} : \boldsymbol{A} = \frac{\partial \det \boldsymbol{F}}{\partial \boldsymbol{F}} : \boldsymbol{A} = \lim_{\alpha \to 0} \frac{\det (\boldsymbol{F} + \alpha \boldsymbol{A}) - \det \boldsymbol{F}}{\alpha}$$
$$= \lim_{\alpha \to 0} \frac{\det [\alpha \boldsymbol{F} (\alpha^{-1} \mathbf{I} + \boldsymbol{F}^{-1} \boldsymbol{A})] - \det \boldsymbol{F}}{\alpha} \tag{18-19}$$

在上式中，由于

$$\det [\alpha \boldsymbol{F} (\alpha^{-1} \mathbf{I} + \boldsymbol{F}^{-1} \boldsymbol{A})] = \det (\alpha \boldsymbol{F}) \cdot \det (\alpha^{-1} \mathbf{I} + \boldsymbol{F}^{-1} \boldsymbol{A}) \tag{18-20}$$

由 (18-17) 和 (18-18) 两式，有

$$\begin{cases} \det (\alpha \boldsymbol{F}) = \alpha^3 \det \boldsymbol{F} = \alpha^3 J \\ \det (\boldsymbol{F}^{-1} \boldsymbol{A} + \alpha^{-1} \mathbf{I}) = \alpha^{-3} + \mathrm{I}_1 (\boldsymbol{F}^{-1} \boldsymbol{A}) \alpha^{-2} + \mathrm{I}_2 (\boldsymbol{F}^{-1} \boldsymbol{A}) \alpha^{-1} + \mathrm{I}_3 (\boldsymbol{F}^{-1} \boldsymbol{A}) \end{cases} \tag{18-21}$$

将 (18-21) 式代入 (18-19) 式，得到

$$\begin{aligned} \frac{\partial J}{\partial \boldsymbol{F}} : \boldsymbol{A} &= \lim_{\alpha \to 0} \frac{\det [\alpha \boldsymbol{F} (\alpha^{-1} \mathbf{I} + \boldsymbol{F}^{-1} \boldsymbol{A})] - \det \boldsymbol{F}}{\alpha} \\ &= \lim_{\alpha \to 0} \frac{\alpha^3 J [\alpha^{-3} + \mathrm{I}_1 (\boldsymbol{F}^{-1} \boldsymbol{A}) \alpha^{-2} + \mathrm{I}_2 (\boldsymbol{F}^{-1} \boldsymbol{A}) \alpha^{-1} + \mathrm{I}_3 (\boldsymbol{F}^{-1} \boldsymbol{A})] - J}{\alpha} \\ &= \lim_{\alpha \to 0} \frac{\mathrm{I}_1 (\boldsymbol{F}^{-1} \boldsymbol{A}) \alpha + \mathrm{I}_2 (\boldsymbol{F}^{-1} \boldsymbol{A}) \alpha^2 + \mathrm{I}_3 (\boldsymbol{F}^{-1} \boldsymbol{A}) \alpha^3}{\alpha} J \\ &= \mathrm{I}_1 (\boldsymbol{F}^{-1} \boldsymbol{A}) J \end{aligned} \tag{18-22}$$

式中，$\mathrm{I}_1 (\boldsymbol{F}^{-1} \boldsymbol{A})$ 表示点积 $\boldsymbol{F}^{-1} \boldsymbol{A}$ 的第一不变量，也就是该点积的迹 (trace)，再由于 (11-9) 式：$\boldsymbol{S} : \boldsymbol{T} = \mathrm{tr} (\boldsymbol{S}^{\mathrm{T}} \boldsymbol{T})$，将其代入 (18-22) 式，有

$$\frac{\partial J}{\partial \boldsymbol{F}} : \boldsymbol{A} = \mathrm{I}_1 (\boldsymbol{F}^{-1} \boldsymbol{A}) J = \mathrm{tr} (\boldsymbol{F}^{-1} \boldsymbol{A}) J = J \boldsymbol{F}^{-\mathrm{T}} : \boldsymbol{A} \tag{18-23}$$

通过上式便证明了连续介质力学的常用关系式：$\dfrac{\partial J}{\partial \boldsymbol{F}} = \dfrac{\partial \det \boldsymbol{F}}{\partial \boldsymbol{F}} = J \boldsymbol{F}^{-\mathrm{T}} = \mathbf{Cof} \boldsymbol{F}$。

事实上，(18-23) 式还可由 (11-12) 式来证明：

$$\frac{\partial J}{\partial \boldsymbol{F}} : \boldsymbol{A} = \mathrm{I}_1 (\boldsymbol{F}^{-1} \boldsymbol{A}) J = J \mathbf{I} : (\boldsymbol{F}^{-1} \boldsymbol{A}) = J \boldsymbol{F}^{-\mathrm{T}} : \boldsymbol{A}$$

18.2.3　张量的标量函数的全微分

设 $f(\boldsymbol{A})$ 是二阶张量 $\boldsymbol{A}$ 的标量函数，则该标量函数的全微分为

$$\begin{aligned} \mathrm{d} f(\boldsymbol{A}) = \frac{\partial f}{\partial \boldsymbol{A}} : \mathrm{d} \boldsymbol{A} &= \frac{\partial f}{\partial A_{11}} \mathrm{d} A_{11} + \frac{\partial f}{\partial A_{12}} \mathrm{d} A_{12} + \cdots + \frac{\partial f}{\partial A_{21}} \mathrm{d} A_{21} + \cdots \\ &\quad + \frac{\partial f}{\partial A_{31}} \mathrm{d} A_{31} + \frac{\partial f}{\partial A_{32}} \mathrm{d} A_{32} + \frac{\partial f}{\partial A_{33}} \mathrm{d} A_{33} \end{aligned} \tag{18-24}$$

18.2.4 张量的标量函数对张量的偏微分

设 $f(\boldsymbol{A})$ 是二阶张量 $\boldsymbol{A}$ 的标量函数, 则该标量函数对 $\boldsymbol{A}$ 的一阶导数为

$$\frac{\partial f(\boldsymbol{A})}{\partial \boldsymbol{A}} = \frac{\partial f}{\partial A_{ij}} \boldsymbol{e}_i \otimes \boldsymbol{e}_j \tag{18-25}$$

事实上, 在 (18-2) 式的运算中, 已经应用了上式. (18-25) 式说明, 一个标量函数对一个二阶张量的偏微分将获得一个二阶张量, 从这里可以得出在求导过程中的升阶规律.

该标量函数对 $\boldsymbol{A}$ 的二阶导数可表示为

$$\frac{\partial^2 f(\boldsymbol{A})}{\partial \boldsymbol{A} \partial \boldsymbol{A}} = \frac{\partial f}{\partial A_{ij} \partial A_{kl}} \boldsymbol{e}_i \otimes \boldsymbol{e}_j \otimes \boldsymbol{e}_k \otimes \boldsymbol{e}_l \tag{18-26}$$

上式表明, 一个标量函数对一个二阶张量的二阶导数将获得一个四阶张量.

例 18.5 设 $\boldsymbol{A}(t)$ 为一和时间有关的二阶张量, 求: $\dfrac{\mathrm{d}}{\mathrm{d}t} \det \boldsymbol{A}(t) = \overline{\det \boldsymbol{A}(t)}^{\boldsymbol{\cdot}}$.

解: 由链式法则, 有

$$\overline{\det \boldsymbol{A}(t)}^{\boldsymbol{\cdot}} = \frac{\partial \det \boldsymbol{A}}{\partial \boldsymbol{A}} : \dot{\boldsymbol{A}}(t) \tag{18-27}$$

再由 (18-15) 式: $\dfrac{\partial \det \boldsymbol{A}}{\partial \boldsymbol{A}} = (\det \boldsymbol{A}) \boldsymbol{A}^{-\mathrm{T}}$, (18-27) 式变为

$$\overline{\det \boldsymbol{A}(t)}^{\boldsymbol{\cdot}} = \frac{\partial \det \boldsymbol{A}}{\partial \boldsymbol{A}} : \dot{\boldsymbol{A}}(t) = (\det \boldsymbol{A}(t)) \boldsymbol{A}^{-\mathrm{T}}(t) : \dot{\boldsymbol{A}}(t) \tag{18-28}$$

再由 (11-12) 式, $\boldsymbol{I} \boldsymbol{A}^{-\mathrm{T}}(t) : \dot{\boldsymbol{A}}(t) = \boldsymbol{I} : \left(\dot{\boldsymbol{A}}(t) \boldsymbol{A}^{-1}(t) \right)$, 由迹的基本关系 (11-11) 式, 得到

$$\begin{aligned}
\overline{\det \boldsymbol{A}(t)}^{\boldsymbol{\cdot}} &= \frac{\partial \det \boldsymbol{A}}{\partial \boldsymbol{A}} : \dot{\boldsymbol{A}}(t) = (\det \boldsymbol{A}(t)) \boldsymbol{A}^{-\mathrm{T}}(t) : \dot{\boldsymbol{A}}(t) \\
&= (\det \boldsymbol{A}(t)) \operatorname{tr} \left[\boldsymbol{A}^{-1}(t) \dot{\boldsymbol{A}}(t) \right]
\end{aligned} \tag{18-29}$$

例 18.6 验证关系式: $\dfrac{\partial \operatorname{tr}(\boldsymbol{A}\boldsymbol{X})}{\partial \boldsymbol{X}} = \boldsymbol{A}^{\mathrm{T}}$, 这里, $\boldsymbol{A}$ 和 $\boldsymbol{X}$ 均为二阶张量.

证明: 按照张量迹运算的定义, 由 (11-11) 和 (11-12) 两式, 有

$$\operatorname{tr}(\boldsymbol{A}\boldsymbol{X}) = \boldsymbol{I} : (\boldsymbol{A}\boldsymbol{X}) = \left(\boldsymbol{A}^{\mathrm{T}} \boldsymbol{I} \right) : \boldsymbol{X} = \boldsymbol{A}^{\mathrm{T}} : \boldsymbol{X} \tag{18-30}$$

按照张量的标量函数微分的定义, 对于任一个二阶张量 $\boldsymbol{S}$ 有

$$\frac{\partial \operatorname{tr}(\boldsymbol{A}\boldsymbol{X})}{\partial \boldsymbol{X}} : \boldsymbol{S} = \lim_{\alpha \to 0} \frac{\boldsymbol{A}^{\mathrm{T}} : (\boldsymbol{X} + \alpha \boldsymbol{S}) - \boldsymbol{A}^{\mathrm{T}} : \boldsymbol{X}}{\alpha} = \boldsymbol{A}^{\mathrm{T}} : \boldsymbol{S} \tag{18-31}$$

故
$$\frac{\partial \text{tr}(\boldsymbol{AX})}{\partial \boldsymbol{X}} = \boldsymbol{A}^{\mathrm{T}} \tag{18-32}$$

证毕.

(18-32) 式再一次验证了张量微分运算的升阶规律, 一个标量函数 $\text{tr}(\boldsymbol{AX})$ 对一个二阶张量 $\boldsymbol{X}$ 的微分将升阶为二阶张量.

18.3 物质时间导数和空间时间导数之间的关系

对于一个用拉格朗日描述的变量的物质场, 如位移场 $\boldsymbol{U}(\boldsymbol{X},t)$, 物质点 X 固定 (holding the material point $\boldsymbol{X}$ fixed) 时, 对时间 t 的导数称为物质时间导数 (material time derivative):

$$\dot{\boldsymbol{U}} = \left(\frac{\partial \boldsymbol{U}(\boldsymbol{X},t)}{\partial t}\right)_{\boldsymbol{X}} = \boldsymbol{V}(\boldsymbol{X},t) \tag{18-33}$$

同理, 一个用欧拉描述的变量空间场, 如位移场 $\boldsymbol{u}(\boldsymbol{x},t)$, 物质点 $\boldsymbol{x}$ 固定时, 对时间 t 的导数称为空间时间导数:

$$\boldsymbol{u}' = \left(\frac{\partial \boldsymbol{u}(\boldsymbol{x},t)}{\partial t}\right)_{\boldsymbol{x}} = \boldsymbol{v}(\boldsymbol{x},t) \tag{18-34}$$

且 (18-33) 式和式 (18-34) 式可通过变换得到

$$\boldsymbol{V}(\boldsymbol{X},t) = \boldsymbol{V}\left[\boldsymbol{\chi}^{-1}(\boldsymbol{x},t), t\right] = \boldsymbol{v}(\boldsymbol{x},t) \tag{18-35}$$

若对于一个光滑的空间场 $f(\boldsymbol{x},t)$, 我们也可以求其物质时间导数 $\dot{f}$. 首先我们将其用拉格朗日描述, 然后求其物质时间导数, 并将结果变换为空间描述:

$$\dot{f}(\boldsymbol{x},t) = \left(\frac{\partial f[\boldsymbol{\chi}(\boldsymbol{X},t),t]}{\partial t}\right)_{\boldsymbol{X}=\boldsymbol{\chi}^{-1}(\boldsymbol{x},t)} \tag{18-36}$$

设 φ 和 $\boldsymbol{u}$ 分别是光滑空间标量场和向量场, 则有物质时间导数和空间时间导数的重要关系式:

$$\begin{cases} \dot{\varphi} = \varphi' + \boldsymbol{v} \cdot \mathbf{grad}\,\varphi \\ \dot{\boldsymbol{u}} = \boldsymbol{u}' + (\mathbf{grad}\,\boldsymbol{u})\,\boldsymbol{v} \end{cases} \tag{18-37}$$

证明: 由莱布尼兹链式法则:

$$\begin{aligned} \dot{\varphi}(\boldsymbol{x},t) &= \left(\frac{\partial \varphi(\boldsymbol{x},t)}{\partial t}\right)_{\boldsymbol{x}} + \left(\frac{\partial \varphi(\boldsymbol{x},t)}{\partial \boldsymbol{x}}\right)_{t} \cdot \left(\frac{\partial \boldsymbol{\chi}(\boldsymbol{X},t)}{\partial t}\right)_{\boldsymbol{X}=\boldsymbol{\chi}^{-1}(\boldsymbol{x},t)} \\ &= \varphi' + (\mathbf{grad}\,\varphi) \cdot \boldsymbol{v} \end{aligned} \tag{18-38}$$

同理

$$\dot{\boldsymbol{u}}(\boldsymbol{x},t) = \left(\frac{\partial \boldsymbol{u}(\boldsymbol{x},t)}{\partial t}\right)_{\boldsymbol{x}} + \left(\frac{\partial \boldsymbol{u}(\boldsymbol{x},t)}{\partial \boldsymbol{x}}\right)_t \cdot \left(\frac{\partial \boldsymbol{\chi}(\boldsymbol{X},t)}{\partial t}\right)_{\boldsymbol{X}=\boldsymbol{\chi}^{-1}(\boldsymbol{x},t)}$$
$$= \boldsymbol{u}' + (\operatorname{grad}\boldsymbol{u})\,\boldsymbol{v} \tag{18-39}$$

§19. 四阶单位张量、对称的四阶单位张量、四阶投影张量

19.1 四阶单位张量、对称的四阶单位张量

用空芯黑色正体符号 $\mathbb{I}$ 表示的四阶单位张量 (fourth-order identity tensor) 定义为

$$\mathbb{I} = \delta_{ik}\delta_{jl}\boldsymbol{e}_i \otimes \boldsymbol{e}_j \otimes \boldsymbol{e}_k \otimes \boldsymbol{e}_l \tag{19-1}$$

而两个传统的二阶单位张量 $\mathbf{I} = \delta_{ij}\boldsymbol{e}_i \otimes \boldsymbol{e}_j$ 的并矢为

$$\mathbf{I} \otimes \mathbf{I} = \delta_{ij}\delta_{kl}\boldsymbol{e}_i \otimes \boldsymbol{e}_j \otimes \boldsymbol{e}_k \otimes \boldsymbol{e}_l \tag{19-2}$$

所以, 通过比较 (19-1) 和 (19-2) 两式, 必须区分:

$$\mathbb{I} \neq \mathbf{I} \otimes \mathbf{I} \tag{19-3}$$

四阶单位张量的定义式 (19-1) 一方面是要满足:

$$\mathbb{I} : \mathbf{I} = \mathbf{I} \tag{19-4}$$

证明上式:

$$\begin{aligned}
\mathbb{I} : \mathbf{I} &= (\delta_{ik}\delta_{jl}\boldsymbol{e}_i \otimes \boldsymbol{e}_j \otimes \boldsymbol{e}_k \otimes \boldsymbol{e}_l) : (\delta_{op}\boldsymbol{e}_o \otimes \boldsymbol{e}_p) \\
&= \delta_{ik}\delta_{jl}\delta_{op}\delta_{ko}\delta_{lp}\boldsymbol{e}_i \otimes \boldsymbol{e}_j = \underbrace{\delta_{ik}\delta_{ko}}_{\delta_{io}}\underbrace{\delta_{jl}\delta_{lp}}_{\delta_{jp}}\delta_{op}\boldsymbol{e}_i \otimes \boldsymbol{e}_j \\
&= \delta_{io}\underbrace{\delta_{jp}\delta_{op}}_{\delta_{jo}}\boldsymbol{e}_i \otimes \boldsymbol{e}_j = \delta_{io}\delta_{jo}\boldsymbol{e}_i \otimes \boldsymbol{e}_j \\
&= \delta_{ij}\boldsymbol{e}_i \otimes \boldsymbol{e}_j = \mathbf{I}
\end{aligned} \tag{19-5}$$

对 (19-4) 式推而广之, 对一个二阶张量 $\boldsymbol{A} = A_{mn}\boldsymbol{e}_m \otimes \boldsymbol{e}_n$ 而言需要满足:

第 3 章 张量代数和微积分

$$\mathbb{I} : \boldsymbol{A} = \boldsymbol{A} \tag{19-6}$$

亦即，另一方面四阶单位张量 $\mathbb{I}$ 需要满足如下关系式：

$$\frac{\partial \boldsymbol{A}}{\partial \boldsymbol{A}} = \mathbb{I} \tag{19-7}$$

上式说明一个二阶张量对其本身的偏导数为四阶单位张量，满足张量微分的升阶规律.

例 19.1 证明 (19-7) 式.

证明：二阶张量 $\boldsymbol{A}$ 为自变量的二阶张量值函数 $\boldsymbol{F}(\boldsymbol{A})$ 的微分形式定义为

$$\lim_{h \to 0} \frac{\boldsymbol{F}(\boldsymbol{A} + h\boldsymbol{D}) - \boldsymbol{F}(\boldsymbol{A})}{h} = \frac{\partial \boldsymbol{F}}{\partial \boldsymbol{A}} : \boldsymbol{D} \tag{19-8}$$

其中 $\boldsymbol{D}$ 为任意二阶张量，h 为一实数. 则对于函数 $\boldsymbol{F}(\boldsymbol{A}) = \boldsymbol{A}$ 有

$$\lim_{h \to 0} \frac{(\boldsymbol{A} + h\boldsymbol{D}) - \boldsymbol{A}}{h} = \frac{\partial \boldsymbol{A}}{\partial \boldsymbol{A}} : \boldsymbol{D} \tag{19-9}$$

即

$$\boldsymbol{D} = \frac{\partial \boldsymbol{A}}{\partial \boldsymbol{A}} : \boldsymbol{D} \tag{19-10}$$

因而，(19-7) 式得证.

我们将 (19-1) 式中的四阶单位张量 $\mathbb{I} = \delta_{ik}\delta_{jl}\boldsymbol{e}_i \otimes \boldsymbol{e}_j \otimes \boldsymbol{e}_k \otimes \boldsymbol{e}_l$ 的指标简称为 1324，是因为它与任意二阶张量 $\boldsymbol{A} = A_{op}\boldsymbol{e}_o \otimes \boldsymbol{e}_p$ 双点积之后，满足 (19-6) 式，亦即

$$\mathbb{I} : \boldsymbol{A} = (\delta_{ik}\delta_{jl}\boldsymbol{e}_i \otimes \boldsymbol{e}_j \otimes \boldsymbol{e}_k \otimes \boldsymbol{e}_l) : (A_{op}\boldsymbol{e}_o \otimes \boldsymbol{e}_p) = \delta_{ik}\delta_{jl}\delta_{ko}\delta_{lp}A_{op}\boldsymbol{e}_i \otimes \boldsymbol{e}_j$$

$$= \underbrace{\delta_{ik}\delta_{ko}}_{\delta_{io}}\underbrace{\delta_{jl}\delta_{lp}}_{\delta_{jp}}A_{op}\boldsymbol{e}_i \otimes \boldsymbol{e}_j = \delta_{io}\delta_{jp}A_{op}\boldsymbol{e}_i \otimes \boldsymbol{e}_j = A_{op}\boldsymbol{e}_o \otimes \boldsymbol{e}_p = \boldsymbol{A} \tag{19-11}$$

换句话说，四阶单位张量的作用就是和任意二阶张量的双点积后还是该任意二阶张量，四阶单位张量还可写成：

$$\mathbb{I} = \boldsymbol{e}_i \otimes \boldsymbol{e}_j \otimes \boldsymbol{e}_i \otimes \boldsymbol{e}_j \tag{19-12}$$

将四阶单位张量的指标顺序由 1324 改为 2314，将得到一个新的四阶张量：$\delta_{jk}\delta_{il}\boldsymbol{e}_i \otimes \boldsymbol{e}_j \otimes \boldsymbol{e}_k \otimes \boldsymbol{e}_l$，我们来看看它和任意二阶张量 $\boldsymbol{A} = A_{op}\boldsymbol{e}_o \otimes \boldsymbol{e}_p$ 双点积之后的结果：

$$(\delta_{jk}\delta_{il}\boldsymbol{e}_i \otimes \boldsymbol{e}_j \otimes \boldsymbol{e}_k \otimes \boldsymbol{e}_l) : (A_{op}\boldsymbol{e}_o \otimes \boldsymbol{e}_p) = \delta_{jk}\delta_{il}\delta_{ko}\delta_{lp}A_{op}\boldsymbol{e}_i \otimes \boldsymbol{e}_j$$

$$= \underbrace{\delta_{jk}\delta_{ko}}_{\delta_{jo}}\underbrace{\delta_{il}\delta_{lp}}_{\delta_{ip}} A_{op} \boldsymbol{e}_i \otimes \boldsymbol{e}_j$$
$$= \delta_{jo}\delta_{ip} A_{op} \boldsymbol{e}_i \otimes \boldsymbol{e}_j = A_{op}\boldsymbol{e}_p \otimes \boldsymbol{e}_o = \boldsymbol{A}^{\mathrm{T}} \tag{19-13}$$

上面的运算表明，$\delta_{jk}\delta_{il}\boldsymbol{e}_i \otimes \boldsymbol{e}_j \otimes \boldsymbol{e}_k \otimes \boldsymbol{e}_l$ 不是四阶单位张量，此张量与任意一个二阶张量 $\boldsymbol{A}$ 双点乘作用后，将得到张量 $\boldsymbol{A}$ 的转置 $\boldsymbol{A}^{\mathrm{T}}$. 因为此张量的 3、4 基矢量与 $\boldsymbol{A}$ 作用后，剩余 21 基矢量，所以结果为张量 $\boldsymbol{A}^{\mathrm{T}}$.

定义：
$$\mathbb{I}^{\mathrm{T}} = \delta_{jk}\delta_{il} \boldsymbol{e}_i \otimes \boldsymbol{e}_j \otimes \boldsymbol{e}_k \otimes \boldsymbol{e}_l \quad \Leftrightarrow \quad \mathbb{I}^{\mathrm{T}} : \boldsymbol{A} = \boldsymbol{A}^{\mathrm{T}} \tag{19-14}$$

上式，也就是四阶单位张量的转置，亦可写为
$$\mathbb{I}^{\mathrm{T}} = \boldsymbol{e}_i \otimes \boldsymbol{e}_j \otimes \boldsymbol{e}_j \otimes \boldsymbol{e}_i \tag{19-15}$$

四阶单位张量和其转置可组成对称的四阶单位张量 (symmetrical fourth-order identity tensor)：
$$\mathbb{I}^s = \frac{\mathbb{I} + \mathbb{I}^{\mathrm{T}}}{2} = \frac{1}{2}\left(\delta_{ik}\delta_{jl} + \delta_{jk}\delta_{il}\right) \boldsymbol{e}_i \otimes \boldsymbol{e}_j \otimes \boldsymbol{e}_k \otimes \boldsymbol{e}_l \tag{19-16}$$

对称的四阶单位张量的作用是，对于任意一个对称的二阶张量 $\boldsymbol{A}$ 满足：
$$\frac{\partial \boldsymbol{A}}{\partial \boldsymbol{A}} = \mathbb{I}^s \tag{19-17}$$

对称的四阶单位张量 (19-16) 式的指标形式为
$$\mathbb{I}^s_{ijkl} = \frac{1}{2}\left(\delta_{ik}\delta_{jl} + \delta_{jk}\delta_{il}\right) \tag{19-18}$$

对称的四阶单位张量满足如下小的对称性条件 (minor symmetries)：
$$\mathbb{I}^s_{ijkl} = \mathbb{I}^s_{jikl}, \quad \mathbb{I}^s_{ijkl} = \mathbb{I}^s_{ijlk} \tag{19-19}$$

以及如下大的对称性条件 (major symmetries)：
$$\mathbb{I}^s_{ijkl} = \mathbb{I}^s_{klij} \tag{19-20}$$

19.2 四阶投影张量

一个任意的二阶张量 $\boldsymbol{A}$ 可分解为球形分量 (spherical part) 和偏量 (deviatoric part)：
$$\boldsymbol{A} = \underbrace{\alpha \mathbf{I}}_{\text{球量}} + \underbrace{\mathrm{dev}\boldsymbol{A}}_{\text{偏量}} \tag{19-21}$$

由于偏量部分是无迹的 (traceless)，对上式求迹，有

$$\alpha = \frac{1}{3}\mathrm{tr}\boldsymbol{A} = \frac{1}{3}(\boldsymbol{I}:\boldsymbol{A}) \tag{19-22}$$

则二阶张量 $\boldsymbol{A}$ 的偏量 $\mathrm{dev}\boldsymbol{A}$ 可通过四阶单位张量 $\mathbb{I}$ 表示为

$$\begin{aligned}\mathrm{dev}\boldsymbol{A} &= \boldsymbol{A} - \frac{1}{3}\boldsymbol{I}(\boldsymbol{I}:\boldsymbol{A}) = \boldsymbol{A} - \frac{1}{3}(\boldsymbol{I}\otimes\boldsymbol{I}):\boldsymbol{A} \\ &= \mathbb{P}:\boldsymbol{A} = \left[\mathbb{I} - \frac{1}{3}(\boldsymbol{I}\otimes\boldsymbol{I})\right]:\boldsymbol{A}\end{aligned} \tag{19-23}$$

式中，四阶投影张量为

$$\mathbb{P} = \mathbb{I} - \frac{1}{3}(\boldsymbol{I}\otimes\boldsymbol{I}) \tag{19-24}$$

思考题和补充材料

3.1 本书中所涉及的终身未婚的著名科学家一览表.

题表 3.1 本书中所涉及的终身未婚的著名科学家 (按照出生年月排序)

科学家姓名	主要贡献	轶事
达·芬奇 Leonardo da Vinci 1452—1519 终年 67 岁	最早明确地提出要以新的态度对待知识的人之一是达·芬奇. 他在体力上和精神上有不可思议的天资，使他成为杰出的语言学家、植物学家、动物学家、解剖学家、地质学家、音乐家、雕塑家、绘画家、建筑学家、发明家和工程师. 达·芬奇约于 1500 年最早使用了意大利文 "la turbolenza (湍流)" 一词. 英文 "turbulence (湍流)" 一词于 1887 年由开尔文勋爵 (Lord Kelvin, 1824—1907) 引入. 达·芬奇对湍流的论述: "doue la turbolenza (turbulence) dell'acqua si genera; doue la turbolenza dell'acqua si mantiene; doue la turbolenza dell'acqua si posa (产生两个水流湍流；维持两个水流湍流；形成两个水流湍流)."	达·芬奇说: "一个人如喜欢没有理论的实践，他就像水手上船而没有舵和罗盘，永远不知道驶向何方 (He who loves practice without theory is like the sailor who boards ship without a rudder and compass and never knows where he may be cast)." 另一方面: "理论好比统帅，实践则是战士 (Theory is the general; experiments are the soliders)." 自 1496 年起，达·芬奇跟着意大利数学家帕乔利 (Luca Pacioli, 1447—1517) 在米兰学了三年几何学. 在遗留的手稿中，达·芬奇多次提到如何将学来的透视法和比例学运用到绘画创作中. 达·芬奇指出: "透视法是数学研究中最微妙的发现，因为它通过线条使近处显得远，小处显得大 (Perspective is a most subtle discovery in mathematical studies, for by means of lines it causes to appear distant that which is near, and large that which is small)."

续表

科学家姓名	主要贡献	轶事
哥白尼 Nicolaus Copernicus 1473—1543 终年 70 岁	文艺复兴时期的波兰数学家、天文学家，他提倡日心说模型，提到太阳为宇宙的中心. 1543 年 3 月临终前发表了《天体运行论》一般认为是现代天文学的起步点. 该书开启了哥白尼革命，并对推动科学革命作出了重要贡献.	据说他闭目的时候，还用冰冷的双手抚摸着刚刚印好的《天体运动论》样书.
笛卡儿 René Descartes 1596—1650 终年 54 岁	被广泛认为是西方现代哲学的奠基人，他第一个创立了一套完整的哲学体系. 对数学最重要的贡献是创立了解析几何和坐标系.	1649 年受瑞典克里斯蒂娜女王之邀来到斯德哥尔摩担任女王的私人教师，但不幸在这片"熊、冰雪与岩石的土地"上得了肺炎，在 1650 年 2 月去世，终年 54 岁.
波义耳 Robert Boyle 1627—1691 终年 64 岁	"把化学确立为科学"的开创者创立气体的波义耳-马略特定律.	1680 年波义耳被选为英国皇家学会会长，但是由于誓言的问题，他拒绝了这一职务.
惠更斯 Christiaan Huygens 1629—1695 终年 66 岁	在数学、力学、物理学、天文学等多个领域都有所建树，许多重要著作是在他逝世后才发表的，《惠更斯全集》共有 22 卷.	相信外星生命的存在. 1665 年 2 月，在病床上发现了两只钟的同步现象.
胡克 Robert Hooke 1635—1703 终年 67 岁	1676 年发表了一条拉丁语字谜：ceiiinosssttuv，两年后公布了谜底 ut tensio sic vis，中文是"力如伸长"，亦即力与伸长量成正比，后称胡克定律.	1703 年 3 月 3 日在伦敦去世，死后在他房间中发现高达 1 万英镑的积蓄，相当于当时一个成功的银行家的财产.
牛顿 Isaac Newton 1643—1727 终年 84 岁	牛顿力学、微积分. 被朗道评价为零级科学家、爱因斯坦为 1/2 级.	1697 年，牛顿时年 54 岁，一天下午 4 点筋疲力尽地从他任职的造币厂回到家中，收到了约翰·伯努利的有关最速下降线的征文，他当晚深夜 4 点就解决了这个世界难题，并匿名寄回. 伯努利看后，敬畏地放下答案，说"从利爪中我认出了雄狮".
莱布尼兹 Gottfried Wilhelm Leibniz 1646—1716 终年 70 岁	和牛顿独立地发明了微积分，其符号应用的更广.	于 1684 年发表第一篇微分论文，定义了微分概念，采用了微分符号 dy 或 dx，1686 年他又发表了积分论文，讨论了微分与积分，使用 $\int$ 作为积分符号.

续表

科学家姓名	主要贡献	轶事
华伦海特 Daniel Gabriel Fahrenheit 1686—1736 终年 50 岁	参见题表 1.5.	1736 年,华伦海特死,瓦特生.
伏尔泰 Voltaire 1694—1778 终年 83 岁	被称为"法兰西思想之父". 与卢梭、孟德斯鸠合称"法兰西启蒙运动三剑侠".	他曾逃至女友夏特莱侯爵夫人在西雷村的庄园,隐居十五年. 他是牛顿主义者,首次告诉了世人有关牛顿与苹果的故事,为夏特莱侯爵夫人所翻译的法文版《自然哲学的数学原理》写了序言.
克莱罗 Alexis Claude Clairaut 1713—1765 终年 52 岁	克莱罗定理.	18 岁当选为法兰西科学院院士.
达朗贝尔 Jean Le Rond d'Alembert 1717—1783 终年 65 岁	达朗贝尔原理,达朗贝尔佯谬.	D'Alembert never married, although he lived for a number of years with Julie de Lespinasse.
卡文迪许 Henry Cavendish 1731—1810 终年 78 岁	被认为是牛顿之后英国最伟大的科学家之一.	Jean Baptiste Biot called him "the richest of all the savants and the most knowledgeable of the rich." At his death, he was the largest depositor in the Bank of England. 法国科学家毕奥称卡文迪许是"有学问的人中最富有的,是富有的人中最有学问的." 卡文迪许去世时,他是英格兰银行最大的储户 早年卡文迪许从叔伯那里承接了大宗遗赠,1783 年他父亲逝世,又给他留下大笔遗产. 这样他的资产超过了 130 万英镑.
热尔曼 Sophie Germain 1776—1831 终年 55 岁	弹性力学和数论.	她常用男性假名与拉格朗日和高斯在内的数学家通信联系.

续表

科学家姓名	主要贡献	轶事
魏尔斯特拉斯 Karl Weierstrass 1815—1897 终年 81 岁	被誉为"现代分析之父".	魏尔斯特拉斯的姐姐说,当她弟弟是一个年轻的中学教师时,要是在他的视线之内有一平方英尺干净的贴墙纸或者一个干净的袖头,就不能放心地把一支铅笔交给他. Weierstrass may have had an illegitimate child named Franz with the widow of his friend Carl Wilhelm Borchardt. At the age of fifty-five, Weierstrass met Sonya Kovalevsky whom he tutored privately after failing to secure her admission to the University. They had a fruitful intellectual, but troubled personal relationship that "far transcended the usual teacher-student relationship".
傅科 Léon Foucault 1819—1868 终年 48 岁	傅科摆.	1851 年傅科在巴黎先贤祠悬挂一枚很长的重摆,藉以显示地球的自转,此举吸引了大批游客,这是首次动力学地证明地球的自转.
兰金 William John Macquorn Rankine 1820—1872 终年 52 岁	与开尔文一起是热力学第一定律的提出者.	见思考题 11.7.
斯宾塞 Herbert Spencer 1820—1903 终年 83 岁	社会达尔文主义之父、进化论的先驱 1851 年出版《社会静力学》(见本书 §65.2.1 小节)	1862年,出版《第一原理》(*First Principles*),该著作是现实中所有领域的根本准则的演变理论的展示. 根据斯宾塞的定义,演变是个不断延续的过程,事物不断改进为复杂和连贯的形式. 这是斯宾塞哲学的主炮—是演变的一个已发展连贯架构的定义.
切比雪夫 Pafnuty Chebyshev 1821—1894 终年 73 岁	切比雪夫不等式、切比雪夫多项式、切比雪夫函数、切比雪夫连杆机构.	见思考题 1.13. 被认为是俄罗斯数学之父 (founding father of Russian mathematics),培养了李雅普诺夫、马尔科夫等杰出数学家.

第 3 章 张量代数和微积分

续表

科学家姓名	主要贡献	轶事
克里斯托费尔 Elwin Bruno Christoffel 1829—1900 终年 70 岁	爱因斯坦在其 1916 年广义相对论一文中对克里斯托费尔给予了很高评价,见本书第 128 页.	教学上的完美主义者,见本书第 129 页.
吉布斯 Josiah Willard Gibbs 1839—1903 终年 64 岁	被爱因斯坦称赞为"美国历史上最伟大的头脑 (the greatest mind in American history)". 诺贝尔物理学奖得主罗伯特·密立根曾这样评价吉布斯:"吉布斯对于统计力学和热力学来说,就如同拉普拉斯之于天体力学,麦克斯韦之于电动力学.他为自己所研究的领域构造了几近完整的理论体系.(Willard Gibbs did for statistical mechanics and for thermodynamics what Laplace did for celestial mechanics and Maxwell did for electrodynamics, namely, made his field a well-nigh finished theoretical structure)"	他有关热力学的开创性的著作多数发表在《康涅狄格学会学报》上.这部期刊由他当图书馆员的姐夫担任主编,在美国读者甚少,而欧洲的读者则更为寥寥.
亥维赛 Oliver Heaviside 1850—1925 终年 74 岁	在本书思考题 2.10—2.13 中介绍过亥维赛的杰出成就和"第一等古怪"的性格.	亥维赛终身未娶.隐居、越来越严重的贫困、耳聋和孤独是亥维赛晚年生活的写照.
特斯拉 Nikola Tesla 1856—1943 终年 86 岁	主要设计了现代交流电力系统而最为人知.他的多项相关的专利以及电磁学的理论研究工作是现代的无线通信和无线电的基石.	国际单位制中,用来衡量磁感应强度的单位,是以特斯拉的名字命名,符号为 T.
乐甫 Augustus Edward Hough Love 1863—1940 终年 77 岁	在《弹性的数学理论教程》(*A Treatise on Mathematical Theory of Elasticity*) 中总结了弹性力学的成就,发展了薄壳理论.该著作仍然是弹性力学的必备参考书.	1911 年在《地球动力学的若干问题》(*Some Problems of Geodynamics*) 中总结了从弹性力学来研究地球的一系列成果,成为地球物理研究的基础.提出固体界面间传播的乐甫波 (Love wave).

续表

科学家姓名	主要贡献	轶事
莱特兄弟 The Wright brothers Wilbur 1867—1912 终年 45 岁 Orville 1871—1948 终年 76 岁	1903 年 12 月 17 日，莱特兄弟驾驶自行研制的固定翼飞机飞行者一号飞行了 12 秒，飞行距离为 36.5 米，实现了人类史上首次重于空气的航空器持续而且受控的动力飞行，被广泛誉为现代飞机的发明者。	Little wonder then that neither felt the need to get married since anyway they occupied themselves for more than 20 years with designing and building the world's first heavier-than-air aircraft, an endeavor that culminated in the Wright flyer's first flight in 1906.
哈代 Godfrey Harold Hardy 1877—1947 终年 70 岁	哈代是数学界公认的将欧洲大陆（法国、瑞士、德国）的严谨数学风格引入英国的数学家. 与另一位英国数学家利特尔伍德进行了长达 35 年的合作，在数学分析与解析数论上有重要贡献.	见思考题 1.6. 哈代被问到什么是他自己对数学最大的贡献，他不加思索的回答是发现了拉马努扬.
冯·卡门 Theodore von Kármán 1881—1963 终年 81 岁	匈牙利裔美国力学家，主要从事航空航天力学方面的工作，是工程力学和航空技术的权威，对于 20 世纪流体力学、空气动力学理论与应用的发展，尤其是在超声速和高超声速气流表征方面，以及亚声速与超声速航空、航天器的设计，产生了重大影响. 　　鉴于冯·卡门的卓著贡献，他被美国授予 1962 年度第一枚"国家科学勋章(National Medal of Science)". 1963 年，2 月 18 日上午，白宫玫瑰园名人聚集宾客如云，授勋仪式即将举行. 当 81 岁的冯·卡门走下台阶时，他因患有严重的关节炎而步履不稳，险些摔倒. 年轻的约翰·肯尼迪总统赶紧走上前，一把将他扶住. 冯·卡门抬头报以感激的微笑，继而轻轻推开总统伸出的手，淡淡地说："总统先生，下坡而行者无须搀扶，惟独举足高攀者才求一臂之力 (Mr. President, one does not need help going down, only going up)."	赫尔曼·外尔 (Hermann Weyl, 1885—1955) 刚去哥廷根的时候，被拒之"圈"外. 所谓的"圈"，是指特普利兹 (Otto Toeplitz, 1881—1940)，施密特 (Erhard Schmidt, 1876—1959)，赫克 (Erich Hecke, 1887—1947) 和哈尔 (Alfréd Haar, 1885—1933) 等一群年轻人一起谈论数学物理，很有贵族的感觉. 一次，大家在等待希尔伯特来上课，特普利兹指着远处的外尔说："看那边的那个家伙，他就是外尔先生. 他也是那种考虑数学的人." 就这样子，外尔就不属于"圈"，这个集合了. 这个故事是柯朗 (Richard Courant, 1888—1972) 讲的，哈尔当时是希尔伯特的助手，哥廷根当时的人们无一不认为他将是那种不朽的数学家. 但是事实证明，外尔的伟大无人能比，尽管哈尔在测度论上贡献突出，但是柯朗还是说他和外尔"根本没法相比". 　　冯·卡门通过哈尔的介绍来到哥廷根，等到哈尔去了匈牙利后，他很快成为"圈"内的领袖. 圈外人外尔再次证明了其优秀，他和冯·卡门同时爱上了一个才貌双全的女孩，并且展开了一场竞争. 最终圈内人都感到特别的沮丧，因为那个女孩子选择了外尔，冯·卡门却终身未娶.

续表

科学家姓名	主要贡献	轶事
诺特 Emmy Noether 1882—1935 终年 53 岁	埃米·诺特是 20 世纪初一个才华洋溢的德国数学家,研究领域为抽象代数和理论物理学. 她善于藉透彻的洞察建立优雅的抽象概念,再将之漂亮地形式化.	被爱因斯坦、外尔等形容为数学史上最重要的女人. 她彻底改变了环,域和代数的理论. 在物理学方面,诺特定理解释了对称性和守恒定律之间的根本联系.
德布罗意 Louis de Broglie 1892—1987 终年 94 岁	于 1924 年完成了博士论文《量子理论研究》,创立了物质的波粒二象性:任何物质同时具备波动和粒子的性质. 参见本书第 19 页有关德布罗意关系和波矢空间的讨论. 于 1927 年提出导航波理论,见本书 §65.3.4 有关金融导航波理论.	爱因斯坦在 1924 年 12 月 16 日给朗之万的信中称赞德布罗意"他已经揭开了大面纱的一角 (He has lifted a corner of the great veil)". 德布罗意于 1929 年获得诺贝尔物理学奖.

3.2 针对题表 3.1 中 1963 年冯·卡门被美国总统肯尼迪 (John Fitzgerald Kennedy, 1917—1963, 终年 46 岁) 授予美国首枚国家科学奖章, 相关情景见题图 3.2.

题图 3.2 冯·卡门被美国肯尼迪总统授予美国首枚国家科学奖章

3.3 在《力学讲义》[1.11] 中, 曾讨论过不少物理学家, 像是爱因斯坦、狄拉克和杨振宁等, 都相信大自然具有数学式的简洁美. 詹姆斯·金斯 (James Jeans, 1877—1946) 曾说过, 上帝是个纯粹数学家! 而诺贝尔奖得主温伯格 (Steven Weinberg, 1933—2021, 1979 年诺贝尔物理学奖获得者) 一边写下一个带 13 个指标的张量一边表示反对, 说自然界是复杂而丑陋的 —— 事实上, 有一整个领域, 就是基于这个信念建立起来的.

3.4 有四位数学家和物理学家对张量分析的早期发展具有重要的推动作用, 如题图 3.4 所示.

3.5 验证 (9-15) 式: $(\boldsymbol{f} \cdot \boldsymbol{n}) \boldsymbol{n} = \boldsymbol{f} \cdot (\boldsymbol{n} \otimes \boldsymbol{n})$.

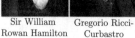

题图 3.4　左起：哈密顿、里奇、列维·奇维塔、爱因斯坦

证明： (9-15) 式左端为

$$(\boldsymbol{f}\cdot\boldsymbol{n})\,\boldsymbol{n} = (f_i\boldsymbol{e}_i\cdot n_j\boldsymbol{e}_j)\,n_k\boldsymbol{e}_k$$
$$= f_i n_j \delta_{ij} n_k \boldsymbol{e}_k = f_i n_i n_k \boldsymbol{e}_k \tag{s3-1}$$

(9-15) 式右端为

$$\boldsymbol{f}\cdot(\boldsymbol{n}\otimes\boldsymbol{n}) = f_i\boldsymbol{e}_i\cdot(n_j n_k \boldsymbol{e}_j\otimes\boldsymbol{e}_k)$$
$$= f_i n_j n_k \delta_{ij}\boldsymbol{e}_k = f_i n_i n_k \boldsymbol{e}_k \tag{s3-2}$$

对比 (s3-1) 和 (s3-2) 两式，则 (9-15) 式得证.

3.6 在例 18.3 的解法二中，有一推导的细节问题必须予以关注，那就是下一步的运算：

$$\mathbf{I}:(\boldsymbol{A}+\alpha\boldsymbol{B})^2 = \mathbf{I}:(\boldsymbol{A}+\alpha\boldsymbol{B})(\boldsymbol{A}+\alpha\boldsymbol{B})$$
$$= \mathbf{I}:\left[\boldsymbol{A}^2+\alpha(\boldsymbol{AB}+\boldsymbol{BA})+\alpha^2\boldsymbol{B}^2\right] \tag{s3-3}$$

而不是写成

$$\mathbf{I}:(\boldsymbol{A}+\alpha\boldsymbol{B})^2 = \mathbf{I}:\left(\boldsymbol{A}^2+2\alpha\boldsymbol{AB}+\boldsymbol{B}^2\right) \tag{s3-4}$$

其原因是，对于二阶张量 $\boldsymbol{A}$ 和 $\boldsymbol{B}$，其点积一般地不满足 $\boldsymbol{AB}\neq\boldsymbol{BA}$. 但和二阶单位张量 $\mathbf{I}$ 点积，也就是求迹的特殊性，确实下列等式存在：

$$\mathbf{I}:(\boldsymbol{AB}+\boldsymbol{BA}) = \mathbf{I}:(2\boldsymbol{AB}) \tag{s3-5}$$

3.7 精细结构常数 (fine-structure constant) 表示电子在第一玻尔轨道上的运动速度和真空中光速的比值，计算公式为 $\alpha = e^2/(4\pi\varepsilon_0\hbar c)$，其中，$e$ 是电子的电荷，ε_0 是真空介电常数，$\hbar$ 是约化普朗克常数，c 是真空中的光速. 精细结构常数是一个无量纲量，$1/\alpha \approx 137.03599976$.

这个不寻常的常数的最不寻常之处在于它是无量纲的. 如果人类要往一颗位于几百光年外的行星发射信号，假设该行星绕着跟太阳一样的恒星转，并且拥有高度的外星文明. 人类要传达的唯一信息就是 137，因为不管外星球的科学家是用什么单位来表示电荷、速度和普朗克常数，他们也会得到 137. 因此，当住在那里的外星人接收到我们的信号时，也会知道我们是一个科学发达并且拥有高科技的文明.

正如费曼 (Richard Feynman, 1918—1988, 1965 年诺贝尔物理学奖获得者) 于 1985 年所指出的: "这个数字自五十多年前发现以来一直是个谜. 所有优秀的理论物理学家都将这个数贴在墙上，为它大伤脑筋 (and all good theoretical physicists put this number up on their wall and worry about it) …… 它是物理学中最大的谜之一，一个该死的谜 (It's one of the greatest damn mysteries of physics): 一个魔数来到我们身边，可没人能理解它. 你也许会说 '上帝之手 (hand of God)' 写下了该数字，而 '我们不知道他是怎样下的笔' ".

3.8 泡利 (Wolfgang Pauli, 1900—1958, 1945 年诺贝尔物理学奖获得者) 与精细结构常数的传说. 在于 1946 年 12 月 13 日在瑞典斯德哥尔摩发表的诺贝尔获奖演讲中，泡利表示他的目标是建立一个理论 "将决定精细结构常数的值从而解释电的原子的结构". 泡利有句名言:"当我死后，我对魔鬼的第一个问题将是: 精细结构常数的意义是什么？" 不幸的是，泡利没有能完成他的目标就于 1958 年 12 月 15 日，在苏黎世红十字医院的 137 号病房去世了 —— 在他去世之前，他就意识到了这种巧合的讽刺.

泡利来到了天堂后，因其在物理学界地位显赫，他被允许同上帝交谈. 上帝说:"泡利，你可以提一个问题，你想知道什么呢？" 泡利立刻就问了那个在他生命的最后十年里一直努力探寻但却没能找到答案的问题:"为什么 $\alpha \approx 1/137$？" 上帝笑了，他拿起粉笔开始在黑板上写公式. 过了几分钟他转向泡利，这时泡利正挥舞着他的手臂嚷道: "Das ist falsch (太荒谬了)！"

题图 3.9　索末菲于 1915 年引入了精细结构常数

3.9 精细结构常数是德国理论物理学家阿诺德·索末菲 (Arnold Sommerfeld, 1868—1951) 于 1915 年引入量子物理学的，因此也被称为索莫菲常数. 2005 年，慕尼黑大学物理系为了纪念索莫菲这位昔日量子力学慕尼黑学派的掌门人，成立了索莫菲理论物理学中心，并在办公楼内树立了他的铜像 (题图 3.9)，上面刻着精细结构常数的表达式. 在高斯单位制中，精细结构常数可由电子的电荷、光速和约化的普朗克常量这三个非常基本的物理学参数表达出来，它描述的是带电粒子之间电磁相互作用的强度.

除本思考题外，本书还在思考题 1.5、§22.2、§55.2、§59.1 中讨论或提及过索末菲这位伟大的理论物理学家. 2013 年，埃克特 (Michael Eckert) 在他为索末菲所做的传记中评价道 [3.17]:"普朗克是权威，爱因斯坦是天才，索末菲是导师 (Planck was the authority, Einstein the genius, and Sommerfeld the teacher)". 是呀，索末菲的学生中有八位获得了诺贝尔奖.

3.10 堪称 20 世纪物理界最后一位全才的大物理学家恩里科·费米 (Enrico Fermi, 1901—1954, 1938 年诺贝尔物理学奖获得者) 有一张在黑板前讲课的照片 (题图 3.10)，里

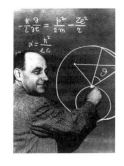

题图 3.10　费米与精细结构常数

面赫然写着十分错误的精细结构常数的表达式，但这丝毫不影响费米作为大物理学家的"光辉形象"，这也只能作为费米的一则"轶事".

3.11 粒子实验物理学家利昂·马克斯·莱德曼 (Leon Max Lederman, 1922—2018, 享年 96 岁，1988 年诺贝尔物理学奖获得者) 于 1979—1989 年曾任费米国家加速器实验室主任，并主持设计了超导超级对撞机建造计划，他将其在费米实验室附近的家的门牌号命名为精细结构常数 137 (Physicist Leon M. Lederman numbered his home near Fermilab 137 based on the significance of the number to those in his profession).

关于精细结构常数 137，莱德曼这样写道："它犹如赤子一般展现在世人面前 (shows up naked all over the place)"，这也就是思考题 3.7 所谈及的，如果宇宙其他行星上有高度文明的话，即便他们对电荷、光速，或普朗克常数采用不同的量纲，他们也会得到 137，因为它是个纯数字.

3.12 给出曲线坐标系中拉梅矢量和黎曼度规张量之间的关系式.

3.13 证明 (18-15) 和 (18-16) 两式中的余子式矩阵 $\mathbf{Cof}\,\boldsymbol{A}$ 满足如下关系式：

$$\boldsymbol{A}\,(\mathbf{Cof}\,\boldsymbol{A})^{\mathrm{T}} = (\mathbf{Cof}\,\boldsymbol{A})^{\mathrm{T}}\,\boldsymbol{A} = (\det\boldsymbol{A})\,\mathbf{I} \tag{s3-6}$$

从而有余子式矩阵的表达式：

$$\mathbf{Cof}\,\boldsymbol{A} = (\det\boldsymbol{A})\,\boldsymbol{A}^{-\mathrm{T}} \tag{s3-7}$$

3.14 在 (18-16) 式的基础上，求证：

$$\frac{\partial J^{-1}}{\partial \boldsymbol{F}} = -J^{-1}\boldsymbol{F}^{-\mathrm{T}} \tag{s3-8}$$

证明：注意到 $J^{-1} = \det^{-1}(\boldsymbol{F})$，利用莱布尼兹链式法则和 (18-16) 式，有

$$\frac{\partial J^{-1}}{\partial \boldsymbol{F}} = \frac{\partial J^{-1}}{\partial J}\frac{\partial J}{\partial \boldsymbol{F}} = -J^{-2}J\boldsymbol{F}^{-\mathrm{T}} = -J^{-1}\boldsymbol{F}^{-\mathrm{T}}$$

故 (s3-8) 式得证. 推而广之，有如下一般关系式：

$$\frac{\partial J^{-n}}{\partial \boldsymbol{F}} = -nJ^{-n}\boldsymbol{F}^{-\mathrm{T}} \tag{s3-9}$$

3.15 对测地线方程 (10-6) 式和 (16-33) 式的进一步解读见图 3.15.

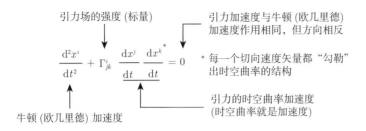

注：'质量' 包含在克里斯托费尔符号中

$$\Gamma^{i}_{jk} = \frac{1}{2} g^{im} (g_{mj,k} + g_{km,j} - g_{jk,m})$$

这里，度规张量 $g_{\alpha\beta}$ 便是由引力质量引起的引力势的表达式

题图 3.15　测地线方程中诸项的进一步解读

参 考 文 献

[3.1] Euler L. Theoria Motus Corporum Solidorum Seu Rigidorum. Greifswald, 1765.

[3.2] Eddington A S. The Mathematical Theory of Relativity. Cambridge: Cambridge University Press, 1923.

[3.3] Ciarlet P G. Mathematical Elasticity: Volume I: three-dimensional elasticity. Amsterdam: North-Holland, 1988. (中译本: P. G. 希亚雷. 数学弹性理论 (卷 I 三维弹性理论. 石钟慈、王烈衡译). 北京: 科学出版社, 1991.)

[3.4] Korn A. Über einige Ungleichungen, welche in der Theorie der elastischen und elektrischen Schwingungen eine Rolle spielen. Bulletin Internationale, Cracovie Akademie Umiejet, Classe de sciences mathematiques et naturelles, 1909, 3: 705-724.

[3.5] Neff P, Pauly D, Witsch K J. Poincaré meets Korn via Maxwell: extending Korn's first inequality to incompatible tensor fields. Journal of Differential Equations, 2015, 258: 1267-1302.

[3.6] Desvillettes L, Villani C. On a variant of Korn's inequality arising in statistical mechanics. ESAIM: Control, Optimisation and Calculus of Variations, 2002, 8: 603-619.

[3.7] Gurtin M E. An Introduction to Continuum Mechanics. New York: Academic Press, 1982. (中译本: 连续介质力学引论 (郭仲衡、郑百哲译). 北京: 高等教育出版社, 1992.)

[3.8] Descartes R. Discours de la methode pour bien conduire sa raison, & chercher la verité dans les sciences. Plus la dioptrique. Les meteores. Et la geometrie, Leiden, Jan Marie, 1637.

[3.9] Gauss C F. Disquisitiones generales circa superficies curvas. Commentationes Societatis Regiae Scientiarum Gottingensis Recentioris, Vol. 6, Gottingen, 1827.

[3.10] Riemann B. Über die Hypothesen, welche der Geometrie zugrunde liegen. Abh. Kgl. Ges. Wiss., Göttingen, 1868.

[3.11] Schwarzschild K. Über das Gravitationsfeld eines Massenpunktes nach der Einstein'schen Theorie. Sitzungsberichte der Königlich Preussischen Akademie der Wissenschaften, 1916, 1: 189–196.

[3.12] Arnold V I. Mathematical Methods of Classical Mechanics. New York: Springer, 2013. (中译本：阿诺尔德. 经典力学的数学方法 (第四版，齐民友译). 北京：高等教育出版社，2006.)

[3.13] Lamé G. Leçons sur la théorie mathématique de l'élasticité des corps solides (In English: Lectures on the Mathematical Theory of Elasticity of Solid Bodies). Paris: Bachelier, 1852.

[3.14] Christoffel E B. Über die Transformation der homogenen Differentialausdrücke zweiten Grades. Journal für die reine und angewandte Mathematik, 1869, 70: 46-70.

[3.15] Einstein A. Die grundlage der allgemeinen relativitätstheorie. Annalen der Physik, 1916, 354: 769-822.

[3.16] Butzer P L. An outline of the life and work of EB Christoffel (1829–1900). Historia Mathematics, 1981, 8: 243-276.

[3.17] Eckert M. Arnold Sommerfeld: Science, Life and Turbulent Times 1868-1951. New York: Springer Science & Business Media, 2013.

第4章　旋转群，拓扑，微分流形上的张量分析

§20. 理性力学中常用的群论

为了消除读者对"群论"的畏惧心理，首先给大家讲一则有关群论的故事.

有位美国老师到法国的一所小学进行数学教学的考察. 美国老师拦住一位七八岁的小学生问道："1+2 等于几？"

法国小学生摇摇头说："我不知道."

美国老师一听，第一反应是都说法国人数学特别厉害，怎么这么大的小学生还不知道这个？

没等美国老师继续想下去，小学生又继续说到："不过我可以告诉你 $1+2 = 2+1$，因为自然数的加法是个阿贝尔交换群."

20.1　对称与群

全体非零实数的乘法构成一个群 (group). 但这个群不是离散的了，是由无限多个实数元素组成的连续群，因为它的所有元素可以看成是由某个参数连续变化而形成. 两个实数相乘可以互相交换，因而这是一个"无限"、"连续"的阿贝尔群 (Abelian group).

可逆方形矩阵在矩阵乘法下也能构成无限的连续群. 矩阵乘法一般不对易，所以构成的是非阿贝尔群 (non-Abelian group).

连续群和离散群的性质大不相同，就像盒子里装的是一堆玻璃弹子，或装的是一堆玻璃细沙不同一样，因而专门有理论研究连续群. 因为连续群是 n 个连续变量之变化而生成的，这 n 个变量同时也张成一个 n 维空间. 如果一个由 n 个变量生成的连续群既有群的结构，又是一个 n 维微分流形，便称之为"李群 (Lie group)"，是以挪威数学家索菲斯·李 (Sophus Lie, 1842—1899，如图 20.1 所示) 的姓氏而命名.

图 20.1　索菲斯·李

群有如下四个基本要求 (群公理)，不妨将它们简称为"群四点"：

(1) 封闭性：两元素相乘后，结果仍然是群中的元素；

(2) 结合律：$(a \cdot b) \cdot c = a \cdot (b \cdot c)$；

(3) 单位元: 存在单位元, 也就是幺元, 与任何元素相乘, 结果不变;

(4) 逆元: 每个元素都存在逆元, 元素与其逆元相乘, 得到幺元.

美丽的对称无处不在, 它在我们的世界中扮演着重要的角色. 自然界遍布虫草花鸟, 人类社会处处有标志性的艺术和建筑, 这些事物无一不体现出对称的和谐与美妙. 几何图形的对称不难理解, 大家都知道 "故宫是左右对称的", "地球是球对称的", "雪花是六角形对称的". 从数学的角度来看待刚才的几个例子, 对称意味着几何图形在某种变换下保持不变. 比如说, 故宫和泰姬陵的左右对称意味着在镜像反射变换下不变 (如图 20.2 所示); 球对称是说在三维旋转变换下的不变性 (如图 20.3); 雪花六角形对称则是说将雪花的图形转动 60°、120°、180°、240°、300° 时图形不变 (如图 20.4). 所以, 对称实际上表达的是事物具有的一种冗余性.

图 20.3 球对称

图 20.2 对称美: 天坛 (左) 和泰姬陵 (右)

上帝设计世界时利用镜像对称, 他只需要设计一半! 利用六角形对称, 他的雪花图案只需画出六分之一! 球对称的天体就更好办了, 画出了一个方向的景色, 就让它们去绕着一个固定点不停地转圈.

对称是一种美. 群论便是描述对称的一种最好的语言.

用数学语言定义对称的优越性之一在于容易推广. 如果将对称概念从几何推广到物理研究中的一般情形, 便被表述为: 如果某种变换能够保持系统的拉格朗日量 (Lagrangian) 不变, 从而保持物理规律不变的话, 就说系统对此变换是对称的.

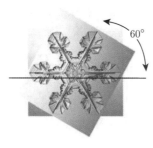

图 20.4 雪花的六角形对称

物理规律应该在变换中保持不变, 这应该是显而易见的. 试想, 如果今天的某个定律明天就不适用了, 或者是麦克斯韦方程组只在剑桥大学适用, 搬到中国科学院大学就不适用了, 那还叫做自然规律吗? 研究它还有任何意义吗? 前面涉及时间平移的不变性, 后面则涉及空间平移的不变性. 前者也叫做时间平移变换, 后者也叫做空间平移变换. 但是, 除了平移变换之外, 还有许多别的种类的变换, 物理定律难道对所有的变换都要保持不变吗? 物理规律有很多, 至少应该不是每一个规律对每一个变换都将保持不变. 那么, 这其中有些什么样的关系呢?

我们所讨论的变换也可以用数学上的"群"来加以分类. 所以, 变换用来描述对称, 群用来描述变换, 因此, 群和对称, 便如此关联起来了.

20.2 外尔的小册子——《对称》

外尔 (Hermann Weyl, 1885—1955) 的《对称》(*Symmetry*), 乃举世闻名的大手笔小册子, 是作者从普林斯顿高等研究院退休前"唱出的一支天鹅曲", 它由普林斯顿大学出版社将外尔退休前的四次系列讲座汇编而成书.

对称, 在我们的日常生活用语中, 就是匀称、平衡、和谐之意, 是指事物各部分相互协调, 构成一个整体. 外尔则把这个概念加以抽象, 从最简单的双边对称 (bilateral symmetry) 开始, 步步深入到旋转对称 (rotational symmetry)、平移对称 (translational symmetry)、全等 (congruence)、相似 (similarity)、正常旋转 (proper rotation)、反常旋转 (improper rotation)、循环群 (cyclic group)、二面角群 (dihedral group)······ 外尔的抽象思维似乎可以无限伸展, 最后进入到 N 维空间.

外尔在进行他的抽象思维的时候, 采用的是现实世界中生动的具体事例. 循着他的诱导, 我们也可以借助"对称"这个概念, 体验和认识丰富多彩的世界: 不管是自然的还是艺术的; 生命的还是非生命的.

外尔指出: "人类自古以来就试图用对称的概念来理解和创造秩序、美和完美 (Man through the ages has tried to comprehend and create order, beauty and perfection)." 古代的艺术, 从苏美尔拉加什古城 (City of Lagash) 的纹章图案 (如图 20.5 所示), 到波斯的彩釉司芬克斯, 从古希腊波利克莱托斯 (Polykleitos) 的人体雕塑, 到意大利拉斐尔的宗教壁画, 无不体现对称的原理. 对称, 是人类审美的最基本的原理; 对称, 它所表现出来是公正、中庸和权威; 对称, 是上帝创造世界所遵守的基本法则.

对称不是绝对的. 生命体要发育生长, 就必须打破一些对称. 人体就是一个很好的例子. 在外表上, 人体是左右对称的. 但是体内的器官就不是这样了. 人体要吸收营养, 肠子就不得不长得很长, 空间布局就顾不上对称了. 还有人体的心脏, 绝大多数是长在左胸, 本身也不是对称的.

但是, 对称的破坏也不是绝对的. 心脏的形状虽然看起来不对称, 但却符合更广意义上的对称. 外尔把它称作"螺旋"(screw), 背着壳的蜗牛就是一个典型的螺旋. 空间中左右并没有本质的区别, 所以左旋转和右旋转本来应该是等同概率发生的. 但是不知什么原因导致生命体的"螺旋"产生了偏向, 人类人体内包含了右

旋的 (dextro-rotatory) 葡萄糖和左旋的 (laevo-rotatory) 果糖. 要是人体内的新陈代谢把这种基因性左右旋搞错了, 那就会产生非常可怕的疾病, 叫做 "苯丙酮酸尿症"(phenyketonuria).

图 20.5 拉加什古城的纹章, 苏美尔, 约公元前 2700 年

在有机自然界, 对称原理也是无处不在起作用, 它对动植物种系发生和个体发生都发生作用. 植物界多彩美丽的鲜花, 展示的是各种循环的对称, 如鸢尾属植物花朵的三对称, 还有更普遍的五对称. 低等生命也展示旋转对称, 如图 20.6 所示的水母.

图 20.6 具有旋转对称性的水母

在无机的自然界，雪花、冰花、晶体，展示了各种各样的对称. 开普勒认定宇宙是和谐的，所以要用对称的几何体解释我们的宇宙.

外尔试图用对称原理来理解当时物理学上最新理论：相对论和量子力学.

相对论与对称有什么关系？要研究空间中几何形状的对称性就必须先研究空间本身的对称性.

空间是高度对称的，空间中的任何一处与它处没有本质的区别. 如莱布尼兹所说，所谓相似，就是在各自本身的框架下来看，两样东西没有区别. 如，平面上的两个正方形，如果考虑它们之间的关系，那确实有很多不同，比方说它们的某一边的朝向就不同. 但就每个正方形本身来说，它们完全相似，不能把一个与另一个区分开来. 那客观性从何谈起呢？

外尔采用亥姆霍兹的做法，在几何相似群中引入一个亚群：全等群. 在普通几何中，长度是相对的，扩张就包含在形状的自同构变换中. 但是原子和基本粒子的结构中已经包含了绝对的长度标准. 所以，两个参照系，只有当其中的所有的几何关系和物理定律都能用同样的代数来表达时，这两个参照系才是等同的. 实现在两个同样可以接受的参照系之间的变换构成了"物理自同构"群.

爱因斯坦相对论的贡献在于：他找到了包括时间维度在内的四维"空间"结构的"物理自同构"群，也就是所谓的"洛伦兹群".

至于量子力学，在外尔看来，对称在原子和分子谱线的分析上起了很大的作用，而解释原子的光谱是量子力学最大的成功. 量子力学成功地解释了氢原子光谱的巴尔末系 (Balmer series)，并且证明其中主要的物理常数是与电子的质量、电荷以及普朗克常数相关. 此后量子力学在光谱理论上取得一个又一个的成功，并导致了重大的发现：电子的自旋和泡利不相容原理.

外尔向我们证明，一旦物理上的基础确定下来，他就可以用对称的原理构造出描写这个物理的数学. 数学和物理之间好像天生就是相通的. 数学上的可能，说不定就成为物理学上的真实.

20.3 旋转李群——SO(n) 和 SU(n)

物理学与各种旋转结下不解之缘，从力学中研究的刚体转动，到量子理论中的粒子自旋. 地球绕太阳转，月亮绕地球转，滚珠在轴承滚道中转，电子绕原子核转，每一层次的实验和理论中似乎都少不了旋转. 物理中的旋转除了在真实时空中的旋转之外，还有一大部分是在假想的、抽象的空间中的旋转，比如动量空间，希尔伯特空间，自旋空间、同位旋空间等.

空间中的旋转也构成群, 并且, 旋转群 (rotation group) 是物理中非常重要的一类群. 旋转群有离散的和连续的之分. 连续旋转群具有天然的流形结构, 是一种李群, 理论物理, 特别是统一理论中所感兴趣的旋转李群有 SO(3)、SO(2)、U(1)、SU(2)、SU(3) 等.

旋转可以用大家熟知的矩阵来表示. 因此, 我们首先用矩阵的语言, 解释一下上面所列的一串符号是什么意思: 括号中的数目字 (3, 2, 1) 等是表示旋转的矩阵空间的维数; 大写字母 O 代表正交矩阵 (orthogonal matrix); U 代表酉矩阵 (unitary matrix); S 是特殊的 (special) 意思, 表示矩阵的行列式为 1.

比如, 举三维空间的旋转群 O(3) 为例. 此处 3 是指旋转空间的维数, O 对应于保持长度和角度不变的正交变换矩阵. 具体一点说, 正交矩阵 O(3) 是一个由 $3 \times 3 = 9$ 个实数组成的矩阵, 它的三个列向量或者三个行向量, 都构成三维空间中三个正交的单位矢量. 一般来说, 正交矩阵 O(3) 的行列式可为 1 或 −1. 当行列式为 −1 时, 正交矩阵表示的变换是旋转再加反演, 这里的负号便来自反演. 上述的 O(3) 旋转群如果加上字母 S, 指的便是特殊旋转群 SO(3), 那意味着, 矩阵行列式被限制为 1. 所以, SO(3) 表示的是三维空间中无反演的纯粹旋转.

观察我们周围的世界: 人的左脸并不完全等同于右脸; 大多数人的心脏长在左边, 大多数的 DNA 分子是右旋的; 地球并不是一个完全规则的球形······ 正是因为对称中有了这些不对称的元素, 对称与不对称的和谐交汇, 创造了我们的世界.

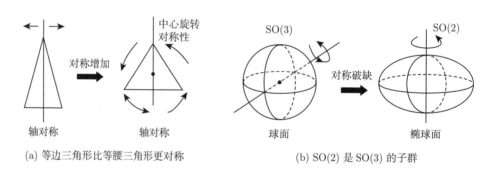

图 20.7 对称性的不同等级

不妨深究一下, 何谓对称? 何谓不对称? 可以说, 对称中有不对称, 不对称中又有对称. 并且, 对称有多种多样, 就几何图像而言, 具有某种变换下的对称, 但对另一种变换便可能不对称. 即使是同一类型的对称, 也有对称程度的高低. 比如说, 一个正三角形, 和一个等腰三角形比较, 正三角形应该更为对称一些, 如图 20.7(a). 再举旋转群为例: 一个球面是三维旋转对称的, 在 SO(3) 群作用下不变, 而椭球面只能看作是在二维旋转群 SO(2) 的作用下不变了. 用不很严格地说法, SO(2) 是 SO(3) 的子群, 因此, 球面比椭球面具有更多的对称性. 如果从对称性的高低等级来定义的话, 系统从对称性高的状态, 演化到对称性更低的状态, 被称为 "对称破缺", 反之, 则可称为 "对称建立". 例如, 当正三角形变形为等腰三角形, 或者当球面变成椭球面, 我们便说 "对称破缺了". 从李群的观点来看, SO(3) 是三阶的, 有三个生成元, SO(2) 只有一个生成元, 从球面到椭球面, 两个对称性被破缺. 因此, 可以从群论的观点来研究对称破缺.

例 20.1 SO(3) 在诺特定理中的应用.

解:《力学讲义》[1.11] §30 中已经详细地推导出诺特定理. SO(3) 主要在诺特定理中针对旋转对称, 如图 20.8 所示.

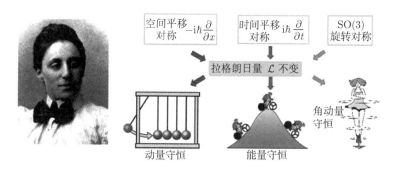

图 20.8 诺特定理与对称性

例 20.2 仔细说明例 20.1 中的一维动量算子 (momentum operator) $\hat{p}$ 和能量算子 (energy operator) $\hat{E}$ 分别为

$$\begin{cases} \hat{p} = -\mathrm{i}\hbar \dfrac{\partial}{\partial x} \\ \hat{E} = \mathrm{i}\hbar \dfrac{\partial}{\partial t} \end{cases} \tag{20-1}$$

帕斯库尔·约尔当 (Pascual Jordan, 1902—1980) 参与了矩阵力学的建立, 从 $[\hat{x}, \hat{p}] = \mathrm{i}\hbar$ 得出表达式 $\hat{p}_x = -\mathrm{i}\hbar\partial_x$ 这是经典力学方程算符化的基础.

证明: 请参阅例 1.7. 单自由粒子的薛定谔方程的解为一个平面波 (plane wave):

$$\psi(x, t) = \exp\left[\frac{\mathrm{i}}{\hbar}(px - Et)\right] \tag{20-2}$$

式中, $\hbar$ 为约化普朗克常数, p 为自由粒子沿 x 方向的动量, E 为自由粒子的能量.

波函数 $\psi(x, t)$ 对 x 的一阶偏导为

$$\frac{\partial \psi(x,t)}{\partial x} = \frac{\mathrm{i}p}{\hbar} \exp\left[\frac{\mathrm{i}}{\hbar}(px - Et)\right] = \frac{\mathrm{i}p}{\hbar}\psi(x,t) \tag{20-3}$$

可从 (20-3) 式简单地推导出：

$$p\psi(x,t) = -\mathrm{i}\hbar \frac{\partial \psi(x,t)}{\partial x} \quad \Rightarrow \quad \hat{p} = -\mathrm{i}\hbar \frac{\partial}{\partial x} \tag{20-4}$$

式中，$\hat{p} = -\mathrm{i}\hbar \dfrac{\partial}{\partial x}$ 即为一维情况时的动量算子. 由于偏导数为线性算子，则，动量算子亦为线性算子.

同理，在三维时，薛定谔方程的平面波的解可从 (20-2) 式类比为

$$\psi(\boldsymbol{r},t) = \exp\left[\frac{\mathrm{i}}{\hbar}(\boldsymbol{p}\cdot\boldsymbol{r} - Et)\right] \tag{20-5}$$

式中，$\boldsymbol{p}$ 为动量矢量，$\boldsymbol{r}$ 为位矢，则波函数 $\psi(\boldsymbol{r},t)$ 的梯度为

$$\begin{aligned}
\boldsymbol{\nabla}\psi(\boldsymbol{r},t) &= \frac{\partial \psi(\boldsymbol{r},t)}{\partial x}\boldsymbol{e}_x + \frac{\partial \psi(\boldsymbol{r},t)}{\partial y}\boldsymbol{e}_y + \frac{\partial \psi(\boldsymbol{r},t)}{\partial z}\boldsymbol{e}_z \\
&= \frac{\mathrm{i}}{\hbar}(p_x\boldsymbol{e}_x + p_y\boldsymbol{e}_y + p_z\boldsymbol{e}_z)\exp\left[\frac{\mathrm{i}}{\hbar}(\boldsymbol{p}\cdot\boldsymbol{r} - Et)\right] \\
&= \frac{\mathrm{i}}{\hbar}\hat{\boldsymbol{p}}\psi(\boldsymbol{r},t)
\end{aligned} \tag{20-6}$$

从 (20-6) 式可容易地导出：

$$\hat{\boldsymbol{p}}\psi(\boldsymbol{r},t) = -\mathrm{i}\hbar\boldsymbol{\nabla}\psi(\boldsymbol{r},t) \tag{20-7}$$

从而有三维的动量算子为

$$\hat{\boldsymbol{p}} = -\mathrm{i}\hbar\boldsymbol{\nabla} \tag{20-8}$$

下面简要证明 (20-1) 式中的能量算子 $\hat{E}$ 的表达式. 式 (20-5) 对时间求偏导，有

$$\frac{\partial \psi(\boldsymbol{r},t)}{\partial t} = -\frac{\mathrm{i}}{\hbar}E\exp\left[\frac{\mathrm{i}}{\hbar}(\boldsymbol{p}\cdot\boldsymbol{r} - Et)\right] = -\frac{\mathrm{i}}{\hbar}E\psi(\boldsymbol{r},t) \tag{20-9}$$

从上式推得

$$\hat{E}\psi(\boldsymbol{r},t) = \mathrm{i}\hbar\frac{\partial \psi(\boldsymbol{r},t)}{\partial t} \quad \Rightarrow \quad \hat{E} = \mathrm{i}\hbar\frac{\partial}{\partial t} \tag{20-10}$$

式中的 $\hat{E} = \mathrm{i}\hbar\dfrac{\partial}{\partial t}$ 即为能量算子. 量子力学中的算子可总结于表 20.1 中.

表 20.1 量子力学中的常用算子

算子名称	表达式	说明
动量算子 Momentum operator	$\hat{p}_x = -\mathrm{i}\hbar\dfrac{\partial}{\partial x}$；$\hat{\boldsymbol{p}} = -\mathrm{i}\hbar\boldsymbol{\nabla}$	诺特定理中的动量守恒与空间平移不变性
角动量算子 Angular momentum operator	$\hat{L}_z = -\mathrm{i}\hbar\dfrac{\partial}{\partial\varphi}$	诺特定理中的角动量守恒与旋转不变性
能量算子 Energy operator	$\hat{E} = \mathrm{i}\hbar\dfrac{\partial}{\partial t}$	诺特定理中的能量守恒与时间平移不变性
哈密顿算子 Hamiltonian operator	$\hat{H} = -\dfrac{\hbar^2}{2m}\dfrac{\partial^2}{\partial x^2} + V(x)$； $\hat{H} = -\dfrac{\hbar^2}{2m}\nabla^2 + V(\boldsymbol{r})$	类比于经典力学的哈密顿量

20.4 理性力学中常用的正交群

如图 20.9 所示，正交群 $\boldsymbol{Q}$，在连续介质力学中常被称为正交张量，是指对于任意两个矢量 $\boldsymbol{u}$ 和 $\boldsymbol{v}$ 满足如下线性变换：

$$\boldsymbol{Qu} \cdot \boldsymbol{Qv} = \boldsymbol{u}\boldsymbol{Q}^\mathrm{T}\boldsymbol{Qv} = \boldsymbol{u} \cdot \mathbf{I}\boldsymbol{v} = \boldsymbol{u} \cdot \boldsymbol{v} \tag{20-11}$$

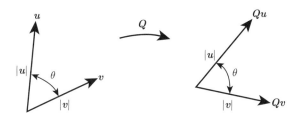

图 20.9 正交群作用于两个矢量间的点积

所以，正交群或正交张量满足：

$$\boldsymbol{Q}^\mathrm{T}\boldsymbol{Q} = \boldsymbol{Q}\boldsymbol{Q}^\mathrm{T} = \mathbf{I} \tag{20-12}$$

满足 (20-2) 式的矩阵 $\boldsymbol{Q}$ 称为正交变换 (orthogonal mapping)，所有正交变换的集合构成一个群，这就是完全正交变换群 (complete orthogaonal transformation group)，正交变换群简称正交群. (20-2) 式等价于 (3-7) 式中第二式

$$\boldsymbol{Q}^{-1} = \boldsymbol{Q}^\mathrm{T} \tag{20-13}$$

(20-11) 式表明，两个矢量 $\boldsymbol{u}$ 和 $\boldsymbol{v}$ 之间的点积 $\boldsymbol{u}\cdot\boldsymbol{v}$ 在两个矢量均进行变换 $\boldsymbol{Qu}$ 和 $\boldsymbol{Qv}$ 后保持不变，充分说明两者的长度和夹角均未发生改变，事实上，有

$$|\boldsymbol{Qu}| = \sqrt{\boldsymbol{Qu}\cdot\boldsymbol{Qu}} = \sqrt{\boldsymbol{u}\boldsymbol{Q}^\mathrm{T}\cdot\boldsymbol{Qu}} = \sqrt{\boldsymbol{u}\cdot\mathbf{I}\boldsymbol{u}} = \sqrt{\boldsymbol{u}\cdot\boldsymbol{u}} = |\boldsymbol{u}| \tag{20-14}$$

这一性质在例 1.5 中被称为等距映射.

正交群或正交张量 Q 的另外一个重要性质是满足:

$$\det\left(QQ^{\mathrm{T}}\right) = (\det Q)\left(\det Q^{\mathrm{T}}\right) = (\det Q)^2 = 1 \quad \Rightarrow \quad \det Q = \pm 1 \qquad (20\text{-}15)$$

式中, 如果 $\det Q = 1$, 则 Q 表示纯转动 (rotation), 为正常正交张量 (proper orthogonal tensor), 也就是 SO(3). 一般地, 正常正交张量在连续介质力学中进一步简称为转动张量 (rotation tensor) 用 R 表示; 反之, 如果 $\det Q = -1$, 则 Q 是非正常正交张量 (improper orthogonal tensor), 表示反射 (reflection). 行列式取正值的正交变换的集合也构成一个群, 这是真正交群 (true orthogonal group).

举例说明, 容易证明, 张量 $[A] = \begin{bmatrix} \cos\theta & \sin\theta & 0 \\ -\sin\theta & \cos\theta & 0 \\ 0 & 0 & 1 \end{bmatrix}$ 为一直角坐标系绕着 z 轴的纯转动, 因而为一转动张量; 而张量 $[B] = \begin{bmatrix} -1 & 0 & 0 \\ 0 & 1 & 0 \\ 0 & 0 & 1 \end{bmatrix}$ 则为一个反射张量.

例 20.3 给出正交群 (正交张量) 关系式 (20-12) 的指标表示形式.

解: 正交张量 Q 将在后续连续介质力学客观性研究中发挥重要作用, 所以弄懂、弄通其基本性质十分关键. 对于 $QQ^{\mathrm{T}} = I$, 由于 $Q = Q_{ij}e_i \otimes e_j$, $Q^{\mathrm{T}} = Q_{sr}e_r \otimes e_s$, 则两者间的点积运算有

$$(Q_{ij}e_i \otimes e_j) \cdot (Q_{sr}e_r \otimes e_s) = Q_{ij}Q_{sr}\delta_{jr}e_i \otimes e_s = Q_{ij}Q_{sj}e_i \otimes e_s \qquad (20\text{-}16)$$

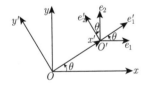

图 20.10 笛卡儿坐标系的旋转

由于二阶单位张量可表示为 $I = \delta_{is}e_i \otimes e_s$, 故, 正交张量基本关系式 (20-12) 的指标形式为

$$Q_{ij}Q_{sj} = \delta_{is} \qquad (20\text{-}17)$$

例 20.4 给出坐标旋转时基矢量的转换关系.

解: 如图 20.10 所示, 由初等几何关系可直接给出基矢量间的转换关系:

$$\begin{bmatrix} e'_1 \\ e'_2 \end{bmatrix} = \begin{bmatrix} \cos\theta & \sin\theta \\ -\sin\theta & \cos\theta \end{bmatrix} \begin{bmatrix} e_1 \\ e_2 \end{bmatrix} = [Q] \begin{bmatrix} e_1 \\ e_2 \end{bmatrix} \qquad (20\text{-}18)$$

式中，$[\boldsymbol{Q}]$ 的元素分别为

$$\begin{cases} Q_{11} = \cos\theta = \cos\langle \boldsymbol{e}_1', \boldsymbol{e}_1\rangle \\ Q_{12} = \sin\theta = \cos\langle \boldsymbol{e}_1', \boldsymbol{e}_2\rangle \\ Q_{21} = -\sin\theta = \cos\langle \boldsymbol{e}_2', \boldsymbol{e}_1\rangle \\ Q_{22} = \cos\theta = \cos\langle \boldsymbol{e}_2', \boldsymbol{e}_2\rangle \end{cases} \tag{20-19}$$

(20-18) 式亦可用指标形式简洁地表示为

$$\boldsymbol{e}_i' = Q_{ij}\boldsymbol{e}_j \tag{20-20}$$

利用矩阵求逆的公式：

$$\boldsymbol{Q}^{-1} = \frac{1}{\det \boldsymbol{Q}}\mathrm{adj}\boldsymbol{Q} = \frac{1}{\det \boldsymbol{Q}}(\mathbf{Cof}\boldsymbol{Q})^{\mathrm{T}} \tag{20-21}$$

式中，$\mathrm{adj}\boldsymbol{Q}$ 为 $\boldsymbol{Q}$ 的伴随矩阵 (adjoint matrix)，$\mathbf{Cof}\boldsymbol{Q}$ 则为其余子式矩阵 (cofactor matrix). 由 (20-21) 式所求得的 $\boldsymbol{Q}^{-1}$ 完全满足 (20-13) 式 $\boldsymbol{Q}^{-1} = \boldsymbol{Q}^{\mathrm{T}}$，则 (20-18) 式的逆为

$$\begin{bmatrix} \boldsymbol{e}_1 \\ \boldsymbol{e}_2 \end{bmatrix} = [\boldsymbol{Q}^{-1}]\begin{bmatrix} \boldsymbol{e}_1' \\ \boldsymbol{e}_2' \end{bmatrix} = \begin{bmatrix} \cos\theta & -\sin\theta \\ \sin\theta & \cos\theta \end{bmatrix}\begin{bmatrix} \boldsymbol{e}_1' \\ \boldsymbol{e}_2' \end{bmatrix} \tag{20-22}$$

例 20.5 最简单的连续群 —— 轴转动群 SO(2) 性质的进一步讨论.

解：$\forall \theta \in \mathbb{R}$，可以按如下形式构成 (20-22) 式中的平面旋转变换矩阵：

$$\theta \mapsto [T(\theta)] = \begin{bmatrix} \cos\theta & -\sin\theta \\ \sin\theta & \cos\theta \end{bmatrix} \in \mathrm{SO}(2) \tag{20-23}$$

这类矩阵的全体集合记为 SO(2). 下面将证明，SO(2) 满足交换性，具有单位元和逆元的结构.

SO(2) 是阿贝尔群的证明. $\forall \alpha, \beta \in \mathbb{R}$，两个旋转的合成等同于分别旋转，且可以交换旋转顺序：

$$\begin{aligned} \begin{bmatrix} \cos(\alpha+\beta) & -\sin(\alpha+\beta) \\ \sin(\alpha+\beta) & \cos(\alpha+\beta) \end{bmatrix} &= \begin{bmatrix} \cos\alpha & -\sin\alpha \\ \sin\alpha & \cos\alpha \end{bmatrix}\begin{bmatrix} \cos\beta & -\sin\beta \\ \sin\beta & \cos\beta \end{bmatrix} \\ &= \begin{bmatrix} \cos\beta & -\sin\beta \\ \sin\beta & \cos\beta \end{bmatrix}\begin{bmatrix} \cos\alpha & -\sin\alpha \\ \sin\alpha & \cos\alpha \end{bmatrix} \\ &= \begin{bmatrix} \cos(\beta+\alpha) & -\sin(\beta+\alpha) \\ \sin(\beta+\alpha) & \cos(\beta+\alpha) \end{bmatrix} \end{aligned} \tag{20-24}$$

(20-24) 式说明，阿贝尔群所具有的交换性: $T(\alpha+\beta) = T(\beta+\alpha)$，在 SO(2) 中得到了保持.

有关 SO(2) 的单位元性，在 (20-23) 式中当 $\theta = 0$ 时，记：

$$[T(0)] = \begin{bmatrix} \cos\theta & -\sin\theta \\ \sin\theta & \cos\theta \end{bmatrix}_{\theta=0} = \begin{bmatrix} 1 & 0 \\ 0 & 1 \end{bmatrix} = [\mathbf{I}] \tag{20-25}$$

这是平面单位矩阵，作用于向量时是恒同的，换句话说，对向量进行了角度为 $\theta = 0$ 的平面旋转，亦即单位元 $\mathbf{0}$.

再考虑 SO(2) 反向旋转，即当角度为 $-\theta$ 时，(20-23) 式变为

$$\begin{aligned}[T(-\theta)] &= \begin{bmatrix} \cos(-\theta) & -\sin(-\theta) \\ \sin(-\theta) & \cos(-\theta) \end{bmatrix} = \begin{bmatrix} \cos\theta & \sin\theta \\ -\sin\theta & \cos\theta \end{bmatrix} \\ &= \begin{bmatrix} \cos\theta & -\sin\theta \\ \sin\theta & \cos\theta \end{bmatrix}^{-1} = [T^{-1}(\theta)] = [T(2\pi-\theta)]\end{aligned} \tag{20-26}$$

(20-26) 式对应了逆元的性质.

综上，轴转动群，亦称平面旋转群 SO(2) 为单参数、连续、联通、阿贝尔、紧致李群.

例 20.6 仿射变换、仿射群、仿射量.

解 如图 20.11 所示，典型的仿射变换包括：转动 (rotation)、平移 (translation)、均匀缩放 (uniform scaling)、非均匀缩放 (nonuniform scaling)、反射 (reflection)、剪切 (shearing).

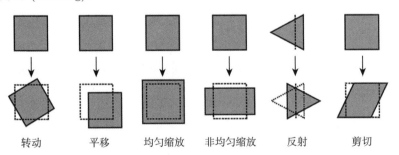

图 20.11 几种典型的仿射变换示意图

仿射变换是任何保持共线性和距离比的变换 (An affine transformation is any transformation that preserves collinearity and ratios of distances). 通俗地讲，所谓"仿射变换"就是"线性变换"+"平移"(any affine transformation is equivalent to a linear transformation of position vectors followed by a translation). 如果 $\mathbb{X}$ 和 $\mathbb{Y}$ 是两个仿射空间, 则任意仿射变换可表示为

$$f: \mathbb{X} \to \mathbb{Y}$$
$$x \mapsto y = f(x) = Mx + b \tag{20-27}$$

式中, M 为仿射空间 $\mathbb{X}$ 中的线性变换 (映射), M 亦称为仿射变换矩阵, x 为 $\mathbb{X}$ 中的向量, b 则为仿射空间 $\mathbb{Y}$ 中的向量.

用增广矩阵 (augmented matrix) 和增广向量, (20-27) 式可优雅地等价表示为

$$\begin{bmatrix} y \\ 1 \end{bmatrix} = \begin{bmatrix} M & b \\ 0 & 1 \end{bmatrix} \begin{bmatrix} x \\ 1 \end{bmatrix} \tag{20-28}$$

普通矩阵向量乘法总将原点映射至原点, 因此无法呈现平移, 也就是原点必须映射至其他点. 借由于所有向量上扩增一坐标 "1", 即将原空间映至更高维空间的一个子集合以进行变换.

仿射群就是全体仿射变换的集合.

所谓 "仿射量" 是指由向量向向量的线性变换:

$$v = Tu \tag{20-29}$$

式中, u 和 v 为向量空间中的任意向量, 仿射量 T 则显然是一个二阶张量, Tu 表示 T 和 u 之间的点积.

20.5 阿贝尔、伽罗瓦在建立群论过程中的曲折经历

柯西 (Augustin-Louis Cauchy, 1789—1857) 是他那个年代法国最主要的数学家. 可以理解, 巴黎科学院在收到数学论文的投稿需要权威的意见时, 柯西是最佳的人选之一. 一般情况下, 柯西是一个迅速而公正的评阅人 (a prompt and just referee). 但是, 柯西也有失误的时候. 在数学史上, 两次主要的灾难 (two of the major disasters) 均是和群论的论文相关, 是由于柯西的疏忽而造成的, 那就是对伽罗瓦的忽略和对阿贝尔的不公正的对待. 对于阿贝尔 (如图 20.12 左图所示), 柯西负部分责任 (Cauchy was only partly to blame); 但是对伽罗瓦 (如图 20.12 右图所示) 的不可原谅的疏忽, 柯西要负全部责任 (For the inexcusable laxity in Galois' case Cauchy alone is responsible).

是呀, 因为柯西的疏忽, 群论晚问世了半个世纪之久!

1826 年 10 月 10 日, 阿贝尔被称为 "永恒的纪念碑" 的有关群论的论文被呈交给巴黎科学院. 时年 74 岁的勒让德和 39 岁的柯西被任命为阿贝尔论文的评阅人.

图 20.12　阿贝尔 (左) 和伽罗瓦 (右)

柯西将阿贝尔的论文带回家, 不知放在什么地方, 完全将其忘了 (Cauchy took the memoir home, mislaid it, and forgot all about it). 多产的柯西 "忙着孵他自己的蛋, 咯咯地叫着, 顾不上谦虚的阿贝尔放在他窝里的确实是大鹏的蛋." (The prolific Cauchy was so busy laying eggs of his own and cackling about them that he had no time to examine the veritable roc's egg which the modest Abel had deposited in the nest.)

两年半后, 1829 年 4 月 8 日, 勒让德在写给雅可比的信中说道: "我们发觉这篇论文很难辨认, 该论文是用淡的几乎是白色的墨水写的, 字写得很糟, 我们两人认为应该要求作者送一份写得整齐易读的来." 而阿贝尔则于 1829 年 4 月 6 日凌晨去世了, 终年仅仅 26 岁!

阿贝尔的祖国在关心着他. 挪威驻巴黎的领事就阿贝尔这份遗失的手稿提出了外交抗议后, 柯西才于 1830 年将它翻了出来. 11 年后的 1841 年, 阿贝尔的这篇史诗般的论文才得以正式发表. 从 1826 年提交到 1841 年发表, 是漫长的 15 年的时间.

伽罗瓦将其积累至 17 岁时的重大发现写成了一篇文章, 准备呈交巴黎科学院. 柯西答应送交这篇论文, 但是他忘记了, 最后柯西甚至竟然不称职地将作者的摘要也遗失了!

刘维尔 (Joseph Liouville, 1809—1882) 在 1836 年 1 月创办了《纯粹与应用数学杂志》(*Journal de Matématiques pures et appliquées*). 1832 年 5 月 30 日, 伽罗瓦在决斗中被杀, 年仅 20 岁. 刘维尔整理了他的部分遗稿并刊登在 1846 年的《纯粹与应用数学杂志》上, 伽罗瓦在代数方面的独创性工作才得以为世人所知.

难道伽罗瓦自己对他论文的遗失和屡遭否定一点儿责任都没有嘛? 或者说后人从中有什么教训可以汲取的吗? 刘维尔在编辑伽罗瓦论文的前言中还善意地谈到, 导致巴黎科学院的审稿专家们拒绝伽罗瓦的论文的原因是 "因为它们模糊费解 (on account of their obscurity)". 刘维尔进而告诫说: "产生这个缺点的原因是过分追求简练，这个缺点是在对待抽象而神秘的纯代数问题时，我们应该首先竭力避免的. 当你试图引导读者远远离开老路，步入更广阔的领域时，确实需要清楚明了. 正如笛卡儿说的'在论述超常的问题时，应该超常地清楚'伽罗瓦常常过于忽略这个告诫了 (An exaggerated desire for conciseness was the cause of this defect which one should strive above all else to avoid when treating the abstract and mysterious matters of pure Algebra. Clarity is, indeed, all the more necessary when one essays to lead the reader farther from the beaten path and into wilder territory. As Descartes said, 'When transcendental questions are under discussion be transcendentally clear.' Too often Galois neglected this precept)."

§21. 微分流形

首先，让我们来总结下人类对空间的认识过程：

• 约公元前 300 年，古希腊数学家欧几里得的《几何原本》奠定了欧氏几何的基础. 欧几里得使用了公理化的方法，这一方法后来成了建立任何知识体系的典范，在差不多二千年间，被奉为必须遵守的严密思维的范例.《几何原本》是欧几里得几何的基础，在西方是仅次于《圣经》而流传最广的书籍.

• 1637 年，法国哲学家、数学家笛卡儿发表《几何学》奠定了解析几何的基础，从而把分析和几何图形联系起来. 笛卡儿在他的《几何学》中第一次出现变量与函数的思想，从而完成了数学史上一项划时代的变革. 对此恩格斯给予了极高的评价："数学中转折点是笛卡儿的变数，有了变数，运动进入了数学，有了变数，辩证法进入了数学，有了变数，微分和积分也就立刻成为必要的了."

• 1788 年，法国数学力学家拉格朗日出版了《分析力学》[1.12]，引进了高维空间来描述动力系统的状态. 哈密顿将拉格朗日的《分析力学》描述为 "一种科学的诗篇 (a kind of scientific poem)"[1.11].

• 德国数学家高斯的研究工作推进了曲面微分几何的发展.

• 德国数学家黎曼系统发展了微分几何，奠定了流形与黎曼空间的基础.

• 20 世纪初，意大利数学家里奇和列维-奇维塔发展了一套流形上的张量分析方法，系统地进行了微分不变量的研究.

• 1899 年，法国数学家嘉当 (Élie Cartan, 1869—1951) 创立了外形式和外微分方法.

再让我们回顾下微分流形的发展历史.

21.1 高斯与内蕴微分几何

高斯首次清楚地将曲线和曲面本身构想为空间. 1827 年,高斯他以其"绝妙定律 (theorema egregium)"[4.1] 创立了内蕴微分几何 (intrinsic differential geometry). 高斯绝妙定理表明,如果一个曲面弯曲而没有拉伸,表面的高斯曲率就不会改变. 换句话说,高斯曲率可以通过测量表面本身上的角度,距离和速率来完全确定,而无需进一步参考表面嵌入环境三维欧几里得空间中的特定方式. 因此,高斯曲率是表面的固有不变量,也就是内蕴几何量. 高斯曲率就是两个主曲率的乘积.

我们不禁要问:为什么高斯将该定理命名为"绝妙定理"?其绝妙之处何在?答案就在于该定理提出并在数学上证明了内蕴几何这个几何史上全新的概念,它说明曲面并不仅仅是嵌入三维欧氏空间中的一个子图形,曲面本身就是一个空间,这个空间有它自身内在的几何学,独立于外界三维空间而存在.

高斯于 1827 年发表的题为"曲面的一般理论研究"的论文[4.1] 标志着微分几何的发展进入了一个新的时代. 陈省身曾对高斯评价道:"历史上最有价值的文献当是高斯的'曲面论'[4.1],这绝非是偶然的,它是'高斯风格'的结晶."

陈省身还曾说道:"高斯是微分几何的始祖,他的曲面论建立了只依据曲面的第一基本形式的几何学,并把欧氏几何推广到了曲面上'弯曲'的几何."

爱因斯坦曾评论高斯的这篇论文[4.1] 道:"高斯为我们提供的最好的东西,好像是独一无二的,如果他没有创造曲面几何,那么黎曼理论就失去了基础,我们就很难想象其他任何人能发现这一理论""高斯对于近代物理理论的发展,尤其是对于相对论理论的数学基础所作出的贡献,它的重要性是超越一切,无与伦比的……"

所谓"内蕴",是相对于"外嵌"而言. 内蕴指的是曲面或曲线不依赖于它在三维空间中嵌入方式的某些性质."内蕴"的概念也可以被解释得更为物理一些:一个观察者在自己生活的物理空间中所能够观察和测量到的几何性质就是这个空间的内蕴性质.

也有人比喻说:外嵌是机械设计工程师看待曲面的方法,将曲面看成为他的三维机械零件的表面;而内蕴几何则是地球上的测地员测量地球表面测量到的几何性质. 比如说,内蕴几何量的最简单例子就是弧长. 一条直线可以在三维空间中看起来转弯抹角地任意弯曲,即随意改变它的曲率和挠率 (torsion),但生活在直线上的"点状蚂蚁"观察不到这些"弯来绕去",只能测量到它爬过的弧长. 因此,空间曲线的曲率和挠率,是三维空间的生物观察这条曲线时得到的重要性质,但却并不是内蕴几何量. 对曲面来说也是如此,弧长并不因为平面卷成了柱面或锥面而改变. 弧长与曲线嵌入空间中的弯曲情况无关,因而是个内蕴几何量.

黎曼比他的老师高斯刚好小 50 岁,于 1826 年生于德国的一个小村庄,有趣的是,按时间算起来,兼任测量局官员的高斯那时候正好在这个地区进行土地测量(如图 21.1 所示). 时间的巧合,给人一种异想天开神话式的联想:上帝是否就在那时候,将非欧几何——黎曼几何的思想种子,植根到了那片被丈量的土地上.

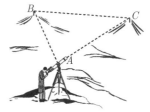

图 21.1 高斯在黎曼的家乡丈量土地,在检验三座山顶点组成的三角形的内角和是否为$180°$

21.2 黎曼与流形

1854 年,28 岁的黎曼在哥廷根大学发表就职演讲. 这个职位是所谓无薪讲师,他的收入完全来自于听课的学生所缴纳的学费. 即使是争取这样一个职位,也需要提供一篇就职论文以及发表一个就职演讲. 1853 年他提交了就职论文,其中讨论了什么样的函数可以展开成三角级数的问题,并导致对定积分的第一个严格数学定义. 之后的就职演讲要求候选人准备三个演讲课题,委员会从中挑选一个作为正式演讲题目.

黎曼挑选了两个思虑多时的课题,外加一个还未及考虑的课题 —— 关于几何学的基本假设. 他几乎确信委员会将挑选前面两个题目之一. 然而,委员会的高斯偏偏就看中了第三个题目. 当时黎曼正沉浸于电、磁、光、引力之间的相互关系问题,从这样的深沉思考中抽身转而研究新的问题无疑是一种巨大的压力,再加上长期的贫穷,一度让黎曼崩溃. 但不久他就重新振作起来,用七个星期时间准备了关于几何学基本假设的演讲. 为了让数学系以外的委员会成员理解他的演讲,黎曼只用了一个公式,并且忽略了所有计算细节. 尽管如此,估计在场鲜有人能理解这次演讲的内容. 只有高斯为黎曼演讲中蕴含的深邃思想激动不已.

黎曼在演讲中提出了 "弯曲空间" 的概念,并给出怎样研究这些空间的建议. "弯曲空间" 正是后来拓扑学研究的主要对象. 在这些对象上,除了可以运用代数拓扑的工具,还可以运用微积分工具,这就形成了 "微分拓扑学".

黎曼认为,几何学的对象缺乏先验的定义,欧几里得的公理只是假设了未定义的几何对象之间的关系,而我们却不知道这些关系怎么来的,甚至不知道为什么几何对象之间会存在关系. 黎曼认为,几何对象应该是一些多度延展的量,体现出各种可能的度量性质. 而我们生活的空间只是一个特殊的三度延展的量,因此欧几里得的公理只能从经验导出,而不是几何对象基本定义的推论. 欧氏几何的公理和定理根本就只是假设而已. 但是,我们可以考察这些定理成立的可能性,然后再试图把它们推广到我们日常观察的范围之外的几何，比如大到不可测的几何，以及小到不可测的几何. 接着，黎曼开始了关于延展性，维数，以及将延展性数量化的讨论. 他给了这些多度延展的量 (几何对象) 一个名称，德文写作 mannigfaltigkeit，英国数学家克利福德 (William Kingdon Clifford, 1845—1879, 见本书题表 1.5) 将其英文翻译为 manifoldness (多层).

中国首个拓扑学家江泽涵将黎曼的 mannigfaltigkeit 这个数学词汇翻译为 "流形"[4.2]，取自文天祥《正气歌》："天地有正气，杂然赋流形"，而其原始出处为《易经 (彖)》："大哉乾元，万物资始，乃统天. 云行雨施，品物流形." 这个翻译比英文翻译更加符合黎曼的原意，即多样化的形体.

21.3　流形的定义

在表 1.3 豪斯多夫空间的基础上，下面给出流形的定义：

设 $\mathcal{M}$ 是豪斯多夫空间，若对于该空间的任意一个点 $x \in \mathcal{M}$，都有 x 在 $\mathcal{M}$ 的一个邻域 $\mathcal{U}$ 同胚于 n 维欧氏空间 $\mathcal{E}^n$ 中的一个开集，则称 $\mathcal{M}$ 是一个 n 维流形.

上述流形定义中的关键词有三个：豪斯多夫空间、邻域和开集. 具体解释如下：

(1) 强调流形是一个豪斯多夫空间，就相当于强调了流形的可分性，这使得流形中连接任意两点的连线可无限细分.

(2) 在表 1.3 中已经指出，所有度量空间 (metric space) 都是豪斯多夫空间，可用度量空间中的距离来定义邻域：在度量空间中，一点的邻域定义为空间中到该点的距离小于某一给定的正实数的点的集合.

(3) 开集 $\mathcal{A}$ 是指空间的子集合，且 $\mathcal{A}$ 中每一点的某一邻域均完全包含在集合 $\mathcal{A}$ 中.

两个空间同胚是指存在一个一一的连续映射，且其逆映射也是连续的. 同胚映射中的的连续性是指，"映射像集" 中的任意开集的原像也是开集.

连续映射保持了流形上各点的临近性，而同胚映射则进一步要求在映射过程中不同的点映射为不同的点，而且不会产生新的点. 对同胚映射的一个直观类比是，将流形想象为橡皮做成的，可任意拉伸、弯曲，但不允许发生撕裂成一分为二或一分为多的情形，也不允许将不同的点黏合在一起，亦即不能发生合二为一的情形.

从几何方面考虑，拉格朗日力学和哈密尔顿力学本质上也都是流形理论.

庞加莱研究了三维流形，并提出一个问题，就是现在所谓的 "庞加莱猜想"：所有闭简单连通的三维流形同胚于三维球吗？这个问题已经于 2003 年完全得以解决，其中最重要的工作是由俄罗斯数学怪才佩雷尔曼做出的，详见思考题 1.2.

外尔 (Hermann Weyl, 1885—1955) 在 1912 年给出了微分流形的一个内在的定义. 该课题的基础性方面在 1930 年代被惠特尼 (Hassler Whitney, 1907—1989) 等人运用从 19 世纪下半叶就开始发展的精确的直觉厘清，并通过微分几何和李群理论得到了发展.

1915 年，爱因斯坦运用黎曼几何和张量分析创立了新的引力理论 —— 广义相对论. 使黎曼几何及其里奇算法成为广义相对论研究的有效数学工具. 而相对论的发展则受到整体微分几何的强烈影响. 例如矢量丛和联络论构成规范场 (杨–米尔斯场) 的数学基础.

庞加莱曾经问了如下的有趣的问题：如果一只具有高度智慧的蚂蚁从出生起就一直生活在一张曲面上，它没有任何三维的概念，那么这只蚂蚁如何判定这张曲面是否存在 "孔洞"？换言之，蚂蚁如何理解这张曲面的拓扑？为此庞加莱发明了代数拓扑，用同伦群的概念完美地加以解决.

后来，爱因斯坦的小女儿问他："为什么你那么有名？"爱因斯坦给她讲了蚂蚁的故事，然后说："别的蚂蚁都以为这张曲面是平直的，只有我这只蚂蚁看出来空间是弯曲的."

§22. 拓扑学与拓扑相变

22.1 拓扑学与亏格

很多欧美人在休闲吃点心时，右手端着一只咖啡杯，左手拿着一个面包圈，这两样东西的形状看上去完全不一样，但它们的拓扑性质是一样的，面包圈可以通过一系列形变，变成咖啡杯，如图 22.1 所示.

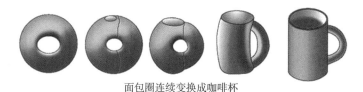

面包圈连续变换成咖啡杯

图 22.1 水杯和面包圈同胚

1736 年，29 岁的莱昂哈德·欧拉 (Leonhard Euler, 1707—1783, 如图 22.2 左图所示) 向圣彼得堡科学院递交了《哥尼斯堡的七座桥》(*The Seven Bridges of*

Königsberg, 如图 22.2 右图所示) 的论文 [4.3], 圆满地回答了哥尼斯堡居民提出的七桥问题, 证明了不可能在所有桥都只走一遍的情况下, 走遍连接河中心两个小岛和两岸的所有七座桥.

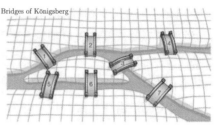

图 22.2 左: 欧拉; 右: 哥尼斯堡的七座桥问题

事实上, 欧拉的解决方法是忽略了桥的长度和岛的大小, 将岛和桥简化成了平面上的点与线. 欧拉的发现为后来的数学新分支 —— 拓扑学的建立奠定了基础.

1847 年, 德国数学家李斯亭 (Johann Benedict Listing, 1808—1882) 将欧拉的上述发现进一步发展, 对于这一新的数学领域, 引入了 "拓扑学 (topology)" 的新概念.

直到 20 世纪 90 年代, 拓扑学的应用终于开始真正的发展. 现在, 几乎所有领域离不开拓扑学了: 生物学家通过扭结理论理解 DNA 的结构; 计算机学家通过扭结在一起的同轴电缆制造量子计算机; 机器人科学家也用相同的理论使机器人走路; 医生以同调论 (homology) 为基础为病人做大脑扫描; 宇宙学家以此来理解银河系的形成; 通信公司运用拓扑学来决定如何布置基站进行网络覆盖; 手机的照相功能也是通过拓扑学原理实现的, 等等.

姜立夫 (1890—1978) 将 "topology" 这个数学名词确立为中文 "拓扑学"[4.2].

在拓扑学家中流传着这么一句俏皮话: 一个拓扑学家分不清面包圈和咖啡杯的差别 (a topologist as someone who doesn't know the difference between a doughnut and a teacup), 如图 22.1 所示. 更为详细的讨论见下一节中有关同胚和微分同胚的讨论.

拓扑学是近代发展起来的数学领域中一个重要的、基础的分支, 研究的是几何图形在连续形变下保持不变的性质. 所谓几何图形的连续形变, 就是允许将几何图形进行伸缩和扭曲等变形, 但不能割断和粘合, 所以拓扑学又被称为 "橡皮膜上的几何学". 比如, 我们在橡皮膜上画一个三角形, 然后随便拉扯甚至扭曲, 只要橡皮膜不破, 所画图形就是在做连续形变. 科幻电影《终结者 2》里, 那个液态机器人杀手的每次变化都可以看作连续形变. 只要图形的闭合性质不被破坏, 在拓扑学上它们就都是等价图形. 所以, 对于拓扑学家来说, 咖啡杯和面包圈没什么区别, 二者是等价的, 因为咖啡杯可以通过连续形变成为面包圈.

拓扑本来是一个数学概念, 是指物体在连续变化下保持不变的性质. 连续变化是指拉伸、扭曲以及变形等等, 但是不能有撕裂. 比如, 一个球和一个椭球, 甚至一个任意形状、没有洞的物体, 在拓扑上都是一样的. 一个面包圈和有一个手柄的茶杯, 甚至任何有一个穿透的洞的物体在拓扑上是一样的. 因此, 洞的个数在数学上叫做亏格 (genus), 是个拓扑不变量, 是整数, 如图 22.3 所示.

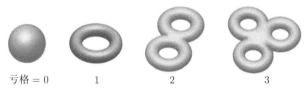

图 22.3 孔洞与亏格

拓扑学只研究图形的等价性, 不考虑研究对象的形状、大小、位置等因素, 从而使很多复杂而抽象的问题大大简化, 这也决定了拓扑学在物理学、生物学、化学、经济学等领域都具有广泛的应用价值.

22.2 拓扑相变

2016 年的诺贝尔物理学奖授予大卫·索利斯 (David J. Thouless, 1934—2019)、邓肯·霍尔丹 (F. Duncan M. Haldane, 1951—) 和迈克·科斯特里兹 (J. Michael Kosterlitz, 1943—), 以表彰他们 "关于拓扑相变和物质拓扑相方面的理论发现 (for theoretical discoveries of topological phase transitions and topological phases of matter)".

拓扑学是数学中的一个分支, 研究阶梯式变化的性质, 比如以上物体的洞的数量, 如图 22.4 所示. 拓扑学是三位得奖者能做出这一成就的关键, 它解释了为什么薄层物质的的导电率会以整数倍发生变化.

如图 22.5 所示, 相变是指由同样的微观粒子组成的宏观体系在不同温度下表现出截然不同的性质. 比如随着温度的下降, 气体变成液体, 液体变成固体. 再比如, 随着温度的下降, 液态氦可以变成超流 —— 也就是说, 变成一种没有黏滞的流体 (类似于超导). 不同宏观性质的表现叫做 "相 (phase)", 比如水的气相、液相、固相, 或者液氦的超流相、正常相.

之所以有相变 (phase transition), 是因为存在两种因素, 即能量与混乱程度 (称作熵) 的互相竞争. 一方面不同的相能量不同, 比如简单来说, 液相比气相能量低，固相又比液相能量低. 而另一方面，液相比固相混乱，气相又比液相混乱. 对于液氦来说，超流相比正常相能量低，正常相比超流相混乱. 混乱程度 (熵) 乘以温度以后可以直接与能量定量比较. 为了稳定，系统既希望能量尽量低，又希望混乱程度尽量高. 最后的结果是，存在某个温度，在这温度之上，系统处于某个相；在这温度之下，系统处于另一个相，这就是相变. 如图 22.5 所示，由温度从高到低，将发生等离子体、气态、液态、固态和量子凝聚等相变.

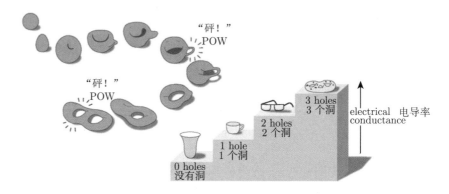

图 22.4　对拓扑学和 2016 年诺贝尔物理学奖的形象比喻

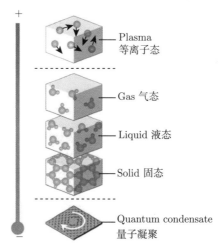

图 22.5　温度从高到低所发生的相变

1972 年以前，物理学家普遍认为，正常相到超流相的相变只能存在于三维系统中. 对于二维系统，当时人们认为在非零温度，不存在相变. 也就是说，任何一个非零温度下，总是正常相赢，因为它在混乱程度上的优势总能战胜在能量上的逆势. 因此，人们说，在二维或一维系统中，在任何非零温度下，热涨落破坏有序，没有相变.

1972 年, 在英国伯明翰大学, 数学物理学教授索利斯和博士后科斯特里兹发现, 通过拓扑的途径, 在二维可以发生一种新的相变, 即拓扑相变.

科斯特里兹和他的老师索利斯的相遇就很有趣. 1962 年, 19 岁的科斯特里兹进入剑桥大学冈维尔与凯斯学院 (Gonville and Caius College) 学习, 他一年级第一次上 "数学物理方法" 课时, 课前发现有个孩子模样的人走进教室, 觉得这孩子太年轻了, 不适合听这个课. 结果这孩子走到讲台上讲起课来! 原来, 这个孩子模样的人就是任课老师索利斯, 时年 28 岁, 长相很年轻. 他 1961 年已经从伯明翰来到剑桥大学做讲师.

具体来说, 这个拓扑的途径是通过涡旋. 涡旋是指某个区域中绕着一个轴旋转的液体 (或者某个物理特性随着绕轴的角度而变), 这是一个拓扑结构, 因为不管怎么旋转, 转 1 圈总归是 360 度, 与没有涡旋的情况截然不同. 表征一个涡旋的量是它的缠绕数, 即绕轴的圈数. 索利斯和科斯特里兹发现, 在二维系统中, 涡旋有两种形态, 一个是旋转方向相反的涡旋两两束缚在一起, 另一个是它们没有互相束缚. 这两种形态有能量与混乱度的竞争, 导致在一个非零温度发生相变. 低于这个温度时, 正反涡旋形成束缚对. 高于这个温度时, 涡旋可以自由运动. 这个相变被称作拓扑相变或者 KT 相变 —— 科斯特里兹-索利斯相变 (Kosterlitz–Thouless transition). 索利斯和科斯特最初讨论的超流薄膜的相变, 但是类似的 KT 相变也存在与其他系统, 如超导薄膜、平面磁系统等等.

苏联物理学家别列津斯基 (Vadim L'vovich Berezinskii, 1935—1980, 终年 44 岁) 也完全独立地对理解这一概念作出了重大贡献, 因此这一相变也被称为是别列津斯基–科斯特利茨–索利斯相变 (Berezinskii-Kosterlitz-Thouless transition, BKT transition). 三位学者的头像如图 22.6 所示.

如图 22.7 所示, 科斯特利茨和索利斯使用拓扑描述了一个超低温下的、薄薄的一层物质上发生的拓扑相变. 在这种极端的寒冷下, 涡旋对形成, 然后在达到相变温度时, 突然分开. 这是在凝聚态物理 20 世纪最重要的发现之一.

索利斯在康奈尔大学的导师是贝特 (Hans Bethe, 1906—2005). 贝特的一位好朋友是同为犹太人的派尔斯 (Rudolf Peierls, 1907—1995), 他后来成为索利斯的博士后导师. 贝特和派尔斯都是索末菲 (Arnold Sommerfeld, 1868—1951) 的学生. 索末菲是德国近代理论物理鼻祖之一, 另一位是玻恩 (Max Born, 1882—1970). 索末菲是历史上获得诺贝尔物理学奖提名次数最多的人, 一生获得 84 次提名, 但终究没有得奖. 不过, 索莫菲培养了很多人才, 其中有几位诺贝尔奖得主, 包括量子力学的创始人海森堡 (Werner Heisenberg, 1901—1976) 和泡利 (Wolfgang Pauli,

1900—1958)，以及德拜 (Peter Debye, 1884—1966) 和贝特.

图 22.6　左起：别列津斯基、科斯特利茨、索利斯

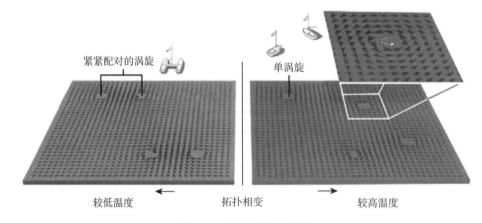

图 22.7　拓扑相变示意图

索利斯的博士后导师派尔斯还有很多关于索利斯的有趣回忆："他很害羞和笨拙，动手能力不是很强. 一个典型例子是他搬来伯明翰时的经历. 他和妻子从他岳父母处借了些旧家具，放在一个旧拖车上，用小汽车拖. 在路上拖车翻了，他好不容易采取紧急措施收好家具. 还有，他的妻子生孩子前，他将汽车停在房子前，准备随时送她去医院. 但是这个时刻到来时，车子却发动不了，他只好叫醒邻居开车." 索利斯伯明翰时期的老同事斯汀康比 (Robin Stinchcombe) 认为，这些描写都是索利斯生活中的典型事例.

§23. 微分同胚，坐标图册，切空间，余切空间

23.1　微分同胚

微分同胚 (diffeomorphism) 是指微分流形之间存在可逆映射，映射及其逆映射

均为光滑的, 即无穷可微.

在拓扑学中, 同胚 (homeomorphism) 是两个拓扑空间之间的双连续函数 (bicontinuous function). 同胚是拓扑空间范畴中的同构; 也就是说, 它们是保持给定空间的所有拓扑性质的映射. 如果两个空间之间存在同胚, 那么这两个空间就称为同胚的, 从拓扑学的观点来看, 两个空间是相同的.

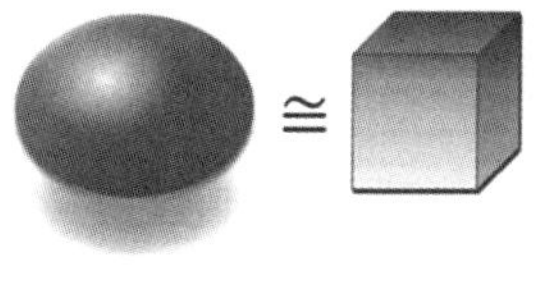

图 23.1 球 (椭球) 和正方形是同胚的

同胚一言以蔽之就是孪生双胞胎! 虽然长得不完全一样, 但是遗传基因一样.

大致地说, 拓扑空间是一个几何物体, 同胚就是把物体连续延展和弯曲, 使其成为一个新的物体. 因此, 如图 23.1 所示, 正方形和球 (椭球) 是同胚的 (homeomorphic), 但球面和环面就不是. 图 23.2 给出是同胚的面包圈和茶杯.

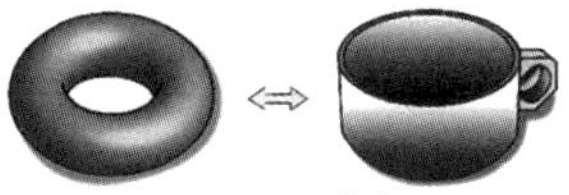

图 23.3 是另外两则典型的同胚的例子.

图 23.3 同胚的典型例子

图 23.2 面包圈和茶杯同胚

从图 23.4 可以领会同胚和微分同胚之间的关系.

通过同胚将 n 维空间的曲线和欧氏空间的直线联系在一起, 将 n 维空间的曲面和欧氏空间的平面联系在一起, 转化为直线和平面上的研究.

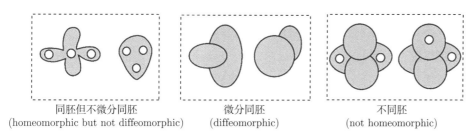

图 23.4 同胚和微分同胚之间的关系

流形上每一点的一个邻域都可以同胚于欧氏空间中的 n 维图形, 则流形的维数为 n. 如图 23.5 所示:

(1) 圆是一个一维流形, 与直线同胚;

(2) 二维平面所有带有交点的曲线不是一维流形;

(3) 球面是一个二维流形, 与平面同胚;

(4) 三维空间中的锥面 $\mathcal{M} = \{(x, y, z): x^2 + y^2 = z^2\}$ 不是流形, 因为在原点附近它不同胚于一个平面. 但半锥面 $\mathcal{M}^+ = \{(x, y, z): x^2 + y^2 = z^2, z \geqslant 0\}$ 则是

一个二维流形.

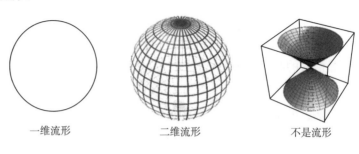

一维流形　　　　　二维流形　　　　　不是流形

图 23.5　流形的典型例子

23.2　坐标图册

映射、坐标图和坐标图册是微分流形中最基本的概念, 坐标图和坐标图册也分别称为坐标卡和坐标卡集. 流形的性质不应与坐标图的选取有关, 所以两个坐标图在相交的部分的复合映射应该光滑.

1943 年, 法国数学家夏尔·埃雷斯曼 (Charles Ehresmann, 1905—1979) 率先以图册 (atlas) 来定义流形.

理性力学为什么需要引入流形? 这自然是读者十分关心的问题. 因为我们需要用流形描述时空, 用定义在流形上的张量描述力学量. 经典力学其实也是这样, 牛顿力学用三维的平直的流形描述空间, 用实数和三维矢量描述所有力学量. 分析力学也只不过是把维度扩张到了自由度的数目的位形空间或相空间.

所以理性力学需要流形. 虽然在 §21.3 已经给出过流形的定义, 但看后依然觉得不够通俗. 到底什么是流形? 为了方便理解, 流形是一个在无限小的区域内看非常像欧氏空间的空间. 首先欧氏空间自己就是流形的一种, 它当然处处看上去都是自己. n 维球 (sphere) 也是流形, 它可以定义成在 $n+1$ 维欧氏空间里到某点距离为定值的点的集合, 比如在二维平面上的一个圆周, 如果看到无限小, 它很像一根直线; 或者三维空间里的一个球壳, 如果看到无限小, 它很像一个平面, 所以人们以为地球是平的.

事实上, 我们可以通过如下顺序来定义流形: 定义开集 (open set) → 定义坐标系 (chart, coordinate system) → 定义地图册 (atlas) → 定义流形 (manifold).

定义开集需要先定义 n 维开球 (open ball). 一个 $\mathbb{R}^n$ 上的开球是满足 $|x-y| < r$ 点 x 的集合, $|x-y| = \sqrt{\sum_i (x_i - y_i)^2}$ 表示两点间的距离. 一个 $\mathbb{R}^n$ 上的开集是任意一些开球的并集, 甚至可以是无限多个开球的并集, 这样做就是要去掉边界的

集合.

坐标系 (chart) 是一个映射 $\varphi: \mathcal{U} \to \mathbb{R}^n$，$\mathcal{U}$ 是 $\mathcal{M}$ 的子集，这个映射要使得像 $\varphi(\mathcal{U})$ 是开集. 这个名字还是挺形象的，就是把子集里的点都标上坐标.

如果 $\mathcal{U}$ 也是 n 维实数集，那 φ 就是普通的函数，我们可以对它求导，p 阶可导的函数叫做 C^p 的，无穷阶可导的话它就是光滑的 (smooth).

地图册这个名字也是很形象的，就是一堆坐标系联在一起，一般看的中国地图都是一个大图，右下角再加上南海诸岛，就是含有两个坐标系的地图册，集合 $\{\mathcal{U}_\alpha, \phi_\alpha\}$ 还要满足：(1) $\mathcal{U}_\alpha$ 的并集等于 $\mathcal{M}$——两张地图拼起来要覆盖整个中国；(2) 两个不同的坐标系重合的部分要平滑地连接：映射 $\phi_\alpha \circ \phi_\beta^{-1}$ 要把 $\phi_\beta(\mathcal{U}_\beta)$ 上的点映射到 $\phi_\alpha(\mathcal{U}_\alpha)$ 上，而且对 C^p 的地图册这个映射是 C^p 的.

最后，n 维 C^p 的流形就是定义了最大 C^p 地图册的点集 $\mathcal{M}$.

流形可以局部映射到欧氏空间的拓扑图形，这些局部映射可以覆盖整个流形. 流形的性质可以转换到欧氏空间进行研究，而这个转换还需要借助切空间、张量和黎曼度量.

各物理量在欧氏空间的分析方法与经典的张量分析一致，微分流形的主要内容在于如何将流形和欧氏空间联系起来，如图 23.6 所示. 流形上的张量转换到欧氏空间的操作叫做推前 (push-forward) 在欧氏空间进行分析之后可以通过拉回 (pull-back) 操作转换到流形上. 虚线表示流形在数学上是开集.

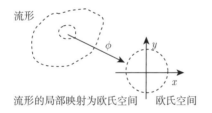

图 23.6 流形和欧氏空间的推前和拉回操作

不同的流形有不同的坐标图册，一个流形也有很多坐标图册. 一般分析过程中并不考虑如何获得映射，即默认映射已知. 圆在一维空间的映射时，分别挖去两个点，并向直线的投影. 图 23.7 为一维流形圆映射到一维欧氏空间，映射分为两个坐标图

图 23.7 中的红点表示映射时挖空的点，这个点发出的射线把圆上的点映射到直线上，如 $X_1 \to x_1$. 在两个映射相交的部分

$$x = \phi_1 \circ \phi_2^{-1}(y), \quad y = \phi_2 \circ \phi_1^{-1}(x) \tag{23-1}$$

$f \circ g$ 表示 f 和 g 的复合映射. g 的值域可以作为 f 的定义域. 微分流形要求这两个复合映射都是无限次光滑映射.

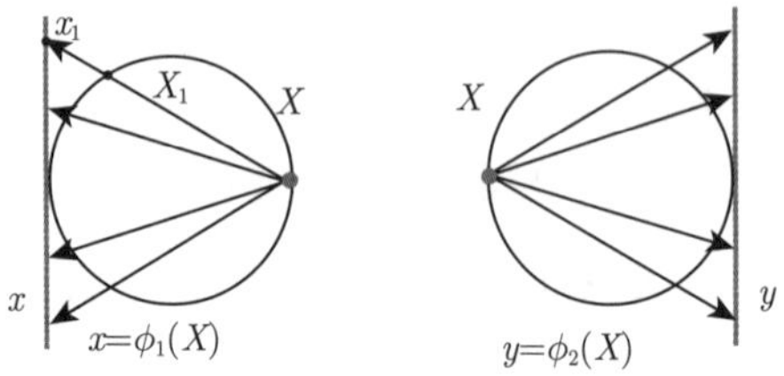

图 23.7　一维流形圆映射到一维欧氏空间

23.3　切空间, 余切空间

流形上矢量和张量都通过切空间定义, 某一点的切矢量通过欧氏空间刻画, 每一点处都存在切空间, 切空间由这一点所有切矢量组成.

流形一点处的切矢量必须通过经过该点的曲线、流形上的光滑函数 (物理量) 来定义.

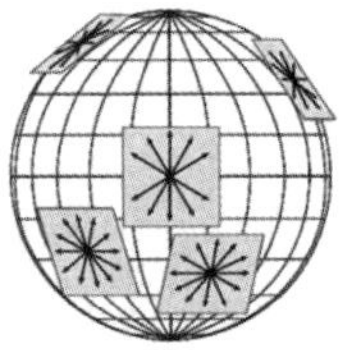

图 23.8　作为流形的地球

例如, 地球是一个流形 (如图 23.8 所示), 在某时刻我们把地球上的每一点处的风向记下来, 画成一张全球风向图. 一点处的风向就是切向量, 整张风向图就是切向量场. 一个著名的定理表明, 地球上任何时刻的风向图中, 必有一处的风速为零, 就是没有风.

切空间的维度与流形的维度一致, 切丛是流形上的点和对应的切空间的集合, 其维度是流形的维度加上切空间的维度, 等于两倍的流形维度.

流形的切空间和欧氏空间的切空间都可以看做流形, 也可以定义两者的映射.

流形上一点处的切矢量需要通过该点的曲线和流形上的物理量确定, 切矢量是一个任意物理量的线性函数, 在欧氏空间的描述是一个方向导数.

切空间的维度与流形的维度一致, 切丛是流形上的点和对应的切空间的集合, 其维度是流形的维度加上切空间的维度, 等于两倍的流形维度.

如图 23.9 所示, 欧氏坐标系中曲线方程为 $c = c(t), t_P = 0$, f 是流形上的可微函数

$$\begin{aligned}\left.\frac{\mathbf{d}f[c(t)]}{\mathbf{d}t}\right|_{t=0} &= \left.\frac{\partial}{\partial x^i}\left(f \circ \varphi^{-1}\right)\right|_{\varphi[c(0)]} \left.\frac{\mathbf{d}x^i[c(t)]}{\mathbf{d}t}\right|_{t=0} \\ &= \xi^i \left(\frac{\partial f}{\partial x^i}\right)_P \end{aligned} \tag{23-2}$$

切矢量为
$$\boldsymbol{T}_P = \xi^i \left(\frac{\partial}{\partial x^i}\right)_P \quad (23\text{-}3)$$
切矢量的基矢量为
$$\left(\frac{\partial}{\partial x^i}\right)_P \quad (23\text{-}4)$$

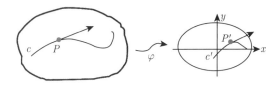

图 23.9 流形上一点沿曲线的切矢量

余切空间是切空间的对偶空间 (dual space), 定义对偶空间可以简化计算 (比如求矢量的坐标) 和深入理解问题的本质 (比如力矢量组成的空间和位移组成的空间是对偶空间).

相当于切空间的基: $\left(\frac{\partial}{\partial x^i}\right)_P$, 对矢量进行分解: $\boldsymbol{A} = A^i \frac{\partial}{\partial x^i} = A_i \mathrm{d}x^i$, 则余切空间的基为
$$\left(\mathrm{d}x^i\right)_P \quad (23\text{-}5)$$
因此, 余切空间基矢和切空间基矢可以类比协变基矢和逆变基矢.

任何函数在 P 点的微分定义为
$$\mathrm{d}f_P(\boldsymbol{T}_P) = \langle \boldsymbol{T}_P, \mathrm{d}f_P \rangle = \boldsymbol{T}_P f \quad (23\text{-}6)$$
则称 $\mathrm{d}f_P$ 为 P 点的一个余切矢量. 由于取 $T_P = \left(\frac{\partial}{\partial x^i}\right)_P$, $f = x^j$ 时, 有
$$\left\langle \left(\frac{\partial}{\partial x^i}\right)_P, \mathrm{d}x^j_p \right\rangle = \left(\frac{\partial}{\partial x^i} x^j\right)_P = \delta_i^j \quad (23\text{-}7)$$
所以, 切空间基矢和余切空间基矢可以分别定义为 $\left(\frac{\partial}{\partial x^i}\right)_P$ 和 $\left(\mathrm{d}x^i\right)_P$.

§24. 微分形式, 外微分运算

24.1 微分形式的引入

对于 n 维欧氏空间 $\mathcal{E}^n$, 空间内一点的坐标为 $(x_1, x_2, \cdots, x_n)$, 可以生成一个以 $\mathrm{d}x_1, \mathrm{d}x_2, \cdots, \mathrm{d}x_n$ 为基向量的 n 维向量空间 $\mathcal{V}$. 以外微分 (exterior derivative),

亦称楔积 (wedge product)，来定义 k 次微分形式为基元素：

$$\mathrm{d}x_{i_1} \wedge \mathrm{d}x_{i_2} \wedge \cdots \wedge \mathrm{d}x_{i_k} \tag{24-1}$$

式中，$1 \leqslant i_1 < i_2 < \cdots < i_k \leqslant n$，基元素前的系数为坐标 $(x_1, x_2, \cdots, x_n)$ 的函数.

对于非零次形微分形式，在线积分、面积分和体积分中经常见到.

1) 0 次形微分形式 (0-form)：

n 维空间上，0 次形微分形式即为一个函数：$f(x_1, x_2, \cdots, x_n)$；

2) 1 次形的微分形式 (1-form)：

如三维空间中的线积分，$\int_l A\mathrm{d}x + B\mathrm{d}y + C\mathrm{d}z$，积分号后即为 1 次形的微分形式：

$$A\mathrm{d}x + B\mathrm{d}y + C\mathrm{d}z \tag{24-2}$$

3) 2 次形的微分形式 (2-form)：

面积分 $\iint_{\partial\Omega} P\mathrm{d}x\mathrm{d}y + Q\mathrm{d}y\mathrm{d}z + R\mathrm{d}z\mathrm{d}x$ 积分号后即为 2 次形的微分形式：

$$P\mathrm{d}x \wedge \mathrm{d}y + Q\mathrm{d}y \wedge \mathrm{d}z + R\mathrm{d}z \wedge \mathrm{d}x \tag{24-3}$$

4) 3 次形的微分形式 (3-form)：

三维空间中的体积分 $\iiint_\Omega H\mathrm{d}x\mathrm{d}y\mathrm{d}z$ 积分号后即为 3 次形的微分形式：

$$H\mathrm{d}x \wedge \mathrm{d}y \wedge \mathrm{d}z \tag{24-4}$$

5) k 次形的微分形式 (k-form)：

由于 1 次形的微分形式可记为

$$w = w_1\mathrm{d}x_1 + w_2\mathrm{d}x_2 + \cdots + w_n\mathrm{d}x_n \tag{24-5}$$

式中，$w_i\,(i = 1, 2, \cdots, n)$ 为 $(x_1, x_2, \cdots, x_n)$ 的函数. 同理，k 次形的微分形式记为

$$\alpha = \sum_{1 \leqslant i_1 < i_2 < \cdots < i_k \leqslant n} \alpha_{i_1 i_2 \cdots i_k} \mathrm{d}x_{i_1} \wedge \mathrm{d}x_{i_2} \wedge \cdots \wedge \mathrm{d}x_{i_k} \tag{24-6}$$

式中，$\alpha_{i_1 i_2 \cdots i_k}$ 为 $(x_1, x_2, \cdots, x_n)$ 的函数.

24.2 外微分运算，庞加莱引理

外微分 (楔积 $\wedge$) 运算满足如下基本性质：

(1) 结合律：$(\mathrm{d}x_1 \wedge \mathrm{d}x_2) \wedge \mathrm{d}x_3 = \mathrm{d}x_1 \wedge (\mathrm{d}x_2 \wedge \mathrm{d}x_3)$.

(2) 分配律: $(w_1 dx_1 + w_2 dx_2) \wedge w_3 dx_3 = w_1 w_3 dx_1 \wedge dx_3 + w_2 w_3 dx_2 \wedge dx_3$.

(3) 反对称 (反交换) 律: $dx_1 \wedge dx_2 = -dx_2 \wedge dx_1$.

除上述性质外, 还有以下推论:

推论一、$dx_1 \wedge dx_1 = 0$, 即对于 k 次微分形式: $dx_{i_1} \wedge dx_{i_2} \wedge \cdots \wedge dx_{i_k}$, 当 $i_p = i_q$ 时, 其中 $1 \leqslant p < q \leqslant k$, 基元素 $dx_{i_1} \wedge dx_{i_2} \wedge \cdots \wedge dx_{i_k} = 0$.

推论二、由推论一知, 当 $i_k > n$, n 维空间上的 k 次形等于零. 原因是上述条件下, 有 $i_k - n$ 个重复的基元素.

推论三、n 维空间上的 k 次形 $dx_{i_1} \wedge dx_{i_2} \wedge \cdots \wedge dx_{i_k}$, 任意 dx_{i_p} 和 dx_{i_q} 交换位置, 其结果反号, 亦即

$$dx_{i_1} \wedge \cdots \wedge dx_{i_p} \wedge \cdots \wedge dx_{i_q} \wedge \cdots \wedge dx_{i_k} = -dx_{i_1} \wedge \cdots \wedge dx_{i_q} \wedge \cdots \wedge dx_{i_p} \wedge \cdots \wedge dx_{i_k}$$

对于基矢量的外微分运算满足性质: $d(dx_i) = d^2 x_i = 0$, 在此基础上即可对微分形式进行外微分运算.

首先令 f 与 g 分别为 n 维空间上 0 次微分形式 $f(x_1, x_2, \cdots, x_n)$ 与 $g(x_1, x_2, \cdots, x_n)$; α_1 与 β_1 分别为 k 次微分形式的基元素 $\alpha_1 = dx_{i_1} \wedge \cdots \wedge dx_{i_k}$ 与 p 次微分形式的基元素 $\beta_1 = dx_{j_1} \wedge \cdots \wedge dx_{j_p}$ $(1 \leqslant k \leqslant n, 1 \leqslant p \leqslant n)$; 令

$$\alpha = f \alpha_1 = f(x_1, x_2, \cdots, x_n) dx_{i_1} \wedge \cdots \wedge dx_{i_k} \tag{24-7}$$

为 k 次微分形式;

$$\beta = g \beta_1 = g(x_1, x_2, \cdots, x_n) dx_{j_1} \wedge \cdots \wedge dx_{j_p} \tag{24-8}$$

为 p 次微分形式. 则对微分形式进行外微分运算:

1) 0 次形, $df(x_1, x_2, \cdots, x_n)$

对于 n 维空间 0 次形 $f(x_1, x_2, \cdots, x_n)$ 的外微分运算就是通常的全微分:

$$df(x_1, x_2, \cdots, x_n) = \sum_{i=1}^{n} \frac{\partial f}{\partial x_i} dx_i \tag{24-9}$$

外微分运算后, 由 0 次形升为 1 次形.

2) k 次形, $d\alpha = d(f dx_{i_1} \wedge \cdots \wedge dx_{i_k}) = \sum_{i=1}^{n} \frac{\partial f}{\partial x_i} dx_i \wedge dx_{i_1} \wedge \cdots \wedge dx_{i_k}$

利用楔积的反对称性质, 则对 α 的外微分运算有

$$d\alpha = d(f\alpha_1) = d(f)\alpha_1 + f d(dx_{i_1} \wedge \cdots \wedge dx_{i_{p-1}} \wedge dx_{i_p} \wedge dx_{i_{p+1}} \wedge \cdots \wedge dx_{i_k})$$

$$
\begin{aligned}
&= \sum_{i=1}^{n} \frac{\partial f}{\partial x_i} \mathrm{d}x_i \wedge \alpha_1 + f \sum_{p=1}^{k} \mathrm{d}\left(\mathrm{d}x_{i_p}\right) \wedge \mathrm{d}x_{i_1} \wedge \cdots \wedge \mathrm{d}x_{i_{p-1}} \wedge \mathrm{d}x_{i_{p+1}} \wedge \cdots \wedge \mathrm{d}x_{i_k} \\
&= \sum_{i=1}^{n} \frac{\partial f}{\partial x_i} \mathrm{d}x_i \wedge \alpha_1 + f \sum_{p=1}^{k} (-1)^{p-1} \wedge \mathrm{d}x_{i_1} \wedge \cdots \wedge \mathrm{d}x_{i_{p-1}} \wedge \underbrace{\mathrm{d}^2 x_{i_p}}_{0} \wedge \mathrm{d}x_{i_{p+1}} \wedge \cdots \wedge \mathrm{d}x_{i_k} \\
&= \sum_{i=1}^{n} \frac{\partial f}{\partial x_i} \mathrm{d}x_i \wedge \alpha_1
\end{aligned}
\tag{24-10}
$$

外微分运算后, 由 k 次形升阶为 $k+1$ 次形.

3) 庞加莱 (Poincaré) 引理, $\mathrm{d}(\mathrm{d}\alpha) = 0$

利用运算 2) 的结论并进一步对结果进行外微分运算, 有

$$
\begin{aligned}
\mathrm{d}(\mathrm{d}\alpha) &= \sum_{j=1}^{n} \sum_{i=1}^{n} \frac{\partial f}{\partial x_i \partial x_j} \mathrm{d}x_j \wedge \mathrm{d}x_i \wedge \alpha_1 \\
&= \frac{1}{2} \sum_{i,j=1}^{n} \left(\frac{\partial f}{\partial x_i \partial x_j} - \frac{\partial f}{\partial x_j \partial x_i} \right) \mathrm{d}x_j \wedge \mathrm{d}x_i \wedge \alpha_1 \\
&= 0
\end{aligned}
\tag{24-11}
$$

此结论即为庞加莱引理, 即对任意 k $(k > 0)$ 次微分形式 α 都满足 $\mathrm{d}(\mathrm{d}\alpha) = 0$.

4) $\mathrm{d}(\alpha \wedge \beta) = \mathrm{d}(\alpha) \wedge \beta + (-1)^k \alpha \wedge \mathrm{d}(\beta)$

对 α 与 β 的楔积 $\alpha \wedge \beta$ 的外微分运算有

$$
\begin{aligned}
\mathrm{d}(\alpha \wedge \beta) &= \mathrm{d}\left(f(x_1, x_2, \cdots, x_n) \mathrm{d}x_{i_1} \wedge \cdots \wedge \mathrm{d}x_{i_k} \wedge g(x_1, x_2, \cdots, x_n) \mathrm{d}x_{j_1} \wedge \cdots \wedge \mathrm{d}x_{j_p}\right) \\
&= \mathrm{d}(fg\alpha_1 \wedge \beta_1) = \frac{\partial(fg)}{\partial x_i} \mathrm{d}x_i \wedge \alpha_1 \wedge \beta_1 = \left(\frac{\partial f}{\partial x_i}g + \frac{\partial g}{\partial x_i}f\right) \mathrm{d}x_i \wedge \alpha_1 \wedge \beta_1 \\
&= g\frac{\partial f}{\partial x_i} \mathrm{d}x_i \wedge \alpha_1 \wedge \beta_1 + f\frac{\partial g}{\partial x_i} \mathrm{d}x_i \wedge \alpha_1 \wedge \beta_1 \\
&= \left(\frac{\partial f}{\partial x_i} \mathrm{d}x_i \wedge \alpha_1\right) \wedge g\beta_1 + (-1)^k f\alpha_1 \wedge \left(\frac{\partial g}{\partial x_i} \mathrm{d}x_i \wedge \beta_1\right) \\
&= \mathrm{d}(\alpha) \wedge \beta + (-1)^k \alpha \wedge \mathrm{d}(\beta)
\end{aligned}
\tag{24-12}
$$

下面给出一些具体的例子.

(1) 0 次形函数 f, 其外微分为 $\mathrm{d}f = \sum_i \frac{\partial f}{\partial x_i} \mathrm{d}x_i$

(2) 三维空间的 1 次形 $\alpha = a_1 \mathrm{d}x_1 + a_2 \mathrm{d}x_2 + a_3 \mathrm{d}x_3$

则根据 $\mathrm{d}x_i \wedge \mathrm{d}x_i = 0$ 和 $\mathrm{d}x_i \wedge \mathrm{d}x_j = -\mathrm{d}x_j \wedge \mathrm{d}x_i$, 有

$$
\mathrm{d}\alpha = \frac{\partial a_1}{\partial x_i} \mathrm{d}x_i \wedge \mathrm{d}x_1 + \frac{\partial a_2}{\partial x_i} \mathrm{d}x_i \wedge \mathrm{d}x_2 + \frac{\partial a_3}{\partial x_i} \mathrm{d}x_i \wedge \mathrm{d}x_3
$$

$$\begin{aligned}
&= \underbrace{\frac{\partial a_1}{\partial x_1}\mathrm{d}x_1 \wedge \mathrm{d}x_1}_{0} + \frac{\partial a_1}{\partial x_2}\mathrm{d}x_2 \wedge \mathrm{d}x_1 + \frac{\partial a_1}{\partial x_3}\mathrm{d}x_3 \wedge \mathrm{d}x_1 \\
&\quad + \frac{\partial a_2}{\partial x_1}\mathrm{d}x_1 \wedge \mathrm{d}x_2 + \underbrace{\frac{\partial a_2}{\partial x_2}\mathrm{d}x_2 \wedge \mathrm{d}x_2}_{0} + \frac{\partial a_2}{\partial x_3}\mathrm{d}x_3 \wedge \mathrm{d}x_2 \\
&\quad + \frac{\partial a_3}{\partial x_1}\mathrm{d}x_1 \wedge \mathrm{d}x_3 + \frac{\partial a_3}{\partial x_2}\mathrm{d}x_2 \wedge \mathrm{d}x_3 + \underbrace{\frac{\partial a_3}{\partial x_3}\mathrm{d}x_3 \wedge \mathrm{d}x_3}_{0} \\
&= \left(\frac{\partial a_2}{\partial x_1} - \frac{\partial a_1}{\partial x_2}\right)\mathrm{d}x_1 \wedge \mathrm{d}x_2 + \left(\frac{\partial a_3}{\partial x_2} - \frac{\partial a_2}{\partial x_3}\right)\mathrm{d}x_2 \wedge \mathrm{d}x_3 \\
&\quad + \left(\frac{\partial a_1}{\partial x_3} - \frac{\partial a_3}{\partial x_1}\right)\mathrm{d}x_3 \wedge \mathrm{d}x_1
\end{aligned} \tag{24-13}$$

上式说明，三维空间 1 次形的外微分就是旋度运算．

(3) 三维空间中的 2 次形 $\beta = a_1 \mathrm{d}x_3 \wedge \mathrm{d}x_1 + a_2 \mathrm{d}x_1 \wedge \mathrm{d}x_2 + a_3 \mathrm{d}x_2 \wedge \mathrm{d}x_3$

令 $\alpha_1 = \mathrm{d}x_3 \wedge \mathrm{d}x_1$, $\alpha_2 = \mathrm{d}x_1 \wedge \mathrm{d}x_2$, $\alpha_3 = \mathrm{d}x_2 \wedge \mathrm{d}x_3$. 则根据外微分遵从的规则 $\mathrm{d}(f\alpha_1) = \sum \frac{\partial f}{\partial x_i}\mathrm{d}x_i \wedge \alpha_1$，有

$$\begin{aligned}
\mathrm{d}\beta &= \frac{\partial a_1}{\partial x_i}\mathrm{d}x_i \wedge \alpha_1 + \frac{\partial a_2}{\partial x_i}\mathrm{d}x_i \wedge \alpha_2 + \frac{\partial a_3}{\partial x_i}\mathrm{d}x_i \wedge \alpha_3 \\
&= \frac{\partial a_1}{\partial x_1}\mathrm{d}x_1 \wedge \mathrm{d}x_2 \wedge \mathrm{d}x_3 + \underbrace{\frac{\partial a_1}{\partial x_2}\mathrm{d}x_2 \wedge \mathrm{d}x_2 \wedge \mathrm{d}x_3}_{0} + \underbrace{\frac{\partial a_1}{\partial x_3}\mathrm{d}x_3 \wedge \mathrm{d}x_2 \wedge \mathrm{d}x_3}_{0} \\
&\quad + \underbrace{\frac{\partial a_2}{\partial x_1}\mathrm{d}x_1 \wedge \mathrm{d}x_3 \wedge \mathrm{d}x_1}_{0} + \frac{\partial a_2}{\partial x_2}\mathrm{d}x_2 \wedge \mathrm{d}x_3 \wedge \mathrm{d}x_1 + \underbrace{\frac{\partial a_2}{\partial x_3}\mathrm{d}x_3 \wedge \mathrm{d}x_3 \wedge \mathrm{d}x_1}_{0} \\
&\quad + \underbrace{\frac{\partial a_3}{\partial x_1}\mathrm{d}x_1 \wedge \mathrm{d}x_1 \wedge \mathrm{d}x_2}_{0} + \underbrace{\frac{\partial a_3}{\partial x_2}\mathrm{d}x_2 \wedge \mathrm{d}x_1 \wedge \mathrm{d}x_2}_{0} + \frac{\partial a_3}{\partial x_3}\mathrm{d}x_3 \wedge \mathrm{d}x_1 \wedge \mathrm{d}x_2 \\
&= \left(\frac{\partial a_1}{\partial x_1} + \frac{\partial a_2}{\partial x_2} + \frac{\partial a_3}{\partial x_3}\right)\mathrm{d}x_1 \wedge \mathrm{d}x_2 \wedge \mathrm{d}x_3
\end{aligned} \tag{24-14}$$

上式说明，三维空间 2 次形的外微分就是散度运算．

(4) 证明在 n 维空间中，任意高于 n 次形的微分形式都为 0．

对 n 维空间有 n 次形，$\omega = f(x_1, \cdots, x_n)\mathrm{d}x_1 \wedge \mathrm{d}x_2 \wedge \cdots \wedge \mathrm{d}x_n$，有

$$\mathrm{d}\omega = \sum_i \frac{\partial f(x_1, \cdots, x_n)}{\partial x_i}\mathrm{d}x_i \wedge \mathrm{d}x_1 \wedge \cdots \wedge \mathrm{d}x_i \wedge \cdots \wedge \mathrm{d}x_n = 0 \tag{24-15}$$

(5) 对 n 维空间有 n 次形，$\omega = f(x_1, \cdots, x_n)\mathrm{d}x_1 \wedge \mathrm{d}x_2 \wedge \cdots \wedge \mathrm{d}x_n$．

采取变数替换: $x_i = x_i(y_1, \cdots, y_n)$, 则由 $\mathrm{d}x_i \wedge \mathrm{d}x_i = 0$ 和 $\mathrm{d}x_i \wedge \mathrm{d}x_j = -\mathrm{d}x_j \wedge \mathrm{d}x_i$, 有如下变换:

$$\omega = f(x_1(y_1,\cdots,y_n),\cdots,x_n(y_1,\cdots,y_n))\left(\sum_{i=1}^n \frac{\partial x_1}{\partial y_i}\mathrm{d}y_i\right)$$

$$\wedge \left(\sum_{i=1}^n \frac{\partial x_2}{\partial y_i}\mathrm{d}y_i\right) \wedge \cdots \wedge \left(\sum_{i=1}^n \frac{\partial x_n}{\partial y_i}\mathrm{d}y_i\right)$$

$$= \varphi(y_1,\cdots,y_n) \frac{\mathrm{D}(x_1,\cdots,x_n)}{\mathrm{D}(y_1,\cdots,y_n)} \mathrm{d}y_1 \wedge \cdots \wedge \mathrm{d}y_n \tag{24-16}$$

式中,

$$J = \frac{\mathrm{D}(x_1,\cdots,x_n)}{\mathrm{D}(y_1,\cdots,y_n)} \tag{24-17}$$

为雅可比行列式 (Jacobian).

24.3 斯托克斯定理

讨论 n 维空间的 k 次微分形式 $v_k = \mathrm{d}x_1 \wedge \mathrm{d}x_2 \wedge \cdots \wedge \mathrm{d}x_k$ 的积分, 按照外微分与积分的关系, 可以将其理解为一个 k 维小微元的体积, 并且具有方向. 该微元的边界由 $k-1$ 维的边界微元 $\partial_i v_k = \mathrm{d}x_1 \wedge \cdots \wedge \mathrm{d}x_{i-1} \wedge \mathrm{d}x_{i+1} \wedge \cdots \wedge \mathrm{d}x_k = \mathrm{d}x_1 \wedge \cdots \wedge \widehat{\mathrm{d}x_i} \wedge \cdots \wedge \mathrm{d}x_k$ 组成, 其中戴帽子的符号 $\widehat{\mathrm{d}x_i}$ 表示没有这个元素. 由于具有方向性, 上述的面元总是成对出现, 并符号相反.

对于一个微分形式 $\omega = \sum_i \omega_i \mathrm{d}x_1 \wedge \cdots \wedge \widehat{\mathrm{d}x_i} \wedge \cdots \wedge \mathrm{d}x_k$, 在上述的微元边界 ∂v_k 上求和有

$$\sum_{\partial v_k} \omega = \sum_i [\omega_i(x_1,\cdots,x_i+\mathrm{d}x_i,\cdots,\mathrm{d}x_k)$$

$$-\omega_i(x_1,\cdots,x_i,\cdots,\mathrm{d}x_k)]\mathrm{d}x_1 \wedge \cdots \wedge \widehat{\mathrm{d}x_i} \wedge \cdots \wedge \mathrm{d}x_k$$

$$= \sum_i \frac{\partial \omega_i}{\partial x_i}\mathrm{d}x_i \wedge \mathrm{d}x_1 \wedge \cdots \wedge \widehat{\mathrm{d}x_i} \wedge \cdots \wedge \mathrm{d}x_k$$

$$= \sum_i (-1)^{i-1} \frac{\partial \omega_i}{\partial x_i}\mathrm{d}x_1 \wedge \cdots \wedge \mathrm{d}x_i \wedge \cdots \wedge \mathrm{d}x_k \tag{24-18}$$

对微分形在整个 $\mathcal{D}$ 上积分, 将 $\mathcal{D}$ 按照坐标平面剖分为小体积元, 用上式最后一项对这些小体积元求和. 同时这些小体积为 $\mathcal{D}$ 内的小体积, 相邻两个小体积共边界面, 边界面对两个小体积一正一负, 相互抵消, 所以对边界面的求和只剩下了边界面, 于是有

$$\int_{\partial \mathcal{D}} \omega = \int_{\mathcal{D}} \sum_i (-1)^{i-1} \frac{\partial w_i}{\partial x_i}\mathrm{d}x_1 \wedge \cdots \wedge \mathrm{d}x_i \wedge \cdots \wedge \mathrm{d}x_k = \int_{\mathcal{D}} \mathrm{d}\omega \tag{24-19}$$

第 4 章 旋转群，拓扑，微分流形上的张量分析

上式即为斯托克斯定理 (Stokes theorem)：$\int_{\partial \mathcal{D}} \omega = \int_{\mathcal{D}} \mathrm{d}\omega$.

下面给出斯托克斯定理的几个特例.

(1) 联系线积分和面积分的格林定理 (Green's theorem). 令

$$\omega = P\mathrm{d}x + Q\mathrm{d}y \tag{24-20}$$

在平面单连通区域 $\mathcal{D}$ 有

$$\oint_{\partial \mathcal{D}} \omega = \oint_{\partial \mathcal{D}} P\mathrm{d}x + Q\mathrm{d}y = \iint_{\mathcal{D}} \mathrm{d}\omega = \iint_{\mathcal{D}} \left(\frac{\partial Q}{\partial x} - \frac{\partial P}{\partial y} \right) \mathrm{d}x\mathrm{d}y \tag{24-21}$$

(2) 开尔文–斯托克斯定理 (Kelvin–Stokes theorem) (旋度定理). 令

$$\omega = P\mathrm{d}x + Q\mathrm{d}y + R\mathrm{d}z \tag{24-22}$$

在空间单连通区域曲面段 $\mathcal{D}$ 有

$$\oint_{\partial \mathcal{D}} \omega = \oint_{\partial \mathcal{D}} P\mathrm{d}x + Q\mathrm{d}y + R\mathrm{d}z = \iint_{\mathcal{D}} \mathrm{d}\omega$$
$$= \iint_{\mathcal{D}} \left(\frac{\partial R}{\partial y} - \frac{\partial Q}{\partial z} \right) \mathrm{d}y\mathrm{d}z + \left(\frac{\partial P}{\partial z} - \frac{\partial R}{\partial x} \right) \mathrm{d}z\mathrm{d}x + \left(\frac{\partial Q}{\partial x} - \frac{\partial P}{\partial y} \right) \mathrm{d}x\mathrm{d}y \tag{24-23}$$

(3) 高斯–奥斯特罗格拉德斯基定理 (Gauss–Ostrogradsky theorem)，亦称：散度定理. 令

$$\omega = P\mathrm{d}y\mathrm{d}z + Q\mathrm{d}z\mathrm{d}x + R\mathrm{d}x\mathrm{d}y \tag{24-24}$$

在空间单连通区域 $\mathcal{D}$ 有

$$\oiint_{\partial \mathcal{D}} \omega = \oiint_{\partial \mathcal{D}} P\mathrm{d}y\mathrm{d}z + Q\mathrm{d}z\mathrm{d}x + R\mathrm{d}x\mathrm{d}y = \iiint_{\mathcal{D}} \left(\frac{\partial P}{\partial x} + \frac{\partial Q}{\partial y} + \frac{\partial R}{\partial z} \right) \mathrm{d}x\mathrm{d}y\mathrm{d}z \tag{24-25}$$

24.4 霍奇星算子和对偶

24.4.1 欧氏空间与闵可夫斯基空间的度量张量

欧氏空间满足：对 $\forall \boldsymbol{x} \neq \boldsymbol{0}$，有 $\boldsymbol{x} \cdot \boldsymbol{x} > 0$，即度量张量 g_{ij} 对应的矩阵为对称正定矩阵. 接下来只讨论拥有标准正交基的 n 维欧氏空间 $\mathcal{E}^n$，此时度量张量的分量可表示为 $g_{ij} = \delta_{ij}$，所对应的矩阵为单位矩阵：$\det(g_{ij}) = 1$.

闵可夫斯基空间中，只讨论四维时空标准正交基 (x_0, x_1, x_2, x_3)，其中 $x_0 = ct$ 此时线段长度的平方满足 (15-5) 式，度量张量的分量为 (15-6) 式，其矩阵表示为 (15-7) 式.

24.4.2 置换符号 $\mathrm{sgn}(\sigma)$

考虑一组按照顺序排列的数列 $1,2,\cdots,n$, 其相应的任意排列可描述为 $\sigma = (i_1, i_2, \cdots, i_n)$. 若排列为偶排列, 则置换符号 $\mathrm{sgn}(\sigma) = 1$; 若排列为奇排列, 则置换符号 $\mathrm{sgn}(\sigma) = -1$. 对于三维数列 $(1,2,3)$ 或者四维数列 $(1,2,3,4)$, 也可等价于列维–奇维塔置换符号 $\varepsilon_{123}, \varepsilon_{1234}$. 以四维数列 $(1,2,3,4)$ 为例, 有

$$\begin{cases} \varepsilon_{1243} = -\varepsilon_{1234} = -1 \\ \varepsilon_{1342} = -\varepsilon_{1324} = -(-\varepsilon_{1234}) = 1 \\ \varepsilon_{4123} = -\varepsilon_{1423} = -(-\varepsilon_{1243}) = -[-(-\varepsilon_{1234})] = -1 \end{cases} \tag{24-26}$$

24.4.3 霍奇星号算子

霍奇星号算子 (Hodge star operator), 简称霍奇星号 (Hodge star), 是英国数学家霍奇 (William Vallance Douglas Hodge, 1903—1975) 在 1931—1932 年访问美国普林斯顿大学期间的新发现.

霍奇星号运算为一种将 n 维向量空间中的 k 维向量投影为 $n-k$ 维向量的运算, 反之亦然. 在微分形式中, 一组 $\mathrm{d}x_i$ 表示的 n 维的基矢量 $(\mathrm{d}x_1, \cdots, \mathrm{d}x_n)$ 按照外微分运算表示为 $\mathrm{d}x_1 \wedge \cdots \wedge \mathrm{d}x_n$, 如度量张量对应的矩阵为正定矩阵, 且相应的排列 $\sigma = (i_1, i_2, \cdots, i_n)$ 为偶排列, 即 $\mathrm{sgn}\,\sigma = 1$, 则 k 维向量相应的霍奇星号为

$$*(\mathrm{d}x_{i_1} \wedge \cdots \wedge \mathrm{d}x_{i_k}) = \mathrm{d}x_{i_{k+1}} \wedge \cdots \wedge \mathrm{d}x_{i_n} \tag{24-27}$$

特别地,

$$*(\mathrm{d}x_1 \wedge \cdots \wedge \mathrm{d}x_k) = \mathrm{d}x_{k+1} \wedge \cdots \wedge \mathrm{d}x_n \tag{24-28}$$

一般情况下, 霍奇星号定义为

$$*(\mathrm{d}x_{i_1} \wedge \cdots \wedge \mathrm{d}x_{i_k}) = \frac{\sqrt{|\det(g_{ij})|}}{(n-k)!} \mathrm{sgn}(\sigma) g_{i_1} \cdots g_{i_k} \mathrm{d}x_{i_{k+1}} \wedge \cdots \wedge \mathrm{d}x_{i_n} \tag{24-29}$$

对于标准正交基 $\det(g_{ij}) = \pm 1$ 时, 且不考虑 $n-k$ 的排序时, 上式退化为

$$*(\mathrm{d}x_{i_1} \wedge \cdots \wedge \mathrm{d}x_{i_k}) = \mathrm{sgn}(\sigma) g_{i_1} \cdots g_{i_k} \mathrm{d}x_{i_{k+1}} \wedge \cdots \wedge \mathrm{d}x_{i_n} \tag{24-30}$$

下面分别以二维、三维欧氏空间与四维闵氏空间为例, 说明该式.

(1) 二维空间欧氏空间 (x, y)

置换符号满足 $\varepsilon_{xy} = 1$, $\varepsilon_{yx} = -1$, 对 k 次形的霍奇星号运算则容易有

$$0\text{-form: } \varepsilon_{xy} = 1, \quad \varepsilon_{yx} = -1 \tag{24-31}$$

第 4 章　旋转群，拓扑，微分流形上的张量分析

$$\text{1-form:} \quad *\mathrm{d}x = \varepsilon_{xy}\mathrm{d}y = \mathrm{d}y; \quad *\mathrm{d}y = \varepsilon_{yx}\mathrm{d}x = -\mathrm{d}x \tag{24-32}$$

$$\text{2-form:} \quad *(\mathrm{d}x \wedge \mathrm{d}y) = 1 \tag{24-33}$$

(2) 三维空间欧氏空间 (x,y,z)

置换符号 ε_{xyz} 对 k 次形的霍奇星号运算有

$$\text{0-form:} \quad *1 = \mathrm{d}x \wedge \mathrm{d}y \wedge \mathrm{d}z \tag{24-34}$$

$$\text{1-form:} \quad *\mathrm{d}x = \mathrm{d}y \wedge \mathrm{d}z; \quad *\mathrm{d}y = \mathrm{d}z \wedge \mathrm{d}x; \quad *\mathrm{d}z = \mathrm{d}x \wedge \mathrm{d}y \tag{24-35}$$

$$\text{2-form:} \quad *(\mathrm{d}x \wedge \mathrm{d}y) = \mathrm{d}z; \quad *(\mathrm{d}z \wedge \mathrm{d}x) = \mathrm{d}y; \quad *(\mathrm{d}y \wedge \mathrm{d}z) = \mathrm{d}x \tag{24-36}$$

$$\text{3-form:} \quad *(\mathrm{d}x \wedge \mathrm{d}y \wedge \mathrm{d}z) = 1 \tag{24-37}$$

(3) 四维闵氏空间 (t,x,y,z)

置换符号为 ε_{txyz}，且度量张量的非零分量为 $g_t = -g_x = -g_y = -g_z = 1$，对 k 次形的霍奇星号运算有如下关系式：

$$\text{0-form:} \quad *1 = \mathrm{d}t \wedge \mathrm{d}x \wedge \mathrm{d}y \wedge \mathrm{d}z \tag{24-38}$$

$$\text{1-form:} \begin{cases} *\mathrm{d}t = \mathrm{d}x \wedge \mathrm{d}y \wedge \mathrm{d}z \\ *\mathrm{d}x = (-1)\cdot(-1)\mathrm{d}t \wedge \mathrm{d}y \wedge \mathrm{d}z = \mathrm{d}t \wedge \mathrm{d}y \wedge \mathrm{d}z \\ *\mathrm{d}y = (-1)\cdot(-1)^2 \mathrm{d}t \wedge \mathrm{d}x \wedge \mathrm{d}z = -\mathrm{d}t \wedge \mathrm{d}x \wedge \mathrm{d}z \\ *\mathrm{d}z = (-1)\cdot(-1)^3 \mathrm{d}t \wedge \mathrm{d}x \wedge \mathrm{d}y = \mathrm{d}t \wedge \mathrm{d}x \wedge \mathrm{d}y \end{cases} \tag{24-39}$$

$$\text{2-form:} \begin{cases} *(\mathrm{d}t \wedge \mathrm{d}x) = -\mathrm{d}y \wedge \mathrm{d}z \\ *(\mathrm{d}t \wedge \mathrm{d}y) = (-1)\cdot(-1)\mathrm{d}x \wedge \mathrm{d}z \\ *\mathrm{d}t \wedge \mathrm{d}z = (-1)\cdot(-1)^2 \mathrm{d}x \wedge \mathrm{d}y = -\mathrm{d}x \wedge \mathrm{d}y \\ *(\mathrm{d}x \wedge \mathrm{d}y) = (-1)^2 \cdot (-1)^2 \mathrm{d}t \wedge \wedge \mathrm{d}z = \mathrm{d}t \wedge \mathrm{d}z \\ *(\mathrm{d}x \wedge \mathrm{d}z) = (-1)^2 \cdot (-1)^3 \mathrm{d}t \wedge \mathrm{d}y = -\mathrm{d}t \wedge \mathrm{d}y \\ *(\mathrm{d}y \wedge \mathrm{d}z) = (-1)^2 \cdot (-1)^4 \mathrm{d}t \wedge \mathrm{d}x = \mathrm{d}t \wedge \mathrm{d}x \end{cases} \tag{24-40}$$

$$\text{3-form:} \begin{cases} *(\mathrm{d}x \wedge \mathrm{d}y \wedge \mathrm{d}z) = (-1)^3 \cdot (-1)^3 \mathrm{d}t = \mathrm{d}t \\ *(\mathrm{d}t \wedge \mathrm{d}y \wedge \mathrm{d}z) = (-1)^2 \cdot (-1)^2 \mathrm{d}x = \mathrm{d}x \\ *(\mathrm{d}t \wedge \mathrm{d}x \wedge \mathrm{d}z) = (-1) \cdot (-1)^2 \mathrm{d}y = -\mathrm{d}y \\ *(\mathrm{d}t \wedge \mathrm{d}x \wedge \mathrm{d}y) = (-1)^2 \mathrm{d}z = \mathrm{d}z \end{cases} \tag{24-41}$$

$$\text{4-form:} \quad *(\mathrm{d}t \wedge \mathrm{d}x \wedge \mathrm{d}y \wedge \mathrm{d}z) = (-1)^3 = -1 \tag{24-42}$$

24.5 霍奇星号在麦克斯韦方程组中的应用

在例 8.4 中的 (8-21) 式中，已经给出了麦克斯韦方程组. 而在思考题 2.5 中已经定义了电磁张量：

$$[F] = \begin{bmatrix} F_{tt} & F_{tx} & F_{ty} & F_{tz} \\ F_{xt} & F_{xx} & F_{xy} & F_{xz} \\ F_{yt} & F_{yx} & F_{yy} & F_{yz} \\ F_{zt} & F_{zx} & F_{zy} & F_{zz} \end{bmatrix} = \begin{bmatrix} 0 & E_x & E_y & E_z \\ -E_x & 0 & -B_z & B_y \\ -E_y & B_z & 0 & -B_x \\ -E_z & -B_y & B_x & 0 \end{bmatrix} \tag{24-43}$$

由此可见，电磁张量是一个 2 次形式 (2-form). 特别地，令电磁张量分解为

$$F = E + B \tag{24-44}$$

式中，电场和磁场的 2 次形式 (2-form) 分别为

$$E = E_x \mathrm{d}t \wedge \mathrm{d}x + E_y \mathrm{d}t \wedge \mathrm{d}y + E_z \mathrm{d}t \wedge \mathrm{d}z \tag{24-45}$$

$$B = B_x \mathrm{d}z \wedge \mathrm{d}y + B_y \mathrm{d}x \wedge \mathrm{d}z + B_z \mathrm{d}y \wedge \mathrm{d}x \tag{24-46}$$

霍奇对偶 (Hodge duals) 关系为

$$*(\mathrm{d}t \wedge \mathrm{d}x) = \mathrm{d}z \wedge \mathrm{d}y, \quad *(\mathrm{d}z \wedge \mathrm{d}y) = -\mathrm{d}t \wedge \mathrm{d}x \tag{24-47}$$

$$*(\mathrm{d}t \wedge \mathrm{d}y) = \mathrm{d}x \wedge \mathrm{d}z, \quad *(\mathrm{d}x \wedge \mathrm{d}z) = -\mathrm{d}t \wedge \mathrm{d}y \tag{24-48}$$

$$*(\mathrm{d}t \wedge \mathrm{d}z) = \mathrm{d}y \wedge \mathrm{d}x, \quad *(\mathrm{d}y \wedge \mathrm{d}x) = -\mathrm{d}t \wedge \mathrm{d}z \tag{24-49}$$

观察上面三式，易知对于任意的 $v \in \mathbb{R}^3$ 满足下列关系式：

$$*E(v) = B(v), \quad *B(v) = -E(v) \tag{24-50}$$

$E(v)$ 和 $B(v)$ 分别是 (24-45) 和 (24-46) 两式定义的 2 次形式 (2-form)，其系数 v. (24-50) 式中的对偶性和麦克斯韦方程组 (8-21) 式的下列对偶性相同：

$$\boldsymbol{E} \mapsto \boldsymbol{B}, \quad \boldsymbol{B} \mapsto -\boldsymbol{E} \tag{24-51}$$

也就是说，将 (24-51) 式的变换代入麦克斯韦方程组 (8-21) 式后，方程组的形式不变.

k 次形式 (k-form) 的外微分 (exterior derivative) 定义为

$$\mathrm{d}(f \mathrm{d}x_{i_1} \wedge \mathrm{d}x_{i_2} \wedge \cdots \wedge \mathrm{d}x_{i_k}) = \sum_{j=1}^{n} \frac{\partial f}{\partial x_j} \mathrm{d}x_j \wedge (\mathrm{d}x_{i_1} \wedge \mathrm{d}x_{i_2} \wedge \cdots \wedge \mathrm{d}x_{i_k}) \tag{24-52}$$

第 4 章 旋转群，拓扑，微分流形上的张量分析

外微分的上述运算使 k 次形式映射到 $(k+1)$ 次形式.

首先来求 $E(\boldsymbol{v})$ 的外微分：

$$\begin{aligned}
\mathrm{d}E(\boldsymbol{v}) &= \mathrm{d}\left[-(v_x\mathrm{d}x + v_y\mathrm{d}y + v_z\mathrm{d}z)\right] \wedge \mathrm{d}t \\
&= -\left[\left(\frac{\partial v_x}{\partial y}\mathrm{d}y + \frac{\partial v_x}{\partial z}\mathrm{d}z\right) \wedge \mathrm{d}x + \left(\frac{\partial v_y}{\partial x}\mathrm{d}x + \frac{\partial v_y}{\partial z}\mathrm{d}z\right) \wedge \mathrm{d}y \right. \\
&\quad \left. + \left(\frac{\partial v_z}{\partial x}\mathrm{d}x + \frac{\partial v_z}{\partial y}\mathrm{d}y\right) \wedge \mathrm{d}z\right] \wedge \mathrm{d}t
\end{aligned} \tag{24-53}$$

对 (24-53) 式进行重新整理后，得到

$$\begin{aligned}
\mathrm{d}E(\boldsymbol{v}) &= \mathrm{d}t \wedge \left[\left(\frac{\partial v_z}{\partial y} - \frac{\partial v_y}{\partial z}\right)\mathrm{d}z \wedge \mathrm{d}y + \left(\frac{\partial v_x}{\partial z} - \frac{\partial v_z}{\partial x}\right)\mathrm{d}x \wedge \mathrm{d}z \right. \\
&\quad \left. + \left(\frac{\partial v_y}{\partial x} - \frac{\partial v_x}{\partial y}\right)\mathrm{d}y \wedge \mathrm{d}x\right] \\
&= \mathrm{d}t \wedge B(\boldsymbol{\nabla} \times \boldsymbol{v})
\end{aligned} \tag{24-54}$$

类似地，对于 $B(\boldsymbol{v})$，我们有

$$\begin{aligned}
\mathrm{d}B(\boldsymbol{v}) &= \mathrm{d}(v_x\mathrm{d}z \wedge \mathrm{d}y + v_y\mathrm{d}x \wedge \mathrm{d}z + v_z\mathrm{d}y \wedge \mathrm{d}x) \\
&= \left[\left(\frac{\partial v_x}{\partial t}\mathrm{d}t + \frac{\partial v_x}{\partial x}\mathrm{d}x\right) \wedge \mathrm{d}z \wedge \mathrm{d}y + \left(\frac{\partial v_y}{\partial t}\mathrm{d}t + \frac{\partial v_y}{\partial y}\mathrm{d}y\right) \wedge \mathrm{d}x \wedge \mathrm{d}z \right. \\
&\quad \left. + \left(\frac{\partial v_z}{\partial t}\mathrm{d}t + \frac{\partial v_z}{\partial z}\mathrm{d}z\right)\right] \wedge \mathrm{d}y \wedge \mathrm{d}x
\end{aligned} \tag{24-55}$$

对上式进行整理，得到

$$\begin{aligned}
\mathrm{d}B(\boldsymbol{v}) &= \mathrm{d}t \wedge \left(\frac{\partial v_x}{\partial t}\mathrm{d}z \wedge \mathrm{d}y + \frac{\partial v_y}{\partial t}\mathrm{d}x \wedge \mathrm{d}z + \frac{\partial v_z}{\partial t}\mathrm{d}y \wedge \mathrm{d}x\right) \\
&\quad - \left(\frac{\partial v_x}{\partial x} + \frac{\partial v_y}{\partial y} + \frac{\partial v_z}{\partial z}\right)\mathrm{d}x \wedge \mathrm{d}y \wedge \mathrm{d}z \\
&= \mathrm{d}t \wedge B\left(\frac{\partial \boldsymbol{v}}{\partial t}\right) - (\boldsymbol{\nabla} \cdot \boldsymbol{v})\mathrm{d}x \wedge \mathrm{d}y \wedge \mathrm{d}z
\end{aligned} \tag{24-56}$$

(24-44) 式中的电磁张量的外微分为

$$\begin{aligned}
\mathrm{d}F &= \mathrm{d}\left[E(\boldsymbol{E}) + B(\boldsymbol{B})\right] \\
&= \mathrm{d}t \wedge B(\boldsymbol{\nabla} \times \boldsymbol{E}) + \mathrm{d}t \wedge B\left(\frac{\partial \boldsymbol{B}}{\partial t}\right) - (\boldsymbol{\nabla} \cdot \boldsymbol{B})\mathrm{d}x \wedge \mathrm{d}y \wedge \mathrm{d}z \\
&= \mathrm{d}t \wedge B\left(\frac{\partial \boldsymbol{B}}{\partial t} + \boldsymbol{\nabla} \times \boldsymbol{E}\right) - (\boldsymbol{\nabla} \cdot \boldsymbol{B})\mathrm{d}x \wedge \mathrm{d}y \wedge \mathrm{d}z
\end{aligned} \tag{24-57}$$

然后计算电磁张量霍奇对偶的外微分：

$$\mathrm{d} * F = \mathrm{d}\left[B(\boldsymbol{E}) - E(\boldsymbol{B})\right]$$

$$= \mathrm{d}t \wedge B\left(\frac{\partial \boldsymbol{E}}{\partial t}\right) - (\boldsymbol{\nabla} \cdot \boldsymbol{E})\,\mathrm{d}x \wedge \mathrm{d}y \wedge \mathrm{d}z - \mathrm{d}t \wedge B\,(\boldsymbol{\nabla} \times \boldsymbol{B})$$

$$= \mathrm{d}t \wedge B\left(\frac{\partial \boldsymbol{E}}{\partial t} - \boldsymbol{\nabla} \times \boldsymbol{B}\right) - (\boldsymbol{\nabla} \cdot \boldsymbol{E})\,\mathrm{d}x \wedge \mathrm{d}y \wedge \mathrm{d}z \tag{24-58}$$

通过对比麦克斯韦方程组 (8-21) 和 (24-57) 及 (24-58) 两式，得出结论：用外微分形式，麦克斯韦方程组可以极为简洁地表达为

$$\mathrm{d}F = \mathrm{d} * F = 0 \tag{24-59}$$

§25. 流形上的矢量和张量分析

25.1 推前和拉回映射

经典张量分析将矢量场定义为有大小、有方向、相等矢量的模和方向相同且满足平行四边形法则的实体. 流形中将矢量定义为线性微分算子且满足莱布尼兹法则 (Leibniz rule).

微分流形中矢量场是切丛的一个截面，流形上的可微矢量场可以看作是作用在流形全体光滑函数的线性微分算子，且满足莱布尼兹法则：

$$\boldsymbol{V}\,[af + bg] = a\boldsymbol{V}\,[f] + b\boldsymbol{V}\,[g] \tag{25-1}$$

$$\boldsymbol{V}\,[fg] = g\boldsymbol{V}\,[f] + f\boldsymbol{V}\,[g] \tag{25-2}$$

经典张量分析通过坐标变换来定义张量概念. 而流形中将矢量看做线性函数，将张量看做多重线性函数，在基向量上的坐标称为该张量的分量.

经典的张量分析将张量定义在三维欧氏空间上. 当坐标系改变时，满足坐标转换关系的有序数组成的集合称为张量. 自由指标的个数与所乘坐标转换系数的次数一致，称为张量的阶数. 流形上的张量定义在 n 维、甚至无穷维的空间.

切映射定义了流形上的切丛对欧氏空间的切丛的映射，是流形对欧氏空间的映射和切丛对基点的投影的复合函数.

推前映射和拉回映射是复合映射，没有定义独立的概念. 研究流形上某点的性质时，需要使用推前映射在欧氏空间进行；流形局部对应的欧氏空间的张量可以通过拉回映射返回给流形. 比如物质加速度和空间加速度.

研究流形上某点的性质时，需要使用推前映射在欧氏空间进行；流形局部对应的欧氏空间的张量可以通过拉回映射返回给流形.

对映射 $x = \Phi(X)$, x 是欧氏空间的坐标, X 是流形上的标号, 推前映射为

$$\Phi_* Y = T\Phi \circ \left[Y \circ \Phi^{-1}(x) \right] \tag{25-3}$$

拉回映射为

$$\Phi^* v = T\Phi^{-1} \circ [v(x)] \tag{25-4}$$

25.2 李导数

张量场沿矢量场的李导数 (Lie derivative) 代表流形沿矢量场的形变下张量的改变率. 李导数是力学中的物质导数的推广. 矢量场的李导数关于两个矢量场反对称.

张量场的物质导数可以写为李导数形式:

$$\frac{\mathrm{d} T}{\mathrm{d} t} = \frac{\partial T}{\partial t} + L_{\boldsymbol{w}} T \tag{25-5}$$

式中泊松括号为

$$[\boldsymbol{w}, \boldsymbol{v}] = L_{\boldsymbol{w}} \boldsymbol{v} \tag{25-6}$$

李导数具有如下性质:

$$[\boldsymbol{w}, \boldsymbol{v}](f+g) = [\boldsymbol{w}, \boldsymbol{v}] f + [\boldsymbol{w}, \boldsymbol{v}] g \tag{25-7}$$

$$[\boldsymbol{w}, \boldsymbol{v}](fg) = f [\boldsymbol{w}, \boldsymbol{v}] g + g [\boldsymbol{w}, \boldsymbol{v}] f \tag{25-8}$$

$$[\boldsymbol{w}, \boldsymbol{v}] = -[\boldsymbol{v}, \boldsymbol{w}] \tag{25-9}$$

$$[\boldsymbol{u} + \boldsymbol{v}, \boldsymbol{w}] = [\boldsymbol{u}, \boldsymbol{w}] + [\boldsymbol{v}, \boldsymbol{w}] \tag{25-10}$$

$$[f\boldsymbol{v}, g\boldsymbol{w}] = f(L_{\boldsymbol{v}} g) \boldsymbol{w} - g(L_{\boldsymbol{w}} g) \boldsymbol{v} + fg [\boldsymbol{v}, \boldsymbol{w}] \tag{25-11}$$

$$[\boldsymbol{u}, [\boldsymbol{v}, \boldsymbol{w}]] + [\boldsymbol{v}, [\boldsymbol{w}, \boldsymbol{u}]] + [\boldsymbol{w}, \boldsymbol{u}, \boldsymbol{v}] = 0 \tag{25-12}$$

对于度量张量场 g_{ij}, 关于位移矢量场 $\boldsymbol{u}$ 的李导数为

$$(L_{\boldsymbol{u}} \boldsymbol{g})_{ij} = u^k \frac{\partial g_{ij}}{\partial x^k} + g_{kj} \frac{\partial u^k}{\partial x^i} + g_{ik} \frac{\partial u^k}{\partial x^j} = 2\varepsilon_{ij} \tag{25-13}$$

式中, $\varepsilon_{ij} = \frac{1}{2} \left(\frac{\partial u^i}{\partial x^j} + \frac{\partial u^j}{\partial x^i} \right)$ 是小应变条件下的柯西应变张量. 说明: 流形沿位移场变形时, 度量张量的改变率正是小应变条件下的应变张量.

表 25.1 给出了经典张量分析和流形上张量分析的联系和区别.

表 25.1　经典和流形上的张量分析比较

相关内容	经典张量分析	流形上的张量分析
坐标	$\{x^a\}$	$\{x^a\}$
坐标基矢量	$\boldsymbol{e}_a = \dfrac{\partial z^i}{\partial x^a}\hat{\boldsymbol{i}}_i$	$\dfrac{\partial}{\partial x^a} = \boldsymbol{e}_a$
对偶基	$\boldsymbol{e}^a = g^{ab}\boldsymbol{e}_b$	$\mathbf{d}x^a = \boldsymbol{e}^a$
对偶关系	$\boldsymbol{e}^a \cdot \boldsymbol{e}_b = \delta^a_b$	$\mathbf{d}x^a \dfrac{\partial}{\partial x^b} = \delta^a_b$
坐标变换	$\tilde{\boldsymbol{e}}_a = \dfrac{\partial x^b}{\partial \tilde{x}^a}\boldsymbol{e}_b,\quad \tilde{\boldsymbol{e}}^a = \dfrac{\partial \tilde{x}^a}{\partial x^b}\boldsymbol{e}^b$	$\dfrac{\partial}{\partial \tilde{x}^a} = \dfrac{\partial x^b}{\partial \tilde{x}^a}\dfrac{\partial}{\partial x^b},\quad \mathbf{d}\tilde{x}^a = \dfrac{\partial \tilde{x}^a}{\partial x^b}\mathbf{d}x^b$
矢量的坐标表示	$\boldsymbol{v} = v^a \boldsymbol{e}_a,\quad v^a = \boldsymbol{e}^a \cdot \boldsymbol{v}$	$\boldsymbol{v} = v^a \dfrac{\partial}{\partial x^a},\quad v^a = \mathbf{d}x^a(\boldsymbol{v})$
1 次形的坐标表示	$\boldsymbol{\alpha} = \alpha_a \boldsymbol{e}^a,\quad \alpha_a = \boldsymbol{e}_\alpha \cdot \boldsymbol{\alpha}$	$\boldsymbol{\alpha} = \alpha_a \mathbf{d}x^a,\quad \alpha_a = \boldsymbol{\alpha}\dfrac{\partial}{\partial x^a}$
方向导数	$\boldsymbol{v}[f] = v^a \dfrac{\partial f}{\partial x^a}$	$\boldsymbol{v}[f] = v^a \dfrac{\partial f}{\partial x^a}$

经典张量分析可以将研究对象整体用一个坐标系描述, 而在流形上只有局部坐标, 不存在整体的坐标系; 流形上的矢量定义为物理量的线性函数, 所以坐标系的基矢量不同, 切空间的基矢量由沿曲线的方向导数引出, 余切空间基矢量由全微分引出.

流形上的坐标变换规则和方向导数与经典张量分析一致.

25.3　黎曼度量与黎曼流形

定义了黎曼度量张量的流形称为黎曼流形, 光滑的流形上一定可以定义黎曼度量. 流形上有了度量张量就可以定义切矢量的长度、一点处两个矢量的夹角和曲线段的弧长. 经典张量分析对欧氏空间的讨论就可以在流形上进行, 引入标架的微分, 与度量张量对应的联络系数, 以及流形上的物质导数.

用黎曼度量张量表示的克里斯托费尔符号, 在欧氏空间为

$$\gamma^c_{ab} = \frac{1}{2}g^{cp}(g_{bp,a} + g_{pa,b} - g_{ab,p}) \tag{25-14}$$

在流形上:

$$\Gamma^c_{ab} = \frac{1}{2}G^{cp}(G_{bp,a} + G_{pa,b} - G_{ab,p}) \tag{25-15}$$

流形绝对运动的物质导数由克里斯托费尔符号决定, 进而由黎曼度量张量唯一地决定.

类比流体力学中的流线,从一点出发在每一点处都对速度积分的积分曲线,曲线在每一点处都与速度相切. 稳态流场的流线与时间无关,数学上称为自治系统. 整个空间的积分曲线的集合称作流,如图 25.1 所示.

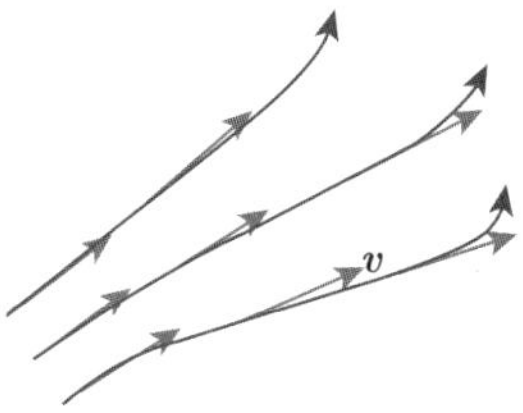

连续介质力学中的运动要求所有点的运动都不会产生重合、穿透或消失. 根据定义,微分流形都可以满足这些要求.

图 25.1 流

思考题和补充材料

4.1 所谓的纤维丛 (fiber bundle),就是定义在一个几何体 (数学上叫底流形) 的函数 (数学上叫纤维). 比如题图 4.1 所示的这把梳子,就可以看成是一个纤维丛,底流形是一个柱面,而纤维则是上面的一根根梳齿 (线段). 更进一步,若将人的头作为基底,头发作为纤维,则长满了头发的脑袋亦可视为纤维丛.

题图 4.1 生活中处处可见 "纤维丛"

4.2 陈省身 (Shiing-Shen Chern, 1911—2004, 如题图 4.2 所示) 鲜明地指出并类比道[4.4]:"在几何学研究中有了坐标这个工具之后,我们现在希望摆脱它的束缚. 这引出了流形这一重要概念. 一个流形在局部上可用坐标刻画,但这个坐标系是可以任意变换的. 换句话说,流形是一个具有可变的或相对的坐标 (相对性原理) 的空间. 或许我可以用人类穿着衣服做个比喻. '人开始穿着衣服' 是一件极端重要的历史事件. '人会改变衣服' 的能力也有着同样重要的意义. 如果把几何看作人体,坐标看作衣服,那么可以像下面这样描写几何进化史:

题图 4.2 陈省身 (右)、杨振宁 (左)

综合几何	裸体人
坐标几何	原始人
流形	现代人

流形这个概念即使对于数学家来说也不是简单的. 例如 J. 阿达玛这样一位大数学家,在讲到以流形这概念为基础的李群理论时就说: '要想对李群理论保持着不只是初等的、肤浅的,而是更多一些的理解,感到有着不可克服的困难.'"

4.3 陈省身在文章[4.4]中进而对杨振宁 (Chen-Ning Yang, 1922— ,如题图 4.2 所示) 的观点进行了如下评价:"丛、联络、上同调和示性类都是艰深的概念,在几何中它们都经过长期的探索和试验才定形下来. 物理学家杨振宁说: '非交换的规范场与纤维丛这个美妙理论 —— 数学家们发展它时并没有参考物理世界 —— 在概念上的一致,对我来说是一大奇迹.' 1975 年他对我讲: '这既是使人震惊的,又是使人迷惑不解的,因为你们数学家是没有依据地虚构出这些概念来的.' 这种迷惑是双方都有的. 事实上, E. 威格纳说起数学在物理中的作用时,曾谈到数学的超乎常理的有效性. 如果一定要找一个理由的话,那么也许可用 '科学的整体性' 这个含糊的词儿来表达."

题图 4.4 斯梅尔

4.4 史蒂芬·斯梅尔 (Stephen Smale, 1930—　, 1966 年菲尔兹奖获得者, 2007 年沃尔夫奖获得者, 见题图 4.4) 在 1970 年发表过一篇题为《拓扑和力学》的文章[4.5,4.6], 很遗憾地是, 该文在力学界的反响不大.

4.5 有关物质流形概念的论述可参阅郭仲衡的相关著作[4.7].

4.6 结合题图 4.6, 理解流形和各种空间之间的关系.

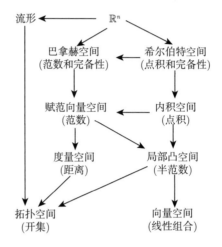

题图 4.6　流形和各类空间的关系图

4.7 为了增进对麦克斯韦方程组的理解, 这里给出一个物理学家文小刚教授给出的故事: "我每次从加拿大开车入境美国, 在边境上, 检察官都这样问: 你是干什么的？我: 搞物理的. 检察官: 你知道麦克斯韦方程有几个吗？我想: 如果检察官是研究生水平, 我应当回答一个. 如果检察官是大学生水平, 我应回答四个. 如果检察官是高中生水平, 我就不知道回答几个了？最后我试着说: 四个. 他就放我过境了."

　　一个麦克斯韦方程是指 (24-59) 式; 而四个麦克斯韦方程是指 (8-21) 式或 (s6-5) 式.

4.8 普朗克对麦克斯韦高度评价道:"他的名字巍然屹立在古典物理学的大门之上, 我们可以这样评价他: 詹姆斯·克拉克·麦克斯韦的出生属于爱丁堡, 他的品格属于剑桥, 他的工作属于全世界."(His name stands magnificently over the portal of classical physics, and we can say this of him; by his birth James Clerk Maxwell belongs to Edinburgh, by his personality he belongs to Cambridge, by his work he belongs to the whole world.)

4.9 斯蒂芬·霍金 (Stephen Hawking, 1942—2018) 对麦克斯韦高度评价道:"麦克斯韦是物理学家中的物理学家 (Maxwell is the physicist's physicist)."

4.10 为了在变形几何学中应用的方便, 这里给出一个非奇异线性映射 (a nonsingular linear map) $\boldsymbol{A}: \mathbb{R}^n \to \mathbb{R}^n$ 保持方向或保持定向 (orientation-preserving) 的条件为

det $\boldsymbol{A} > 0$.

4.11 数学大师阿诺尔德 (Vladimir Igorevich Arnold, 1937—2010) 于 1998 年出版了第一部拓扑流体力学的专著[4.8]. 拓扑流体力学之目的在于研究有复杂轨迹的流动的拓扑特征, 并对其加以广泛应用. 拓扑流体力学的交叉学科有: 流体稳定性, 黎曼几何与辛几何, 磁流体力学, 李代数与李群, 纽结理论, 动力学系统等.

4.12 韦尔切克 (1951— , 2004 年诺贝尔物理学奖获得者) 在 2019 年 5 月发表的《物理学的统一愿景》一文中对于 "从流体中的涡旋到拓扑物质" 做出如下评述: "拓扑学首次大规模侵入物理学是通过涡旋的流体动力学. 亥姆霍兹于 1858 关于理想流体中涡旋的开创性数学研究表明了涡旋环 (vortex loop) 的稳定性, 以及涡旋环间相互作用本质上颇为简单的类磁体特性. 泰特大约于 1867 的烟圈实验又增强了这些认识. 现实的宏观物理中的这些发现激发了开尔文勋爵的灵感, 他提出原子或许是以太中的涡旋扭结 (knot). 拓扑上不同的扭结的离散性, 以及涡旋环的稳定性似乎抓住了一个主要的化学事实: 存在不同的稳定元素. 至于分子则可以被想象成是相互连接的环. 受到这些想法的启发, 泰特对扭结和链环的分类做了非常严肃且具有原创性的数学工作. 涡旋已经成为我们理解磁性材料、超流体、超导体和许多其他物态的主要成分." 拓扑流体动力学新近的实验进展见题图 4.12.

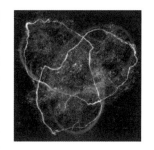

题图 4.12 实验中观察到的三叶结型涡环结构[4.9]

参 考 文 献

[4.1] Gauss C F. Disquisitiones generales circa superficies curvas. Ostwald's Klassiker der Exakten Wissenschaften, 1827, 5.

[4.2] 江泽涵. 我国数学名词的早期工作. 数学通报, 1980, 12: 23-24.

[4.3] Euler L. Solutio problematis ad geometriam situs pertinensis. Comm. Acad. Sci. Imper. Petropol., 1736, 8: 128-140.

[4.4] Chern S S. From triangles to manifolds. The American Mathematical Monthly, 1979, 86: 339-349. (中文版: 陈省身. 从三角形到流形 (尤承业译). 自然杂志, 1979, 8: 473-478.)

[4.5] Smale S. Topology and mechanics. I. Inventiones mathematicae, 1970, 10: 305-331.

[4.6] Smale S. Topology and mechanics. II. Inventiones mathematicae, 1970, 11: 45-64.

[4.7] 郭仲衡, 梁浩云. 变形体非协调理论. 重庆: 重庆出版社, 1989.

[4.8] Arnold V I, Khesin B A. Topological Methods in Hydrodynamics. New York: Springer, 1998.

[4.9] Kleckner D, Irvine W T M. Creation and dynamics of knotted vortices. Nature Physics, 2013, 9: 253-258.

第 5 章 变形运动学、功共轭

连续介质力学主要包括三大部分：(1) 变形几何学，也就是变形运动学，此部分主要是要通过变形几何场，引出相关的应变度量 (strain measures) 和与之相功共轭 (work conjugate) 的应力度量 (stress measures)；(2) 场方程；(3) 联系应力和应变的本构方程或本构关系.

§26. 变形梯度 F 及其极分解

26.1 变形梯度张量 F 及其转置、逆、逆的转置的详细推导

连续介质力学在研究可变形体的力学行为时，可测量的基本量是距离和时间间隔. 故，为了描述和研究物体的变形，需要一个带有时钟的参考标价——时空系 (Σ). Σ 是被赋予刚性标架的称为物理空间的三维欧几里得 (欧氏) 空间 ($\mathcal{E}^3$) 和一个称为时间的一维欧氏空间 $\mathbb{R}$ 的积空间 (product space)：

$$\Sigma = \mathcal{E}^3 \times \mathbb{R}$$

为了确定在欧氏空间 $\mathcal{E}^3$ 中的一个位置点 $\boldsymbol{x}$，可用一个笛卡儿直角坐标系 $\boldsymbol{x} = x^i \boldsymbol{e}_i$ 来表示，相当于这个笛卡儿直角坐标系，$\mathcal{E}^3$ 就与一个有向的内积空间 $\mathbb{R}^3$ 建立了微分同胚的关系.

可变形体在数学语言上可用物质流形来表示. 物质流形的力学 (mechanics on the material manifold) 常常被简称为 M^3. 那什么是物质流形？所谓物质流形 ($\mathcal{B}$) 是一个可以赋予整体坐标系的三维有向微分流形. 利用在欧氏空间 $\mathcal{E}^3$ 上的笛卡儿坐标系 (x^i) 我们可认为在物质流形 $\mathcal{B}$ 上的整体坐标系给出了从 $\mathcal{B}$ 到 $\mathcal{E}^3$ 的一个保持定向的微分同胚：

$$\mathcal{C}: \mathcal{B} \to \mathbb{R}^3 \overset{(x^i)}{\leftrightarrow} \mathcal{E}^3$$

这样的微分同胚称为物质流形 $\mathcal{B}$ 的一个构形 (configuration). 可以指定由每一个物质点 $X \in \mathcal{B}$ 所占据的位置 $\boldsymbol{x} = \mathcal{C}(X) \in \mathcal{E}^3$ 的坐标 (x^i) 来说明物体 $\mathcal{B}$ 的一个构形. 因此，构形也可以简单地理解为物体 $\mathcal{B}$ 在欧氏空间 $\mathcal{E}^3$ 中占据的位置 $\boldsymbol{x}$ 的整体.

一族具有单个参变量——时间 $t \in \mathbb{R}$ ——的构形 $\mathcal{C}_t$ 称为物体 $\mathcal{B}$ 的运动,就是物体所处的当前构形 (current configuration). 选定某一特定的构形 $\mathcal{K}$,通常是时间 $t=0$ 时即 $\mathcal{K} = \mathcal{C}_0$,通过比较 $\mathcal{K}$ 和 $\mathcal{C}_t$ 来描述物体 $\mathcal{B}$ 的运动,则这个特定的构形称为参考构形 (reference configuration). 如果用 (X^A) 来表示在参考构形 $\mathcal{K}$ 上的笛卡儿坐标,亦即

$$\mathcal{K}: \mathcal{B} \to \mathbb{R}^3, \quad \mathcal{K}(X) = \boldsymbol{X} = (X^A) = (X^A(X))$$

式中,$X^A(X)$ 被称为参考点的坐标. 如图 26.1 所示,有一个映射 χ 联系参考构形中的点 $\boldsymbol{X}$ 经过时间 t 到当前构形的点 $\boldsymbol{x}$:

$$\boldsymbol{x} = \boldsymbol{\chi}(\boldsymbol{X}, t) \tag{26-1}$$

χ 亦称为变形函数. 参考构形又经常被称为初始构形 (initial configuration) 和未变形构形 (undeformed configuration);当前构形也被称为变形后的构形 (deformed configuration).

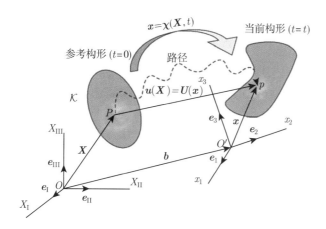

图 26.1 参考构形和当前构形之间的关系:单射和保持方向映射

对线段微元的关系进行一阶泰勒展开:

$$\begin{aligned} \mathrm{d}\boldsymbol{x} &= \boldsymbol{\chi}(\boldsymbol{X} + \mathrm{d}\boldsymbol{X}, t) - \boldsymbol{\chi}(\boldsymbol{X}, t) \\ &= \boldsymbol{F}\mathrm{d}\boldsymbol{X} = \mathrm{d}\boldsymbol{X}\boldsymbol{F}^{\mathrm{T}} \end{aligned} \tag{26-2}$$

式中,$\boldsymbol{F}\mathrm{d}\boldsymbol{X}$ 表示二阶张量 $\boldsymbol{F}$ 和矢量 $\mathrm{d}\boldsymbol{X}$ 的点积. 一般地,将二阶张量和矢量点积中的"·"略去. 上式中,$\boldsymbol{F}^{\mathrm{T}}$ 是变形梯度 $\boldsymbol{F}$ 的转置.

由于 x 是当前构形中的点，在三维欧几里得空间中，其可表示为

$$\boldsymbol{x} = x_1\boldsymbol{e}_1 + x_2\boldsymbol{e}_2 + x_3\boldsymbol{e}_3 = x_i\boldsymbol{e}_i \quad (i=1,2,3) \tag{26-3}$$

式中，i 被称为哑标 (dummy index). 而参考构形中的点 $\boldsymbol{X}$ 在笛卡儿坐标系中表示为

$$\boldsymbol{X} = X_\mathrm{I}\boldsymbol{e}_\mathrm{I} + X_\mathrm{II}\boldsymbol{e}_\mathrm{II} + X_\mathrm{III}\boldsymbol{e}_\mathrm{III} = X_K\boldsymbol{e}_K \quad (K=\mathrm{I,II,III}) \tag{26-4}$$

这样，(26-2) 中的张量 $\boldsymbol{F}$ 可表示为

$$\boldsymbol{F} = \frac{\partial \boldsymbol{x}}{\partial \boldsymbol{X}} = \boldsymbol{x} \otimes \nabla_X = \frac{\partial x_i}{\partial X_K}\boldsymbol{e}_i \otimes \boldsymbol{e}_K \tag{26-5}$$

上式定义了一个线性切映射场：$\boldsymbol{F}(\boldsymbol{X}):\mathcal{T}\to\mathcal{T},\boldsymbol{X}\in\mathcal{K}(\mathcal{B})$，称为变形梯度场，这里的 $\mathcal{T}$ 表示切空间. 从 (26-2) 和 (26-5) 两式可以看出，$\boldsymbol{F}$ 是联系当前构形和参考构形变形的二阶张量，因而被称为变形梯度张量 (deformation gradient tensor)，它的前基 $\boldsymbol{e}_i$ 在当前构形，后基 $\boldsymbol{e}_K$ 在参考构形，故尔，变形梯度 $\boldsymbol{F}$ 往往被称为"两点张量 (two-point tensor)"，也被称为"混合欧拉–拉格朗日型张量 (mixed Euler–Lagrange tensor)". 为了活跃上课的气氛，也可将两点张量称为"脚踩两只船"，这两只船分别是当前构形和参考构形.

变形梯度张量 (26-5) 式为何被称为"混合欧拉–拉格朗日型张量"？这里，我们先将二阶张量做如下的划分：

▷ 两个基均在参考构形的二阶张量被称为"拉格朗日型张量"；

▷ 两个基均在当前构形的二阶张量被称为"欧拉型张量"；

▷ 前基在当前构形，后基在参考构形的二阶张量被称为"混合欧拉–拉格朗日型张量"；

▷ 前基在参考构形，后基在当前构形的二阶张量被称为"混合拉格朗日–欧拉型张量".

事实上，变形梯度 $\boldsymbol{F} = \boldsymbol{\chi} \otimes \nabla_X$ 只不过是一个矩阵：

$$[\boldsymbol{F}] = \frac{\partial(x,y,z)}{\partial(X,Y,Z)} = \begin{bmatrix} \dfrac{\partial x}{\partial X} & \dfrac{\partial x}{\partial Y} & \dfrac{\partial x}{\partial Z} \\ \dfrac{\partial y}{\partial X} & \dfrac{\partial y}{\partial Y} & \dfrac{\partial y}{\partial Z} \\ \dfrac{\partial z}{\partial X} & \dfrac{\partial z}{\partial Y} & \dfrac{\partial z}{\partial Z} \end{bmatrix} \tag{26-6}$$

它表示映射向量 $\boldsymbol{\chi}$ 的弗雷歇导数 (Fréchet derivative), 见附录 D.2. 上式的转置为

$$\left[\boldsymbol{F}^{\mathrm{T}}\right] = \begin{bmatrix} \dfrac{\partial x}{\partial X} & \dfrac{\partial y}{\partial X} & \dfrac{\partial z}{\partial X} \\ \dfrac{\partial x}{\partial Y} & \dfrac{\partial y}{\partial Y} & \dfrac{\partial z}{\partial Y} \\ \dfrac{\partial x}{\partial Z} & \dfrac{\partial y}{\partial Z} & \dfrac{\partial z}{\partial Z} \end{bmatrix} \tag{26-7}$$

因此, (26-5) 式的转置, 亦即变形梯度的转置可表示为

$$\boldsymbol{F}^{\mathrm{T}} = \frac{\partial x_i}{\partial X_K} \boldsymbol{e}_K \otimes \boldsymbol{e}_i \tag{26-8}$$

映射 (26-1) 式还可以有逆映射:

$$\boldsymbol{X} = \boldsymbol{\chi}^{-1}(\boldsymbol{x}, t) \tag{26-9}$$

同样其变形关系为

$$\begin{aligned} \mathrm{d}\boldsymbol{X} &= \boldsymbol{\chi}^{-1}(\boldsymbol{x} + \mathrm{d}\boldsymbol{x}, t) - \boldsymbol{\chi}^{-1}(\boldsymbol{x}, t) \\ &= \boldsymbol{F}^{-1} \mathrm{d}\boldsymbol{x} = \mathrm{d}\boldsymbol{x} \boldsymbol{F}^{-\mathrm{T}} \end{aligned} \tag{26-10}$$

上式已经进行了一阶泰勒展开, 式中

$$\boldsymbol{F}^{-1} = \frac{\partial X_A}{\partial x_b} \boldsymbol{e}_A \otimes \boldsymbol{e}_b \tag{26-11}$$

显然, 上述张量的前基和后基分别在参考构形和当前构形, 也是一个典型的"混合拉格朗日-欧拉型张量". 对 (26-5) 式和 (26-11) 式中的两个张量进行点积 (相邻的基矢量基进行缩并), 有

$$\begin{aligned} \boldsymbol{F}\boldsymbol{F}^{-1} &= \left(\frac{\partial x_i}{\partial X_K} \boldsymbol{e}_i \otimes \boldsymbol{e}_K\right) \cdot \left(\frac{\partial X_L}{\partial x_m} \boldsymbol{e}_L \otimes \boldsymbol{e}_m\right) = \frac{\partial x_i}{\partial X_K} \frac{\partial X_L}{\partial x_m} \delta_{KL} \boldsymbol{e}_i \otimes \boldsymbol{e}_m \\ &= \frac{\partial x_i}{\partial x_m} \boldsymbol{e}_i \otimes \boldsymbol{e}_m = \delta_{im} \boldsymbol{e}_i \otimes \boldsymbol{e}_m = \mathbf{I} \end{aligned} \tag{26-12}$$

$\boldsymbol{F}^{-1}$ 则称为变形梯度张量 $\boldsymbol{F}$ 的逆 (inverse), 用矩阵表示为

$$\left[\boldsymbol{F}^{-1}\right] = \frac{\partial(X, Y, Z)}{\partial(x, y, z)} = \begin{bmatrix} \dfrac{\partial X}{\partial x} & \dfrac{\partial X}{\partial y} & \dfrac{\partial X}{\partial z} \\ \dfrac{\partial Y}{\partial x} & \dfrac{\partial Y}{\partial y} & \dfrac{\partial Y}{\partial z} \\ \dfrac{\partial Z}{\partial x} & \dfrac{\partial Z}{\partial y} & \dfrac{\partial Z}{\partial z} \end{bmatrix} \tag{26-13}$$

上式的转置 (Transpose) 为

$$\left[\boldsymbol{F}^{-\mathrm{T}} \right] = \left[\left(\boldsymbol{F}^{-1} \right)^{\mathrm{T}} \right] = \left[\left(\boldsymbol{F}^{\mathrm{T}} \right)^{-1} \right] = \begin{bmatrix} \dfrac{\partial X}{\partial x} & \dfrac{\partial Y}{\partial x} & \dfrac{\partial Z}{\partial x} \\ \dfrac{\partial X}{\partial y} & \dfrac{\partial Y}{\partial y} & \dfrac{\partial Z}{\partial y} \\ \dfrac{\partial X}{\partial z} & \dfrac{\partial Y}{\partial z} & \dfrac{\partial Z}{\partial z} \end{bmatrix} \tag{26-14}$$

所以有

$$\boldsymbol{F}^{-\mathrm{T}} = \left(\boldsymbol{F}^{-1} \right)^{\mathrm{T}} = \left(\boldsymbol{F}^{\mathrm{T}} \right)^{-1} = \frac{\partial X_A}{\partial x_b} \boldsymbol{e}_b \otimes \boldsymbol{e}_A \tag{26-15}$$

这里先谈谈我们经常用到的几个点积.

例 26.1 求点积: $\boldsymbol{C} = \boldsymbol{F}^{\mathrm{T}} \boldsymbol{F}$, $\boldsymbol{C}$ 被称为 "右柯西–格林变形张量 (right Cauchy-Green deformation tensor)".

解: 由 (26-8) 式和 (26-5) 式, 有 (初学者要特别注意更换指标的事情):

$$\boldsymbol{F}^{\mathrm{T}} \boldsymbol{F} = \left(\frac{\partial x_i}{\partial X_K} \boldsymbol{e}_K \otimes \boldsymbol{e}_i \right) \cdot \left(\frac{\partial x_j}{\partial X_L} \boldsymbol{e}_j \otimes \boldsymbol{e}_L \right)$$

$$= \frac{\partial x_i}{\partial X_K} \frac{\partial x_j}{\partial X_L} \delta_{ij} \boldsymbol{e}_K \otimes \boldsymbol{e}_L = \frac{\partial x_i}{\partial X_K} \frac{\partial x_i}{\partial X_L} \boldsymbol{e}_K \otimes \boldsymbol{e}_L \tag{26-16}$$

显然, 右柯西–格林变形张量 $\boldsymbol{C} = \boldsymbol{F}^{\mathrm{T}} \boldsymbol{F}$ 是一个在参考构形中的拉格朗日型张量, 满足对称性: $\left(\boldsymbol{F}^{\mathrm{T}} \boldsymbol{F} \right)^{\mathrm{T}} = \boldsymbol{F}^{\mathrm{T}} \boldsymbol{F}$, 将主要用于构建基于参考构形的应变张量.

例 26.2 求点积: $\boldsymbol{B} = \boldsymbol{F} \boldsymbol{F}^{\mathrm{T}}$, $\boldsymbol{B}$ 被称为 "左柯西–格林变形张量 (left Cauchy-Green deformation tensor)".

解: 由 (26-5) 式和 (26-8) 式, 有

$$\boldsymbol{F} \boldsymbol{F}^{\mathrm{T}} = \left(\frac{\partial x_i}{\partial X_K} \boldsymbol{e}_i \otimes \boldsymbol{e}_K \right) \cdot \left(\frac{\partial x_j}{\partial X_L} \boldsymbol{e}_L \otimes \boldsymbol{e}_j \right)$$

$$= \frac{\partial x_i}{\partial X_K} \frac{\partial x_j}{\partial X_L} \delta_{KL} \boldsymbol{e}_i \otimes \boldsymbol{e}_j = \frac{\partial x_i}{\partial X_K} \frac{\partial x_j}{\partial X_K} \boldsymbol{e}_i \otimes \boldsymbol{e}_j \tag{26-17}$$

显然, 左柯西–格林变形张量 $\boldsymbol{B} = \boldsymbol{F} \boldsymbol{F}^{\mathrm{T}}$ 是一个在当前构形中的欧拉型张量, 且满足对称性: $\left(\boldsymbol{F} \boldsymbol{F}^{\mathrm{T}} \right)^{\mathrm{T}} = \boldsymbol{F} \boldsymbol{F}^{\mathrm{T}}$. $\boldsymbol{B}$ 也被广泛地称为芬格变形张量 (Finger deformation tensor), 主要是基于约瑟夫·芬格 (Josef Finger, 1841—1925) 于 1894 年的工作.

例 26.3 求点积: $\boldsymbol{c} = \boldsymbol{F}^{-\mathrm{T}} \boldsymbol{F}^{-1}$, $\boldsymbol{c}$ 被称为 "柯西变形张量 (Cauchy deformation tensor)".

解: 由 (26-15) 式和 (26-11) 式, 有

$$\boldsymbol{F}^{-\mathrm{T}} \boldsymbol{F}^{-1} = \left(\frac{\partial X_A}{\partial x_b} \boldsymbol{e}_b \otimes \boldsymbol{e}_A \right) \cdot \left(\frac{\partial X_B}{\partial x_c} \boldsymbol{e}_B \otimes \boldsymbol{e}_c \right)$$

$$=\frac{\partial X_A}{\partial x_b}\frac{\partial X_B}{\partial x_c}\delta_{AB}\boldsymbol{e}_b\otimes\boldsymbol{e}_c=\frac{\partial X_A}{\partial x_b}\frac{\partial X_A}{\partial x_c}\boldsymbol{e}_b\otimes\boldsymbol{e}_c \tag{26-18}$$

显然，柯西变形张量 $\boldsymbol{c}=\boldsymbol{F}^{-\mathrm{T}}\boldsymbol{F}^{-1}$ 是一个在当前构形中的欧拉型张量. 显然, 柯西变形张量和左柯西-格林变形张量之间满足如下基本关系式:

$$\boldsymbol{c}=\boldsymbol{B}^{-1} \tag{26-19}$$

上述关系式将主要用于当前构形中阿尔曼西应变 (Almansi strain) 的构建.

例 26.4 求点积: $\boldsymbol{F}^{-1}\boldsymbol{F}^{-\mathrm{T}}$, 显然, 该张量是右柯西-格林变形张量 $\boldsymbol{C}=\boldsymbol{F}^{\mathrm{T}}\boldsymbol{F}$ 的逆.

解: 由 (26-11) 式和 (26-15) 式, 有

$$\boldsymbol{F}^{-1}\boldsymbol{F}^{-\mathrm{T}}=\left(\frac{\partial X_A}{\partial x_b}\boldsymbol{e}_A\otimes\boldsymbol{e}_b\right)\cdot\left(\frac{\partial X_B}{\partial x_c}\boldsymbol{e}_c\otimes\boldsymbol{e}_B\right)$$

$$=\frac{\partial X_A}{\partial x_b}\frac{\partial X_B}{\partial x_c}\delta_{bc}\boldsymbol{e}_A\otimes\boldsymbol{e}_B=\frac{\partial X_A}{\partial x_b}\frac{\partial X_B}{\partial x_b}\boldsymbol{e}_A\otimes\boldsymbol{e}_B \tag{26-20}$$

显然, $\boldsymbol{C}^{-1}=\boldsymbol{F}^{-1}\boldsymbol{F}^{-\mathrm{T}}$ 是一个在参考构形中的拉格朗日型张量, 也经常性地被称为芬格张量 (Finger tensor).

26.2 变形梯度的极分解

极分解定理 (polar decomposition theorem) 表明, 任意一个可逆的非奇异的二阶张量 $\boldsymbol{F}$ 可被唯一的分解为

$$\boldsymbol{F}=\boldsymbol{R}\boldsymbol{U}=\boldsymbol{V}\boldsymbol{R} \tag{26-21}$$

式中, $\boldsymbol{R}$ 为正交张量中的转动张量, 满足 $\boldsymbol{R}^{\mathrm{T}}=\boldsymbol{R}^{-1}$, 表示纯转动. $\boldsymbol{U}$ 和 $\boldsymbol{V}$ 分别为对应于右、左极分解的正定对称的右、左伸长张量 (right and left stretch tensors):

$$\boldsymbol{U}=\boldsymbol{U}^{\mathrm{T}},\quad \boldsymbol{V}=\boldsymbol{V}^{\mathrm{T}} \tag{26-22}$$

26.2.1 右极分解

由 (26-21) 式中第一个等号和 (26-22) 式中第一式, 有 $\boldsymbol{F}^{\mathrm{T}}=\boldsymbol{U}\boldsymbol{R}^{\mathrm{T}}$, 因此有

$$\boldsymbol{F}^{\mathrm{T}}\boldsymbol{F}=\boldsymbol{C}=\boldsymbol{U}^2 \tag{26-23}$$

由谱分解定理知, 右柯西-格林变形张量 $\boldsymbol{C}$ 可以对角化为如下形式:

$$\boldsymbol{C}=\boldsymbol{U}^2=\sum_{\Gamma}\lambda_{\Gamma}^2\boldsymbol{N}_{\Gamma}\otimes\boldsymbol{N}_{\Gamma}\quad(\Gamma=\mathrm{I},\mathrm{II},\mathrm{III}) \tag{26-24}$$

式中 λ_Γ^2 和 $\boldsymbol{N}_\Gamma$ 分别为 $\boldsymbol{C}$ 的本征值和本征向量，$\boldsymbol{N}_\Gamma$ 亦为参考构形中的基矢. 再应用平方根定理 (square-root theorem)，右伸长张量 $\boldsymbol{U}$ 则可表示为

$$\boldsymbol{U} = \sqrt{\boldsymbol{C}} = \sum_\Gamma \lambda_\Gamma \boldsymbol{N}_\Gamma \otimes \boldsymbol{N}_\Gamma \quad (\Gamma = \text{I}, \text{II}, \text{III}) \tag{26-25}$$

λ_Γ 被称为主伸长 (principal stretch)，它们均为正.

26.2.2 左极分解

由 (26-21) 式中第二个等号和 (26-22) 式中第二式，有 $\boldsymbol{F}^\text{T} = \boldsymbol{R}^\text{T} \boldsymbol{V}$，则有

$$\boldsymbol{F}\boldsymbol{F}^\text{T} = \boldsymbol{B} = \boldsymbol{V}^2 \tag{26-26}$$

由 (26-17) 式知，$\boldsymbol{F}\boldsymbol{F}^\text{T} = \boldsymbol{B}$ 为一欧拉型张量，则左伸长张量 $\boldsymbol{V}$ 可通过主轴表示为

$$\boldsymbol{V} = \sum_{\gamma=\Gamma} \lambda_\gamma \boldsymbol{n}_\gamma \otimes \boldsymbol{n}_\gamma \quad (\lambda_\gamma = \lambda_\Gamma) \tag{26-27}$$

式中，$\boldsymbol{n}_\gamma$ 为当前构形中的基矢. 则左柯西–格林变形张量 $\boldsymbol{B}$ 可通过主轴表示为

$$\boldsymbol{B} = \boldsymbol{V}^2 = \boldsymbol{c}^{-1} = \boldsymbol{F}\boldsymbol{F}^\text{T} = \sum_{\gamma=\Gamma} \lambda_\gamma^2 \boldsymbol{n}_\gamma \otimes \boldsymbol{n}_\gamma \quad (\lambda_\gamma = \lambda_\Gamma) \tag{26-28}$$

式中，

$$\boldsymbol{c} = \boldsymbol{F}^{-\text{T}} \boldsymbol{F}^{-1} = \sum_{\gamma=\Gamma} \lambda_\gamma^{-2} \boldsymbol{n}_\gamma \otimes \boldsymbol{n}_\gamma \quad (\lambda_\gamma = \lambda_\Gamma) \tag{26-29}$$

是柯西变形张量.

§27. 拉格朗日描述下的格林应变与欧拉描述下的阿尔曼西应变

27.1 拉格朗日描述下有限变形的格林应变

业已表明，$\boldsymbol{C} = \boldsymbol{F}^\text{T}\boldsymbol{F}$ 是一个在参考构形中的拉格朗日型张量. 因此，可据此构造一个在参考构形中的应变张量，该张量被称为格林应变张量，也有书中将其称为拉格朗日应变的.

例 27.1 构造用变形梯度张量表示的格林应变张量.

解：由变形梯度张量的定义 (26-2) 式，有变形后和变形前线段微元的平方差为

$$\text{d}\boldsymbol{x}^2 - \text{d}\boldsymbol{X}^2 = \boldsymbol{F}\text{d}\boldsymbol{X} \cdot \boldsymbol{F}\text{d}\boldsymbol{X} - \text{d}\boldsymbol{X}^2 = \text{d}\boldsymbol{X}\boldsymbol{F}^\text{T} \cdot \boldsymbol{F}\text{d}\boldsymbol{X} - \text{d}\boldsymbol{X} \cdot \boldsymbol{I}\text{d}\boldsymbol{X}$$

第 5 章　变形运动学、功共轭

$$= \mathrm{d}\boldsymbol{X} \cdot \left(\boldsymbol{F}^{\mathrm{T}}\boldsymbol{F} - \mathbf{I}\right) \mathrm{d}\boldsymbol{X} = \mathrm{d}\boldsymbol{X} \cdot 2\boldsymbol{E}\mathrm{d}\boldsymbol{X} \tag{27-1}$$

因此可得，张量形式的格林应变为

$$\boldsymbol{E} = \frac{1}{2}\left(\boldsymbol{C} - \mathbf{I}\right) = \frac{1}{2}\left(\boldsymbol{F}^{\mathrm{T}}\boldsymbol{F} - \mathbf{I}\right) \tag{27-2}$$

事实上，著者在为研究生授课时，几乎每届都会有同学提问下列问题：为何要通过当前构形和参考构形中微分长度的平方差 $\mathrm{d}\boldsymbol{x}^2 - \mathrm{d}\boldsymbol{X}^2$ 来定义应变张量？

回答是在经典的连续介质力学中，有四种定义应变的方法：

(1) 长度变化量与原长的比，称为工程应变 (engineering strain)，主要适用于小变形情形，常见于材料力学和弹性力学教材. 考虑一维情形，设试件的原长 (original length) 和现长 (current length) 分别为 L 和 l，则工程应变：

$$\varepsilon = \frac{l - L}{L} = \frac{l}{L} - 1 = \lambda - 1 \tag{27-3}$$

式中，λ 称为 "伸长比 (stretch ratio)"，当试件拉伸时，$\lambda > 1$，从而工程应变大于零；反之，当试件压缩时，$\lambda < 1$，从而工程应变小于零.

例 27.2　如图 27.1 所示，详细推导二维的工程应变的表达式.

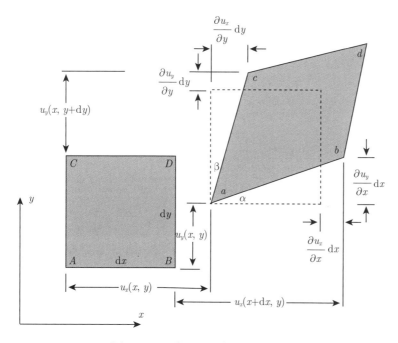

图 27.1　二维工程应变推导示意图

解: 这是一个繁琐但不难的问题, 通过仔细考察该问题, 可对工程应变的深入理解有很大好处.

首先推导沿 x 方向的应变. 我们知道线段 AB 的原长为 $\mathrm{d}x$, 其现长为 $\widehat{ab}$:

$$\widehat{ab} = \sqrt{\left(\mathrm{d}x + \frac{\partial u_x}{\partial x}\mathrm{d}x\right)^2 + \left(\frac{\partial u_y}{\partial x}\mathrm{d}x\right)^2} \tag{27-4}$$

则根据定义, 沿 x 方向的工程应变为

$$\varepsilon_x = \frac{\widehat{ab} - \widehat{AB}}{\widehat{AB}} = \frac{\sqrt{\left(\mathrm{d}x + \frac{\partial u_x}{\partial x}\mathrm{d}x\right)^2 + \left(\frac{\partial u_y}{\partial x}\mathrm{d}x\right)^2} - \mathrm{d}x}{\mathrm{d}x}$$

$$= \sqrt{\left(1 + \frac{\partial u_x}{\partial x}\right)^2 + \left(\frac{\partial u_y}{\partial x}\right)^2} - 1 = \underbrace{\sqrt{1 + 2\frac{\partial u_x}{\partial x} + \left(\frac{\partial u_x}{\partial x}\right)^2 + \left(\frac{\partial u_y}{\partial x}\right)^2}}_{\text{忽略掉二阶项}} - 1 \approx \frac{\partial u_x}{\partial x} \tag{27-5}$$

同理, 沿 y 方向的正应变为

$$\varepsilon_y = \frac{\widehat{ac} - \widehat{AC}}{\widehat{AC}} = \frac{\sqrt{\left(\mathrm{d}y + \frac{\partial u_y}{\partial y}\mathrm{d}y\right)^2 + \left(\frac{\partial u_x}{\partial y}\mathrm{d}y\right)^2} - \mathrm{d}y}{\mathrm{d}y} \approx \frac{\partial u_y}{\partial y} \tag{27-6}$$

工程切应变 (engineering shear strain) γ_{xy} 定义为角度的改变, 但必须应用小变形假设

$$\gamma_{xy} = \alpha + \beta \approx \tan\alpha + \tan\beta = \frac{(\partial u_y/\partial x)\mathrm{d}x}{\mathrm{d}x + \frac{\partial u_x}{\partial x}\mathrm{d}x} + \frac{(\partial u_x/\partial y)\mathrm{d}y}{\mathrm{d}y + \frac{\partial u_y}{\partial y}\mathrm{d}y}$$

$$= \frac{\partial u_y/\partial x}{1 + \underbrace{\partial u_x/\partial x}_{\text{和 1 比为小量}}} + \frac{\partial u_x/\partial y}{1 + \underbrace{\partial u_y/\partial y}_{\text{和 1 比为小量}}} \approx \frac{\partial u_y}{\partial x} + \frac{\partial u_x}{\partial y} \tag{27-7}$$

连续介质力学的切应变和工程切应变之间的关系为 $\varepsilon_{xy} = \frac{1}{2}\gamma_{xy}$.

值得特别关注的是, 工程应变不是张量的分量, 见本章思考题 5.14.

(2) 当前长度与原长比的对数, 称为亨奇应变 (Hencky strain)、对数应变 (logarithmic strain)[1.6] 或真应变 (true strain), 适用于大变形情形. 此时, 应变可表示为如下增量形式:

$$\mathrm{d}e = \frac{\mathrm{d}l}{l} \tag{27-8}$$

对上式从原长 L 积分到现长 l，则得到亨奇对数应变 e 的表达式：

$$e = \int_L^l \frac{\mathrm{d}l}{l} = \ln\frac{l}{L} = \ln\lambda = \ln(1+\varepsilon) \tag{27-9}$$

将上式亨奇对数应变 e 对柯西应变 ε 进行泰勒级数展开：

$$e = \ln(1+\varepsilon) = \varepsilon - \frac{\varepsilon^2}{2} + \frac{\varepsilon^3}{3} - \cdots \tag{27-10}$$

可见只有当柯西应变很小时，二阶及其以上的高阶项才可被忽略，两者才可近似等价.

(3) 现长度平方与原长度平方的差 $\mathrm{d}\boldsymbol{x}^2 - \mathrm{d}\boldsymbol{X}^2$ 与原长平方 $\mathrm{d}\boldsymbol{X}^2$ 比的一半，称为格林应变 (Green strain)，这是由英国自学成才的数学力学家格林 (George Green, 1793—1841) 于 1839 年引入的方法. 格林应变满足对称性：$\boldsymbol{E} = \boldsymbol{E}^{\mathrm{T}}$.

例 27.3 用最简单的一维杆的变形 (如图 27.2 所示) 来说明格林应变定义的有效性.

图 27.2 一维杆的变形

解：取如图所示的微元体，位移增量为 $\mathrm{d}u(x) = \mathrm{d}x' - \mathrm{d}x$，则现长与原长的平方差为

$$(\mathrm{d}x')^2 - (\mathrm{d}x)^2 = (\mathrm{d}x + \mathrm{d}u)^2 - (\mathrm{d}x)^2 = \underbrace{2\mathrm{d}x\mathrm{d}u}_{\text{保留 }\mathrm{d}u\text{ 一阶项}} + \underbrace{\mathrm{d}u\mathrm{d}u}_{\text{忽略掉二阶项}}$$

$$\approx 2\mathrm{d}x\mathrm{d}u = 2\mathrm{d}x\frac{\partial u}{\partial x}\mathrm{d}x = 2\mathrm{d}x\underbrace{\varepsilon_{xx}}_{\text{一维应变}}\mathrm{d}x \tag{27-11}$$

在忽略掉二阶项 $\mathrm{d}u\mathrm{d}u$ 时，则得到柯西应变为

$$\varepsilon_{xx} = \frac{1}{2}\frac{(\mathrm{d}x')^2 - (\mathrm{d}x)^2}{(\mathrm{d}x)^2} = \frac{\partial u}{\partial x} \tag{27-12}$$

如果进一步保留二阶项 $\mathrm{d}u\mathrm{d}u$ 时，则一维的格林应变为

$$(\mathrm{d}x')^2 - (\mathrm{d}x)^2 = (\mathrm{d}x + \mathrm{d}u)^2 - (\mathrm{d}x)^2 = 2\mathrm{d}x\mathrm{d}u + \underbrace{\mathrm{d}u\mathrm{d}u}_{\text{保留 }\mathrm{d}u\text{ 二阶项}}$$

$$= 2\mathrm{d}x\frac{\partial u}{\partial x}\mathrm{d}x + \mathrm{d}x\left(\frac{\partial u}{\partial x}\right)^2\mathrm{d}x = 2\mathrm{d}x\underbrace{\varepsilon_{xx}}_{\text{大变形}}\mathrm{d}x \tag{27-13}$$

则在大变形时，一维的格林应变为

$$\varepsilon_{xx} = \frac{1}{2}\frac{(\mathrm{d}x')^2 - (\mathrm{d}x)^2}{(\mathrm{d}x)^2} = \frac{\partial u}{\partial x} + \frac{1}{2}\left(\frac{\partial u}{\partial x}\right)^2 \tag{27-14}$$

例 27.4 再用二维板的面内变形 (如图 27.3 所示) 来说明格林应变定义的有效性.

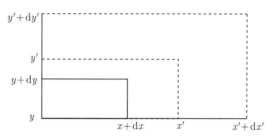

图 27.3　二维板的面内变形

解：类似于上例的一维情形，取如图所示的二维微元体，位移增量为

$$\mathrm{d}x' = \mathrm{d}x + \mathrm{d}u_x, \quad \mathrm{d}y' = \mathrm{d}y + \mathrm{d}u_y \tag{27-15}$$

则现长与原长的平方差为

$$(\mathrm{d}x')^2 + (\mathrm{d}y')^2 - (\mathrm{d}x)^2 - (\mathrm{d}y)^2 = (\mathrm{d}x + \mathrm{d}u_x)^2 - (\mathrm{d}x)^2 + (\mathrm{d}y + \mathrm{d}u_y)^2 - (\mathrm{d}y)^2$$

$$= \underbrace{2\mathrm{d}x\mathrm{d}u_x + 2\mathrm{d}y\mathrm{d}u_y + (\mathrm{d}u_x)^2 + (\mathrm{d}u_y)^2}_{\text{保留 } \mathrm{d}u_x \text{ 和 } \mathrm{d}u_y \text{ 一阶项}} \approx 2\mathrm{d}x\mathrm{d}u_x + 2\mathrm{d}y\mathrm{d}u_y$$

$$= 2\left[\mathrm{d}x\left(\frac{\partial u_x}{\partial x}\mathrm{d}x + \frac{\partial u_x}{\partial y}\mathrm{d}y\right) + \mathrm{d}y\left(\frac{\partial u_y}{\partial x}\mathrm{d}x + \frac{\partial u_y}{\partial y}\mathrm{d}y\right)\right]$$

$$= 2\mathrm{d}x\underbrace{\frac{\partial u_x}{\partial x}}_{\varepsilon_{xx}}\mathrm{d}x + 2\mathrm{d}x\underbrace{\left(\frac{\partial u_x}{\partial y} + \frac{\partial u_y}{\partial x}\right)}_{2\varepsilon_{xy}=2\varepsilon_{yx}}\mathrm{d}y + 2\mathrm{d}y\underbrace{\frac{\partial u_y}{\partial y}}_{\varepsilon_{yy}}\mathrm{d}y$$

$$= 2\mathrm{d}x\frac{\partial u_x}{\partial x}\mathrm{d}x + \mathrm{d}x\left(\frac{\partial u_x}{\partial y} + \frac{\partial u_y}{\partial x}\right)\mathrm{d}y + \mathrm{d}y\left(\frac{\partial u_y}{\partial x} + \frac{\partial u_x}{\partial y}\right)\mathrm{d}x + 2\mathrm{d}y\frac{\partial u_y}{\partial y}\mathrm{d}y$$

$$= 2\begin{bmatrix}\mathrm{d}x & \mathrm{d}y\end{bmatrix}\begin{bmatrix}\varepsilon_{xx} & \varepsilon_{xy} \\ \varepsilon_{yx} & \varepsilon_{yy}\end{bmatrix}\begin{bmatrix}\mathrm{d}x \\ \mathrm{d}y\end{bmatrix} \tag{27-16}$$

式中，在忽略掉二阶项 $(\mathrm{d}u_x)^2 + (\mathrm{d}u_y)^2$ 时，柯西应变的各分量为

$$\varepsilon_{xx} = \frac{\partial u_x}{\partial x}, \quad \varepsilon_{yy} = \frac{\partial u_y}{\partial y}, \quad \varepsilon_{xy} = \varepsilon_{yx} = \frac{1}{2}\left(\frac{\partial u_y}{\partial x} + \frac{\partial u_x}{\partial y}\right) \tag{27-17}$$

如果不忽略掉二阶项，将 $(\mathrm{d}u_x)^2 = \left(\dfrac{\partial u_x}{\partial x}\mathrm{d}x + \dfrac{\partial u_x}{\partial y}\mathrm{d}y\right)^2$ 和 $(\mathrm{d}u_y)^2 = \left(\dfrac{\partial u_y}{\partial x}\mathrm{d}x + \dfrac{\partial u_y}{\partial y}\mathrm{d}y\right)^2$ 代入 (27-16) 式，则 (27-17) 式将变为

$$\begin{cases} \varepsilon_{xx} = \dfrac{\partial u_x}{\partial x} + \dfrac{1}{2}\left[\left(\dfrac{\partial u_x}{\partial x}\right)^2 + \left(\dfrac{\partial u_y}{\partial x}\right)^2\right] \\[2mm] \varepsilon_{yy} = \dfrac{\partial u_y}{\partial y} + \dfrac{1}{2}\left[\left(\dfrac{\partial u_x}{\partial y}\right)^2 + \left(\dfrac{\partial u_y}{\partial y}\right)^2\right] \\[2mm] \varepsilon_{xy} = \varepsilon_{yx} = \dfrac{1}{2}\left(\dfrac{\partial u_y}{\partial x} + \dfrac{\partial u_x}{\partial y} + \dfrac{\partial u_x}{\partial x}\dfrac{\partial u_x}{\partial y} + \dfrac{\partial u_y}{\partial x}\dfrac{\partial u_y}{\partial y}\right) \end{cases} \quad (27\text{-}18)$$

(4) 现长度平方与原长度平方的差 $\mathrm{d}\boldsymbol{x}^2 - \mathrm{d}\boldsymbol{X}^2$ 与现长平方 $\mathrm{d}\boldsymbol{x}^2$ 比的一半，称为阿尔曼西应变 (Almansi strain)，见 §27.2 的讨论.

例 27.5 朗道和栗弗席兹教程第 7 卷《弹性理论》[1.8] 中在不区分两种构形的情况下，如何构造格林应变张量？

解：朗道十卷理论物理教程主要是为物理学家撰写的. 《弹性理论》[1.8] 虽然只有区区一百多页，但却包括了物理学家感兴趣的几乎所有的弹性形变的问题，包括液晶力学的这种软物质力学的内容. 该书十分深刻，但行文简约，对问题的讨论往往是单刀直入，作为研究生教材的难度也是举世公认的.

设变形前质点的矢径为 $\boldsymbol{r}$，发生位移 $\boldsymbol{u}$ 后，质点新的矢径为 $\boldsymbol{r}'$，分量形式为

$$\mathrm{d}u_i = \mathrm{d}x'_i - \mathrm{d}x_i \quad (i = x, y, z) \tag{27-19}$$

变形前后线段的平方差为

$$\mathrm{d}l'^2 - \mathrm{d}l^2 = (\mathrm{d}x_i + \mathrm{d}u_i)^2 - \mathrm{d}x_i^2 = 2\mathrm{d}x_i\mathrm{d}u_i + \mathrm{d}u_i^2 \tag{27-20}$$

由于全微分为

$$\mathrm{d}u_i = \dfrac{\partial u_i}{\partial x_k}\mathrm{d}x_k \tag{27-21}$$

式中，$i = x, y, z$ 为自由指标 (free index)，而 $k = x, y, z$ 为哑标 (dummy index). 例如，对于 $i = x$，(27-21) 式为 $\mathrm{d}u_x = \dfrac{\partial u_x}{\partial x}\mathrm{d}x + \dfrac{\partial u_x}{\partial y}\mathrm{d}y + \dfrac{\partial u_x}{\partial z}\mathrm{d}z$.

将 (27-21) 式代入 (27-20) 式，并注意到 (27-20) 式最右端项 $\mathrm{d}u_i^2$ 可更换哑标为 $\mathrm{d}u_l \cdot \mathrm{d}u_l$，则 (27-20) 式可进一步表示为

$$\mathrm{d}l'^2 - \mathrm{d}l^2 = 2\mathrm{d}x_i\mathrm{d}u_i + \mathrm{d}u_l^2 = 2\mathrm{d}x_i\dfrac{\partial u_i}{\partial x_k}\mathrm{d}x_k + \mathrm{d}x_i\dfrac{\partial u_l}{\partial x_i}\dfrac{\partial u_l}{\partial x_k}\mathrm{d}x_k \tag{27-22}$$

式中，考虑到指标的对称性，$2\mathrm{d}x_i \dfrac{\partial u_i}{\partial x_k} \mathrm{d}x_k$ 可进一步表示为

$$\begin{aligned} \mathrm{d}l'^2 - \mathrm{d}l^2 &= 2\mathrm{d}x_i \dfrac{\partial u_i}{\partial x_k} \mathrm{d}x_k + \mathrm{d}x_i \dfrac{\partial u_l}{\partial x_i} \dfrac{\partial u_l}{\partial x_k} \mathrm{d}x_k \\ &= \mathrm{d}x_i \left(\dfrac{\partial u_i}{\partial x_k} + \dfrac{\partial u_k}{\partial x_i} \right) \mathrm{d}x_k + \mathrm{d}x_i \dfrac{\partial u_l}{\partial x_i} \dfrac{\partial u_l}{\partial x_k} \mathrm{d}x_k \\ &= \mathrm{d}x_i \cdot 2 E_{ik} \cdot \mathrm{d}x_k \end{aligned} \tag{27-23}$$

则应变 E_{ij} 的指标形式为

$$E_{ik} = \dfrac{1}{2} \left(\dfrac{\partial u_i}{\partial x_k} + \dfrac{\partial u_k}{\partial x_i} + \dfrac{\partial u_l}{\partial x_i} \dfrac{\partial u_l}{\partial x_k} \right) \tag{27-24}$$

上式的张量形式为

$$\boldsymbol{E} = \dfrac{1}{2} \left(\boldsymbol{u} \otimes \boldsymbol{\nabla} + \boldsymbol{\nabla} \otimes \boldsymbol{u} + \boldsymbol{\nabla} \otimes \boldsymbol{u} \cdot \boldsymbol{u} \otimes \boldsymbol{\nabla} \right) \tag{27-25}$$

在 (27-24) 式或 (27-25) 式中，由于 $\dfrac{\partial u_l}{\partial x_i} \dfrac{\partial u_l}{\partial x_k}$ 或 $\boldsymbol{\nabla} \otimes \boldsymbol{u} \cdot \boldsymbol{u} \otimes \boldsymbol{\nabla}$ 和前两项相比为二阶小量，在小变形时可略去，此时可表示为

$$E_{ik} \approx \dfrac{1}{2} \left(\dfrac{\partial u_i}{\partial x_k} + \dfrac{\partial u_k}{\partial x_i} \right), \quad \boldsymbol{E} \approx \dfrac{1}{2} \left(\boldsymbol{u} \otimes \boldsymbol{\nabla} + \boldsymbol{\nabla} \otimes \boldsymbol{u} \right) \tag{27-26}$$

(27-26) 式所表示的应变就是柯西应变张量 (12-24) 式.

值得注意的是，朗道第 7 卷的推导方法和例 27.1 的区别在于 (27-21) 式：

$$\mathrm{d}u_i = \dfrac{\partial u_i}{\partial X_K} \mathrm{d}X_K = F_{iK} \mathrm{d}X_K \tag{27-27}$$

亦即，未明确地区分参考构形和当前构形，也未引入变形梯度张量. 但是，朗道教程中隐含了两种构形，所以殊途同归.

图 27.4 左：慕尼黑工业大学设立的弗普尔奖章；右：弗普尔像

例 27.6 给出薄平板的大挠度 (large deflections of thin flat plates) 的弗普尔–冯·卡门应变 (Föppl–von Kármán strain, FvK strain) 的详细推导过程.

解：弗普尔–冯·卡门应变主要基于奥古斯特·奥托·弗普尔 (August Otto Föppl, 1854—1924，见图 27.4；本书经常提及的普朗特的博士导师和岳父，见图

55.4) 于 1907 年 [5.1] 和冯·卡门于 1910 年 [5.2] 的贡献. 设板的挠度为 ξ, 在上例的基础上, 有

$$\mathrm{d}l'^2 - \mathrm{d}l^2 = (\mathrm{d}x_i + \mathrm{d}u_i)^2 + \mathrm{d}\xi^2 - \mathrm{d}x_i^2 = 2\mathrm{d}x_i\mathrm{d}u_i + \mathrm{d}u_i^2 + \mathrm{d}\xi^2 \tag{27-28}$$

在上式中, 忽略掉面内变形的二阶项 $\mathrm{d}u_i^2$, 这样 (27-28) 式变为

$$\mathrm{d}l'^2 - \mathrm{d}l^2 = 2\mathrm{d}x_i\mathrm{d}u_i + \mathrm{d}\xi^2 = 2\mathrm{d}x_i\frac{\partial u_i}{\partial x_k}\mathrm{d}x_k + \mathrm{d}x_i\frac{\partial \xi}{\partial x_i}\frac{\partial \xi}{\partial x_k}\mathrm{d}x_k$$

$$= \mathrm{d}x_i\left(\frac{\partial u_i}{\partial x_k} + \frac{\partial u_k}{\partial x_i}\right)\mathrm{d}x_k + \mathrm{d}x_i\frac{\partial \xi}{\partial x_i}\frac{\partial \xi}{\partial x_k}\mathrm{d}x_k \tag{27-29}$$

故, 弗普尔–冯·卡门应变张量为

$$\varepsilon_{ik} = \underbrace{\frac{1}{2}\left(\frac{\partial u_i}{\partial x_k} + \frac{\partial u_k}{\partial x_i}\right)}_{\text{平板面内小变形}} + \underbrace{\frac{1}{2}\frac{\partial \xi}{\partial x_i}\frac{\partial \xi}{\partial x_k}}_{\text{板横向大变形}} \quad \text{或} \quad \boldsymbol{E} = \frac{1}{2}\left(\boldsymbol{u}\otimes\boldsymbol{\nabla} + \boldsymbol{\nabla}\otimes\boldsymbol{u}\right) + \frac{1}{2}(\boldsymbol{\nabla}\xi)\otimes(\xi\boldsymbol{\nabla}) \tag{27-30}$$

例 27.7 给出用算子表示的格林应变张量的详细推导过程.

解: 由 (27-25) 式所给出的格林应变未区分两种构形. 本例的目的是要给出区分两种构形的格林应变表达式.

位移场可在当前构形 (基矢量 $\boldsymbol{e}_i$, 指标用小写) 和参考构形 (基矢量 $\boldsymbol{e}_J$, 指标用大写) 下分别表示为

$$\boldsymbol{u}(\boldsymbol{x}) = \boldsymbol{u}(\boldsymbol{X}) \tag{27-31}$$

在欧拉描述下, 位移场可以用矢量和分量分别表示为

$$\begin{cases} \boldsymbol{u}(\boldsymbol{x},t) = \boldsymbol{x} - \boldsymbol{X}(\boldsymbol{x},t) = u_i\boldsymbol{e}_i \\ u_i = x_i - \delta_{iJ}X_J \end{cases} \tag{27-32}$$

在拉格朗日描述下, 位移场可以用矢量和分量分别表示为

$$\begin{cases} \boldsymbol{u}(\boldsymbol{X},t) = \boldsymbol{x}(\boldsymbol{X},t) - \boldsymbol{X} = u_J\boldsymbol{e}_J \\ u_J = \delta_{Ji}x_i - X_J \end{cases} \tag{27-33}$$

且 (27-32) 和 (27-33) 两式是等价的, 满足转换关系 $\boldsymbol{e}_i \cdot \boldsymbol{e}_J = \delta_{iJ} = \delta_{Ji}$, $\boldsymbol{e}_i = \delta_{iJ}\boldsymbol{e}_J$, 和 $u_i = \delta_{iJ}u_J$, $u_J = \delta_{Ji}u_i$, 则其等价关系可表示为

$$\boldsymbol{u}(\boldsymbol{x}) = u_i\boldsymbol{e}_i = u_i\delta_{iJ}\boldsymbol{e}_J = u_J\boldsymbol{e}_J = \boldsymbol{u}(\boldsymbol{X}) \tag{27-34}$$

在拉格朗日描述下，可以得到位移梯度张量（$H = u \otimes \nabla_X$ 或 $\text{Grad} u$）与变形梯度张量 $F(X,t)$ 的关系，可分别用张量和分量表示为

$$\begin{cases} H = u \otimes \nabla_X = x(X,t) \otimes \nabla_X - X \otimes \nabla_X = F(X,t) - I \\ H_{iK} = \dfrac{\partial u_i}{\partial X_K} = F_{iK} - \delta_{iK} \end{cases} \quad (27\text{-}35)$$

其中，$F = F_{iK} e_i \otimes e_K$，$F_{iK} = \partial x_i / \partial X_K$；由 (27-34) 式知，(27-35) 第一式左端可表示为

$$u \otimes \nabla_X = \dfrac{\partial u_J}{\partial X_K} E_J \otimes E_K = \dfrac{\partial u_i}{\partial X_K} e_i \otimes E_K. \quad (27\text{-}36)$$

同理，在欧拉描述下，我们可以得到的位移梯度张量（$u \otimes \nabla_x$ 或 $\text{grad} u$）与变形梯度张量的逆 $F^{-1}(x,t)$ 的关系：

$$\begin{cases} u \otimes \nabla_x = x \otimes \nabla_x - X(x,t) \otimes \nabla_x = I - F^{-1}(x,t) \\ \dfrac{\partial u_J}{\partial x_k} = \delta_{Jk} - F_{Jk}^{-1} \end{cases} \quad (27\text{-}37)$$

由 (27-34) 式知，(27-37) 式左端可表示为

$$u \otimes \nabla_x = \dfrac{\partial u_i}{\partial x_k} e_i \otimes e_k = \dfrac{\partial u_J}{\partial x_k} e_J \otimes e_k. \quad (27\text{-}38)$$

由 (27-35) 式以及右柯西–格林变形张量 C 和格林应变张量 E 的定义：

$$E = \dfrac{1}{2}(C - I) = \dfrac{1}{2}\left(F^T F - I\right) = \dfrac{1}{2}\left[(u \otimes \nabla_X + I)^T \cdot (u \otimes \nabla_X + I) - I\right]$$

$$= \dfrac{1}{2}(u \otimes \nabla_X + \nabla_X \otimes u + \nabla_X \otimes u \cdot u \otimes \nabla_X) \quad (27\text{-}39)$$

为证明格林应变张量在直角坐标系下的分量形式，可以从两种形式出发：

形式一：由 F 的张量形式，有

$$F^T F = F_{iK} e_K \otimes e_i \cdot F_{mL} e_m \otimes e_L = F_{iK} F_{mL} \delta_{im} e_K e_L = F_{iK} F_{iL} e_K \otimes e_L \quad (27\text{-}40)$$

其中 $F_{iK} F_{iL} = \dfrac{\partial x_i}{\partial X_K} \dfrac{\partial x_i}{\partial X_L}$，则有 $E_{KL} = \dfrac{1}{2}\left(\dfrac{\partial x_i}{\partial X_K} \dfrac{\partial x_i}{\partial X_L} - \delta_{KL}\right)$.

形式二：由 F 与 $u \otimes \nabla_X$ 的关系式出发，根据 (27-35) 式和 (27-40) 式，有

$$F^T = F_{iK} e_K \otimes e_i = \left(\dfrac{\partial u_i}{\partial X_K} e_i \otimes e_K + \delta_{iK} e_i \otimes e_K\right)^T = \dfrac{\partial u_i}{\partial X_K} e_K \otimes e_i + \delta_{iK} e_K \otimes e_i \quad (27\text{-}41)$$

$$F = F_{mL} e_m \otimes e_L = \dfrac{\partial u_m}{\partial X_L} e_m \otimes e_L + \delta_{mL} e_m \otimes e_L \quad (27\text{-}42)$$

根据位移关系 (27-36) 式可知

$$\frac{\partial u_i}{\partial X_K}\boldsymbol{e}_K \otimes \boldsymbol{e}_i = \frac{\partial u_J}{\partial X_K}\boldsymbol{e}_K \otimes \boldsymbol{e}_J, \quad \frac{\partial u_m}{\partial X_L}\boldsymbol{e}_m \otimes \boldsymbol{e}_L = \frac{\partial u_N}{\partial X_L}\boldsymbol{e}_N \otimes \boldsymbol{e}_L \tag{27-43}$$

同时有

$$\delta_{iK}\boldsymbol{e}_K \otimes \boldsymbol{e}_i = \boldsymbol{e}_K \otimes \boldsymbol{e}_K, \delta_{mL}\boldsymbol{e}_m \otimes \boldsymbol{e}_L = \boldsymbol{e}_L \otimes \boldsymbol{e}_L \tag{27-44}$$

由 (27-41)—(27-44) 诸式, 可知 (27-39) 式中的右柯西-格林变形张量为

$$\begin{aligned}
\boldsymbol{F}^\mathrm{T}\boldsymbol{F} &= \left(\frac{\partial u_i}{\partial X_K}\boldsymbol{e}_K \otimes \boldsymbol{e}_i + \delta_{iK}\boldsymbol{e}_K \otimes \boldsymbol{e}_i\right) \cdot \left(\frac{\partial u_m}{\partial X_L}\boldsymbol{e}_m \otimes \boldsymbol{e}_L + \delta_{mL}\boldsymbol{e}_m \otimes \boldsymbol{e}_L\right) \\
&= \left(\frac{\partial u_J}{\partial X_K}\boldsymbol{e}_K \otimes \boldsymbol{e}_J + \boldsymbol{e}_K \otimes \boldsymbol{e}_K\right) \cdot \left(\frac{\partial u_N}{\partial X_L}\boldsymbol{e}_N \otimes \boldsymbol{e}_L + \boldsymbol{e}_L \otimes \boldsymbol{e}_L\right) \\
&= \frac{\partial u_J}{\partial X_K}\frac{\partial u_N}{\partial X_L}\delta_{JN}\boldsymbol{e}_K \otimes \boldsymbol{e}_L + \frac{\partial u_J}{\partial X_K}\delta_{JL}\boldsymbol{e}_K \otimes \boldsymbol{e}_L + \frac{\partial u_N}{\partial X_L}\delta_{KN}\boldsymbol{e}_K \otimes \boldsymbol{e}_L + \delta_{KL}\boldsymbol{e}_K \otimes \boldsymbol{e}_L \\
&= \left(\frac{\partial u_L}{\partial X_K} + \frac{\partial u_K}{\partial X_L} + \frac{\partial u_J}{\partial X_K}\frac{\partial u_J}{\partial X_L} + \delta_{KL}\right)\boldsymbol{e}_K \otimes \boldsymbol{e}_L
\end{aligned} \tag{27-45}$$

将 (27-45) 式代回 (27-39) 式, 则格林应变张量为

$$\boldsymbol{E} = \frac{1}{2}\left(\boldsymbol{F}^\mathrm{T}\boldsymbol{F} - \boldsymbol{I}\right) = \frac{1}{2}\left(\frac{\partial u_L}{\partial X_K} + \frac{\partial u_K}{\partial X_L} + \frac{\partial u_J}{\partial X_K}\frac{\partial u_J}{\partial X_L}\right)\boldsymbol{e}_K \otimes \boldsymbol{e}_L \tag{27-46}$$

亦即格林应变的分量形式为

$$E_{KL} = \frac{1}{2}\left(\frac{\partial u_L}{\partial X_K} + \frac{\partial u_K}{\partial X_L} + \frac{\partial u_J}{\partial X_K}\frac{\partial u_J}{\partial X_L}\right) \tag{27-47}$$

27.2 欧拉描述下有限变形的阿尔曼西应变

正如 §27.1 中所指出的, 现长度平方与原长度平方的差 $\mathrm{d}\boldsymbol{x}^2 - \mathrm{d}\boldsymbol{X}^2$ 与现原长平方 $\mathrm{d}\boldsymbol{x}^2$ 比之半, 称为阿尔曼西应变 (Almansi strain)[5.3]. 注意到 $\mathrm{d}\boldsymbol{X} = \boldsymbol{F}^{-1}\mathrm{d}\boldsymbol{x}$, 则有

$$\begin{aligned}
\mathrm{d}\boldsymbol{x}^2 - \mathrm{d}\boldsymbol{X}^2 &= \mathrm{d}\boldsymbol{x}^2 - \boldsymbol{F}^{-1}\mathrm{d}\boldsymbol{x} \cdot \boldsymbol{F}^{-1}\mathrm{d}\boldsymbol{x} = \mathrm{d}\boldsymbol{x}^2 - \mathrm{d}\boldsymbol{x} \cdot \boldsymbol{F}^{-\mathrm{T}}\boldsymbol{F}^{-1}\mathrm{d}\boldsymbol{x} \\
&= \mathrm{d}\boldsymbol{x} \cdot \left(\boldsymbol{I} - \boldsymbol{F}^{-\mathrm{T}}\boldsymbol{F}^{-1}\right)\mathrm{d}\boldsymbol{x} = \mathrm{d}\boldsymbol{x} \cdot 2\boldsymbol{e}\mathrm{d}\boldsymbol{x}
\end{aligned} \tag{27-48}$$

则阿尔曼西应变张量为

$$\boldsymbol{e} = \frac{1}{2}\left(\boldsymbol{I} - \boldsymbol{F}^{-\mathrm{T}}\boldsymbol{F}^{-1}\right) = \frac{1}{2}\sum_{\gamma=1}^{3}\left(1 - \frac{1}{\lambda_\gamma^2}\right)\boldsymbol{n}_\gamma \otimes \boldsymbol{n}_\gamma \tag{27-49}$$

由 (26-17)—(26-19) 诸式，可见阿尔曼西应变可通过左柯西–格林变形张量 $\boldsymbol{B} = \boldsymbol{F}\boldsymbol{F}^{\mathrm{T}}$ 和柯西变形张量 $\boldsymbol{c} = \boldsymbol{F}^{-\mathrm{T}}\boldsymbol{F}^{-1}$ 来表示为

$$\boldsymbol{e} = \frac{1}{2}\left(\boldsymbol{I} - \boldsymbol{c}\right) = \frac{1}{2}\left(\boldsymbol{I} - \boldsymbol{B}^{-1}\right) \tag{27-50}$$

§28. 赛斯–希尔应变度量

28.1 希尔应变度量

理性连续介质力学中应变有十几种，不要说作为应用者的工程师们被各种应变概念弄得头晕目眩，就是专门从事理性力学研究的学者很多时候也有莫名其妙的感觉.

在应变定义门派林立，众说纷纭之时，事实上是最能体现钱学森称之的理性力学 "把关的工作" 作用的是希尔应变度量所做的 "统一" 或 "抽象"，真可谓 "沧海横流，方显英雄本色".

罗德尼·希尔 (Rodney Hill, 1921—2011) 于 1968 年，在右伸长张量 $\boldsymbol{U} = \sqrt{\boldsymbol{C}} = \sqrt{\boldsymbol{F}^{\mathrm{T}}\boldsymbol{F}}$ 这一拉格朗日型张量的基础上，定义了通类 (a general class) 应变度量函数 [5.4,5.5]：

$$\boldsymbol{E}_{\mathrm{Hill}} = \boldsymbol{f}(\boldsymbol{U}) = \sum_{\Gamma} f(\lambda_{\Gamma})\boldsymbol{N}_{\Gamma} \otimes \boldsymbol{N}_{\Gamma} \tag{28-1}$$

式中，$f(\lambda_{\Gamma})$ 为一光滑、严格递增的标量函数，说明二阶张量函数 $\boldsymbol{f}(\boldsymbol{U})$ 同右伸长张量 $\boldsymbol{U}$ 共轴. (28-1) 式中的标量函数 $f(\lambda_{\Gamma})$ 满足如下三个条件：

(1) 伸长比为 1 时，没有发生变形，应变为零:

$$f(1) = 0 \tag{28-2}$$

(2) 应变是递增函数，换句话说应变随伸长比的递增而递增:

$$\frac{\mathrm{d}f}{\mathrm{d}\lambda_{\Gamma}} > 0 \tag{28-3}$$

(3) 小应变的退化条件:

$$\frac{\mathrm{d}f}{\mathrm{d}\lambda_{\Gamma}}\bigg|_{\lambda_{\Gamma}=1} = f'(1) = 1 \tag{28-4}$$

亦即，在小应变时满足

$$\mathrm{d}f = \mathrm{d}\lambda_{\Gamma} \tag{28-5}$$

例 28.1 进一步透彻地解释 (28-4) 和 (28-5) 两式.

解：由于 $\lambda_\Gamma = 1$ 对应于未发生变形，故对伸长比 λ_Γ 在 1 附近进行小参量展开：

$$\lambda_\Gamma \approx 1 + \Delta\lambda, \quad \Delta\lambda \ll 1 \tag{28-6}$$

此时，由于原长为 1，所以所对应的小应变为

$$\varepsilon = \frac{\Delta\lambda}{1} = \Delta\lambda \tag{28-7}$$

将 (28-1) 式中的应变标量函数 $f(\lambda_\Gamma)$ 的参数在 1 附近进行一阶泰勒展开，注意到 (28-2) 式，有

$$f(\lambda_\Gamma) \approx f(1 + \Delta\lambda) = f(1) + f'(1)\Delta\lambda + O\left((\Delta\lambda)^2\right)$$
$$\approx f'(1)\Delta\lambda \tag{28-8}$$

上式说明，只有当 (28-4) 式成立，亦即 $f'(1) = 1$ 时，才有

$$f(\lambda_\Gamma) \approx \Delta\lambda \tag{28-9}$$

换句话说，只有满足 $f'(1) = 1$ 时，才能保证希尔应变 (28-1) 式在小应变时，与已有柯西应变的定义相一致。再由于

$$\mathrm{d}f = f(1 + \mathrm{d}\lambda) - f(1) = f(1 + \mathrm{d}\lambda) - 0 \approx \mathrm{d}\lambda \tag{28-10}$$

故，(28-5) 式得证。以上三个条件可谓对何种参量能作为应变度量做出了诠释。

(28-1) 式给出的是拉格朗日描述时的希尔应变度量的主轴形式，同理，可通过左伸长张量 $\boldsymbol{V} = \sqrt{\boldsymbol{B}} = \sqrt{\boldsymbol{F}\boldsymbol{F}^{\mathrm{T}}}$ 这一欧拉型张量给出如下欧拉描述时的希尔应变度量的主轴形式：

$$\boldsymbol{e}_{\mathrm{Hill}} = \boldsymbol{f}(\boldsymbol{V}) = \sum_\gamma f(\lambda_\gamma)\boldsymbol{n}_\gamma \otimes \boldsymbol{n}_\gamma \tag{28-11}$$

拉格朗日和欧拉描述的应变度量之间满足如下关系：

$$\boldsymbol{E}_{\mathrm{Hill}} = \boldsymbol{R}^{\mathrm{T}}\boldsymbol{e}_{\mathrm{Hill}}\boldsymbol{R}, \quad \boldsymbol{e}_{\mathrm{Hill}} = \boldsymbol{R}\boldsymbol{E}_{\mathrm{Hill}}\boldsymbol{R}^{\mathrm{T}}, \quad \boldsymbol{R} = \sum_{\gamma,\Gamma}\boldsymbol{n}_\gamma \otimes \boldsymbol{N}_\Gamma \tag{28-12}$$

希尔于 1950 年在时年 28 岁时，出版了名著《塑性数学理论》(*The Mathematical Theory of Plasticity*)[5.6]，该书很快奠定了其该领域国际权威学者的地位 (This book very rapidly established Hill as an international authority on the subject)，该书迄今已经被引用了一万多次。希尔因为理性力学等方面的工作，被广泛认为是 20 世纪下半叶理性力学和固体力学基础的最重要贡献者，在行业内俗称"第一把交椅"。

28.2 赛斯应变度量

印度理性力学家、印度理论与应用力学学会 (ISTAM) 的创始人赛斯 (Bhoj Raj Seth, 1907—1979) 创造性地将 (28-1) 式中的特征值 λ_α 的度量函数表示为[5.7-5.9]

$$f(\lambda_\alpha) = \begin{cases} \dfrac{1}{2m}\left(\lambda_\alpha^{2m} - 1\right), & \text{对于 } m \neq 0 \\ \ln \lambda_\alpha, & \text{对于 } m = 0 \end{cases} \tag{28-13}$$

式中，$2m$ 为正或负的整数，函数 $f(\lambda_\alpha)$ 满足 (28-2)—(28-4) 式的三个限制条件，其所对应的函数关系可用图 28.1 表示出。图中 L 和 l 分别对应参考构形和当前构形中微小线段的长度。图右下角小框中是 $\lambda_\alpha = 1$ 邻域内 $f(\lambda_\alpha)$ 的近似直线段情形.

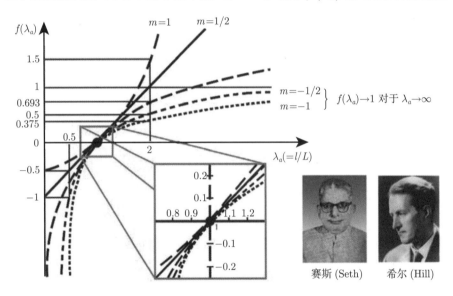

图 28.1 $f(\lambda_\alpha)$ 和 λ_α 关系图

事实上，可用罗必塔法则 (l'Hôpital's rule, 1696 年) 来证明 (28-13) 式中第一式当 $m \to 0$ 时，可得到第二式的对数函数：

$$\begin{aligned}
\lim_{m \to 0} \frac{1}{2m}\left(\lambda_\alpha^{2m} - 1\right) &= \lim_{m \to 0} \frac{\exp(2m \ln \lambda_\alpha) - 1}{2m} \\
&= \lim_{m \to 0} \frac{\partial\left[\exp(2m \ln \lambda_\alpha) - 1\right]/\partial m}{\partial(2m)/\partial m} \\
&= \lim_{m \to 0} \frac{\exp(2m \ln \lambda_\alpha)(2 \ln \lambda_\alpha)}{2} = \ln \lambda_\alpha
\end{aligned} \tag{28-14}$$

将 (28-13) 式代入 (28-1) 式，得到拉格朗日描述下的赛斯应变度量：

$$\begin{cases} \boldsymbol{E}^{(m)} = \dfrac{1}{2m} \sum_{\Gamma} \left(\lambda_{\Gamma}^{2m} - 1 \right) \boldsymbol{N}_{\Gamma} \otimes \boldsymbol{N}_{\Gamma} = \dfrac{1}{2m} \left(\boldsymbol{U}^{2m} - \boldsymbol{I} \right), & \text{如果 } m \neq 0 \\ \boldsymbol{E}^{(0)} = \ln \boldsymbol{U} = \sum_{\Gamma} (\ln \lambda_{\Gamma}) \boldsymbol{N}_{\Gamma} \otimes \boldsymbol{N}_{\Gamma}, & \text{如果 } m = 0 \end{cases} \quad (28\text{-}15)$$

在保证 $2m$ 为正或负的整数的情况下，对于不同的 m 取值，可得到如表 28.1 所示的几种常用的应变情形. 表中所谓 "物质 (material)" 是指拉格朗日描述.

表 28.1　在拉格朗日描述下赛斯应变度量的几种退化形式

m 取值	赛斯应变度量表达式	应变名称
$m = 1$	$\boldsymbol{E}^{(1)} = \boldsymbol{E} = \dfrac{1}{2}(\boldsymbol{U}^2 - \boldsymbol{I}) = \dfrac{1}{2}(\boldsymbol{C} - \boldsymbol{I})$	格林应变
$m = 1/2$	$\boldsymbol{E}^{(1/2)} = \boldsymbol{U} - \boldsymbol{I} = \sum_{\Gamma}(\lambda_{\Gamma} - 1)\boldsymbol{N}_{\Gamma} \otimes \boldsymbol{N}_{\Gamma}$	物质毕奥应变
$m = 0$	$\boldsymbol{E}^{(0)} = \ln \boldsymbol{U} = \dfrac{1}{2}\ln \boldsymbol{C} = \dfrac{1}{2}\ln(\boldsymbol{F}^{\mathrm{T}}\boldsymbol{F})$	物质亨奇应变, 右亨奇应变
$m = -1$	$\boldsymbol{E}^{(-1)} = \dfrac{1}{2}(\boldsymbol{I} - \boldsymbol{U}^{-2}) = \dfrac{1}{2}(\boldsymbol{I} - \boldsymbol{C}^{-1}) = \dfrac{1}{2}(\boldsymbol{I} - \boldsymbol{F}^{-1}\boldsymbol{F}^{-\mathrm{T}})$	物质皮奥拉应变

同理，将 (28-13) 式代入 (28-11) 式，则得到欧拉描述下赛斯应变度量：

$$\begin{cases} \boldsymbol{e}^{(m)} = \dfrac{1}{2m} \sum_{\gamma} \left(\lambda_{\gamma}^{2m} - 1 \right) \boldsymbol{n}_{\gamma} \otimes \boldsymbol{n}_{\gamma} = \dfrac{1}{2m} \left(\boldsymbol{V}^{2m} - \boldsymbol{I} \right), & \text{如果 } m \neq 0 \\ \boldsymbol{e}^{(0)} = \ln \boldsymbol{V} = \sum_{\gamma} (\ln \lambda_{\gamma}) \boldsymbol{n}_{\gamma} \otimes \boldsymbol{n}_{\gamma}, & \text{如果 } m = 0 \end{cases} \quad (28\text{-}16)$$

对于不同的 m 取值，可得到如表 28.2 所示的几种常用的应变情形. 表中所谓 "空间 (spatial)" 是指欧拉描述.

表 28.2　在欧拉描述下赛斯应变度量的几种退化形式

m 取值	赛斯应变度量表达式	应变名称
$m = 1$	$\boldsymbol{e}^{(1)} = \dfrac{1}{2}(\boldsymbol{V}^2 - \boldsymbol{I}) = \dfrac{1}{2}(\boldsymbol{B} - \boldsymbol{I}) = \dfrac{1}{2}(\boldsymbol{F}\boldsymbol{F}^{\mathrm{T}} - \boldsymbol{I})$	芬格应变
$m = 0$	$\boldsymbol{e}^{(0)} = \ln \boldsymbol{V} = \dfrac{1}{2}\ln \boldsymbol{B} = \dfrac{1}{2}\ln(\boldsymbol{F}\boldsymbol{F}^{\mathrm{T}})$	空间亨奇应变, 左亨奇应变
$m = -1/2$	$\boldsymbol{e}^{(-1/2)} = \boldsymbol{I} - \boldsymbol{V}^{-1} = \boldsymbol{I} - \boldsymbol{B}^{-1/2} = \boldsymbol{I} - \sqrt{\boldsymbol{F}\boldsymbol{F}^{\mathrm{T}}}$	空间毕奥应变
$m = -1$	$\boldsymbol{e}^{(-1)} = \boldsymbol{e} = \dfrac{1}{2}(\boldsymbol{I} - \boldsymbol{V}^{-2}) = \dfrac{1}{2}(\boldsymbol{I} - \boldsymbol{B}^{-1}) = \dfrac{1}{2}(\boldsymbol{I} - \boldsymbol{F}^{-\mathrm{T}}\boldsymbol{F}^{-1})$	阿尔曼西应变

值得注意的是，赛斯的三篇经典文章均发表在会议文集上，这再一次说明，好的文章无论发表在那里都是好文章. 在思考题 1.2 中，佩雷尔曼有关庞加莱猜想的

文章不投给任何杂志, 而只在网络上发表.

§29. 功 共 轭

29.1 面元变换的南森公式

根据定义, 反映两种构形中线元变化的变形梯度的混合欧拉–拉格朗日型张量为 $\boldsymbol{F} = \dfrac{\partial \boldsymbol{x}}{\partial \boldsymbol{X}}$, 而反映两种构形中体元变化的雅可比 (Jacobian) 为

$$J = \det \boldsymbol{F} = \frac{\mathrm{d}v}{\mathrm{d}V} = \frac{\mathrm{d}\boldsymbol{x} \cdot \mathrm{d}\boldsymbol{a}}{\mathrm{d}\boldsymbol{X} \cdot \mathrm{d}\boldsymbol{A}} \quad \Leftrightarrow \quad \mathrm{d}v = J\mathrm{d}V \tag{29-1}$$

式中, $\boldsymbol{x}$, $\boldsymbol{a}$ 和 v 对应当前构形; $\boldsymbol{X}$, $\boldsymbol{A}$ 和 V 对应参考构形. 将 (29-1) 式进一步展开, 有

$$\begin{aligned} \mathrm{d}v &= \mathrm{d}\boldsymbol{x} \cdot \mathrm{d}\boldsymbol{a} = \boldsymbol{F}\mathrm{d}\boldsymbol{X} \cdot \mathrm{d}\boldsymbol{a} = \mathrm{d}\boldsymbol{X}\boldsymbol{F}^{\mathrm{T}} \cdot \mathrm{d}\boldsymbol{a} \\ &= J\mathrm{d}V = J\mathrm{d}\boldsymbol{X} \cdot \mathrm{d}\boldsymbol{A} \end{aligned} \tag{29-2}$$

即 $\mathrm{d}\boldsymbol{X} \cdot \left(\boldsymbol{F}^{\mathrm{T}}\mathrm{d}\boldsymbol{a} - J\mathrm{d}\boldsymbol{A}\right) = 0$, 根据线段矢量 $\mathrm{d}\boldsymbol{X}$ 的任意性, 可得到关系式:

$$\boldsymbol{F}^{\mathrm{T}}\mathrm{d}\boldsymbol{a} = J\mathrm{d}\boldsymbol{A} \quad \Rightarrow \quad \mathrm{d}\boldsymbol{a} = J\boldsymbol{F}^{-\mathrm{T}}\mathrm{d}\boldsymbol{A} = (\mathbf{Cof}\boldsymbol{F})\mathrm{d}\boldsymbol{A} \tag{29-3}$$

上式中两个面积微元转换的关系式被称为南森关系式, 如图 29.1 所示.

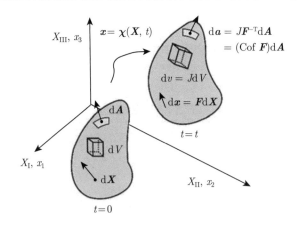

图 29.1 线元、面元和体元在两个不同构形间的转换关系式

南森关系式在定义应力度量时作用十分基础. 著者在《近代连续介质力学》中的第 86 页, 对爱德华·约翰·南森 (Edward John Nanson, 1850—1936) 做了较为详细的介绍, 他在选举制度改革数学模型方面的经典文献发表于 1882 年 [5.10].

南森共有两段婚姻，第一任妻子为他生育了一个儿子和五个女儿，第二任妻子为他生育了四个女儿. 因此，南森共有十个孩子，一男九女.

例 29.1 详细证明连续介质力学常用关系式——(29-1) 式中的 $J = \det \boldsymbol{F} = \dfrac{\mathrm{d}v}{\mathrm{d}V}$.

证明：如图 6.1 所示，由三个不共面的矢量 $\boldsymbol{a}$, $\boldsymbol{b}$ 和 $\boldsymbol{c}$ 组成的平行六面体的体积由 (6-4) 式给出：

$$\det(\boldsymbol{a},\boldsymbol{b},\boldsymbol{c}) = \boldsymbol{a} \cdot (\boldsymbol{b} \times \boldsymbol{c}) = \boldsymbol{b} \cdot (\boldsymbol{c} \times \boldsymbol{a}) = \boldsymbol{c} \cdot (\boldsymbol{a} \times \boldsymbol{b}) = \begin{vmatrix} a_x & a_y & a_z \\ b_x & b_y & b_z \\ c_x & c_y & c_z \end{vmatrix} \tag{29-4}$$

则当前构形中体元的体积为 $\mathrm{d}v = \det(\mathrm{d}\boldsymbol{x},\mathrm{d}\boldsymbol{y},\mathrm{d}\boldsymbol{z})$，参考构形中体元的体积为 $\mathrm{d}V = \det(\mathrm{d}\boldsymbol{X},\mathrm{d}\boldsymbol{Y},\mathrm{d}\boldsymbol{Z})$，从而有

$$\mathrm{d}v = \det(\mathrm{d}\boldsymbol{x},\mathrm{d}\boldsymbol{y},\mathrm{d}\boldsymbol{z}) = \mathrm{d}\boldsymbol{x} \cdot (\mathrm{d}\boldsymbol{y} \times \mathrm{d}\boldsymbol{z}) \tag{29-5}$$

由于有如下全微分的基本关系式：

$$\begin{cases} \mathrm{d}\boldsymbol{x} = \dfrac{\partial x}{\partial X}\mathrm{d}\boldsymbol{X} + \dfrac{\partial x}{\partial Y}\mathrm{d}\boldsymbol{Y} + \dfrac{\partial x}{\partial Z}\mathrm{d}\boldsymbol{Z} \\ \mathrm{d}\boldsymbol{y} = \dfrac{\partial y}{\partial X}\mathrm{d}\boldsymbol{X} + \dfrac{\partial y}{\partial Y}\mathrm{d}\boldsymbol{Y} + \dfrac{\partial y}{\partial Z}\mathrm{d}\boldsymbol{Z} \\ \mathrm{d}\boldsymbol{z} = \dfrac{\partial z}{\partial X}\mathrm{d}\boldsymbol{X} + \dfrac{\partial z}{\partial Y}\mathrm{d}\boldsymbol{Y} + \dfrac{\partial z}{\partial Z}\mathrm{d}\boldsymbol{Z} \end{cases} \tag{29-6}$$

将 (29-6) 式代入 (29-5) 式，则有当前构形微元体的体积：

$$\begin{aligned} \mathrm{d}\boldsymbol{x} \cdot (\mathrm{d}\boldsymbol{y} \times \mathrm{d}\boldsymbol{z}) =& \dfrac{\partial x}{\partial X}\dfrac{\partial y}{\partial Y}\dfrac{\partial z}{\partial Z}\mathrm{d}\boldsymbol{X} \cdot (\mathrm{d}\boldsymbol{Y} \times \mathrm{d}\boldsymbol{Z}) + \dfrac{\partial x}{\partial X}\dfrac{\partial y}{\partial Z}\dfrac{\partial z}{\partial Y}\mathrm{d}\boldsymbol{X} \cdot (\mathrm{d}\boldsymbol{Z} \times \mathrm{d}\boldsymbol{Y}) \\ &+ \dfrac{\partial x}{\partial Y}\dfrac{\partial y}{\partial Z}\dfrac{\partial z}{\partial X}\mathrm{d}\boldsymbol{Y} \cdot (\mathrm{d}\boldsymbol{Z} \times \mathrm{d}\boldsymbol{X}) + \dfrac{\partial x}{\partial Y}\dfrac{\partial y}{\partial X}\dfrac{\partial z}{\partial Z}\mathrm{d}\boldsymbol{Y} \cdot (\mathrm{d}\boldsymbol{X} \times \mathrm{d}\boldsymbol{Z}) \\ &+ \dfrac{\partial x}{\partial Z}\dfrac{\partial y}{\partial X}\dfrac{\partial z}{\partial Y}\mathrm{d}\boldsymbol{Z} \cdot (\mathrm{d}\boldsymbol{X} \times \mathrm{d}\boldsymbol{Y}) + \dfrac{\partial x}{\partial Z}\dfrac{\partial y}{\partial Y}\dfrac{\partial z}{\partial X}\mathrm{d}\boldsymbol{Z} \cdot (\mathrm{d}\boldsymbol{Y} \times \mathrm{d}\boldsymbol{X}) \\ =& \begin{vmatrix} \dfrac{\partial x}{\partial X} & \dfrac{\partial x}{\partial Y} & \dfrac{\partial x}{\partial Z} \\ \dfrac{\partial y}{\partial X} & \dfrac{\partial y}{\partial Y} & \dfrac{\partial y}{\partial Z} \\ \dfrac{\partial z}{\partial X} & \dfrac{\partial z}{\partial Y} & \dfrac{\partial z}{\partial Z} \end{vmatrix} \mathrm{d}\boldsymbol{X} \cdot (\mathrm{d}\boldsymbol{Y} \times \mathrm{d}\boldsymbol{Z}) = |\boldsymbol{F}|\mathrm{d}V = J\mathrm{d}V \end{aligned} \tag{29-7}$$

故得证.

29.2 基尔霍夫应力、第一类和第二类皮奥拉–基尔霍夫应力 (PK1 和 PK2)

变形体上的受力 $\mathrm{d}\boldsymbol{f}$，既可用参考构形又可用当前构形的面积微元来进行分解：

$$\mathrm{d}\boldsymbol{f} = \boldsymbol{\sigma}\mathrm{d}\boldsymbol{a} = \boldsymbol{P}\mathrm{d}\boldsymbol{A} \tag{29-8}$$

其中，$\boldsymbol{\sigma}$ 是对应当前构形的柯西应力，$\boldsymbol{P}$ 是对应参考构形的第一类 PK 应力. 根据 (29-3) 式，可进一步得到两者之间的关系：

$$\boldsymbol{\sigma} \cdot \boldsymbol{F}^{-\mathrm{T}} J \mathrm{d}\boldsymbol{A} = \boldsymbol{P} \cdot \mathrm{d}\boldsymbol{A}$$

通过上式，可定义基尔霍夫应力 (Kirchhoff stress):

$$\boldsymbol{\tau} = J\boldsymbol{\sigma} = \boldsymbol{P}\boldsymbol{F}^{\mathrm{T}} \tag{29-9}$$

和第一类皮奥拉–基尔霍夫应力 (first Piola–Kirchhoff stress, PK1):

$$\boldsymbol{P} = J\boldsymbol{\sigma}\boldsymbol{F}^{-\mathrm{T}} = \boldsymbol{\tau}\boldsymbol{F}^{-\mathrm{T}} = \boldsymbol{\sigma}\mathrm{Cof}\boldsymbol{F} \tag{29-10}$$

显然 PK1 不满足对称性要求. 可通过如下指标运算

$$\begin{aligned}\boldsymbol{P} = J\boldsymbol{\sigma}\boldsymbol{F}^{-\mathrm{T}} &= J\left(\sigma_{ab}\boldsymbol{e}_a \otimes \boldsymbol{e}_b\right) \cdot \left(\frac{\partial X_C}{\partial x_d}\boldsymbol{e}_d \otimes \boldsymbol{e}_C\right) \\ &= J\sigma_{ab}\frac{\partial X_C}{\partial x_d}\delta_{bd}\boldsymbol{e}_a \otimes \boldsymbol{e}_C = J\sigma_{ab}\frac{\partial X_C}{\partial x_b}\boldsymbol{e}_a \otimes \boldsymbol{e}_C \end{aligned} \tag{29-11}$$

来说明 PK1 是一个混合欧拉–拉格朗日型张量，即两点张量.

第二类皮奥拉–基尔霍夫应力 (second Piola–Kirchhoff stress, PK2) 就是构造出来的满足对称性的拉格朗日型的张量：

$$\boldsymbol{T} = \boldsymbol{F}^{-1}\boldsymbol{P} = \boldsymbol{F}^{-1}\boldsymbol{\tau}\boldsymbol{F}^{-\mathrm{T}} \tag{29-12}$$

首先验证 (29-12) 式的对称性：

$$\boldsymbol{T}^{\mathrm{T}} = \left(\boldsymbol{F}^{-1}\boldsymbol{\tau}\boldsymbol{F}^{-\mathrm{T}}\right)^{\mathrm{T}} = \left(\boldsymbol{F}^{-\mathrm{T}}\right)^{\mathrm{T}}\boldsymbol{\tau}^{\mathrm{T}}\left(\boldsymbol{F}^{-1}\right)^{\mathrm{T}} = \boldsymbol{F}^{-1}\boldsymbol{\tau}\boldsymbol{F}^{-\mathrm{T}} = \boldsymbol{T} \tag{29-13}$$

然后再通过指标运算来验证 (29-12) 式，也就是 PK2，是拉格朗日型张量：

$$\begin{aligned}\boldsymbol{T} = \boldsymbol{F}^{-1}\boldsymbol{\tau}\boldsymbol{F}^{-\mathrm{T}} &= J\left(\frac{\partial X_A}{\partial x_b}\boldsymbol{e}_A \otimes \boldsymbol{e}_b\right) \cdot \left(\sigma_{cd}\boldsymbol{e}_c \otimes \boldsymbol{e}_d\right) \cdot \left(\frac{\partial X_E}{\partial x_f}\boldsymbol{e}_f \otimes \boldsymbol{e}_E\right) \\ &= J\frac{\partial X_A}{\partial x_b}\sigma_{cd}\frac{\partial X_E}{\partial x_f}\delta_{bc}\delta_{df}\boldsymbol{e}_A \otimes \boldsymbol{e}_E = J\frac{\partial X_A}{\partial x_b}\frac{\partial X_E}{\partial x_f}\sigma_{bf}\boldsymbol{e}_A \otimes \boldsymbol{e}_E\end{aligned} \tag{29-14}$$

听课同学们经常问道：在实际的大变形的力学问题的研究中，到底用 PK1 还是用 PK2? 回答是没有一个一致的答案，有的学者偏爱 PK1，有的则偏爱 PK2，其原因是：

(1) PK1 的优点是对于参考构形单位面积的力，也就是满足 (29-8) 式，而 PK2 不具有这个特点，也就是说，PK2 是为了满足对称性构造出来的，不对应具体的面积微元.

(2) PK2 满足对称性，PK1 则不满足对称性.

PK1 和 PK2 的理论架构是由意大利数学家卡布里奥·皮奥拉 (Gabrio Piola, 1794—1850, 图 29.2 左) 和德国物理学家古斯塔夫·罗伯特·基尔霍夫 (Gustav Robert Kirchhoff, 1824—1887, 图 29.2 右) 所共同创立的，更为详细的历史可参阅《近代连续介质力学》[1.6] 中 §7.1 的内容.

基尔霍夫曾对力学做了如下定义：“力学是研究运动的科学；她的任务是以最完备和最简单的方式来描述自然界发生的运动 (Mechanics is the science of motion; its object may be stated to be to describe in the most complete and simple way the motion that takes place in nature)”，基尔霍夫的定义特别强调了“最完备和最简单的方式”. 该定义得到了马赫主义者的赞同.

图 29.2　皮奥拉 (左)；基尔霍夫 (右)

29.3　功共轭

根据定义及 (29-1) 式，当前构形与参考构形的应变能的变化率为

$$\dot{W} = \int \dot{w} \mathrm{d}v = \int_v \boldsymbol{\sigma} : \boldsymbol{d} \mathrm{d}v = \int_V J\boldsymbol{\sigma} : \boldsymbol{d} \mathrm{d}V = \int_V \boldsymbol{\tau} : \boldsymbol{d} \mathrm{d}V \qquad (29\text{-}15)$$

式中 $\boldsymbol{d} = \dfrac{1}{2}\left(\boldsymbol{l} + \boldsymbol{l}^\mathrm{T}\right)$ 为应变率，是一个二阶对称张量；$\boldsymbol{l}$ 为速度矢量的右梯度，简称为"速度梯度 (velocity gradient)". 速度梯度张量可分解为 $\boldsymbol{l} = \boldsymbol{d} + \boldsymbol{\omega}$, $\boldsymbol{\omega} = \dfrac{1}{2}\left(\boldsymbol{l} - \boldsymbol{l}^\mathrm{T}\right)$ 为旋率 (spin), 它为一个二阶反对称张量. 根据定义，可得到 $\boldsymbol{l}$ 与变形

梯度 F 的关系:

$$l = v \otimes \nabla_x = \frac{\partial v}{\partial x} = \frac{\partial v}{\partial X}\frac{\partial X}{\partial x} = \frac{\partial}{\partial X}\left(\frac{\mathrm{d}x}{\mathrm{d}t}\right)\frac{\partial X}{\partial x} = \frac{\mathrm{d}}{\mathrm{d}t}\left(\frac{\partial x}{\partial X}\right)\frac{\partial X}{\partial x} = \dot{F}F^{-1} \quad (29\text{-}16)$$

则应变能密度的变化率可表示为

$$\dot{w} = \boldsymbol{\tau} : \boldsymbol{d} = \boldsymbol{\tau} : (\boldsymbol{l} - \boldsymbol{\omega}) = \boldsymbol{\tau} : \boldsymbol{l} = \boldsymbol{\tau} : \dot{F}F^{-1} \quad (29\text{-}17)$$

由常用关系式 (11-12), $\boldsymbol{A} : (\boldsymbol{BC}) = \boldsymbol{B}^{\mathrm{T}}\boldsymbol{A} : \boldsymbol{C} = \boldsymbol{AC}^{\mathrm{T}} : \boldsymbol{B}$, 可将 (29-17) 式简化为

$$\dot{w} = \boldsymbol{\tau} F^{-\mathrm{T}} : \dot{F} = \boldsymbol{P} : \dot{F} \quad (29\text{-}18)$$

上式说明 PK1 和变形梯度张量 F 之间满足功共轭关系, 但也有学者认为, 由于变形梯度 F 不是应变, 所以 PK1 和变形梯度张量之间的功共轭关系不是真正意义上的功共轭.

下面继续详细证明 PK2 和格林应变之间的功共轭关系. 由 (29-18) 式:

$$\begin{aligned}\dot{w} &= \boldsymbol{\tau} F^{-\mathrm{T}} : \dot{F} = J\boldsymbol{\sigma} F^{-\mathrm{T}} : \dot{F} \\ &= JFF^{-1}\boldsymbol{\sigma} F^{-\mathrm{T}} : \dot{F} = F^{-1}J\boldsymbol{\sigma} F^{-\mathrm{T}} : \left(F^{\mathrm{T}}\dot{F}\right) \\ &= \boldsymbol{T} : \left(F^{\mathrm{T}}\dot{F}\right)\end{aligned} \quad (29\text{-}19)$$

又格林应变张量 $\boldsymbol{E} = \frac{1}{2}\left(F^{\mathrm{T}}F - \boldsymbol{I}\right)$ 的率可表示为

$$\dot{\boldsymbol{E}} = \frac{1}{2}\left(\dot{F}^{\mathrm{T}}F + F^{\mathrm{T}}\dot{F}\right) \quad (29\text{-}20)$$

由于 (29-19) 式中 PK2 的对称性, 基于 (12-14) 式, (29-19) 式可进一步写为

$$\begin{aligned}\dot{w} &= \boldsymbol{T} : \left(F^{\mathrm{T}}\dot{F}\right) = \boldsymbol{T} : \frac{F^{\mathrm{T}}\dot{F} + \left(F^{\mathrm{T}}\dot{F}\right)^{\mathrm{T}}}{2} \\ &= \boldsymbol{T} : \dot{\boldsymbol{E}}\end{aligned} \quad (29\text{-}21)$$

综合上述推导的结果, 则有如下连续介质力学的功共轭的关系:

$$\dot{w} = J\boldsymbol{\sigma} : \boldsymbol{d} = \boldsymbol{\tau} : \boldsymbol{d} = \boldsymbol{P} : \dot{F} = \boldsymbol{T} : \dot{\boldsymbol{E}} \quad (29\text{-}22)$$

值得指出的是, 功共轭关系 (29-22) 式是连续介质力学中最为重要的关系式之一, 其推导过程应熟练掌握并能熟练应用.

例 29.2 变形梯度和速度梯度均为连续介质力学中的重要张量. 证明下列常用关系式:

$$\overline{\dot{F^{-1}}} = -F^{-1}l \tag{29-23}$$

证明: 由基本关系式: $FF^{-1} = I$, 则有

$$\overline{\dot{FF^{-1}}} = \dot{F}F^{-1} + F\overline{\dot{F^{-1}}} = 0$$

将速度梯度张量所满足的 (29-16) 式 $l = \dot{F}F^{-1}$ 代入上式, 有

$$F\overline{\dot{F^{-1}}} = -l$$

上式两端左侧点积 F^{-1}, 则 (29-23) 式得证.

例 29.3 证明下列常用关系式:

$$\overline{\dot{F^{-T}}} = -l^T F^{-T} \tag{29-24}$$

证明: 由于对时间的导数和转置操作顺序可互换, 则有

$$\overline{\dot{F^T}} = \left(\dot{F}\right)^T = \dot{F}^T \tag{29-25}$$

则由 (29-23) 式可得

$$\overline{\dot{F^{-T}}} = \left(\overline{\dot{F^{-1}}}\right)^T = \left(-F^{-1}l\right)^T = -l^T F^{-T} \tag{29-26}$$

例 29.4 证明格林应变率为

$$\dot{E} = F^T dF \tag{29-27}$$

证明: 由格林应变的定义 (27-2) 式: $E = \frac{1}{2}(C - I) = \frac{1}{2}(F^T F - I)$, 对其求时间导数, 则有 (29-20) 式: $\dot{E} = \frac{1}{2}\left(\dot{F}^T F + F^T \dot{F}\right)$. 再由速度梯度的定义式 (29-16): $l = \dot{F}F^{-1}$, 因而有下列有关变形梯度率的基本关系式:

$$\dot{F} = lF, \quad \dot{F}^T = F^T l^T \tag{29-28}$$

将 (29-28) 式代入 (29-20) 式, 得到格林应变率满足:

$$\dot{E} = \frac{1}{2}\left(\dot{F}^T F + F^T \dot{F}\right) = \frac{1}{2}\left(F^T l^T F + F^T l F\right)$$

$$= \boldsymbol{F}^{\mathrm{T}}\frac{\boldsymbol{l}^{\mathrm{T}}+\boldsymbol{l}}{2}\boldsymbol{F} = \boldsymbol{F}^{\mathrm{T}}\boldsymbol{dF}$$

亦即 (29-27) 式成立. (29-27) 式说明, 通过左端点积 $\boldsymbol{F}^{\mathrm{T}}$ 和右端点积 $\boldsymbol{F}$ 的变换, 将位于当前构形中的应变率 $\boldsymbol{d} = \frac{1}{2}\left(\boldsymbol{l}+\boldsymbol{l}^{\mathrm{T}}\right)$ "拉回 (pull-back)" 到参考构形中的应变率 $\dot{\boldsymbol{E}}$. 其逆变换 $\boldsymbol{d} = \boldsymbol{F}^{-\mathrm{T}}\dot{\boldsymbol{E}}\boldsymbol{F}^{-1}$ 则称为 "推前 (push-forward)" 操作, 亦即将参考构形中的格林应变率推前到当前构形中的应变率.

(29-27) 式说明, 右柯西–格林变形张量 $\boldsymbol{C} = \boldsymbol{F}^{\mathrm{T}}\boldsymbol{F}$ 的率为

$$\dot{\boldsymbol{C}} = 2\dot{\boldsymbol{E}} = 2\boldsymbol{F}^{\mathrm{T}}\boldsymbol{dF} \tag{29-29}$$

上式为连续介质力学重要关系式, 在建立超弹性本构关系时, 经常会用到.

例 29.5 求阿尔曼西应变率 $\dot{\boldsymbol{e}}$.

解: 由阿尔曼斯应变的定义式 (27-49): $\boldsymbol{e} = \frac{1}{2}\left(\mathbf{I}-\boldsymbol{F}^{-\mathrm{T}}\boldsymbol{F}^{-1}\right)$, 则其率为

$$\dot{\boldsymbol{e}} = -\frac{1}{2}\left(\overline{\boldsymbol{F}^{-\mathrm{T}}}\,\boldsymbol{F}^{-1}+\boldsymbol{F}^{-\mathrm{T}}\,\overline{\boldsymbol{F}^{-1}}\right) \tag{29-30}$$

将 (29-26) 和 (29-23) 两式代入 (29-30) 式, 再代入由阿尔曼西应变的定义式所得到的 $\boldsymbol{F}^{-\mathrm{T}}\boldsymbol{F}^{-1} = \mathbf{I}-2\boldsymbol{e}$, 则阿尔曼西应变率的表达式为

$$\dot{\boldsymbol{e}} = \frac{1}{2}\left(\boldsymbol{l}^{\mathrm{T}}\boldsymbol{F}^{-\mathrm{T}}\boldsymbol{F}^{-1}+\boldsymbol{F}^{-\mathrm{T}}\boldsymbol{F}^{-1}\boldsymbol{l}\right) = \frac{1}{2}\left[\boldsymbol{l}^{\mathrm{T}}\left(\mathbf{I}-2\boldsymbol{e}\right)+\left(\mathbf{I}-2\boldsymbol{e}\right)\boldsymbol{l}\right]$$
$$= \boldsymbol{d}-\boldsymbol{l}^{\mathrm{T}}\boldsymbol{e}-\boldsymbol{el} \tag{29-31}$$

例 29.6 求左柯西–格林变形张量 $\boldsymbol{B} = \boldsymbol{FF}^{\mathrm{T}}$ 的率 $\dot{\boldsymbol{B}}$.

解: 由定义, 有

$$\dot{\boldsymbol{B}} = \dot{\boldsymbol{F}}\boldsymbol{F}^{\mathrm{T}}+\boldsymbol{F}\dot{\boldsymbol{F}}^{\mathrm{T}} \tag{29-32}$$

将 (29-28) 式代入上式, 整理得到

$$\dot{\boldsymbol{B}} = \dot{\boldsymbol{F}}\boldsymbol{F}^{\mathrm{T}}+\boldsymbol{F}\dot{\boldsymbol{F}}^{\mathrm{T}} = \boldsymbol{lFF}^{\mathrm{T}}+\boldsymbol{FF}^{\mathrm{T}}\boldsymbol{l}^{\mathrm{T}}$$
$$= \boldsymbol{lB}+\boldsymbol{Bl}^{\mathrm{T}} \tag{29-33}$$

例 29.7 给出如图 29.3 所示的剪切流 (shear flow) 的应变率和旋率.

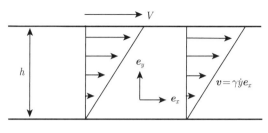

图 29.3 剪切流示意图

解：对于如图 29.3 所示的三角形线性分布的流动，其剪切应变率为 $\dot{\gamma} = V/h$，则速度矢量为

$$\boldsymbol{v} = \dot{\gamma} y \boldsymbol{e}_x \tag{29-34}$$

这是一个定常流动 (steady flow)，也就是满足：

$$\frac{\partial \boldsymbol{v}}{\partial t} = \boldsymbol{0} \tag{29-35}$$

由 (29-34) 式所示，该流动的速度梯度张量为

$$\boldsymbol{l} = \boldsymbol{v} \otimes \boldsymbol{\nabla} = \dot{\gamma} \boldsymbol{e}_x \otimes \boldsymbol{e}_y \tag{29-36}$$

则该流动的应变率和旋率分别为

$$\boldsymbol{d} = \frac{1}{2}\left(\boldsymbol{l} + \boldsymbol{l}^{\mathrm{T}}\right) = \frac{1}{2}\begin{bmatrix} 0 & \dot{\gamma} & 0 \\ \dot{\gamma} & 0 & 0 \\ 0 & 0 & 0 \end{bmatrix}, \quad \boldsymbol{\omega} = \frac{1}{2}\left(\boldsymbol{l} - \boldsymbol{l}^{\mathrm{T}}\right) = \frac{1}{2}\begin{bmatrix} 0 & \dot{\gamma} & 0 \\ -\dot{\gamma} & 0 & 0 \\ 0 & 0 & 0 \end{bmatrix} \tag{29-37}$$

例 29.8 求证连续介质力学基本关系式：

$$\dot{J} = J \mathrm{div} \boldsymbol{v} = J \mathrm{tr} \boldsymbol{d} \tag{29-38}$$

证明：由雅可比和变形梯度之间的关系式 (29-1)：$J = \det \boldsymbol{F}$，链式法则给出：

$$\dot{J} = \frac{\partial J}{\partial \boldsymbol{F}} : \dot{\boldsymbol{F}} \tag{29-39}$$

由例 18.5 中的 (18-29) 式和 $\boldsymbol{l} = \dot{\boldsymbol{F}} \boldsymbol{F}^{-1}$，可得

$$\dot{J} = \frac{\partial J}{\partial \boldsymbol{F}} : \dot{\boldsymbol{F}} = J \mathrm{tr}\left(\boldsymbol{F}^{-1} \dot{\boldsymbol{F}}\right) = J \mathrm{tr}\left(\dot{\boldsymbol{F}} \boldsymbol{F}^{-1}\right)$$
$$= J \mathrm{tr} \boldsymbol{l} \tag{29-40}$$

由 $\boldsymbol{l} = \boldsymbol{d} + \boldsymbol{\omega}$，而作为反对称张量的旋率的迹为零：$\mathrm{tr}\boldsymbol{\omega} = 0$，则 (29-40) 式为

$$\dot{J} = J \mathrm{tr} \boldsymbol{l} = J \mathrm{tr}\left(\boldsymbol{d} + \boldsymbol{\omega}\right) = J \mathrm{tr} \boldsymbol{d} \tag{29-41}$$

再由

$$\mathrm{tr}\boldsymbol{l} = \mathrm{tr}\left(\boldsymbol{v}\otimes\boldsymbol{\nabla}\right) = \mathrm{tr}\left(\mathbf{grad}^{\mathrm{T}}\boldsymbol{v}\right)$$
$$= \mathrm{tr}\left(\mathbf{grad}\boldsymbol{v}\right) = \mathbf{grad}\boldsymbol{v}:\mathbf{I} = \mathrm{div}\boldsymbol{v} \tag{29-42}$$

则 (29-38) 式得证.

例 29.9 求证连续介质力学基本关系式:

$$\dot{\overline{\mathrm{d}v}} = \frac{\mathrm{d}}{\mathrm{d}t}\left(\mathrm{d}v\right) = \left(\mathrm{div}\boldsymbol{v}\right)\mathrm{d}v \tag{29-43}$$

证明: 由于当前构形中的体积微元 $\mathrm{d}v$ 和参考构形中的体积微元 $\mathrm{d}V$ 之间通过雅可比来联系: $\mathrm{d}v = J\mathrm{d}V$, 且 $\mathrm{d}V$ 不随时间变化, 再应用 (29-38) 式, 则有

$$\dot{\overline{\mathrm{d}v}} = \frac{\mathrm{d}}{\mathrm{d}t}\left(\mathrm{d}v\right) = \frac{\mathrm{d}}{\mathrm{d}t}\left(J\mathrm{d}V\right) = \dot{J}\mathrm{d}V$$
$$= \left(\mathrm{div}\boldsymbol{v}\right)J\mathrm{d}V = \left(\mathrm{div}\boldsymbol{v}\right)\mathrm{d}v \tag{29-44}$$

例 29.10 求证介质不可压缩的条件可以等价地表示为

$$\begin{cases} J = \det\boldsymbol{F} = 1 \\ \dot{J} = 0 \\ \mathrm{div}\boldsymbol{v} = 0 \\ \mathrm{tr}\boldsymbol{d} = 0 \\ \mathrm{tr}\boldsymbol{l} = 0 \\ \mathrm{d}v = \mathrm{const} \\ \boldsymbol{F}^{-\mathrm{T}}:\dot{\boldsymbol{F}} = 0 \end{cases} \tag{29-45}$$

证明: 由例 29.8 和例 29.9, (29-45) 式中的前六个条件已经自明. 而 (29-45) 式的最后一个条件, 则可由 (18-28) 式可得

$$\dot{J} = \frac{\partial J}{\partial\boldsymbol{F}}:\dot{\boldsymbol{F}} = J\boldsymbol{F}^{-\mathrm{T}}:\dot{\boldsymbol{F}} \tag{29-46}$$

则得证.

例 29.11 为何图 26.1 强调两种构形之间的映射是单射的 (injective) 且保持方向的 (orientation-preserving)? 这两种要求的条件是什么?

答: 要求从参考构形到当前构形间的映射是单射的或 "一对一的 (one-to-one)" 是基于如下原因: (1) 从数学上讲, 严格递增或严格递减函数是单射的, 如果这里

所述的关系是应力—应变关系或者应变能密度和应变之间的关系的话，单射 (injection) 可从数学上保证问题解的唯一性 (uniqueness of solution)；(2) 从物理上讲，单射保证了在变形过程中，如图 29.4 所示的材料的相互贯穿 (interpenetration) 的情况可以得到避免；(3) 单射所需要满足的具体条件限制是

$$\int_V J \mathrm{d}V = \int_v \mathrm{d}v \leqslant \mathrm{vol}\chi \tag{29-47}$$

上式说明，若物质流形变形后发生了穿透，那么当前构形中的一个体积微元对应参考构形中的多个体积微元，则当前构形的体积小于通过对参考构形微元积分计算得到的体积.

在 §26.1 中，曾谈到非奇异线性映射 "有向" "保持定向的" 和 "保持方向映射 (orientation-preserving mapping)"，其数学要求是 (详见思考题 4.10)：$J = \det \boldsymbol{F} > 0$，这事实上是要求 $\boldsymbol{F} \in \mathbb{M}_+^3$，也就是 $\boldsymbol{F}$ 属于正定的 3×3 实矩阵的集合.

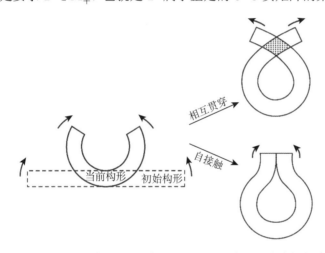

图 29.4　构形间的单射性可避免材料的相互贯穿但允许自接触情况

思考题和补充材料

5.1 对于面积微元矢量 $\mathrm{d}\boldsymbol{a}$，求证其时间变化率满足 $\dot{\overline{\mathrm{d}\boldsymbol{a}}} = \left[(\mathrm{div}\boldsymbol{v})\boldsymbol{I} - \boldsymbol{l}^\mathrm{T}\right]\mathrm{d}\boldsymbol{a}$.

5.2 我们已经在 §20.5 中提到过，如题图 5.2 所示的柯西是他那个年代法国最著名的数学家，27 岁便当选为法国科学院院士. 柯西十分高产，他一生一共 "孵出的蛋"(著作) 包括 789 篇论文和几本书，仅次于大数学家欧拉. 他那个年代的期刊很少，出版周期又很长，我们不禁要问：他的近 800 篇论文是如何发表的？这里的答案涉及了婚姻和姻亲的重要性!

题图 5.2　柯西的邮票和首日封 (first day cover)

柯西在 1818 年时年 28 岁上娶了比他小 5 岁的 Aloïse de Bure 为妻. de Bure 家族是巴黎著名的出版商和书商，该家族出版了柯西的大部分著作.

由于柯西投给每周出版的法国科学院院刊 (*Comptes Rendu Acad. Sci. Paris*) 的论文实在太多，几乎每周两篇长文，一方面是法国科学院院刊不可能刊登一位院士这么多的文章，另一方面是法国科学院要负担很大的印刷费用，这也超出科学院的预算. 因此，法国科学院规定了柯西投给法国科学院院刊的论文篇数的上限和页数的限制.

这种出版的限制当然难不倒柯西！他经常性地动用他妻子 Aloïse de Bure 娘家是出版商和书商的"出书和卖书一条龙"的便利条件，他们不但为柯西出版了大量的论文和书籍，有一段时间还为柯西创刊了只有柯西一个作者的题为《数学练习》(*Exercices de Mathématiques*) 的数学刊物.

5.3 幼年的柯西受到过 3L (见思考题 1.4) 中的 2L (拉普拉斯和拉格朗日) 的重要影响. 柯西的父亲是波旁政府里的一名议会律师，同时也是巴黎警署的一名中尉，这样的家庭背景在法国大革命时期能够保住性命就已经是很幸运的了. 老柯西带着全家逃到了乡下一个村庄，并开始了长达十余年的隐居生活. 柯西的不幸福童年是时代造成的，家里吃不饱穿不暖，外面所有的学校都关闭了，有文化的人被大批的送上断头台，看起来柯西即不可能也没有必要接受任何教育. 幸运的是，柯西的爸爸自己编写了教材来完成对孩子们的教育，特别是为了培养孩子们的兴趣，有几本教材是用诗歌体写成. 在学习自然科学知识的同时，小柯西也领略到了文学和诗歌的美，并且此后一生都痴迷于法文诗和拉丁文诗.

此外，柯西一家还有一个值得一提的邻居，就是在法国大革命里如鱼得水的拉普拉斯. 拉普拉斯的优点是喜欢交际而且平易近人. 很快拉普拉斯就在他贫穷的邻居家里看到了营养不良的柯西，并且发现了他的数学才能. 拉普拉斯没有想到的是这种才能在几年以后会让他惊出一身冷汗，差点心脏病发作.

18 世纪的最后一年，法国大革命终于结束了，老柯西得以重新回到巴黎工作，在拿破仑 1799 年执政后，老柯西被晋升为参议院秘书长 (Secretary-General of the Senate) 直接在拉普拉斯的领导下工作 (working directly under Laplace). 老柯西在卢森堡宫里有一间自己的办公室,而少年柯西则在父亲的办公室的一角学习,于是他会经常见到当时的参议员 —— 巴黎综合工科学校 (École Polytechnique) 的数学教授拉格朗日,柯西是一位名副其实的"官员子弟". 和拉普拉斯一样,没过多久,拉格朗日就发现了柯西在数学方面的与众不同,并且当着拉普拉斯等人的面夸奖柯西道:"你们看到那位瘦小的年轻人了吗? 嘿! 作为数学家的我们迟早要被他所取代 (You see that little young man? Well! He will supplant all of us in so far as we are mathematicians)."

拉格朗日年轻的时候曾因为过度劳累,饱受疾病的困扰,看到柯西瘦弱的身体,他给了老柯西一个很特别的忠告:"在 17 岁之前不要让他接触任何数学书籍 (Don't let him touch a mathematical book till he is seventeen)." 这里拉格朗日当然指的是高等数学 (higher mathematics).

拉格朗日进一步对老柯西忠告道:"如果你不赶快给奥古斯汀 (柯西) 一点可靠的文学教育,他的趣味就会使他冲昏头脑; 他将成为一名伟大的数学家,但是他不知道如何用自己的语言来写作 (If you don't hasten to give Augustin a solid literary education his tastes will carry him away; he will be a great mathematician but he won't know how to write his own language)."

老柯西很好地遵从了他那个年代最伟大的数学力学家拉格朗日的忠告,从 13 岁进入学校开始,柯西就用法语、希腊语、拉丁语在五花八门诗歌和作文比赛里屡屡获奖.

拉格朗日真是一位慧眼识才的伯乐呀! 数学力学大师拉格朗日针对少年柯西的那些忠告仍然值得中国那些望子成龙的家长回味.

5.4 题表 1.5 中业已提及的大数学家和物理学家布莱士·帕斯卡 (Blaise Pascal, 如题图 5.4),可怜的布莱士在与他聪明的大脑一起继承了一个极差的体格 (But poor Blaise inherited a wretched physique along with his brilliant mind). 老帕斯卡又是如何对神童小帕斯卡如何培养的呢?

题图 5.4 帕斯卡的代表性邮票

根据年轻的天才可能会用脑太多而过于劳累的理论,数学是禁忌的 (Mathematics was taboo, on the theory that the young genius might overstrain himself by using his head)! 老帕斯卡是一位极好的教练,但又是一位蹩脚的心理学家. 老帕斯卡对小帕斯卡学习数学的禁令很自然地激发了小帕斯卡对数学的好奇心 (Pascal senior was an excellent drillmaster but a poor psychologist. His ban on mathematics naturally excited the boy's curiosity).

少年的帕斯卡到底数学有多强? 帕斯卡的姐姐吉尔伯特声称: 她的弟弟独自重新发现了欧几里得的前三十二个命题,而且是以欧几里得的同样次序发现它们的 (Gilberte declared that her brother had rediscovered for himself the first thirty two propositions of Euclid, and that he had found them *in the same order* as that in which Euclid sets them forth).

帕斯卡还发明了世界上第一个机械计算器 (mechanical calculator),如题图 5.4 右图所示.

5.5 力学大师冯·卡门耐人寻味的作息时间表和他的特殊才能 [5.11]: "他的日程表令人惊异: 6 点起床, 7 点或 8 点和客人 (有时多达 12 位) 共进早餐. 餐后, 口述各种信件, 一直到中午时分. 接下来再抽样浏览一下世界各地科研人员寄来的大量书和论文. 午餐以喝烈性威士忌酒开始. 通常, 这常常又是一段工作时间, 不是接待来访的要人, 就是会见从前的学生. 下午 3 点午睡, 一般 5 点起床, 为晚上约会做准备. 晚餐照例先喝几杯上好的威士忌, 才吃丰盛的正餐, 吃完饭再喝酒, 一直要喝到半夜. 晚上, 他谈天说地兴致最佳, 通常总有几位年轻漂亮的女客在座. 他善于把享乐和事业结合在一起, 使两者都不偏废; 能办到这一点的人不多. 他还有一种特殊本领: 表面上从事别的活动, 头脑里却在进行科学思考. 他从不一段段地划分时间; 但他会离开餐桌一小时, 去推导一个方程或起草一份文件, 然后回到客人们中间, 重新捡起他离席时的话题. (His schedule was incredible. He rose at six and breakfasted at seven or eight with a guest — sometimes with as many as twelve guests. He dictated correspondence until about noon and then glanced through samples of the vast numbers of books and papers which usually came to him from scientific workers from all over the world. Lunch, preceded by a strong whisky, was usually another business session, or a time for greeting visiting dignitaries or former students. He napped at three in the afternoon and usually arose at five to prepare for the evening engagement, which started typically with several Jack Daniels and ran through a large meal followed by after-dinner drinks until midnight. It was in the evening that von Kármán was in his top form as he told stories, usually in the company of the prettiest young ladies present. He was one of the few people who knew how to successfully mix pleasure with business, without sacrificing either, and he had the extraordinary knock of carrying on his scientific thinking while apparently conducting other activities. He never compartmentalized his time, but he would disappear for an one hour from a party to note down an equation or work up a paper and then return to his guests to take up where he left off.)"

5.6 大作业一: 有关生物力学中的生长张量 (growth tensor) 的深入调研. 生物体的典型特征是伴随着生长或凋亡. 针对此类问题, 文献 [5.12] 指出, 变形梯度张量 $\boldsymbol{F}$ 可进一步如下乘法分解为

$$\boldsymbol{F} = \boldsymbol{F}_e \boldsymbol{F}_g \tag{s5-1}$$

式中, $\boldsymbol{F}_g$ 为生长张量 (growth tensor), $\boldsymbol{F}_e$ 则为变形梯度的弹性部分, 简称弹性张量或协调张量 (accommodation tensor). 如题图 5.6 所示, 初始构形 $\mathcal{B}_0$ 和当前构形 $\mathcal{B}_t$ 之间的中间构形 $\mathcal{B}_g$, 则可称为生长构形 (growth configuration), $\mathcal{B}_g$ 是一个无应力构形 (unstressed configuration). 问题:

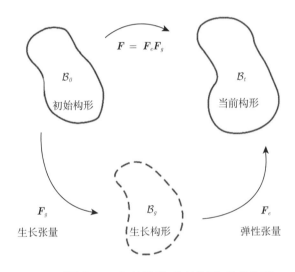

题图 5.6 初始构形–生长构形–当前构形

(1) 这种对构形的乘法分解源于弹塑性力学中对弹性构形 $\boldsymbol{F}_e$ 和塑性构形 $\boldsymbol{F}_p$ 的乘法分解:

$$\boldsymbol{F} = \boldsymbol{F}_e \boldsymbol{F}_p \tag{s5-2}$$

讨论该乘法分解是否和顺序有关, 或者说 $\boldsymbol{F}_e \boldsymbol{F}_g$ 或者 $\boldsymbol{F}_g \boldsymbol{F}_e$ 是否相同? 请参阅文献 [1.6] 中的 23.6 节中有关 $\boldsymbol{F}_e \boldsymbol{F}_p$ 和 $\boldsymbol{F}_p \boldsymbol{F}_e$ 的相关讨论.

(2) 引入位移梯度张量 $\boldsymbol{H}$, 也就是变形梯度和单位张量之差:

$$\boldsymbol{H} = \boldsymbol{F} - \boldsymbol{I} \tag{s5-3}$$

从而，生长张量和弹性张量均可进行类似的分解：

$$\boldsymbol{H}_g = \boldsymbol{F}_g - \boldsymbol{I}, \quad \boldsymbol{H}_e = \boldsymbol{F}_e - \boldsymbol{I} \tag{s5-4}$$

则变形梯度 (s5-1) 式可表示为

$$\boldsymbol{F} = \boldsymbol{F}_e \boldsymbol{F}_g = (\boldsymbol{H}_e + \boldsymbol{I})(\boldsymbol{H}_g + \boldsymbol{I}) \tag{s5-5}$$

验证：在微小变形时，亦即当 $\|\boldsymbol{H}_g\| \ll 1$、$\|\boldsymbol{H}_e\| \ll 1$ 时，格林应变张量 (27-2) 式可表示为

$$\boldsymbol{E} = \frac{1}{2}\left(\boldsymbol{F}^{\mathrm{T}}\boldsymbol{F} - \boldsymbol{I}\right) \approx \frac{1}{2}\left(\boldsymbol{H}_g + \boldsymbol{H}_g^{\mathrm{T}} + \boldsymbol{H}_e + \boldsymbol{H}_e^{\mathrm{T}}\right) + o\left(\|\boldsymbol{H}\|\right) \tag{s5-6}$$

亦即，在微小变形时，应变可以进行和分解 (additive decomposition)：

$$\begin{aligned} \boldsymbol{E} = \boldsymbol{e} &\approx \boldsymbol{e}_g + \boldsymbol{e}_e \\ &= \frac{1}{2}\left(\boldsymbol{H}_g + \boldsymbol{H}_g^{\mathrm{T}}\right) + \frac{1}{2}\left(+\boldsymbol{H}_e + \boldsymbol{H}_e^{\mathrm{T}}\right) \end{aligned} \tag{s5-7}$$

(3) 进一步调研生长张量在不同类型生物体中的应用[5.13,5.14].

(4) "accommodation tensor" 一词还未有相对应的得到一致认可的中文术语，目前对应的中文术语有 "协调张量" 或 "相容张量".

5.7 大作业二：中间构形在确定残余应力中的应用.

5.8 给出罗德尼·希尔被连续介质力学界公认为 "第一把交椅" 的理由和你对他的学术贡献的体会.

5.9 用狄拉克符号表示格林应变 (27-39) 式或 (27-25) 式的演算过程.

解：应用 §5 中引入的狄拉克符号，变形后和变形前线段的平方差可表示为

$$\begin{aligned} \langle \mathrm{d}\boldsymbol{x}|\mathrm{d}\boldsymbol{x}\rangle - \langle \mathrm{d}\boldsymbol{X}|\mathrm{d}\boldsymbol{X}\rangle &= \langle \mathrm{d}\boldsymbol{u} + \mathrm{d}\boldsymbol{X}|\mathrm{d}\boldsymbol{u} + \mathrm{d}\boldsymbol{X}\rangle - \langle \mathrm{d}\boldsymbol{X}|\mathrm{d}\boldsymbol{X}\rangle \\ &= \langle (\boldsymbol{u} \otimes \boldsymbol{\nabla}_{\boldsymbol{X}} + \boldsymbol{I})\mathrm{d}\boldsymbol{X}|(\boldsymbol{u} \otimes \boldsymbol{\nabla}_{\boldsymbol{X}} + \boldsymbol{I})\mathrm{d}\boldsymbol{X}\rangle - \langle \mathrm{d}\boldsymbol{X}|\mathrm{d}\boldsymbol{X}\rangle \\ &= \langle \mathrm{d}\boldsymbol{X}|(\boldsymbol{u} \otimes \boldsymbol{\nabla}_{\boldsymbol{X}} + \boldsymbol{I})^{\mathrm{T}}(\boldsymbol{u} \otimes \boldsymbol{\nabla}_{\boldsymbol{X}} + \boldsymbol{I})|\mathrm{d}\boldsymbol{X}\rangle - \langle \mathrm{d}\boldsymbol{X}|\mathrm{d}\boldsymbol{X}\rangle \\ &= 2\langle \mathrm{d}\boldsymbol{X}|\frac{1}{2}(\boldsymbol{u} \otimes \boldsymbol{\nabla}_{\boldsymbol{X}} + \boldsymbol{\nabla}_{\boldsymbol{X}} \otimes \boldsymbol{u} + \boldsymbol{\nabla}_{\boldsymbol{X}} \otimes \boldsymbol{u} \cdot \boldsymbol{u} \otimes \boldsymbol{\nabla}_{\boldsymbol{X}})|\mathrm{d}\boldsymbol{X}\rangle \\ &= 2\langle \mathrm{d}\boldsymbol{X}|\boldsymbol{E}|\mathrm{d}\boldsymbol{X}\rangle \end{aligned} \tag{s5-8}$$

故 (27-39) 式或 (27-25) 式得证.

5.10 用狄拉克符号表示阿尔曼西应变 (27-48) 式的演算过程.

解: 应用狄拉克符号, 有

$$
\begin{aligned}
\langle \mathrm{d}\boldsymbol{x}|\,\mathrm{d}\boldsymbol{x}\rangle - \langle \mathrm{d}\boldsymbol{X}|\,\mathrm{d}\boldsymbol{X}\rangle &= \langle \mathrm{d}\boldsymbol{x}|\,\mathrm{d}\boldsymbol{x}\rangle - \langle \mathrm{d}\boldsymbol{x} - \mathrm{d}\boldsymbol{u}|\,\mathrm{d}\boldsymbol{x} - \mathrm{d}\boldsymbol{u}\rangle \\
&= \langle \mathrm{d}\boldsymbol{x}|\,\mathrm{d}\boldsymbol{x}\rangle - \langle (\mathbf{I} - \boldsymbol{u}\otimes\boldsymbol{\nabla}_{\boldsymbol{x}})\,\mathrm{d}\boldsymbol{x}|\,(\mathbf{I} - \boldsymbol{u}\otimes\boldsymbol{\nabla}_{\boldsymbol{x}})\,\mathrm{d}\boldsymbol{x}\rangle \\
&= \langle \mathrm{d}\boldsymbol{x}|\,\mathrm{d}\boldsymbol{x}\rangle - \langle \mathrm{d}\boldsymbol{x}|\,(\mathbf{I} - \boldsymbol{u}\otimes\boldsymbol{\nabla}_{\boldsymbol{x}})^{\mathrm{T}}\,(\mathbf{I} - \boldsymbol{u}\otimes\boldsymbol{\nabla}_{\boldsymbol{x}})|\,\mathrm{d}\boldsymbol{x}\rangle \\
&= \langle \mathrm{d}\boldsymbol{x}|\,\mathbf{I} - (\mathbf{I} - \boldsymbol{u}\otimes\boldsymbol{\nabla}_{\boldsymbol{x}})^{\mathrm{T}}\,(\mathbf{I} - \boldsymbol{u}\otimes\boldsymbol{\nabla}_{\boldsymbol{x}})|\,\mathrm{d}\boldsymbol{x}\rangle \\
&= 2\,\langle \mathrm{d}\boldsymbol{x}|\,\tfrac{1}{2}(\boldsymbol{u}\otimes\boldsymbol{\nabla}_{\boldsymbol{x}} + \boldsymbol{\nabla}_{\boldsymbol{x}}\otimes\boldsymbol{u} - \boldsymbol{\nabla}_{\boldsymbol{x}}\otimes\boldsymbol{u}\cdot\boldsymbol{u}\otimes\boldsymbol{\nabla}_{\boldsymbol{x}})|\,\mathrm{d}\boldsymbol{x}\rangle
\end{aligned}
\tag{s5-9}
$$

因此, 阿尔曼西应变可表示为

$$
\boldsymbol{e} = \frac{1}{2}\left(\boldsymbol{u}\otimes\boldsymbol{\nabla}_{\boldsymbol{x}} + \boldsymbol{\nabla}_{\boldsymbol{x}}\otimes\boldsymbol{u} - \boldsymbol{\nabla}_{\boldsymbol{x}}\otimes\boldsymbol{u}\cdot\boldsymbol{u}\otimes\boldsymbol{\nabla}_{\boldsymbol{x}}\right) \tag{s5-10}
$$

5.11 大作业三: 柱坐标 (cylindrical polar coordinates, 参见图 15.3) 下的变形梯度张量. 设参考构形和当前构形的正交但非单位的基矢量分别为 $(\boldsymbol{g}_R, \boldsymbol{g}_\Theta, \boldsymbol{g}_Z)$ 和 $(\boldsymbol{g}_r, \boldsymbol{g}_\theta, \boldsymbol{g}_z)$. 变形前后的点分别为

$$
\begin{cases}
\boldsymbol{X} = R\boldsymbol{g}_R + \Theta\boldsymbol{g}_\Theta + Z\boldsymbol{g}_Z \\
\boldsymbol{x} = r(R,\Theta,Z)\,\boldsymbol{g}_r + \theta(R,\Theta,Z)\,\boldsymbol{g}_\theta + z(R,\Theta,Z)\,\boldsymbol{g}_z
\end{cases}
\tag{s5-11}
$$

梯度算子表示为

$$
\boldsymbol{\nabla}_X = \frac{\partial}{\partial X^A}\boldsymbol{g}^A = \frac{\partial}{\partial R}\boldsymbol{g}^R + \frac{\partial}{\partial \Theta}\boldsymbol{g}^\Theta + \frac{\partial}{\partial Z}\boldsymbol{g}^Z \tag{s5-12}
$$

由 (26-5) 式, 变形梯度可通过正交非单位基矢量表示为

$$
\begin{aligned}
\boldsymbol{F} = \boldsymbol{x}\otimes\boldsymbol{\nabla}_X &= \frac{\partial x^i}{\partial X^A}\boldsymbol{g}_i\otimes\boldsymbol{g}^A \\
&= \frac{\partial r}{\partial R}\boldsymbol{g}_r\otimes\boldsymbol{g}^R + \frac{\partial r}{\partial \Theta}\boldsymbol{g}_r\otimes\boldsymbol{g}^\Theta + \frac{\partial r}{\partial Z}\boldsymbol{g}_r\otimes\boldsymbol{g}^Z \\
&\quad + \frac{\partial \theta}{\partial R}\boldsymbol{g}_\theta\otimes\boldsymbol{g}^R + \frac{\partial \theta}{\partial \Theta}\boldsymbol{g}_\theta\otimes\boldsymbol{g}^\Theta + \frac{\partial \theta}{\partial Z}\boldsymbol{g}_\theta\otimes\boldsymbol{g}^Z \\
&\quad + \frac{\partial z}{\partial R}\boldsymbol{g}_z\otimes\boldsymbol{g}^R + \frac{\partial z}{\partial \Theta}\boldsymbol{g}_z\otimes\boldsymbol{g}^\Theta + \frac{\partial z}{\partial Z}\boldsymbol{g}_z\otimes\boldsymbol{g}^Z
\end{aligned}
\tag{s5-13}
$$

其中, (s5-13) 式中基矢量的模长满足柱坐标系的黎曼度规张量 (15-11) 式:

$$
[g_{mn}] = [\boldsymbol{g}_m\cdot\boldsymbol{g}_n] = \begin{bmatrix} 1 & 0 & 0 \\ 0 & r^2 & 0 \\ 0 & 0 & 1 \end{bmatrix} \tag{s5-14}
$$

为方便起见, 引入参考和当前两种构形中的正交且单位的基矢量 $(\boldsymbol{e}_R, \boldsymbol{e}_\Theta, \boldsymbol{e}_Z)$ 和 $(\boldsymbol{e}_r, \boldsymbol{e}_\theta, \boldsymbol{e}_z)$, 根据度规张量单位化的基矢量, 对应的协变基矢量转换涉及:

$$
\boldsymbol{e}_\theta = \frac{1}{r}\boldsymbol{g}_\theta \tag{s5-15}
$$

逆变基矢量的转化关系为

$$e^\Theta = Rg^\Theta \qquad (\text{s5-16})$$

相对应的梯度算子 (s5-12) 式用正交且单位化的基矢量表示为

$$\nabla_R = \frac{\partial}{\partial R}e^R + \frac{1}{R}\frac{\partial}{\partial \Theta}e^\Theta + \frac{\partial}{\partial Z}e^Z \qquad (\text{s5-17})$$

从而，柱坐标系下的变形梯度可用正交且单位化的基矢量表示如下：

$$\begin{aligned}
F = x \otimes \nabla_X &= \frac{\partial x^i}{\partial X^A}g_i \otimes g^A \\
&= \frac{\partial r}{\partial R}e_r \otimes e^R + \frac{1}{R}\frac{\partial r}{\partial \Theta}e_r \otimes e^\Theta + \frac{\partial r}{\partial Z}e_r \otimes e^Z \\
&\quad + r\frac{\partial \theta}{\partial R}e_\theta \otimes e^R + \frac{r}{R}\frac{\partial \theta}{\partial \Theta}e_\theta \otimes e^\Theta + r\frac{\partial \theta}{\partial Z}e_\theta \otimes e^Z \\
&\quad + \frac{\partial z}{\partial R}e_z \otimes e^R + \frac{1}{R}\frac{\partial z}{\partial \Theta}e_z \otimes e^\Theta + \frac{\partial z}{\partial Z}e_z \otimes e^Z
\end{aligned} \qquad (\text{s5-18})$$

亦即，柱坐标系下的变形梯度可表示为如下矩阵形式：

$$[F] = \begin{bmatrix} \dfrac{\partial r}{\partial R} & \dfrac{1}{R}\dfrac{\partial r}{\partial \Theta} & \dfrac{\partial r}{\partial Z} \\ r\dfrac{\partial \theta}{\partial R} & \dfrac{r}{R}\dfrac{\partial \theta}{\partial \Theta} & r\dfrac{\partial \theta}{\partial Z} \\ \dfrac{\partial z}{\partial R} & \dfrac{1}{R}\dfrac{\partial z}{\partial \Theta} & \dfrac{\partial z}{\partial Z} \end{bmatrix} \qquad (\text{s5-19})$$

5.12 大作业四：球坐标 (spherical coordinates, 参见图 15.4) 下的变形梯度张量. 参考和当前两种构形中的正交但非单位基矢量分别为 (g_R, g_Θ, g_ψ) 和 (g_r, g_θ, g_ϕ). 变形前和变形后的点的坐标分别为

$$\begin{cases} X = Rg_R + \Theta g_\Theta + \psi g_\psi \\ x = r(R,\Theta,\psi)g_r + \theta(R,\Theta,\psi)g_\theta + \varphi(R,\Theta,\psi)g_\phi \end{cases} \qquad (\text{s5-20})$$

球坐标系的梯度算子为

$$\nabla_X = \frac{\partial}{\partial X^A}g^A = \frac{\partial}{\partial R}g^R + \frac{\partial}{\partial \Theta}g^\Theta + \frac{\partial}{\partial \psi}g^\psi \qquad (\text{s5-21})$$

球坐标系下用正交且非单位基矢量表示的变形梯度张量为

$$\begin{aligned}
F = x \otimes \nabla_X &= \frac{\partial x^i}{\partial X^A}g_i \otimes g^A \\
&= \frac{\partial r}{\partial R}g_r \otimes g^R + \frac{\partial r}{\partial \Theta}g_r \otimes g^\Theta + \frac{\partial r}{\partial \psi}g_r \otimes g^\psi \\
&\quad + \frac{\partial \theta}{\partial R}g_\theta \otimes g^R + \frac{\partial \theta}{\partial \Theta}g_\theta \otimes g^\Theta + \frac{\partial \theta}{\partial \psi}g_\theta \otimes g^\psi \\
&\quad + \frac{\partial \phi}{\partial R}g_\phi \otimes g^R + \frac{\partial \phi}{\partial \Theta}g_\phi \otimes g^\Theta + \frac{\partial \phi}{\partial \psi}g_\phi \otimes g^\psi
\end{aligned} \qquad (\text{s5-22})$$

式中，基矢量的模长为满足球坐标系黎曼度规张量 (15-15) 式：

$$[g_{mn}] = [\boldsymbol{g}_m \cdot \boldsymbol{g}_n] = \begin{bmatrix} 1 & 0 & 0 \\ 0 & r^2 & 0 \\ 0 & 0 & r^2 \sin^2 \theta \end{bmatrix} \tag{s5-23}$$

为方便起见，仍引入参考和当前两种构形中的正交且单位化的基矢量 $(\boldsymbol{e}_R, \boldsymbol{e}_\Theta, \boldsymbol{e}_\psi)$ 和 $(\boldsymbol{e}_r, \boldsymbol{e}_\theta, \boldsymbol{e}_\phi)$. 根据度规张量单位化的基矢量，对应的协变基矢量和逆变基矢量的转换涉及如下关系：

$$\boldsymbol{e}_\theta = \frac{1}{r} \boldsymbol{g}_\theta, \quad \boldsymbol{e}_\phi = \frac{1}{r \sin \theta} \boldsymbol{g}_\varphi$$
$$\boldsymbol{e}^\Theta = R \boldsymbol{g}^\Theta, \quad \boldsymbol{e}^\psi = R \sin \Theta \boldsymbol{g}^\psi \tag{s5-24}$$

相对应的用正交且单位化的基矢量表示的梯度算子为

$$\boldsymbol{\nabla}_X = \frac{\partial}{\partial R} \boldsymbol{e}^R + \frac{1}{R} \frac{\partial}{\partial \Theta} \boldsymbol{e}^\Theta + \frac{1}{R \sin \Theta} \frac{\partial}{\partial \psi} \boldsymbol{e}^\psi \tag{s5-25}$$

则球坐标系下用正交且单位化的基矢量表示的变形梯度张量为

$$\begin{aligned}
\boldsymbol{F} &= \boldsymbol{x} \otimes \boldsymbol{\nabla}_X = \frac{\partial x^i}{\partial X^A} \boldsymbol{g}_i \otimes \boldsymbol{g}^A \\
&= \frac{\partial r}{\partial R} \boldsymbol{e}_r \otimes \boldsymbol{e}^R + \frac{1}{R} \frac{\partial r}{\partial \Theta} \boldsymbol{e}_r \otimes \boldsymbol{e}^\Theta + \frac{1}{R \sin \Theta} \frac{\partial r}{\partial \psi} \boldsymbol{e}_r \otimes \boldsymbol{e}^\psi \\
&\quad + r \frac{\partial \theta}{\partial R} \boldsymbol{e}_\theta \otimes \boldsymbol{e}^R + \frac{r}{R} \frac{\partial \theta}{\partial \Theta} \boldsymbol{e}_\theta \otimes \boldsymbol{e}^\Theta + \frac{r}{R \sin \Theta} \frac{\partial \theta}{\partial \psi} \boldsymbol{e}_\theta \otimes \boldsymbol{e}^\psi \\
&\quad + r \sin \theta \frac{\partial \phi}{\partial R} \boldsymbol{e}_\phi \otimes \boldsymbol{e}^R + \frac{r \sin \theta}{R} \frac{\partial \phi}{\partial \Theta} \boldsymbol{e}_\phi \otimes \boldsymbol{e}^\Theta + \frac{r \sin \theta}{R \sin \Theta} \frac{\partial \phi}{\partial \psi} \boldsymbol{e}_\phi \otimes \boldsymbol{e}^\psi
\end{aligned} \tag{s5-26}$$

亦即，球坐标系下的变形梯度张量的矩阵形式为

$$[\boldsymbol{F}] = \begin{bmatrix} \dfrac{\partial r}{\partial R} & \dfrac{1}{R} \dfrac{\partial r}{\partial \Theta} & \dfrac{1}{R \sin \Theta} \dfrac{\partial r}{\partial \psi} \\ r \dfrac{\partial \theta}{\partial R} & \dfrac{r}{R} \dfrac{\partial \theta}{\partial \Theta} & \dfrac{r}{R \sin \Theta} \dfrac{\partial \theta}{\partial \psi} \\ r \sin \theta \dfrac{\partial \phi}{\partial R} & \dfrac{r \sin \theta}{R} \dfrac{\partial \phi}{\partial \Theta} & \dfrac{r \sin \theta}{R \sin \Theta} \dfrac{\partial \phi}{\partial \psi} \end{bmatrix} \tag{s5-27}$$

5.13 对于小变形，亦即当位移梯度满足 $\boldsymbol{H} \to \boldsymbol{0}$ 时，验证：

$$J^{-1} = \det{}^{-1}(\boldsymbol{F}) \approx 1 - \mathrm{tr} \boldsymbol{H} \tag{s5-28}$$

证明：当位移梯度满足 $\boldsymbol{H} \to \boldsymbol{0}$ 时，由 (s3-6) 式，有如下一阶泰勒展开：

$$\begin{aligned}
J^{-1} &= \det{}^{-1}(\boldsymbol{F}) = \det{}^{-1}(\boldsymbol{I} + \boldsymbol{H}) \approx \det{}^{-1}(\boldsymbol{I}) + \left. \frac{\partial J^{-1}}{\partial \boldsymbol{F}} \right|_{\boldsymbol{F} = \boldsymbol{I}} : \boldsymbol{H} \\
&= 1 - \left[\det{}^{-1}(\boldsymbol{F}) \boldsymbol{F}^{-\mathrm{T}} \right]\Big|_{\boldsymbol{F} = \boldsymbol{I}} : \boldsymbol{H} = 1 - \det{}^{-1}(\boldsymbol{I}) \, \mathrm{tr}\left(\boldsymbol{F}^{-1} \boldsymbol{H}\right)\Big|_{\boldsymbol{F} = \boldsymbol{I}} \\
&= 1 - \mathrm{tr} \boldsymbol{H}
\end{aligned} \tag{s5-29}$$

5.14 大作业五：验证由例 27.2 所给出的工程应变不是张量的分量.

证明：由例 20.2 可知，在坐标系发生旋转后，旧和新坐标系的关系为

$$\begin{cases} x = x'\cos\theta - y'\sin\theta \\ y = x'\sin\theta + y'\cos\theta \end{cases} \text{或者} \begin{bmatrix} x \\ y \end{bmatrix} = \boldsymbol{Q}^{\mathrm{T}} \begin{bmatrix} x' \\ y' \end{bmatrix} \tag{s5-30}$$

新位移 (u', v') 和旧位移 (u, v) 之间满足类似 (20-18) 式的关系：

$$\begin{cases} u' = u\cos\theta + v\sin\theta \\ v' = -u\sin\theta + v\cos\theta \end{cases} \text{或者} \begin{bmatrix} u' \\ v' \end{bmatrix} = \boldsymbol{Q} \begin{bmatrix} u \\ v \end{bmatrix} \tag{s5-31}$$

按照例 27.2 所给出的工程应变的定义，坐标旋转后工程应变的分量为

$$[\boldsymbol{\gamma}'] = \begin{bmatrix} \varepsilon_x' & \gamma_{xy}' \\ \gamma_{xy}' & \varepsilon_y' \end{bmatrix} = \begin{bmatrix} \dfrac{\partial u'}{\partial x'} & \dfrac{\partial v'}{\partial x'} + \dfrac{\partial u'}{\partial y'} \\ \dfrac{\partial v'}{\partial x'} + \dfrac{\partial u'}{\partial y'} & \dfrac{\partial v'}{\partial y'} \end{bmatrix} \tag{s5-32}$$

为了简洁起见，仅以 ε_x' 为例说明，其表达式为

$$\begin{aligned} \varepsilon_x' &= \frac{\partial u'}{\partial x'} = \frac{\partial u'}{\partial u}\left(\frac{\partial u}{\partial x}\frac{\partial x}{\partial x'} + \frac{\partial u}{\partial y}\frac{\partial y}{\partial x'}\right) + \frac{\partial u'}{\partial v}\left(\frac{\partial v}{\partial x}\frac{\partial x}{\partial x'} + \frac{\partial v}{\partial y}\frac{\partial y}{\partial x'}\right) \\ &= \frac{\partial u}{\partial x}\cos^2\theta + \left(\frac{\partial u}{\partial y} + \frac{\partial v}{\partial x}\right)\cos\theta\sin\theta + \frac{\partial v}{\partial y}\sin^2\theta \\ &= \varepsilon_x\cos^2\theta + \gamma_{xy}\cos\theta\sin\theta + \varepsilon_y\sin^2\theta \end{aligned} \tag{s5-33}$$

按照正交矩阵变换所得到的工程应变的分量为

$$\begin{aligned} [\boldsymbol{\gamma}''] &= [\boldsymbol{Q}\boldsymbol{\gamma}\boldsymbol{Q}^{\mathrm{T}}] = \begin{bmatrix} \varepsilon_x'' & \gamma_{xy}'' \\ \gamma_{xy}'' & \varepsilon_y'' \end{bmatrix} \\ &= \begin{bmatrix} \cos\theta & \sin\theta \\ -\sin\theta & \cos\theta \end{bmatrix} \begin{bmatrix} \varepsilon_x & \gamma_{xy} \\ \gamma_{xy} & \varepsilon_y \end{bmatrix} \begin{bmatrix} \cos\theta & -\sin\theta \\ \sin\theta & \cos\theta \end{bmatrix} \end{aligned} \tag{s5-34}$$

仅以 ε_x'' 为例说明，其表达式为

$$\varepsilon_x'' = \varepsilon_x\cos^2\theta + 2\gamma_{xy}\cos\theta\sin\theta + \varepsilon_y\sin^2\theta \tag{s5-35}$$

对比以上两种方法所得到了分量 (s5-33) 和 (s5-35) 两式，易知 $\varepsilon_x' \neq \varepsilon_x''$，亦即 $\boldsymbol{\gamma}' \neq \boldsymbol{Q}\boldsymbol{\gamma}\boldsymbol{Q}^{\mathrm{T}}$. 因此，这就证明了工程应变不是张量的分量. 这里还提醒读者参阅 (37-2) 式有关二阶张量客观性的定义.

参 考 文 献

[5.1] Föppl A. Vorlesungen über technische Mechanik. B. G. Teubner, Bd. 5: 132-144, Leipzig, Germany, 1907.

[5.2] von Kármán T. Festigkeitsproblem im Maschinenbau. Encyclopädie der mathematischen Wissenschaften, IV: 311-385, 1910.

[5.3] Almansi E. Sulledeformazioni finite dei solidi elastici isotropi, I. Rendiconti della Reale Accademia dei Lincei. Classe di scienze fisiclie, matematiclie e naturali, 1911, 20: 705-714.

[5.4] Hill R. On constitutive inequalities for simple materials—I. Journal of the Mechanics and Physics of Solids, 1968, 16: 229-242.

[5.5] Hill R. Aspects of invariance in solid mechanics. Advances in Applied Mechanics, 18: 1–75, New York: Academic Press, 1978.

[5.6] Hill R. The Mathematical Theory of Plasticity. Oxford: Oxford University Press, 1950. (中译本: 希尔. 塑性数学理论 (王仁译). 北京: 科学出版社, 1966.)

[5.7] Seth B R. Generalized strain measure with applications to physical problems. MRC Technical Summary Report #248 (Mathematics Research Center, United States Army, University of Wisconsin), 1961, 1-18.

[5.8] Seth B R. Generalized strain measure with applications to physical problems. IUTAM Symposium on Second Order Effects in Elasticity, Plasticity and Fluid Mechanics, Haifa, 1962.

[5.9] Seth B R. Generalized strain measure with applications to physical problems. In: Second-order Effects in Elasticity, Plasticity and Fluid Dynamics (Reiner M, Abir D eds.), Oxford: Pergamon, pp. 162-172, 1964.

[5.10] Nanson E J. Methods of Election. Transactions and Proceedings of the Royal Society of Victoria, 1882, 18: 197-240.

[5.11] von Kármán T, Edson L. The Wind and Beyond: Theodore von Kármán, Pioneer in Aviation and Pathfinder in Space. Little: Brown, 1967 (中译本: 冯·卡门, 李·爱特生. 冯·卡门——钱学森的导师 (王克仁译). 西安: 西安交通大学出版社, 2015.)

[5.12] Rodriguez E K, Hoger A, McCulloch A D. Stress-dependent finite growth in soft elastic tissues. Journal of Biomechanics, 1994, 27: 455-467.

[5.13] Jones G W, Chapman S J. Modeling growth in biological materials. SIAM Review, 2012, 54: 52-118.

[5.14] Goriely A. The Mathematics and Mechanics of Biological Growth. New York: Springer, 2017.

第 6 章 守恒律与场方程

§30. 雷诺输运定理

雷诺输运定理 (Reynolds' transport theorem, RTT) 是进行连续介质力学场方程推导的重要基础,本节将给出其详细证明.

在例 29.10 中,我们业已证明了当前构形下微元体积的时间变化率,由 (29-43) 式或 (29-44) 式给出. 本节中我们将用另外一种方法进行证明,其目的是为了从不同侧面出发,进一步加深对该基本问题的理解.

当前构形中的线段 $\mathrm{d}\boldsymbol{r}$ 和参考构形中的线段 $\mathrm{d}\boldsymbol{R}$ 通过两点张量或称混合欧拉–拉格朗日型张量——变形梯度张量 $\boldsymbol{F}$——来联系:

$$\mathrm{d}\boldsymbol{r} = \boldsymbol{F}\mathrm{d}\boldsymbol{R} \tag{30-1}$$

对上式对时间进行求导:

$$\frac{\mathrm{d}}{\mathrm{d}t}(\mathrm{d}\boldsymbol{r}) = \frac{\mathrm{d}}{\mathrm{d}t}(\boldsymbol{F}\mathrm{d}\boldsymbol{R}) = \frac{\mathrm{d}}{\mathrm{d}t}\left(\frac{\partial \boldsymbol{r}}{\partial \boldsymbol{R}}\mathrm{d}\boldsymbol{R}\right) = \frac{\mathrm{d}}{\mathrm{d}t}\left(\frac{\partial \boldsymbol{r}}{\partial \boldsymbol{R}}\right)\mathrm{d}\boldsymbol{R}$$

$$= \frac{\partial}{\partial \boldsymbol{R}}\left(\frac{\mathrm{d}\boldsymbol{r}}{\mathrm{d}t}\right)\mathrm{d}\boldsymbol{R} = \frac{\partial \boldsymbol{v}}{\partial \boldsymbol{R}}\mathrm{d}\boldsymbol{R} = \frac{\partial \boldsymbol{v}}{\partial \boldsymbol{r}}\frac{\partial \boldsymbol{r}}{\partial \boldsymbol{R}}\mathrm{d}\boldsymbol{R}$$

$$= \frac{\partial \boldsymbol{v}}{\partial \boldsymbol{r}}\boldsymbol{F}\mathrm{d}\boldsymbol{R} = (\boldsymbol{v} \otimes \boldsymbol{\nabla})\mathrm{d}\boldsymbol{r} = \boldsymbol{l}\mathrm{d}\boldsymbol{r} \tag{30-2}$$

式中, $\boldsymbol{v}$ 为速度矢量, $\boldsymbol{l} = \boldsymbol{v} \otimes \boldsymbol{\nabla}$ 为速度梯度. 体积微元可通过标量三重积表示为

$$\mathrm{d}v = [\mathrm{d}\boldsymbol{r}_1 \quad \mathrm{d}\boldsymbol{r}_2 \quad \mathrm{d}\boldsymbol{r}_3] \tag{30-3}$$

利用标量三重积的性质 (6-4) 式,体积微元的时间变化率为

$$\frac{\mathrm{d}}{\mathrm{d}t}(\mathrm{d}v) = \frac{\mathrm{d}}{\mathrm{d}t}[\mathrm{d}\boldsymbol{r}_1 \cdot (\mathrm{d}\boldsymbol{r}_2 \times \mathrm{d}\boldsymbol{r}_3)] = \frac{\mathrm{d}}{\mathrm{d}t}(\mathrm{d}\boldsymbol{r}_1) \cdot (\mathrm{d}\boldsymbol{r}_2 \times \mathrm{d}\boldsymbol{r}_3) + \mathrm{d}\boldsymbol{r}_1 \cdot \left[\frac{\mathrm{d}}{\mathrm{d}t}(\mathrm{d}\boldsymbol{r}_2) \times \mathrm{d}\boldsymbol{r}_3\right]$$

$$+ \mathrm{d}\boldsymbol{r}_1 \cdot \left[\mathrm{d}\boldsymbol{r}_2 \times \frac{\mathrm{d}}{\mathrm{d}t}(\mathrm{d}\boldsymbol{r}_3)\right]$$

$$= \boldsymbol{l} \cdot \mathrm{d}\boldsymbol{r}_1 \cdot (\mathrm{d}\boldsymbol{r}_2 \times \mathrm{d}\boldsymbol{r}_3) + \mathrm{d}\boldsymbol{r}_1 \cdot [\boldsymbol{l} \cdot \mathrm{d}\boldsymbol{r}_2 \times \mathrm{d}\boldsymbol{r}_3] + \mathrm{d}\boldsymbol{r}_1 \cdot [\mathrm{d}\boldsymbol{r}_2 \times \boldsymbol{l} \cdot \mathrm{d}\boldsymbol{r}_3]$$

$$= \boldsymbol{l} \cdot \underbrace{\mathrm{d}\boldsymbol{r}_1 \cdot (\mathrm{d}\boldsymbol{r}_2 \times \mathrm{d}\boldsymbol{r}_3)}_{\mathrm{d}v} + \boldsymbol{l} \cdot \underbrace{\mathrm{d}\boldsymbol{r}_2 \cdot \mathrm{d}\boldsymbol{r}_3 \times \mathrm{d}\boldsymbol{r}_1}_{\mathrm{d}v} + \boldsymbol{l} \cdot \underbrace{\mathrm{d}\boldsymbol{r}_3 \cdot \mathrm{d}\boldsymbol{r}_1 \times \mathrm{d}\boldsymbol{r}_2}_{\mathrm{d}v}$$

$$= l_{ij}\boldsymbol{e}_i \otimes \boldsymbol{e}_j \cdot \boldsymbol{e}_1 \mathrm{d}r_1 \mathrm{d}r_2 \mathrm{d}r_3 \cdot \boldsymbol{e}_1 + (l_{ij}\boldsymbol{e}_i \otimes \boldsymbol{e}_j \cdot \boldsymbol{e}_2 \mathrm{d}r_2) \cdot \mathrm{d}r_1 \mathrm{d}r_3 \boldsymbol{e}_2$$

$$+ (l_{ij}\boldsymbol{e}_i \otimes \boldsymbol{e}_j \cdot \boldsymbol{e}_3 \mathrm{d}r_3) \cdot \mathrm{d}r_1 \mathrm{d}r_2 \boldsymbol{e}_3$$

$$= l_{ij}\boldsymbol{e}_i \delta_{j1} \cdot \boldsymbol{e}_1 \mathrm{d}r_1 \mathrm{d}r_2 \mathrm{d}r_3 + l_{ij}\boldsymbol{e}_i \delta_{j2} \cdot \boldsymbol{e}_2 \mathrm{d}r_1 \mathrm{d}r_2 \mathrm{d}r_3 + l_{ij}\boldsymbol{e}_i \delta_{j3} \cdot \boldsymbol{e}_3 \mathrm{d}r_1 \mathrm{d}r_2 \mathrm{d}r_3$$

$$= l_{ij}\delta_{i1}\delta_{j1}\mathrm{d}r_1 \mathrm{d}r_2 \mathrm{d}r_3 + l_{ij}\delta_{i2}\delta_{j2}\mathrm{d}r_1 \mathrm{d}r_2 \mathrm{d}r_3 + l_{ij}\delta_{i3}\delta_{j3}\mathrm{d}r_1 \mathrm{d}r_2 \mathrm{d}r_3$$

$$= (l_{11} + l_{22} + l_{33})\mathrm{d}r_1 \mathrm{d}r_2 \mathrm{d}r_3 = (\boldsymbol{l} : \mathbf{I})\,\mathrm{d}v = \mathrm{tr}\,(\boldsymbol{v} \otimes \boldsymbol{\nabla})\,\mathrm{d}v = (\boldsymbol{\nabla} \cdot \boldsymbol{v})\,\mathrm{d}v \quad (30\text{-}4)$$

上式的结果和 (29-44) 式 $\dot{J} = J \mathrm{div}\boldsymbol{v}$ 完全相同.

下面计算某个积分量的导数:

$$\frac{\mathrm{d}}{\mathrm{d}t}\int \Phi \mathrm{d}v = \int_v \frac{\mathrm{d}}{\mathrm{d}t}(\Phi \cdot \mathrm{d}v) = \int_v \left[\frac{\mathrm{d}\Phi}{\mathrm{d}t}\mathrm{d}v + \Phi\frac{\mathrm{d}}{\mathrm{d}t}(\mathrm{d}v)\right]$$

$$= \int_v \left[\frac{\mathrm{d}\Phi(\boldsymbol{x},t)}{\mathrm{d}t} + (\boldsymbol{\nabla} \cdot \boldsymbol{v})\,\Phi\right]\mathrm{d}v = \int_v \left[\frac{\partial \Phi}{\partial t} + \frac{\partial \Phi}{\partial \boldsymbol{x}} \cdot \frac{\partial \boldsymbol{x}}{\partial t} + (\boldsymbol{\nabla} \cdot \boldsymbol{v})\,\Phi\right]\mathrm{d}v$$

$$= \int_v \left[\frac{\partial \Phi}{\partial t} + \boldsymbol{v} \cdot (\boldsymbol{\nabla}\Phi) + (\boldsymbol{\nabla} \cdot \boldsymbol{v})\,\Phi\right]\mathrm{d}v = \int_v \left[\frac{\partial \Phi}{\partial t} + \boldsymbol{\nabla} \cdot (\Phi\boldsymbol{v})\right]\mathrm{d}v \quad (30\text{-}5)$$

上式所给出的积分量的时间导数的关系式: $\dfrac{\mathrm{d}}{\mathrm{d}t}\int \Phi \mathrm{d}v = \int_v \left[\dfrac{\partial \Phi}{\partial t} + \boldsymbol{\nabla} \cdot (\Phi\boldsymbol{v})\right]\mathrm{d}v$, 便称为雷诺输运定理. 该定理之所以用著名流体力学家奥斯鲍恩·雷诺 (Osborne Reynolds, 1842—1912, 如图 30.1 所示) 来命名, 主要是基于他于 1903 年的工作 [6.1].

图 30.1 奥斯鲍恩·雷诺: 1868 年到欧文学院任职 (左); 1895 年 (中); 1904 年退休肖像 (右)

雷诺在曼彻斯特欧文学院授课的轶事在本书前言中已经给出, 其英文原文见思考题 11.6. 该授课的个例再一次说明，好的授课并没有统一的模式.

§31. 质 量 守 恒

31.1 欧拉描述下的质量守恒方程

将雷诺输运定理 (30-5) 式中的变量 Φ 用密度 ρ 来代替的话，将得到欧拉描述下质量守恒定律或质量连续性方程 (mass continuity equation) 的表达式为

$$\frac{\partial \rho}{\partial t} + \boldsymbol{\nabla} \cdot (\rho \boldsymbol{v}) = \frac{\partial \rho}{\partial t} + \mathrm{div}\,(\rho \boldsymbol{v}) = 0 \tag{31-1}$$

引入质量流密度 (mass flux density) 矢量：

$$\boldsymbol{j} = \rho \boldsymbol{v} \tag{31-2}$$

其方向沿着流体流动的方向，其大小为单位时间垂直于速度方向的单位面积所流过的质量. 则欧拉描述下质量连续性方程 (31-1) 式还可表示为

$$\frac{\partial \rho}{\partial t} + \boldsymbol{\nabla} \cdot \boldsymbol{j} = 0 \tag{31-3}$$

再由散度的运算法则，一个标量函数 ρ 和一个矢量函数的乘积 $\rho \boldsymbol{v}$ 的散度可以进一步分解为

$$\mathrm{div}\,(\rho \boldsymbol{v}) = \rho \mathrm{div}\,\boldsymbol{v} + \boldsymbol{v} \cdot \mathbf{grad}\,\rho \tag{31-4}$$

从而，欧拉描述下的质量连续性方程可等价地表示为 [6.2]

$$\frac{\partial \rho}{\partial t} + \rho \mathrm{div}\,\boldsymbol{v} + \boldsymbol{v} \cdot \mathbf{grad}\,\rho = 0 \quad \text{或} \quad \frac{\mathrm{d}\rho}{\mathrm{d}t} + \rho \mathrm{div}\,\boldsymbol{v} = 0 \tag{31-5}$$

当流体不可压缩时，$\rho = \mathrm{const}$，亦即 $\frac{\mathrm{d}\rho}{\mathrm{d}t} = 0$，从而有体积的时间变化率为零

$$\boldsymbol{\nabla} \cdot \boldsymbol{v} = 0 \quad \text{或} \quad \mathrm{div}\,\boldsymbol{v} = 0 \tag{31-6}$$

例 31.1 给出电荷守恒方程.

答：电荷守恒方程又称为电荷连续性方程 (charge continuity equation)，在形式上和 (31-3) 式完全相同，其中，ρ 为电荷密度 (charge density)，而 $\boldsymbol{j}$ 为电流密度 (current density).

例 31.2 用微元体的概念说明 (31-6) 式的物理意义.

解：取如图 31.1 所示的体积为 $\mathrm{d}V = \mathrm{d}x_1 \mathrm{d}x_2 \mathrm{d}x_3$ 的微元体，设介质的密度 ρ 为常量.

第 6 章 守恒律与场方程

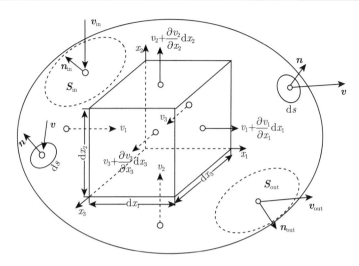

图 31.1 输运问题的微元体

沿 x_1 轴的质量的流入随时间的变化率为

$$\left.\frac{\mathrm{d}m_1}{\mathrm{d}t}\right|_{\mathrm{in}} = \mathrm{d}x_2\mathrm{d}x_3\rho v_1 \tag{31-7}$$

而沿 x_1 轴的质量的流出随时间的变化率为

$$\left.\frac{\mathrm{d}m_1}{\mathrm{d}t}\right|_{\mathrm{out}} = \mathrm{d}x_2\mathrm{d}x_3\rho\left(v_1 + \frac{\partial v_1}{\partial x_1}\mathrm{d}x_1\right) \tag{31-8}$$

则沿 x_1 轴的净质量随时间的变化率为

$$\frac{\mathrm{d}m_1}{\mathrm{d}t} = \left.\frac{\mathrm{d}m_1}{\mathrm{d}t}\right|_{\mathrm{out}} - \left.\frac{\mathrm{d}m_1}{\mathrm{d}t}\right|_{\mathrm{in}} = \mathrm{d}x_1\mathrm{d}x_2\mathrm{d}x_3\rho\frac{\partial v_1}{\partial x_1} \tag{31-9}$$

同理，沿 x_2 和 x_3 轴的净质量随时间的变化率分别为

$$\frac{\mathrm{d}m_2}{\mathrm{d}t} = \mathrm{d}x_1\mathrm{d}x_2\mathrm{d}x_3\rho\frac{\partial v_2}{\partial x_2}, \quad \frac{\mathrm{d}m_3}{\mathrm{d}t} = \mathrm{d}x_1\mathrm{d}x_2\mathrm{d}x_3\rho\frac{\partial v_3}{\partial x_3} \tag{31-10}$$

则整个微元体的净质量随时间的变化率为

$$\begin{aligned}\frac{\mathrm{d}m}{\mathrm{d}t} &= \frac{\mathrm{d}m_1}{\mathrm{d}t} + \frac{\mathrm{d}m_2}{\mathrm{d}t} + \frac{\mathrm{d}m_3}{\mathrm{d}t} \\ &= \mathrm{d}x_1\mathrm{d}x_2\mathrm{d}x_3\rho\left(\frac{\partial v_1}{\partial x_1} + \frac{\partial v_2}{\partial x_2} + \frac{\partial v_3}{\partial x_3}\right) = \rho\mathrm{d}V\boldsymbol{\nabla}\cdot\boldsymbol{v}\end{aligned} \tag{31-11}$$

由于微元体积 $\mathrm{d}V$ 的任意性，故，介质的不可压缩性条件可以表示为

$$\boldsymbol{\nabla}\cdot\boldsymbol{v} = 0 \tag{31-12}$$

例 31.3 证明 (31-4) 式的散度运算准则.

证明：让我们来看一个任意的标量场 φ 和一个矢量场 $\boldsymbol{v}$ 的乘积 $\varphi\boldsymbol{v}$ 的散度，由于

$$\boldsymbol{\nabla} = \frac{\partial}{\partial x_i}\boldsymbol{e}_i, \quad \boldsymbol{v} = v_j\boldsymbol{e}_j \tag{31-13}$$

则有

$$\mathrm{div}\,(\varphi\boldsymbol{v}) = \frac{\partial}{\partial x_i}\boldsymbol{e}_i \cdot (\varphi v_j\boldsymbol{e}_j) = \varphi\frac{\partial v_j}{\partial x_i}\delta_{ij} + \left(\frac{\partial \varphi}{\partial x_i}\boldsymbol{e}_i\right) \cdot (v_j\boldsymbol{e}_j)$$

$$= \varphi\frac{\partial v_i}{\partial x_i} + \left(\frac{\partial \varphi}{\partial x_i}\boldsymbol{e}_i\right) \cdot \boldsymbol{v} = \varphi\,\mathrm{div}\,\boldsymbol{v} + \boldsymbol{v} \cdot \mathbf{grad}\,\varphi \tag{31-14}$$

值得说明的是，上述证明是在笛卡儿坐标系下给出的，即未考虑基矢量随坐标的变化.

例 31.4 给出正交曲纹坐标系下的连续性方程.

解：在曲纹坐标系下，梯度算子 (nabla) 定义为

$$\boldsymbol{\nabla} = \boldsymbol{g}^i\frac{\partial}{\partial \xi^i} \tag{31-15}$$

从而在正交曲纹坐标系下，速度的散度为

$$\boldsymbol{\nabla} \cdot \boldsymbol{v} = \left(\boldsymbol{g}^i\frac{\partial}{\partial \xi^i}\right) \cdot (v^j\boldsymbol{g}_j) = \boldsymbol{g}^i \cdot \left(\frac{\partial v^j}{\partial \xi^i}\boldsymbol{g}_j + \frac{\partial \boldsymbol{g}_j}{\partial \xi^i}v^j\right)$$

$$= v^i_{,i} + v^j\Gamma^i_{ij} \tag{31-16}$$

因此，正交曲纹坐标系下的连续性方程 (31-1) 式可表示为

$$\frac{\partial \rho}{\partial t} + \boldsymbol{\nabla} \cdot (\rho\boldsymbol{v}) = \frac{\partial \rho}{\partial t} + (\rho v^i)_{,i} + (\rho v^j)\Gamma^i_{ij} = 0 \tag{31-17}$$

在正交曲纹坐标系下，介质的不可压缩性条件可以相应地表示为

$$\boldsymbol{\nabla} \cdot \boldsymbol{v} = v^i_{,i} + v^j\Gamma^i_{ij} = 0 \tag{31-18}$$

31.2 拉格朗日描述下的质量守恒方程

质量守恒意味着在参考构形和当前构形下质量相等：

$$m = \int \rho_0\mathrm{d}V = \int \rho\mathrm{d}v \tag{31-19}$$

由两种构形下体积的变化关系：$\mathrm{d}v = J\mathrm{d}V$，因此 (31-19) 式可进一步表示为

$$m = \int \rho_0\mathrm{d}V = \int \rho J\mathrm{d}V = \int \rho\det\boldsymbol{F}\mathrm{d}V \tag{31-20}$$

上式等价于：$\int (\rho_0 - \rho J) \mathrm{d}V = \int (\rho_0 - \rho \det \boldsymbol{F}) \mathrm{d}V = 0$，亦即在拉格朗日描述下，连续性方程为

$$\rho_0 = \rho J = \rho \det \boldsymbol{F} \tag{31-21}$$

例 31.5 结合思考题 5.6，给出考虑生长的连续性方程.

答：当考虑材料生长或凋亡时，需要给出单位当前体积质量的增长率 (rate of increase of mass per unit current volume) $\gamma(\boldsymbol{r}, t)$，此时对于考虑生长的问题，连续性方程需要改写为

$$\begin{cases} \text{当前构形}: \dfrac{\partial \rho}{\partial t} + \boldsymbol{\nabla} \cdot (\rho \boldsymbol{v}) = \gamma \\ \text{初始构形}: \dfrac{\mathrm{d}(J\rho)}{\mathrm{d}t} = J\gamma \end{cases} \tag{31-22}$$

31.3 笛卡儿坐标系、柱坐标系和球坐标系下的质量守恒方程

为了便于初学者的理解和应用上的方便，本小节将给出连续性方程 (31-1) 式或 (31-6) 式在几个常用坐标系下的分量形式.

在笛卡儿坐标系 (x, y, z) 下，连续性方程为

$$\frac{\partial \rho}{\partial t} + \left(v_x \frac{\partial \rho}{\partial x} + v_y \frac{\partial \rho}{\partial y} + v_z \frac{\partial \rho}{\partial z} \right) + \rho \left(\frac{\partial v_x}{\partial x} + \frac{\partial v_y}{\partial y} + \frac{\partial v_z}{\partial z} \right) = 0 \tag{31-23}$$

在柱坐标系 (r, θ, z) 下，连续性方程为

$$\frac{\partial \rho}{\partial t} + \frac{1}{r} \frac{\partial (\rho r v_r)}{\partial r} + \frac{1}{r} \frac{\partial (\rho v_\theta)}{\partial \theta} + \frac{\partial (\rho v_z)}{\partial z} = 0 \tag{31-24}$$

在球坐标系 (r, θ, φ) 下，连续性方程为

$$\frac{\partial \rho}{\partial t} + \frac{1}{r^2} \frac{\partial (\rho r^2 v_r)}{\partial r} + \frac{1}{r \sin \theta} \frac{\partial (\rho v_\theta \sin \theta)}{\partial \theta} + \frac{1}{r \sin \theta} \frac{\partial (\rho v_\phi)}{\partial \phi} = 0 \tag{31-25}$$

§32. 动 量 守 恒

32.1 动量守恒与空间平移不变性

根据图 20.8 所示的诺特定理，线动量守恒是空间平移不变性的必然结果. 以下给出其证明的主要步骤.

空间平移不变性意味着空间的均匀性 (homogeneity of space). 此时，可假设一个封闭的力学系统均经历了一个微小的位移矢量 $\boldsymbol{\varepsilon}$，由于该系统的速度没有发生变

化，从而拉格朗日量的变分为

$$\delta L = \sum_\alpha \frac{\partial L}{\partial \boldsymbol{r}_\alpha} \cdot \delta \boldsymbol{r}_\alpha = \boldsymbol{\varepsilon} \cdot \sum_\alpha \frac{\partial L}{\partial \boldsymbol{r}_\alpha} \tag{32-1}$$

由拉格朗日方程 (4-2) 式，有 $\frac{\partial L}{\partial \boldsymbol{r}_\alpha} = \frac{\mathrm{d}}{\mathrm{d}t}\frac{\partial L}{\partial \dot{\boldsymbol{r}}_\alpha} = \frac{\mathrm{d}}{\mathrm{d}t}\frac{\partial L}{\partial \boldsymbol{v}_\alpha}$，由 $\delta L = 0$，则由于 (32-1) 式中 $\boldsymbol{\varepsilon}$ 的任意性，显然有

$$\sum_\alpha \frac{\partial L}{\partial \boldsymbol{r}_\alpha} = \sum_\alpha \frac{\mathrm{d}}{\mathrm{d}t}\frac{\partial L}{\partial \boldsymbol{v}_\alpha} = \frac{\mathrm{d}}{\mathrm{d}t} \sum_\alpha \frac{\partial L}{\partial \boldsymbol{v}_\alpha} = \boldsymbol{0} \tag{32-2}$$

由于 $\boldsymbol{p}_\alpha = \frac{\partial L}{\partial \boldsymbol{v}_\alpha}$ 为质点 α 的广义动量，(32-2) 式 $\frac{\mathrm{d}}{\mathrm{d}t}\sum_\alpha \boldsymbol{p}_\alpha = \boldsymbol{0}$ 等价于整个系统的广义动量守恒：$\boldsymbol{P} = \sum_\alpha \boldsymbol{p}_\alpha = \mathrm{const}$，亦即，空间平移不变性是动量守恒的根源.

32.2　欧拉描述下流体的动量守恒方程

动量守恒意味着动量矢量的时间变化率等于合力矢量. 将雷诺输运方程中的积分核换为 $\Phi = \rho \boldsymbol{v}$，则系统的动量变化率为 $\frac{\mathrm{d}}{\mathrm{d}t}\int \rho \boldsymbol{v} \mathrm{d}v$，从而 (30-5) 式变为

$$\frac{\mathrm{d}}{\mathrm{d}t}\int \rho \boldsymbol{v} \mathrm{d}v = \int_v \left[\frac{\partial (\rho \boldsymbol{v})}{\partial t} + \boldsymbol{\nabla} \cdot (\rho \boldsymbol{v} \otimes \boldsymbol{v})\right] \mathrm{d}v \tag{32-3}$$

对于流体而言，合力由三部分组成：体力 (如重力等)、正应力 (静水压强 $-p\mathbf{I}$) 和反映黏性的切应力 $\boldsymbol{\tau}$ 组成，应用动量守恒，亦即令动量矢量的时间变化率等于合力矢量，则有

$$\frac{\mathrm{d}}{\mathrm{d}t}\int \rho \boldsymbol{v} \mathrm{d}v = \int_v \left[\frac{\partial (\rho \boldsymbol{v})}{\partial t} + \boldsymbol{\nabla} \cdot (\rho \boldsymbol{v} \otimes \boldsymbol{v})\right] \mathrm{d}v = \int_v \rho \boldsymbol{f} \mathrm{d}v + \oiint_a (-p\mathbf{I} + \boldsymbol{\tau})\boldsymbol{n}\mathrm{d}a$$

$$= \int_v \rho \boldsymbol{f} \mathrm{d}v + \int_v \boldsymbol{\nabla} \cdot (-p\mathbf{I} + \boldsymbol{\tau})\mathrm{d}v \tag{32-4}$$

式中，$(-p\mathbf{I} + \boldsymbol{\tau})\boldsymbol{n}$ 表示单位外法线矢量 $\boldsymbol{n}$ 和二阶张量 $(-p\mathbf{I} + \boldsymbol{\tau})$ 之间的点积. 对上式合并同类项，再由于体积微元的任意性，则得到流体的动量方程为

$$\frac{\partial (\rho \boldsymbol{v})}{\partial t} + \boldsymbol{\nabla} \cdot (\rho \boldsymbol{v} \otimes \boldsymbol{v} + p\mathbf{I}) = \rho \boldsymbol{f} + \boldsymbol{\nabla} \cdot \boldsymbol{\tau} \tag{32-5}$$

(32-4) 和 (32-5) 两式中，$\boldsymbol{f}$ 为单位质量的体力，如对于重力而言，$\boldsymbol{f} = \boldsymbol{g}$.

32.3　欧拉描述下固体的动量守恒方程

固体和流体的重要区别在于，不用再将应力区分为正应力 (静水压强) 和黏性切应力. 因此，固体力学的动量方程从形式上比流体力学的动量方程要简单些.

由连续性方程，在积分核中 $\rho \mathrm{d}v$ 可看做守恒量，则有

$$\frac{\mathrm{d}}{\mathrm{d}t}\int_v \rho \boldsymbol{v}\mathrm{d}v = \int_v \rho\frac{\mathrm{d}\boldsymbol{v}}{\mathrm{d}t}\mathrm{d}v \tag{32-6}$$

应用动量守恒定律，有

$$\frac{\mathrm{d}}{\mathrm{d}t}\int_v \rho \boldsymbol{v}\mathrm{d}v = \int_v \rho\frac{\mathrm{d}\boldsymbol{v}}{\mathrm{d}t}\mathrm{d}v = \int_v \rho \boldsymbol{f}\mathrm{d}v + \oint_a \boldsymbol{\sigma}\boldsymbol{n}\mathrm{d}a$$

$$= \int_v \rho \boldsymbol{f}\mathrm{d}v + \int_v \boldsymbol{\sigma}\cdot\boldsymbol{\nabla}\mathrm{d}v \tag{32-7}$$

式中，$\boldsymbol{\sigma}\boldsymbol{n}$ 表示的是柯西应力张量和单位外法线矢量间的点积。由上式可得到在当前构形中欧拉描述的运动方程:

$$\boldsymbol{\sigma}\cdot\boldsymbol{\nabla} + \rho \boldsymbol{f} = \rho \boldsymbol{a} \tag{32-8}$$

式中，$\frac{\mathrm{d}\boldsymbol{v}}{\mathrm{d}t} = \boldsymbol{a}(\boldsymbol{x},t)$ 为加速度。上式也称为柯西第一运动定律，亦称柯西动量方程。其中 $\boldsymbol{\nabla}\cdot = \mathrm{div}$ 为当前构形中的散度算子。如果惯性力为零 ($\rho\boldsymbol{a}=\boldsymbol{0}$) 同时也不考虑体力时，(32-8) 式则退化为欧拉描述下的平衡方程：

$$\mathrm{div}\boldsymbol{\sigma} = \boldsymbol{0} \quad \text{或} \quad \boldsymbol{\sigma}\cdot\boldsymbol{\nabla}_{\boldsymbol{x}} = \boldsymbol{0} \tag{32-9}$$

(32-9) 式当然可用微元体的方法导出，请参阅 §49 的相关内容。

32.4 拉格朗日描述下固体的动量守恒方程

在参考构形中，设单位质量 ρ_0 所受的体力矢量为 $\boldsymbol{f}_0(\boldsymbol{X},t)$，则相应的动量守恒定律可写为

$$\oint_{\partial V_0} \boldsymbol{PN}\mathrm{d}A + \int_{V_0} \rho_0 \boldsymbol{f}_0 \mathrm{d}V = \frac{\mathrm{d}}{\mathrm{d}t}\int_{V_0} \rho_0 \boldsymbol{V}\mathrm{d}V = \int_{V_0} \rho_0 \boldsymbol{A}\mathrm{d}V \tag{32-10}$$

式中，$\boldsymbol{P}$ 为由 (29-7) 式所定义的第一类皮奥拉–基尔霍夫应力 (简称 PK1)，$\boldsymbol{N}$ 为参考构形中面积微元 $\mathrm{d}A$ 的外法线，$\boldsymbol{PN}$ 表示 $\boldsymbol{P}$ 和 $\boldsymbol{N}$ 两者间的点积。$\frac{\mathrm{d}\boldsymbol{V}}{\mathrm{d}t} = \boldsymbol{a}_0(\boldsymbol{X},t)$ 则为参考构形中质点的加速度，$\boldsymbol{V}$ 为参考构形中质点的速度。一般地，$\boldsymbol{V}$ 和 $\boldsymbol{a}_0$ 还被称为拉格朗日速度和加速度场。对 (32-10) 式应用格林定理，则得到用非对称的两点张量 PK1 $\boldsymbol{P}$ 所表示的参考构形中拉格朗日描述下的运动方程：

$$\mathrm{Div}\boldsymbol{P} + \rho_0 \boldsymbol{f}_0 = \rho_0 \boldsymbol{a}_0 \tag{32-11}$$

上式亦称为布辛涅斯克动量方程 (Boussinesq momentum equation). 式中, $\text{Div}\boldsymbol{P} = \boldsymbol{P} \cdot \nabla_X$ 表示参考构形中的散度.

再将 PK2 和 PK1 之间的关系 (29-12) 式代入上式, 则可给出通过 PK2 $\boldsymbol{T}$ 表示的运动方程:

$$\text{Div}\left(\boldsymbol{F}\boldsymbol{T}\right) + \rho_0 \boldsymbol{f}_0 = \rho_0 \boldsymbol{a}_0 \tag{32-12}$$

上式通常称为基尔霍夫动量方程 (Kirchhoff momentum equation).

§33. 动量矩守恒

33.1 角动量守恒与空间旋转不变性

根据图 20.8 所示的诺特定理, 角动量守恒是空间旋转不变性的必然结果. 以下给出其证明的主要步骤.

空间旋转不变性意味着空间的各向同性 (isotropy of space). 各向同性是指和旋转方向无关, 设一个封闭的力学系统有一个微小的旋转角度矢量 $\delta\boldsymbol{\varphi}$, 此时由该旋转角度所引起的位移和速度矢量分别为

$$\delta\boldsymbol{r} = \delta\boldsymbol{\varphi} \times \boldsymbol{r}, \quad \delta\boldsymbol{v} = \delta\boldsymbol{\varphi} \times \boldsymbol{v} \tag{33-1}$$

此时, 拉格朗日量的变分为

$$\delta L = \sum_{\alpha} \left(\frac{\partial L}{\partial \boldsymbol{r}_{\alpha}} \cdot \delta\boldsymbol{r}_{\alpha} + \frac{\partial L}{\partial \boldsymbol{v}_{\alpha}} \cdot \delta\boldsymbol{v}_{\alpha} \right) = 0 \tag{33-2}$$

由于 $\boldsymbol{p}_{\alpha} = \dfrac{\partial L}{\partial \boldsymbol{v}_{\alpha}}$ 为广义动量, $\dot{\boldsymbol{p}}_{\alpha} = \dfrac{\partial L}{\partial \boldsymbol{r}_{\alpha}}$ 为广义力, 则通过代入 (33-1) 式, 并考虑到矢量混合积的符号的轮换性规则 (6-4) 式, (33-2) 式可表示为

$$\begin{aligned}\delta L &= \sum_{\alpha} (\dot{\boldsymbol{p}}_{\alpha} \cdot \delta\boldsymbol{r}_{\alpha} + \boldsymbol{p}_{\alpha} \cdot \delta\boldsymbol{v}_{\alpha}) = \sum_{\alpha} (\dot{\boldsymbol{p}}_{\alpha} \cdot \delta\boldsymbol{\varphi} \times \boldsymbol{r}_{\alpha} + \boldsymbol{p}_{\alpha} \cdot \delta\boldsymbol{\varphi} \times \boldsymbol{v}_{\alpha}) \\ &= \sum_{\alpha} (\delta\boldsymbol{\varphi} \cdot \boldsymbol{r}_{\alpha} \times \dot{\boldsymbol{p}}_{\alpha} + \delta\boldsymbol{\varphi} \cdot \boldsymbol{v}_{\alpha} \times \boldsymbol{p}_{\alpha}) = \delta\boldsymbol{\varphi} \cdot \frac{\mathrm{d}}{\mathrm{d}t} \sum_{\alpha} \boldsymbol{r}_{\alpha} \times \boldsymbol{p}_{\alpha} = \delta\boldsymbol{\varphi} \cdot \frac{\mathrm{d}\boldsymbol{M}}{\mathrm{d}t} = 0\end{aligned} \tag{33-3}$$

由于 $\delta\boldsymbol{\varphi}$ 的任意性, (33-2) 式表明 $\dfrac{\mathrm{d}\boldsymbol{M}}{\mathrm{d}t} = \boldsymbol{0}$, 意味着系统的动量矩或角动量守恒:

$$\boldsymbol{M} = \sum_{\alpha} \boldsymbol{r}_{\alpha} \times \boldsymbol{p}_{\alpha} = \text{const} \tag{33-4}$$

33.2 柯西应力的对称性

两个矢量的叉积，以力矩为例，可用矩阵形式表示为

$$\boldsymbol{r} \times \boldsymbol{F} = \begin{vmatrix} \hat{\boldsymbol{i}} & \hat{\boldsymbol{j}} & \hat{\boldsymbol{k}} \\ x & y & z \\ F_x & F_y & F_z \end{vmatrix} \tag{33-5}$$

按照牛顿第三定律，在弹性体的内部的相互作用在求和时会相互抵消掉. 因此，根据高斯散度定理，对于固体的任一体积，其内应力合力的三个分量中的每一个 $\int F_i \mathrm{d}V$ 都可以变换为体积表面的面积分，此时，矢量 F_i 应该是柯西应力张量 σ_{ik} 的散度：

$$F_i = \frac{\partial \sigma_{ik}}{\partial x_k} \tag{33-6}$$

从而由高斯散度定理，有

$$\int F_i \mathrm{d}v = \int \frac{\partial \sigma_{ik}}{\partial x_k} \mathrm{d}v = \int \boldsymbol{\sigma} \cdot \boldsymbol{\nabla} \mathrm{d}v = \oint \sigma_{ik} \mathrm{d}a_k \tag{33-7}$$

若无体力和惯性力，则由上式可直接获得欧拉描述下的平衡方程 (32-9) 式.

下面我们按照 (33-5) 式来确定作用在整个体积上的力矩：

$$\begin{aligned} M_{ik} &= \int (x_i F_k - x_k F_i) \mathrm{d}v = \int \left(x_i \frac{\partial \sigma_{kl}}{\partial x_l} - x_k \frac{\partial \sigma_{il}}{\partial x_l} \right) \mathrm{d}v \\ &= \int \frac{\partial (x_i \sigma_{kl} - x_k \sigma_{il})}{\partial x_l} \mathrm{d}v + \int \left(\sigma_{il} \frac{\partial x_k}{\partial x_l} - \sigma_{kl} \frac{\partial x_i}{\partial x_l} \right) \mathrm{d}v \\ &= \oint (x_i \sigma_{kl} - x_k \sigma_{il}) \mathrm{d}a_l + \int (\sigma_{il} \delta_{kl} - \sigma_{kl} \delta_{il}) \mathrm{d}v \\ &= \oint (x_i \sigma_{kl} - x_k \sigma_{il}) \mathrm{d}a_l + \int (\sigma_{ik} - \sigma_{ki}) \mathrm{d}v \end{aligned} \tag{33-8}$$

由于体积分项必须为零，因此，柯西应力必须满足对称性：

$$\sigma_{ik} = \sigma_{ki} \quad \Leftrightarrow \quad \boldsymbol{\sigma} = \boldsymbol{\sigma}^{\mathrm{T}} \tag{33-9}$$

33.3 用 PK1 表示应力的对称性条件

由柯西应力 $\boldsymbol{\sigma}$ 和 PK1 $\boldsymbol{P}$ 的关系式 (29-10) 知：$\boldsymbol{\sigma} = \frac{1}{J} \boldsymbol{P} \boldsymbol{F}^{\mathrm{T}}$，将该式代入 $\boldsymbol{\sigma} = \boldsymbol{\sigma}^{\mathrm{T}}$ 中，得到用 PK1 所表示的等价对称性条件：

$$\boldsymbol{P} \boldsymbol{F}^{\mathrm{T}} = \boldsymbol{F} \boldsymbol{P}^{\mathrm{T}} \tag{33-10}$$

§34. 能量守恒

34.1 能量守恒与时间平移不变性

根据图 20.8 所示的诺特定理，能量守恒是时间平移不变性的必然结果. 以下给出其证明的主要步骤.

时间平移不变性意味着时间的均匀性 (homogeneity of time)，换句话说，也就是封闭力学系统的拉格朗日量不显含时间，此时拉格朗日量的全时间导数中不含有 $\dfrac{\partial L}{\partial t}$ 项：

$$\frac{\mathrm{d}L}{\mathrm{d}t} = \sum_i \frac{\partial L}{\partial q_i}\dot{q}_i + \sum_i \frac{\partial L}{\partial \dot{q}_i}\ddot{q}_i \tag{34-1}$$

将拉格朗日方程 $\dfrac{\partial L}{\partial q_i} = \dfrac{\mathrm{d}}{\mathrm{d}t}\dfrac{\partial L}{\partial \dot{q}_i}$ 代入 (34-1) 式，有

$$\frac{\mathrm{d}L}{\mathrm{d}t} = \sum_i \frac{\mathrm{d}}{\mathrm{d}t}\frac{\partial L}{\partial \dot{q}_i}\dot{q}_i + \sum_i \frac{\partial L}{\partial \dot{q}_i}\ddot{q}_i = \sum_i \frac{\mathrm{d}}{\mathrm{d}t}\left(\frac{\partial L}{\partial \dot{q}_i}\dot{q}_i\right) = \frac{\mathrm{d}}{\mathrm{d}t}\sum_i\left(\frac{\partial L}{\partial \dot{q}_i}\dot{q}_i\right) \tag{34-2}$$

将上式移项，得到

$$\frac{\mathrm{d}}{\mathrm{d}t}\left[\sum_i\left(\frac{\partial L}{\partial \dot{q}_i}\dot{q}_i\right) - L\right] = 0 \tag{34-3}$$

由欧拉齐次函数定理，$\sum_i\left(\dfrac{\partial L}{\partial \dot{q}_i}\dot{q}_i\right) = 2T$，这里 T 为动能，由 (34-3) 式可得出

$$E = 2T - L = 2T - (T - U) = T + U = \mathrm{const} \tag{34-4}$$

上式说明，时间平移不变性意味着能量守恒.

连续介质力学中的能量守恒定律 (Energy Conservation Law) 表述为：动能加内能的时间变化率等于外力的功率与单位时间进入或流出物体的所有其他能量之和.

34.2 欧拉描述下流体力学的能量守恒

热力学第一定律，也就是能量守恒定律表明，一个系统内能加动能的时间变化率等于外部向系统传热的时间变化率减去系统向外所做的功率：

$$\frac{\mathrm{d}\mathcal{E}}{\mathrm{d}t} = \frac{\mathrm{d}Q}{\mathrm{d}t} - \frac{\mathrm{d}W}{\mathrm{d}t} \tag{34-5}$$

第 6 章 守恒律与场方程

式中, $\mathcal{E}$ 为系统的内能 U 和动能 K 之和, 该广延量 (extensive quantity) 可表示为

$$\mathcal{E} = \underbrace{\rho v}_{\text{质量}} \underbrace{(u+k)}_{\text{比内能和比动能之和}} = \rho v \left(u + \frac{1}{2} \boldsymbol{v} \cdot \boldsymbol{v} \right) = \rho v \left(u + \frac{1}{2} v_k v_k \right) \tag{34-6}$$

式中, u 和 k 分别为单位质量的比内能和比动能, 它们均为强度量 (intensive quantity). (34-5) 式中的系统向外所做的功的时间变化率即功率可表示为

$$\mathrm{d}W = \boldsymbol{F} \cdot \mathrm{d}\boldsymbol{x} = \boldsymbol{F} \cdot \boldsymbol{v}\mathrm{d}t \quad \Rightarrow \quad \frac{\mathrm{d}W}{\mathrm{d}t} = \boldsymbol{F} \cdot \boldsymbol{v} \tag{34-7}$$

应用 (34-5) 式, 能量守恒则可表示为

$$\underbrace{\frac{\mathrm{d}}{\mathrm{d}t} \int_v \rho \left(u + \frac{1}{2} v_k v_k \right) \mathrm{d}v}_{\frac{\mathrm{d}\mathcal{E}}{\mathrm{d}t}} = \underbrace{\int_a (-q_i) n_i \mathrm{d}a}_{\frac{\mathrm{d}Q}{\mathrm{d}t}} - \underbrace{\left[\int_a (-n_i \sigma_{ij} v_j) \mathrm{d}a + \int_v (-\rho f_i v_i) \mathrm{d}v \right]}_{\frac{\mathrm{d}W}{\mathrm{d}t}}$$
$$\tag{34-8}$$

再应用雷诺输运定理 (RTT), 则 (34-8) 式可进一步表示为如下黑体形式

$$\frac{\partial}{\partial t} \left[\rho \left(u + \frac{1}{2} \boldsymbol{v} \cdot \boldsymbol{v} \right) \right] + \boldsymbol{\nabla} \cdot \left[\rho \boldsymbol{v} \left(u + \frac{1}{2} \boldsymbol{v} \cdot \boldsymbol{v} \right) \right] = -\boldsymbol{\nabla} \cdot \boldsymbol{q} + \boldsymbol{\nabla} \cdot (\boldsymbol{\sigma} \cdot \boldsymbol{v}) + \rho \boldsymbol{v} \cdot \boldsymbol{f} \tag{34-9}$$

由 §32.2, 对于流体力学, 应力可分解为静水压强和切应力: $\boldsymbol{\sigma} = -p\mathbf{I} + \boldsymbol{\tau}$, 则 (34-9) 式可进一步表示为

$$\frac{\partial}{\partial t} \left[\rho \left(u + \frac{1}{2} \boldsymbol{v} \cdot \boldsymbol{v} \right) \right] + \boldsymbol{\nabla} \cdot \left[\rho \boldsymbol{v} \left(u + \frac{1}{2} \boldsymbol{v} \cdot \boldsymbol{v} \right) \right] = -\boldsymbol{\nabla} \cdot \boldsymbol{q} - \boldsymbol{\nabla} \cdot (p\boldsymbol{v}) + \boldsymbol{\nabla} \cdot (\boldsymbol{\tau} \cdot \boldsymbol{v}) + \rho \boldsymbol{v} \cdot \boldsymbol{f}$$
$$\tag{34-10}$$

将上式右端的第二项移到左端的第二项中, 则 (34-10) 式可等价地表示为

$$\frac{\partial}{\partial t} \left[\rho \left(u + \frac{1}{2} \boldsymbol{v} \cdot \boldsymbol{v} \right) \right] + \boldsymbol{\nabla} \cdot \left[\rho \boldsymbol{v} \left(u + \frac{p}{\rho} + \frac{1}{2} \boldsymbol{v} \cdot \boldsymbol{v} \right) \right] = -\boldsymbol{\nabla} \cdot \boldsymbol{q} + \boldsymbol{\nabla} \cdot (\boldsymbol{\tau} \cdot \boldsymbol{v}) + \rho \boldsymbol{v} \cdot \boldsymbol{f} \tag{34-11}$$

由热力学, 可引入比焓 (specific enthalpy) 的表达式:

$$h = u + \frac{p}{\rho} \tag{34-12}$$

将 (34-12) 式代入 (34-11) 式, 则 (34-11) 式可等价地表示为

$$\frac{\partial}{\partial t} \left[\rho \left(u + \frac{1}{2} \boldsymbol{v} \cdot \boldsymbol{v} \right) \right] + \boldsymbol{\nabla} \cdot \left[\rho \boldsymbol{v} \left(h + \frac{1}{2} \boldsymbol{v} \cdot \boldsymbol{v} \right) \right] = -\boldsymbol{\nabla} \cdot \boldsymbol{q} + \boldsymbol{\nabla} \cdot (\boldsymbol{\tau} \cdot \boldsymbol{v}) + \rho \boldsymbol{v} \cdot \boldsymbol{f} \tag{34-13}$$

(34-13) 式还可用指标形式表示为

$$\frac{\partial}{\partial t} \left[\rho \left(u + \frac{1}{2} v_k v_k \right) \right] + \frac{\partial}{\partial x_i} \left[\rho v_i \left(h + \frac{1}{2} v_k v_k \right) \right] = -\frac{\partial q_i}{\partial x_i} + \frac{\partial (\tau_{ik} v_k)}{\partial x_i} + \rho v_i f_i \tag{34-14}$$

34.3 固体力学中的动能定理

在任意时刻 t 和当前构形中任意体积 V，动能为 $K = \dfrac{1}{2}\int_V \rho \boldsymbol{v} \cdot \boldsymbol{v} \mathrm{d}v = \dfrac{1}{2}\int_V \rho \boldsymbol{v}^2 \mathrm{d}v$，注意到 $\rho \mathrm{d}v$ 对时间的守恒性，则动能的物质时间导数为

$$\dot{K} = \frac{1}{2}\int_V \frac{\mathrm{d}}{\mathrm{d}t}\left(\boldsymbol{v}^2\right)\rho \mathrm{d}v = \int_V \rho \boldsymbol{a} \cdot \boldsymbol{v} \mathrm{d}v \tag{34-15}$$

将 (32-8) 式中的运动方程 $\rho \boldsymbol{a} = \mathrm{div}\boldsymbol{\sigma} + \rho \boldsymbol{f}$ 代入上式，并利用张量分析公式：$(\boldsymbol{v}\boldsymbol{\sigma})\mathrm{div} = \boldsymbol{v}\cdot(\boldsymbol{\sigma}\mathrm{div}) + (\boldsymbol{v}\otimes\boldsymbol{\nabla}):\boldsymbol{\sigma}$，则 (34-15) 式可写作如下积分形式：

$$\dot{K} = \int_V \boldsymbol{v}\cdot(\mathrm{div}\boldsymbol{\sigma} + \rho\boldsymbol{f})\mathrm{d}v = \underbrace{\int_V (\boldsymbol{v}\boldsymbol{\sigma})\mathrm{div}\,\mathrm{d}v}_{\text{用散度定理体积分变面积分}} + \int_V \rho\boldsymbol{f}\cdot\boldsymbol{v}\mathrm{d}v - \int_V \boldsymbol{\sigma}:\boldsymbol{l}\mathrm{d}v$$

$$= \int_{\partial V} \boldsymbol{t_n}\cdot\boldsymbol{v}\mathrm{d}a + \int_V \rho\boldsymbol{f}\cdot\boldsymbol{v}\mathrm{d}v - \int_V \boldsymbol{\sigma}:\frac{\boldsymbol{l}+\boldsymbol{l}^{\mathrm{T}}}{2}\mathrm{d}v$$

$$= \underbrace{\int_{\partial V} \boldsymbol{t_n}\cdot\boldsymbol{v}\mathrm{d}a}_{\text{面力功率}} + \underbrace{\int_V \rho\boldsymbol{f}\cdot\boldsymbol{v}\mathrm{d}v}_{\text{体力功率}} - \underbrace{\int_V \boldsymbol{\sigma}:\boldsymbol{d}\mathrm{d}v}_{\text{内约束力功率}} \tag{34-16}$$

上式右端中的内约束力功率刚好和变形功率差一符号。利用积分体积域的任意性，上式可写为如下微分（局部）形式的动能方程：

$$\frac{1}{2}\rho\frac{\mathrm{d}}{\mathrm{d}t}\left(\boldsymbol{v}^2\right) = (\boldsymbol{v}\boldsymbol{\sigma})\mathrm{div} + \rho\boldsymbol{f}\cdot\boldsymbol{v} - \boldsymbol{\sigma}:\boldsymbol{d} \tag{34-17}$$

再应用当前构形和初始构形体积转换的关系式 (29-1)：$\mathrm{d}v = J\mathrm{d}V$，内约束力功率项 $\int_V \boldsymbol{\sigma}:\boldsymbol{d}\mathrm{d}v = \int_{V_0} J\boldsymbol{\sigma}:\boldsymbol{d}\mathrm{d}V$，再利用 (29-22) 式：$\dot{w} = J\boldsymbol{\sigma}:\boldsymbol{d} = \boldsymbol{T}:\dot{\boldsymbol{E}}$，因此，积分形式的动能方程 (34-16) 式可通过 PK2 和格林应变率表示为另一等价形式：

$$\dot{K} = \int_{\partial V} \boldsymbol{t_n}\cdot\boldsymbol{v}\mathrm{d}a + \int_V \rho\boldsymbol{f}\cdot\boldsymbol{v}\mathrm{d}v - \int_{V_0} \boldsymbol{T}:\dot{\boldsymbol{E}}\mathrm{d}V \tag{34-18}$$

34.4 固体力学中的能量守恒律

如果每单位质量的内能记为 u，则总的内能可表示为

$$U = \int_V u\rho \mathrm{d}v \tag{34-19}$$

总能量 P 为动能与内能组成，即 $P = K + U$。热力学第一定律，也就是能量守恒定律，告诉我们：总能量的物质时间导数等于作用在该体积域的外力功率与每单位时间从该体积域外部所加的热，

$$\dot{P} = \dot{K} + \dot{U} = \int_{\partial V} \boldsymbol{t_n}\cdot\boldsymbol{v}\mathrm{d}a + \int_V \rho\boldsymbol{f}\cdot\boldsymbol{v}\mathrm{d}v - \int_{\partial V} \boldsymbol{q}\cdot\boldsymbol{n}\mathrm{d}a + \int_V \lambda\rho\mathrm{d}v \tag{34-20}$$

第 6 章 守恒律与场方程

式中，q 为热流通量 $(\text{Jm}^{-2}\text{s}^{-1})$，也就是每单位面积和每单位时间的热流，$\lambda$ 为每单位质量的热源.

应用动能定理 (34-16) 式，则可得到内能对时间的变化率为

$$\dot{U} = \int_V \dot{u}\rho\mathrm{d}v = -\underbrace{\int_{\partial V} \boldsymbol{q} \cdot \boldsymbol{n}\mathrm{d}a}_{\text{热流}} + \underbrace{\int_V \lambda\rho\mathrm{d}v}_{\text{热源}} + \underbrace{\int_V \boldsymbol{\sigma} : \boldsymbol{d}\mathrm{d}v}_{\text{变形功率转化}} \tag{34-21}$$

上式即为积分形式的热力学第一定律，也就是连续介质力学的能量守恒定律.

对 (34-21) 式面积分项应用散度定理，得到微分形式的热力学第一定律：

$$\rho\dot{u} = -\mathrm{div}\boldsymbol{q} + \rho\lambda + \boldsymbol{\sigma} : \boldsymbol{d}$$

$$= -\boldsymbol{\nabla}_{\boldsymbol{x}} \cdot \boldsymbol{q} + \rho\lambda + \boldsymbol{\sigma} : \boldsymbol{d} \tag{34-22}$$

上式即为当前构形中的能量方程，也就是能量方程的欧拉描述.

下面给出能量方程的拉格朗日描述，也就是参考构形中的能量方程：

$$\rho_0\dot{u}(\boldsymbol{X},t) = -\mathrm{Div}\boldsymbol{q}_0 + \rho_0\lambda(\boldsymbol{X},t) + \boldsymbol{T} : \dot{\boldsymbol{E}}$$

$$= -\boldsymbol{\nabla}_{\boldsymbol{X}} \cdot \boldsymbol{q}_0 + \rho_0\lambda(\boldsymbol{X},t) + \boldsymbol{T} : \dot{\boldsymbol{E}} \tag{34-23}$$

§35. 熵守恒和热力学不等式

35.1 熵平衡方程和熵不等式

对于非耗散过程的绝热等熵运动，应用雷诺输运定理 (RTT)，单位质量的比熵 (specific entropy) s 的连续性方程为

$$\frac{\partial(\rho s)}{\partial t} + \boldsymbol{\nabla} \cdot (\rho s\boldsymbol{v}) = 0 \tag{35-1}$$

经典热力学的熵不等式为

$$\mathrm{d}S \geqslant \frac{\mathrm{d}Q}{\theta} \tag{35-2}$$

式中，θ 为绝对温度，熵和热的增量可分别表示为

$$\begin{cases} \mathrm{d}S = \rho s\mathrm{d}v \\ \mathrm{d}Q = -\boldsymbol{q} \cdot \boldsymbol{n}\mathrm{d}a\mathrm{d}t \end{cases} \tag{35-3}$$

将 (35-3) 式代入 (35-2) 式，积分后再对时间求导得到

$$\frac{\mathrm{d}}{\mathrm{d}t} \int_v \rho s\mathrm{d}v \geqslant \int_{\partial v} -\frac{\boldsymbol{q}}{\theta} \cdot \boldsymbol{n}\mathrm{d}a \tag{35-4}$$

对上式左端应用雷诺输运定理,再对上式右端应用散度定理,稍加整理便可得到如下熵不等式:

$$\frac{\partial(\rho s)}{\partial t} + \boldsymbol{\nabla} \cdot \left(\rho s \boldsymbol{v} + \frac{\boldsymbol{q}}{\theta}\right) \geqslant 0 \tag{35-5}$$

35.2 热力学第二定律在固体力学中的应用

如 §35.1,以 s 代表单位质量的熵 (熵密度),则体积域 V 内的总熵为

$$S = \int_V s\rho \mathrm{d}v \tag{35-6}$$

热力学第二定律表明:

$$\theta \dot{S} \geqslant \int_V \lambda \rho \mathrm{d}v - \int_{\partial V} \boldsymbol{q} \cdot \boldsymbol{n} \mathrm{d}a \tag{35-7}$$

式中 ">" 号对应于不可逆过程 (irreversible process),而 "=" 则对应于可逆过程 (reversible process). 上式亦可通过每单位质量的熵表示为

$$\dot{S} = \int_V \dot{s}\rho \mathrm{d}v \geqslant \int_V \underbrace{\frac{\lambda}{\theta}}_{\text{熵源}} \rho \mathrm{d}v - \int_{\partial V} \underbrace{\frac{\boldsymbol{q}}{\theta}}_{\text{熵流}} \cdot \boldsymbol{n} \mathrm{d}a \tag{35-8}$$

上式即为连续介质力学中积分形式的热力学第二定律,也就是 "克劳修斯–迪昂不等式 (Clausius–Duhem inequality, CDI)".

再定义 (35-8) 不等式两端之差为体积域 V 内的熵的生成率 Γ,若以 γ 为单位质量的熵生成率,亦即

$$\Gamma = \int_V \gamma \rho \mathrm{d}v \quad (\gamma \geqslant 0, \quad \Gamma \geqslant 0) \tag{35-9}$$

则克劳修斯–迪昂不等式可改写为如下积分形式的熵平衡方程:

$$\int_V \dot{s}\rho \mathrm{d}v = \int_V \frac{\lambda}{\theta}\rho \mathrm{d}v - \int_{\partial V} \frac{\boldsymbol{q}}{\theta} \cdot \boldsymbol{n} \mathrm{d}a + \int_V \gamma \rho \mathrm{d}v \tag{35-10}$$

对上式右端的面积分项应用散度定理,则有如下微分形式的熵平衡方程:

$$\rho \dot{s} = \rho\left(\frac{\lambda}{\theta} + \gamma\right) - \left(\frac{\boldsymbol{q}}{\theta}\right) \cdot \boldsymbol{\nabla}_{\boldsymbol{x}} \tag{35-11}$$

利用矢量分析的恒等式:$(c\boldsymbol{q}) \cdot \boldsymbol{\nabla}_{\boldsymbol{x}} = c(\boldsymbol{q} \cdot \boldsymbol{\nabla}_{\boldsymbol{x}}) + (c\boldsymbol{\nabla}_{\boldsymbol{x}}) \cdot \boldsymbol{q}$,令 $c = \dfrac{1}{\theta}$,则上式可改写为

$$\theta\dot{s} = \lambda - \frac{1}{\rho}\left(\boldsymbol{\nabla}_{\boldsymbol{x}}\cdot\boldsymbol{q}\right) + \frac{1}{\rho\theta}\left(\boldsymbol{\nabla}_{\boldsymbol{x}}\theta\right)\cdot\boldsymbol{q} + \theta\gamma \qquad (35\text{-}12)$$

由 (35-12) 式可给出连续介质力学中微分形式的克劳修斯–迪昂不等式:

$$\rho\dot{s} \geqslant \frac{\rho\lambda}{\theta} - \frac{1}{\theta}\left(\boldsymbol{\nabla}_{\boldsymbol{x}}\cdot\boldsymbol{q}\right) + \frac{1}{\theta^2}\left(\boldsymbol{\nabla}_{\boldsymbol{x}}\theta\right)\cdot\boldsymbol{q} \qquad (35\text{-}13)$$

由 (34-12) 式得到 $-\boldsymbol{\nabla}_{\boldsymbol{x}}\cdot\boldsymbol{q} = \rho\dot{u} - \rho\lambda - \boldsymbol{\sigma}:\boldsymbol{d}$, 代入 (35-13) 式, 可得到与 (35-13) 式等价的在当前构形中的克劳修斯–迪昂不等式:

$$\rho\dot{s}(\boldsymbol{x},t) + \frac{1}{\theta}\boldsymbol{\sigma}:\boldsymbol{d} - \frac{1}{\theta}\rho\dot{u}(\boldsymbol{x},t) - \frac{1}{\theta^2}\left(\boldsymbol{\nabla}_{\boldsymbol{x}}\theta\right)\cdot\boldsymbol{q} \geqslant 0 \qquad (35\text{-}14)$$

可对应地给出在参考构形中相应的克劳修斯–迪昂不等式为

$$\rho_0\dot{s}(\boldsymbol{X},t) + \frac{1}{\theta}\boldsymbol{T}:\dot{\boldsymbol{E}} - \frac{1}{\theta}\rho_0\dot{u}(\boldsymbol{X},t) - \frac{1}{\theta^2}\left(\boldsymbol{\nabla}_{\boldsymbol{X}}\theta\right)\cdot\boldsymbol{q}_0 \geqslant 0 \qquad (35\text{-}15)$$

(35-9) 式所定义的每单位质量的熵生成率 γ 可分两部分: 一是由热传导所产生的热传导熵生成率 γ_{th}; 二是由于熵增率 $\dot{s}$ 超过每单位质量从邻域及从外部吸收的热的率所产生的, 该部分也称为内禀熵生成率 γ_{int}, 以及

$$\begin{cases} \text{总耗散率:} \theta\gamma = \theta\left(\gamma_{\text{th}} + \gamma_{\text{int}}\right) \geqslant 0 \\ \text{热耗散率:} \theta\gamma_{\text{th}} = -\dfrac{1}{\rho\theta}\left(\boldsymbol{\nabla}_{\boldsymbol{x}}\theta\right)\cdot\boldsymbol{q} \\ \text{内禀耗散率:} \theta\gamma_{\text{int}} = \theta\dot{s} - \left[-\dfrac{1}{\rho}\left(\boldsymbol{\nabla}_{\boldsymbol{x}}\cdot\boldsymbol{q}\right) + \lambda\right] \end{cases} \qquad (35\text{-}16)$$

(35-9) 式中的限制条件是每单位质量的总熵生成率 $\gamma \geqslant 0$, 而为了简化分析, 一般地学界更强地要求热传导和内禀两部分熵生成率均大于等于零:

$$\gamma_{\text{th}} \geqslant 0 \quad \text{和} \quad \gamma_{\text{int}} \geqslant 0 \qquad (35\text{-}17)$$

式中的第二个不等式即为 "克劳修斯–普朗克不等式". 由 (34-22) 式可得 $\rho\lambda = \rho\dot{u} + \boldsymbol{\nabla}_{\boldsymbol{x}}\cdot\boldsymbol{q} - \boldsymbol{\sigma}:\boldsymbol{d}$, 将其代入 (35-16) 式中第三式, 得到内禀耗散率为

$$\theta\gamma_{\text{int}} = \theta\dot{s} - \left(\dot{u} - \frac{\boldsymbol{\sigma}:\boldsymbol{d}}{\rho}\right) \qquad (35\text{-}18)$$

上式右端的第二项表示每单位质量内能增加率超过每单位质量变形率的部分. 由 (35-17) 式, 在当前构形下, 克劳修斯–普朗克不等式可表示为

$$\rho \dot{s}(\boldsymbol{x},t) + \frac{1}{\theta}\boldsymbol{\sigma}:\boldsymbol{d} - \frac{1}{\theta}\rho \dot{u}(\boldsymbol{x},t) \geqslant 0 \tag{35-19}$$

在参考构形时，克劳修斯–普朗克不等式则可相应地表示为

$$\rho_0 \dot{s}(\boldsymbol{X},t) + \frac{1}{\theta}\boldsymbol{T}:\dot{\boldsymbol{E}} - \frac{1}{\theta}\rho_0 \dot{u}(\boldsymbol{X},t) \geqslant 0 \tag{35-20}$$

在上面的讨论中，均是以每单位质量的熵 s 为自变量，但在建立本构关系时，往往是以熵的共轭对温度 θ 作为自变量，这需要用到勒让德变换 (Legendre transform)，也就是每单位质量的亥姆霍兹自由能 (Helmholtz free energy) f、内能 u 和束缚能 θs 的关系为

$$f = u - \theta s \tag{35-21}$$

以对应于针对体积积分后的亥姆霍兹自由能的表达式：$F = U - \theta S$. 这样，上式的率形式为 $\dot{f} + s\dot{\theta} = \dot{u} - \theta \dot{s}$，从而 (35-18) 式中的内熵耗散率可用温度作为自变量表示为

$$\theta \gamma_{\text{int}} = -s\dot{\theta} - \left(\dot{f} - \frac{\boldsymbol{\sigma}:\boldsymbol{d}}{\rho}\right) \tag{35-22}$$

亦即用亥姆霍兹自由能表示的当前构形中的克劳修斯–普朗克不等式为

$$\boldsymbol{\sigma}:\boldsymbol{d} - \rho\left(s\dot{\theta} + \dot{f}\right) \geqslant 0 \tag{35-23}$$

相应地，还给出用亥姆霍兹自由能表示的参考构形中的克劳修斯–普朗克不等式为

$$\boldsymbol{T}:\dot{\boldsymbol{E}} - \rho_0\left(s\dot{\theta} + \dot{f}\right) \geqslant 0 \tag{35-24}$$

同理，通过应用 $\dot{f} + s\dot{\theta} = \dot{u} - \theta\dot{s}$，当前构形中的克劳修斯–迪昂不等式 (35-14) 亦可用亥姆霍兹自由能等价地表示为

$$\boldsymbol{\sigma}:\boldsymbol{d} - \rho\left(\dot{f} + s\dot{\theta}\right) - \frac{1}{\theta}(\boldsymbol{\nabla}_{\boldsymbol{x}}\theta)\cdot\boldsymbol{q} \geqslant 0 \tag{35-25}$$

而参考构形的中的克劳修斯–迪昂不等式 (35-15) 则可用亥姆霍兹自由能等价地表示为

$$\boldsymbol{T}:\dot{\boldsymbol{E}} - \rho_0\left(\dot{f} + s\dot{\theta}\right) - \frac{1}{\theta}(\boldsymbol{\nabla}_{\boldsymbol{X}}\theta)\cdot\boldsymbol{q}_0 \geqslant 0 \tag{35-26}$$

思考题和补充材料

6.1 应用散度定理,雷诺输运定理 (RTT) 亦可表示为如下形式:

$$\frac{\mathrm{d}}{\mathrm{d}t}\int_{\mathrm{CV}}\Phi\mathrm{d}v = \int_{\mathrm{CV}}\left[\frac{\partial\Phi}{\partial t} + \boldsymbol{\nabla}\cdot(\Phi\boldsymbol{v})\right]\mathrm{d}v = \int_{\mathrm{CV}}\frac{\partial\Phi}{\partial t}\mathrm{d}v + \int_{\mathrm{CS}}\Phi\boldsymbol{v}\cdot\boldsymbol{n}\mathrm{d}a \qquad \text{(s6-1)}$$

式中, CV 为控制体积, CS 为控制表面, $\boldsymbol{n}$ 为面元 $\mathrm{d}a$ 的外法线.

6.2 守恒律的主方程. 质量守恒、动量守恒、动量矩守恒和机械能守恒在当前构形下可以写成如下统一的形式:

$$\frac{\mathrm{d}}{\mathrm{d}t}\int_{\Omega}f(\boldsymbol{x},t)\,\mathrm{d}v = \int_{\partial\Omega}\varphi(\boldsymbol{x},t,\boldsymbol{n})\,\mathrm{d}a + \int_{\Omega}\Psi(\boldsymbol{x},t)\,\mathrm{d}v \qquad \text{(s6-2)}$$

式中等号右端的第一项的面积分项是应用了散度定理的结果. 上述方程被称为连续介质力学守恒律的主方程 (master equation), 亦称为 "主平衡原理 (master balance principle)".

题表 6.2 守恒律在主方程中所对应的项

守恒律	f	φ	Ψ
质量守恒	ρ	0	0
动量守恒 (32-7)	$\rho\boldsymbol{v}$	$\boldsymbol{t} = \boldsymbol{\sigma}\boldsymbol{n}$	$\boldsymbol{b} = \rho\boldsymbol{f}$
动量矩守恒	$\boldsymbol{x}\times(\rho\boldsymbol{v})$	$\boldsymbol{x}\times\boldsymbol{t} = \boldsymbol{x}\times(\boldsymbol{\sigma}\boldsymbol{n})$	$\boldsymbol{x}\times\boldsymbol{b} = \boldsymbol{x}\times(\rho\boldsymbol{f})$
机械能守恒 (34-11)	$\frac{1}{2}\rho\boldsymbol{v}^2 + \underbrace{\rho u}_{\text{内能}}$	$\boldsymbol{\sigma}\boldsymbol{n}\cdot\boldsymbol{v} + \underbrace{q_n}_{\text{热流}}$	$\rho\boldsymbol{f}\cdot\boldsymbol{v} + \underbrace{\rho\lambda}_{\text{热源}}$

6.3 结合思考题 5.6 和例 31.5, 考虑生长时, 由变形梯度张量分解的 (s5-2) 式, 雅可比可进一步分解为

$$J = \det\boldsymbol{F} = \det(\boldsymbol{F}_e\boldsymbol{F}_g)$$

$$= \det(\boldsymbol{F}_e)\cdot\det(\boldsymbol{F}_g) = J_e J_g \qquad \text{(s6-3)}$$

式中, $J_e = \det\boldsymbol{F}_e$ 为弹性或相容张量 (accommodation tensor), $J_g = \det\boldsymbol{F}_g$ 为生长梯度张量的雅可比. 简要证明此时拉格朗日描述的连续性方程为

$$\frac{\mathrm{d}(J\rho)}{\mathrm{d}t} = J_e J_g \gamma \qquad \text{(s6-4)}$$

6.4 麦克斯韦方程组可表示为

$$\begin{cases} \boldsymbol{\nabla} \cdot \boldsymbol{D} = \rho \\ \boldsymbol{\nabla} \cdot \boldsymbol{B} = 0 \\ \boldsymbol{\nabla} \times \boldsymbol{E} = -\dfrac{\partial \boldsymbol{B}}{\partial t} \\ \boldsymbol{\nabla} \times \boldsymbol{H} = \boldsymbol{J} + \dfrac{\partial \boldsymbol{D}}{\partial t} \end{cases} \tag{s6-5}$$

式中，$\boldsymbol{J}$ 为电流密度矢量，$\boldsymbol{E}$ 和 $\boldsymbol{B}$ 分别代表电场和磁场矢量. 另外两个场为位移场 $\boldsymbol{D}$ 和磁场 $\boldsymbol{H}$ 矢量，它们通过常数与 $\boldsymbol{E}$ 和 $\boldsymbol{B}$ 相关联. 问题：利用矢量的旋度的散度为零的性质，证明电荷守恒方程 (31-3) 式.

6.5 证明：麦克斯韦方程组 (s6-5) 是线性的.

6.6 将各类连续性方程总结于题表 6.6.

题表 6.6 各种类型的连续性方程一览表

守恒量	守恒方程	符号说明	等价形式
质量	$\dfrac{\partial \rho}{\partial t} + \boldsymbol{\nabla} \cdot \boldsymbol{J} = 0$	$\boldsymbol{J} = \rho \boldsymbol{v}$ 为质量流密度或质量通量	$\dfrac{\mathrm{d}\rho}{\mathrm{d}t} + \rho \boldsymbol{\nabla} \cdot \boldsymbol{v} = 0$
电荷	$\dfrac{\partial \rho}{\partial t} + \boldsymbol{\nabla} \cdot \boldsymbol{J} = 0$	$\boldsymbol{J} = \rho \boldsymbol{v}_d$ 为电流密度 ρ 为电荷密度 $\boldsymbol{v}_d$ 为带电粒子的平均漂移速度	$\dfrac{\mathrm{d}\rho}{\mathrm{d}t} + \rho \boldsymbol{\nabla} \cdot \boldsymbol{v}_d = 0$
量子力学概率密度	$\dfrac{\partial \rho}{\partial t} + \boldsymbol{\nabla} \cdot \boldsymbol{J} = 0$	$\rho = \|\psi\|^2$ 为概率密度 $\boldsymbol{J} = \dfrac{\hbar}{2mi}(\psi^* \boldsymbol{\nabla}\psi - \psi \boldsymbol{\nabla}\psi^*)$ $= \dfrac{\hbar}{m} \mathrm{Im}(\psi^* \boldsymbol{\nabla}\psi)$ 为概率流	
扩散	$\dfrac{\partial c}{\partial t} + \boldsymbol{\nabla} \cdot \boldsymbol{J} = 0$	$\boldsymbol{J} = -D\boldsymbol{\nabla}c$ 为扩散通量 D 为扩散系数 $\boldsymbol{\nabla}c$ 为浓度梯度	$\dfrac{\partial c}{\partial t} = D\nabla^2 c$
热传导	$\rho c_p \dfrac{\partial \theta}{\partial t} + \boldsymbol{\nabla} \cdot \boldsymbol{q} = 0$	$\boldsymbol{q} = -\kappa \boldsymbol{\nabla}\theta$ 为热流通量 c_p 为定压比热 κ 为热导率 $\boldsymbol{\nabla}\theta$ 为温度梯度	$\dfrac{\partial \theta}{\partial t} = \alpha \nabla^2 \theta$ $\alpha = \dfrac{\kappa}{\rho c_p}$ 为温导率
动量扩散	$\dfrac{\partial (\rho \boldsymbol{v})}{\partial t} + \boldsymbol{\nabla} \cdot \boldsymbol{\tau} = \boldsymbol{0}$	$\boldsymbol{\tau} = -\mu\left(\boldsymbol{\nabla} \otimes \boldsymbol{v} + \boldsymbol{v} \otimes \boldsymbol{\nabla} - \dfrac{2}{3}(\boldsymbol{\nabla} \cdot \boldsymbol{v})\mathbf{I}\right)$ 不可压缩条件： $\boldsymbol{\nabla} \cdot \boldsymbol{v} = 0, \quad \boldsymbol{\nabla} \cdot (\boldsymbol{v} \otimes \boldsymbol{\nabla}) = \boldsymbol{0}$	$\dfrac{\partial \boldsymbol{v}}{\partial t} = \nu \nabla^2 \boldsymbol{v}$

6.7 证明：柯西应力的对称性条件 (33-9) 还可用置换符号方便地表示为

$$\varepsilon_{ijk}\sigma_{ji} = 0 \tag{s6-6}$$

参 考 文 献

[6.1] Reynolds O. Papers on Mechanical and Physical Subjects. Vol. 3, The Sub-Mechanics of the Universe. Cambridge: Cambridge University Press, 1903.

[6.2] Landau L D, Lifshitz E M. Fluid Mechanics. Oxford: Pergamon Press, 1987. (中译本：朗道, 栗弗席兹. 流体动力学 (李植译). 北京：高等教育出版社, 2013.)

第 7 章 连续介质力学中的客观性

沃尔特·诺尔

2017 年 6 月 6 日去世的著名理性力学权威沃尔特·诺尔 (Walter Noll, 1925—2017, 享年 92 岁) 曾指出: "我对于 1958 年所引入的过时的术语负责, 我现在很后悔我误导过很多人 (I was responsible for introducing the obsolete term in 1958 and now regret that I misled a lot of people)", 诺尔这里所指的过时的术语是 "材料客观性原理 (principle of material objectivity)", 该术语后来被 "材料标架无差异性原理 (principle of material frame-indifference)" 所替代, 简称 MFI.

理性力学大师克利福德·特鲁斯德尔 (Clifford Ambrose Truesdell III, 1919—2000) 和诺尔于 1965 年出版的理性力学经典著作《力学的非线性场论》(*The Non-Linear Field Theories of Mechanics, NLFT*)[7.1] 中, 将 MFI 陈述为: "材料的响应对所有观察者而言均相同 (the response of a material is the same for all observers)". MFI 还可简单地陈述为材料性质在观察者发生变化时的不变性 (invariant under changes of observers)[7.2].

§36. 标量, 位移、速度、加速度矢量的欧几里得客观性

36.1 欧几里得变换

首先定义观察者 (observer), 一个三维欧几里得空间中的观察者可度量不同点的相对位置和时段 (relative position and intervals of time). 这里的空间位置 x 是一个有三个分量的位矢, 而时间 t 则为标量.

一个 "事件 (event)" 可用位矢和时间对 (pair) (x_0, t_0) 和 (x, t) 来刻画. 一个事件映射到另一事件, 用上标 * 来表示. 引入一个和时间相关的二阶正交张量 $Q(t)$, 该映射使得两点之间的距离 $u = x - x_0$ 保持不变:

$$u^* = x^* - x_0^* = Q(t)(x - x_0) = Q(t)u \tag{36-1}$$

则联系 (x, t) 到 (x^*, t^*) 一对一映射的欧几里得变换 (Euclidean transformation) 表示为

$$\begin{cases} x^* = c(t) + Q(t)x \\ \alpha = t^* - t = t_0^* - t_0 \end{cases} \tag{36-2}$$

第 7 章 连续介质力学中的客观性

式中，$c(t) = x_0^* - Q(t)x_0$, α 则为时差.

有关等距映射和正交张量的内容详见例 1.5 和 §20.4.

36.2 标量和位移矢量的欧几里得客观性

和 (36-2) 式中第二式中作为标量的时间一样，诸如密度、反映体积变化的雅可比行列式 (Jacobian)、温度等标量亦满足欧几里得客观性 (Euclidean objectivity).

(36-1) 式已经清楚地表明，位移矢量为满足欧几里得客观性的矢量.

36.3 速度矢量的欧几里得客观性

由 (36-1) 式两端对时间求导，得到速度的变化率为

$$\dot{x}^* = \dot{x}_0^* + Q\dot{x} + \dot{Q}(x - x_0) \tag{36-3}$$

再由 (36-1) 式得到 $x - x_0 = Q^T(x^* - x_0^*)$，将其代入上式，得到

$$\dot{x}^* = \dot{x}_0^* + Q\dot{x} + \dot{Q}Q^T(x^* - x_0^*)$$
$$= Q\dot{x} + \dot{x}_0^* + \Omega(x^* - x_0^*) \tag{36-4}$$

式中，

$$\Omega = \dot{Q}Q^T \tag{36-5}$$

对 (20-2) 式 $QQ^T = I$ 两端对时间求导，有

$$\dot{Q}Q^T + Q\dot{Q}^T = 0 \quad \Rightarrow \quad \Omega = -\Omega^T \tag{36-6}$$

由于 Ω 满足反对称性 (skew symmetric)，说明 Ω 为一个旋率张量 (spin tensor)，事实上，一般地 Q 为正常正交张量，此时 $\Omega(t) = \dot{R}(t)(R(t))^T = \dot{R}(t)R^T(t)$.

由 (36-4) 式可知，速度矢量一般不满足欧几里得客观性：$\dot{x}^* = Q\dot{x}$，只有同时满足 $\dot{x}_0^* = \Omega = 0$，也就是，当 $x_0^* = \text{const}$, $Q = \text{const}$, 亦即对应于刚体运动的场合，速度才是客观矢量.

36.4 加速度矢量的欧几里得客观性

对 (36-4) 式进一步对时间求导，有

$$\ddot{x}^* = Q\ddot{x} + \ddot{x}_0^* + \dot{Q}\dot{x} + \dot{\Omega}(x^* - x_0^*) + \Omega(\dot{x}^* - \dot{x}_0^*) \tag{36-7}$$

再由 (36-4) 式，有

$$\dot{x} = Q^T(\dot{x}^* - \dot{x}_0^*) - Q^T\Omega(x^* - x_0^*) \tag{36-8}$$

将 (36-8) 式代回 (36-7) 式，并稍加整理得到

$$\ddot{\boldsymbol{x}}^* = \boldsymbol{Q}\ddot{\boldsymbol{x}} + \underbrace{\ddot{\boldsymbol{x}}_0^*}_{\text{牵连加速度}} + \underbrace{2\boldsymbol{\Omega}\left(\dot{\boldsymbol{x}}^* - \dot{\boldsymbol{x}}_0^*\right)}_{\text{科里奥利加速度}} + \underbrace{\dot{\boldsymbol{\Omega}}\left(\boldsymbol{x}^* - \boldsymbol{x}_0^*\right)}_{\text{欧拉角加速度}} - \underbrace{\boldsymbol{\Omega}^2\left(\boldsymbol{x}^* - \boldsymbol{x}_0^*\right)}_{\text{离心加速度}} \tag{36-9}$$

只有当 $\ddot{\boldsymbol{x}}_0^* = \boldsymbol{0}$ 且 $\boldsymbol{\Omega} = \boldsymbol{0}$ 时，加速度才为客观矢量，此时等价于要求：$\dot{\boldsymbol{x}}_0^* =$ const 且 $\boldsymbol{Q} = $ const，也就是做 "匀速直线运动" 的情形，此时伽利略变换 (Galilean transformation) 成立.

伽利略变换是经典力学的基本内容. 在著者为国科大一年级本科新生所讲授的普物力学中，"伽利略变换" 就是其中的重点内容之一，可参阅《力学讲义》[1.11]. 可从伽利略变换群的角度出发予以理解. 对伽利略变换可做如下三部分内容的分解. 第一步，是关于匀速直线运动速度 v 的变换：

$$g_1(t, \boldsymbol{x}) = (t, \boldsymbol{x} + \boldsymbol{v}t) \quad \forall t \in \mathbb{R},\ \boldsymbol{x} \in \mathbb{R}^3 \tag{36-10}$$

式中，$\mathbb{R}$ 表示所有实数的集合，$\mathbb{R}^3$ 则表示三维矢量空间 (vector space) 或线性空间 (linear space)，可见第一步操作共有 3 个生成元 (generators).

第二步，是关于原点的平移变换：

$$g_2(t, \boldsymbol{x}) = (t + \tau, \boldsymbol{x} + \boldsymbol{s}) \quad \forall t \in \mathbb{R},\ \boldsymbol{x} \in \mathbb{R}^3 \tag{36-11}$$

式中，τ 为两个坐标系时钟的时差，而 $\boldsymbol{s}$ 则为坐标系原点的空间平移矢量，可见第二步操作共有 $1 + 3 = 4$ 个生成元.

最后一步，是关于坐标轴的旋转变换：

$$g_3(t, \boldsymbol{x}) = (t, \boldsymbol{R}\boldsymbol{x}) \quad \forall t \in \mathbb{R},\ \boldsymbol{x} \in \mathbb{R}^3 \tag{36-12}$$

式中，$\boldsymbol{R}\boldsymbol{x}$ 表示的是旋转张量 $\boldsymbol{R}$ 和位置矢量 $\boldsymbol{x}$ 之间的点积，该点积操作后仍然为一个矢量，可见第三步操作共有 3 个生成元.

综上所述，伽利略变换群共有 $3 + 4 + 3 = 10$ 个生成元.

事实上，所谓 "经典力学" 也就是在伽利略变换下所有的不变的性质. 而所谓 "狭义相对论力学" 其实也就是在洛伦兹变换下所有不变的性质. 见本章思考题 7.3.

例 36.1 证明连续性方程 (31-1) 的客观性.

证明：方程 (31-1) 的时空转换关系为

$$\frac{\partial \rho^*}{\partial t^*} + \boldsymbol{\nabla}^* \cdot (\rho \boldsymbol{v})^* = 0 \tag{36-13}$$

运用指标法, 有

$$\frac{\partial \rho^*}{\partial t^*} + \frac{\partial (\rho^* Q_{lk} v_k)}{\partial x_l^*} = \frac{\partial \rho^*}{\partial t^*} + Q_{lk}\frac{\partial (\rho^* v_k)}{\partial x_p}\frac{\partial x_p}{\partial x_l^*} = 0 \tag{36-14}$$

由 (36-1) 式可知: $\dfrac{\partial x_p}{\partial x_l^*} = Q_{lp}$, 将其代入 (36-14) 式, 有

$$\frac{\partial \rho^*}{\partial t^*} + Q_{lk}\frac{\partial (\rho^* v_k)}{\partial x_p}\frac{\partial x_p}{\partial x_l^*} = \frac{\partial \rho^*}{\partial t^*} + Q_{lk}Q_{lp}\frac{\partial (\rho^* v_k)}{\partial x_p} = 0 \tag{36-15}$$

由于正交张量满足:

$$Q_{lk}Q_{lp} = \delta_{kp} \tag{36-16}$$

将 (36-16) 式代入 (36-15) 式, 得到

$$\begin{aligned}\frac{\partial \rho^*}{\partial t^*} + Q_{lk}Q_{lp}\frac{\partial (\rho^* v_k)}{\partial x_p} &= \frac{\partial \rho^*}{\partial t^*} + \delta_{kp}\frac{\partial (\rho^* v_k)}{\partial x_p} \\ &= \frac{\partial \rho^*}{\partial t^*} + \frac{\partial (\rho^* v_k)}{\partial x_k} \\ &= \frac{\partial \rho^*}{\partial t^*} + \boldsymbol{\nabla} \cdot (\rho^* \boldsymbol{v}) = 0\end{aligned} \tag{36-17}$$

再由于标量满足的时空转换关系为 $\rho^* = \rho$, $t^* = t$, 则 (36-17) 式完全退化为 (31-1) 式, 亦即, 连续性方程 (31-1) 满足客观性要求.

§37. 张量的欧几里得客观性和客观率

首先定义 n 阶张量的客观性. 一个 n 阶张量 $\boldsymbol{u}_1 \otimes \boldsymbol{u}_2 \cdots \otimes \boldsymbol{u}_n$, 当更换观察者时, 所对应的变换满足下式:

$$(\boldsymbol{u}_1 \otimes \boldsymbol{u}_2 \cdots \otimes \boldsymbol{u}_n)^* = (\boldsymbol{Q}\boldsymbol{u}_1) \otimes (\boldsymbol{Q}\boldsymbol{u}_2) \cdots \otimes (\boldsymbol{Q}\boldsymbol{u}_n) \tag{37-1}$$

对于二阶张量, 上式退化为

$$(\boldsymbol{u}_1 \otimes \boldsymbol{u}_2)^* = (\boldsymbol{Q}\boldsymbol{u}_1) \otimes (\boldsymbol{Q}\boldsymbol{u}_2) = (\boldsymbol{Q}\boldsymbol{u}_1) \otimes \left(\boldsymbol{u}_2 \boldsymbol{Q}^{\mathrm{T}}\right) = \boldsymbol{Q}(\boldsymbol{u}_1 \otimes \boldsymbol{u}_2)\boldsymbol{Q}^{\mathrm{T}} \tag{37-2}$$

亦即, 满足 (37-2) 式的二阶张量就是满足欧几里得客观性.

37.1 变形梯度张量的欧几里得客观性

可从两种途径判断变形梯度的客观性. 其一, 从其基本定义出发: $\boldsymbol{F} = \partial \boldsymbol{x}/\partial \boldsymbol{X}$, 对于另外一个观察者而言, 有

$$\boldsymbol{F}^* = \frac{\partial \boldsymbol{x}^*}{\partial \boldsymbol{X}} = \frac{\partial (\boldsymbol{Q}\boldsymbol{x})}{\partial \boldsymbol{x}}\frac{\partial \boldsymbol{x}}{\partial \boldsymbol{X}} = \boldsymbol{Q}\boldsymbol{F} \tag{37-3}$$

上式反映出，变形梯度的变换就像一个矢量一样，它作为两点张量场，不满足欧几里得客观性的要求. 其原因之一, 是参考构形中的矢量 $\boldsymbol{X}$ 是不变的, 亦即 $\boldsymbol{X}^* = \boldsymbol{X}$, 换句话说, $\boldsymbol{X}$ 和观察者无关.

判断变形梯度是否满足客观性要求的途径是由当前构形和参考构形中线段的联系: $\mathrm{d}\boldsymbol{x} = \boldsymbol{F}\mathrm{d}\boldsymbol{X}$, 由于 $(\mathrm{d}\boldsymbol{x})^* = \boldsymbol{Q}\mathrm{d}\boldsymbol{x}$, $(\mathrm{d}\boldsymbol{x})^* = (\boldsymbol{F}\mathrm{d}\boldsymbol{X})^* = \boldsymbol{F}^*\mathrm{d}\boldsymbol{X}$, 因此, (37-3) 式成立.

由于雅可比 $J = \det \boldsymbol{F}$ 作为一个反映体积变化的标量毫无疑问是满足欧几里得客观性的, 现利用 (37-3) 式证明如下:

$$J^* = \det \boldsymbol{F}^* = \det(\boldsymbol{QF}) = (\det \boldsymbol{Q})(\det \boldsymbol{F}) = \det \boldsymbol{F} = J \tag{37-4}$$

下面我们来看 "右柯西–格林" 和 "左柯西–格林"(芬格) 变形张量的客观性:

$$\begin{cases} \boldsymbol{C}^* = \left(\boldsymbol{F}^\mathrm{T}\boldsymbol{F}\right)^* = (\boldsymbol{QF})^\mathrm{T} \boldsymbol{QF} = \boldsymbol{F}^\mathrm{T}\boldsymbol{Q}^\mathrm{T}\boldsymbol{QF} = \boldsymbol{F}^\mathrm{T}\boldsymbol{F} = \boldsymbol{C} \\ \boldsymbol{B}^* = \left(\boldsymbol{FF}^\mathrm{T}\right)^* = \boldsymbol{QF}(\boldsymbol{QF})^\mathrm{T} = \boldsymbol{QFF}^\mathrm{T}\boldsymbol{Q}^\mathrm{T} = \boldsymbol{QBQ}^\mathrm{T} \end{cases} \tag{37-5}$$

上式说明, 右柯西–格林变形张量 $\boldsymbol{C}$ 由于是参考构形中的张量, 其变换关系和标量一样, 是不满足张量的欧几里得客观性要求的; 而芬格张量 $\boldsymbol{B}$ 作为当前构形中的欧拉型张量, 满足张量客观性的要求.

37.2 柯西应力的欧几里得客观性

柯西面力 (traction) 矢量 $\boldsymbol{t}$ 是柯西应力 $\boldsymbol{\sigma}$ 和截面外法线 $\boldsymbol{n}$ 的点积: $\boldsymbol{t} = \boldsymbol{\sigma n}$. 两个矢量 $\boldsymbol{t}$ 和 $\boldsymbol{n}$ 均满足客观性: $\boldsymbol{t}^* = \boldsymbol{Qt}$, $\boldsymbol{n}^* = \boldsymbol{Qn}$, 因此有

$$\boldsymbol{t}^* = \boldsymbol{Qt} = (\boldsymbol{\sigma n})^* = \boldsymbol{\sigma}^*\boldsymbol{n}^* = \boldsymbol{\sigma}^*\boldsymbol{Qn} \Rightarrow \boldsymbol{Qt} = \boldsymbol{Q\sigma n} = \boldsymbol{\sigma}^*\boldsymbol{Qn} \Rightarrow \boldsymbol{\sigma}^* = \boldsymbol{Q\sigma Q}^\mathrm{T} \tag{37-6}$$

上式说明, 柯西应力作为欧拉型的当前构形中的张量, 是满足欧几里得客观性要求的.

基尔霍夫应力由于和柯西应力之间满足 $\boldsymbol{\tau} = J\boldsymbol{\sigma}$, 显然也是客观的应力张量.

37.3 PK1 和 PK2 应力张量的欧几里得客观性

PK1 应力的定义式: $\boldsymbol{P} = J\boldsymbol{\sigma F}^{-\mathrm{T}}$, 应用 (37-6) 和 (12-13) 两式, 故有

$$\boldsymbol{P}^* = J^*\boldsymbol{\sigma}^*\left(\boldsymbol{F}^*\right)^{-\mathrm{T}} = J\boldsymbol{Q\sigma Q}^\mathrm{T}(\boldsymbol{QF})^{-\mathrm{T}} = J\boldsymbol{Q\sigma Q}^\mathrm{T}\boldsymbol{Q}^{-\mathrm{T}}\boldsymbol{F}^{-\mathrm{T}}$$

$$= J\boldsymbol{Q\sigma F}^{-\mathrm{T}} = \boldsymbol{QP} \tag{37-7}$$

第 7 章　连续介质力学中的客观性

上式说明, 作为两点张量的 PK1 的变换关系和另外一个两点张量——变形梯度张量的变换关系一致, 不满足二阶张量的欧几里得客观性要求.

PK2 应力的定义式: $\boldsymbol{T} = \boldsymbol{F}^{-1}\boldsymbol{P}$, 变换观察者时, 其变换关系为

$$\boldsymbol{T}^* = (\boldsymbol{QF})^{-1}\boldsymbol{QP} = \boldsymbol{F}^{-1}\boldsymbol{Q}^{-1}\boldsymbol{QP} = \boldsymbol{F}^{-1}\boldsymbol{P} = \boldsymbol{T} \qquad (37\text{-}8)$$

和右柯西-格林变形张量 $\boldsymbol{C}$ 类似, PK2 作为参考构形中的应力张量, 其变换关系和标量一样, 不满足张量的欧几里得客观性要求.

37.4　速度梯度、应变率、旋率张量的欧几里得客观性

首先看速度梯度张量的客观性. 由于 $\boldsymbol{l} = \dot{\boldsymbol{F}}\boldsymbol{F}^{-1}$, 故, 速度梯度的变换关系为

$$\boldsymbol{l}^* = \overline{\boldsymbol{QF}}(\boldsymbol{QF})^{-1} = \left(\dot{\boldsymbol{Q}}\boldsymbol{F} + \boldsymbol{Q}\dot{\boldsymbol{F}}\right)\boldsymbol{F}^{-1}\boldsymbol{Q}^{-1} = \dot{\boldsymbol{Q}}\boldsymbol{Q}^{\mathrm{T}} + \boldsymbol{Q}\boldsymbol{l}\boldsymbol{Q}^{\mathrm{T}} = \boldsymbol{\Omega} + \boldsymbol{Q}\boldsymbol{l}\boldsymbol{Q}^{\mathrm{T}} \quad (37\text{-}9)$$

式中, 已经利用了正交张量的基本性质 (20-13) 式. 利用 (36-6) 式中的 $\boldsymbol{\Omega} = -\boldsymbol{\Omega}^{\mathrm{T}}$, 有

$$\begin{cases} \boldsymbol{d}^* = \dfrac{\boldsymbol{l}^* + \boldsymbol{l}^{*\mathrm{T}}}{2} = \boldsymbol{Q}\dfrac{\boldsymbol{l} + \boldsymbol{l}^{\mathrm{T}}}{2}\boldsymbol{Q}^{\mathrm{T}} = \boldsymbol{Q}\boldsymbol{d}\boldsymbol{Q}^{\mathrm{T}} \\ \boldsymbol{\omega}^* = \dfrac{\boldsymbol{l}^* - \boldsymbol{l}^{*\mathrm{T}}}{2} = \boldsymbol{\Omega} + \boldsymbol{Q}\boldsymbol{\omega}\boldsymbol{Q}^{\mathrm{T}} \end{cases} \qquad (37\text{-}10)$$

上式表明, 应变率作为当前构形的欧拉型张量是满足欧几里得客观性要求的; 而旋率的转换关系和速度梯度相同, 均不满足张量的客观性要求.

从 (37-10) 式中第二式: $\boldsymbol{\omega}^* = \dot{\boldsymbol{Q}}\boldsymbol{Q}^{\mathrm{T}} + \boldsymbol{Q}\boldsymbol{\omega}\boldsymbol{Q}^{\mathrm{T}}$, 再利用旋率张量的反对称性质: $\boldsymbol{\omega}^{\mathrm{T}} = -\boldsymbol{\omega}$ 和 $\boldsymbol{\omega}^{*\mathrm{T}} = -\boldsymbol{\omega}^*$ 容易得到

$$\begin{cases} \dot{\boldsymbol{Q}} = \boldsymbol{\omega}^*\boldsymbol{Q} - \boldsymbol{Q}\boldsymbol{\omega} \\ \dot{\boldsymbol{Q}}^{\mathrm{T}} = -\boldsymbol{Q}^{\mathrm{T}}\boldsymbol{\omega}^* + \boldsymbol{\omega}\boldsymbol{Q}^{\mathrm{T}} \end{cases} \qquad (37\text{-}11)$$

37.5　客观矢量率的定义

由客观矢量的定义 (36-1) 式, 对时间求导, 有 $\dot{\boldsymbol{u}}^* = \overline{\boldsymbol{Qu}} = \dot{\boldsymbol{Q}}\boldsymbol{u} + \boldsymbol{Q}\dot{\boldsymbol{u}}$, 将 (37-11) 式中第一式代入, 得到如下关系:

$$\begin{aligned} \dot{\boldsymbol{u}}^* &= \dot{\boldsymbol{Q}}\boldsymbol{u} + \boldsymbol{Q}\dot{\boldsymbol{u}} = (\boldsymbol{\omega}^*\boldsymbol{Q} - \boldsymbol{Q}\boldsymbol{\omega})\boldsymbol{u} + \boldsymbol{Q}\dot{\boldsymbol{u}} \\ &= \boldsymbol{\omega}^*\boldsymbol{u}^* + \boldsymbol{Q}\dot{\boldsymbol{u}} - \boldsymbol{Q}\boldsymbol{\omega}\boldsymbol{u} \end{aligned} \qquad (37\text{-}12)$$

由上式整理得到

$$(\dot{\boldsymbol{u}} - \boldsymbol{\omega}\boldsymbol{u})^* = \boldsymbol{Q}(\dot{\boldsymbol{u}} - \boldsymbol{\omega}\boldsymbol{u}) \qquad (37\text{-}13)$$

通过上式可定义满足矢量客观性要求的共旋率 (co-rotational rate)：

$$\overset{\circ}{\boldsymbol{u}} = \dot{\boldsymbol{u}} - \boldsymbol{\omega}\boldsymbol{u} \qquad (37\text{-}14)$$

37.6 客观张量率的定义

对于任意满足客观性要求的二阶张量 $\boldsymbol{A}$，有 $\boldsymbol{A}^* = \boldsymbol{Q}\boldsymbol{A}\boldsymbol{Q}^{\mathrm{T}}$，其时间导数：

$$\dot{\boldsymbol{A}}^* = \dot{\boldsymbol{Q}}\boldsymbol{A}\boldsymbol{Q}^{\mathrm{T}} + \boldsymbol{Q}\dot{\boldsymbol{A}}\boldsymbol{Q}^{\mathrm{T}} + \boldsymbol{Q}\boldsymbol{A}\dot{\boldsymbol{Q}}^{\mathrm{T}} \qquad (37\text{-}15)$$

将 (37-11) 式代入上式，整理得到

$$\left(\dot{\boldsymbol{A}} - \boldsymbol{\omega}\boldsymbol{A} + \boldsymbol{A}\boldsymbol{\omega}\right)^* = \boldsymbol{Q}\left(\dot{\boldsymbol{A}} - \boldsymbol{\omega}\boldsymbol{A} + \boldsymbol{A}\boldsymbol{\omega}\right)\boldsymbol{Q}^{\mathrm{T}} \qquad (37\text{-}16)$$

上式便定义了连续介质力学中最为常用的尧曼-扎伦巴 (Jaumann-Zaremba) 客观率

$$\overset{\triangle}{\boldsymbol{A}} = \dot{\boldsymbol{A}} - \boldsymbol{\omega}\boldsymbol{A} + \boldsymbol{A}\boldsymbol{\omega} \qquad (37\text{-}17)$$

特别地，当 $\boldsymbol{A}$ 取柯西应力 $\boldsymbol{\sigma}$，得到连续介质力学中常用的柯西应力的尧曼-扎伦巴客观率：

$$\overset{\triangle}{\boldsymbol{\sigma}} = \dot{\boldsymbol{\sigma}} - \boldsymbol{\omega}\boldsymbol{\sigma} + \boldsymbol{\sigma}\boldsymbol{\omega} \qquad (37\text{-}18)$$

例 37.1 证明无体力时的弹性力学平衡方程的客观性.

证明：柯西应力的散度项的时空变换关系为

$$\text{左散度} \quad \boldsymbol{\nabla}^* \cdot \boldsymbol{\sigma}^* = \frac{\partial}{\partial x_l^*}\boldsymbol{e}_l \cdot \sigma_{mk}^*\boldsymbol{e}_m \otimes \boldsymbol{e}_k = \frac{\partial \sigma_{lk}^*}{\partial x_l^*}\boldsymbol{e}_k \qquad (37\text{-}19)$$

$$\text{右散度} \quad \boldsymbol{\sigma}^* \cdot \boldsymbol{\nabla}^* = \sigma_{km}^*\boldsymbol{e}_k \otimes \boldsymbol{e}_m \cdot \frac{\partial}{\partial x_l^*}\boldsymbol{e}_l = \frac{\partial \sigma_{kl}^*}{\partial x_l^*}\boldsymbol{e}_k \qquad (37\text{-}20)$$

由于柯西应力满足 $(\boldsymbol{\sigma}^*)^{\mathrm{T}} = \boldsymbol{\sigma}^*$，由 (37-19) 和 (37-20) 两式可知，柯西应力左散度与右散度的时空变换关系相等.

由 (37-6) 式中最后一个式子，柯西应力的时空变换关系为

$$\begin{aligned}\boldsymbol{\sigma}^* &= Q_{lm}\boldsymbol{e}_l \otimes \boldsymbol{e}_m \cdot \sigma_{rs}\boldsymbol{e}_r \otimes \boldsymbol{e}_s \cdot Q_{kn}\boldsymbol{e}_n \otimes \boldsymbol{e}_k \\ &= Q_{lm}\sigma_{rs}\delta_{mr}\delta_{sn}Q_{kn}\boldsymbol{e}_l \otimes \boldsymbol{e}_k = Q_{lm}\sigma_{mn}Q_{kn}\boldsymbol{e}_l \otimes \boldsymbol{e}_k\end{aligned} \qquad (37\text{-}21)$$

同时考虑到 $\dfrac{\partial x_p}{\partial x_l^*} = Q_{lp}$ 和 $Q_{lp}Q_{lm} = \delta_{mp}$，则有

$$\boldsymbol{\nabla}^* \cdot \boldsymbol{\sigma}^* = \frac{\partial}{\partial x_s^*}\boldsymbol{e}_s \cdot (Q_{lm}\sigma_{mn}Q_{kn}\boldsymbol{e}_l \otimes \boldsymbol{e}_k) = \frac{\partial x_p}{\partial x_s^*}\boldsymbol{e}_s \cdot \frac{\partial Q_{lm}\sigma_{mn}Q_{kn}\boldsymbol{e}_l \otimes \boldsymbol{e}_k}{\partial x_p}$$

$$= Q_{sp}\boldsymbol{e}_s \cdot \frac{\partial Q_{lm}\sigma_{mn}Q_{kn}\boldsymbol{e}_l \otimes \boldsymbol{e}_k}{\partial x_p} = Q_{sp}\delta_{sl}Q_{lm}Q_{kn}\frac{\partial \sigma_{mn}}{\partial x_p}\boldsymbol{e}_k$$

$$= Q_{lp}Q_{lm}Q_{kn}\frac{\partial \sigma_{mn}}{\partial x_p}\boldsymbol{e}_k = \delta_{mp}Q_{kn}\frac{\partial \sigma_{mn}}{\partial x_p}\boldsymbol{e}_k$$

$$= Q_{kn}\frac{\partial \sigma_{mn}}{\partial x_m}\boldsymbol{e}_k = \boldsymbol{Q}\left(\nabla \cdot \boldsymbol{\sigma}\right) \tag{37-22}$$

上式说明，$\nabla \cdot \boldsymbol{\sigma}$ 从而 $\boldsymbol{\sigma} \cdot \nabla$ 均满足客观矢量的时空变换关系. 因此，平衡方程 (32-9) 满足客观性要求.

§38. 流体动力学的客观性

通过研究纳维–斯托克斯方程 (Navier-Stokes equations) 在各种变换下的行为，我们可以推导出它们所描述的流体流动的重要性质. 这些性质中最重要的是雷诺数相似性 (Reynolds number similarity) [7.3]、固定旋转和坐标轴反射下的不变性、伽利略不变性以及坐标系旋转下不变性的丢失.

如图 38.1 所示，考虑一个在实验室中进行的特殊流体力学实验，再考虑第二个实验，它与第一个实验相似，但在某些方面有所不同. 例如：第二次实验可以在不同的时间进行；该装置可以放置在不同的位置；它可以有不同的方向；它可以放在一个移动的平台上；可以使用不同的液体；或者可以构造出第二种仪器，它在几何上与第一种类似，但规模不同. 对于这些不同，我们可以问两个实验中的速度场是否相似？也就是说，当速度场被适当地缩放并引用到适当的坐标系中时，它们是相同的吗？这些问题可以通过研究 N-S 方程变换性质来回答 (亦被称为不变性或对称性).

这些都是湍流模型的重要考虑因素. 除非模型的变换性质与 N-S 方程的性质相一致，否则模型在定性上是不正确的.

图 38.1(a) 是第一个参考实验中考虑的仪器示意图. 该仪器的尺寸由其长度 $\mathcal{L}$ 来表征，用速度标度 $\mathcal{U}$ 表征速度的初始条件和边界条件. 坐标系 (用 E 表示，标准正交基向量为 $\boldsymbol{e}_i$) 的原点和坐标轴相对于惯性坐标系中静止的装置是固定的.

38.1 基本方程组

首先给出相关的基本方程. 第一个方程是质量守恒方程中的不可压缩条件：

$$\nabla \cdot \boldsymbol{U} = 0 \tag{38-1}$$

第二个方程是由动量方程和牛顿流体导出的 N-S 方程：

$$\frac{DU}{Dt} = \frac{\partial U}{\partial t} + (U \cdot \nabla) U = -\frac{1}{\rho} \nabla p + \nu \nabla^2 U \tag{38-2}$$

首先讨论 N-S 方程 (38-2) 式左端加速度项取散度，其张量形式为

$$\begin{aligned}
\nabla \cdot \frac{DU}{Dt} &= \nabla \cdot \left[\frac{\partial U}{\partial t} + (U \cdot \nabla) U \right] = \frac{\partial (\nabla \cdot U)}{\partial t} \\
&\quad + (\nabla \otimes U) : (U \otimes \nabla) + U \cdot [\nabla (\nabla \cdot U)] \\
&= \frac{\partial (\nabla \cdot U)}{\partial t} + U \cdot [\nabla (\nabla \cdot U)] + (\nabla \otimes U) : (U \otimes \nabla) \\
&= \frac{D (\nabla \cdot U)}{Dt} + (\nabla \otimes U) : (U \otimes \nabla)
\end{aligned} \tag{38-3}$$

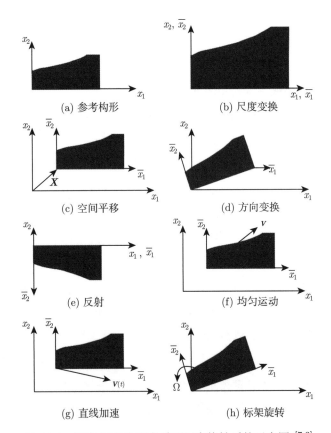

图 38.1　研究 N-S 方程各种不同变换性质的示意图 [7.3]

用分量形式验证 (38-3) 式：

$$\frac{\partial}{\partial x_i} \left(\frac{\partial U_i}{\partial t} + U_j \frac{\partial U_i}{\partial x_j} \right) = \frac{\partial}{\partial t} \frac{\partial U_i}{\partial x_i} + \frac{\partial U_j}{\partial x_i} \frac{\partial U_i}{\partial x_j} + U_j \frac{\partial^2 U_i}{\partial x_i \partial x_j}$$

第 7 章 连续介质力学中的客观性

$$=\frac{D}{Dt}\left(\frac{\partial U_i}{\partial x_i}\right)+\frac{\partial U_j}{\partial x_i}\frac{\partial U_i}{\partial x_j}$$
$$=\frac{D}{Dt}\left(\boldsymbol{\nabla}\cdot\boldsymbol{U}\right)+\frac{\partial U_j}{\partial x_i}\frac{\partial U_i}{\partial x_j} \tag{38-4}$$

再对 (38-2) 式右端取散度，在密度均匀的情况下，有

$$\boldsymbol{\nabla}\cdot\left(-\frac{1}{\rho}\boldsymbol{\nabla}p+\nu\nabla^2\boldsymbol{U}\right)=-\frac{1}{\rho}\nabla^2 p+\nu\nabla^2\left(\boldsymbol{\nabla}\cdot\boldsymbol{U}\right) \tag{38-5}$$

合并 (38-4) 式和 (38-5) 式，则有

$$\frac{D}{Dt}\left(\boldsymbol{\nabla}\cdot\boldsymbol{U}\right)+\frac{\partial U_j}{\partial x_i}\frac{\partial U_i}{\partial x_j}=-\frac{1}{\rho}\nabla^2 p+\nu\nabla^2\left(\boldsymbol{\nabla}\cdot\boldsymbol{U}\right) \tag{38-6}$$

当介质不可压缩时，亦即 (38-1) 式：$\boldsymbol{\nabla}\cdot\boldsymbol{U}=0$，上式变为如下泊松方程：

$$\nabla^2 p=-\rho\frac{\partial U_j}{\partial x_i}\frac{\partial U_i}{\partial x_j} \tag{38-7}$$

因此，可得出结论：泊松方程 (38-7) 的满足是一个无源速度场保持其无源性的充分必要条件.

38.2 基本方程组的无量纲化

特征长度 $\mathcal{L}$ 和特征速度 $\mathcal{U}$ 的引入可用于定义无量纲自变量：

$$\hat{\boldsymbol{x}}=\frac{\boldsymbol{x}}{\mathcal{L}},\quad \hat{t}=\frac{t\mathcal{U}}{\mathcal{L}} \tag{38-8}$$

和无量纲因变量：

$$\hat{\boldsymbol{U}}\left(\hat{\boldsymbol{x}},\hat{t}\right)=\frac{\boldsymbol{U}\left(\boldsymbol{x},t\right)}{\mathcal{U}},\quad \hat{p}\left(\hat{\boldsymbol{x}},\hat{t}\right)=\frac{p\left(\boldsymbol{x},t\right)}{\rho\mathcal{U}^2} \tag{38-9}$$

首先讨论不可压缩条件 (38-1) 式的无量纲化：

$$\left(\boldsymbol{\nabla}^*\mathcal{L}^{-1}\right)\cdot\left(\hat{\boldsymbol{U}}\mathcal{U}\right)=0 \tag{38-10}$$

上式可进一步简化为如下无量纲形式：

$$\boldsymbol{\nabla}^*\cdot\hat{\boldsymbol{U}}=0 \quad \text{或} \quad \frac{\partial \hat{U}_k}{\partial \hat{x}_k}=0 \tag{38-11}$$

然后给出 N-S 方程的无量纲过程：

$$\frac{\partial\left(\hat{\boldsymbol{U}}\mathcal{U}\right)}{\partial\left(\hat{t}\mathcal{L}/\mathcal{U}\right)}+\left[\left(\hat{\boldsymbol{U}}\mathcal{U}\right)\cdot\frac{\boldsymbol{\nabla}^*}{\mathcal{L}}\right]\left(\hat{\boldsymbol{U}}\mathcal{U}\right)=-\frac{1}{\rho}\frac{\boldsymbol{\nabla}^*}{\mathcal{L}}\left(\hat{p}\rho\mathcal{U}^2\right)+\nu\left(\frac{\boldsymbol{\nabla}^*}{\mathcal{L}}\right)^2\left(\hat{\boldsymbol{U}}\mathcal{U}\right) \tag{38-12}$$

整理上式，得到无量纲形式的 N-S 方程：
$$\frac{\partial \hat{U}}{\partial \hat{t}} + \left(\hat{U} \cdot \nabla^*\right)\hat{U} = -\nabla^*\hat{p} + \frac{1}{Re}\nabla^{*2}\hat{U} \tag{38-13}$$

或表示为如下分量形式：
$$\frac{\partial \hat{U}_i}{\partial \hat{t}} + \hat{U}_j \frac{\partial \hat{U}_i}{\partial \hat{x}_j} = -\frac{\partial \hat{p}}{\partial \hat{x}_i} + \frac{1}{Re}\frac{\partial^2 \hat{U}_i}{\partial \hat{x}_j \hat{x}_j} \tag{38-14}$$

最后给出压强所需要满足的泊松方程 (38-7) 的无量纲形式：
$$\frac{\partial^2 \hat{p}}{\partial \hat{x}_i \partial \hat{x}_i} = -\frac{\partial \hat{U}_i}{\partial \hat{x}_j}\frac{\partial \hat{U}_j}{\partial \hat{x}_i} \tag{38-15}$$

进行总结，无量纲方程组 (连续性方程 + N-S 方程 + 泊松方程) 为

$$\begin{cases} \dfrac{\partial \hat{U}_k}{\partial \hat{x}_k} = 0 \\ \dfrac{\partial \hat{U}_i}{\partial \hat{t}} + \hat{U}_j \dfrac{\partial \hat{U}_i}{\partial \hat{x}_j} = -\dfrac{\partial \hat{p}}{\partial \hat{x}_i} + \dfrac{1}{Re}\dfrac{\partial^2 \hat{U}_i}{\partial \hat{x}_j \hat{x}_j} \\ \dfrac{\partial^2 \hat{p}}{\partial \hat{x}_i \partial \hat{x}_i} = -\dfrac{\partial \hat{U}_i}{\partial \hat{x}_j}\dfrac{\partial \hat{U}_j}{\partial \hat{x}_i} \end{cases} \tag{38-16}$$

在上式无量纲方程组中，雷诺数
$$Re = \frac{\mathcal{U}\mathcal{L}}{\nu} \tag{38-17}$$

是唯一的无量纲数.

38.3 雷诺数相似性

图 38.1(b) 所示的实验具有不同的特征长度 $\mathcal{L}_b$，特征速度 $\mathcal{U}_b$，和流体性质，ν_b 和 ρ_b. 若定义无量纲量：

$$\hat{x} = \frac{x}{\mathcal{L}_b}, \quad \hat{U} = \frac{U}{\mathcal{U}_b}, \quad \hat{t} = \frac{t\mathcal{U}_b}{\mathcal{L}_b}, \quad Re_b = \frac{\mathcal{U}_b \mathcal{L}_b}{\nu_b} \tag{38-18}$$

且用 $\hat{U}\left(\hat{x}, \hat{t}\right)$ 表示的两个实验的边界条件相同. 在雷诺数相同的情况下，则无量纲速度场 $\hat{U}\left(\hat{x}, \hat{t}\right)$ 也将相同，原因是它们由具有相同初始条件和边界条件的相同方程控制. 这就是雷诺数相似的性质.

38.4 时空不变性

N-S 方程最简单的不变性是它们关于时间和空间平移的不变性. 空间和时间的平移由方程可表示为

$$\begin{cases} \hat{x} = \dfrac{x - X}{\mathcal{L}} \\ \hat{t} = \dfrac{(t - T)\mathcal{U}}{\mathcal{L}} \end{cases} \tag{38-19}$$

第 7 章　连续介质力学中的客观性

由于上式中 X 和 T 均为定值，求导后变为零，所以连续性方程 (38-16) 式中第一式、N-S 方程 (38-16) 式中第二式、泊松方程 (38-16) 式中第三式在经过 (38-19) 式的空间和时间变换后，均保持形式的不变性，因此满足时空不变性.

38.5　时间反演不变性

时间反演意味着时间和速度变号：

$$\begin{cases} \hat{t} = -\dfrac{t\mathcal{U}}{\mathcal{L}} \\ \hat{\boldsymbol{U}}(\hat{\boldsymbol{x}},\hat{t}) = -\dfrac{\boldsymbol{U}(\boldsymbol{x},t)}{\mathcal{U}} \end{cases} \tag{38-20}$$

方程 (38-16) 式中第一式 $\dfrac{\partial \hat{U}_k}{\partial \hat{x}_k} = 0$，由于右端为零，则满足时间反演不变性；

方程 (38-16) 式中第三式 $\dfrac{\partial^2 \hat{p}}{\partial \hat{x}_i \partial \hat{x}_i} = -\dfrac{\partial \hat{U}_i}{\partial \hat{x}_j}\dfrac{\partial \hat{U}_j}{\partial \hat{x}_i}$，由于右端有两个 $\hat{U}$，亦满足时间反演不变性.

再让我们来看 N-S 方程 (38-16) 式中第二式，$\dfrac{\partial \hat{U}_i}{\partial \hat{t}} + \hat{U}_j \dfrac{\partial \hat{U}_i}{\partial \hat{x}_j} = -\dfrac{\partial \hat{p}}{\partial \hat{x}_i} + \dfrac{1}{Re}\dfrac{\partial^2 \hat{U}_i}{\partial \hat{x}_j \hat{x}_j}$，左端第一项分子和分母均变号，满足不变性；左端第二项，由于有两个 $\hat{U}$ 亦满足时间反演不变性；右端第一项满足反演不变性；关键是黏性项，也就是动量扩散项，由于只有一个 $\hat{U}$，该项不满足时间反演不变性，因此，方程 N-S 不满足时间反演不变性.

38.6　旋转和反射不变性

图 38.1(d) 为与参考实验方向不同的仪器. 通过旋转参考坐标轴 (E) 得到合适的坐标系 ($\bar{E}$). 图 38.1(e) 显示了一个不同的设备，构造为参考仪器的镜像. 在这种情况下，适当的 $\bar{E}$ 坐标系是通过坐标轴的反射得到的.

这些坐标变换——轴的旋转和反射——正是笛卡儿张量中考虑的那些. 设参考坐标系的单位基矢量为 $\boldsymbol{e}_i$，而旋转或反射坐标系的单位基矢量为 $\bar{\boldsymbol{e}}_j$，两者之间的点积为方向余弦：$a_{ij} = \boldsymbol{e}_i \cdot \bar{\boldsymbol{e}}_j$，从而无量纲化的坐标和速度为

$$\begin{cases} \hat{x}_i = \dfrac{\bar{x}_i}{\mathcal{L}} = \dfrac{\bar{\boldsymbol{e}}_i \cdot (x_j \boldsymbol{e}_j)}{\mathcal{L}} = \dfrac{a_{ji} x_j}{\mathcal{L}} \\ \hat{U}_i = \dfrac{\bar{U}_i}{\mathcal{U}} = \dfrac{\bar{\boldsymbol{e}}_i \cdot (U_j \boldsymbol{e}_j)}{\mathcal{U}} = \dfrac{a_{ji} U_j}{\mathcal{U}} \end{cases} \tag{38-21}$$

从 N-S 方程可以用笛卡儿张量符号表示的事实可以直接得出，变换后的方程与参照系中的方程 (38-16) 是相同的. 因此，N-S 方程对于坐标轴的旋转和反射是不变的.

在上述的分析中，重要的是要区分两种"旋转". 这里我们考虑的是通过 E 坐标轴的固定旋转得到的 $\bar{E}$ 坐标系. 固定的意思是方向余弦 a_{ij} 不依赖于时间.

反射不变性具有重要的物理和数学意义. 其物理意义是，N-S 方程不包含对右手或左手运动的偏爱. 当然，这种偏差可以发生在气流中 —— 尤其是龙卷风中 —— 但它是由初始条件或边界条件引起的，或者是由标架旋转引起的，而不是由运动方程 (用惯性系表示) 引起的.

任何用笛卡儿张量符号表示的方程都能保证在旋转和坐标轴反射下的不变性. 相反，用矢量符号写的方程，包含赝矢量 (如涡度)，或用置换符号 ε_{ijk} 的指标符号写的方程，不能保证这些旋转或反射不变性.

38.7 伽利略不变性

在四维仿射空间 $\mathbb{A}^4$ 中，进行如下伽利略变换：

$$\begin{cases} \bar{\boldsymbol{x}} = \boldsymbol{x} - \boldsymbol{V}t \\ \bar{t} = t \\ \bar{\boldsymbol{U}}(\bar{\boldsymbol{x}}, \bar{t}) = \boldsymbol{U}(\boldsymbol{x}, t) - \boldsymbol{V} \end{cases} \quad (38\text{-}22)$$

则有如下关系式：

$$\begin{cases} \dfrac{\partial \bar{U}_i}{\partial \bar{x}_j} = \dfrac{\partial (U_i - V_i)}{\partial (x_j - V_j t)} = \dfrac{\partial U_i}{\partial x_j} \\ \dfrac{\partial \bar{U}_i}{\partial \bar{t}} = \dfrac{\partial U_i}{\partial t} + \dfrac{\partial U_i}{\partial x_j}\dfrac{\partial x_j}{\partial t} = \dfrac{\partial U_i}{\partial t} + \dfrac{\partial (\bar{x}_j + V_j t)}{\partial t}\dfrac{\partial U_i}{\partial x_j} = \dfrac{\partial U_i}{\partial t} + V_j \dfrac{\partial U_i}{\partial x_j} \\ \dfrac{D\bar{U}_i}{D\bar{t}} = \dfrac{\partial \bar{U}_i}{\partial \bar{t}} + \bar{U}_j \dfrac{\partial \bar{U}_i}{\partial \bar{x}_j} = \dfrac{DU_i}{Dt} \end{cases} \quad (38\text{-}23)$$

(38-23) 式中第一式和 (38-23) 式中第三式表明速度梯度和流体加速度是伽利略不变量；而 (38-22) 式中第三式和 (38-23) 式中第二式两式则表明速度和它的时间偏导数则不具有伽利略不变性.

结果表明，转换后的 N-S 方程与 (38-16) 式是一致的. 因而具有伽利略不变性.

本小节的重要结论是：就像经典力学中描述的所有现象一样，流体流动在所有惯性系中的行为是相同的.

38.8 扩展伽利略不变性

N-S 方程的一个特殊性质是，它们在标架的直线加速度下是不变的. 让我们来考虑如图 38.1(g) 所示的在变速 $\boldsymbol{V}(t)$ 平台上所进行的第二次实验，但是没有坐标系的旋转，因此坐标方向 (单位基矢量 $\boldsymbol{e}_i$ 和 $\bar{\boldsymbol{e}}_j$) 仍然是平行的.

由 (38-22 式所定义的变换后的变量 $\bar{x}$, $\bar{t}$ 和 $\bar{U}$, 转换后的 N-S 方程为

$$\frac{\partial \bar{U}_i}{\partial \bar{t}} + \bar{U}_j \frac{\partial \bar{U}_i}{\partial \bar{x}_j} = \nu \frac{\partial^2 \bar{U}_i}{\partial \bar{x}_j \partial \bar{x}_j} - \frac{1}{\rho} \frac{\partial p}{\partial \bar{x}_i} - A_i \qquad (38\text{-}24)$$

式中, 方程右端的附加项为标架的加速度 $\boldsymbol{A} = \dfrac{\mathrm{d}\boldsymbol{V}}{\mathrm{d}t}$, 方程右端的最后两项可以改写为

$$\frac{1}{\rho} \frac{\partial p}{\partial \bar{x}_i} + A_i = \frac{1}{\rho} \frac{\partial}{\partial \bar{x}_i} (p + \rho \bar{x}_j A_j) \qquad (38\text{-}25)$$

上式表明标架加速度可以被修正后的压强吸收. 在无量纲化的 N-S 方程 (38-16) 式中第二式中只需引入如下无量纲量:

$$\hat{\boldsymbol{U}} = \frac{\bar{U}}{\mathcal{U}}, \quad \hat{p} = \frac{p + \rho \bar{\boldsymbol{x}} \cdot \boldsymbol{A}}{\rho \mathcal{U}^2} \qquad (38\text{-}26)$$

此时无量纲的 N-S 方程在形式上不变, 则 $\hat{\boldsymbol{U}}$ 和 $\hat{p}$ 在具有任意直线加速度的坐标系中与惯性坐标系中相应量相同, 该性质被称为扩展的伽利略不变性.

38.9 标架旋转

最后, 让我们考虑如图 38.1(h) 所示的在非惯性旋转坐标系上所进行的第二次实验. 在 $\bar{E}$ 坐标系, 依赖于时间的基矢量 $\bar{e}_i(t)$ 满足下列关系:

$$\frac{\mathrm{d}\bar{\boldsymbol{e}}_i}{\mathrm{d}t} = \tilde{\Omega}_{ij} \bar{\boldsymbol{e}}_j \qquad (38\text{-}27)$$

式中, $\tilde{\Omega}_{ij}(t) = -\tilde{\Omega}_{ji}(t)$ 为反对称的旋转率张量, 此时, 方向余弦 $a_{ij} = \boldsymbol{e}_i \cdot \bar{\boldsymbol{e}}_j$ 具有时间相关性. 转换后的 N-S 方程 (38-24) 中的加速度将由离心加速度、科里奥利加速度、角加速度三部分组成:

$$A_i = \underbrace{\bar{x}_j \tilde{\Omega}_{jk} \tilde{\Omega}_{ki}}_{\text{离心加速度}} + \underbrace{2\bar{U}_j \tilde{\Omega}_{ji}}_{\text{科氏加速度}} + \underbrace{\bar{x}_j \frac{\mathrm{d}\tilde{\Omega}_{ji}}{\mathrm{d}\bar{t}}}_{\text{旋转加速度}} \qquad (38\text{-}28)$$

式中的三个加速度分别代表的虚拟力为: 离心力、科里奥利力和角加速度力. 离心力可以被吸收成一个修正的压强, 但剩下的两个力却不能. 众所周知, 在气象学和叶轮机械, 科里奥利力可以对旋转标架下的流动有着重要的影响.

在旋转坐标系和非旋转坐标系中相同的量被称为具有物质标架无差异性 (material-frame indifference). 显然, N-S 方程不具备这种性质.

38.10 关于虚拟力的进一步讨论

(38-28) 式中的离心加速度沿外径向、垂直于旋转轴，与粒子在旋转坐标系的运动无关，其可进一步表示为

$$-\boldsymbol{\omega} \times (\boldsymbol{\omega} \times \boldsymbol{r}) = -\boldsymbol{\nabla}\left[\frac{1}{2}(\boldsymbol{\omega} \times \boldsymbol{r})^2\right] \tag{38-29}$$

亦即，和离心加速度相关的离心虚拟力为无旋的保守力，参照 (28-25) 式，离心力可以被吸收成一个修正的压强：

$$-\boldsymbol{\nabla} p - \rho \boldsymbol{\omega} \times (\boldsymbol{\omega} \times \boldsymbol{r}) = -\boldsymbol{\nabla}\left[p - \frac{1}{2}\rho(\boldsymbol{\omega} \times \boldsymbol{r})^2\right] \tag{38-30}$$

换句话说，离心加速度和其虚拟力满足扩展伽利略不变性. (38-29) 式的一个演算细节为

$$\begin{aligned}
\boldsymbol{\nabla}\left[\frac{1}{2}(\boldsymbol{\omega} \times \boldsymbol{r})^2\right] &= [\boldsymbol{\nabla} \otimes (\boldsymbol{\omega} \times \boldsymbol{r})] \cdot (\boldsymbol{\omega} \times \boldsymbol{r}) \\
&= [(\boldsymbol{\nabla} \otimes \boldsymbol{\omega}) \times \boldsymbol{r} - (\boldsymbol{\nabla} \otimes \boldsymbol{r}) \times \boldsymbol{\omega}] \cdot (\boldsymbol{\omega} \times \boldsymbol{r}) \\
&= -(\mathbf{I} \times \boldsymbol{\omega}) \cdot (\boldsymbol{\omega} \times \boldsymbol{r}) = -\boldsymbol{\omega} \times (\boldsymbol{\omega} \times \boldsymbol{r})
\end{aligned} \tag{38-31}$$

事实上，在上式的运算中，还应用到下列等式：

$$\begin{aligned}
(\mathbf{I} \times \boldsymbol{a}) \cdot \boldsymbol{b} &= \delta_{ij} a_m b_n (\boldsymbol{e}_i \otimes \boldsymbol{e}_j \times \boldsymbol{e}_m) \cdot \boldsymbol{e}_n \\
&= a_m b_n \boldsymbol{e}_i \otimes \varepsilon_{imk}\boldsymbol{e}_k \cdot \boldsymbol{e}_n = a_m b_n \varepsilon_{imn}\boldsymbol{e}_i \\
&= \boldsymbol{a} \times \boldsymbol{b}
\end{aligned} \tag{38-32}$$

(38-28) 式中的科氏加速度 $-2\boldsymbol{\omega} \times \boldsymbol{v}$ 与粒子在旋转坐标系的运动相关，因此，其虚拟力，亦即科里奥利力，为有旋的非保守力.

(38-28) 式中的角加速度项 $-\frac{\mathrm{d}\boldsymbol{\omega}}{\mathrm{d}t} \times \boldsymbol{r}$ 由于沿 $\boldsymbol{r}$ 的切线方向，所以和其相对应的虚拟力，也就是欧拉力或称角加速度力，为有旋的非保守力.

故而，科里奥利力和角加速度力这两种虚拟力作为非保守力，不能被吸收成修正的压强，不满足扩展伽利略不变性.

让著者以爱因斯坦的如下名言来作为本章的结束语："如果上帝满足于惯性系，他就不会创造出引力 (If God had been satisfied with inertial frames, he would not have created gravitation)."

思考题和补充材料

7.1 理性力学大师克利福德·特鲁斯德尔 (Clifford Ambrose Truesdell III, 1919—2000) 的珍贵照片.

题图 7.1　特鲁斯德尔于 1975 年 7 月和他的妻子夏洛特 (Charlotte) 在美国马里兰州巴尔的摩市

7.2 美国自然哲学学会第一次会议的合影 (1963 年 3 月 25—26 日). 多位著名理性力学家参会,包括:Truesdell 夫妇、Coleman、Ericksen、Sternberg、Noll、Toupin、Rivlin 等.

题图 7.2　美国自然哲学学会的第一次合影. 第三排右三为 M. E. Gurtin;第六排左二为 A. C. Eringen;最后一排左六为 A. C. Pipkin

7.3 证明四维仿射空间 ($\mathbb{A}^4$) 中的伽利略变换:

$$\begin{cases} x' = x - vt \\ y' = y' \\ z' = z \\ t' = t \end{cases} \tag{s7-1}$$

是如下闵可夫斯基时空 (Minkowski spacetime) 中洛伦兹变换的特例 [1.6]:

$$\begin{cases} x' = \gamma(x - vt) \\ y' = y' \\ z' = z \\ t' = \gamma\left(t - \dfrac{vx}{c^2}\right) \end{cases} \tag{s7-2}$$

(s7-1) 和 (s7-2) 两式中,v 是坐标系间相对的匀速直线运动速度,c 为光速,洛伦兹因子 γ 的表达式为

$$\gamma = \dfrac{1}{\sqrt{1 - \dfrac{v^2}{c^2}}} \tag{s7-3}$$

7.4 在 §1.1 中,曾陈述道 "理性连续介质力学中十分关键的 '客观性' 的概念可上溯至爱因斯坦及其更早期的物理学家的相对论思想中",请找到这种论述的依据.

7.5 所谓 "理想流体 (ideal fluid)" 是一种没有黏滞性 (viscosity) 亦不可压缩 (incompressible) 的流体,所以它没有剪应力存在,也无能量损耗,它流动的行为正如无黏滞流 (inviscid flow) 的特性. 一般状况是指此流体存在于远离任何物体几何形状影响之区域,而来描述此流体流动特性时,可视此流体即为理想流体,以别与在物体周围附近流场因黏滞力的作用而存在速度梯度. 理想流体由欧拉方程 (56-11) 式描述. 证明如下定理 [3.7]: 理想流体的响应与观察者无关.

参 考 文 献

[7.1] Truesdell C, Noll W. The Non-Linear Field Theories of Mechanics. In: Encyclopedia of Physics (Flügge S, Ed., Vol. 3), Berlin: Springer-Verlag, 1965.

[7.2] Holzapfel G A. Nonlinear Solid Mechanics: A Continuum Approach for Engineering Science. Chichester: John Wiley & Sons, 2000.

[7.3] Pope S B. Turbulent Flows. Cambridge: Cambridge University Press, 2001.

第8章 本构关系

§39. 理性力学中的公理

39.1 本构公理的提出与建立

公理是理性力学的基石 [8.1]. 詹姆斯·奥尔德罗伊德 (James Gardner Oldroyd, 1921—1982) 于 1950 年发表了被誉为 "或许是理论流变学领域最为重要的单篇论文 (probably the most important single paper in theoretical rheology)" 的题为《关于流变状态方程的推导》(On the Formulation of Rheological Equations of State) 的经典之作 [8.2]，开启了有关理性力学本构公理的研究，其内容详见表 39.1.

表 39.1 理性力学中的奥尔德罗伊德公理和诺尔三公理一览表

理性力学公理名称	公理内容	提出人和年代
奥尔德罗伊德本构公理	流变状态方程必须具有正确的不变性性质. The right invariance properties which must be satisfied by a rheological equation of state.	奥尔德罗伊德, 1950
应力的决定性原理 principle of determinism for the stress	粒子 X 在时刻 t 的应力 $S(t)$ 是由过去任意小的 X 邻域运动的历史决定. The stress $S(t)$ at a particle X and at time t is determined by the past history of the motion of an arbitrarily small neighborhood of X.	诺尔, 1958
局部作用原理 principle of local action	在确定给定粒子 X 处的应力时，可以忽略 X 任意邻域外的运动. In determing the stress at a given particle X, the motion outside an arbitrary neighborhood of X may be disregarded.	
材料性质的客观性原理 principle of objectivity of material properties	如果一个过程{运动—θ, 应力—S} 和一个本构方程相容，同样，与它等价的所有过程{θ', S'} 都必须与相同的本构方程相容. If a process $\{\theta, S\}$ is compatible with a constitutive equation, then also all processes $\{\theta', S'\}$ equivalent to it must be compatible with the same constitutive equation.	

沃尔特·诺尔 (Walter Noll, 1925—2017) 继而于 1958 年提出的 "确定性公理、局部作用公理和客观性公理" 是构造本构理论的基础 [8.3]，诺尔三公理迄今仍被理性力学或连续介质力学教材所广泛引用. 其内容亦详见表 39.1.

美国工程科学学会的创始人 (founder of the Society of Engineering Science) 埃林根 (Ahmed Cemal Eringen, 1921—2009)[8.4,8.5] 进一步扩充了诺尔的公理结构，使之成为工程科学学派的理论基石. 作为现代理性力学核心内容的力学公理化体系的建立，奠定了现代连续介质力学体系的基础. 埃林根的公理体系见表 39.2.

表 39.2 埃林根的理性力学公理一览表

公理名称	公理内容
因果性公理 axiom of causality	在物体的每一个热力学状态中，将物体的物质点的运动、温度、电荷看成是自明的可测效应. 而将进入到克劳修斯–迪昂不等式中的其余的量看成是运动、温度、电荷等这个 "原因" 所产生的结果，这些量称为 "响应函数" 或者 "本构依赖变量". The motions, temperatures and charges of the material points of a body are the cause of all physical phenomena. The remaining variables (other than those derivable from motion, temperature and charges) that enter the expressions of the Clausius–Duhem (C–D) inequality are the response functions (or constitutive-dependent variables).
确定性公理 axiom of determinism	物体中的物质点在时刻 t 的热力学本构泛函以及应力状态由物体中所有物质点的运动和温度历史所决定. The constitutive-dependent variables at a material point X, at time t, are functionals of the independent variables over the entire material points X' of the body, at all past times t' up to and including the present time t.
等存在公理 axiom of equipresence	一开始，所有的本构泛函都应该用同样的独立本构变量来表示，直到推出相反的结果为止. At the outset, all constitutive-dependent variables must be expressed as functionals of the same list of independent constitutive variables until the contrary is deduced.
客观性公理 axiom of objectivity	本构方程对于空间参照系的刚体运动必须是形式不变的. Constitutive equations must be form-invariant with respect to rigid motions of the spatial frame of reference.
物质不变性公理 axiom of material invariance	本构方程必须具有关于物质点对称群的形式不变量. Constitutive equations must be form-invariant with respect to the symmetry group of the material points.

公理名称	公理内容
邻域公理 axiom of neighborhood	物体中物质点的应力状态与离开该物质点有限距离的其他物质点的运动无关. The values of the independent constitutive variables at distant material points X' from the reference point X do not appreciably affect the value of the constitutive-dependent variables at X.
记忆公理 axiom of memory	本构变量在远离现在的过去时刻的值, 不明显地影响本构函数的值. The values of the constitutive-independent variables at distants past the present do not appreciably affect the values of the constitutive functionals at the present time.
相容性公理 axiom of admissibility	所有本构方程必须与守恒定律和熵不等式相一致. All constitutive equations must be consistent with the balance laws and the entropy inequality.

埃林根还在八条公理的基础上, 添加了坐标不变性公理和对因次 (单位) 系统不变性公理. 由于埃林根本构公理比较复杂且广泛, 诺尔三公理仍然发挥着不可替代的作用.

爱因斯坦的如下著名论断针对理性力学而言亦成立: "所有科学的宏大目标是从最小数目的假说或公理出发, 通过逻辑演绎来覆盖最大数目的经验事实 (The grand aim of all science is to cover the greatest number of empirical facts by logical deduction from the smallest number of hypotheses or axioms)."

39.2 里夫林等学者对连续介质力学公理的批评

理性力学大师克利福德·特鲁斯德尔 (Clifford Truesdell, 1919—2000) 学派 (包括其学生诺尔) 对理性连续介质力学的最主要贡献就是使其公理化. 翻开任何一本理性力学或连续介质力学的专著或教材, 都或多或少地包含有公理化的内容, 因此, 可以说特鲁斯德尔学派在连续介质力学中的地位是不可撼动的.

但是, 由于特鲁斯德尔四面出击 (如他对 1968 年诺贝尔化学奖得主拉斯·昂萨格的攻击), 其工作也招致了一些学者的尖锐批评. 1981 年, 伍兹 (Leslie Colin Woods, 1922—2007) 就直接撰写了题为 "连续介质力学臆造的公理 (The bogus axioms of continuum mechanics)" 的文章直接针对特鲁斯德尔. 还有就是拉文达 (Bernard Howard Lavenda, 1945—) 说 "理性热力学有理性的话, 那么它一定隐藏的很好 (If there is something rational in rational thermodynamics it is well-hidden)", 讽刺特鲁斯德尔及其追随者并没有找到理性或公理.

具有讽刺意味的是, 理性热力学在特鲁斯德尔所擅长的领域失效: 当人们发现理性热力学不能用于非牛顿流体时,理性热力学就承受了更多的毁誉 (More damage was suffered by rational thermodynamics when it was found that the theory could not be applied to non-Newtonian fluids).

另外一位理性力学大师里夫林 (Ronald Rivlin, 1915—2005),作为非线性连续介质力学的开创者,在其著名的题为 Red herrings and sundry unidentified fish in nonlinear continuum mechanics 文章中 [8.6],以明显挑衅的 (evidently provocative) 美式幽默以不点名的方式,来猛烈地攻击特鲁斯德尔的公理化的工作并没有像希尔伯特的公理化在纯数学领域一样带给我们什么新的发现,而且还造成了很多错误,让人误入歧途. 里夫林指出:"实际上,就内容而言,它(即理性连续介质力学) 仅仅是将人们熟知的连续介质运动学和力学的一些概念用基础的集合论语言翻译了一下 (Indeed, in content, it appears to be little more than a translation of a few familiar concepts in the kinematics and mechanics of continua into the language of elementary set theory)."

里夫林文章题目中的 "red herring" 原意是撒了盐烟熏了的鲱鱼,可引申为分散注意力而提出的不相干事实或论点,这里指的是特鲁斯德尔让人误入歧途的理论;标题第二部分 "sundry unidentified fish" 直译是各类叫不出名的鱼的鱼干. 因此,里夫林的讽刺文章 [8.6] 的题目可大致地意译为 "非线性连续介质里让人误入歧途的理论和长久以来没有意识到的问题".

§40. 线弹性本构关系——广义胡克定律

1678 年,罗伯特·胡克 (Robert Hooke, 1635—1703) 建立了弹簧的线弹性定律 [8.7],被称为胡克定律.

1782 年,意大利学者佐丹奴·黎卡提 (Giordano Riccati, 1709—1790) 首次在实验上确定了材料的弹性模量,黎卡提用弯曲振动 (flexural vibrations) 的方法 [8.8],所给出的钢和黄铜的弹性模量之比为 2.06 倍,即使在当代,这个数据也是十分准确的. 黎卡提有关材料弹性模量的研究比托马斯·杨早了 25 年.

可变形体的弹性模量又被广泛地称为杨氏模量,其原因是英国科学家托马斯·杨 (Thomas Young, 1773—1829) 于 1807 年引入了该概念 [8.9].

1821 年,纳维 (C. L. M. H. Navier, 1785—1836) 在其《论弹性体的平衡与运动》论文中最早提出弹性体运动的一般方程 [8.10].

1828 年,柯西 (Augustin-Louis Cauchy, 1789—1857) 在《弹性或非弹性固体的

运动方程》[8.11] 一文中建立了弹性力学平衡与运动的普遍方程.

1829 年, 泊松 (Siméon Denis Poisson, 1781—1840) 在《弹性体的平衡与运动》[8.12] 中最早提出弹性体变形的泊松比 (Poisson's ratio).

1852 年, 法国数学家和弹性力学家 —— 拉梅 (Gabriel Lamé, 1795—1870, 如图 40.1 所示) 出版了 *Leçons sur la Théorie Mathématique de l'Élasticité des Corps Solides*[3.8], 这是第一本系统陈述弹性力学的专著.

图 40.1　加布里埃尔·拉梅: 戴帽像 (左); 脱帽像 (中); 雕塑 (右)

1892 年, 乐甫 (Augustus Edward Hough Love, 1863—1940) 在《弹性的数学理论教程》[8.13] 中全面地总结了弹性力学的成就, 发展了薄壳理论. 至此, 弹性力学的范式得以形成.

绝大部分的工程研究课题仅仅用到小变形、各向同性的线弹性本构关系. 因此, 将其深入浅出地弄通、弄懂、会熟练地应用, 十分关键.

40.1　材料力学和弹性力学中的广义胡克定律 (应变—应力关系式)

考虑一个三维的弹性体, 只在 x 方向施加应力 σ_{xx}, 将产生三个应变: 沿 x 方向的伸长 $\varepsilon_{xx} = \dfrac{\sigma_{xx}}{E}$, 由于泊松效应, 沿 y 和 z 方向的收缩: $\varepsilon_{yy} = \varepsilon_{zz} = -\nu\varepsilon_{xx} = -\nu\dfrac{\sigma_{xx}}{E}$.

同理, 分别沿 y 和 z 方向施加应力 σ_{yy} 和 σ_{zz} 的话, 沿三个方向分别所产生的应变如表 40.1 所示.

表 40.1　在三个方向施加应力分量后所对应的应变分量

所施加的应力分量	在 x 方向产生的应变	在 y 方向产生的应变	在 z 方向产生的应变
σ_{xx}	$\varepsilon_{xx} = \dfrac{\sigma_{xx}}{E}$	$\varepsilon_{yy} = -\nu\dfrac{\sigma_{xx}}{E}$	$\varepsilon_{zz} = -\nu\dfrac{\sigma_{xx}}{E}$
σ_{yy}	$\varepsilon_{xx} = -\nu\dfrac{\sigma_{yy}}{E}$	$\varepsilon_{yy} = \dfrac{\sigma_{yy}}{E}$	$\varepsilon_{zz} = -\nu\dfrac{\sigma_{yy}}{E}$
σ_{zz}	$\varepsilon_{xx} = -\nu\dfrac{\sigma_{zz}}{E}$	$\varepsilon_{yy} = -\nu\dfrac{\sigma_{zz}}{E}$	$\varepsilon_{zz} = \dfrac{\sigma_{zz}}{E}$

由于线弹性这个前提，我们可用叠加原理. 将上表中的三个不同方向的应变进行叠加，得到如下三个正应变–正应力关系式：

$$\begin{cases} \varepsilon_{xx} = \dfrac{\sigma_{xx}}{E} - \dfrac{\nu}{E}(\sigma_{yy} + \sigma_{zz}) = \dfrac{1+\nu}{E}\sigma_{xx} - \dfrac{\nu}{E}(\sigma_{xx} + \sigma_{yy} + \sigma_{zz}) \\ \varepsilon_{yy} = \dfrac{\sigma_{yy}}{E} - \dfrac{\nu}{E}(\sigma_{zz} + \sigma_{xx}) = \dfrac{1+\nu}{E}\sigma_{yy} - \dfrac{\nu}{E}(\sigma_{xx} + \sigma_{yy} + \sigma_{zz}) \\ \varepsilon_{zz} = \dfrac{\sigma_{zz}}{E} - \dfrac{\nu}{E}(\sigma_{xx} + \sigma_{yy}) = \dfrac{1+\nu}{E}\sigma_{zz} - \dfrac{\nu}{E}(\sigma_{xx} + \sigma_{yy} + \sigma_{zz}) \end{cases} \quad (40\text{-}1)$$

三个切应力和切应变的关系式为

$$\varepsilon_{xy} = \dfrac{1}{2G}\sigma_{xy}, \quad \varepsilon_{yz} = \dfrac{1}{2G}\sigma_{yz}, \quad \varepsilon_{zx} = \dfrac{1}{2G}\sigma_{zx} \quad (40\text{-}2)$$

注意到材料力学材料常数的常用关系式：

$$G = \dfrac{E}{2(1+\nu)} \quad (40\text{-}3)$$

故 (40-1) 式和 (40-2) 式可统一地表示为

$$\begin{cases} \varepsilon_{xx} = \dfrac{1+\nu}{E}\sigma_{xx} - \dfrac{\nu}{E}(\sigma_{xx} + \sigma_{yy} + \sigma_{zz}) \\ \varepsilon_{yy} = \dfrac{1+\nu}{E}\sigma_{yy} - \dfrac{\nu}{E}(\sigma_{xx} + \sigma_{yy} + \sigma_{zz}) \\ \varepsilon_{zz} = \dfrac{1+\nu}{E}\sigma_{zz} - \dfrac{\nu}{E}(\sigma_{xx} + \sigma_{yy} + \sigma_{zz}) \\ \varepsilon_{xy} = \dfrac{1+\nu}{E}\sigma_{xy} \\ \varepsilon_{yz} = \dfrac{1+\nu}{E}\sigma_{yz} \\ \varepsilon_{zx} = \dfrac{1+\nu}{E}\sigma_{zx} \end{cases} \quad (40\text{-}4)$$

上式可以十分简洁地用张量形式表示为

$$\varepsilon_{ij} = \dfrac{1+\nu}{E}\sigma_{ij} - \dfrac{\nu}{E}\sigma_{kk}\delta_{ij} \quad \text{或} \quad \boldsymbol{\varepsilon} = \dfrac{1+\nu}{E}\boldsymbol{\sigma} - \dfrac{\nu}{E}(\mathrm{tr}\boldsymbol{\sigma})\mathbf{I} \quad (40\text{-}5)$$

上式就是用张量表示的应变和应力的线弹性本构关系.

对 (40-5) 式求迹，有

$$\varepsilon_{kk} = \dfrac{1+\nu}{E}\sigma_{kk} - 3\dfrac{\nu}{E}\sigma_{kk} = \dfrac{1-2\nu}{E}\sigma_{kk} \quad \text{或} \quad \mathrm{tr}\boldsymbol{\varepsilon} = \dfrac{1+\nu}{E}\mathrm{tr}\boldsymbol{\sigma} - 3\dfrac{\nu}{E}\mathrm{tr}\boldsymbol{\sigma} = \dfrac{1-2\nu}{E}\mathrm{tr}\boldsymbol{\sigma}$$
$$(40\text{-}6)$$

第 8 章 本 构 关 系

静水压强 σ_m 定义为 $\sigma_m = \sigma_{kk}/3$, 由静水压强 σ_m 和体积应变 ε_{kk} 可定义体模量 (bulk modulus) K, 利用 (40-6) 式, 有

$$K = \frac{\sigma_m}{\varepsilon_{kk}} = \frac{E}{3(1-2\nu)} \tag{40-7}$$

由 (40-3) 和 (40-7) 两式, 有

$$\frac{1}{9K} - \frac{1}{6G} = \frac{3(1-2\nu)}{9E} - \frac{2(1+\nu)}{6E} = -\frac{\nu}{E} \tag{40-8}$$

对比 (40-5) 式、(40-8) 式和 (40-3) 式, 线弹性用应力表示的广义胡克定律还十分普遍地表示为

$$\varepsilon_{ij} = \frac{1}{2G}\sigma_{ij} + \left(\frac{1}{9K} - \frac{1}{6G}\right)\sigma_{kk}\delta_{ij} \quad \text{或} \quad \boldsymbol{\varepsilon} = \frac{1}{2G}\boldsymbol{\sigma} + \left(\frac{1}{9K} - \frac{1}{6G}\right)(\text{tr}\boldsymbol{\sigma})\mathbf{I} \tag{40-9}$$

上式即为《近代连续介质力学》[1.6] 中的 (11-63) 式以及 (33-1) 式的前半段, 在工程中应用十分广泛. (40-9) 式还可用四阶柔度张量 (compliance tensor) 更为简洁地表示为

$$\begin{aligned}
\boldsymbol{\varepsilon} &= \frac{1}{2G}\boldsymbol{\sigma} + \left(\frac{1}{9K} - \frac{1}{6G}\right)(\text{tr}\boldsymbol{\sigma})\mathbf{I} = \frac{1}{2G}\boldsymbol{\sigma} + \left(\frac{1}{9K} - \frac{1}{6G}\right)\mathbf{I}(\mathbf{I}:\boldsymbol{\sigma}) \\
&= \frac{1}{2G}\mathbb{I}^s : \boldsymbol{\sigma} + \left(\frac{1}{9K} - \frac{1}{6G}\right)(\mathbf{I} \otimes \mathbf{I}) : \boldsymbol{\sigma} \\
&= \boldsymbol{S} : \boldsymbol{\sigma}
\end{aligned} \tag{40-10}$$

式中的四阶柔度张量为

$$\boldsymbol{S} = \frac{1}{2G}\mathbb{I}^s + \left(\frac{1}{9K} - \frac{1}{6G}\right)(\mathbf{I} \otimes \mathbf{I}) \tag{40-11}$$

式中,

$$\mathbb{I}^s = \frac{\mathbb{I} + \mathbb{I}^{\text{T}}}{2} = \frac{1}{2}(\delta_{ik}\delta_{jl} + \delta_{jk}\delta_{il})\boldsymbol{e}_i \otimes \boldsymbol{e}_j \otimes \boldsymbol{e}_k \otimes \boldsymbol{e}_l \tag{40-12}$$

为对称的四阶单位张量. 请注意口诀: 1324、2314, 1234 $(\mathbf{I} \otimes \mathbf{I})$, 它们之间的区别.

用指标形式, (40-10) 式和 (40-11) 式表示为

$$\begin{cases} \varepsilon_{ij} = S_{ijkl}\sigma_{kl} \\ S_{ijkl} = \frac{1}{2G}\frac{\delta_{ik}\delta_{jl} + \delta_{il}\delta_{jk}}{2} + \left(\frac{1}{9K} - \frac{1}{6G}\right)\delta_{ij}\delta_{kl} \end{cases} \tag{40-13}$$

40.2 应力-应变关系式

在本构关系求逆时，最主要的是求迹. 将 (40-6) 式代回 (40-5) 式，可得到 (40-5) 式的逆形式：

$$\sigma_{ij} = \frac{E}{1+\nu}\varepsilon_{ij} + \frac{\nu E}{(1+\nu)(1-2\nu)}\varepsilon_{kk}\delta_{ij} \quad \text{或} \quad \boldsymbol{\sigma} = \frac{E}{1+\nu}\boldsymbol{\varepsilon} + \frac{\nu E}{(1+\nu)(1-2\nu)}(\text{tr}\boldsymbol{\varepsilon})\mathbf{I} \tag{40-14}$$

为了使上式更为简洁，引入拉梅常数 (Lamé constants)，使得

$$\sigma_{ij} = 2\mu\varepsilon_{ij} + \lambda\varepsilon_{kk}\delta_{ij} \quad \text{或} \quad \boldsymbol{\sigma} = 2\mu\boldsymbol{\varepsilon} + \lambda(\text{tr}\boldsymbol{\varepsilon})\mathbf{I} \tag{40-15}$$

式中，拉梅常数和其他弹性常数之间的关系为

$$\mu = G = \frac{E}{2(1+\nu)}, \quad \lambda = \frac{\nu E}{(1+\nu)(1-2\nu)} \tag{40-16}$$

(40-15) 式还可更为简洁地用四阶刚度张量 (stiffness tensor) 表示为

$$\boldsymbol{\sigma} = 2\mu\mathbb{I}^s : \boldsymbol{\varepsilon} + \lambda\mathbf{I}(\mathbf{I} : \boldsymbol{\varepsilon}) = 2\mu\mathbb{I}^s : \boldsymbol{\varepsilon} + \lambda(\mathbf{I}\otimes\mathbf{I}) : \boldsymbol{\varepsilon} = \boldsymbol{C} : \boldsymbol{\varepsilon} \tag{40-17}$$

则四阶刚度张量为

$$\boldsymbol{C} = 2\mu\mathbb{I}^s + \lambda(\mathbf{I}\otimes\mathbf{I}) \tag{40-18}$$

用指标形式，(40-17) 和 (40-18) 两式还可表示为

$$\begin{cases} \sigma_{ij} = C_{ijkl}\varepsilon_{kl} \\ C_{ijkl} = \mu(\delta_{ik}\delta_{jl} + \delta_{il}\delta_{jk}) + \lambda\delta_{ij}\delta_{kl} \end{cases} \tag{40-19}$$

式中，四阶刚度张量满足如下小对称性 (minor symmetry)：

$$C_{ijkl} = C_{jikl} = C_{ijlk} \tag{40-20}$$

也满足如下大对称性 (major symmetry)：

$$C_{ijkl} = C_{klij} \tag{40-21}$$

有关弹性刚度张量小对称性 (40-20) 式成立的原因可见 (49-6)′ 式，亦可见 §47.2. 因此，小对称性来自于柯西应力和应变的对称性，而正如本书 §33 所讨论过的，柯西应力的对称性来自于诺特定理的空间旋转对称性所导致的角动量守恒.

有关弹性刚度张量大对称性 (40-21) 式成立的原因可见 §47、(49-6)″ 式和 (49-18) 式.

值得注意的是，(40-11) 式和 (40-18) 式中为何是对称的四阶单位张量，而不是四阶单位张量，其原因是，和其进行二次缩并的是对称的应力张量或应变张量.

40.3 对弹性常数的限制

出发点一、一般性常识的角度

一维情况，施加应力后，材料一般要伸长，所以要求杨氏模量 $E > 0$.

材料的体模量需为正，也就是静水压强使材料压缩，由 (40-7) 式得到

$$K = \frac{E}{3(1-2\nu)} > 0 \quad \Rightarrow \quad \nu < \frac{1}{2} \tag{40-22}$$

当 $\nu = \frac{1}{2}$ 时，体模量为无穷大：$K \to \infty$，此时材料为不可压缩.

材料的剪切模量需要为正，由 (40-3) 式：

$$G = \frac{E}{2(1+\nu)} > 0 \quad \Rightarrow \quad \nu > -1 \tag{40-23}$$

综上所述，对各向同性材料常数的限制条件为如下两个：

$$E > 0, \quad -1 < \nu < \frac{1}{2} \tag{40-24}$$

出发点二、应变能的正定性的角度

由于体应变为 ε_{kk}，偏应变为 $\varepsilon_{ij} - \frac{1}{3}\varepsilon_{kk}\delta_{ij}$，因此，应变能密度可写为

$$w = \frac{1}{2}K\varepsilon_{kk}^2 + G\left(\varepsilon_{ij} - \frac{1}{3}\varepsilon_{kk}\delta_{ij}\right)^2 \tag{40-25}$$

应变能密度的正定性要求：(40-22) 式和 (40-23) 式成立，亦即 (40-24) 式成立.

例 40.1 美国布朗大学皮普金 (Allen Compere Pipkin, 1931—1994, 见题图 7.2) 曾对软物质 (soft matter) 定义如下：介于固体和流体之间的一种物质形式 (A form of matters between solids and fluids). 原子处于一种部分有序的状态 (Atoms are in a partially ordered form). 当时间足够长时 $(t \to \infty)$，剪切弹性模量 G 和体积模量 K 之间满足

$$\frac{G}{K} \to O(\varepsilon) \tag{40-26}$$

对剪切的弱抵抗使得我们可以获得其本构公式的渐近展开 (Weak resistance to shear enables us to get an asymptotic expansion for the constitutive formulaties).

解：由 (40-3) 和 (40-7) 两式，得到

$$\frac{G}{K} = \frac{3(1-2\nu)}{2(1+\nu)} \tag{40-27}$$

当 $\frac{G}{K} \to 0$ 时，泊松比要满足 $\nu \to 1/2$，这是固体材料不可压缩的条件，亦即，大多

数软物质材料的泊松比都满足接近 1/2 的条件.

再让我们看看美国哈佛大学詹姆斯·罗伯特·赖斯 (James Robert Rice, 1940—) 在《大英百科全书》(Britannica) 中, 对 "固体" 所下的定义: "一种材料被称为固体, 而不是流体, 如果它在某些自然过程或技术应用感兴趣的时间尺度上也能承受巨大的剪切力的话 (A material is called solid rather than fluid if it can also support a substantial shearing force over the time scale of some natural process or technological application of interest).''

同时, 赖斯也指出: "固体和流体的界限并不明确, 而且在很多情况下依赖于时间尺度 (The distinction between solids and fluids is not precise and in many cases will depend on the time scale).'' 请参阅《力学讲义》[1.11] 中 §48 中应用德博拉数 (Deborah number) 对固体和流体等介质的划分.

40.4 固体力学材料常数常用关系的简单证明

让我们首先讨论应变能密度的表达式, 所谓 "应变能密度" 是指单位体积的应变能. 由于应力和应变之间的线弹性关系, 通过应力–应变关系中的三角形, 所以应变能密度为

$$w = \frac{1}{2}\left(\sigma_{xx}\varepsilon_{xx} + \sigma_{yy}\varepsilon_{yy} + \sigma_{zz}\varepsilon_{zz} + 2\sigma_{xy}\varepsilon_{xy} + 2\sigma_{yz}\varepsilon_{yz} + 2\sigma_{zx}\varepsilon_{zx}\right) \tag{40-28}$$

式中, 右端后三个式子中的 2 是由于剪应力互等的原因, 即, $2\sigma_{xy}\varepsilon_{xy}$ 是 $\sigma_{xy}\varepsilon_{xy}$ 和 $\sigma_{yx}\varepsilon_{yx}$ 两项之和. 如果用工程应变, 则无此 2 的倍数. 将广义胡克定律 (40-1) 和 (40-2) 两式代入 (40-28) 式, 得到

$$w = \frac{1}{2E}\left[\left(\sigma_{xx}^2 + \sigma_{yy}^2 + \sigma_{zz}^2\right) - 2\nu\left(\sigma_{xx}\sigma_{yy} + \sigma_{yy}\sigma_{zz} + \sigma_{zz}\sigma_{xx}\right)\right] + \frac{1}{2G}\left(\sigma_{xy}^2 + \sigma_{yz}^2 + \sigma_{zx}^2\right) \tag{40-29}$$

当应用三个主应力 (principal stress) 时, (40-29) 式简化为

$$w = \frac{1}{2E}\left[\left(\sigma_1^2 + \sigma_2^2 + \sigma_3^2\right) - 2\nu\left(\sigma_1\sigma_2 + \sigma_2\sigma_3 + \sigma_3\sigma_1\right)\right] \tag{40-30}$$

式中, σ_1、σ_2 和 σ_3 分别为第一、第二和第三主应力, 满足 $\sigma_1 \geqslant \sigma_2 \geqslant \sigma_3$.

纯剪切 (pure shear) 时, 剪切应变能密度可简单地表示为

$$w = \frac{\tau\gamma}{2} \tag{40-31}$$

由剪切胡克定律 $\tau = G\gamma$, 上式可写为

$$w = \frac{\tau^2}{2G} \tag{40-32}$$

第 8 章 本构关系

由应力的莫尔圆 (Mohr's circle) 知, 纯剪切的两个主应力 (principal stresses) 分别为

$$\sigma_1 = \tau, \quad \sigma_2 = -\tau \tag{40-33}$$

平面应力状态下的应变能密度为

$$w = \frac{1}{2E}\left(\sigma_1^2 + \sigma_2^2 - 2\nu\sigma_1\sigma_2\right) \tag{40-34}$$

将纯剪切时的主应力 (40-33) 式代入式 (40-34), 得应变能密度为

$$w = \frac{\tau^2(1+\nu)}{E} \tag{40-35}$$

由 (40-32) 和 (40-35) 两式相等, 则证得常用关系式 (40-3) 式: $G = \dfrac{E}{2(1+\nu)}$.

§41. 流体力学本构关系

流体是液体和气体的总称. 本节从全局的角度讨论流体的本构关系.

固体和流体的主要区别是, 固体在静止状态能承受剪力, 这主要是固体的分子位置相对固定的缘故; 而流体在静止状态下不承受剪力, 一旦受到剪力作用就处于流动状态, 发生无休止的变形.

固体的本构关系是应力-应变之间的关系; 而流体的本构关系是应力-应变率 (速度梯度) 之间的关系.

液体一般情况下很难被压缩, 密度基本为常数; 而气体非常容易被压缩, 密度容易变化. 按照我们在生活上的经验, 液体的黏度系数会随着温度的升高而降低, 亦即, 温度越高液体越易流动; 气体的黏度系数会随着温度的升高而升高, 如在蒸桑拿时, 气体的黏度系数会随着温度的升高而愈发地黏稠.

流体可分为两大类:

一、理想流体 (ideal fluids), 亦称无黏流体 (inviscid fluids), 是一种设想的没有黏性的流体, 只能承受法向、压缩应力, 在流动时各层之间没有相互作用的切应力, 即没有内摩擦力. 指无黏性而不可压的流体, 流体的密度为常数, 速度散度为零.

二、实际流体 (real fluids), 亦称黏性流体 (viscous fluids), 即实际流动间均有内摩擦力.

什么是黏性的微观机制? 从物理力学概念上来讲, 黏性是相邻流体之间动量交换过程的体现. 微观上看, 根据流体性质的不同, 导致流体分子间动量交换的方式

也会不同. 固体分子相对其固定位置振动，动量交换主要通过相邻分子间相互作用力；气体分子作无规则热运动，动量交换以非紧邻分子间相互碰撞方式为主；液体分子间的动量交换则介于这两者之间，两种方式都可以有.

41.1 帕斯卡定律和帕斯卡水桶实验

法国数学家布莱士·帕斯卡 (Blaise Pascal, 1623—1662, 见题表 1.5 和思考题 5.4) 于 1647—1648 年间, 研究了受限的不可压缩的静止流体 (confined, incompressible fluids at rest) 所产生的正应强 (normal pressure), 创建了帕斯卡定律 [8.14].

帕斯卡定律 (Pascal's law): 在受限静止液体的内部，在一个给定的点，压强在所有方向上均相等 (In a confined fluid at rest, pressure acts equally in all directions at a given point).

上述的 "在所有方向上均相等" 就是 "各向同性 (isotropy)".

帕斯卡定律可用公式表示为

$$\boldsymbol{\sigma} = -p_0\mathbf{I}, \quad \sigma_{ij} = -p_0\delta_{ij} \tag{41-1}$$

图 41.1 帕斯卡正在进行桶裂实验

式中, p_0 为静水压强 (hydrostatic pressure).

如图 41.1 所示, 帕斯卡在 1646 年表演了一个著名的实验, 简称为 "帕斯卡水桶实验 (Pascal's barrel experiment)": 帕斯卡用一个密闭的装满水的木桶, 在桶盖上插入一根细长的管子, 从楼房的阳台上向细管子里灌水. 结果只用了几杯水, 就把木桶压裂了, 桶里的水就从木条的裂缝中流了出来. 原来由于细管子的底面积较小, 几杯水灌进去, 其深度 h 很大, 所以根据阿基米德定律, 作用在水桶木板上静水压强 $\sigma = \rho g h$ 就很大. 这就是历史上有名的帕斯卡桶裂 (barrel-buster) 实验.

事实上, 近年来在石油和页岩气工业中所广泛采用的 "水力压裂 (hydraulic fracturing)" 其实就可以追溯到帕斯卡的桶裂实验.

41.2 流体本构关系的一般形式

在帕斯卡的水静力学 (hydrostatics) 定律的基础上, 我们可以给出水动力学 (hydrodynamics) 情形时, 流体的热力学本构关系的张量和分量的一般形式分别为

$$\begin{cases} \boldsymbol{\sigma} = -p_0\mathbf{I} + \boldsymbol{\Psi}(\rho, \boldsymbol{S}, \theta) \\ \sigma_{ij} = -p_0\delta_{ij} + \Psi_{ij}(\rho, S_{ij}, \theta) \end{cases} \tag{41-2}$$

第 8 章 本构关系 · 287 ·

式中，ρ 为流体的密度，为一个强度量，见附录 B；θ 为绝对温度，也为一个强度量. 对称的速度梯度 (应变率) 张量 $\boldsymbol{S}$ 和速度矢量 $\boldsymbol{v}$ 之间的关系为

$$\boldsymbol{S} = \frac{\boldsymbol{v} \otimes \boldsymbol{\nabla} + \boldsymbol{\nabla} \otimes \boldsymbol{v}}{2} = \frac{(\boldsymbol{\nabla} \otimes \boldsymbol{v})^{\mathrm{T}} + \boldsymbol{\nabla} \otimes \boldsymbol{v}}{2} \tag{41-3}$$

式中，$\boldsymbol{\nabla} \otimes \boldsymbol{v}$ 称为速度的左梯度，$\boldsymbol{v} \otimes \boldsymbol{\nabla}$ 为速度梯度的右梯度，两者互为转置关系，均为二阶张量.

根据张量函数 $\boldsymbol{\Psi}(\rho, \boldsymbol{S}, \theta)$ 的性质，流体还可以划分为表 41.1 中的三种情形.

表 41.1 流体的三种类型

流体类型	$\boldsymbol{\Psi}(\rho, \boldsymbol{S}, \theta)$ 性质	说明
理想流体 (perfect fluids)	$\boldsymbol{\Psi}(\rho, \boldsymbol{S}, \theta) = \boldsymbol{0}$	(41-2) 式退化为 (41-1) 式
牛顿流体 (Newtonnian fluids)	$\boldsymbol{\Psi}(\rho, \boldsymbol{S}, \theta)$ 是速度梯度 $\boldsymbol{S}$ 的线性函数	见本书 §53.2
斯托克斯流体 (Stokesian fluids)	$\boldsymbol{\Psi}(\rho, \boldsymbol{S}, \theta)$ 是其参量的非线性函数	

41.3 牛顿流体本构关系的一般形式

类似于胡克弹性固体的本构关系 (41-16) 式，牛顿流体本构关系的一般形式为

$$\begin{cases} \boldsymbol{\sigma} = -p\mathbf{I} + \mathbb{C} : \boldsymbol{S} \\ \sigma_{ij} = -p\delta_{ij} + \mathbb{C}_{ijkl} S_{kl} \end{cases} \tag{41-4}$$

式中，$\mathbb{C}$ 为四阶黏性张量，它满足类似于线弹性固体的四阶刚度张量所满足的 (40-20) 式和 (40-21) 式所给出的对称性条件. 类比于线弹性固体的 (40-17) 式，牛顿流体的四阶黏性张量可相应地表示为

$$\begin{cases} \mathbb{C} = 2\mu \mathbb{I}^s + \lambda (\mathbf{I} \otimes \mathbf{I}) \\ \mathbb{C}_{ijkl} = \mu (\delta_{ik}\delta_{jl} + \delta_{il}\delta_{jk}) + \lambda \delta_{ij}\delta_{kl} \end{cases} \tag{41-5}$$

类比于线弹性胡克固体的 (40-15) 式，牛顿流体的本构关系还可写为

$$\begin{cases} \boldsymbol{\sigma} = -p\mathbf{I} + 2\mu \boldsymbol{S} + \lambda (\mathrm{tr} \boldsymbol{S}) \mathbf{I} \\ \sigma_{ij} = -p\delta_{ij} + 2\mu S_{ij} + \lambda S_{kk} \delta_{ij} \end{cases} \tag{41-6}$$

(41-5) 和 (41-6) 两式中，μ 为动力黏性系数，λ 为第二黏性系数，这两个参数不必为常数，它们可能依赖于流体的密度 ρ 和温度 θ.

§42. 不可压缩超弹性材料的新胡克本构模型

新胡克体 (Neo-Hookean solid) 是一种基于统计力学的超弹性材料模型,可用于预测经历大变形的材料的非线性应力-应变行为. 该模型由罗纳德·塞缪尔·里夫林 (Ronald Samuel Rivlin, 1915—2005) 于 1948 年提出. 新胡克材料的应力-应变曲线不再是线性的.

为求得一个超弹性体的本构关系, 我们从其分子结构出发, 将分子构型通过熵增与外力做功联系起来, 实现微观到宏观的计算. 对于一个由长链分子组成的, 不可压缩且各向同性的超弹性一维杆, 长度为 l_0, 分子总数 N, 如图 42.1 所示, 单个分子首尾两端间距 x 满足高斯分布, 即满足概率:

$$p(x)\,\mathrm{d}x = \frac{\beta}{\sqrt{\pi}}\mathrm{e}^{-\beta^2 x^2}\mathrm{d}x \tag{42-1}$$

即 $x \sim N\left(0, \dfrac{1}{2\beta^2}\right)$. 式中 β 是只和分子链本身参数有关的常量: $\dfrac{1}{\beta^2} = \dfrac{2}{3}l_e^2 Z\dfrac{1+\cos\theta}{1-\cos\theta}$. 其中, l_e 为 C—C 键长, Z 是长链分子中节点数, θ 为化学键键角的补角.

单轴拉伸变形之后, 杆伸长率为 $\alpha = l/l_0$, 此时分子两端间距满足概率分布 $x \sim N\left(0, \dfrac{\alpha^2}{2\beta^2}\right)$:

$$p'(x)\,\mathrm{d}x = \frac{\beta}{\alpha\sqrt{\pi}}\mathrm{e}^{-\frac{\beta^2}{\alpha^2}x^2}\mathrm{d}x \tag{42-2}$$

拉伸前后分别称两个状态为 S 态与 S' 态, 如图 42.2 所示.

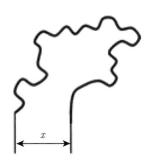

图 42.1 分子首尾间距

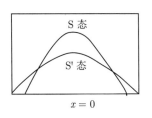

图 42.2 拉伸前后的两个状态

为计算处于两个状态的微观状态数 Ω, 得到其最概然分布 ($\delta\ln\Omega = 0$), 我们取长度为 Δx 的微元, 长链分子处于该微元的概率为 p_i, 设初始时, 处于该微元的分子数目为 $n_{i,}$, 对应的微观状态数 $\Omega_0 = \dfrac{N!}{\prod\limits_i n_i!}\prod\limits_i p_i^{n_i}$, 其中 n_i 满足 $n_i = N \times p_i \times \Delta x_i$, N 为总分子数. 不妨假设另一状态中同样概率下, 处于该微元的分子数目为 s_i, 则此时微观状态数 $\Omega = \dfrac{N!}{\prod\limits_i s_i!}\prod\limits_i p_i^{s_i}$, 其中 $s_i = N \times p_i' \times \Delta x_i$. 根据斯特林近似 (Stirling approximation) $\left(n! = \sqrt{2\pi n}\left(\dfrac{n}{\mathrm{e}}\right)^n\right)$, 二者的比例可表示为

$$\frac{\Omega}{\Omega_0} = \prod_i \left(\frac{n_i}{s_i}\right)^{s_i} \quad \text{或者} \quad \ln\frac{\Omega}{\Omega_0} = \sum_i s_i \ln\frac{n_i}{s_i} \tag{42-3}$$

第 8 章 本构关系

则根据初始态 S 态的分布, 满足关系式 $n_i = N \times p(x) = \frac{N\beta}{\sqrt{\pi}}\mathrm{e}^{-\beta^2 x_i^2}\Delta x_i$, 同理 S′ 满足 $s' = N \times p'(x) = \frac{N\beta}{\alpha\sqrt{\pi}}\mathrm{e}^{-\frac{\beta^2}{\alpha^2}x^2}\Delta x_i$. 则有 $\frac{n_i}{s_i} = \alpha \mathrm{e}^{-\beta^2 x_i^2(\alpha^2-1)/\alpha^2}$, 将之代入 (42-3) 式:

$$\ln\frac{\Omega}{\Omega_0} = \sum_i s_i \left[\ln\alpha - \frac{\beta^2 x_i^2(\alpha^2-1)}{\alpha^2}\right]$$

$$= \sum_i s_i \ln\alpha - \frac{\beta^2(\alpha^2-1)}{\alpha^2}\sum_i s_i x_i^2$$

$$= N\ln\alpha - N\frac{(\alpha^2-1)}{2} \tag{42-4}$$

利用近似 $\sum_i s_i x_i^2 = \overline{x^2}\sum_i s_i = N \cdot \frac{\alpha^2}{2\beta^2}(\overline{x^2} = \sigma^2 + \mu^2)$, 显然, 当 $\alpha = 1$ 时, Ω 达到最大值, 即 Ω_0.

熵由微观状态数通过玻尔兹曼常数 k_B 表示为: $S = k_\mathrm{B}\ln\Omega$. 由 (42-4) 式, 拉伸之后熵增为

$$S - S_0 = k_\mathrm{B}\ln\frac{\Omega}{\Omega_0}$$

$$= \frac{k_\mathrm{B}N}{2}\left[2\ln\alpha - \alpha^2 + 1\right] \tag{42-5}$$

将 $\alpha = l/l_0$ 代入, 得到

$$S - S_0 = \frac{k_\mathrm{B}N}{2}\left[2\ln\frac{l}{l_0} - \left(\frac{l}{l_0}\right)^2 + 1\right] \tag{42-6}$$

在拉伸前后, 自由能变化为零, 由 (42-6) 式以及自由能与熵增以及外力做功的微分关系式: $\mathrm{d}U = T\mathrm{d}S + F\mathrm{d}l = 0$, 可得到外力与拉伸率的关系:

$$F = -T\left(\frac{\partial S}{\partial l}\right)_U = \frac{Nk_\mathrm{B}T}{l_0}\left(\frac{l}{l_0} - \frac{l_0}{l}\right) = \frac{Nk_\mathrm{B}T}{l_0}\left(\alpha - \frac{1}{\alpha}\right) \tag{42-7}$$

进一步考虑三维单轴拉伸问题的本构关系, 依然从微观状态分布着手. 拉伸前, x, y, z 方向的长度均为 l_0, 分子间距分布满足:

$$p(x,y,z)\mathrm{d}x\mathrm{d}y\mathrm{d}z = \frac{\beta^3}{\pi^{3/2}}\mathrm{e}^{-\beta^2(x^2+y^2+z^2)}\mathrm{d}x\mathrm{d}y\mathrm{d}z \tag{42-8}$$

对 x 方向单轴拉伸, 伸长率为 $\alpha = l/l_0$, 则根据体积不变 $\delta V = 0$, y 和 z 方向的伸长率为 $\alpha^{-1/2}$, 且拉伸后, 分子间距分布满足:

$$p'(x,y,z)\mathrm{d}x\mathrm{d}y\mathrm{d}z = \frac{\beta^3}{\pi^{3/2}}\mathrm{e}^{-\beta^2\left(\frac{x^2}{\alpha^2}+\alpha y^2+\alpha z^2\right)}\mathrm{d}x\mathrm{d}y\mathrm{d}z \tag{42-9}$$

由一维的微观状态数, 可得到

$$\begin{cases} \Omega = \dfrac{N!}{\prod_i s_i!} \prod_i p_i^{s_i} \\ \Omega_0 = \dfrac{N!}{\prod_i n_i!} \prod_i p_i^{n_i} \\ n_i = N \times p_i \times \Delta x_i \\ s_i = N \times p_i' \times \Delta x_i \end{cases} \qquad (42\text{-}10)$$

进一步得到熵增为

$$\begin{aligned} S - S_0 &= k_{\text{B}} \ln \frac{\Omega}{\Omega_0} \\ &= k_{\text{B}} \sum_i s_i \left[-\frac{\beta^2 x_i^2 \left(\alpha^2 - 1 \right)}{\alpha^2} - y_i^2 (1-\alpha) - z_i^2 (1-\alpha) \right] \\ &= k_{\text{B}} N \left(-\frac{\alpha^2 - 1}{2} + \frac{1-\alpha}{\alpha} \right) \\ &= -\frac{k_{\text{B}} N}{2} \left(\alpha^2 - \frac{2}{\alpha} - 3 \right) \end{aligned} \qquad (42\text{-}11)$$

其中用到近似为 $\sum_i s_i x_i^2 = \overline{x^2} \sum_i s_i = N \cdot \dfrac{\alpha^2}{2\beta^2}$, $\sum_i s_i y_i^2 = \overline{y^2} \sum_i s_i = N \cdot \dfrac{1}{2\alpha\beta^2}$ 和 $\sum_i s_i z_i^2 = \overline{z^2} \sum_i s_i = N \cdot \dfrac{1}{2\alpha\beta^2}$.

将 $\alpha = l/l_0$ 代入 (42-11) 式, 并根据自由能与熵增的关系, 得到外力大小为

$$F = -T \left(\frac{\partial S}{\partial l} \right)_U = N k_{\text{B}} T \left(\frac{l}{l_0^2} + \frac{l_0}{l^2} \right) = \frac{\rho A R T}{M} \left(\frac{l}{l_0^3} + \frac{1}{l^2} \right) \qquad (42\text{-}12)$$

当 $l_0 = 1$ 时, 得到 $E = \dfrac{F}{A} \cdot \dfrac{1}{l-1} = \dfrac{\rho R T}{M(l-1)} \left(l + \dfrac{1}{l^2} \right)$.

对于一个标准立方体, 即 x, y, z 方向的长度 $l_0 = 1$, 则三个方向的伸长率分别为 λ_1, λ_2, 和 λ_3. 则变形前后, 分子首尾间距分布规律满足:

$$p(x, y, z) \mathrm{d}x \mathrm{d}y \mathrm{d}z = \frac{\beta^3}{\pi^{3/2}} \mathrm{e}^{-\beta^2 \left(x^2 + y^2 + z^2 \right)} \mathrm{d}x \mathrm{d}y \mathrm{d}z \qquad (42\text{-}13)$$

$$p'(x, y, z) \mathrm{d}x \mathrm{d}y \mathrm{d}z = \frac{\beta^3}{\pi^{3/2}} \mathrm{e}^{-\beta^2 \left(\frac{x^2}{\lambda_1^2} + \frac{y^2}{\lambda_2^2} + \frac{z^2}{\lambda_3^2} \right)} \mathrm{d}x \mathrm{d}y \mathrm{d}z \qquad (42\text{-}14)$$

第 8 章 本构关系

由微观状态数, 可得到

$$\begin{cases} \Omega = \dfrac{N!}{\prod\limits_i s_i!} \prod\limits_i p_i^{s_i} \\ \Omega_0 = \dfrac{N!}{\prod\limits_i n_i!} \prod\limits_i p_i^{n_i} \\ n_i = N \times p_i \times \Delta x_i \\ s_i = N \times p_i' \times \Delta x_i \end{cases} \quad (42\text{-}15)$$

则熵增为

$$\begin{aligned} S - S_0 &= k_\text{B} \ln \frac{\Omega}{\Omega_0} \\ &= -k_\text{B} \beta^2 \sum_i s_i \left[x_i^2 \left(1 - \frac{1}{\lambda_1^2}\right) + y_i^2 \left(1 - \frac{1}{\lambda_2^2}\right) + z_i^2 \left(1 - \frac{1}{\lambda_3^2}\right) \right] \\ &= -k_\text{B} \beta^2 N \left[\frac{\lambda_1^2}{2\beta^2}\left(1 - \frac{1}{\lambda_1^2}\right) + \frac{\lambda_2^2}{2\beta^2}\left(1 - \frac{1}{\lambda_2^2}\right) + \frac{\lambda_3^2}{2\beta^2}\left(1 - \frac{1}{\lambda_3^2}\right) \right] \\ &= -\frac{k_\text{B} N}{2} \left(\lambda_1^2 + \lambda_2^2 + \lambda_3^2 - 3 \right) \end{aligned} \quad (42\text{-}16)$$

则外力做功用熵增表示为

$$\begin{aligned} W = F \Delta l &= -T \Delta S = -k_\text{B} T \ln \frac{\Omega}{\Omega_0} \\ &= \frac{N k_\text{B} T}{2} \left(\lambda_1^2 + \lambda_2^2 + \lambda_3^2 - 3 \right) \\ &= \frac{1}{2} \frac{\rho R T}{M} \left(\lambda_1^2 + \lambda_2^2 + \lambda_3^2 - 3 \right) \\ &= \frac{1}{2} \mu \left(\lambda_1^2 + \lambda_2^2 + \lambda_3^2 - 3 \right) \end{aligned} \quad (42\text{-}17)$$

上式即新胡克本构模型 (neo-Hooken constitutive model) 的关系.

下面求应力与应变的关系. 因为 $\lambda_1 \lambda_2 \lambda_3 = 1$, x、y、z 方向的的拉力分别为 f_1、f_2、f_3, 则全微分表示为

$$\begin{aligned} \mathrm{d}W &= \frac{1}{2} \mu \left[\left(2\lambda_1 - \frac{2}{\lambda_1^3 \lambda_2^2} + \right) \mathrm{d}\lambda_1 + \left(2\lambda_2 - \frac{2}{\lambda_1^2 \lambda_2^3} \right) \mathrm{d}\lambda_2 \right] \\ &= \mu \left[\left(\lambda_1^2 - \frac{1}{\lambda_1^2 \lambda_2^2} \right) \frac{\mathrm{d}\lambda_1}{\lambda_1} + \left(\lambda_2^2 - \frac{1}{\lambda_1^2 \lambda_2^2} \right) \frac{\mathrm{d}\lambda_2}{\lambda_2} \right] \end{aligned} \quad (42\text{-}18)$$

微元的应力可表示为

$$S_1 = \frac{f_1}{\lambda_2 \lambda_3} = f_1 \lambda_1, \quad S_2 = \frac{f_2}{\lambda_1 \lambda_3} = f_2 \lambda_2, \quad S_3 = \frac{f_3}{\lambda_1 \lambda_2} = f_3 \lambda_3 \tag{42-19}$$

应变可分别表示为

$$\frac{\mathrm{d}\lambda_1}{\lambda_1}, \frac{\mathrm{d}\lambda_2}{\lambda_2}, \frac{\mathrm{d}\lambda_3}{\lambda_3} \quad \text{且} \quad \frac{\mathrm{d}\lambda_1}{\lambda_1} + \frac{\mathrm{d}\lambda_2}{\lambda_2} + \frac{\mathrm{d}\lambda_3}{\lambda_3} = 0 \tag{42-20}$$

则微功可表示为

$$\begin{aligned} \mathrm{d}W &= S_1 \frac{\mathrm{d}\lambda_1}{\lambda_1} + S_2 \frac{\mathrm{d}\lambda_2}{\lambda_2} + S_3 \frac{\mathrm{d}\lambda_3}{\lambda_3} \\ &= (S_1 - S_3) \frac{\mathrm{d}\lambda_1}{\lambda_1} + (S_2 - S_3) \frac{\mathrm{d}\lambda_2}{\lambda_2} \end{aligned} \tag{42-21}$$

对比 (42-18) 式和 (42-21) 式, 得到

$$S_1 - S_3 = \mu \left(\lambda_1^2 - \frac{1}{\lambda_1^2 \lambda_2^2} \right) = \mu \left(\lambda_1^2 - \lambda_3^2 \right)$$

$$S_2 - S_3 = \mu \left(\lambda_2^2 - \frac{1}{\lambda_1^2 \lambda_2^2} \right) = \mu \left(\lambda_2^2 - \lambda_3^2 \right) \tag{42-22}$$

采用张量法:

由 (42-17) 式, 得到 $W = \frac{1}{2}\mu(\mathrm{I}_1 - 3)$, $J = 1$ 时, 根据静水压 p 的性质, 将应变能函数改写为 $W = W(\mathrm{I}_1, \mathrm{I}_2) - \frac{1}{2}p(\mathrm{I}_3 - 1)$

则 PK2 应力可表示为

$$\boldsymbol{T} = \frac{\partial W}{\partial \boldsymbol{E}} = 2\frac{\partial W}{\partial \boldsymbol{C}} = 2\left(\frac{\partial W}{\partial \mathrm{I}_1}\frac{\partial \mathrm{I}_1}{\partial \boldsymbol{C}} + \frac{\partial W}{\partial \mathrm{I}_2}\frac{\partial \mathrm{I}_2}{\partial \boldsymbol{C}} + \frac{\partial W}{\partial \mathrm{I}_3}\frac{\partial \mathrm{I}_3}{\partial \boldsymbol{C}} \right) \tag{42-23}$$

其中, $\frac{\partial W}{\partial \mathrm{I}_1} = \frac{\mu}{2}$, $\frac{\partial \mathrm{I}_1}{\partial \boldsymbol{C}} = \mathbf{I}$, $\frac{\partial W}{\partial \mathrm{I}_3} = -\frac{1}{2}pJ$, $\frac{\partial \mathrm{I}_3}{\partial \boldsymbol{C}} = \mathrm{I}_3 C^{-1}$, 且 $\mathrm{I}_3 = J^2 = 1$, 则

$$\boldsymbol{T} = \mu \mathbf{I} - p\boldsymbol{C}^{-1} \tag{42-24}$$

则柯西应力可表示为

$$\boldsymbol{\sigma} = \boldsymbol{F}\boldsymbol{T}\boldsymbol{F}^{\mathrm{T}} = \mu \boldsymbol{B} - p\mathbf{I} \tag{42-25}$$

§43. 超弹性材料的本构方程和应力

43.1 用 PK1 表示的不可压缩 ($J = 1$) 和可压缩 ($J \neq 1$) 的超弹性本构关系

设应变能密度 (也就是单位体积的应变能) 函数是两点张量 (也就是混合的欧拉–拉格朗日型张量, mixed Euler–Lagrange tensor)——变形梯度 $\boldsymbol{F}$ 的标量函数

$w(\boldsymbol{F})$，则由于连续介质力学基石的功共轭关系：$\dot{w} = \boldsymbol{P} : \dot{\boldsymbol{F}}$，所以有如下张量和 $\boldsymbol{F}$ 指标形式：

$$\boldsymbol{P} = \frac{\partial w}{\partial \boldsymbol{F}} \quad \Rightarrow \quad P_{iK} = \frac{\partial w}{\partial F_{iK}} \tag{43-1}$$

式中的指标形式 (index form) 形象地反映出，PK1 $\boldsymbol{P}$ 和 $\boldsymbol{F}$ 均为两点张量. (43-1) 式为可压缩的 (compressible) 用 PK1 表示的超弹性本构关系，如果 $w(\boldsymbol{F})$ 有用 $\boldsymbol{F}$ 表示的显函数的话，则可给出 (43-1) 式的具体形式.

如果超弹性材料为不可压缩的 (incompressible)，则由于雅可比 (Jacobian) $J = \det \boldsymbol{F}$ 需要满足条件：

$$J = \det \boldsymbol{F} = 1 \tag{43-2}$$

对于可压缩性超弹时材料，在限制条件 (43-2) 式下，用拉格朗日乘子法，此时的超弹性材料的应变能密度函数可表示为

$$w' = w(\boldsymbol{F}) - p(J - 1) \tag{43-3}$$

式中，p 为拉格朗日乘子，由于 $p(J-1)$ 的量纲是单位体积的能量，$(J-1)$ 是体积的相对变化，则拉格朗日乘子 p 只能是静水压强 (hydrostatic pressure). 再由于如下基本关系式：

$$\frac{\partial J}{\partial \boldsymbol{F}} = J\boldsymbol{F}^{-\mathrm{T}} \tag{43-4}$$

故而，用 PK1 表示的可压缩的超弹性本构关系为

$$\boldsymbol{P} = \frac{\partial w(\boldsymbol{F})}{\partial \boldsymbol{F}} - pJ\boldsymbol{F}^{-\mathrm{T}} \tag{43-5}$$

在不可压缩时，上式右端中的最后一项变为 $-p\boldsymbol{F}^{-\mathrm{T}}$.

43.2 用 PK2 和 PK1 表示的可压缩和不可压缩的超弹性本构关系

设应变能密度函数是满足对称性要求的拉格朗日型张量——格林应变 $\boldsymbol{E}$ 的标量函数 $w(\boldsymbol{E})$，则用 PK2 $\boldsymbol{T}$ 表示的超弹性本构关系可写为

$$\boldsymbol{T} = \frac{\partial w}{\partial \boldsymbol{E}} \tag{43-6}$$

由 PK1 $\boldsymbol{P}$ 和 PK2 $\boldsymbol{T}$ 的关系：$\boldsymbol{T} = \boldsymbol{F}^{-1}\boldsymbol{P}$，则由 (43-1) 和 (43-6) 两式，得到

$$\boldsymbol{T} = \boldsymbol{F}^{-1}\frac{\partial w}{\partial \boldsymbol{F}} \quad \Rightarrow \quad T_{IJ} = F_{Ik}^{-1}\frac{\partial w}{\partial F_{kJ}} \tag{43-7}$$

或者等价地，

$$\boldsymbol{P} = \boldsymbol{F}\frac{\partial w}{\partial \boldsymbol{E}} \quad \Rightarrow \quad P_{iK} = F_{iL}\frac{\partial w}{\partial E_{LK}} \tag{43-8}$$

由格林应变张量的表达式：$E = \frac{1}{2}(C - I) = \frac{1}{2}\left(F^T F - I\right)$，则有

$$T = 2\frac{\partial w}{\partial C} \tag{43-9}$$

对于可压缩超弹性材料，(43-3) 式可改写为

$$w' = w(E) - p(J-1) \quad \text{或} \quad w' = w(C) - p(J-1) \tag{43-10}$$

再注意到右柯西-格林张量 C 的对称性，有 (43-11) 式增加推导细节为

$$\begin{aligned}\frac{\partial J}{\partial C} &= \frac{\partial \sqrt{\det C}}{\partial C} = \frac{1}{2}\frac{1}{\sqrt{\det C}}\frac{\partial(\det C)}{\partial C} = \frac{1}{2}\frac{\det C}{J}C^{-T} \\ &= \frac{1}{2}JC^{-1} = \frac{1}{2}JF^{-1}F^{-T}\end{aligned} \tag{43-11}$$

则用 PK2 表示的不可压缩超弹性材料的本构关系为

$$\begin{aligned}T &= \frac{\partial w}{\partial E} = 2\frac{\partial w}{\partial C} = 2\frac{\partial w}{\partial C} - 2p\frac{\partial(J-1)}{\partial C} \\ &= 2\frac{\partial w}{\partial C} - pJC^{-1} = 2\frac{\partial w}{\partial C} - pJF^{-1}F^{-T}\end{aligned} \tag{43-12}$$

(43-12) 式亦可用 PK1 表示为

$$P = FT = 2F\frac{\partial w}{\partial C} - pJFF^{-1}F^{-T} = 2F\frac{\partial w}{\partial C} - pJF^{-T} \tag{43-13}$$

43.3 用柯西应力表示的可压缩和不可压缩的超弹性本构关系

由 PK1 和柯西应力的基本关系式：

$$\sigma = \frac{1}{J}PF^T \quad \Rightarrow \quad \sigma_{ij} = \frac{1}{J}P_{iK}F^T_{Kj} \tag{43-14}$$

由 (43-1) 和 (43-14) 两式可得用柯西应力表示的可压缩超弹性材料的本构关系：

$$\sigma = \frac{1}{J}\frac{\partial w}{\partial F}F^T \quad \Rightarrow \quad \sigma_{ij} = \frac{1}{J}\frac{\partial w}{\partial F_{iK}}F^T_{Kj} \tag{43-15}$$

由 (43-13) 和 (43-14) 两式，可得用柯西应力表示的不可压缩的超弹性本构关系 ($J = 1$)：

$$\sigma = PF^T = 2F\frac{\partial w}{\partial C}F^T - pI \tag{43-16}$$

以及基尔霍夫应力的表达式：

$$\tau = J\sigma = \frac{\partial w}{\partial F}F^T = 2\frac{\partial w}{\partial B}B \tag{43-17}$$

第 8 章 本构关系

例 43.1 证明 (43-17) 式.

证明方法一： 由 (18-13) 式知，左柯西-格林变形张量 (芬格变形张量) $\boldsymbol{B}$ 的三个不变量为

$$\begin{cases} \mathrm{I}_1 = \mathrm{tr}\boldsymbol{B} \\ \mathrm{I}_2 = \dfrac{1}{2}\left[\mathrm{tr}^2(\boldsymbol{B}) - \mathrm{tr}(\boldsymbol{B}^2)\right] = \mathrm{tr}(\mathbf{Cof}\boldsymbol{B}) = \det\boldsymbol{B}\,\mathrm{tr}(\boldsymbol{B}^{-1}) \\ \mathrm{I}_3 = \det\boldsymbol{B} = \dfrac{1}{6}\left[\mathrm{tr}^3(\boldsymbol{B}) - 3\mathrm{tr}(\boldsymbol{B})\mathrm{tr}(\boldsymbol{B}^2) + 2\mathrm{tr}(\boldsymbol{B}^3)\right] \end{cases} \tag{43-18}$$

对左柯西-格林变形张量 $\boldsymbol{B}$ 的偏导为

$$\begin{cases} \dfrac{\partial \mathrm{I}_1}{\partial \boldsymbol{B}} = \dfrac{\partial \mathrm{tr}\boldsymbol{B}}{\partial \boldsymbol{B}} = \mathbf{I} \\ \dfrac{\partial \mathrm{I}_2}{\partial \boldsymbol{B}} = \dfrac{1}{2}\left(2\mathrm{I}_1\mathbf{I} - 2\boldsymbol{B}^{\mathrm{T}}\right) = \mathrm{I}_1\mathbf{I} - \boldsymbol{B} \\ \dfrac{\partial \mathrm{I}_3}{\partial \boldsymbol{B}} = \mathrm{I}_3\boldsymbol{B}^{-\mathrm{T}} \end{cases} \tag{43-19}$$

在 (43-19) 式的运算中，用到了以下的微分关系：

对第一不变量：

$$\begin{aligned} \dfrac{\partial \mathrm{tr}\boldsymbol{B}}{\partial \boldsymbol{B}} : \boldsymbol{A} &= \lim_{\varepsilon \to 0} \dfrac{\mathbf{I}:(\boldsymbol{B}+\varepsilon\boldsymbol{A}) - \mathbf{I}:\boldsymbol{B}}{\varepsilon} \\ &= \lim_{\varepsilon \to 0} \dfrac{\mathbf{I}:\varepsilon\boldsymbol{A} + \mathbf{I}:\boldsymbol{B} - \mathbf{I}:\boldsymbol{B}}{\varepsilon} = \lim_{\varepsilon \to 0} \dfrac{\mathbf{I}:\varepsilon\boldsymbol{A}}{\varepsilon} = \mathbf{I}:\boldsymbol{A} \end{aligned}$$

即

$$\dfrac{\partial \mathrm{tr}(\boldsymbol{B})}{\partial \boldsymbol{B}} = \mathbf{I} \tag{43-20}$$

对第二不变量 (43-18) 式中的右端第一项，有

$$\dfrac{\partial \mathrm{tr}^2(\boldsymbol{B})}{\partial \boldsymbol{B}} = 2\mathrm{tr}(\boldsymbol{B})\mathbf{I} = 2\mathrm{I}_1\mathbf{I} \tag{43-21}$$

对第二不变量 (43-18) 式中的右端第二项，有

$$\begin{aligned} \dfrac{\partial \mathrm{tr}\boldsymbol{B}^2}{\partial \boldsymbol{B}} : \boldsymbol{A} &= \lim_{\varepsilon \to 0} \dfrac{\mathbf{I}:(\boldsymbol{B}+\varepsilon\boldsymbol{A})^2 - \mathbf{I}:\boldsymbol{B}^2}{\varepsilon} = \lim_{\varepsilon \to 0} \dfrac{\mathbf{I}:(\boldsymbol{B}+\varepsilon\boldsymbol{A})(\boldsymbol{B}+\varepsilon\boldsymbol{A}) - \mathbf{I}:\boldsymbol{B}^2}{\varepsilon} \\ &= \lim_{\varepsilon \to 0} \dfrac{\mathbf{I}:\boldsymbol{B}^2 + \mathbf{I}:\varepsilon(\boldsymbol{B}\boldsymbol{A}+\boldsymbol{A}\boldsymbol{B}) + \mathbf{I}:(\varepsilon\boldsymbol{A})^2 - \mathbf{I}:\boldsymbol{B}^2}{\varepsilon} \\ &= \lim_{\varepsilon \to 0} \dfrac{\mathbf{I}:\varepsilon(\boldsymbol{B}\boldsymbol{A}+\boldsymbol{A}\boldsymbol{B}) + \mathbf{I}:(\varepsilon\boldsymbol{A})^2}{\varepsilon} = \mathbf{I}:(\boldsymbol{B}\boldsymbol{A}+\boldsymbol{A}\boldsymbol{B}) \\ &= \boldsymbol{B}^{\mathrm{T}}\mathbf{I}:\boldsymbol{A} + \mathbf{I}\boldsymbol{B}^{\mathrm{T}}:\boldsymbol{A} = 2\boldsymbol{B}^{\mathrm{T}}:\boldsymbol{A} \end{aligned}$$

即
$$\frac{\partial \mathrm{tr}\boldsymbol{B}^2}{\partial \boldsymbol{B}} = 2\boldsymbol{B}^{\mathrm{T}} = 2\boldsymbol{B} \tag{43-22}$$

对第三不变量的右端第一项，有

$$\frac{\partial \mathrm{tr}\boldsymbol{B}^3}{\partial \boldsymbol{B}} : \boldsymbol{A}$$

$$= \lim_{\varepsilon \to 0} \frac{\mathbf{I} : (\boldsymbol{B} + \varepsilon \boldsymbol{A})^3 - \mathbf{I} : \boldsymbol{B}^3}{\varepsilon} = \lim_{\varepsilon \to 0} \frac{\mathbf{I} : (\boldsymbol{B} + \varepsilon \boldsymbol{A})(\boldsymbol{B} + \varepsilon \boldsymbol{A})(\boldsymbol{B} + \varepsilon \boldsymbol{A}) - \mathbf{I} : \boldsymbol{B}^3}{\varepsilon}$$

$$= \lim_{\varepsilon \to 0} \frac{\mathbf{I} : \left[\boldsymbol{B}^3 + \varepsilon\left(\boldsymbol{B}^2\boldsymbol{A} + \boldsymbol{A}\boldsymbol{B}^2 + \boldsymbol{B}\boldsymbol{A}\boldsymbol{B}\right) + \varepsilon^2\left(\boldsymbol{A}\boldsymbol{B}\boldsymbol{A} + \boldsymbol{B}\boldsymbol{A}^2 + \boldsymbol{A}^2\boldsymbol{B}\right) + \varepsilon^3\boldsymbol{A}^3\right] - \mathbf{I} : \boldsymbol{B}^3}{\varepsilon}$$

$$= \mathbf{I} : \left(\boldsymbol{B}^2\boldsymbol{A} + \boldsymbol{A}\boldsymbol{B}^2 + \boldsymbol{B}\boldsymbol{A}\boldsymbol{B}\right) = \left(\boldsymbol{B}^{\mathrm{T}}\right)^2 \mathbf{I} : \boldsymbol{A} + \mathbf{I}\left(\boldsymbol{B}^{\mathrm{T}}\right)^2 : \boldsymbol{A} + \boldsymbol{B}^{\mathrm{T}}\mathbf{I}\boldsymbol{B}^{\mathrm{T}} : \boldsymbol{A}$$

$$= 3\left(\boldsymbol{B}^{\mathrm{T}}\right)^2 : \boldsymbol{A}$$

即
$$\frac{\partial \mathrm{tr}\boldsymbol{B}^3}{\partial \boldsymbol{B}} = 3\left(\boldsymbol{B}^{\mathrm{T}}\right)^2 \tag{43-23}$$

则第三不变量对 $\boldsymbol{B}$ 的偏导为

$$\frac{\partial \mathrm{I}_3}{\partial \boldsymbol{B}} = \frac{1}{6}\frac{\partial}{\partial \boldsymbol{B}}\left[\mathrm{tr}^3(\boldsymbol{B}) - 3\mathrm{tr}(\boldsymbol{B})\mathrm{tr}(\boldsymbol{B}^2) + 2\mathrm{tr}(\boldsymbol{B}^3)\right]$$

$$= \frac{1}{6}\left[3\mathrm{tr}^2(\boldsymbol{B})\frac{\partial \mathrm{tr}(\boldsymbol{B})}{\partial \boldsymbol{B}} - 3\frac{\partial \mathrm{tr}(\boldsymbol{B})}{\partial \boldsymbol{B}}\mathrm{tr}(\boldsymbol{B}^2) - 3\mathrm{tr}(\boldsymbol{B})\frac{\partial \mathrm{tr}(\boldsymbol{B}^2)}{\partial \boldsymbol{B}} + 2\frac{\partial \mathrm{tr}(\boldsymbol{B}^3)}{\partial \boldsymbol{B}}\right]$$

$$= \frac{1}{6}\left[3\mathrm{I}_1^2\mathbf{I} - 3\mathrm{tr}(\boldsymbol{B}^2)\mathbf{I} - 6\mathrm{I}_1\boldsymbol{B}^{\mathrm{T}} + 6\left(\boldsymbol{B}^{\mathrm{T}}\right)^2\right]$$

$$= \frac{1}{2}\mathrm{I}_1^2\mathbf{I} - \frac{1}{2}\mathrm{tr}(\boldsymbol{B}^2)\mathbf{I} - \mathrm{I}_1\boldsymbol{B}^{\mathrm{T}} + \left(\boldsymbol{B}^{\mathrm{T}}\right)^2$$

$$= \mathrm{I}_2\mathbf{I} - \mathrm{I}_1\boldsymbol{B}^{\mathrm{T}} + \left(\boldsymbol{B}^{\mathrm{T}}\right)^2 \tag{43-24}$$

根据凯莱-哈密顿定理 (Cayley-Hamilton theorem)：

$$\left(\boldsymbol{B}^{\mathrm{T}}\right)^3 - \mathrm{I}_1\left(\boldsymbol{B}^{\mathrm{T}}\right)^2 + \mathrm{I}_2\boldsymbol{B}^{\mathrm{T}} - \mathrm{I}_3\mathbf{I} = \boldsymbol{0} \tag{43-25}$$

用 $\boldsymbol{B}^{-\mathrm{T}}$ 乘以上式，得到

$$\left(\boldsymbol{B}^{-\mathrm{T}}\right)^3\left(\boldsymbol{B}^{\mathrm{T}}\right) - \mathrm{I}_1\left(\boldsymbol{B}^{\mathrm{T}}\right)^2\boldsymbol{B}^{-\mathrm{T}} + \mathrm{I}_2\boldsymbol{B}^{\mathrm{T}}\boldsymbol{B}^{-\mathrm{T}} = \mathrm{I}_3\mathbf{I}\boldsymbol{B}^{-\mathrm{T}}$$

亦即

$$\left(\boldsymbol{B}^{\mathrm{T}}\right)^2 - \mathrm{I}_1\boldsymbol{B}^{\mathrm{T}} + \mathrm{I}_2\mathbf{I} = \mathrm{I}_3\boldsymbol{B}^{-\mathrm{T}} \tag{43-26}$$

将 (43-26) 式代入 (43-24) 式的结果中, 即可得到

$$\frac{\partial I_3}{\partial \boldsymbol{B}} = I_2\mathbf{I} - I_1\boldsymbol{B}^{\mathrm{T}} + \left(\boldsymbol{B}^{\mathrm{T}}\right)^2 = I_3\boldsymbol{B}^{-\mathrm{T}} \tag{43-27}$$

第三不变量与相对体积变化 J 满足 $I_3 = \det\boldsymbol{B} = \det\left(\boldsymbol{F}\boldsymbol{F}^{\mathrm{T}}\right) = J^2$, 则 J 对左柯西–格林变形张量的偏导:

$$\frac{\partial J}{\partial \boldsymbol{B}} = \frac{\partial \det\boldsymbol{F}}{\partial \boldsymbol{B}} = \frac{\partial\sqrt{\det\boldsymbol{B}}}{\partial \boldsymbol{B}} = \frac{1}{2}J^{-1}\frac{\partial I_3}{\partial \boldsymbol{B}} = \frac{1}{2}J\boldsymbol{B}^{-\mathrm{T}} = \frac{1}{2}J\boldsymbol{B}^{-1} \tag{43-28}$$

由莱布尼兹链式法则, 应变能密度对左柯西–格林变形张量 $\boldsymbol{B}$ 的偏导数为

$$\begin{aligned}\frac{\partial w}{\partial \boldsymbol{B}} &= \frac{\partial w}{\partial I_1}\frac{\partial I_1}{\partial \boldsymbol{B}} + \frac{\partial w}{\partial I_2}\frac{\partial I_2}{\partial \boldsymbol{B}} + \frac{\partial w}{\partial I_3}\frac{\partial I_3}{\partial \boldsymbol{B}}\\ &= \frac{\partial w}{\partial I_1}\mathbf{I} + \frac{\partial w}{\partial I_2}\left(I_1\mathbf{I} - \boldsymbol{B}\right) + \frac{\partial w}{\partial I_3}I_3\boldsymbol{B}^{-1}\end{aligned} \tag{43-29}$$

右柯西–格林应变张量 $\boldsymbol{C} = \boldsymbol{F}^{\mathrm{T}}\boldsymbol{F}$ 与左柯西–格林应变张量 $\boldsymbol{B} = \boldsymbol{F}\boldsymbol{F}^{\mathrm{T}}$ 满足:

$$I_1\left(\boldsymbol{B}\right) = I_1\left(\boldsymbol{C}\right), \quad I_2\left(\boldsymbol{B}\right) = I_2\left(\boldsymbol{C}\right), \quad I_3\left(\boldsymbol{B}\right) = I_3\left(\boldsymbol{C}\right) \tag{43-30}$$

因此, 应变能密度对于右柯西–格林变形张量 $\boldsymbol{C}$ 的偏导数可相应地表示为

$$\begin{aligned}\frac{\partial w}{\partial \boldsymbol{C}} &= \frac{\partial w}{\partial I_1}\frac{\partial I_1}{\partial \boldsymbol{C}} + \frac{\partial w}{\partial I_2}\frac{\partial I_2}{\partial \boldsymbol{C}} + \frac{\partial w}{\partial I_3}\frac{\partial I_3}{\partial \boldsymbol{C}}\\ &= \frac{\partial w}{\partial I_1}\mathbf{I} + \frac{\partial w}{\partial I_2}\left(I_1\mathbf{I} - \boldsymbol{C}\right) + \frac{\partial w}{\partial I_3}I_3\boldsymbol{C}^{-1}\end{aligned} \tag{43-31}$$

对 (43-29) 式进行对左柯西–格林变形张量 $\boldsymbol{B}$ 的点积操作, 有

$$\frac{\partial w}{\partial \boldsymbol{B}}\boldsymbol{B} = \boldsymbol{B}\frac{\partial w}{\partial \boldsymbol{B}} = \left(\frac{\partial w}{\partial I_1} + \frac{\partial w}{\partial I_2}I_1\right)\boldsymbol{B} - \frac{\partial w}{\partial I_2}\boldsymbol{B}^2 + \frac{\partial w}{\partial I_3}I_3\mathbf{I} \tag{43-32}$$

由 (43-31) 式, 有如下关系式:

$$\begin{aligned}\boldsymbol{F}\frac{\partial w}{\partial \boldsymbol{C}}\boldsymbol{F}^{\mathrm{T}} &= \left(\frac{\partial w}{\partial I_1} + \frac{\partial w}{\partial I_2}I_1\right)\boldsymbol{F}\boldsymbol{F}^{\mathrm{T}} - \frac{\partial w}{\partial I_2}\boldsymbol{F}\boldsymbol{C}\boldsymbol{F}^{\mathrm{T}} + \frac{\partial w}{\partial I_3}I_3\boldsymbol{F}\boldsymbol{C}^{-1}\boldsymbol{F}^{\mathrm{T}}\\ &= \left(\frac{\partial w}{\partial I_1} + \frac{\partial w}{\partial I_2}I_1\right)\boldsymbol{B} - \frac{\partial w}{\partial I_2}\boldsymbol{F}\boldsymbol{F}^{\mathrm{T}}\boldsymbol{F}\boldsymbol{F}^{\mathrm{T}} + \frac{\partial w}{\partial I_3}I_3\boldsymbol{F}\left(\boldsymbol{F}^{\mathrm{T}}\boldsymbol{F}\right)^{-1}\boldsymbol{F}^{\mathrm{T}}\\ &= \left(\frac{\partial w}{\partial I_1} + \frac{\partial w}{\partial I_2}I_1\right)\boldsymbol{B} - \frac{\partial w}{\partial I_2}\boldsymbol{B}^2 + \frac{\partial w}{\partial I_3}I_3\mathbf{I}\end{aligned} \tag{43-33}$$

通过比较 (43-32) 和 (43-33) 两式, 可知:

$$\frac{\partial w}{\partial \boldsymbol{B}}\boldsymbol{B} = \boldsymbol{B}\frac{\partial w}{\partial \boldsymbol{B}} = \boldsymbol{F}\frac{\partial w}{\partial \boldsymbol{C}}\boldsymbol{F}^{\mathrm{T}} \tag{43-34}$$

由于 PK2 可通过应变能密度表示为

$$T = 2\frac{\partial w}{\partial C} = F^{-1}\tau F^{-T} = F^{-1}P = F^{-1}\frac{\partial w}{\partial F} \tag{43-35}$$

则基尔霍夫应力可表示为

$$\tau = FTF^{T} = \frac{\partial w}{\partial F}F^{T} = 2\frac{\partial w}{\partial B}B \tag{43-36}$$

证明方法二：根据定义，应变能密度的变化率可以用基尔霍夫应力与其共轭的应变率表示为

$$\dot{w} = \tau : d$$

其中 $d = \frac{1}{2}\left(l + l^{T}\right)$ 为应变率；l 为速度矢量的右梯度. 又根据链式法则，应变能密度的变化率可以用左柯西–格林应变张量表示为

$$\dot{w} = \frac{\partial w}{\partial B} : \dot{B} \tag{43-37}$$

由 (29-16) 式：$l = \dot{F}F^{-1}$，则有

$$\dot{F} = lF \quad \text{与} \quad \dot{F}^{T} = F^{T}l^{T} \tag{43-38}$$

则根据 $\dot{B}$ 可以用 l 与 B 表示为

$$\dot{B} = \overline{FF^{T}} = \dot{F}F^{T} + F\dot{F}^{T} = lFF^{T} + FF^{T}l^{T} = lB + Bl^{T} \tag{43-39}$$

将 (43-39) 式代入 (43-37) 式，根据对称性，有

$$\frac{\partial w}{\partial B} : lB = \frac{\partial w}{\partial B} : \left(\frac{lB + Bl^{T}}{2} + \frac{lB - Bl^{T}}{2}\right)$$

$$= \frac{\partial w}{\partial B} : \frac{lB + Bl^{T}}{2} + \frac{\partial w}{\partial B} : \left(\frac{lB - Bl^{T}}{2}\right)$$

$$= \frac{\partial w}{\partial B} : \frac{lB + Bl^{T}}{2}$$

得到

$$\dot{w} = \frac{\partial w}{\partial B} : \dot{B} = \frac{\partial w}{\partial B} : \left(lB + Bl^{T}\right) = 2\frac{\partial w}{\partial B} : \left(\frac{lB + Bl^{T}}{2}\right) = 2\frac{\partial w}{\partial B} : lB \tag{43-40}$$

根据对称性以及恒等式：$A : (BC) = B^{T}A : C = AC^{T} : B$，有

$$\dot{w} = 2\frac{\partial w}{\partial B} : lB = 2\frac{\partial w}{\partial B}B : l = 2\frac{\partial w}{\partial B}B : d \tag{43-41}$$

对比 (43-36) 和 (43-41) 两式，即可得证：$\tau = 2\frac{\partial w}{\partial B}B$.

§44. 可压缩超弹性体材料的穆尼–里夫林本构模型

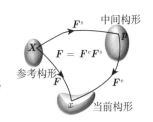

图 44.1 参考构形、中间构形和当前构形

如图 44.1 所示，考虑介质的可压缩性时，除了参考构形和当前构形外，还必须要引入中间构形.

变形梯度张量 $\boldsymbol{F}$ 可通过链式法则，实现下列乘法分解 [8.15]：

$$\boldsymbol{F} = \frac{\partial \boldsymbol{x}}{\partial \boldsymbol{X}} = \frac{\partial x_i}{\partial X_K}\boldsymbol{e}_i \otimes \boldsymbol{e}_K = \frac{\partial \boldsymbol{x}}{\partial \boldsymbol{p}}\frac{\partial \boldsymbol{p}}{\partial \boldsymbol{X}} = \boldsymbol{F}^{\mathrm{e}}\boldsymbol{F}^{\mathrm{s}} \tag{44-1}$$

式中，$\boldsymbol{p}$ 为中间构形，$\boldsymbol{F}^{\mathrm{e}}$ 为保体积弹性部分变形梯度，$\boldsymbol{F}^{\mathrm{s}}$ 为体胀部分变形梯度.

保体积部分的相对体积变化为 $J_{\mathrm{e}} = \det\boldsymbol{F}^{\mathrm{e}} = 1$，则整体的相对体积变化：

$$J = \det\boldsymbol{F} = \det\left(\boldsymbol{F}^{\mathrm{e}}\boldsymbol{F}^{\mathrm{s}}\right) = \det\left(\boldsymbol{F}^{\mathrm{e}}J_s^{1/3}\mathbf{I}\right) = \left(J_s^{1/3}\right)^3 \det\boldsymbol{F}^{\mathrm{e}} = J_s \tag{44-2}$$

体胀部分的变形梯度为

$$\boldsymbol{F}^{\mathrm{s}} = J_s^{1/3}\mathbf{I} = J^{1/3}\mathbf{I} \tag{44-3}$$

值得注意的是，上式只针对三维问题. 对于 n 维问题，任意二阶张量与一个标量的乘积满足 $\det(a\boldsymbol{F}) = a^n \det\boldsymbol{F}$，即 $\boldsymbol{F}^{\mathrm{s}} = J_s^{1/n}\mathbf{I} = J^{1/n}\mathbf{I}$. 则 $\boldsymbol{F}^{\mathrm{e}} = \bar{\boldsymbol{F}}$，有 $\boldsymbol{F} = \boldsymbol{F}^{\mathrm{e}}\boldsymbol{F}^{\mathrm{s}} = J^{1/3}\bar{\boldsymbol{F}}$.

左柯西–格林变形张量 $\boldsymbol{B} = \boldsymbol{F}\boldsymbol{F}^{\mathrm{T}} = J^{2/3}\bar{\boldsymbol{B}}$，其中 $\bar{\boldsymbol{B}} = \bar{\boldsymbol{F}}\bar{\boldsymbol{F}}^{\mathrm{T}}$，仅考虑保体积部分，左柯西–格林变形变形张量 $\bar{\boldsymbol{B}}$ 的三个不变量与 $\boldsymbol{B}$ 的三个不变量有如下关系：

$$\begin{cases} \bar{\mathrm{I}}_1 = \mathrm{tr}\bar{\boldsymbol{B}} = J^{-2/3}\mathrm{I}_1 \\ \bar{\mathrm{I}}_2 = \dfrac{1}{2}\left[\mathrm{tr}^2\left(\bar{\boldsymbol{B}}\right) - \mathrm{tr}\left(\bar{\boldsymbol{B}}^2\right)\right] = J^{-4/3}\mathrm{I}_2 \\ \bar{\mathrm{I}}_3 = \det\bar{\boldsymbol{B}} = J^{-6/3}\mathrm{I}_3 = 1 \end{cases} \tag{44-4}$$

考虑保体积部分的应变能密度，采用六参数的奥格登超弹性模型 (Ogden hyperelasic model)：

$$\bar{w} = \sum_{i=1}^{N}\frac{c_i}{\alpha_i}\left(\bar{\lambda}_1^{\alpha_i} + \bar{\lambda}_2^{\alpha_i} + \bar{\lambda}_3^{\alpha_i} - 3\right) \tag{44-5}$$

式中，$\bar{w}$ 和 $\bar{\lambda}_i$ 分别为应变能密度和主伸长比. 当参数取 $N = 2$，$\alpha_1 = 2$，$\alpha_2 = -2$ 时，(44-5) 式的奥格登超弹性模型退化为穆尼–里夫林模型 (Mooney-Rivlin model)：

$$\bar{w} = \frac{1}{2}c_1\left(\bar{\lambda}_1^2 + \bar{\lambda}_2^2 + \bar{\lambda}_3^2 - 3\right) + \frac{1}{2}c_2\left(\bar{\lambda}_1^{-2} + \bar{\lambda}_2^{-2} + \bar{\lambda}_3^{-2} - 3\right) \tag{44-6}$$

即
$$\bar{w} = \frac{1}{2}c_1\left(\bar{I}_1 - 3\right) + \frac{1}{2}c_2\left(\bar{I}_2 - 3\right) \tag{44-7}$$

式中，系数 c_1、c_2 和剪切模量 μ 间满足 $c_1 + c_2 = \frac{1}{2}\mu$.

考虑可压缩性时的穆尼-里夫林本构模型为
$$w = \frac{1}{2}c_1\left(\bar{I}_1 - 3\right) + \frac{1}{2}c_2\left(\bar{I}_2 - 3\right) + p\left(J - J^s\right) \tag{44-8}$$

式中，p 为静水压. 将 (44-4) 式代入上式，亦可将之表示为
$$w = \frac{1}{2}c_1\left(J^{-2/3}I_1 - 3\right) + \frac{1}{2}c_2\left(J^{-4/3}I_2 - 3\right) + p\left(J - J^s\right) \tag{44-9}$$

则基尔霍夫应力为
$$\boldsymbol{\tau} = \frac{\partial w}{\partial \boldsymbol{F}} \cdot \boldsymbol{F}^{\mathrm{T}} = 2\frac{\partial w}{\partial \boldsymbol{B}} \cdot \boldsymbol{B} = 2\left(\frac{\partial w}{\partial I_1}\frac{\partial I_1}{\partial \boldsymbol{B}} + \frac{\partial w}{\partial I_2}\frac{\partial I_2}{\partial \boldsymbol{B}} + \frac{\partial w}{\partial J}\frac{\partial J}{\partial \boldsymbol{B}}\right) \cdot \boldsymbol{B} \tag{44-10}$$

其中，
$$\frac{\partial I_1}{\partial \boldsymbol{B}} = \boldsymbol{I}, \qquad \frac{\partial I_2}{\partial \boldsymbol{B}} = I_1 \boldsymbol{I} - \boldsymbol{B}, \qquad \frac{\partial J}{\partial \boldsymbol{B}} = \frac{1}{2}J\boldsymbol{B}^{-1}$$

将之代入 (44-10) 式，则得到
$$\begin{aligned}\boldsymbol{\tau} &= 2\left(\frac{\partial w}{\partial I_1}\frac{\partial I_1}{\partial \boldsymbol{B}} + \frac{\partial w}{\partial I_2}\frac{\partial I_2}{\partial \boldsymbol{B}} + \frac{\partial w}{\partial J}\frac{\partial J}{\partial \boldsymbol{B}}\right) \cdot \boldsymbol{B} \\ &= 2\left[\frac{\partial w}{\partial I_1}\boldsymbol{I} + \frac{\partial w}{\partial I_2}\left(I_1\boldsymbol{I} - \boldsymbol{B}\right) + \frac{\partial w}{\partial J}\frac{1}{2}J\boldsymbol{B}^{-1}\right] \cdot \boldsymbol{B} \\ &= 2\left(\frac{\partial w}{\partial I_1} + I_1\frac{\partial w}{\partial I_2}\right)\boldsymbol{B} - 2\frac{\partial w}{\partial I_2}\boldsymbol{B}^2 + J\frac{\partial w}{\partial J}\boldsymbol{I}\end{aligned} \tag{44-11}$$

又由 (44-9) 式，得到
$$\begin{cases}\dfrac{\partial w}{\partial I_1} = \dfrac{1}{2}c_1 J^{-2/3} \\ \dfrac{\partial w}{\partial I_2} = \dfrac{1}{2}c_2 J^{-4/3} \\ \dfrac{\partial w}{\partial J} = -\dfrac{1}{3}c_1 J^{-5/3}I_1 - \dfrac{2}{3}c_2 J^{-7/3}I_2 + p\end{cases} \tag{44-12}$$

将上式代入 (44-11) 式，得到基尔霍夫应力与左柯西-格林应变张量的关系：
$$\begin{aligned}\boldsymbol{\tau} &= 2\left(\frac{\partial w}{\partial I_1} + I_1\frac{\partial w}{\partial I_2}\right)\boldsymbol{B} - 2\frac{\partial w}{\partial I_2}\boldsymbol{B}^2 + J\frac{\partial w}{\partial J}\boldsymbol{I} \\ &= \left(c_1 J^{-2/3} + c_2 I_1 J^{-4/3}\right)\boldsymbol{B} - c_2 J^{-4/3}\boldsymbol{B}^2 - \left(\frac{1}{3}c_1 J^{-5/3}I_1 + \frac{2}{3}c_2 J^{-7/3}I_2 - p\right)J\boldsymbol{I} \\ &= \left(c_1 + c_2\bar{I}_1\right)\bar{\boldsymbol{B}} - c_2\bar{\boldsymbol{B}}^2 - \left(\frac{1}{3}c_1\bar{I}_1 + \frac{2}{3}c_2\bar{I}_2 - p\right)J\boldsymbol{I}\end{aligned} \tag{44-13}$$

思考题和补充材料

8.1 给出式 (43-23) 式的另外一种证明方法.

解: 由于 $\mathrm{tr}\boldsymbol{B}^3 = \boldsymbol{I}:\boldsymbol{B}^3 = \boldsymbol{I}:(\boldsymbol{B}\boldsymbol{B}^2) = (\boldsymbol{B}^\mathrm{T}\boldsymbol{I}):\boldsymbol{B}^2 = \boldsymbol{B}^\mathrm{T}:\boldsymbol{B}^2$, 则有

$$\frac{\partial \mathrm{tr}\boldsymbol{B}^3}{\partial \boldsymbol{B}}:\boldsymbol{A} = \lim_{\varepsilon\to 0}\frac{(\boldsymbol{B}+\varepsilon\boldsymbol{A})^\mathrm{T}:(\boldsymbol{B}+\varepsilon\boldsymbol{A})^2 - \boldsymbol{B}^\mathrm{T}:\boldsymbol{B}^2}{\varepsilon}$$

$$= \lim_{\varepsilon\to 0}\frac{(\boldsymbol{B}^\mathrm{T}+\varepsilon\boldsymbol{A}^\mathrm{T}):\left[\boldsymbol{B}^2+\varepsilon(\boldsymbol{B}\boldsymbol{A}+\boldsymbol{A}\boldsymbol{B})+\varepsilon^2\boldsymbol{A}^2\right] - \boldsymbol{B}^\mathrm{T}:\boldsymbol{B}^2}{\varepsilon}$$

$$= \boldsymbol{B}^\mathrm{T}:(\boldsymbol{B}\boldsymbol{A}+\boldsymbol{A}\boldsymbol{B}) + \boldsymbol{A}^\mathrm{T}:\boldsymbol{B}^2$$

应用 (11-12) 式, 有

$$\boldsymbol{B}^\mathrm{T}:(\boldsymbol{B}\boldsymbol{A}+\boldsymbol{A}\boldsymbol{B}) = 2\left(\boldsymbol{B}^2\right)^\mathrm{T}:\boldsymbol{A}$$

和

$$\boldsymbol{A}^\mathrm{T}:\boldsymbol{B}^2 = \left(\boldsymbol{A}^\mathrm{T}\boldsymbol{I}\right):\boldsymbol{B}^2 = \boldsymbol{I}:(\boldsymbol{A}\boldsymbol{B}^2) = \left(\boldsymbol{B}^2\right)^\mathrm{T}:\boldsymbol{A}$$

有

$$\frac{\partial \mathrm{tr}\boldsymbol{B}^3}{\partial \boldsymbol{B}}:\boldsymbol{A} = 2\boldsymbol{B}^\mathrm{T}:(\boldsymbol{B}\boldsymbol{A}) + \boldsymbol{A}^\mathrm{T}:\boldsymbol{B}^2 = 2\left(\boldsymbol{B}^2\right)^\mathrm{T}:\boldsymbol{A} + \left(\boldsymbol{B}^2\right)^\mathrm{T}:\boldsymbol{A}$$

$$= 3\left(\boldsymbol{B}^2\right)^\mathrm{T}:\boldsymbol{A}$$

则 $\dfrac{\partial \mathrm{tr}\boldsymbol{B}^3}{\partial \boldsymbol{B}} = 3\left(\boldsymbol{B}^2\right)^\mathrm{T}$ 得证, 这也是 (18-12) 式的一个特例.

8.2 给出式 (43-23) 式的第三种证明方法.

解: 由于 $\mathrm{tr}\boldsymbol{B}^3 = \boldsymbol{I}:\boldsymbol{B}^3 = \boldsymbol{I}:(\boldsymbol{B}^2\boldsymbol{B}) = \left(\boldsymbol{B}^2\right)^\mathrm{T}:\boldsymbol{B}$, 则有

$$\frac{\partial \mathrm{tr}\boldsymbol{B}^3}{\partial \boldsymbol{B}}:\boldsymbol{A} = \lim_{\varepsilon\to 0}\frac{\left[(\boldsymbol{B}+\varepsilon\boldsymbol{A})^2\right]^\mathrm{T}:(\boldsymbol{B}+\varepsilon\boldsymbol{A}) - \left(\boldsymbol{B}^2\right)^\mathrm{T}:\boldsymbol{B}}{\varepsilon}$$

$$= \lim_{\varepsilon\to 0}\frac{\left[\boldsymbol{B}^2+\varepsilon(\boldsymbol{B}\boldsymbol{A}+\boldsymbol{A}\boldsymbol{B})+\varepsilon^2\boldsymbol{A}^2\right]^\mathrm{T}:(\boldsymbol{B}+\varepsilon\boldsymbol{A}) - \left(\boldsymbol{B}^2\right)^\mathrm{T}:\boldsymbol{B}}{\varepsilon}$$

$$= \left(\boldsymbol{B}^2\right)^\mathrm{T}:\boldsymbol{A} + \left(\boldsymbol{A}^\mathrm{T}\boldsymbol{B}^\mathrm{T}+\boldsymbol{B}^\mathrm{T}\boldsymbol{A}^\mathrm{T}\right):\boldsymbol{B} = \left(\boldsymbol{B}^2\right)^\mathrm{T}:\boldsymbol{A} + 2\left(\boldsymbol{B}^2\right)^\mathrm{T}:\boldsymbol{A}$$

$$= 3\left(\boldsymbol{B}^2\right)^\mathrm{T}:\boldsymbol{A}$$

则 $\dfrac{\partial \mathrm{tr}\boldsymbol{B}^3}{\partial \boldsymbol{B}} = 3\left(\boldsymbol{B}^2\right)^\mathrm{T}$ 得证.

8.3 穆尼–里夫林本构模型 (44-7) 式是由梅尔文·穆尼 (Melvin Mooney, 1893—1968, 题图 8.3 左图) 于 1940 年和罗纳德·塞缪尔·里夫林 (Ronald Samuel Rivlin, 1915—2005, 题图 8.3 右图) 于 1948 年建立的.

题图 8.3　左：穆尼；右：里夫林

8.4　里夫林和特鲁斯德尔是理性力学领域的两位顶级大师，本书已经在 §39.2 中介绍过里夫林对特鲁斯德尔的猛烈攻击，本题则继续讨论他们的学术风格、品味和对学术界的影响．为使读者能原汁原味地体会，著者对本题的英文不做翻译．

Rivlin is one of a few giants who created the branch of theoretical physics and engineering now known as nonlinear continuum mechanics. Much of his work deals with fundamental problems that were being investigated by specialists in mechanics, such as William Prager and Daniel Drucker at Brown University; however, Prager and Drucker had little or no influence on Rivlin's work.

Clifford Truesdell, on the contrary, had a great deal of influence on Rivlin. Truesdell was the leader of a group of talented mechanicians and mathematicians who called their work "rational mechanics." In the early days, Rivlin and Truesdell had cordial relations based on mutual respect. Later, their relationship deteriorated, although many have traced their conflicts back to the 1960s. Truesdell's followers, especially Walter Noll, were encouraged to develop mechanics as an axiomatic subject in the spirit of David Hilbert. Rivlin's approach was to combine deep theoretical analysis with concrete practicality.

Rivlin was one of the creators of the theory of large elastic deformation, based on the theory of neo-Hookean and Mooney-Rivlin solids. Rivlin's work on the deformations and strengths of rubber and rubber-like materials, which began early on in his career under the influence of Treloar, had tremendous implications for real-world applications. Whereas the classical mechanics approach of applying linear theory of elasticity did not have practical applications, Rivlin, with unsurpassed elegance, overcame the formidable difficulties in understanding the finite deformations characteristic of these materials. His exact solutions are universally recognized as a tipping point that demonstrated to the engineering community that

第 8 章 本构关系

rigorous mathematical results for nonlinear and large deformations could guide engineering design. Today, Rivlin's work provides a mathematical foundation for a wide variety of applications.

8.5 小应变时，圣维南–基尔霍夫材料 (St Venant-Kirchhoff materials) 的应变能密度可表示为

$$w(\boldsymbol{E}) = \mu \mathrm{tr} \boldsymbol{E}^2 + \frac{\lambda}{2} (\mathrm{tr} \boldsymbol{E})^2 + o(\|\boldsymbol{E}\|^2) \tag{s8-1}$$

证明：圣维南–基尔霍夫材料的本构关系可通过 PK2 $\boldsymbol{T}$ 以及格林应变 $\boldsymbol{E}$ 表示为

$$\boldsymbol{T} = 2\mu \boldsymbol{E} + \lambda (\mathrm{tr} \boldsymbol{E}) \boldsymbol{I} + o(\|\boldsymbol{E}\|) \tag{s8-2}$$

提示：由 (18-6) 式可得 $\dfrac{\partial (\mathrm{tr} \boldsymbol{E})^2}{\partial \boldsymbol{E}} = 2 (\mathrm{tr} \boldsymbol{E}) \boldsymbol{I}$；再由 (18-11) 式可得 $\dfrac{\partial \mathrm{tr} \boldsymbol{E}^2}{\partial \boldsymbol{E}} = 2 \boldsymbol{E}^{\mathrm{T}} = 2\boldsymbol{E}$；将上两式代入不可压缩超弹性本构关系式 (43-6)，则 (s8-2) 式得证.

8.6 力–化耦合的胡克定律. 当考虑化学的扩散时，广义胡克定律可以表示为 [8.16]

$$\varepsilon_{ij} = \frac{1}{E} \left[(1+\nu) \sigma_{ij} - \nu \sigma_{kk} \delta_{ij} \right] + \frac{C\Omega}{3} \delta_{ij} \tag{s8-3}$$

式中，Ω 是扩散物质摩尔体积 ($\mathrm{m}^3/\mathrm{mol}$)，$C$ 是扩散物质的浓度 ($\mathrm{mol}/\mathrm{m}^3$). 验证：(s8-3) 式的逆形式为

$$\sigma_{ij} = 2\mu \left(\varepsilon_{ij} - \frac{C\Omega}{3} \delta_{ij} \right) + \lambda (\varepsilon_{kk} - C\Omega) \delta_{ij} \tag{s8-4}$$

或者

$$\boldsymbol{\sigma} = 2\mu \left(\boldsymbol{\varepsilon} - \frac{C\Omega}{3} \boldsymbol{I} \right) + \lambda (\mathrm{tr} \boldsymbol{\varepsilon} - C\Omega) \boldsymbol{I} \tag{s8-5}$$

8.7 大作业：(1) 进一步分析、归纳和总结固体、流体、流变体、软物质各自作为连续介质一部分之间的异同点；(2) 具体地分析晶体塑性变形中的位错线和流体中涡丝的类比性 [8.17].

8.8 固体力学的本构关系和流体力学的湍流被认为是连续介质力学的两个主要尚待解决的难题. 在大数据时代，你是否认为人工智能 (AI) 有助于这两个难题的解决？

参 考 文 献

[8.1] Truesdell C, Noll W. The Non-Linear Field Theories of Mechanics (3rd Edition). Berlin: Springer, 2004.

[8.2] Oldroyd J G. On the formulation of rheological equations of state. Proceedings of the Royal Society of London A, 1950, 200: 523-541.

[8.3] Noll W. A mathematical theory of the mechanical behavior of continuous media. Archive for Rational Mechanics and Analysis, 1958, 2: 197-226.

[8.4] Eringen A C. Mechanics of Continua (2nd Edition). New York: Robert E. Krieger, 1980.

[8.5] Eringen A C. Nonlocal Continuum Field Theories. New York: Springer, 2002.

[8.6] Rivlin R S. Red herrings and sundry unidentified fish in nonlinear continuum mechanics//Collected Papers of RS Rivlin. Springer, New York, NY, 1997, pp. 2765-2782.

[8.7] Hooke R. De Potentia Restitutiva, or of Spring Explaining the Power of Springing Bodies. London, 1678.

[8.8] Riccati G. Delle Vibrazione Sonore dei Cilindri. Memoire Matematica e Fisica Società Italiana, 1782, 1: 444-525.

[8.9] Young T. A Course of Lectures on Natural Philosophy and the Mechanical Arts. Johnson, 1807.

[8.10] Navier L. De l'équilibre et du mouvement des corps solides élastiques. Paper read to the Académie des Sciences, 1821, 14.

[8.11] Cauchy A L. Sur les équations qui expriment les conditions d'équilibre ou les lois du mouvement intérieur d'un corps solide, élastique, ou non élastique. Ex. de Math, 1828, 3: 160-187.

[8.12] Poisson S D. Mémoire sur l'Equilibre et le Mouvement des Corps élastiques. Bulletin des Sciences Mathématiques, Astromatiques, Physiques et Chimiques, 1829, 11: 98-111.

[8.13] Love A E H. A Treatise on the Mathematical Theory of Elasticity (1st Edition). Cambridge: The University Press, 1892.

[8.14] Pascal B. Récit de la grande expérience de l'équilibre des liqueurs. Paris, 1648.

[8.15] Weickenmeier J, Saez P, Butler C A M, Young P G, Goriely A, Kuhl E. Bulging brains. Journal of Elasticity, 2017, 129: 197-212.

[8.16] Li J C M. Physical chemistry of some microstructural phenomena. Metallurgical Transactions A, 1978, 9: 1353-1380.

[8.17] Zhao Y P. Explaining the analogy between dislocation line in crystal and vortex filament in fluid. Mechanics Research Communications, 1998, 25: 487-492.

第 9 章 虚功原理在连续介质力学中的应用

§45. 微元长度、面积、体积和雅可比的变分

45.1 知道虚位移后如何确定虚体积？

虚位移 $\delta\boldsymbol{u}$ 在当前构形下所产生的虚体积为

$$\delta(\mathrm{d}v) = \mathrm{div}\delta\boldsymbol{u}(\mathrm{d}v) \tag{45-1}$$

证明方法一：在产生了虚位移 $\delta\boldsymbol{u}$ 后，在当前构形所产生的虚体积应变为

$$\delta\varepsilon_{ii} = \delta u_{i,i} = \mathrm{div}\delta\boldsymbol{u} \tag{45-2}$$

实体积微元 $\mathrm{d}v$ 乘以虚体积应变后，即获得虚体积的表达式 (45-1) 式. 证毕.

证明方法二：由于当前构形下的体积微元为 $\mathrm{d}v = \mathrm{d}x\mathrm{d}y\mathrm{d}z$，则其变分为

$$\begin{aligned}
\delta(\mathrm{d}v) &= (\delta\mathrm{d}x)\mathrm{d}y\mathrm{d}z + \mathrm{d}x(\delta\mathrm{d}y)\mathrm{d}z + \mathrm{d}x\mathrm{d}y(\delta\mathrm{d}z) \\
&= \left(\frac{\delta\mathrm{d}x}{\mathrm{d}x} + \frac{\delta\mathrm{d}y}{\mathrm{d}y} + \frac{\delta\mathrm{d}z}{\mathrm{d}z}\right)\mathrm{d}x\mathrm{d}y\mathrm{d}z \\
&= \mathrm{div}\delta\boldsymbol{u}(\mathrm{d}v)
\end{aligned} \tag{45-3}$$

45.2 知道虚位移后如何确定雅可比的变分？

方法一、直接法：由于雅可比定义为当前构形和参考构形中微元的体积比：

$$J = \frac{\mathrm{d}x\mathrm{d}y\mathrm{d}z}{\mathrm{d}X\mathrm{d}Y\mathrm{d}Z} \tag{45-4}$$

由于参考构形中的体积微元 $\mathrm{d}V = \mathrm{d}X\mathrm{d}Y\mathrm{d}Z$ 的不变性，则对 (45-4) 式进行变分：

$$\begin{aligned}
\delta J &= \frac{\delta(\mathrm{d}x\mathrm{d}y\mathrm{d}z)}{\mathrm{d}X\mathrm{d}Y\mathrm{d}Z} = \frac{(\delta\mathrm{d}x)\mathrm{d}y\mathrm{d}z + \mathrm{d}x(\delta\mathrm{d}y)\mathrm{d}z + \mathrm{d}x\mathrm{d}y(\delta\mathrm{d}z)}{\mathrm{d}V} \\
&= \frac{\mathrm{d}v}{\mathrm{d}V}\left(\frac{\delta\mathrm{d}x}{\mathrm{d}x} + \frac{\delta\mathrm{d}y}{\mathrm{d}y} + \frac{\delta\mathrm{d}z}{\mathrm{d}z}\right) \\
&= J\mathrm{div}\delta\boldsymbol{u}
\end{aligned} \tag{45-5}$$

方法二、严格法：变形梯度 $\boldsymbol{F}$ 与位移梯度 $\boldsymbol{H}$ 的关系为

$$\boldsymbol{F} = \mathrm{Grad}\boldsymbol{u} + \boldsymbol{I} = \boldsymbol{H} + \boldsymbol{I} \tag{45-6}$$

由数学算子和变分符号的可交换性，则变形梯度张量 $\boldsymbol{F}$ 的变分为

$$\delta \boldsymbol{F} = \mathrm{Grad}\,\delta \boldsymbol{u} = \delta \boldsymbol{H} = (\delta \boldsymbol{u}) \otimes \boldsymbol{\nabla}_X \tag{45-7}$$

由于雅可比 $J = \det \boldsymbol{F}$ 为张量的标量函数，则其变分为

$$\delta J = \frac{\partial J}{\partial \boldsymbol{F}} : \delta \boldsymbol{F} = \frac{\partial J}{\partial \boldsymbol{F}} : (\delta \boldsymbol{u}) \otimes \boldsymbol{\nabla}_X \tag{45-8}$$

再注意到常用关系式 (18-16)：$\dfrac{\partial J}{\partial \boldsymbol{F}} = J\boldsymbol{F}^{-\mathrm{T}}$，将该式代回 (45-8) 式有

$$\delta J = J\boldsymbol{F}^{-\mathrm{T}} : (\delta \boldsymbol{u}) \otimes \boldsymbol{\nabla}_X \tag{45-9}$$

由于在当前构形中位移的梯度为

$$\mathrm{grad}\,\boldsymbol{u} = \boldsymbol{u} \otimes \boldsymbol{\nabla}_x = \frac{\partial\,(\boldsymbol{x} - \boldsymbol{X})}{\partial \boldsymbol{x}} = \boldsymbol{I} - \frac{\partial \boldsymbol{X}}{\partial \boldsymbol{x}} = \boldsymbol{I} - \boldsymbol{F}^{-1} \tag{45-10}$$

亦即变形梯度张量的逆可表示为

$$\boldsymbol{F}^{-1} = \boldsymbol{I} - \boldsymbol{u} \otimes \boldsymbol{\nabla}_x = \boldsymbol{I} - \mathrm{grad}\,\boldsymbol{u} \tag{45-11}$$

将 (45-11) 式代回 (45-9) 式，得到 (另一更直观的方法见思考题 9.4)

$$\begin{aligned}
\delta J &= J\left(\boldsymbol{I} - \boldsymbol{u} \otimes \boldsymbol{\nabla}_x\right)^{\mathrm{T}} : (\delta \boldsymbol{u}) \otimes \boldsymbol{\nabla}_X = J\left(\boldsymbol{I} - \boldsymbol{\nabla}_x \otimes \boldsymbol{u}\right) : (\delta \boldsymbol{u}) \otimes \boldsymbol{\nabla}_X \\
&= J\left((\delta \boldsymbol{u}) \cdot \boldsymbol{\nabla}_X - \underbrace{\boldsymbol{\nabla}_x \otimes \boldsymbol{u} : (\delta \boldsymbol{u}) \otimes \boldsymbol{\nabla}_X}_{\text{二阶小量，可略去}} \right) \\
&\approx J\mathrm{Div}\,\delta \boldsymbol{u}
\end{aligned} \tag{45-12}$$

(45-12) 式为大变形的结果，对于小变形，(45-12) 式退化为

$$\delta J = J\mathrm{div}\,\delta \boldsymbol{u} \tag{45-13}$$

值得注意的是，在 (45-12) 式的运算中，运用了如下基本性质：

$$\begin{cases} \boldsymbol{I} : \mathrm{grad}\,\boldsymbol{u} = \boldsymbol{I} : \boldsymbol{u} \otimes \boldsymbol{\nabla}_x = \mathrm{tr}\,(\boldsymbol{u} \otimes \boldsymbol{\nabla}_x) = \mathrm{div}\,\boldsymbol{u} \\ \boldsymbol{I} : \mathrm{Grad}\,\boldsymbol{u} = \boldsymbol{I} : (\delta \boldsymbol{u}) \otimes \boldsymbol{\nabla}_X = \mathrm{tr}\,((\delta \boldsymbol{u}) \otimes \boldsymbol{\nabla}_X) = \mathrm{Div}\,\boldsymbol{u} \end{cases} \tag{45-14}$$

45.3 知道虚位移后如何确定微线段矢量的变分？

由当前构形和参考构形中线段矢量变换的基本关系式 (26-2)：$\mathrm{d}\boldsymbol{x} = \boldsymbol{F}\mathrm{d}\boldsymbol{X}$，则两种构形中对虚位移的梯度为

$$\mathrm{grad}\,(\delta \boldsymbol{u}) = \frac{\partial\,(\delta \boldsymbol{u})}{\partial \boldsymbol{x}} = \frac{\partial\,(\delta \boldsymbol{u})}{\partial \boldsymbol{X}} \frac{\partial \boldsymbol{X}}{\partial \boldsymbol{x}} = \mathrm{Grad}\,(\delta \boldsymbol{u})\,\boldsymbol{F}^{-1} \tag{45-15}$$

由上式得到
$$\mathbf{Grad}(\delta \boldsymbol{u}) = \mathbf{grad}(\delta \boldsymbol{u}) \boldsymbol{F} \tag{45-16}$$

再由变形梯度的定义，其变分为
$$\delta \boldsymbol{F} = \delta(\mathbf{Grad}\,\boldsymbol{u}) = \delta(\boldsymbol{\nabla}_X \otimes \boldsymbol{u}) = \mathbf{Grad}\,\delta \boldsymbol{u} = (\delta \boldsymbol{u}) \otimes \boldsymbol{\nabla}_X \tag{45-17}$$

由 (45-16) 和 (45-17) 两式可知，变形梯度的变分为
$$\delta \boldsymbol{F} = \mathbf{Grad}\,\delta \boldsymbol{u} = \mathbf{grad}(\delta \boldsymbol{u}) \boldsymbol{F} \tag{45-18}$$

对 (26-2) 式 $\mathrm{d}\boldsymbol{x} = \boldsymbol{F}\mathrm{d}\boldsymbol{X}$ 进行变分，并将 (45-18) 式代入，最终得到
$$\delta(\mathrm{d}\boldsymbol{x}) = \delta \boldsymbol{F}\mathrm{d}\boldsymbol{X} = \mathbf{grad}(\delta \boldsymbol{u}) \boldsymbol{F}\mathrm{d}\boldsymbol{X}$$
$$= \mathbf{grad}(\delta \boldsymbol{u})\mathrm{d}\boldsymbol{x} = (\delta \boldsymbol{u}) \otimes \boldsymbol{\nabla}_x \mathrm{d}\boldsymbol{x} \tag{45-19}$$

45.4 知道虚位移后如何确定微面积矢量的变分？

微元面积 $\mathrm{d}\boldsymbol{a}$ 是一个向量，其方向是其外法线的方向. 微元面积 $\mathrm{d}\boldsymbol{a}$ 和微线段 $\mathrm{d}\boldsymbol{x}$ 之间的点积构成了当前构形中的微体积：
$$\mathrm{d}v = \mathrm{d}\boldsymbol{a} \cdot \mathrm{d}\boldsymbol{x} \tag{45-20}$$

对上式进行变分，并利用 (45-1) 式和 (45-19) 式，有
$$\delta(\mathrm{d}v) = \mathrm{div}(\delta \boldsymbol{u})\mathrm{d}v = \delta(\mathrm{d}\boldsymbol{a}) \cdot \mathrm{d}\boldsymbol{x} + \mathrm{d}\boldsymbol{a} \cdot \underbrace{\delta(\mathrm{d}\boldsymbol{x})}_{\mathbf{grad}(\delta \boldsymbol{u})\mathrm{d}\boldsymbol{x}} \tag{45-21}$$

再利用 (45-19) 式，(45-21) 式的后两步变为
$$\mathrm{div}(\delta \boldsymbol{u})\mathrm{d}\boldsymbol{a} \cdot \mathrm{d}\boldsymbol{x} = \delta(\mathrm{d}\boldsymbol{a}) \cdot \mathrm{d}\boldsymbol{x} + \mathrm{d}\boldsymbol{a} \cdot \mathbf{grad}(\delta \boldsymbol{u})\mathrm{d}\boldsymbol{x}$$
$$= \delta(\mathrm{d}\boldsymbol{a}) \cdot \mathrm{d}\boldsymbol{x} + \mathbf{grad}^{\mathrm{T}}(\delta \boldsymbol{u})\mathrm{d}\boldsymbol{a} \cdot \mathrm{d}\boldsymbol{x}$$
$$= \left[\delta(\mathrm{d}\boldsymbol{a}) + \mathbf{grad}^{\mathrm{T}}(\delta \boldsymbol{u})\mathrm{d}\boldsymbol{a}\right] \cdot \mathrm{d}\boldsymbol{x} \tag{45-22}$$

对上式进行移项，有
$$\left\{\delta(\mathrm{d}\boldsymbol{a}) - \left[\mathrm{div}(\delta \boldsymbol{u})\mathbf{I} - \mathbf{grad}^{\mathrm{T}}(\delta \boldsymbol{u})\right]\mathrm{d}\boldsymbol{a}\right\} \cdot \mathrm{d}\boldsymbol{x} = 0 \tag{45-23}$$

由上式中微线段 $\mathrm{d}\boldsymbol{x}$ 的任意性，从而有
$$\delta(\mathrm{d}\boldsymbol{a}) = \left[\mathrm{div}(\delta \boldsymbol{u})\mathbf{I} - \mathbf{grad}^{\mathrm{T}}(\delta \boldsymbol{u})\right]\mathrm{d}\boldsymbol{a}$$
$$= \left[\boldsymbol{\nabla}_x \cdot (\delta \boldsymbol{u})\mathbf{I} - \boldsymbol{\nabla}_x \otimes (\delta \boldsymbol{u})\right]\mathrm{d}\boldsymbol{a} \tag{45-24}$$

§46. 虚功原理在连续介质力学中的应用

46.1 当前构形中的虚位移

如图 46.1 所示，在 $t = 0$ 时，所研究的连续体在初始构形中所占据的体积为 Ω_0，其中的质点 $\boldsymbol{X}$ 经过时间 t 变形到当前构形 Ω 中的质点 $\boldsymbol{x}$，所发生的位移矢量为 $\boldsymbol{u}$.

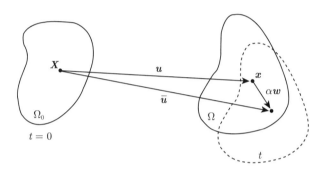

图 46.1　初始构形、当前构形、虚位移

设在时间 t 时刻，质点 $\boldsymbol{x}$ 发生了一个虚位移：

$$\delta \boldsymbol{u} = \bar{\boldsymbol{u}} - \boldsymbol{u} = \alpha \boldsymbol{w} \tag{46-1}$$

式中，α 为一无限小量 $(\alpha \to 0)$. 按照经典力学的定义，虚位移 $\delta \boldsymbol{u}$ 和实位移 $\mathrm{d}\boldsymbol{u}$ 的区别是，实位移需要时间，而虚位移是在某一固定时刻发生的，并不需要时间.

对 (46-1) 式在当前构形中求梯度，有

$$\mathbf{grad}\,(\delta \boldsymbol{u}) = (\delta \boldsymbol{u}) \otimes \boldsymbol{\nabla}_{\boldsymbol{x}} = \mathbf{grad}\,\bar{\boldsymbol{u}} - \mathbf{grad}\,\boldsymbol{u} \tag{46-2}$$

由于按照梯度的定义：$\mathbf{grad}\,\boldsymbol{u} = \dfrac{\partial \boldsymbol{u}}{\partial \boldsymbol{x}}$，其变分为

$$\delta\,(\mathbf{grad}\,\boldsymbol{u}) = \delta\,(\boldsymbol{u} \otimes \boldsymbol{\nabla}_{\boldsymbol{x}}) = \mathbf{grad}\,\bar{\boldsymbol{u}} - \mathbf{grad}\,\boldsymbol{u} \tag{46-3}$$

比较上两式，有

$$\mathbf{grad}\,(\delta \boldsymbol{u}) = \delta\,(\mathbf{grad}\,\boldsymbol{u}) \tag{46-4}$$

上式表明，位移梯度场的变分等于位移变分的梯度，也就是说，梯度算子和变分顺序可以交换. 同理，我们还可以得出变分和定积分的顺序亦可交换的结论.

46.2 和变形梯度张量相关的变分

重要的是，需要区分当前构形的梯度 (grad) 和参考构形中的物质梯度 (Grad). 事实上，通过变形梯度张量 $\boldsymbol{F}$ 来联系 $\mathrm{grad}\,(\delta\boldsymbol{u})$ 和 $\mathrm{Grad}\,(\delta\boldsymbol{u})$：

$$\mathrm{grad}\,(\delta\boldsymbol{u}) = \frac{\partial(\delta\boldsymbol{u})}{\partial\boldsymbol{x}} = \frac{\partial(\delta\boldsymbol{u})}{\partial\boldsymbol{X}}\frac{\partial\boldsymbol{X}}{\partial\boldsymbol{x}} = \mathrm{Grad}\,(\delta\boldsymbol{u})\,\boldsymbol{F}^{-1} \tag{46-5}$$

上式右端表示的是虚位移的物质梯度 $\mathrm{Grad}\,(\delta\boldsymbol{u})$ 和变形梯度张量逆 $\boldsymbol{F}^{-1}$ 之间的点积. (46-5) 式的指标形式为

$$\frac{\partial \delta u_a}{\partial x_b} = \frac{\partial \delta u_a}{\partial X_A} F_{Ab}^{-1} \tag{46-6}$$

变形梯度张量 $\boldsymbol{F}$ 的变分和虚位移的物质梯度之间满足：

$$\delta\boldsymbol{F} = \delta\,(\mathrm{Grad}\,\boldsymbol{u}) = \delta\,(\boldsymbol{u}\otimes\boldsymbol{\nabla}_X) = \mathrm{Grad}\,\delta\boldsymbol{u} = (\delta\boldsymbol{u})\otimes\boldsymbol{\nabla}_X \tag{46-7}$$

对恒等式 $\boldsymbol{F}^{-1}\boldsymbol{F} = \boldsymbol{I}$ 求变分，得到

$$\delta\boldsymbol{F}^{-1}\boldsymbol{F} + \boldsymbol{F}^{-1}\delta\boldsymbol{F} = \boldsymbol{0} \quad\Rightarrow\quad \delta\boldsymbol{F}^{-1}\boldsymbol{F} = -\boldsymbol{F}^{-1}\delta\boldsymbol{F} \tag{46-8}$$

利用 (46-5) 式和 (46-7) 式，由上式得到

$$\delta\boldsymbol{F}^{-1} = -\boldsymbol{F}^{-1}\delta\boldsymbol{F}\boldsymbol{F}^{-1} = -\boldsymbol{F}^{-1}\mathrm{Grad}\,(\delta\boldsymbol{u})\,\boldsymbol{F}^{-1} = -\boldsymbol{F}^{-1}\mathrm{grad}\,\delta\boldsymbol{u} \tag{46-9}$$

虚位移 $\delta\boldsymbol{u}$ 所产生的虚体积为 (45-1) 式：$\delta\,(\mathrm{d}v) = \mathrm{div}\,\delta\boldsymbol{u}\,\mathrm{d}v$，雅可比 (Jacobian) 为当前构形和初始构形的单位体积之比，因此，雅可比的变分可表示为 (45-5) 式：$\delta J = J\mathrm{div}\,\delta\boldsymbol{u}$.

46.3 格林和柯西应变张量的变分

参考构形中的格林应变张量：$\boldsymbol{E} = \dfrac{1}{2}\left(\boldsymbol{F}^{\mathrm{T}}\boldsymbol{F} - \boldsymbol{I}\right)$，对其进行变分，有

$$\begin{aligned}
\delta\boldsymbol{E} &= \frac{1}{2}\left[\delta\left(\boldsymbol{F}^{\mathrm{T}}\right)\boldsymbol{F} + \boldsymbol{F}^{\mathrm{T}}\delta\boldsymbol{F}\right] = \frac{1}{2}\left[(\delta\boldsymbol{F})^{\mathrm{T}}\boldsymbol{F} + \boldsymbol{F}^{\mathrm{T}}\delta\boldsymbol{F}\right] \\
&= \frac{1}{2}\left[(\mathrm{Grad}\,\delta\boldsymbol{u})^{\mathrm{T}}\boldsymbol{F} + \boldsymbol{F}^{\mathrm{T}}\mathrm{Grad}\,\delta\boldsymbol{u}\right] = \frac{1}{2}\left[\left(\boldsymbol{F}^{\mathrm{T}}\mathrm{Grad}\,\delta\boldsymbol{u}\right)^{\mathrm{T}} + \boldsymbol{F}^{\mathrm{T}}\mathrm{Grad}\,\delta\boldsymbol{u}\right] \\
&= \mathrm{sym}\left(\boldsymbol{F}^{\mathrm{T}}\mathrm{Grad}\,\delta\boldsymbol{u}\right)
\end{aligned} \tag{46-10}$$

作为当前构形中欧拉型二阶张量，柯西应变的变分显然可表示为

$$\delta\boldsymbol{\varepsilon} = \frac{1}{2}\left(\mathrm{grad}^{\mathrm{T}}\delta\boldsymbol{u} + \mathrm{grad}\,\delta\boldsymbol{u}\right) = \mathrm{sym}\,(\mathrm{grad}\,\delta\boldsymbol{u}) \tag{46-11}$$

46.4 虚功原理

46.4.1 当前构形中的虚功原理

当前构形中的运动方程 (柯西第一运动方程) 为 (32-6) 式:

$$\text{div}\boldsymbol{\sigma} + \boldsymbol{b} = \rho\ddot{\boldsymbol{u}} \tag{46-12}$$

当前构形下的矢量形式的运动方程和虚位移矢量的点积作用后产生一个新的标量函数, 在当前构形所占有的体积 Ω 下积分, 有

$$f(\boldsymbol{u}, \delta\boldsymbol{u}) = \int_\Omega (-\text{div}\boldsymbol{\sigma} - \boldsymbol{b} + \rho\ddot{\boldsymbol{u}}) \cdot \delta\boldsymbol{u}\text{d}v = 0 \tag{46-13}$$

式中, 由于

$$\text{div}\boldsymbol{\sigma} \cdot \delta\boldsymbol{u} = \text{div}(\boldsymbol{\sigma}\delta\boldsymbol{u}) - \boldsymbol{\sigma} : \text{grad}\delta\boldsymbol{u} \tag{46-14}$$

将 (46-14) 式代回 (46-13) 式, 并由散度定理得到

$$f(\boldsymbol{u}, \delta\boldsymbol{u}) = \int_\Omega [\boldsymbol{\sigma} : \text{grad}(\delta\boldsymbol{u}) - (\boldsymbol{b} - \rho\ddot{\boldsymbol{u}}) \cdot \delta\boldsymbol{u}]\text{d}v - \int_{\partial\Omega} \boldsymbol{\sigma}\delta\boldsymbol{u} \cdot \boldsymbol{n}\text{d}a = 0 \tag{46-15}$$

式中, 由于柯西应力的对称性, 有

$$\boldsymbol{\sigma} : \text{grad}\delta\boldsymbol{u} = \boldsymbol{\sigma} : \text{sym}(\text{grad}\delta\boldsymbol{u}) = \boldsymbol{\sigma} : \delta\boldsymbol{\varepsilon} \tag{46-16}$$

以及由柯西应力原理, 有

$$\boldsymbol{\sigma}\delta\boldsymbol{u} \cdot \boldsymbol{n} = \boldsymbol{\sigma} \cdot \boldsymbol{n} \cdot \delta\boldsymbol{u} = \bar{\boldsymbol{t}} \cdot \delta\boldsymbol{u} \tag{46-17}$$

式中, $\bar{\boldsymbol{t}} = \boldsymbol{\sigma} \cdot \boldsymbol{n}$ 为面力. 将 (46-16) 式和 (46-17) 式代回 (46-15) 式, 得到

$$\underbrace{\int_\Omega \boldsymbol{\sigma} : \delta\boldsymbol{\varepsilon}\text{d}v}_{\text{当前构形内虚功}} = \underbrace{\int_\Omega (\boldsymbol{b} - \rho\ddot{\boldsymbol{u}}) \cdot \delta\boldsymbol{u}\text{d}v}_{\text{当前构形体力和惯性力虚功}} + \underbrace{\int_{\partial\Omega} \bar{\boldsymbol{t}} \cdot \delta\boldsymbol{u}\text{d}a}_{\text{当前构形面力虚功}} \tag{46-18}$$

式中, 柯西应力在虚柯西应变上所做的内机械虚功 (internal mechanical virtual work) 为左端项:

$$\delta W_{\text{int}} = \int_\Omega \boldsymbol{\sigma} : \delta\boldsymbol{\varepsilon}\text{d}v \tag{46-19}$$

而 (46-18) 式中的右端项为体力、惯性力和面力在虚位移上等所做的外机械虚功 (external mechanical virtual work) 为

$$\delta W_{\text{ext}} = \int_\Omega (\boldsymbol{b} - \rho\ddot{\boldsymbol{u}}) \cdot \delta\boldsymbol{u}\text{d}v + \int_{\partial\Omega} \bar{\boldsymbol{t}} \cdot \delta\boldsymbol{u}\text{d}a \tag{46-20}$$

虚功原理 (虚位移原理) 表明, 在当前构形上, 内虚功和外虚功相等:

$$\delta W_{\text{int}} = \delta W_{\text{ext}} \tag{46-21}$$

46.4.2 参考构形中的虚功原理

在参考构形中, 拉格朗日描述下的运动方程由 (32-9) 式给出:

$$\text{Div}\boldsymbol{P} + \boldsymbol{B} = \rho_0 \ddot{\boldsymbol{u}} \tag{46-22}$$

对应于 (46-13) 式, 对于参考构形下体积 Ω_0 的积分, 有

$$\mathcal{F}(\boldsymbol{u}, \delta\boldsymbol{u}) = \int_{\Omega_0} (-\text{Div}\boldsymbol{P} - \boldsymbol{B} + \rho_0 \ddot{\boldsymbol{u}}) \cdot \delta\boldsymbol{u} \, \mathrm{d}V = 0 \tag{46-23}$$

由于

$$\text{Div}\boldsymbol{P} \cdot \delta\boldsymbol{u} = \text{Div}(\boldsymbol{P}\delta\boldsymbol{u}) - \boldsymbol{P} : \text{Grad}\delta\boldsymbol{u}$$
$$= \text{Div}(\boldsymbol{P}\delta\boldsymbol{u}) - \boldsymbol{P} : \delta\boldsymbol{F} \tag{46-24}$$

将 (46-24) 式代回 (46-26) 式, 从而得到

$$\underbrace{\int_{\Omega_0} \boldsymbol{P} : \delta\boldsymbol{F} \mathrm{d}V}_{\text{参考构形内虚功}} = \underbrace{\int_{\Omega_0} (\boldsymbol{B} - \rho_0 \ddot{\boldsymbol{u}}) \cdot \delta\boldsymbol{u} \, \mathrm{d}V + \int_{\partial\Omega_0} \bar{\boldsymbol{T}} \cdot \delta\boldsymbol{u} \, \mathrm{d}A}_{\text{参考构形外虚功}} \tag{46-25}$$

式中, 参考构形上的面力为 $\bar{\boldsymbol{T}} = \boldsymbol{P}\boldsymbol{N}$, 也就是第一类皮奥拉-基尔霍夫应力 $\boldsymbol{P}$ 和外法线 $\boldsymbol{N}$ 之间的点积. (46-25) 式就是参考构形上的虚功原理.

容易验证, 当前构形和参考构形上的虚功原理完全是等价的.

§47. 贝蒂定理与材料弹性模量对称性之间的关系

47.1 积分法

47.1.1 利用弹性模量对称性推导贝蒂定理

同一物体有两种不同的真实状态:

第一状态: 体力 $f_i^{(1)}$, 面力 $P_i^{(1)}$, 应力 $\sigma_{ij}^{(1)}$, 应变 $\varepsilon_{ij}^{(1)}$, 位移 $u_i^{(1)}$;
第二状态: 体力 $f_i^{(2)}$, 面力 $P_i^{(2)}$, 应力 $\sigma_{ij}^{(2)}$, 应变 $\varepsilon_{ij}^{(2)}$, 位移 $u_i^{(2)}$.
由虚功原理可知:

$$\int_V f_i^{(1)} u_i^{(2)} \mathrm{d}V + \int_S P_i^{(1)} u_i^{(2)} \mathrm{d}A = \int_V \sigma_{ij}^{(1)} \varepsilon_{ij}^{(2)} \mathrm{d}V \tag{47-1}$$

$$\int_V f_i^{(2)} u_i^{(1)} \mathrm{d}V + \int_S P_i^{(2)} u_i^{(1)} \mathrm{d}A = \int_V \sigma_{ij}^{(2)} \varepsilon_{ij}^{(1)} \mathrm{d}V \tag{47-2}$$

其中
$$\sigma_{ij}^{(1)}\varepsilon_{ij}^{(2)} = C_{ijkl}\varepsilon_{kl}^{(1)}\varepsilon_{ij}^{(2)} \tag{47-3}$$

在弹性模量满足大对称性的前提之下，有
$$\sigma_{ij}^{(1)}\varepsilon_{ij}^{(2)} = C_{ijkl}\varepsilon_{kl}^{(1)}\varepsilon_{ij}^{(2)} = C_{klij}\varepsilon_{kl}^{(1)}\varepsilon_{ij}^{(2)} = \sigma_{kl}^{(2)}\varepsilon_{kl}^{(1)} \tag{47-4}$$

从而有
$$\int_V f_i^{(1)}u_i^{(2)}\mathrm{d}V + \int_S P_i^{(1)}u_i^{(2)}\mathrm{d}A = \int_V f_i^{(2)}u_i^{(1)}\mathrm{d}V + \int_S P_i^{(2)}u_i^{(1)}\mathrm{d}A \tag{47-5}$$

因此，弹性模量大对称性是贝蒂定理的充分条件.

47.1.2 若已知贝蒂定理

此时有
$$\int_V f_i^{(1)}u_i^{(2)}\mathrm{d}V + \int_S P_i^{(1)}u_i^{(2)}\mathrm{d}A = \int_V f_i^{(2)}u_i^{(1)}\mathrm{d}V + \int_S P_i^{(2)}u_i^{(1)}\mathrm{d}A \tag{47-6}$$

可得
$$\sigma_{ij}^{(1)}\varepsilon_{ij}^{(2)} = \sigma_{ij}^{(2)}\varepsilon_{ij}^{(1)} \tag{47-7}$$

其中
$$\sigma_{ij}^{(1)}\varepsilon_{ij}^{(2)} = C_{ijkl}\varepsilon_{kl}^{(1)}\varepsilon_{ij}^{(2)} \tag{47-8}$$

则有
$$\sigma_{ij}^{(2)}\varepsilon_{ij}^{(1)} = \sigma_{kl}^{(2)}\varepsilon_{kl}^{(1)} = C_{klij}\varepsilon_{kl}^{(2)}\varepsilon_{ij}^{(1)} \tag{47-9}$$

因此，有
$$C_{ijkl} = C_{klij} \tag{47-10}$$

即在已知贝蒂定理的前提条件之下可以推导得到弹性模量的大对称性 (major symmetry)，不能直接得到弹性模量的小对称性 (minor symmetry).

47.2 微分法

47.2.1 应用偏导数对自变量的求导次序无关的性质

在自由能密度函数是应变张量的三次可微的条件下，偏导数对自变量的求导次序无关. 由于应变张量以及应力张量的对称性：$\sigma_{ij} = \sigma_{ji}$，$\varepsilon_{ij} = \varepsilon_{ji}$ 有

$$\sigma_{ij}^{(1)}\varepsilon_{ij}^{(2)} = \frac{\partial \sigma_{ij}^{(1)}}{\partial \varepsilon_{kl}^{(1)}}\varepsilon_{kl}^{(1)}\varepsilon_{ij}^{(2)} = \frac{\partial^2 f}{\partial \varepsilon_{ij}^{(1)}\partial \varepsilon_{kl}^{(1)}}\varepsilon_{kl}^{(1)}\varepsilon_{ij}^{(2)} = C_{ijkl}\varepsilon_{kl}^{(1)}\varepsilon_{ij}^{(2)}$$

第 9 章 虚功原理在连续介质力学中的应用

$$
\begin{aligned}
&= \frac{\partial^2 f}{\partial \varepsilon_{kl}^{(1)} \partial \varepsilon_{ij}^{(1)}} \varepsilon_{kl}^{(1)} \varepsilon_{ij}^{(2)} = C_{klij} \varepsilon_{kl}^{(1)} \varepsilon_{ij}^{(2)} \\
&= \frac{\partial^2 f}{\partial \varepsilon_{ji}^{(1)} \partial \varepsilon_{kl}^{(1)}} \varepsilon_{kl}^{(1)} \varepsilon_{ij}^{(2)} = C_{jikl} \varepsilon_{kl}^{(1)} \varepsilon_{ij}^{(2)} \\
&= \frac{\partial^2 f}{\partial \varepsilon_{ij}^{(1)} \partial \varepsilon_{lk}^{(1)}} \varepsilon_{kl}^{(1)} \varepsilon_{ij}^{(2)} = C_{ijlk} \varepsilon_{kl}^{(1)} \varepsilon_{ij}^{(2)} \\
&= \varepsilon_{kl}^{(1)} \sigma_{kl}^{(2)} = \varepsilon_{ij}^{(1)} \sigma_{ij}^{(2)}
\end{aligned}
\qquad (47\text{-}11)
$$

即可推导得到贝蒂定理, 此时四阶弹性刚度张量 C_{ijkl} 满足大和小对称性.

47.2.2 在已知贝蒂定理的前提之下

此时有

$$
\begin{aligned}
\sigma_{ij}^{(1)} \varepsilon_{ij}^{(2)} &= \frac{\partial \sigma_{ij}^{(1)}}{\partial \varepsilon_{kl}^{(1)}} \varepsilon_{kl}^{(1)} \varepsilon_{ij}^{(2)} = \frac{\partial^2 f}{\partial \varepsilon_{ij}^{(1)} \partial \varepsilon_{kl}^{(1)}} \varepsilon_{kl}^{(1)} \varepsilon_{ij}^{(2)} = C_{ijkl} \varepsilon_{kl}^{(1)} \varepsilon_{ij}^{(2)} \\
\varepsilon_{ij}^{(1)} \sigma_{ij}^{(2)} &= \frac{\partial \sigma_{ij}^{(2)}}{\partial \varepsilon_{kl}^{(2)}} \varepsilon_{kl}^{(2)} \varepsilon_{ij}^{(1)} = \frac{\partial^2 f}{\partial \varepsilon_{ij}^{(2)} \partial \varepsilon_{kl}^{(2)}} \varepsilon_{kl}^{(2)} \varepsilon_{ij}^{(1)} = C_{ijkl} \varepsilon_{kl}^{(2)} \varepsilon_{ij}^{(1)}
\end{aligned}
\qquad (47\text{-}12)
$$

有 $\sigma_{ij}^{(1)} \varepsilon_{ij}^{(2)} = \varepsilon_{ij}^{(1)} \sigma_{ij}^{(2)}$, 亦即 $C_{ijkl} \varepsilon_{kl}^{(1)} \varepsilon_{ij}^{(2)} = C_{ijkl} \varepsilon_{kl}^{(2)} \varepsilon_{ij}^{(1)}$, 即可得到四阶弹性刚度张量 C_{ijkl} 的大对称性, 结合应变张量以及应力张量的对称性 $\sigma_{ij} = \sigma_{ji}, \varepsilon_{ij} = \varepsilon_{ji}$ 可以得出四阶弹性刚度张量具有小对称性.

思考题和补充材料

9.1 恩里科·贝蒂 (Enrico Betti Glaoui, 1823—1892, 如题图 9.1 所示), 意大利数学家. 意大利统一后对数学的复兴起重大作用的人之一. 曾在比萨大学学习数学, 曾参加意大利独立战争, 1865 年获比萨大学教授职位, 一直到去世. 1862 年任国会议员, 1884 年任参议员, 1874 年任短期教育部副部长. 意大利多位重要的数学、力学家为他的博士生, 如: Cesare Arzelà, Luigi Bianchi, Ulisse Dini, Federigo Enriques, Gregorio Ricci-Curbastro, Vito Volterra.

9.2 贝蒂定理 (Betti's theorem), 也被称为麦克斯韦–贝蒂功的互等定理 (Maxwell-Betti reciprocal work theorem), 该定理由贝蒂于 1872 年提出 [9.1]. 奥古斯塔斯·爱德华·霍夫·乐甫 (Augustus Edward Hough Love, 1863—1940, 见题表 3.1) 在其经典教材《弹性的数学理论教程》[8.13] 中将功的互等定理归功于贝蒂 (Love attributed the theorem to E. Betti)[9.2]. 乐甫在书中还指出, 贝蒂互等定理是瑞利勋爵更广形式的一个特例 (It is a special case of a more general theorem due to Lord Rayleigh). 在

题图 9.1 贝蒂

瑞利勋爵的经典著作《声的理论》[9.3] 中，瑞利也将互等定理归功于贝蒂. 事实上，在 1864 年的文章中 [9.4]，麦克斯韦确实包含了互等定理一种方式的陈述 (Maxwell did include a statement of a form of the reciprocal theorem in his paper)，麦克斯韦提出功的互等定理也确实早于贝蒂. 因此，将功的互等定理称为 "麦克斯韦–贝蒂功的互等定理" 是比较公正的.

9.3 经典力学中的功的互等定理是否也可理解成一种 "纠缠态"？

9.4 给出 (45-12) 式另外一种更为巧妙的证明方法.

证明：对 (45-9) 式进行变换，并利用连续介质力学常用关系式 (11-12) 式，即可得

$$\begin{aligned}\delta J &= J\boldsymbol{F}^{-\mathrm{T}} : \frac{\partial \delta \boldsymbol{u}}{\partial \boldsymbol{x}} \frac{\partial \boldsymbol{x}}{\partial \boldsymbol{X}} = J\boldsymbol{F}^{-\mathrm{T}} : ((\delta \boldsymbol{u}) \otimes \boldsymbol{\nabla}_{\boldsymbol{x}})\,\boldsymbol{F} \\ &= J\boldsymbol{F}^{-\mathrm{T}} \boldsymbol{F}^{\mathrm{T}} : ((\delta \boldsymbol{u}) \otimes \boldsymbol{\nabla}_{\boldsymbol{x}}) \\ &= J\mathbf{I} : ((\delta \boldsymbol{u}) \otimes \boldsymbol{\nabla}_{\boldsymbol{x}}) = J\mathrm{div}\delta \boldsymbol{u}\end{aligned} \quad (\text{s9-1})$$

9.5 在参考构形 (reference configuration) 和当前构形 (current configuration) 中，二阶单位张量分别定义为

$$\mathbf{I}_{rc} = \frac{\partial \boldsymbol{X}}{\partial \boldsymbol{X}} = \delta_{JK} \boldsymbol{e}_J \otimes \boldsymbol{e}_K, \quad \mathbf{I}_{cc} = \frac{\partial \boldsymbol{x}}{\partial \boldsymbol{x}} = \delta_{jk} \boldsymbol{e}_j \otimes \boldsymbol{e}_k \quad (\text{s9-2})$$

问题：(1) 在小变形且初始应力比较小甚至可忽略情况下，可以认为 (s9-2) 式中两种不同构形的二阶单位张量是一样的；(2) 在大变形时，两种不同构形的二阶单位张量将存在明显的差别，应予以区分，此时 (45-14) 式将改写为

$$\begin{cases} \mathbf{I}_{cc} : \mathbf{grad}\boldsymbol{u} = \mathbf{I}_{cc} : \boldsymbol{u} \otimes \boldsymbol{\nabla}_{\boldsymbol{x}} = \mathrm{tr}\,(\boldsymbol{u} \otimes \boldsymbol{\nabla}_{\boldsymbol{x}}) = \mathrm{div}\boldsymbol{u} \\ \mathbf{I}_{rc} : \mathbf{Grad}\boldsymbol{u} = \mathbf{I}_{rc} : \boldsymbol{u} \otimes \boldsymbol{\nabla}_{\boldsymbol{X}} = \mathrm{tr}\,(\boldsymbol{u} \otimes \boldsymbol{\nabla}_{\boldsymbol{X}}) = \mathrm{Div}\boldsymbol{u} \end{cases} \quad (\text{s9-3})$$

参 考 文 献

[9.1] Betti E. Teoria della elasticita'. Il Nuovo Cimento (1869-1876), 1872, 7: 69-97.

[9.2] Charlton T M. A historical note on the reciprocal theorem and theory of statically indeterminate frameworks. Nature, 1960, 187: 231-232.

[9.3] Lord Rayleigh. The Theory of Sound (Volume 1, 2$^\mathrm{nd}$ Edition). Macmillan, 1894.

[9.4] Maxwell J C. On the calculation of the equilibrium and stiffness of frames. Philosophical Magazine, 1864, 27: 294-299.

第 10 章　固体力学要义

§48. 材料力学之提纲挈领

如图 48.1 所示，纳维 (Claude-Louis Navier, 1785—1836) 于 1826 年出版了国际上首部《材料力学》[10.1] 教程，并于 1833 和 1838 年实现了重印. 该书对于材料力学范式的形成起到了奠基性的作用.

图 48.1　纳维的《材料力学》封面，左：1826 年首版；中：1833 年重印版；右：1838 年重印版

图 48.2 为对固体力学和强度研究作出杰出贡献的学者及其所生活的年代. 从该图可以看出，1800—1900 年是有关固体力学强度研究人才辈出的年代.

1638 年，74 岁的伽利略 (Galileo Galilei, 1564—1642) 在荷兰莱顿出版的《关于两门新科学的对话》[10.2] 一书被认为是材料力学 (strength of materials, mechanics of materials) 开始形成一门独立学科的标志. 在该书中，伽利略所讨论的第一门新科学便是材料力学，不但提出了固体的强度问题，介绍了他最早进行的梁强度的实验，提出了等强度梁的概念，还给出了简单情形下的虚功原理等等. 图 48.3 是书中的两幅著名插图.

伽利略在参观威尼斯一家兵工厂时，对一些具有几何相似的结构物的强度进行了分析. 对于自身重量作用下具有几何相似的悬臂梁进行了仔细理论分析，得出了如下重要结论：梁的强度会随着尺寸增大而减小. 伽利略指出："无论是艺术或是自然的产物，都不可以无限地扩大其尺寸. 所以，建造无穷巨大的船、宫殿和庙宇是不可能的. 就好像自然界的树木不会超过其尺寸而任意生长，否则它的枝杈会在自重作用下折断."

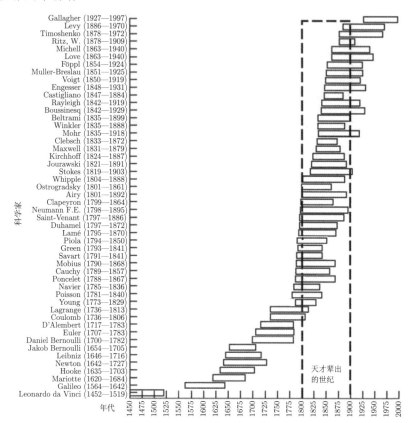

图 48.2　对固体力学和强度学科发展作出重要贡献的学者和所生活的年代

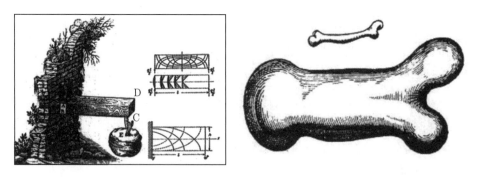

图 48.3　伽利略书中的两幅插图，左：悬臂梁；右：狗骨头的尺寸效应

伽利略对比了轻盈的鸟骨头和粗笨的恐龙骨头，从而得出了结论：随着尺度的增加，承重能力的增加与面积成正比，也就是同尺度成二次方关系；而自重的增

第 10 章　固体力学要义

加与体积成正比, 也就是同尺度成三次方关系. 自重的增加比承重能力的增加要快得多.

因此, 伽利略敏锐地得到如下结论: 动物的尺寸不能太大, 否则骨骼要么需要更强的材质, 要么放大骨骼, 变成巨灵怪物, 如图 48.3 中的右图所示. 他还深刻地指出: "动物形体尺寸减小时, 躯体的强度并不按比例减小. 一只小狗也许可以在它背上驮两三只同样大小的狗, 但我相信一匹马也许连一匹和它同样大小的马也驮不起来." 也就是说, 物体越小, 其相对强度越大. 据测定, 一只蚂蚁可以举起其自身重量 100—400 倍的重物.

铁摩辛柯 (Stephen Prokofyevich Timoshenko, 1878—1972) 系列教材的出版, 对固体力学范式的形成起到了巨大的作用. 见表 48.1.

表 48.1　铁摩辛柯的力学教材

出版年份	书名	合著者
1928	*Vibration Problem in Engineering*（《工程中的振动问题》）	1955 年与 Young 合著; 1974 与 Young 和 Weaver 合著
1930	*Strength of Materials*（《材料力学》）	
1934	*Theory of Elasticity*（《弹性理论》）	与 J. N. Goodier 合著
1936	*Theory of Elastic Stability*（《弹性稳定理论》）	1962 年与 James M. Gere 合著
1940	*Theory of Plates and Shells*（《板壳理论》）	1959 年与 S. Woinowsky-Krieger 合著
1948	*Advanced Dynamics*（《高等动力学》）	与 D. H. Young 合著
1953	*History of Strength of Materials*（《材料力学史》）	

黑板的力量 —— 授课中的铁摩辛柯

材料力学和弹性力学的基本假定为:

连续性假设: 指物体内部任何一点的力学性质都是连续的, 例如密度、位移、应变、应力等力学量除在某些点、线、面上而外都是空间的连续变量, 而且变形后物体上的质点与变形前物体上的质点是一一对应的;

各向同性假设: 物体在每一个点各个方向上的弹性性质均相同;

均匀性假设: 物体各点的弹性性质均相同;

完全弹性假设: 物体在卸载后完全能恢复其初始的形状和尺寸;

无初始应力假设: 解的唯一性要求物体中无初始应力作用;

小变形假设: 可略去位移在应变中的二次项, 此时线弹性规律也就是胡克定律成立. 线弹性意味着叠加原理 (superposition principle) 的成立.

所谓叠加原理是指线性映射 $f(x)$ 同时满足两个性质: (1) 可加性 (additivity):

$f(x+y) = f(x) + f(y)$;(2) 一次齐次性 (homogeneity of degree one):$f(\lambda x) = \lambda f(x)$.

材料力学主要研究杆件的拉伸、扭转，梁的弯曲以及柱体的压缩失稳 (屈曲) 等内容. 其主要特点和公式总结在表 48.2 中.

表 48.2 材料力学的主要内容之梳理一览表

变形类型	图例	主要公式	特征量
拉伸 Tension	图 48.4	应变:$\varepsilon = \delta/L$ (48-1) 应力:$\sigma = P/A$ (48-2) 胡克定律:$\sigma = E\varepsilon$ (48-3) 伸长:$\delta = \dfrac{PL}{EA}$ (48-4)	轴向刚度:EA
扭转 Torsion	图 48.5	最大切应变:$\gamma_{\max} = \dfrac{bb'}{ab} = \dfrac{r\mathrm{d}\varphi}{\mathrm{d}x} = r\theta$ (48-5) 胡克定律:$\tau_{\max} = G\gamma_{\max} = Gr\theta$ (48-6) 总扭转角:$\phi = \dfrac{TL}{GI_P}$ (48-7)	极惯性矩:$I_P = \displaystyle\int_A \rho^2 \mathrm{d}A$ 扭矩:$T = \dfrac{\tau_{\max}}{r} I_P$ 扭转刚度:GI_P
弯曲 Bending	图 48.6	曲率和弯矩的关系:$\dfrac{1}{\rho} = \dfrac{M}{EI}$ (48-8) 胡克定律:$\sigma_x = E\varepsilon_x = -\dfrac{Ey}{\rho}$ (48-9) 弯曲方程:$EI\dfrac{\mathrm{d}^2 y}{\mathrm{d}x^2} = M$ (48-10)	惯性矩:$I = \displaystyle\int_A y^2 \mathrm{d}A$ 弯曲刚度:EI $V = EI\dfrac{\mathrm{d}^3 y}{\mathrm{d}x^3}$ $q = EI\dfrac{\mathrm{d}^4 y}{\mathrm{d}x^4}$
屈曲 Buckling	图 48.7	$\dfrac{\mathrm{d}^2 y}{\mathrm{d}x^2} + \dfrac{P}{EI} y = 0$ (48-11) $P_{\mathrm{cr}} = \dfrac{\pi^2 EI}{L^2}$ (48-12)	

例 48.1 如图 48.8 所示,所生长纳米线的力学行为可用原子力针尖在可视为微悬臂梁的纳米线上施加力使其发生弹性变形来反求,这是一篇被广泛引用的文献[10.3],到本书正式出版,该文已经被 google scholar 引用近 6000 次. 如图所示,求纳米线在均布和集中载荷作用下的挠度变形.

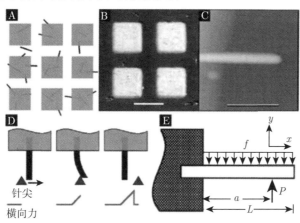

图 48.8 纳米线悬臂梁的弯曲变形

解: 由于是小变形的线弹性问题,可利用叠加原理求解. 由表 48.2,梁截面弯矩:

$$M(x) = \frac{E}{\rho} \int_A y^2 \mathrm{d}A = \frac{EI}{\rho}$$

故梁轴线的曲率可表示为 (48-8) 式:

$$\frac{1}{\rho(x)} = \frac{M(x)}{EI}$$

根据曲率的定义,平面曲线 $w = w(x)$ 的曲率 $1/\rho(x)$ 可进一步表示为

$$\frac{1}{\rho(x)} = \pm \frac{w''}{(1+w'^2)^{3/2}} = \frac{M(x)}{EI}$$

在小变形情况下,w'^2 为高阶小量,可忽略不计,上式可简化为

$$\pm w'' = \frac{M(x)}{EI}$$

符号采用弯矩逆时针为正的约定 (convention),上式可进一步表示为

$$w'' = -\frac{M(x)}{EI}$$

将上式分别对 x 积分一次和二次,便得到梁的转角曲线方程和挠曲线方程:

$$\theta = w' = -\int \frac{M(x)}{EI}\mathrm{d}x + C \tag{48-13}$$

$$w = -\int\left(\int\frac{M(x)}{EI}\mathrm{d}x\right)\mathrm{d}x + Cx + D \tag{48-14}$$

由于在推导挠曲线微分方程时应用了线弹性的曲率和弯矩关系, 同时应用了小变形的限制, 从而将非线性微分方程简化为线性方程. 此时叠加原理成立, 也就是说, 梁上某点在多个荷载同时作用下的变形, 等于每个荷载单独作用时引起同一点变形的代数和.

均布荷载作用下悬臂梁的弯矩为

$$M(x) = -\int_0^{L-x} fx\mathrm{d}x = -\frac{1}{2}f(L-x)^2 \tag{48-15}$$

悬臂梁在固定端处挠度和转角均为零, 即

$$w(0) = 0, \quad w'(0) = \theta(0) = 0 \tag{48-16}$$

将式 (48-15) 代入式 (48-14) 中, 结合式 (48-16), 得到均布荷载作用下的挠度为

$$w_1(x) = \frac{1}{24EI}fx^2(x^2 - 4Lx + 6L^2) \tag{48-17}$$

集中荷载作用下悬臂梁的弯矩为

$$M(x) = P(a-x), \quad 0 \leqslant x \leqslant a \tag{48-18}$$

同样将式 (48-18) 代入式 (48-13) 中, 结合式 (48-16), 得到集中荷载作用下的转角为

$$\theta_2(x) = \frac{1}{2EI}Px(x-2a), \quad 0 \leqslant x \leqslant a \tag{48-19}$$

将式 (48-18) 代入式 (48-14) 中, 结合式 (48-16), 得到集中荷载作用下的挠度为

$$w_{21}(x) = \frac{Px^2}{6EI}(x - 3a), \quad 0 \leqslant x \leqslant a \tag{48-20}$$

$$w_{22}(x) = w_{21}(a) + (x-a)\theta_2(a) = \frac{Pa^2}{6EI}(a - 3x), \quad a \leqslant x \leqslant L \tag{48-21}$$

将均布荷载和集中荷载分别作用下的挠度叠加得到

$$w(x) = \begin{cases} \dfrac{1}{24EI}fx^2(x^2 - 4Lx + 6L^2) + \dfrac{Px^2}{6EI}(x-3a), & 0 \leqslant x \leqslant a, \\ \dfrac{1}{24EI}fx^2(x^2 - 4Lx + 6L^2) + \dfrac{Pa^2}{6EI}(a-3x), & a \leqslant x \leqslant L. \end{cases} \tag{48-22}$$

例 48.2 建立等直杆中扭转的波动方程.

解：取如图 48.9 所示的微元体，其长度为 $\mathrm{d}x$，则下表面 (lower cross-section, 左) 的扭矩和转角分别为 $T(x)$ 和 $\phi(x)$，上表面 (upper cross-section, 右) 的扭矩和转角分别为 $T(x+\mathrm{d}x) \approx T + \dfrac{\partial T}{\partial x}\mathrm{d}x$ 和 $\phi(x+\mathrm{d}x) \approx \phi + \dfrac{\partial \phi}{\partial x}\mathrm{d}x$，这里只做了一阶泰勒展开.

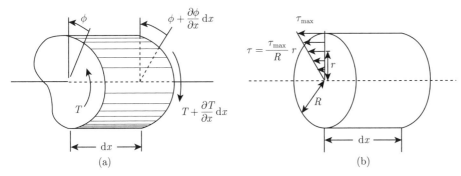

图 48.9 扭转微元体

要建立扭转的波动方程，就需要应用动量矩定理. 其中，扭矩的改变为 $\dfrac{\partial T}{\partial x}\mathrm{d}x$，对于圆轴而言，采用微元体的方法，扭矩和单位转角 $\theta = \dfrac{\partial \phi}{\partial x}$ 之间的关系为

$$T = \int_A \tau r \mathrm{d}A = \frac{\tau_{\max}}{R}\int_A r^2 \mathrm{d}A = \frac{\tau_{\max}}{R} I_P \tag{48-23}$$

式中，$I_P = \displaystyle\int_A r^2 \mathrm{d}A$ 为极惯性矩 (polar moment of inertia). 对于扭转问题，基于小变形假定，在转角很小时 (满足 $\tan\theta \approx \theta$)，剪应力和单位扭转角 $\theta = \dfrac{\partial \phi}{\partial x}$ 之间满足

$$\tau_{\max} = GR\theta = GR\frac{\partial \phi}{\partial x} \tag{48-24}$$

将 (48-24) 式代回 (48-23) 式，得到用单位扭转角表示的扭矩表达式：

$$T = GI_P \frac{\partial \phi}{\partial x} \tag{48-25}$$

微元体的动量矩为

$$\int_A r^2 \frac{\partial \phi}{\partial t}\underbrace{\rho \mathrm{d}x \mathrm{d}A}_{\mathrm{d}m} = \rho \mathrm{d}x \frac{\partial \phi}{\partial t}\int_A r^2 \mathrm{d}A = \rho \mathrm{d}x I_P \frac{\partial \phi}{\partial t} \tag{48-26}$$

上式亦表明，微元体的动量矩还可通过转动惯量 (moment of inertia) $I = \displaystyle\int_A r^2 \mathrm{d}m$ 表示为 $I\dfrac{\partial \phi}{\partial t}\mathrm{d}x$. 动量矩定理表明：

$$\frac{\partial T}{\partial x}\mathrm{d}x = GI_P \frac{\partial^2 \phi}{\partial x^2}\mathrm{d}x$$

$$= \frac{\partial}{\partial t}\left(\rho \mathrm{d}x I_P \frac{\partial \phi}{\partial t}\right) = \rho \mathrm{d}x I_P \frac{\partial^2 \phi}{\partial t^2} \tag{48-27}$$

整理上式，有 $G\dfrac{\partial^2 \phi}{\partial x^2} = \rho \dfrac{\partial^2 \phi}{\partial t^2}$，则杆扭转的波动方程为

$$\frac{\partial^2 \phi}{\partial x^2} = \frac{1}{c_t^2}\frac{\partial^2 \phi}{\partial t^2} \tag{48-28}$$

上式为一个双曲型的二阶偏微分方程. 式中，扭转波速就是横波波速：

$$c_t = \sqrt{\frac{G}{\rho}} \tag{48-29}$$

上式和 (14-20) 式中第二式完全相同.

例 48.3 给出简单剪切 (simple shear) 和纯剪切 (pure shear) 的定义.

答：定义如下：

(1) 如图 48.10，如果一个变形的变形梯度张量 $\boldsymbol{F} \in \mathrm{GL}^+(3)$ 满足下列关系的话，则被称为简单剪切：

$$[\boldsymbol{F}] = \begin{bmatrix} 1 & \gamma & 0 \\ 0 & 1 & 0 \\ 0 & 0 & 1 \end{bmatrix} \tag{48-30}$$

式中，$\gamma \in \mathbb{R}$，$\mathrm{GL}^+(3) = \{\boldsymbol{X} \in \mathbb{R}^{3\times 3} \,|\, \det \boldsymbol{X} > 0\}$ 是行列式大于零的 3×3 可逆矩阵组成的实数域的一般线性群 (general linear group). 更为具体的讨论见思考题 10.4.

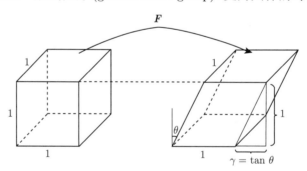

图 48.10 简单剪切示意图

(2) 如图 13.3(b) 所示，如果一个应力场 $\boldsymbol{\sigma} \in \mathrm{Sym}(3)$ 满足如下关系的话，则被称为纯剪切：

$$[\boldsymbol{\sigma}] = \begin{bmatrix} 0 & s & 0 \\ s & 0 & 0 \\ 0 & 0 & 0 \end{bmatrix} = s\left(\boldsymbol{e}_1 \otimes \boldsymbol{e}_2 + \boldsymbol{e}_2 \otimes \boldsymbol{e}_1\right) \tag{48-31}$$

式中，$s \in \mathbb{R}$，$\text{Sym}(3) = \left\{ \boldsymbol{X} \in \mathbb{R}^{3\times 3} \middle| \boldsymbol{X} = \boldsymbol{X}^{\text{T}} \right\}$ 为 3×3 对称矩阵组成的实数域的对称群 (symmetrical group). 更为具体的讨论见思考题 10.5.

圣维南原理 (Saint Venant's principle, SVP): 如图 48.11 所示，如果作用在边界 (端点) 的小面积上 ΔA 的力被作用在这块小面积上的另一组外力所代替，这组外力与原外力是静力等效的 (即合力相等、合力矩相等)，那么在物体内部产生的应力改变随着与 ΔA 距离的增加迅速衰减.

图 48.11 (a) 表明，试件上部的集中载荷使其局部的网格变形发生扭曲，所影响的范围大致和试件的宽度在一个数量级；图 48.11 (a) 和 (b) 均表明，在试件的底部由于固定端的束缚作用，其局部的网格也都发生了扭曲.

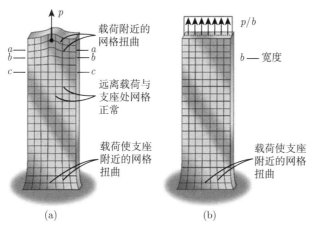

图 48.11　拉伸样品中的圣维南原理示意图

(a) 集中载荷作用；(b) 等效力系的均布载荷作用

SVP 在工程实际中的广泛应用：该原理是从大量工程实际中所总结出来的. SVP 最大的好处是帮助我们在实际问题中把复杂的边界载荷问题加以简化，可以用一个最简单的、在局部静力等效的应力分布来替代边界上的复杂应力.

圣维南原理的要点有两处：(一) 两个力系必须是按照刚化原理的 "静力等效" 力系；(二) 替换所在的表面必须小，并且替换导致在小表面附近失去精确解. 一般对连续体而言，替换所造成显著影响的区域深度与小表面的直径有关.

我们再以如图 48.12 所示的受集中载荷作用的半无限大体为例说明，该例子也被广泛地称为 "布辛涅斯克问题" (Boussinesq problem). 静力等效的两个条件均已经满足要求，即两个分力和虚线等距，且两个分力的距离是个小量. 那么，距离力的作用点足够远的点的应力分布满足圣维南原理.

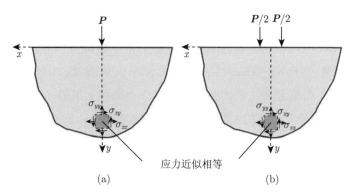

(a)   (b)

图 48.12   半无限大体中的圣维南原理示意图

(a) 集中载荷作用；(b) 等效力系的两个集中载荷作用

§49. 弹性力学提法和方程

49.1 弹性力学平衡方程

在 §32 中，已经应用动量守恒的方法推导出了弹性力学的平衡方程 (32-9) 式. 为了加深对该问题的理解，下面将用微元体的方法从另一侧面进行推导. 选取如图 49.1 所示的微元体，其体积为 $\mathrm{d}V = \mathrm{d}x_1 \mathrm{d}x_2 \mathrm{d}x_3$.

为了简化问题，先只考虑垂直于 x_1 轴的两个表面的应力分布.

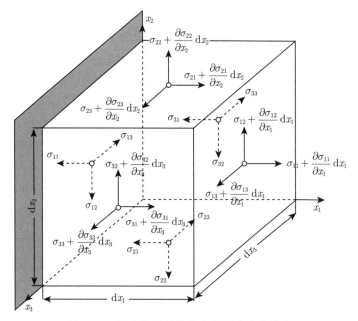

图 49.1   整个微元体的各个表面的应力分布

第 10 章 固体力学要义

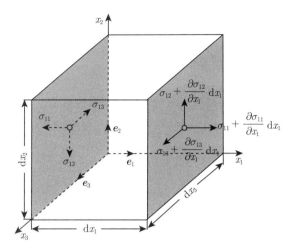

图 49.2 仅仅考虑微元体中垂直于 x_1 两个表面的应力分布

如图 49.2 所示，微元体中垂直于 x_1 两个截面上沿 x_1 方向上力的变化为

$$\mathrm{d}x_2\mathrm{d}x_3\left(\sigma_{11}+\frac{\partial\sigma_{11}}{\partial x_1}\mathrm{d}x_1-\sigma_{11}\right)\bm{e}_1 = \mathrm{d}V\frac{\partial\sigma_{11}}{\partial x_1}\bm{e}_1 \tag{49-1}$$

同理，垂直于 x_1 两个截面上沿 x_2 和 x_3 方向上力的变化分别为

$$\mathrm{d}V\frac{\partial\sigma_{12}}{\partial x_1}\bm{e}_2,\quad \mathrm{d}V\frac{\partial\sigma_{13}}{\partial x_1}\bm{e}_3 \tag{49-2}$$

综上，两个垂直于 x_1 的截面上合力变化为

$$\mathrm{d}\bm{F}_1 = \mathrm{d}V\left(\frac{\partial\sigma_{11}}{\partial x_1}\bm{e}_1 + \frac{\partial\sigma_{12}}{\partial x_1}\bm{e}_2 + \frac{\partial\sigma_{13}}{\partial x_1}\bm{e}_3\right) \tag{49-3}$$

将上述只针对垂直于 x_1 的截面的结果推广至整个微元体，则单位体积总的合力变化为

$$\begin{aligned}\frac{\mathrm{d}\bm{F}}{\mathrm{d}V} &= \left(\frac{\partial\sigma_{11}}{\partial x_1}+\frac{\partial\sigma_{21}}{\partial x_2}+\frac{\partial\sigma_{31}}{\partial x_3}\right)\bm{e}_1 + \left(\frac{\partial\sigma_{12}}{\partial x_1}+\frac{\partial\sigma_{22}}{\partial x_2}+\frac{\partial\sigma_{32}}{\partial x_3}\right)\bm{e}_2 \\ &\quad+ \left(\frac{\partial\sigma_{13}}{\partial x_1}+\frac{\partial\sigma_{23}}{\partial x_2}+\frac{\partial\sigma_{33}}{\partial x_3}\right)\bm{e}_3 \\ &= \bm{\sigma}\cdot\bm{\nabla}\end{aligned} \tag{49-4}$$

显然，由于微元体体积 $\mathrm{d}V$ 任意性，则弹性体的平衡方程为 (32-9) 式：$\bm{\sigma}\cdot\bm{\nabla}=\bm{0}$.

49.2 弹性模量独立分量的个数

对于小变形的线弹性体，在等温情况下本构关系可表示为

$$\bm{\sigma}=\bm{\mathcal{C}}:\bm{\varepsilon},\quad \sigma_{ij}=\mathcal{C}_{ijkl}\varepsilon_{kl} \tag{49-5}$$

其中, $\mathcal{C}$ 为弹性模量, 是一个四阶张量, 原则上应有 $3^4 = 81$ 个独立的分量, 按照指标是否相同分为四类:

第 1 类: 四个指标完全相同, 共有 $\mathrm{C}_3^1 = 3$ 个系数:
$$\mathcal{C}_{1111}, \quad \mathcal{C}_{2222}, \quad \mathcal{C}_{3333};$$

第 2 类: 三个指标相同, 共有 $\mathrm{C}_3^1 \mathrm{C}_2^1 \mathrm{C}_4^1 = 24$ 个系数, 例如:
$$\mathcal{C}_{1112}, \quad \mathcal{C}_{2221}, \quad \mathcal{C}_{1113}, \quad \mathcal{C}_{3331}, \quad \cdots;$$

第 3 类: 指标两两相同, 共有 $\mathrm{C}_3^2 \mathrm{C}_4^2 = 18$ 个系数, 例如:
$$\mathcal{C}_{1122}, \quad \mathcal{C}_{2211}, \quad \mathcal{C}_{1212}, \quad \mathcal{C}_{2121}, \quad \mathcal{C}_{2112}, \quad \mathcal{C}_{1221}, \quad \cdots;$$

第 4 类: 只有两个指标相同, 共有 $\mathrm{C}_3^1 \mathrm{C}_4^2 = 36$ 个系数, 例如:
$$\mathcal{C}_{1123}, \quad \mathcal{C}_{2311}, \quad \mathcal{C}_{1213}, \quad \mathcal{C}_{2131}, \quad \mathcal{C}_{2113}, \quad \mathcal{C}_{1231}, \quad \cdots.$$

然而, 该张量具有完全对称性, 即
$$\mathcal{C}_{ijkl} = \mathcal{C}_{jikl} = \mathcal{C}_{ijlk} = \mathcal{C}_{klij} \tag{49-6}$$

根据应变张量 ε 和应力张量 σ 的对称性, 有
$$\begin{aligned}&\mathcal{C}_{ijkl}\varepsilon_{kl} = \mathcal{C}_{ijkl}\varepsilon_{lk} = \mathcal{C}_{ijlk}\varepsilon_{kl} \\ &\sigma_{ij} = \mathcal{C}_{ijkl}\varepsilon_{kl} = \sigma_{ji} = \mathcal{C}_{jikl}\varepsilon_{kl}\end{aligned} \tag{49-6$'$}$$

则 (49-6) 式前两个等号成立, $\mathcal{C}$ 的前后两组脚标的独立个数分别由 3×3 变为 6, 则 $\mathcal{C}$ 的独立分量只有 36 个.

又由于弹性模量和应变能密度 w 之间满足:
$$\mathcal{C}_{ijkl} = \frac{\partial^2 w}{\partial \varepsilon_{ij} \partial \varepsilon_{kl}} = \frac{\partial^2 w}{\partial \varepsilon_{kl} \partial \varepsilon_{ij}} = \mathcal{C}_{klij} \tag{49-6$''$}$$

则 (49-6) 式中第三个等号成立, $\mathcal{C}$ 的前后两组脚标对称, 则 $\mathcal{C}$ 的独立分量只有 $\frac{1}{2}n(n+1) = \frac{1}{2}\times 6 \times 7 = 21$ 个.

根据 $\mathcal{C}$ 的对称性, 此时 (49-5) 式可以表示为矩阵形式:

$$\begin{bmatrix} \sigma_{11} \\ \sigma_{22} \\ \sigma_{33} \\ \sqrt{2}\sigma_{12} \\ \sqrt{2}\sigma_{23} \\ \sqrt{2}\sigma_{31} \end{bmatrix} = \begin{bmatrix} \mathcal{C}_{1111} & \mathcal{C}_{1122} & \mathcal{C}_{1133} & \sqrt{2}\mathcal{C}_{1112} & \sqrt{2}\mathcal{C}_{1123} & \sqrt{2}\mathcal{C}_{1131} \\ & \mathcal{C}_{2222} & \mathcal{C}_{2233} & \sqrt{2}\mathcal{C}_{2212} & \sqrt{2}\mathcal{C}_{2223} & \sqrt{2}\mathcal{C}_{2231} \\ & & \mathcal{C}_{3333} & \sqrt{2}\mathcal{C}_{3312} & \sqrt{2}\mathcal{C}_{3323} & \sqrt{2}\mathcal{C}_{3331} \\ & & & 2\mathcal{C}_{1212} & 2\mathcal{C}_{1223} & 2\mathcal{C}_{1231} \\ & & & & 2\mathcal{C}_{2323} & 2\mathcal{C}_{2331} \\ & & & & & 2\mathcal{C}_{3131} \end{bmatrix} \begin{bmatrix} \varepsilon_{11} \\ \varepsilon_{22} \\ \varepsilon_{33} \\ \sqrt{2}\varepsilon_{12} \\ \sqrt{2}\varepsilon_{23} \\ \sqrt{2}\varepsilon_{31} \end{bmatrix}$$

即
$$\boldsymbol{\sigma} = \boldsymbol{D} \cdot \boldsymbol{\varepsilon}$$

其中 $\boldsymbol{D}$ 为刚度矩阵. 此时的材料为完全各向异性, 当材料的性质具有某种对称性时, $\mathcal{C}$ 的独立分量还会减少.

对材料进行坐标变换, 其坐标转换矩阵为 $\boldsymbol{A}$, 对应的应变张量 $\boldsymbol{\varepsilon}'$ 和应力张量 $\boldsymbol{\sigma}'$ 变换为

$$\boldsymbol{\varepsilon}' = \boldsymbol{A}\boldsymbol{\varepsilon}\boldsymbol{A}^{\mathrm{T}}, \quad \varepsilon'_{pq} = A_{pk}A_{ql}\varepsilon_{kl}, \quad \varepsilon_{kl} = A_{pk}A_{ql}\varepsilon'_{pq}$$

$$\boldsymbol{\sigma}' = \boldsymbol{A}\boldsymbol{\sigma}\boldsymbol{A}^{\mathrm{T}}, \quad \sigma'_{rt} = A_{ri}A_{tj}\sigma_{ij}, \quad \sigma_{ij} = A_{ri}A_{tj}\sigma'_{rt}$$

又 $\sigma_{ij} = \mathcal{C}_{ijkl}\varepsilon_{kl}$, 则有

$$A_{ri}A_{tj}T'_{rt} = \mathcal{C}_{ijkl}A_{pk}A_{ql}E'_{pq}; \tag{49-7}$$

用 $A_{mi}A_{nj}$ 乘 (49-7) 式两边, 考虑

$$A_{mi}A_{ri} = \delta_{mr}, \quad A_{nj}A_{tj} = \delta_{nt},$$

得

$$A_{mi}A_{nj}A_{ri}A_{tj}\sigma'_{rt} = \delta_{mr}\delta_{nt}\sigma'_{rt} = \sigma'_{mn} = \mathcal{C}_{ijkl}A_{mi}A_{nj}A_{pk}A_{ql}\varepsilon'_{pq} \tag{49-8}$$

新坐标系下的应力应变关系表示为

$$\sigma'_{mn} = \mathcal{C}'_{mnpq}\varepsilon'_{pq}, \tag{49-9}$$

对比 (49-8) 和 (49-9) 两式, 有

$$\left(\mathcal{C}'_{mnpq} - A_{mi}A_{nj}A_{pk}A_{ql}\mathcal{C}_{ijkl}\right)\varepsilon'_{pq} = 0$$

由于 $\mathcal{C}'_{mnpq}$, $\mathcal{C}_{ijkl}$ 的完全对称性, 上式对任意对称张量 ε'_{pq} 都成立并推出

$$\mathcal{C}'_{mnpq} = A_{mi}A_{nj}A_{pk}A_{ql}\mathcal{C}_{ijkl} \tag{49-10}$$

考虑当材料具有对称性时, $\mathcal{C}$ 的独立分量个数. 若变换 $\boldsymbol{A}$ 满足材料的对称性, 此时转换前后, 表示物性的张量 $\mathcal{C}$ 的分量保持不变, 即

$$\mathcal{C}'_{mnpq} = \mathcal{C}_{ijkl} \tag{49-11}$$

当材料具有一个对称面时, 设 $x = 0$ 为对称面, 引进变换如图 49.3 所示. 此时坐标的转换矩阵为

$$[\boldsymbol{A}] = \begin{bmatrix} -1 & 0 & 0 \\ 0 & 1 & 0 \\ 0 & 0 & 1 \end{bmatrix}$$

且变换前后弹性张量 $\mathcal{C}$ 满足 (49-10) 和 (49-11) 两式, 易得

$$C'_{1112} = A_{11}A_{11}A_{11}A_{22}C_{1112} = -C_{1112} = C_{1112} = 0$$

即脚标中含有奇数个 1 的分量为 0, 同理可导出

$$C_{1112} = C_{2212} = C_{3312} = C_{1223} = C_{1131} = C_{2231} = C_{3331} = C_{2331} = 0$$

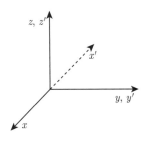

图 49.3 对称面为 $x = 0$

这种情况下, $\mathcal{C}$ 的矩阵形式 ($\boldsymbol{D}$) 便为

$$\begin{bmatrix} C_{1111} & C_{1122} & C_{1133} & 0 & \sqrt{2}C_{1123} & 0 \\ & C_{2222} & C_{2233} & 0 & \sqrt{2}C_{2223} & 0 \\ & & C_{3333} & 0 & \sqrt{2}C_{3323} & 0 \\ & & & 2C_{1212} & 0 & 2C_{1231} \\ & & & & 2C_{2323} & 0 \\ & & & & & 2C_{3131} \end{bmatrix} \quad (49\text{-}12)$$

因此, 当材料具有一个对称面时, $\mathcal{C}$ 的独立分量有 13 个.

当材料为正交各向异性材料, 即同时具有第二个对称面 (相互正交面) 时, 在 (49-12) 式的基础上设 $y = 0$ 为对称面, 同样可以证明脚标中含有奇数个 2 的分量为 0, 即

$$\begin{bmatrix} C_{1111} & C_{1122} & C_{1133} & 0 & 0 & 0 \\ & C_{2222} & C_{2233} & 0 & 0 & 0 \\ & & C_{3333} & 0 & 0 & 0 \\ & & & 2C_{1212} & 0 & 0 \\ & & & & 2C_{2323} & 0 \\ & & & & & 2C_{3131} \end{bmatrix} \quad (49\text{-}13)$$

注意到, 脚标包含奇数个 3 的分量也都为 0 了, 可以看出, 有两个相互正交的对称面时, 就有三个相互正交的对称面. 即正交各向异性材料, $\mathcal{C}$ 的独立分量有 9 个.

如果材料性能关于一根轴对称, 称之为横观各向同性. 设 z 轴为对称轴, 那么将坐标系绕 z 轴顺时针旋转 $90°$ (如图 49.4), 其转换矩阵为

$$[\boldsymbol{A}] = \begin{bmatrix} 0 & 1 & 0 \\ -1 & 0 & 0 \\ 0 & 0 & 1 \end{bmatrix}$$

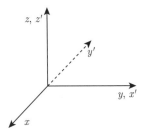

图 49.4 z 轴为对称轴顺时针旋转 $90°$

此时在 (49-13) 式的基础上, 有

$$\mathcal{C}'_{1111} = \mathcal{C}_{1111} = A_{12}A_{12}A_{12}A_{12}\mathcal{C}_{2222} = \mathcal{C}_{2222}$$

$$\mathcal{C}'_{1133} = \mathcal{C}_{1133} = A_{12}A_{12}A_{33}A_{33}\mathcal{C}_{2233} = \mathcal{C}_{2233}$$

$$\mathcal{C}'_{3131} = \mathcal{C}_{3131} = A_{33}A_{12}A_{33}A_{12}\mathcal{C}_{3232} = \mathcal{C}_{2323}$$

若将坐标系绕 z 轴顺时针旋转 $45°$ (如图 49.5), 其转换矩阵为

$$[\boldsymbol{A}] = \begin{bmatrix} \sqrt{2}/2 & \sqrt{2}/2 & 0 \\ -\sqrt{2}/2 & \sqrt{2}/2 & 0 \\ 0 & 0 & 1 \end{bmatrix}$$

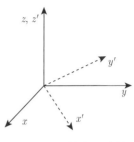

图 49.5 z 轴为对称轴
顺时针旋转 $45°$

在这个变换下, 有

$$\begin{aligned}
\mathcal{C}'_{1111} &= \mathcal{C}_{1111} \\
&= A_{11}A_{11}A_{11}A_{11}\mathcal{C}_{1111} + A_{11}A_{11}A_{12}A_{12}\mathcal{C}_{1122} \\
&\quad + A_{12}A_{12}A_{11}A_{11}\mathcal{C}_{2211} + A_{11}A_{12}A_{11}A_{12}\mathcal{C}_{1212} \\
&\quad + A_{12}A_{11}A_{12}A_{11}\mathcal{C}_{2121} + A_{11}A_{12}A_{12}A_{11}\mathcal{C}_{1221} \\
&\quad + A_{12}A_{11}A_{11}A_{12}\mathcal{C}_{2112} + A_{12}A_{12}A_{12}A_{12}\mathcal{C}_{2222} \\
&= \frac{1}{2}\mathcal{C}_{1111} + \frac{1}{2}\mathcal{C}_{1122} + \frac{1}{2}\mathcal{C}_{1212} + \frac{1}{2}\mathcal{C}_{1221} \\
&= \frac{1}{2}\mathcal{C}_{1111} + \frac{1}{2}\mathcal{C}_{1122} + \mathcal{C}_{1212}
\end{aligned}$$

即 $\mathcal{C}_{1212} = \frac{1}{2}(\mathcal{C}_{1111} - \mathcal{C}_{1122})$.

结合以上各式, 此时 $\mathcal{C}$ 的矩阵形式 $(\boldsymbol{D})$ 便为

$$\begin{bmatrix} \mathcal{C}_{1111} & \mathcal{C}_{1122} & \mathcal{C}_{1133} & 0 & 0 & 0 \\ & \mathcal{C}_{1111} & \mathcal{C}_{1133} & 0 & 0 & 0 \\ & & \mathcal{C}_{3333} & 0 & 0 & 0 \\ & & & \mathcal{C}_{1111} - \mathcal{C}_{1122} & 0 & 0 \\ & & & & 2\mathcal{C}_{2323} & 0 \\ & & & & & 2\mathcal{C}_{2323} \end{bmatrix} \quad (49\text{-}14)$$

因此, 当材料为横观各向同性时, $\mathcal{C}$ 的独立分量有 5 个.

若材料关于两个坐标轴对称, 可以证明它有是三个对称轴, 为各向同性材料. 在 (49-14) 式的基础上, 设 x 轴也为对称轴, 绕其顺时针旋转 $90°$ 和 $45°$ 的旋转矩阵分别为 (图 49.6):

$$[A] = \begin{bmatrix} 1 & 0 & 0 \\ 0 & 0 & 1 \\ 0 & -1 & 0 \end{bmatrix}, \quad [A] = \begin{bmatrix} 1 & 0 & 0 \\ 0 & \sqrt{2}/2 & \sqrt{2}/2 \\ 0 & -\sqrt{2}/2 & \sqrt{2}/2 \end{bmatrix}$$

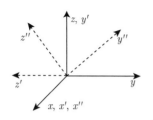

图 49.6 对称轴为 x 轴

在这种情况下, 可证明:

$$\mathcal{C}'_{1122} = \mathcal{C}_{1122} = A_{11}A_{11}A_{23}A_{23}\mathcal{C}_{1133} = \mathcal{C}_{1133}$$

$$\mathcal{C}'_{2222} = \mathcal{C}_{2222} = A_{23}A_{23}A_{23}A_{23}\mathcal{C}_{3333} = \mathcal{C}_{3333}$$

$$\mathcal{C}'_{2323} = \mathcal{C}_{2323} = \mathcal{C}_{1212} = \frac{1}{2}(\mathcal{C}_{1111} - \mathcal{C}_{1122})$$

此时 $\mathcal{C}$ 的矩阵形式 (D) 便为

$$\begin{bmatrix} \mathcal{C}_{1111} & \mathcal{C}_{1122} & \mathcal{C}_{1122} & 0 & 0 & 0 \\ & \mathcal{C}_{1111} & \mathcal{C}_{1122} & 0 & 0 & 0 \\ & & \mathcal{C}_{1111} & 0 & 0 & 0 \\ & & & \mathcal{C}_{1111} - \mathcal{C}_{1122} & 0 & 0 \\ & & & & \mathcal{C}_{1111} - \mathcal{C}_{1122} & 0 \\ & & & & & \mathcal{C}_{1111} - \mathcal{C}_{1122} \end{bmatrix}$$

定义

$$\mathcal{C}_{1111} = \lambda + 2\mu$$
$$\mathcal{C}_{1212} = \mu$$
$$\mathcal{C}_{1122} = \mathcal{C}_{1111} - 2\mathcal{C}_{1212} = \lambda$$

发现经过以上所有的变换, 对各向同性材料, 易有 (40-19) 式中第二式

$$\mathcal{C}_{ijks} = \lambda \delta_{ij}\delta_{ks} + \mu(\delta_{ik}\delta_{js} + \delta_{is}\delta_{jk})$$

将之代入 (49-5) 式有 (40-15) 式中第一式

$$\sigma_{ij} = \lambda \varepsilon_{kk}\delta_{ij} + 2\mu\varepsilon_{ij}$$

各向同性的本构关系可表示为

$$\begin{bmatrix} \sigma_{11} \\ \sigma_{22} \\ \sigma_{33} \\ \sqrt{2}\sigma_{12} \\ \sqrt{2}\sigma_{23} \\ \sqrt{2}\sigma_{31} \end{bmatrix} = \begin{bmatrix} \lambda+2\mu & \lambda & \lambda & 0 & 0 & 0 \\ & \lambda+2\mu & \lambda & 0 & 0 & 0 \\ & & \lambda+2\mu & 0 & 0 & 0 \\ & & & 2\mu & 0 & 0 \\ & & & & 2\mu & 0 \\ & & & & & 2\mu \end{bmatrix} \begin{bmatrix} \varepsilon_{11} \\ \varepsilon_{22} \\ \varepsilon_{33} \\ \sqrt{2}\varepsilon_{12} \\ \sqrt{2}\varepsilon_{23} \\ \sqrt{2}\varepsilon_{31} \end{bmatrix}$$

这样，当材料为各向同性时，$\mathcal{C}$ 的独立分量有 2 个.

49.3 勒让德–阿达玛不等式

第一类皮奥拉–基尔霍夫应力 PK1 定义为

$$\boldsymbol{P} = J\boldsymbol{\sigma}\boldsymbol{F}^{-\mathrm{T}} = \boldsymbol{\sigma}\mathbf{Cof}\boldsymbol{F} \tag{49-15}$$

式中，$\boldsymbol{\sigma} \in \mathbb{S}^3$ 为柯西应力，换句话说，柯西应力属于 3×3 对称矩阵. $\mathbf{Cof}\boldsymbol{F} = (\det \boldsymbol{F})\boldsymbol{F}^{-\mathrm{T}} = J\boldsymbol{F}^{-\mathrm{T}}$ 为变形梯度 $\boldsymbol{F}$ 的余子式矩阵 (cofactor matrix).

对于超弹性材料，应变能密度函数，也就是单位体积的应变能为满足客观性要求的标量函数 $w(\boldsymbol{F}) = w(\boldsymbol{Q}\boldsymbol{F})$，这里 $\boldsymbol{Q}$ 为满足 $\boldsymbol{Q}\boldsymbol{Q}^{\mathrm{T}} = \mathbf{I}$ 的正常正交张量，则 PK1 为

$$\boldsymbol{P} = \frac{\partial w}{\partial \boldsymbol{F}} \tag{49-16}$$

从而低弹性本构关系 (hypoelastic constitutive relation) 可表示为

$$\dot{\boldsymbol{P}} = \boldsymbol{C} : \dot{\boldsymbol{F}} \tag{49-17}$$

式中，四阶弹性张量 (elasticity tensor) 为

$$\boldsymbol{C} = \frac{\partial^2 w}{\partial \boldsymbol{F} \partial \boldsymbol{F}}, \quad C_{iJkL} = \frac{\partial^2 w}{\partial F_{iJ} \partial F_{kL}} = \frac{\partial^2 w}{\partial F_{kL} \partial F_{iJ}} = C_{kLiJ} \tag{49-18}$$

不考虑体力时的用 PK1 表示的平衡方程为

$$\boldsymbol{P} \cdot \boldsymbol{\nabla}_{\boldsymbol{X}} = \boldsymbol{0} \tag{49-19}$$

由于变形梯度 $\boldsymbol{F}$ 和位移矢量 $\boldsymbol{u}(\boldsymbol{X})$ 之间的关系为

$$\boldsymbol{F} = \mathbf{I} + \boldsymbol{u} \otimes \boldsymbol{\nabla}_{\boldsymbol{X}} = \mathbf{I} + \boldsymbol{H} \tag{49-20}$$

式中，$\boldsymbol{H} = \boldsymbol{u} \otimes \boldsymbol{\nabla}_X$ 为位移梯度，也是一个两点张量场. 将 (49-16) 式和 (49-20) 式代入 (49-19) 式，得到用位移表示的平衡方程：

$$C_{iJkL} \frac{\partial^2 u_k}{\partial X_J \partial X_L} = 0 \tag{49-21}$$

事实上，弹性力学用位移所表示的平衡方程 (49-21) 从本质上应该为一个椭圆型偏微分方程. 容易理解，是四阶弹性张量 C_{iJkL} 决定了二阶、拟线性偏微分方程 (49-21) 的类型.

勒让德–阿达玛不等式 (Legendre-Hadamard inequality)[10.4] 就是指对于任意矢量 $\boldsymbol{a}$ 和 $\boldsymbol{b}$，四阶弹性张量 (49-18) 式满足如下本构不等式：

$$(\boldsymbol{a} \otimes \boldsymbol{b}) : \boldsymbol{C} : (\boldsymbol{a} \otimes \boldsymbol{b}) \equiv C_{iJkL} a_i b_J a_k b_L \geqslant 0 \tag{49-22}$$

上式亦被称为椭圆性条件 (ellipticity condition). 当 $\boldsymbol{a} \neq 0$ 和 $\boldsymbol{b} \neq 0$ 不等式 (49-22) 严格成立时，则称为强椭圆性条件 (strong ellipticity condition) 或强勒让德–阿达玛条件 (strong Legendre-Hadamard condition).

对于波动问题，在无体力时，用 PK1 表示的波动方程为如下布辛涅斯克动量方程：

$$\boldsymbol{P} \cdot \boldsymbol{\nabla}_X = \rho_0 \boldsymbol{a}_0 = \rho_0 \ddot{\boldsymbol{u}}_0 = \rho_0 \ddot{\boldsymbol{u}} \tag{49-23}$$

则 (49-21) 式改写为

$$C_{iJkL} \frac{\partial^2 u_k}{\partial X_J \partial X_L} = \rho_0 \frac{\partial^2 u_i}{\partial t^2} \tag{49-24}$$

则 (49-24) 式为一个二阶双曲型偏微分方程.

不失一般性，下面仅讨论小变形的线性化的波动问题，此时可不必再区分指标的大小写，也可令 $\rho_0 = \rho$，从而 (49-24) 式可进一步简化为

$$C_{ijkl} \frac{\partial^2 u_k}{\partial x_j \partial x_l} = \rho \frac{\partial^2 u_i}{\partial t^2}, \quad \boldsymbol{C} \vdots \frac{\partial^2 \boldsymbol{u}}{\partial \boldsymbol{x} \partial \boldsymbol{x}} = \rho \frac{\partial^2 \boldsymbol{u}}{\partial t^2} \tag{49-25}$$

参照 (14-13) 式，位移矢量可表示为

$$\boldsymbol{u}(\boldsymbol{x}, t) = \boldsymbol{a} \varphi[k(\boldsymbol{x} \cdot \boldsymbol{m} - ct)] \tag{49-26}$$

式中，$\boldsymbol{a}$ 为振幅矢量，φ 为一无量纲函数，k 为波数 (量纲为长度的倒数)，$\boldsymbol{m}$ 为满足 $\boldsymbol{m}^2 = 1$ 的单位方向矢量，c 为波速. 由于

$$\ddot{\boldsymbol{u}} = c^2 k^2 \varphi'' \boldsymbol{a}, \quad \boldsymbol{C} \vdots \frac{\partial^2 \boldsymbol{u}}{\partial \boldsymbol{x} \partial \boldsymbol{x}} = k^2 \varphi'' \boldsymbol{C} \vdots (\boldsymbol{m} \otimes \boldsymbol{a} \otimes \boldsymbol{m}) \tag{49-27}$$

阿达玛 (Jacques Hadamard, 1865—1963)

第 10 章　固体力学要义

将 (49-27) 式代入 (49-25) 式, 得到

$$\boldsymbol{C} \vdots (\boldsymbol{k} \otimes \boldsymbol{a} \otimes \boldsymbol{k}) = \rho c^2 \boldsymbol{a}, \quad C_{ijkl} k_j a_k k_l = \rho c^2 a_i \tag{49-28}$$

对于波的传播问题, 由于振幅矢量 $\boldsymbol{a}$ 满足 $\boldsymbol{a}^2 = a^2 > 0$, (49-28) 式中第一式可改写为

$$c^2 = \frac{1}{\rho a^2} (\boldsymbol{a} \otimes \boldsymbol{m}) : \boldsymbol{C} : (\boldsymbol{a} \otimes \boldsymbol{m}) \tag{49-29}$$

实的波速要求 $c^2 > 0$, 即要求:

$$(\boldsymbol{a} \otimes \boldsymbol{m}) : \boldsymbol{C} : (\boldsymbol{a} \otimes \boldsymbol{m}) > 0, \quad C_{ijkl} a_i m_j a_k m_l > 0 \tag{49-30}$$

也就是强椭圆性条件或强勒让德–阿达玛条件的成立. 换言之, 强勒让德–阿达玛条件保证了波动方程 (49-25) 的双曲型, 这样波才能传播.

由于二阶声学张量 (acoustic tensor) $\boldsymbol{A}$ 定义为

$$\boldsymbol{A} = \frac{1}{\rho} \boldsymbol{C} : (\boldsymbol{m} \otimes \boldsymbol{m}), \quad A_{ik} = \frac{1}{\rho} C_{ijkl} m_j m_l \tag{49-31}$$

强椭圆性条件或强勒让德–阿达玛条件保证了声学张量 $\boldsymbol{A}$ 的正定性. 需要特别注意的是, (49-31) 式中第二式清楚地注明了单位矢量 $\boldsymbol{m}$ 和 $\boldsymbol{C}$ 指标的作用顺序是 24 而非 34, 这在下面的运算中十分重要.

对于各向同性材料, 将 (40-18) 式 $\boldsymbol{C} = 2\mu \mathbb{I}^s + \lambda (\mathbf{I} \otimes \mathbf{I})$ 代入 (49-31) 式中第一式, 有

$$\begin{aligned} \boldsymbol{A}(\boldsymbol{m}) &= \frac{1}{\rho} \boldsymbol{C} : (\boldsymbol{m} \otimes \boldsymbol{m}) = [2\mu \mathbb{I}^s + \lambda (\mathbf{I} \otimes \mathbf{I})] : (\boldsymbol{m} \otimes \boldsymbol{m}) \\ &= \frac{\lambda + 2\mu}{\rho} (\boldsymbol{m} \otimes \boldsymbol{m}) + \frac{\mu}{\rho} (\mathbf{I} - \boldsymbol{m} \otimes \boldsymbol{m}) \\ &= c_l^2 (\boldsymbol{m} \otimes \boldsymbol{m}) + c_t^2 (\mathbf{I} - \boldsymbol{m} \otimes \boldsymbol{m}) \end{aligned} \tag{49-32}$$

则 (49-32) 式和 (14-19) 式完全相同. c_l 和 c_t 分别为纵波和横波波速. (49-32) 式是谱分解定理中张量 $\boldsymbol{A}$ 有两个不同本征值时 (13-2) 式的一个应用的具体实例.

例 49.1　给出 (49-32) 式的推导细节.

证明: 将 (40-19) 式中第二式 $\boldsymbol{C} = [\lambda \delta_{ij} \delta_{kl} + \mu (\delta_{ik} \delta_{jl} + \delta_{il} \delta_{jk})] \boldsymbol{e}_i \otimes \boldsymbol{e}_j \otimes \boldsymbol{e}_k \otimes \boldsymbol{e}_l$ 和 $\boldsymbol{m} \otimes \boldsymbol{m} = m_p m_q \boldsymbol{e}_p \otimes \boldsymbol{e}_q$ 代入 (49-32) 式, 在注意到上面已经提醒过的, 单位矢量 $\boldsymbol{m}$ 和 $\boldsymbol{C}$ 指标的作用顺序是 24 而非 34, 有

$$\begin{aligned}
\boldsymbol{A} &= \frac{1}{\rho}\left\{[\lambda\delta_{ij}\delta_{kl}+\mu(\delta_{ik}\delta_{jl}+\delta_{il}\delta_{jk})]\underbrace{\boldsymbol{e}_i\otimes\boldsymbol{e}_j\otimes\boldsymbol{e}_k\otimes\boldsymbol{e}_l}_{\text{第二和第四个基和后面基作用}}\right\}:(m_p m_q \boldsymbol{e}_p\otimes\boldsymbol{e}_q)\\
&=\frac{1}{\rho}[\lambda\delta_{ij}\delta_{kl}+\mu(\delta_{ik}\delta_{jl}+\delta_{il}\delta_{jk})]m_p m_q \delta_{jp}\delta_{lq}\boldsymbol{e}_i\otimes\boldsymbol{e}_k\\
&=\frac{1}{\rho}[\lambda m_i m_k+\mu(\delta_{ik}m_l m_l+m_i m_k)]\boldsymbol{e}_i\otimes\boldsymbol{e}_k\\
&=\frac{1}{\rho}\left[(\lambda+2\mu)m_i m_k+\mu\left(\delta_{ik}\underbrace{m_l m_l}_{=1}-m_i m_k\right)\right]\boldsymbol{e}_i\otimes\boldsymbol{e}_k\\
&=\frac{1}{\rho}[(\lambda+2\mu)m_i m_k+\mu(\delta_{ik}-m_i m_k)]\boldsymbol{e}_i\otimes\boldsymbol{e}_k
\end{aligned} \tag{49-33}$$

从而 (49-32) 式得证.

思考题和补充材料

10.1 圣维南 (Adhémar Jean Claude Barré de Saint-Venant, 1797—1886, 题图 10.1 左) 由柯西推荐申请院士, 到柯西死后才被评上, 46 岁首次参加, 到评上院士时是 71 岁.

圣维南的学生约瑟夫·布辛涅斯克 (Joseph Boussinesq, 1842—1929, 题图 10.1 中) 成为法国科学院院士的经历比较坎坷. 1870 年由圣维南首次推荐失败, 1868、1871、1872 (两次)、1873、1880 以及 1883 年的申请均告失败. 1883 年圣维南已经是力学组的头. 他在 1886 年的选举推荐报告被保存了下来, 可能想到布辛涅斯克会重复自己的老路, 他指出布辛涅斯克是唯一一位应该评上院士的生存者了, 这是圣维南为布辛涅斯克做的最后的努力. 圣维南死于 1886 年 1 月 6 号, 而选举投票日期是 1886 年 1 月 18 号, 以微弱票当选 (29 票, 竞争对手 Deprez 得到 26 票).

圣维南的另外一个学生莫里斯·列维 (Maurice Lévy, 1838—1910, 题图 10.1 右) 于 1883 年评为法兰西科学院院士, 时年 45 岁.

题图 10.1 圣维南、布辛涅斯克、列维

第 10 章　固体力学要义

圣维南工作到生命的最后一刻，他的最后一篇论文于 1886 年 1 月 2 日发表于法国科学院院刊，四天后的 1886 年 1 月 6 日这位伟大的力学家 89 岁时无疾而终.

10.2 对于不可压缩流体：$\boldsymbol{\sigma} = -p\mathbf{I} + \boldsymbol{\tau}$，则 (49-4) 式变为 $\dfrac{\mathrm{d}\boldsymbol{F}}{\mathrm{d}V} = -\boldsymbol{\nabla}p + \boldsymbol{\nabla}\cdot\boldsymbol{\tau}$.

10.3 为了保证位移场为单值性，弹性变形场必须满足协调方程 (compatibility equation). 考虑二维情形，柯西应变的三个分量已经由 (27-17) 式给出. 问题：验证柯西应变三个分量满足如下应变协调方程：

$$2\frac{\partial^2 \varepsilon_{xy}}{\partial x \partial y} = \frac{\partial^2 \varepsilon_{xx}}{\partial y^2} + \frac{\partial^2 \varepsilon_{yy}}{\partial x^2} \tag{s10-1}$$

推而广之，三维情形时，应用紧凑符号：$(\cdot)_{,i} = \partial(\cdot)/\partial x_i$，则应变协调方程为 (12-38) 式并可等价地表示为

$$\boldsymbol{\nabla} \times \boldsymbol{\varepsilon} \times \boldsymbol{\nabla} = \varepsilon_{kim}\varepsilon_{jln}\varepsilon_{ij,kl}\boldsymbol{e}_m \otimes \boldsymbol{e}_n = \boldsymbol{0} \tag{s10-2}$$

运用置换符号的运算规则，则 (s10-2) 式还可更为明确地表示为

$$\varepsilon_{ik,lm} + \varepsilon_{lm,ik} = \varepsilon_{il,km} + \varepsilon_{km,il} \tag{s10-3}$$

事实上，(s10-2) 式或 (s10-3) 式由下列六个方程组成：

$$\begin{cases} \varepsilon_{11,22} + \varepsilon_{22,11} = 2\varepsilon_{12,12} \\ \varepsilon_{22,33} + \varepsilon_{33,22} = 2\varepsilon_{23,23} \\ \varepsilon_{33,11} + \varepsilon_{11,33} = 2\varepsilon_{13,13} \\ \varepsilon_{11,23} + \varepsilon_{23,11} = \varepsilon_{31,21} + \varepsilon_{12,31} \\ \varepsilon_{22,31} + \varepsilon_{23,12} = \varepsilon_{31,22} + \varepsilon_{12,32} \\ \varepsilon_{33,12} + \varepsilon_{23,13} = \varepsilon_{31,23} + \varepsilon_{12,33} \end{cases} \tag{s10-4}$$

问题：结合例 1.3，阐明经典弹性理论所讨论的变形是从三维欧氏空间 ($\mathcal{E}^3$) 到三维欧氏空间 ($\mathcal{E}^3$) 的同构.

10.4 如题图 10.4 所示的简单剪切示意图. 当前构形的坐标 (x_1, x_2, x_3) 和初始构形坐标 (X_1, X_2, X_3) 之间的关系为

$$\begin{cases} x_1 = X_1 + \gamma(t)X_2 \\ x_2 = X_2 \\ x_3 = X_3 \end{cases} \tag{s10-5}$$

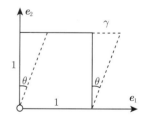

题图 10.4　简单剪切

则由 (26-6) 式和 (26-7) 式得到变形梯度和其转置的如下表达式：

$$[\boldsymbol{F}] = \begin{bmatrix} 1 & \gamma & 0 \\ 0 & 1 & 0 \\ 0 & 0 & 1 \end{bmatrix}, \quad [\boldsymbol{F}^{\mathrm{T}}] = \begin{bmatrix} 1 & 0 & 0 \\ \gamma & 1 & 0 \\ 0 & 0 & 1 \end{bmatrix} \tag{s10-6}$$

由 (27-2) 式可得到简单剪切时的格林应变为

$$[\boldsymbol{E}] = \frac{1}{2}\begin{bmatrix} 0 & \gamma & 0 \\ \gamma & \gamma^2 & 0 \\ 0 & 0 & 0 \end{bmatrix} \qquad (\text{s}10\text{-}7)$$

如果忽略掉格林应变中的高阶项 γ^2，则 (s10-7) 式则退化为小应变情形：

$$[\boldsymbol{\varepsilon}] = \frac{1}{2}\begin{bmatrix} 0 & \gamma & 0 \\ \gamma & 0 & 0 \\ 0 & 0 & 0 \end{bmatrix} \qquad (\text{s}10\text{-}8)$$

在小变形假设的前提下，表 48.2 中的圆轴的扭转则对应 (s10-8) 式的情形.

10.5 如题图 10.5 所示的纯剪切情形，当前构形的坐标 (x_1, x_2, x_3) 和初始构形坐标 (X_1, X_2, X_3) 之间的关系为

$$\begin{cases} x_1 = \dfrac{\lambda + \lambda^{-1}}{2}X_1 + \dfrac{\lambda - \lambda^{-1}}{2}X_2 \\ x_2 = \dfrac{\lambda - \lambda^{-1}}{2}X_1 + \dfrac{\lambda + \lambda^{-1}}{2}X_2 \\ x_3 = X_3 \end{cases} \qquad (\text{s}10\text{-}9)$$

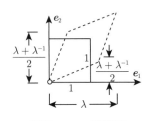

题图 10.5 纯剪切

其变形梯度和其转置为

$$[\boldsymbol{F}] = \frac{1}{2}\begin{bmatrix} \lambda + \lambda^{-1} & \lambda - \lambda^{-1} & 0 \\ \lambda - \lambda^{-1} & \lambda + \lambda^{-1} & 0 \\ 0 & 0 & 2 \end{bmatrix}, \quad [\boldsymbol{F}^{\mathrm{T}}] = [\boldsymbol{F}] \qquad (\text{s}10\text{-}10)$$

由 (27-2) 式可得到纯剪切时的格林应变为

$$[\boldsymbol{E}] = \frac{1}{4}\begin{bmatrix} \lambda^2 + \lambda^{-2} - 2 & \lambda^2 - \lambda^{-2} & 0 \\ \lambda^2 - \lambda^{-2} & \lambda^2 + \lambda^{-2} - 2 & 0 \\ 0 & 0 & 0 \end{bmatrix} \qquad (\text{s}10\text{-}11)$$

10.6 有限变形时的变形协调条件表示为

$$\boldsymbol{F} \times \boldsymbol{\nabla}_X = (F_{iA}\boldsymbol{e}_i \otimes \boldsymbol{e}_A) \times \left(\frac{\partial}{\partial X_B}\boldsymbol{e}_B\right) = \varepsilon_{ABC}F_{iA,B}\boldsymbol{e}_i \otimes \boldsymbol{e}_C = \boldsymbol{0} \qquad (\text{s}10\text{-}12)$$

证明： 利用二阶偏微分的对称性质，有如下关系：

$$F_{iA,B} = \frac{\partial^2 x_i}{\partial X_A \partial X_B} = \frac{\partial^2 x_i}{\partial X_B \partial X_A} = F_{iB,A} \qquad (\text{s}10\text{-}13)$$

因此, 对上式进行移项有

$$F_{iA,B} - F_{iB,A} = 0 \qquad (\text{s}10\text{-}14)$$

对上式两端乘以置换符号 ε_{ABC}，则有

$$\varepsilon_{ABC}(F_{iA,B} - F_{iB,A}) = \varepsilon_{ABC}F_{iA,B} - \varepsilon_{ABC}F_{iB,A} = 0 \qquad (\text{s}10\text{-}15)$$

对于置换符号满足 $\varepsilon_{ABC} = -\varepsilon_{BAC}$，则有

$$-\varepsilon_{ABC}F_{iB,A} = \underbrace{\varepsilon_{BAC}F_{iB,A}}_{B \text{ 和 } A \text{ 均为哑标}\atop\text{可更换顺序}} = \varepsilon_{ABC}F_{iA,B} \tag{s10-16}$$

将 (s10-16) 式代入 (s10-15) 式，得到

$$\varepsilon_{ABC}F_{iA,B} - \varepsilon_{ABC}F_{iB,A} = 2\varepsilon_{ABC}F_{iA,B} = 0 \tag{s10-17}$$

故，(s10-12) 式得证.

10.7 大作业一：弹性力学解的唯一性证明 [10.5].

10.8 大作业二：弹性力学基本假定中为什么包含"无初始应力假设"？

证明： 变形梯度 F 和位移梯度 H 之间的关系为

$$F = \frac{\partial x}{\partial X} = x \otimes \nabla_X = \frac{\partial(X+u)}{\partial X} = I + u \otimes \nabla_X = I + H \tag{s10-18}$$

线弹性体是简单物质 (simple matter)[1.6] 中的一种. 所谓简单物质是指其力学行为仅仅依赖于变形梯度 F. 换言之，第一类皮奥拉–基尔霍夫应力 PK1 P 和柯西应力 σ 仅为变形梯度的函数：$P = \hat{P}(F) = \hat{P}(I+H)$，$\sigma = \hat{\sigma}(F) = \hat{\sigma}(I+H)$. 对于小变形，亦即位移梯度趋于零时 $H \to 0$，应用前言中所述及的三个特别"行之有效"(俗称"好使") 中的一阶泰勒展开，有

$$\begin{cases} P = \hat{P}(F) = \hat{P}(I+H) \approx \hat{P}(I) + \left.\frac{\partial \hat{P}(F)}{\partial F}\right|_{F=I} : H = \hat{P}(I) + C_1 : H \\ \sigma = \hat{\sigma}(F) = \hat{\sigma}(I+H) \approx \hat{\sigma}(I) + \left.\frac{\partial \hat{\sigma}(F)}{\partial F}\right|_{F=I} : H = \hat{\sigma}(I) + C_2 : H \end{cases} \tag{s10-19}$$

式中，C_1 和 C_2 为四阶弹性刚度张量. $\hat{P}(I) \approx \hat{\sigma}(I)$ 为小变形时初始构形中的初始应力 (initial stress). 由 PK1 和柯西应力之间的关系式 (29-10) 式，再利用 (s5-28) 式，舍掉高阶项后，有

$$\begin{aligned} \sigma = \hat{\sigma}(F) &= \det{}^{-1}(F)\hat{P}(F)F^T \\ &= (1 - \text{tr}H)\left(\hat{P}(I) + C_1 : H\right)\left(I + H^T\right) \\ &= \left[(1 - \text{tr}H)I + H^T\right]\hat{P}(I) + C_1 : H \end{aligned} \tag{s10-20}$$

联立 (s10-20) 和 (s10-19) 式中第二式两式，则得到两个四阶弹性刚度张量 C_1 和 C_2 之间的如下关系：

$$\hat{\sigma}(I) + C_2 : H = \left[(1-\text{tr}H)I + H^T\right]\hat{P}(I) + C_1 : H \tag{s10-21}$$

当初始应力为零时，

$$\hat{P}(I) \approx \hat{\sigma}(I) = 0 \tag{s10-22}$$

则有
$$C_1 = C_2 = C \tag{s10-23}$$

式中的 C 则为 (40-17) 式中的四阶弹性刚度张量. 考虑到 C 的大和小对称性, 在小变形和无初始应力的前提下, 本构关系可表示为

$$P = \sigma = C : H = C : \frac{H + H^T}{2} = C : \varepsilon \tag{s10-24}$$

10.9 **大作业三**: "上天、入地、下海" 是力学重大应用的形象概括. 对于 "入地" 问题而言, 由于地应力的永恒存在, 弹性力学就势必存在如下基本难题[10.5]: 具有初始应力场的一般弹性理论的解的唯一性问题.

10.10 **大作业四**: 菲隆 (Louis Napoleon George Filon, 1875—1937, 1912 年当选 FRS) 于 1903 年研究了在梁的中点施加一对儿集中压力的情形[10.6], 如题图 10.10(a) 所示, 菲隆发现当距离力的作用点大于 1.3 倍梁的半厚时 ($|x| > 1.3h$), 梁中将几乎没有应力存在, 菲隆的这个结果从一个方面佐证了圣维南原理的合理性. 铁摩辛柯和古迪尔的《弹性理论》(见表 48.1) 引用了菲隆的结果. 问题: (1) 如图 10.10(b) 所示, 考虑一个理想的二维问题, 一把尖锐的钳子夹住一个板条[10.7], 请应用菲隆的结论确定板条中应力影响的范围, 也就是图中虚线的特征尺度; (2) 如图 10.10(c) 所示, 一端部作用有平衡力系但包含有裂纹的梁, 也就是裂纹尖端存在应力的奇异性时[10.8], 圣维南原理是否依然成立? (3) 圣维南原理在纳米尺度是否依然成立?

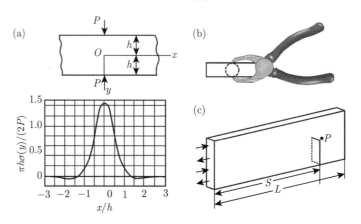

题图 10.10 圣维南原理的三个例子

参 考 文 献

[10.1] Navier C L M H. Résumé des leçons données à l'École royale des ponts et chaussées sur l'application de la mécanique à l'établissement des constructions et des machines. Didot, 1826.

[10.2] Galileo G. Dialogues Concerning Two New Sciences. 1638.

[10.3] Wong E W, Sheehan P E, Lieber C M. Nanobeam mechanics: elasticity, strength, and toughness of nanorods and nanotubes. Science, 1997, 277: 1971-1975.

[10.4] Hadamard J. Leçons sur la propagation des ondes et les équations de Vhydrodynamique. Paris: Hermann, 1903.

[10.5] 高梦霓, 赵亚溥. 若干弹性力学问题解的唯一性定理. 中国科学: 物理学 力学 天文学. 2020, 50: 084601.

[10.6] Filon L N G. On an approximate solution for the bending of a beam of rectangular cross-section under any system of load, with special reference to points of concentrated or discontinuous loading. Philosophical Transactions of the Royal Society of London A, 1903, 201: 63-155.

[10.7] 王敏中, 王炜, 武际可. 弹性力学教程 (修订版). 北京: 北京大学出版社, 2011.

[10.8] Toupin R A. Saint-Venant's principle. Archive for Rational Mechanics and Analysis, 1965, 18: 83-96.

第11章 流体动力学

§50. 从哈维的血液循环学说到血压计的发明

血压 (blood pressure, BP) 是人体健康的最重要的指标之一. 人类对血压特别是高血压的认识是医学发展的一个缩影. 血压计 (sphygmomanometer) 早已成为现代社会每位成员的生活必须品. 血压计的机理和流体力学关系密切. 因此, 本书的流体力学部分, 就让我们从血压计的研发历史谈起.

1628 年, 近代生理学之父威廉·哈维 (William Harvey, 1578—1657, 如图 50.1 所示) 用拉丁文出版的仅仅 72 页的著作《心血运动论》[11.1], 是人类历史上第一次对人体的循环系统做了系统的描述. 从而推翻了 "血液潮汐论", 建立了 "血液循环学说". 他在实验中发现当动脉被割破时, 血液会像被压力驱使一样从血管里喷涌而出, 这种力在触摸脉搏时也可以感受到. 哈维的该经典著作被称为 "宣布并演示血液循环的划时代的著作 (epoch-making treatise announcing and demonstrating the circulation of the blood)".

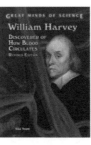

图 50.1 哈维像 (左、中); 哈维正在给查理一世国王和王子演示他对鹿所做的实验 (右)

1733 年, 英国学者斯蒂芬·黑尔斯 (Stephen Hales, 1677—1761) 首次测量了动物的血压. 如图 50.2 所示, 黑尔斯用尾端接有小金属管的长 9 英尺 (274 厘米) 直径六分之一英寸的玻璃管插入一匹马的颈动维持 270 厘米的柱高. 此时血液立即涌入玻璃管内, 高达 8.3 英尺 (270 厘米). 黑尔斯和助手看到玻璃管内血柱的高度并不是稳定不动的, 而是有规律地忽高忽低, 他们瞬间就明白了那就是马儿心脏跳动的节律, 心脏收缩时, 玻璃管内的血柱就升高, 心脏舒张时, 血柱就下降, 这也就是现在常说的收缩压 (systolic pressure) 和舒张压 (diastolic pressure).

1828 年，法国医生和生理学家泊肃叶 (Jean Louis Marie Poiseuille, 1797—1869) 在其医学院的论文中 [11.2]，采用内装水银的玻璃管来测量狗主动脉的血压并引入了 "毫米汞柱 (mmHg)" 的压强计量单位，该单位至今仍在使用. 由于水银的密度是水的 13.6 倍，此法大大减少了所用玻璃管的长度，即使玻璃管内的压力很大，也不至于把管中的水银柱顶起太高. 比起黑尔斯来，泊肃叶的这种血压测量法要简便了许多. 此时，泊肃叶已经对血压之于人体生理的意义进行了一些初步探索.

学者们开始探索无创的方法，既然在体表可以感受到动脉的搏动，那么能否在不割开血管的情况下，直接让脉搏的搏动传导给水银柱呢？

图 50.2　黑尔斯正在测量一匹马的血压

1896 年，意大利内科和儿科医生 (Italian internist and pediatrician) 里瓦罗基 (Scipione Riva-Rocci, 1863—1937)[11.3]，如图 50.3 所示，终于改制成了一种真正意义上的袖带水银血压计 (an easy-to-use cuff-based version mercury-sphygmomanometer). 这种血压计由袖带、压力表和气球三个部分构成. 测量血压时，将袖带平铺缠绕在手臂上部，用手捏压气球，然后观察压力表跳动的高度，以此推测血压的数值. 这种血压计测量血压较之以前是安全得多了，但和血腥的直接测量法相比，在准确性上还是稍逊一筹，它只能测量动脉的收缩压，而且测量出的数值也只是一个推测性的约数.

图 50.3　里瓦罗基 (左) 和他发明的血压计 (右)

1816 年法国名医雷奈克 (René Laennec, 1781—1826) 发明了听诊器. 1851 年，爱尔兰医生 Arthur Leared 发明了双耳听诊器 (binaural stethoscope). 1852 年，英国医师乔治·菲力普·卡门 (George Philip Cammann) 对双耳听诊器进一步进行了改良并实现了商业化.

俄国外科医生 (surgeon) 尼古拉·柯洛特克夫 (Nikolai Sergeyevich Korotkov, 1874—1920) 对里瓦罗基的听诊器进行了改进，在测血压时，加上了双耳听诊器，如图 50.4 所示. 这一点今天被称为 "集成式创新" 的改进使血压测量飞跃到一个全新的水平, 一直到现在仍然是血压测量的基本方法. 1905 年 11 月 8 日, 柯洛特克夫首次就他的方法在皇家军事医学科学院的跨学科会议 (interdisciplinary conference of the Imperial Military Medical Academy) 上进行了报告; 他还于 1905 年 12 月 13 日将实验测量数据对外公开 [11.4].

 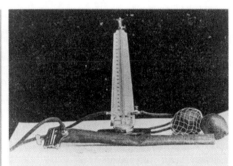

图 50.4 1900 年的柯洛特克夫 (左) 和他发明的血压计 (右)

在测定血压时, 另在袖带里面靠肘窝内侧动脉搏动处放上听诊器. 测量开始时先用气囊向缠缚于上臂的袖带内充气加压, 压力经软组织作用于肱动脉将其压瘪, 阻断其内的血流, 当所加压力高于心脏收缩压力时, 由气球慢慢向外放气, 袖带内的压力即随之下降, 当袖带内的压力等于或稍低于心缩压时, 随着心室收缩射血, 血液即可冲开被阻断的血管, 发出与心脏搏动相应的节律音 —— 柯氏音, 用听诊器听到这一声音的瞬间, 水银柱所指示的压力值即相当于收缩压; 继续缓慢放气, 使袖带内压力继续逐渐降低, 这段时间里, 袖带内压力低于心收缩压, 但高于心舒张压, 因此心脏每收缩一次, 均可听到一次柯氏音. 当袖带压力继续降低达到等于或稍低于舒张压时, 血流复又畅通, 这种声音便突然变弱或消失, 这个声音明显变调时水银柱所指示的压力值即相当于舒张压.

血压计的发明, 从 1733 年黑尔斯对马测量血压的开创性的实验到 1905 年带听诊器的袖带式血压计的完善则前后历时 170 余年. 其中, 蕴含着丰富的生物流体力学原理.

§51. 伯努利方程的建立

51.1 星光灿烂的伯努利家族

伯努利家族 (the Bernoullis) 星光闪耀、人才济济, 在三代中产生了八位世界级的数学家. 很多人一直在争论这是基于先天的基因还是后天的培养? 就这个家族的第二代、第三代从事数学研究的大多数成员来看,最值得注意的事情是他们并不是有意选择数学作为职业,"而是像酒鬼离不开酒那样不由自主地陷入了数学"[1.13]. 事实上,伯努利的家族的一个约定俗成的规矩,就是父辈们总是希望下一代能够经商,或者在医学方面有所建树.

伯努利方程,亦称伯努利定律,是丹尼尔·伯努利 (Daniel Bernoulli, 1700—1782) 创建的,所以本节将主要围绕丹尼尔展开.

丹尼尔的伯伯雅克布·伯努利 (Jacob Bernoulli, 1654—1705) 的座右铭是"我违父意,研究群星",这是对他父亲徒劳地反对他研究数学和天文学的讽刺性的纪念. 丹尼尔固执的爷爷开始也是试图强迫其父约翰·伯努利 (Johann Bernoulli, 1667—1748) 从事家族业务的经营. 约翰最初学医,于 1690 年获医学硕士学位,1694 年又获得博士学位,其论文是关于肌肉的收缩问题. 约翰最终在哥哥雅克布的教育下 (如图 51.1 所示),成为著名的数学家,兄弟俩均为德国大数学家莱布尼兹最好的朋友 (closest friends),即时在牛顿和莱布尼兹展开微积分优先权的论战之时,也坚定地站在莱布尼兹一边. 请特别注意的是,这为后来丹尼尔将莱布尼兹的"活力守恒定律 (law of vis viva conservation)"进一步发展成为"伯努利定律"或"伯努利方程"打下了重要的伏笔!

图 51.1 伯努利兄弟:雅克布在给弟弟约翰讲授数学

1686 年由莱布尼兹发表的《笛卡儿的一个出名错误的简短证明》(*Brevis demonstratio erroris memorabilis Cartesii*) 文章中所提出活力守恒定律表明:

$$\text{altitude} + \text{vis viva} = \text{constant}, \quad \text{vis viva} = mv^2 \tag{51-1}$$

式中的"altitude (高度)"显然对应于后来的势能 (potential energy),活力 (vis viva, living force, mv^2) 后来加上系数 1/2 后发展为动能 (kinetic energy),(51-1) 式也将进一步发展成为机械能守恒定律 (law of mechanical energy conservation).

牛顿的功底到底有多强? 1696 年 6 月,约翰·伯努利在莱布尼兹的杂志《教师学报》上刊登了一个他自己知道答案的挑战性问题——"最速下降线". 约翰原定于 1697 年 1 月 1 日向数学界公布答案. 但遗憾的是,到最后期限截止时,他只收到了莱布尼兹寄来的一份答案,并且,约翰说"莱布尼兹谦恭地请求我延长最后期限到复活节,以便在公布答案时 …… 没有人会抱怨说给的时间太短了. 我不仅同意了他的请求,而且还决定亲自宣布延长期限."

约翰的挑战目标非常明确,他把他的最速下降线问题抄了一份,装进信封,寄往英国的牛顿. 当然,1697 年,牛顿作为厂长正在忙于皇家铸币厂的事务,而且,正如他自己所承认的那样,他的头脑已不似全盛期时那样机敏了. 当时,牛顿与他的外甥女凯瑟琳·康迪特 (Catherine Barton Conduitt, 1679—1739) 一起住在伦敦. 凯瑟琳记述了这样的故事:

"1697 年的一天, 收到伯努利寄来的问题时, 艾萨克·牛顿爵士正在铸币厂里忙着改铸新币的工作, 直到 4 点钟才精疲力尽地回到家里, 但是, 直到解出这道难题, 他才上床休息, 这时, 正是凌晨 4 点钟." 牛顿一个晚上就解决了最速下降线问题!

到 1697 年复活节时挑战的最后期限到了, 最后寄来的答案, 信封上盖着英国的邮戳. 约翰打开后, 发现答案虽然是匿名的, 但却完全正确. 他显然遇到了他的对手牛顿. 答案虽然没有署名, 但却明显地出于一位绝顶天才之手. 约翰半是羞恼, 半是敬畏地说: "我从他的利爪就认出了这头狮子 (I recognize the lion by his paw)! "

51.2 伯努利定律

51.2.1 学生时期的丹尼尔·伯努利

流体力学最初的发展和医学有着很深的渊源.

丹尼尔的父亲约翰由于申请家乡巴塞尔大学的教授职位失败后, 担任了荷兰格罗宁根大学数学教授的职务. 丹尼尔就于 1700 年出生在荷兰的格罗宁根. 父亲对丹尼尔的培养很是重视, 丹尼尔 13 岁时开始学习哲学和逻辑学, 并在 15 岁获得学士学位, 16 岁获得艺术硕士学位. 在这期间, 他的父亲慷慨地教他学习数学和自然科学知识, 他的哥哥尼古拉·伯努利也不例外, 使他受到了数学家庭的熏陶. 让丹尼尔印象十分深刻地是, 他的父亲约翰作为莱布尼兹最好的朋友, 给他讲解莱布尼兹的 "活力 (vis viva, living force)" 和 "活力守恒定律" (51-1) 式 [11.5]. 该原理适用于质点和刚体力学. 这为丹尼尔后来在圣彼得堡科学院将其推广为适用于流体的 "伯努利定律" 奠定了坚实的基础.

当丹尼尔完成学业并在学业上快要取得成功时, 丹尼尔的父亲约翰也要求丹尼尔去当一个商业学徒, 谋一个经商的职业. 不过, 经商这个想法同样没有取得胜利.

但是丹尼尔的父亲约翰并没有放弃让丹尼尔从事数学研究及教学以外的职业, 于是萌发了让丹尼尔去学医, 这一次, 丹尼尔并没有断然拒绝, 反而答应了父亲的要求, 这使丹尼尔的父亲约翰十分高兴.

丹尼尔的学医历程也很曲折, 起初在家乡巴塞尔, 1718 年到了德国的海德堡, 1719 年又到了法国的施特拉斯堡学习, 在 1720 年他又回到了家乡巴塞尔. 在学医期间, 丹尼尔特别对意大利生物学家博雷利的《论动物的运动》和英国生理学家哈维的《心血运动论》[11.1] 表现出浓厚的兴趣. 哈维的血液循环学说使他对流体力学发生了兴趣.

虽然丹尼尔兴趣宽广，但是其他的一切东西怎么也不能让他随意抛开对数学的热爱，在学校学习医学期间，反而让他意外获得了更多的时间去自学数学，钻研数学. 而且在学校可以不受父亲监视和束缚，经济状况不错的家庭背景，不用像其他大多数学生一样为生活担心，丹尼尔一边学医，一边研习数学，很是自由.

1721 年，到了毕业时间了，丹尼尔通过了论文答辩，获得医学博士学位. 那一年他申请巴塞尔大学的两个教授职位，一个是解剖学和植物学 (anatomy and botany)，另一个是逻辑学 (logic)，都进入了最后的答辩，他从未想到过他的这次申请会失败. 但事实是，丹尼尔申请的两个职位却都没有成功，这当然令只有 21 岁的他大失所望，他当时正处于年轻气盛的年龄.

51.2.2 伯努利在意大利的游学

1723 年，申请巴塞尔大学教授职位失败的丹尼尔逃离了家乡，直奔意大利的帕多瓦，但是很不幸的是，刚刚抵达的丹尼尔却因为发烧得起了重病 (deathly ill)，病愈过程几乎持续了一年的时间. 期间，他和他的朋友克里斯蒂安·哥德巴赫 (Christian Goldbach, 1690—1764) 就一些学术思想和科学发现开始密集通信.

丹尼尔还参加了法国科学院的年度竞赛，使他大吃一惊的是，24 岁的他竟然获得了一等奖！他还未从获大奖的兴奋中缓过神来，更令他吃惊的消息便接踵而至了：哥德巴赫是如此地被丹尼尔来信的内容印象深刻，以至于要替丹尼尔出版成书！尽管丹尼尔以书信为非正式的形式，而且很多问题缺乏细节为由反对出书. 在抱怨无济于事的情况下，丹尼尔选择了一个不容易被人注意的书名——《一些数学练习》(Some Mathematical Exercises)，而且将书的作者谦逊地写为"丹尼尔·伯努利，约翰之子"以回报他父亲对他前期的培养. 该书的出版给丹尼尔带来了很大荣誉.

1725 年，25 岁的丹尼尔从意大利信心满满地返回他的故乡巴塞尔. 他收到了大量书的读者来信，称赞他的书及他本人. 最令他吃惊的是，他收到了来自于圣彼得堡的俄国女沙皇叶卡捷琳娜一世 (Catherine I, 1684—1727, 1725—1727 在位) 的邀请信，信中叶卡捷琳娜一世称赞了丹尼尔出众的才华，并邀请他到圣彼得堡皇家科学院担任数学教授.

51.2.3 伯努利定律建立的背景和过程

正如本书 §4.3 中所谈到的，1725—1733 年的 8 年间丹尼尔在圣彼得堡生活和工作. 除了和他父亲的高足欧拉在梁的变形力学方面有合作外，他所感兴趣的重点是流体力学，他开展了大量的实验和理论研究．

值得一说的是，丹尼尔·伯努利研究流体的管道流动最初是从研究血液的流速和血压的关系开始的．丹尼尔·伯努利既然在学校里学了医，他深深被哈维的发现所吸引．他认为血液在血管中流动，就有流动速度，心脏既然是一个血泵，就一定有压力．于是血管内的血液流速和压强也应当存在一定的关系．

如图 51.2 所示，伯努利设计的测量血压的方法，是把一根很细的玻璃管 CR 插入病人的动脉中，并且使它保持垂直．管上读出血液的高度 CT 的压强就相当于该处的血压．同样，当血压为负压时，用铅直向下插入动脉血管的玻璃管 cr 也可以得到血压值，血液高度 ct 对应的压强，也就是血压的负压值．伯努利当时和欧拉都在研究用这种方法测量病人的血压．每次测量血压都要刺破血管．尽管这样，这种测量血压的方法，在伯努利之后还是应用了达 170 年之久．一直到 1896 年，意大利医生里瓦罗基发明的袖带水银血压计，伯努利的测量血压的方法才被淘汰．

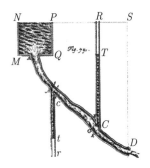

图 51.2 伯努利的测血压的方法

丹尼尔经过大量的实验和分析，并结合哈维的《心血运动论》有关血液循环以及达·芬奇的河流宽窄对流速的影响进行的综合分析，他在莱布尼兹的"活力守恒定律" (51-1) 式的基础上，做出了如下重要类比：

$$\text{altitude} \sim \text{pressure}\ (p) \quad \text{vis viva} \sim \rho v^2 \tag{51-2}$$

从而他给出了后来称为伯努利定律的方程式：

$$p + \rho v^2 = \text{const} \tag{51-3}$$

大约一个世纪后，法国物理学家科里奥利于 1829 年用 $\frac{1}{2}\rho v^2$ 来取代 ρv^2，(51-3) 式才变为其现代形式：

$$p + \frac{1}{2}\rho v^2 = \text{const} \tag{51-4}$$

我们将在 §51.2.5 中继续讨论现代形式的伯努利定律及其重要应用．本节给出的是丹尼尔·伯努利当年研究该问题的"初心"．

51.2.4　丹尼尔·伯努利出版传世之作《流体动力学》

1732 年，丹尼尔终于获得了家乡巴塞尔大学解剖学和植物学教授的职位，加之经他推荐去圣彼得堡工作的欧拉的声誉日隆，也由于他不喜欢圣彼得堡冬天的严寒，他决定返回家乡．

此时的丹尼尔已经将他在圣彼得堡工作期间的主要成果汇集成册，只差一个结束语了．在离开圣彼得堡前，经他推荐，欧拉接替了丹尼尔数学教授的职位．

第 11 章 流体动力学

从圣彼得堡乘坐四轮马车 (horse-drawn coach) 到瑞士巴塞尔的长途跋涉需要大约两个整月的时间. 正是在这个漫长的旅行中, 使丹尼尔一生都感到荣幸的对话发生了. 在四轮马车上, 当一位同行的植物学家 (botanist) 询问他的名字时, 丹尼尔回答道: "我是丹尼尔·伯努利." 那位植物学家认为自己受到了嘲弄, 便以挖苦的口吻说: "那我就是艾萨克·牛顿."

丹尼尔对那位植物学家说他没有撒谎时, 这位植物学家坚持说, 你这么年轻, 怎么可能是著名的丹尼尔·伯努利呢? 当丹尼尔拿出身份证明后, 这位植物学家先是心慌不安, 然后在整个行程都变成了丹尼尔的追星一族. 丹尼尔充分地感到: 他现在的确已经是一位著名人物了! 终于有人拿他和科学巨人牛顿来进行类比了!

1734 年他在家乡的巴塞尔大学安顿下来后, 他原来的对家乡天堂般的梦想也变成了梦魇. 这一年他和父亲同时被遴选为法国科学院年度竞赛的并列第一名. 这彻底地激怒了他的父亲.

在 1734 年底, 丹尼尔完成了他的书稿, 交给了斯特拉斯堡的出版商. 那时的出版业费时又费力, 书的出版和装订花去了三年多的时间.

1738 年, 丹尼尔终于拿到了他正式出版发行的书《流体动力学》[11.6], 当他看到书的封面 (如图 51.3) 上印着 "流体动力学, 丹尼尔·伯努利, 约翰之子" 时, 他激动的热泪盈眶 (his eyes began to water).

图 51.3 丹尼尔·伯努利和他的《流体动力学》

51.2.5 伯努利方程的详细推导过程

理想流体 (ideal fluid) 是具有如下性质的流体:

- 不可压缩 (incompressible)——有如下多种表述方法: (1) 密度为常量: $\rho = \rho_0$; (2) 速度的散度为零: $\mathrm{div}\boldsymbol{v} = 0$; (3) 表征体积比的雅可比满足 $J = \det \boldsymbol{F} = 1$. 可详见 (29-45) 式所给出的七个条件.
- 无旋 (irrotational)——流动是光滑的，处于层流，没有湍流.
- 无黏 (inviscid)——流体无内摩擦，黏度系数 $\mu = 0$.
- 无表面张力 (no surface tension)——表面张力 $\gamma = 0$.

而放弃了以上四条假定的流体为真实流体 (real fluid). 换言之，真实流体是可压缩的、有黏度、有表面张力，其流动很多情况下是湍流.

自然界虽然不存在理想流体，但是对于像一般情况下的空气和水等，由于其黏度系数小，在一些条件下，可视为理想流体，其好处是分析简便.

考虑管道内理想流体的一维流动. 取如图 51.4 所示的一段流体分析. 两截面压力对该段流体做功的功率 (单位时间所做的功) 分别为

$$p_1 A_1 v_1, \quad -p_2 A_2 v_2 \tag{51-5}$$

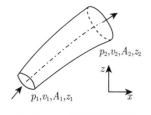

图 51.4 伯努利方程的分析模型

单位时间动能改变量:

$$\frac{1}{2}\rho v_2 A_2 \cdot v_2^2 - \frac{1}{2}\rho v_1 A_1 \cdot v_1^2 \tag{51-6}$$

单位时间重力势能的改变量:

$$\rho g v_2 A_2 z_2 - \rho g v_1 A_1 z_1 \tag{51-7}$$

由能量守恒可得

$$p_1 v_1 A_1 - p_2 v_2 A_2 = \frac{1}{2}\rho v_2 A_2 \cdot v_2^2 - \frac{1}{2}\rho v_1 A_1 \cdot v_1^2 + \rho g v_2 A_2 z_2 - \rho g v_1 A_1 z_1 \tag{51-8}$$

由连续性方程，也就是两截面流量相同:

$$v_2 A_2 = v_1 A_1 \tag{51-9}$$

由以上两式得

$$p_1 + \rho g z_1 + \frac{1}{2}\rho v_1^2 = p_2 + \rho g z_2 + \frac{1}{2}\rho v_2^2 \tag{51-10}$$

由于截面 1、2 是任意取的，故上式也可写成

$$\frac{p}{\rho} + gz + \frac{v^2}{2} = \mathrm{const} \tag{51-11}$$

上式即为伯努利方程 (Bernoulli equation). 其意义为: (1) 伯努利方程是理想流体流动能量守恒的结果; (2) 重力作用下的不可压缩理想流体做定常运动，总能量在流线上守恒. 左端各项分别代表单位质量内的压力能，势能和动能.

例 51.1 文丘里效应 (the Venturi effect). 这种现象以其发现者, 意大利物理学家文丘里 (Giovanni Battista Venturi, 1746—1822) 命名, 该工作发表于 1797 年 [11.7]. 如图 51.5 所示, 这种效应是指在高速流动的气体附近, 静压强会减少, 从而产生吸附作用. 利用这种效应可以制作出文氏管.

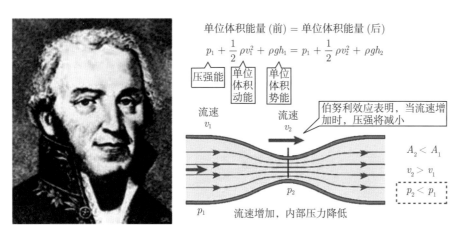

图 51.5 文丘里和文丘里效应示意图

值得注意的是, 文丘里还是一位著名科学史家 (historian of science). 他在 1797 年出版的一个题为 *Essai sur les ouvrages physico-mathématiques de Léonard da Vinci* 的小册子中, 第一次呼吁, 达·芬奇不仅仅是一位艺术家, 而且还是一位科学家.

例 51.2 伯努利夹持 (Bernoulli grip) 和伯努利吸盘 (Bernoulli chuck) 在先进制造中的广泛应用.

如图 51.6 所示, 该夹持利用了伯努利原理, 气流快速流动的地方静压强低. 因此, 可将物体提升至一定高度, 实现非接触式吸附或夹持. 这在微电子加工中的晶圆的运移, 易接触感染样品的运移, 以及软体爬行机器人的设计等方面, 均有着广泛的应用价值.

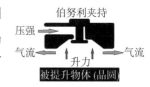

图 51.6 气流造成物体 (晶圆, wafer) 的升力, 从而形成非接触式的吸附

伯努利吸盘如图 51.7 所示, 在工业界也被广泛地称为伯努利机械手指, 其工作原理是, 供给口导入的气体会从吸盘的内部圆筒状侧面的喷口高速喷出, 并在吸盘内部的筒状空间内形成旋转气流, 形成负压, 也就是所谓的旋风效应. 上述气流最终通过机械手指吸附表面和晶圆 (wafer) 表面之间的间隙释放到外部空间. 气流在旋风吸盘和晶圆之间的空间中形成稳定的层流, 并造成晶圆上下表面间的压力差, 最终形成对晶圆向上的吸附力.

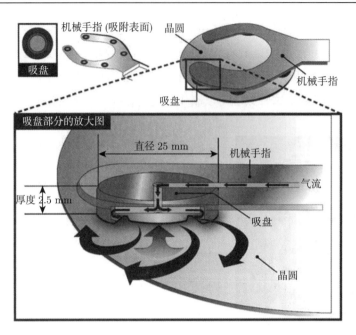

图 51.7　先进制造中的伯努利机械手指示意图

§52. 流体力学势流问题

52.1　势流的特点

当流体的旋度 (涡量) $\boldsymbol{\Omega} = \boldsymbol{\nabla} \times \boldsymbol{v} = \boldsymbol{0}$ 时，可以引入一个势函数 φ，通过其梯度表示速度矢量：

$$\boldsymbol{v} = \boldsymbol{\nabla} \varphi \tag{52-1}$$

势流 (potential flow) 问题也被称为无旋流动 (irrotational flows)，原因是势流问题中的速度场实际上是速度势函数的梯度，而梯度的旋度始终为零. 可简单证明如下：

$$\begin{aligned}
\boldsymbol{\nabla} \times (\boldsymbol{\nabla}\varphi) &= \begin{vmatrix} \hat{\boldsymbol{i}} & \hat{\boldsymbol{j}} & \hat{\boldsymbol{k}} \\ \dfrac{\partial}{\partial x} & \dfrac{\partial}{\partial y} & \dfrac{\partial}{\partial z} \\ \dfrac{\partial \varphi}{\partial x} & \dfrac{\partial \varphi}{\partial y} & \dfrac{\partial \varphi}{\partial z} \end{vmatrix} \\
&= \left(\dfrac{\partial^2 \varphi}{\partial y \partial z} - \dfrac{\partial^2 \varphi}{\partial z \partial y} \right) \hat{\boldsymbol{i}} - \left(\dfrac{\partial^2 \varphi}{\partial x \partial z} - \dfrac{\partial^2 \varphi}{\partial z \partial x} \right) \hat{\boldsymbol{j}} + \left(\dfrac{\partial^2 \varphi}{\partial x \partial y} - \dfrac{\partial^2 \varphi}{\partial y \partial x} \right) \hat{\boldsymbol{k}} = \boldsymbol{0}
\end{aligned} \tag{52-2}$$

第 11 章 流体动力学

因此,速度的旋度 (涡量) 为零:

$$\nabla \times \boldsymbol{v} = \nabla \times (\nabla \varphi) = \boldsymbol{0} \tag{52-3}$$

亦即,势流问题无法应用在尾涡和边界层问题中.

52.2 不可压缩流体的特性与势流方程

对于不可压缩流体,根据连续性方程,有 (29-45) 式中第三式和 (31-12) 式: $\nabla \cdot \boldsymbol{v} = 0$,因此速度势满足经典的拉普拉斯方程 (二阶线性偏微分方程):

$$\nabla^2 \varphi = 0 \tag{52-4}$$

此时流体可由其运动学行为完全确定. 在二维情况下, 势流将简化为一个可以通过复变函数分析的简单系统.

建立二维正交坐标系 (x, y), 则速度的两个分量 $u(x, y)$, $v(x, y)$ 分别为

$$u = \frac{\partial \varphi}{\partial x}, \quad v = \frac{\partial \varphi}{\partial y} \tag{52-5}$$

引入另一个标量函数 $\psi(x, y)$, 使得

$$u = \frac{\partial \psi}{\partial y}, \quad v = -\frac{\partial \psi}{\partial x} \tag{52-6}$$

则 $\psi(x, y)$ 称为流函数 (stream function). 流函数与势函数之间满足

$$\frac{\partial \varphi}{\partial x} = \frac{\partial \psi}{\partial y}, \quad \frac{\partial \varphi}{\partial y} = -\frac{\partial \psi}{\partial x} \tag{52-7}$$

$$\Delta \psi = \frac{\partial^2 \psi}{\partial x^2} + \frac{\partial^2 \psi}{\partial y^2} = -\frac{\partial^2 \varphi}{\partial x \partial y} + \frac{\partial^2 \varphi}{\partial x \partial y} = 0 \tag{52-8}$$

因此流函数亦满足拉普拉斯方程 (52-8) 式: $\nabla^2 \psi = \Delta \psi = 0$.

在二维情况下, 流函数与势函数都是调和函数, 并且二者之间满足柯西–黎曼条件, 因此是一对共轭调和函数.

52.3 势流的分析

不可压缩流体二维势流可以使用保角映射、复平面变换进行分析. 基本思想就是使用解析函数 f, 将其物理域 (x, y) 映射到变换域 (φ, ψ). 虽然 x, y, φ, ψ 都是实数, 但可以很容易地定义其复数量如下:

$$z = x + \mathrm{i}y, \quad \omega = \varphi + \mathrm{i}\psi \tag{52-9}$$

构造复变数函数

$$f(x+\mathrm{i}y) = \varphi + i\psi \quad \text{或} \quad f(z) = \omega \tag{52-10}$$

则 x, y 方向的速度 u, v 可以直接由 f 对 z 求导得到

$$\frac{\mathrm{d}f}{\mathrm{d}z} = u + \mathrm{i}v \tag{52-11}$$

式中，$f(z)$ 称为复速度势.

流函数 ψ 为常数的线就是流线 (streamlines)，而势函数 φ 是常数的线是等势线 (equipotential lines). 势线与流线相互正交，证明如下：

$$\mathrm{d}\psi = \frac{\partial \psi}{\partial x}\mathrm{d}x + \frac{\partial \psi}{\partial y}\mathrm{d}y = -v\mathrm{d}x + u\mathrm{d}y = 0 \tag{52-12}$$

$$\mathrm{d}\varphi = \frac{\partial \varphi}{\partial x}\mathrm{d}x + \frac{\partial \varphi}{\partial y}\mathrm{d}y = u\mathrm{d}x + v\mathrm{d}y = 0 \tag{52-13}$$

$$\boldsymbol{\nabla}\varphi \cdot \boldsymbol{\nabla}\psi = \frac{\partial \varphi}{\partial x}\frac{\partial \psi}{\partial x} + \frac{\partial \varphi}{\partial y}\frac{\partial \psi}{\partial y} = uv - uv = 0 \tag{52-14}$$

对于无旋假设成立的流体区域势流理论是成立的. 比如在飞行器周围的流动、地下水的流动、声学、水波以及电渗流等领域. 目前, 对于自由涡和点源等简单势流问题存在解析解. 这些解的叠加是能够处理满足各种边界条件的更复杂的流动. 此外, 当真实流动与势流之间存在较小的偏差时, 会出现很多有价值的解. 例如, 在计算流体力学中, 一种技术是将边界层外的势流解与边界层内的边界方程的解耦合起来, 边界层效应的缺失意味着任何流线都可以被不改变流场的固体边界所取代, 这是许多气动设计方案所采用的技术.

52.4 基本流

下面所给出的简单解析函数所代表的流动是研究复杂流动的基础. 复杂流动可由简单流叠加或变换得到. 考虑势流时通常结合流函数与势函数一起分析, 下面给出常见势流解析解及相关势函数与流函数对比.

52.4.1 不同坐标系下势函数所满足的拉普拉斯方程

笛卡儿坐标系 (x, y, z) 下:

$$\Delta\varphi = \frac{\partial^2 \varphi}{\partial x^2} + \frac{\partial^2 \varphi}{\partial y^2} + \frac{\partial^2 \varphi}{\partial z^2} = 0 \tag{52-15}$$

圆柱坐标系 (r, θ, z) 下:

$$\Delta\varphi = \frac{1}{r}\frac{\partial}{\partial r}\left(r\frac{\partial \varphi}{\partial r}\right) + \frac{1}{r^2}\frac{\partial^2 \varphi}{\partial \theta^2} + \frac{\partial^2 \varphi}{\partial z^2} = 0 \tag{52-16}$$

球坐标系 (r, θ, α) 下:
$$\Delta\varphi = \frac{1}{r^2\sin\theta}\left[\frac{\partial}{\partial r}\left(r^2\sin\theta\frac{\partial\varphi}{\partial r}\right) + \frac{\partial}{\partial\theta}\left(\sin\theta\frac{\partial\varphi}{\partial\theta}\right) + \frac{\partial}{\partial\alpha}\left(\sin\theta\frac{\partial\varphi}{\partial\alpha}\right)\right] = 0 \quad (52\text{-}17)$$

52.4.2 流函数

前面业已述及, 流函数 ψ 为常数的线就是流线, 由 (52-12) 式, 得到
$$\frac{\mathrm{d}y}{\mathrm{d}x} = \frac{v}{u} \quad (52\text{-}18)$$
上式说明, 流函数 ψ 为常数的线其切线方向即是速度方向.

两个为常数流函数之差代表两条流线之间的体积流量. 证明如下:

如图 52.1 所示, ab 与 cd 代表两条流线, 则体积流量增量为
$$\mathrm{d}Q = \boldsymbol{v}\cdot\boldsymbol{n}\mathrm{d}l = \left[u\frac{\mathrm{d}y}{\mathrm{d}l} + v\left(-\frac{\mathrm{d}x}{\mathrm{d}l}\right)\right]\mathrm{d}l = u\mathrm{d}y - v\mathrm{d}x = \frac{\partial\psi}{\partial y}\mathrm{d}y + \frac{\partial\psi}{\partial x}\mathrm{d}x = \mathrm{d}\psi \quad (52\text{-}19)$$

52.4.3 流函数与势函数之间的关系

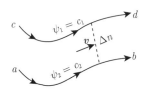

图 52.1

两者之间有如下区别: (1) 速度势函数是无旋的结果, 而流函数是不可压缩流体结果; (2) 势函数可以在广义的三维空间定义, 而流函数只适用于二维空间; (3) 流函数并不局限于无旋场. 若用公式表示的话, 可见表 52.1.

表 52.1　流函数与势函数之间的区别

	$u = \dfrac{\partial\varphi}{\partial x} = \dfrac{\partial\psi}{\partial y}$	$v = \dfrac{\partial\varphi}{\partial y} = -\dfrac{\partial\psi}{\partial x}$
无旋流动	$\dfrac{\partial v}{\partial x} - \dfrac{\partial u}{\partial y} = 0$	$\Delta\varphi = \dfrac{\partial^2\varphi}{\partial x^2} + \dfrac{\partial^2\varphi}{\partial y^2} = 0$
不可压缩流动	$\dfrac{\partial u}{\partial x} + \dfrac{\partial v}{\partial y} = 0$	$\Delta\psi = \dfrac{\partial^2\psi}{\partial x^2} + \dfrac{\partial^2\psi}{\partial y^2} = 0$

下面给出常见简单流动的势函数与流函数的解析解.

应用幂律保角映射, 从 $z = x + \mathrm{i}y$ 到 $\omega = \varphi + \mathrm{i}\psi$:
$$\omega = Az^n \quad (52\text{-}20)$$
将 z 在极坐标情况下表示, $z = x + \mathrm{i}y = \mathrm{e}^{\mathrm{i}\theta}$, 则势函数和流函数可分别表示为
$$\varphi = Ar^n\cos n\theta, \quad \psi = Ar^n\sin n\theta \quad (52\text{-}21)$$
式中, n 的取值情况对应于特定的流动, 常数 A 是标度参数. A 的模表示大小, 转角表示旋转量. 黑色的线表示流动的边界, 深蓝色的线是流线, 浅蓝色的线是势线.

当 $n=1$ 时，为均匀来流情形. 此时有

$$\omega = Az^1 \Rightarrow \frac{\partial \varphi}{\partial x} = u = A, \quad \frac{\partial \varphi}{\partial y} = v = 0 \Rightarrow \frac{\partial \psi}{\partial x} = -v = 0, \quad \frac{\partial \psi}{\partial y} = u = A \tag{52-22}$$

表明流体是 x 方向的均匀来流 (如图 52.2 所示), 其势函数可以简单地表示为

$$\phi = Ax, \quad \psi = Ay \tag{52-23}$$

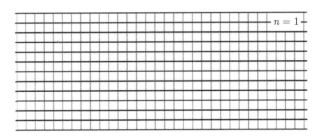

图 52.2　当 $n=1$ 时的流场

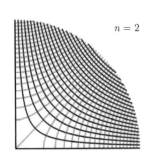

图 52.3　当 $n=2$ 时的流场

当 $n=2$ 时，所表示的情形是一个角或滞点附近的流动，此时有

$$\begin{cases} \omega = Az^2 = A\left(x^2 - y^2 + 2xy\mathrm{i}\right) \\ \dfrac{\partial \varphi}{\partial x} = u = 2Ax, \quad \dfrac{\partial \varphi}{\partial y} = v = -2Ay \\ \dfrac{\partial \psi}{\partial x} = -v = 2Ay, \quad \dfrac{\partial \psi}{\partial y} = u = 2Ax \end{cases} \tag{52-24}$$

此时的势函数和流函数可分别表示为

$$\varphi = A\left(x^2 - y^2\right), \quad \psi = 2Axy \tag{52-25}$$

此时的流线为双曲线，如图 52.3 所示.

当 $n=-1$ 时，所表示的是由源偶极子引起的流动情景，如图 52.4 所示.

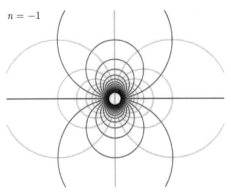

图 52.4　当 $n=-1$ 时的流场

此时有
$$\omega = \frac{A}{z} = \frac{A}{x+\mathrm{i}y} = \frac{A(x-\mathrm{i}y)}{(x+\mathrm{i}y)(x-\mathrm{i}y)} = \frac{A(x-\mathrm{i}y)}{(x^2+y^2)} \tag{52-26}$$

亦即

$$\begin{cases} \dfrac{\partial \varphi}{\partial x} = u = \dfrac{A(y^2-x^2)}{(x^2+y^2)^2} \\ \dfrac{\partial \varphi}{\partial y} = v = \dfrac{2xyA}{(x^2+y^2)^2} \end{cases}, \quad \begin{cases} \dfrac{\partial \psi}{\partial x} = -v = -\dfrac{2xyA}{(x^2+y^2)^2} \\ \dfrac{\partial \psi}{\partial y} = u = \dfrac{A(y^2-x^2)}{(x^2+y^2)^2} \end{cases} \tag{52-27}$$

考虑到其形式的复杂性, 在极坐标系下 $(x = r\cos\theta, y = r\sin\theta)$ 对其进行分析, 此时有

$$u_r = -\frac{A}{r^2}\cos\theta, \quad u_\theta = -\frac{A}{r^2}\sin\theta \tag{52-28}$$

势函数和流函数分别为

$$\varphi = \frac{A}{r}\cos\theta, \quad \psi = -\frac{A}{r}\sin\theta \tag{52-29}$$

当 $n = \dfrac{1}{2}$ 时, 对应于半无限长平板周围的流动, 如图 52.5 所示.

当 $n = \dfrac{2}{3}$ 时, 对应于右直角附近流动情形, 如图 52.6 所示.

下面讨论如图 52.7 所示的源流情形.

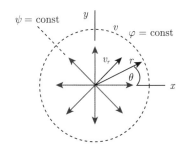

图 52.7 源流

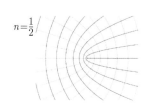

图 52.5 半无限长平板周围的流动

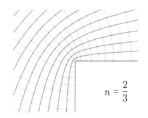

图 52.6 直角附件的流动

其流函数与势函数满足二维拉普拉斯方程:

$$\begin{aligned} \Delta \varphi &= \frac{1}{r}\frac{\partial}{\partial r}\left(r\frac{\partial \varphi}{\partial r}\right) + \frac{1}{r^2}\frac{\partial^2 \varphi}{\partial \theta^2} = 0 \\ \Delta \psi &= \frac{1}{r}\frac{\partial}{\partial r}\left(r\frac{\partial \psi}{\partial r}\right) + \frac{1}{r^2}\frac{\partial^2 \psi}{\partial \theta^2} = 0 \end{aligned} \tag{52-30}$$

由于势函数是对称的, 不随转角变化, 流线沿着半径方向. 因此,

$$\Delta\varphi = \frac{1}{r}\frac{\partial}{\partial r}\left(r\frac{\partial \varphi}{\partial r}\right) = 0 \quad \Rightarrow \quad \varphi = c_1 \ln r + c_2$$
$$\Delta\psi = \frac{1}{r^2}\frac{\partial^2 \psi}{\partial \theta^2} = 0 \quad \Rightarrow \quad \psi = c_3 \theta + c_4 \tag{52-31}$$

速度满足

$$u_r = \frac{\partial \varphi}{\partial r} = \frac{1}{r}\frac{\partial \psi}{\partial \theta} = \frac{c}{r}, \quad u_\theta = \frac{1}{r}\frac{\partial \varphi}{\partial \theta} = -\frac{\partial \psi}{\partial r} = 0 \tag{52-32}$$

此时通过半径为 r 的圆的体积流量可获得 (52-32) 式中的待定常数:

$$Q = 2\pi r \cdot u_r = 2\pi c \quad \Rightarrow \quad c = \frac{Q}{2\pi} \tag{52-33}$$

因此, 势函数与流函数分别为

$$\varphi = \frac{Q}{2\pi}\ln r, \quad \psi = \frac{Q}{2\pi}\theta \tag{52-34}$$

在本节的最后讨论如图 52.8 所示的涡流情形.

涡流的流函数与势函数与源流的相应函数形式恰好相反, 满足二维拉普拉斯方程:

$$\Delta\varphi = \frac{1}{r}\frac{\partial}{\partial r}\left(r\frac{\partial \varphi}{\partial r}\right) + \frac{1}{r^2}\frac{\partial^2 \varphi}{\partial \theta^2} = 0$$
$$\Delta\psi = \frac{1}{r}\frac{\partial}{\partial r}\left(r\frac{\partial \psi}{\partial r}\right) + \frac{1}{r^2}\frac{\partial^2 \psi}{\partial \theta^2} = 0 \tag{52-35}$$

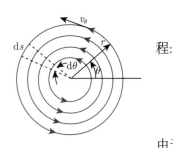

图 52.8 涡流

由于势函数是对称的, 不随转角变化, 流线沿着半径方向. 因此,

$$\Delta\varphi = \frac{1}{r^2}\frac{\partial^2 \varphi}{\partial \theta^2} = 0 \quad \Rightarrow \quad \varphi = c_1 \theta + c_2$$
$$\Delta\psi = \frac{1}{r}\frac{\partial}{\partial r}\left(r\frac{\partial \psi}{\partial r}\right) = 0 \quad \Rightarrow \quad \psi = c_3 \ln r + c_4 \tag{52-36}$$

速度满足

$$u_r = \frac{\partial \varphi}{\partial r} = \frac{1}{r}\frac{\partial \psi}{\partial \theta} = 0, \quad u_\theta = \frac{1}{r}\frac{\partial \varphi}{\partial \theta} = -\frac{\partial \psi}{\partial r} = \frac{c}{r} \tag{52-37}$$

此时通过半径为 r 的环量 (circulation) 为

$$\Gamma = -\oint \boldsymbol{v}\cdot \mathrm{d}\boldsymbol{s} = -\int_0^{2\pi}\frac{c}{r}r\mathrm{d}\theta = -2\pi c \quad \Rightarrow \quad c = -\frac{\Gamma}{2\pi} \tag{52-38}$$

因此, 势函数与流函数分别为

$$\varphi = -\frac{\Gamma}{2\pi}\theta, \quad \psi = \frac{\Gamma}{2\pi}\ln r \tag{52-39}$$

第 11 章　流体动力学

§53. 流变体与牛顿流体

53.1 流变体的定义

图 53.1 给出的是 2018 年 5 月 5 日美国夏威夷州的基拉韦厄火山 (Kilauea volcano) 喷发所产生的熔岩河 (右图), 左图给出的是待喷发状的岩浆, 就像即将出炉的钢水, 火红而炽热. 岩浆的温度一般在 900—1200℃ 之间, 最高可达 1400℃, 足以熔化岩石. 熔岩是一种典型的流变体, 其黏度大约是水的十万倍.

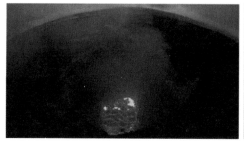

图 53.1　火山口中待喷发的熔岩 (左); 2018 年美国夏威夷火山喷发产生的熔岩河 (右)

黏度是岩浆的重要性质, 它决定着岩浆流动的状态和程度. 岩浆中 SiO_2 的含量对黏度影响最大, 其次是 Al_2O_3、Cr_2O_3, 它们的含量增高, 岩浆黏度会明显增大. 酸性岩中 SiO_2、Al_2O_3 的含量很高, 因此, 黏度也最大; 溶解在岩浆中的挥发份可以降低岩浆的黏度、降低矿物的熔点, 使岩浆容易流动, 结晶时间延长; 此外, 岩浆的温度高, 黏度相应变小; 岩浆承受的压力加大, 岩浆的黏度也增大.

流变体的性质如图 53.2 所示. 横轴是速度梯度 (应变率), 纵轴是切应力. 由于 §51.2.5 中所讨论的理想流体无黏特性, 所以横轴所代表的就是理想流体, 而一般的理想固体的弹性行为和应变率无关, 故, 纵轴代表的就是理想固体.

一般地, 流变体的行为可表示为如下关系:

$$\tau = \mu \dot{\gamma}^n \tag{53-1}$$

式中, 表观黏度 (apparant viscosity) 可定义为

$$\mu_{\text{app}} = \frac{\tau}{\dot{\gamma}} = \mu |\dot{\gamma}|^{n-1} \tag{53-2}$$

流变体可分为如下三种情形: (1) $n = 1$ 为牛顿流体 (Newtonian fluids), 其黏性系数为常数, 和剪切率无关; 所有 $n \neq 1$ 的流体为非牛顿流体 (non-Newtonian fluids); (2) $n > 1$ 为剪切致稠流变体 (shear-thickening), 表观黏度系数将随着剪切

率的增大而增大；(3) $0 < n < 1$ 为剪切致稀流变体 (shear-thinning)，表观黏度系数将随着剪切率的增大而减小.

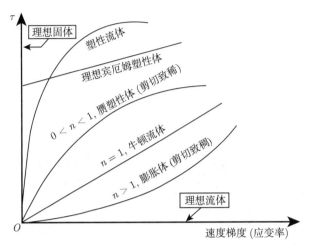

图 53.2　牛顿流体和剪切致稀、剪切致稠流变体的比较

53.2　牛顿流体

牛顿于 1687 年在《自然哲学的数学原理》($Principia$) 中定义了线性黏性定律. 牛顿在该书的第三篇第九章"流体的圆形运动"的一开头，给出了一条假设："由于流体各部分之间缺乏滑润性而产生的阻力在其他情况相同时与流体各部分彼此分开的速度成正比例."

斯托克斯 (George Gabriel Stokes, 1819—1903) 在推导黏性流体运动方程时，引进了流体剪应力与剪应变速度成正比例的关系. 这个关系就是牛顿上述假设的更准确的具体表示.

后人将这种流体称为牛顿流体 (Newtonian fluids)，以区别于那种剪应力与剪应变速度不是线性关系的流体. 后者也称为非牛顿流体 (non-Newtonian fluids).

如图 53.3 所示，对于不可压缩和各向同性牛顿流体，两层流体间的切应力 τ 与其速度梯度 (应变率，单位 s^{-1}) $\dot{\gamma} = \dfrac{\partial v}{\partial y}$ 成正比：

$$\tau = \mu \dot{\gamma} = \mu \frac{\partial v}{\partial y} \tag{53-3}$$

式中，v 为流体沿水平方向的速度，μ 为动力黏度系数 (dynamic viscosity)，由于切应力的单位为 Pa，而速度梯度 (剪切率) $\dot{\gamma} = \dfrac{\partial v}{\partial y}$ 的单位为 s^{-1}，故动力黏度系数的单位为 Pa·s. 作为流体力学中最为重要的一个参量，动力黏度的量纲为

$$[\mu] = \mathrm{MLT^{-2}L^{-2}T = ML^{-1}T^{-1}} \tag{53-4}$$

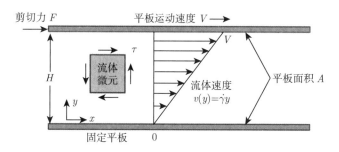

图 53.3 牛顿流体的定义

由于切应力 τ 的量纲为单位面积的力，亦即 $[\tau] = \left[\dfrac{F}{A}\right] = \left[\dfrac{\text{动量}}{\text{面积} \cdot \text{时间}}\right]$，则切应力 τ 具有动量通量 (momentum flux) 的量纲，切应力的变化将引起动量扩散 (momentum diffusion)。

53.3 牛顿流体的本构关系

设流体中一点处的速度 $\bm{v}$ 的三个分量为 (v_x, v_y, v_z)，在空间坐标 (x, y, z) 处取微元体 $(x+\delta x, y+\delta y, z+\delta z)$，则速度 $\bm{v}$ 的增量 $\delta \bm{v}$ 可表示为

$$\begin{cases} \delta v_x = \dfrac{\partial v_x}{\partial x}\delta x + \dfrac{\partial v_x}{\partial y}\delta y + \dfrac{\partial v_x}{\partial z}\delta z \\ \delta v_y = \dfrac{\partial v_y}{\partial x}\delta x + d\dfrac{\partial v_y}{\partial y}\delta y + \dfrac{\partial v_y}{\partial z}\delta z \\ \delta v_z = \dfrac{\partial v_z}{\partial x}\delta x + \dfrac{\partial v_z}{\partial y}\delta y + \dfrac{\partial v_z}{\partial z}\delta z \end{cases} \tag{53-5}$$

式中的速度梯度可用矩阵表示为

$$[\bm{v} \otimes \bm{\nabla}] = \begin{bmatrix} \dfrac{\partial v_x}{\partial x} & \dfrac{\partial v_x}{\partial y} & \dfrac{\partial v_x}{\partial z} \\ \dfrac{\partial v_y}{\partial x} & \dfrac{\partial v_y}{\partial y} & \dfrac{\partial v_y}{\partial z} \\ \dfrac{\partial v_z}{\partial x} & \dfrac{\partial v_z}{\partial y} & \dfrac{\partial v_z}{\partial z} \end{bmatrix} = [\bm{S} + \bm{\Omega}] \tag{53-6}$$

式中，对称矩阵 (symmetrical matrix) 为

$$[\boldsymbol{S}] = [\boldsymbol{S}^{\mathrm{T}}] = \left[\frac{\boldsymbol{v} \otimes \boldsymbol{\nabla} + \boldsymbol{\nabla} \otimes \boldsymbol{v}}{2}\right]$$

$$= \begin{bmatrix} \dfrac{\partial v_x}{\partial x} & \dfrac{1}{2}\left(\dfrac{\partial v_x}{\partial y} + \dfrac{\partial v_y}{\partial x}\right) & \dfrac{1}{2}\left(\dfrac{\partial v_x}{\partial z} + \dfrac{\partial v_z}{\partial x}\right) \\ \dfrac{1}{2}\left(\dfrac{\partial v_x}{\partial y} + \dfrac{\partial v_y}{\partial x}\right) & \dfrac{\partial v_y}{\partial y} & \dfrac{1}{2}\left(\dfrac{\partial v_y}{\partial z} + \dfrac{\partial v_z}{\partial y}\right) \\ \dfrac{1}{2}\left(\dfrac{\partial v_x}{\partial z} + \dfrac{\partial v_z}{\partial x}\right) & \dfrac{1}{2}\left(\dfrac{\partial v_y}{\partial z} + \dfrac{\partial v_z}{\partial y}\right) & \dfrac{\partial v_z}{\partial z} \end{bmatrix} \quad (53\text{-}7)$$

式中，T 表示 "转置 (transpose)". 二阶对称张量

$$\boldsymbol{S} = \frac{1}{2}(\boldsymbol{v} \otimes \boldsymbol{\nabla} + \boldsymbol{\nabla} \otimes \boldsymbol{v}) = \frac{1}{2}\left[(\boldsymbol{\nabla} \otimes \boldsymbol{v})^{\mathrm{T}} + \boldsymbol{\nabla} \otimes \boldsymbol{v}\right] \quad (53\text{-}8)$$

称为应变率张量，它只有六个独立分量. 对角线上的分量 (diagonal components) 表示的是体元的伸缩变形率，而非对角分量则表示体元的剪切变形率. $\boldsymbol{S}$ 的分量形式为

$$S_{ij} = \frac{1}{2}\left(\frac{\partial v_i}{\partial x_j} + \frac{\partial v_j}{\partial x_i}\right) \quad (i, j = 1, 2, 3) \quad (53\text{-}9)$$

式中，v_i 为速度的三个分量 (v_x, v_y, v_z)，x_i 为空间位置的三个分量 (x, y, z).

(53-6) 式中的反对称 (antisymmetrical) 矩阵为

$$[\boldsymbol{\Omega}] = [-\boldsymbol{\Omega}^{\mathrm{T}}] = \left[\frac{\boldsymbol{v} \otimes \boldsymbol{\nabla} - \boldsymbol{\nabla} \otimes \boldsymbol{v}}{2}\right]$$

$$= \begin{bmatrix} 0 & \dfrac{1}{2}\left(\dfrac{\partial v_x}{\partial y} - \dfrac{\partial v_y}{\partial x}\right) & \dfrac{1}{2}\left(\dfrac{\partial v_x}{\partial z} - \dfrac{\partial v_z}{\partial x}\right) \\ -\dfrac{1}{2}\left(\dfrac{\partial v_x}{\partial y} - \dfrac{\partial v_y}{\partial x}\right) & 0 & \dfrac{1}{2}\left(\dfrac{\partial v_y}{\partial z} - \dfrac{\partial v_z}{\partial y}\right) \\ -\dfrac{1}{2}\left(\dfrac{\partial v_x}{\partial z} - \dfrac{\partial v_z}{\partial x}\right) & -\dfrac{1}{2}\left(\dfrac{\partial v_y}{\partial z} - \dfrac{\partial v_z}{\partial y}\right) & 0 \end{bmatrix} \quad (53\text{-}10)$$

该反对称张量 $\boldsymbol{\Omega}$ 和 (s2-4) 式中的涡量 $\boldsymbol{\Omega}$ 通过列维–奇维塔符号联系：$\Omega_{ij} = \dfrac{1}{2}\varepsilon_{ijk}\Omega_k$.

对于不可压缩流体，牛顿流体的定义式 (53-3) 可改写为如下张量形式：

$$\boldsymbol{\tau} = 2\mu \boldsymbol{S} \quad \text{或} \quad \tau_{ij} = \mu\left(\frac{\partial v_i}{\partial x_j} + \frac{\partial v_j}{\partial x_i}\right) \quad (53\text{-}11)$$

如果用全应力 $\boldsymbol{\sigma}$ 来表示的话，则不可压缩牛顿流体的本构关系为

$$\boldsymbol{\sigma} = -p\mathbf{I} + 2\mu \boldsymbol{S} \quad \text{或} \quad \sigma_{ij} = -p\delta_{ij} + \mu\left(\frac{\partial v_i}{\partial x_j} + \frac{\partial v_j}{\partial x_i}\right) \quad (50\text{-}12)$$

式中，p 为静水压强.

§54. 哈根-泊肃叶流动定律

泊肃叶 (Jean Léonard Marie Poiseuille, 1799—1869, 图 54.1 右图所示) 是法国生理学家. 他在巴黎综合工科学校 (École Polytechnique) 毕业后, 又攻读了医学, 长期研究血液在血管内的流动. 在求学时代即已发明血压计用以测量狗主动脉的血压. 他发表过一系列关于血液在动脉和静脉内流动的论文 (最早一篇发表于 1819 年). 其中 1840—1841 年发表的论文《小管径内液体流动的实验研究》[11.8] 对流体力学的发展起了重要作用. 他在文中指出, 流量与单位长度上的压力降并与管径的四次方成正比.

图 54.1 哈根 (左) 和泊肃叶 (右)

现在流体力学中常把黏性流体在圆管道中的流动称为泊肃叶流动. 医学上把小血管管壁近处流速较慢的流层称为泊肃叶层.

德国工程师哈根 (Gotthilf Hagen, 1797—1884, 如图 54.1 左图所示) 在 1839 年曾得到和泊肃叶同样的结果. 考虑到哈根于 1839 年发表的实验结果 [11.9] 相当准确, 而且在发表时间上确实早于泊肃叶 1840—1841 年的相关论文, 德国化学家沃尔夫冈·奥斯特瓦尔德 (Wolfgang Ostwald, 1883—1943)[11.10] 最早使用了 "哈根-泊肃叶定律" (Hagen-Poiseuille law) 这一名称, 流体力学大师普朗特和蒂特延斯在 1934 年的专著中 [11.11] 建议该层流定律采用奥斯特瓦尔德的称法. 值得注意的是, 沃尔夫冈·奥斯特瓦尔德是德国著名化学家, 1909 年诺贝尔化学奖获得者威廉·奥斯特瓦尔德 (Wilhelm Ostwald, 1853—1932) 的儿子.

泊肃叶和哈根的经验定律是斯托克斯于 1845 年建立的关于黏性流体运动基本理论 [11.12] 的重要实验证明 [11.13].

为了方便理解, 下面给出哈根-泊肃叶定律的四种推导方法. 应用纳维-斯托克斯方程来推导哈根-泊肃叶定律的方法可见思考题 11.4.

方法一：从力平衡以及流量公式出发

设管道长度为 L，半径为 R，两端压强差为 Δp，则流动达到稳定状态时，黏性阻力：

$$f = -\underbrace{2\pi rL}_{\text{面积}} \underbrace{\mu \frac{\mathrm{d}v}{\mathrm{d}r}}_{\text{黏性应力}} \tag{54-1}$$

式中 μ 为黏性系数. 静压力为

$$F = \Delta p \pi r^2 \tag{54-2}$$

稳定流动时，

$$\Delta p \pi r^2 = -2\pi L r \mu \frac{\mathrm{d}v}{\mathrm{d}r} \tag{54-3}$$

结合无滑移边界条件，$v|_{r=R} = 0$，(54-3) 式的解为

$$v = \frac{\Delta p}{4\mu L}\left(R^2 - r^2\right) \tag{54-4}$$

此时通过管道内的总流量 (单位时间的体积) 为

$$\mathrm{d}Q = v\mathrm{d}S = v2\pi r\mathrm{d}r \tag{54-5}$$

积分上式，得到总流量的解析表达式：

$$Q = \int_0^R \frac{\Delta p}{4\mu L}\left(R^2 - r^2\right) \cdot 2\pi r\mathrm{d}r = \frac{\Delta p}{8\mu L}\pi R^4 \tag{54-6}$$

由上式得到哈根–泊肃叶定律为

$$\Delta p = \frac{8\mu L Q}{\pi R^4} \tag{54-7}$$

方法二：从层与层间的力平衡以及流量公式出发

设管道长度为 L，半径为 R，两端压强差为 Δp，则流动达到稳定状态时，层间厚度为 $\mathrm{d}r$，快速流动层的黏性阻力为

$$f_{\text{vis,faster}} = -2\pi r L \mu \left.\frac{\mathrm{d}v}{\mathrm{d}r}\right|_r \tag{54-8}$$

慢速流动层的黏性阻力为

$$f_{\text{vis,slower}} = 2\pi \left(r+\mathrm{d}r\right) L \mu \left.\frac{\mathrm{d}v}{\mathrm{d}r}\right|_{r+\mathrm{d}r} \tag{54-9}$$

压力：

$$F = \Delta p 2\pi r \mathrm{d}r \tag{54-10}$$

通过力平衡条件:
$$F + f_{\text{vis,slower}} + f_{\text{vis,faster}} = 0 \tag{54-11}$$

亦即
$$\Delta p 2\pi r \mathrm{d}r - 2\pi r L \mu \left.\frac{\mathrm{d}v}{\mathrm{d}r}\right|_r + 2\pi(r+\mathrm{d}r)L\mu \left.\frac{\mathrm{d}v}{\mathrm{d}r}\right|_{r+\mathrm{d}r} = 0 \tag{54-12}$$

根据泰勒级数展开
$$\left.\frac{\mathrm{d}v}{\mathrm{d}r}\right|_{r+\mathrm{d}r} = \left.\frac{\mathrm{d}v}{\mathrm{d}r}\right|_r + \left.\frac{\mathrm{d}^2 v}{\mathrm{d}r^2}\right|_r \mathrm{d}r \tag{54-13}$$

将上式代入上述力平衡方程 (54-12) 式：
$$\Delta p 2\pi r \mathrm{d}r + 2\pi \mathrm{d}r L \mu \frac{\mathrm{d}v}{\mathrm{d}r} + 2\pi r L \mu \mathrm{d}r \frac{\mathrm{d}^2 v}{\mathrm{d}r^2} + 2\pi (\mathrm{d}r)^2 L \mu \frac{\mathrm{d}^2 v}{\mathrm{d}r^2} = 0 \tag{54-14}$$

在上式中忽略掉高阶项 $(\mathrm{d}r)^2$，有
$$\frac{-\Delta p}{L\mu} = \frac{1}{r}\frac{\mathrm{d}}{\mathrm{d}r}\left(r\frac{\mathrm{d}v}{\mathrm{d}r}\right) \tag{54-15}$$

根据边界条件 $v|_{r=R} = 0$ 和 $\left.\frac{\mathrm{d}v}{\mathrm{d}r}\right|_{r=0} = 0$，上式的解为
$$v = \frac{\Delta p}{4\mu L}\left(R^2 - r^2\right) \tag{54-16}$$

则总流量为
$$Q = \int_0^R \frac{\Delta p}{4\mu L}\left(R^2 - r^2\right) \cdot 2\pi r \mathrm{d}r = \frac{\Delta p}{8\mu L}\pi R^4 \tag{54-17}$$

由上式立即得到哈根-泊肃叶定律的 (54-7) 式.

方法三：从常规的量纲分析出发

设流体在细圆管内做稳定流动，则根据物理分析，其体积流率与管道内压强梯度 $\Delta p/L$、流体黏性系数 μ 以及管道半径 R 有关，则
$$Q = C_1 R^a \mu^b \left(\frac{\Delta p}{L}\right)^c \tag{54-18}$$

其中, C_1, a, b, c 均是常数。上式中各物理量的量纲为
$$[Q] = \mathrm{L}^3\mathrm{T}^{-1}, \quad [R] = \mathrm{L}, \quad [\mu] = \mathrm{ML}^{-1}\mathrm{T}^{-1}, \quad \left[\frac{\Delta p}{L}\right] = \mathrm{ML}^{-2}\mathrm{T}^{-2} \tag{54-19}$$

式中, $[\cdot]$ 表示某物理量的量纲, M、L 和 T 分别表示质量、长度和时间的基本量纲. 根据公式两端量纲一致性原理, 有
$$a = 4, \quad b = -1, \quad c = 1 \tag{54-20}$$

得到哈根–泊肃叶定律的流量和各参数之间的关系：

$$Q = C_1 R^4 \frac{\Delta p}{\mu L} \tag{54-21}$$

其中，常数 C_1 可根据实验测得. 上述过程表明，量纲分析方法除了具体的系数不能确定外，各参量间的标度关系可——确定.

方法四：从量纲分析的快速匹配法出发

著者在《力学讲义》[1.11] 中提出了量纲分析的快速匹配法.

设流体在细圆管内做稳定流动，其体积流率 $[Q] = \mathrm{L}^3\mathrm{T}^{-1}$ 与管道半径 $[R] = \mathrm{L}$、流体黏性系数 $[\mu] = \mathrm{ML}^{-1}\mathrm{T}^{-1}$ 和管道内压强梯度 $\left[\dfrac{\Delta p}{L}\right] = \mathrm{ML}^{-2}\mathrm{T}^{-2}$ 有关，则根据快速匹配法：

第一步：黏性系数 μ 中的质量量纲 M 仅与 $\dfrac{\Delta p}{L}$ 配比：$\left[\dfrac{\Delta p}{\mu L}\right] = \mathrm{L}^{-1}\mathrm{T}^{-1}$；

第二步：通过 R 与 $\dfrac{\Delta p}{\mu L}$ 配比：$\left[\dfrac{R^4 \Delta p}{\mu L}\right] = \mathrm{L}^3\mathrm{T}^{-1}$，得到 Q 的量纲.

第三步：得到 (54-21) 式：$Q = C_1 R^4 \dfrac{\Delta p}{\mu L}$.

通过本例可以看出，只要对问题的物理过程理解透彻的话，则量纲分析的快速匹配法应用起来非常的便利.

§55. 达朗贝尔佯谬

思考题 1.2 中的七个数学猜想——千禧年大奖难题中的第三个便是纳维–斯托克斯方程解的存在性和光滑性，其赏金 100 万美元仍尚未发出.

55.1 马略特有关流体阻力的研究

埃德姆·马略特 (Edmé Mariotte, 1620—1684) 是法国物理学家和植物生理学家. 马略特是创建于 1666 年的法国科学院的创建者之一，并成为该院首批院士 (1666 年). 在 1668 年，法国科学院成立了一个以惠更斯、让·皮卡 (Jean Picard, 1620—1682)、马略特等组成的委员会，任务是用实验的方法来验证托里拆利原理. 这个题目后来被适当扩充去研究流体流动冲击在平面上的效果.

托里拆利 (Evangelista Torricelli, 1608—1647)，意大利数学家、物理学家，曾任伽利略助手. 1644 年，托里拆利在《论重物的运动》(*De motu gravium*) 中证明了孔口出流的速度与液高的平方根成比例，即托里拆利定理.

第 11 章　流体动力学

1670 年，马略特开展了后来被称为 "牛顿的摇篮 (Newton's cradle)"(亦简称 "牛顿摆") 的球摆的撞击实验，如图 55.1 所示，五个球，如果拿起左边一个，击打后，右边起来一个，然后进入循环. 马略特的名字和实验在 1687 年牛顿出版的《自然哲学的数学原理》中被提及. 理想弹性、等质量、正碰撞，速度在两端的球间才能实现交换.

图 55.1　牛顿的摇篮

马略特在 1673 年出版的《论物体的撞击与碰撞》(*Traité de la percussion ou choc des corps*) 中总结了前人碰撞问题的实验，得到了动量守恒定律.

马略特对流体力学进行了深入的实验研究，并将结果写成题为《论水和其他流体的运动》(*Traité du mouvement des eaux et des autres corps fluides*) 的著作，于 1686 年出版. 其中特别讨论了流体的摩擦问题.

在该著作中，马略特研究了流体的阻力，得到流体阻力与速度的平方成比例的结论.

在这本书中，马略特论述了：(1) 液体与浮体的平衡；(2) 讨论了由容器流出的液体射流的摩阻，并且解释了实验与理论的一些差异；(3) 给出了一种后来称为马略特瓶的容器的描述，从它流出的流体可以较长时间保持恒速；(4) 给出了圆管中流体压强分布规律，讨论了水在管中流动阻力以及喷水高度问题.

马略特是第一个研究流体阻力的学者，之后，惠更斯也几乎同时得到了流体阻力与流速的平方成正比的结论.

如图 55.2 所示，在带有瓶塞的玻璃瓶中插入一根开口的玻璃管，就制成了一个马略特瓶. 再在马略特瓶的瓶塞中插入 U 形虹吸管的短管.

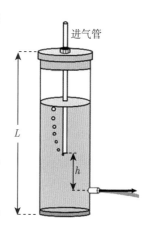

图 55.2　马略特瓶

如图 55.3 所示，当打开虹吸管的阀门时，水就均匀地从虹吸管中流出. 马略特瓶中的水面不断下降但是流量始终保持均匀. 这是因为，马略特瓶中插入的开口玻璃管改变了瓶内的压强分布，使玻璃管下端开口处的压强始终保持与大气压强相同 (这时玻璃管内没有水). 所以在虹吸管的入口处也保持了恒定的压强，使水以均匀的流量流出.

55.2　达朗贝尔佯谬

1749 年，柏林科学院设立了有关流体阻力 (flow drag) 问题的有奖竞赛. 达朗贝尔开展了对该问题的研究，得出的结论是："我承认，我看不出人们如何能够用令人满意的方法在理论上来解释流体的阻力. 据我看来恰好相反：从这个经过深刻研究的理论中，至少在绝大多数的情况下，我们只能得出阻力绝对等于零的结论. 这是一个谜，只好让几何学家们 (数学家们) 来解答 (It seems to me that the

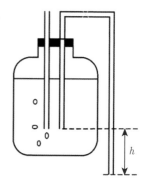

图 55.3　虹吸瓶

theory (potential flow), developed in all possible rigor, gives, at least in several cases, a strictly vanishing resistance, a singular paradox which I leave to future Geometers [i.e. mathematicians — the two terms were used interchangeably at that time] to elucidate)."

图 55.4　普朗特夫妇，其夫人是他导师弗普尔教授的长女

普朗特 (Ludwig Prandtl, 1875—1953, 如图 55.4 所示) 的边界层理论不仅解决了这个疑难, 而且给出了计算物体在流体中运动时阻力的近似方法. 在计算机出现以前, 人们尚无计算黏性流体绕流的途径, 所以边界层理论极大地促进了航空、航天工业的发展. 实际上, 小黏性的作用是无黏方程的奇异摄动, 所以它也是后来发展起来的匹配渐近展开法的原型和物理基础.

1904 年, 当时 29 岁的普朗特在第三次世界数学大会上发表了 *über die Flüssigbewegung bei sehr kleiner Reibung* (《论黏性很小的流体的运动》)[11.14] 的论文. 他根据实验观测, 提出了大雷诺数 (小黏性) 的流体运动边界层的概念, 即黏性仅在固壁附近的薄边界层内起作用, 故层内黏性流体运动方程可以简化, 称为边界层方程, 该层以外可以用理想无黏流体来处理, 从而解决了平板边界层问题. 他还研究了在逆压梯度下的边界层的分离, 注意到流动一旦分离, 边界层便会形成包住尾流的涡面, 改变流动的拓扑结构. 普朗特的这一理论可以应用到所有大雷诺数 (小黏性) 的流体运动上. 普朗特的论文受到了哥廷根大学数学权威菲利克斯·克莱因 (Felix Klein, 1849—1925) 的赏识, 克莱因推荐普朗特担任哥廷根大学应用力学系主任, 后又支持他建立并主持空气动力实验所和威廉皇家流体力学研究所. 值得提及的是, 在哥廷根大学一度被边缘化的西奥多·冯·卡门 (Theodore von Kármán, 1881—1963) 也曾受到过克莱因的提携, 见本书 §59.2.

1908 年, 在罗马召开的第四届世界数学大会上 (the 4$^\text{th}$ International Congress of Mathematicians in Rome), 物理学大师索末菲 (Arnold Sommerfeld, 1868—1951, 如图 55.5 所示) 宣读了一篇有关流动稳定性的论文[11.15], 在其论文中所包含的后来被广泛地称为奥尔–索末菲方程 (Orr-Sommerfeld equation) 中, 索末菲引入了一个重要的无量纲数 —— 雷诺数. 索末菲在文章中写道: "*eine reine Zahl, die wir die Reynolds'sche Zahl nennen wollen* (R is a pure number; we will call it the Reynolds number. R 是一个无量纲的纯数字, 我们称之为雷诺数)."

奥尔–索末菲方程是在流体力学基本方程纳维–斯托克斯方程上叠加微小扰动后线性化所得到的, 描述的是在已知的层流速度剖面上叠加振幅的微小扰动后, 扰动随时间的变化. 如果对于某一特定的雷诺数, 任意扰动都随时间衰减, 那么这个流动就是稳定的. 反之, 则流动是不稳定的.

图 55.5 左: 索末菲和弟子海森堡在一起; 右: 索末菲和弟子泡利在一起

由索末菲引入的雷诺数迄今 (作者写该节的时间为 2018 年 8 月) 已经整整 110 周年, 雷诺数已成为流体力学中最基本的无量纲数之一.

§56. 纳维–斯托克斯方程

56.1 用拉格朗日方程推导纳维–斯托克斯方程

如图 56.1 所示, 纳维–斯托克斯方程 (Navier-Stokes equations)[11.16,11.17] 是在 1757 年建立的不可压缩理想流体的欧拉方程的基础上, 分别由法国数学力学家纳维 (Claude-Louis Navier, 1785—1836) 于 1822 年和英国数学力学家斯托克斯 (George Gabriel Stokes, 1819—1903) 于 1845 年建立的.

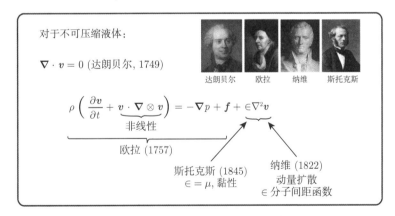

图 56.1 不可压缩流体的纳维–斯托克斯方程的简史

取流体微元 τ, 为了得到系统的拉格朗日量, 首先需要给出动能 K、压强势能

U_p 和体积力势能 U_F 的表达式:

$$\begin{cases} K = \int_\tau \frac{1}{2}\rho \dot{\boldsymbol{u}}^2 \mathrm{d}\tau = \int_\tau \frac{1}{2}\rho \left(\frac{\mathrm{d}\boldsymbol{u}}{\mathrm{d}t}\right)^2 \mathrm{d}\tau \\ U_p = \oint_S \int p \mathrm{d}\boldsymbol{u} \cdot \mathrm{d}\boldsymbol{s} = \int_\tau \mathrm{div}\left(\int p \mathrm{d}\boldsymbol{u}\right)\mathrm{d}\tau \\ U_F = -\int_\tau \mathrm{d}\tau \int \boldsymbol{f} \cdot \mathrm{d}\boldsymbol{u} \end{cases} \quad (56\text{-}1)$$

式中, p 为流体压强. 在 (56-1) 式中第二式, 存在关系式:

$$\mathrm{div}\left(\int p \mathrm{d}\boldsymbol{u}\right) = \int p \mathrm{d}(\mathrm{div}\boldsymbol{u}) + \int \mathbf{grad}p \cdot \mathrm{d}\boldsymbol{u} \quad (56\text{-}2)$$

由于总势能为 $U = U_p + U_F$, 则拉格朗日量为

$$\begin{aligned} L = K - U &= \int_\tau \frac{1}{2}\rho \dot{u}^2 \mathrm{d}\tau - \int_\tau \mathrm{div}\left(\int p \mathrm{d}\boldsymbol{u}\right)\mathrm{d}\tau + \int_\tau \mathrm{d}\tau \int \boldsymbol{f} \cdot \mathrm{d}\boldsymbol{u} \\ &= \int_\tau \mathrm{d}\tau \left[\frac{1}{2}\rho \dot{u}^2 - \mathrm{div}\left(\int p \mathrm{d}\boldsymbol{u}\right) + \int \boldsymbol{f} \cdot \mathrm{d}\boldsymbol{u}\right] \\ &= \int_\tau \mathrm{d}\tau \left[\frac{1}{2}\rho \dot{u}^2 - \int p \mathrm{d}(\mathrm{div}\boldsymbol{u}) - \int \mathbf{grad}p \cdot \mathrm{d}\boldsymbol{u} + \int \boldsymbol{f} \cdot \mathrm{d}\boldsymbol{u}\right] \end{aligned} \quad (56\text{-}3)$$

应力张量 $\boldsymbol{P}$ 写成各项同性部分 $-p\mathbf{I}$ 和各项异性部分 (偏量) $\boldsymbol{P}'$ 之和:

$$\boldsymbol{P} = -p\mathbf{I} + \boldsymbol{P}', \quad P_{ij} = -p\delta_{ij} + P'_{ij} \quad (56\text{-}4)$$

为了得到流体的本构关系, 首先需要给出变形速度 (速度梯度) 张量:

$$S_{ij} = \frac{1}{2}\left(\frac{\partial \dot{u}_i}{\partial x_j} + \frac{\partial \dot{u}_j}{\partial x_i}\right) \quad (56\text{-}5)$$

对于各项同性牛顿流体, 应力偏量可表示为

$$P'_{ij} = \lambda S_{kk}\delta_{ij} + 2\mu S_{ij}$$

由 (56-4) 式, 则应力张量的分量形式为

$$\begin{aligned} P_{ij} &= (-p + \lambda S_{kk})\delta_{ij} + 2\mu S_{ij} \\ &= \left[-p + \left(\lambda + \frac{2}{3}\mu\right)S_{kk}\right]\delta_{ij} + 2\mu\left(S_{ij} - \frac{1}{3}S_{kk}\delta_{ij}\right) \end{aligned}$$

在上式中, 令 $\mu' = \lambda + \frac{2}{3}\mu$, 有

$$P_{ij} = (-p + \mu' S_{kk})\delta_{ij} + 2\mu\left(S_{ij} - \frac{1}{3}S_{kk}\delta_{ij}\right) \quad (56\text{-}6)$$

式中，P_{ij} 为应力张量，p 为流体压强函数，μ 为剪切黏性系数，μ' 为膨胀黏性系数. 由于体积改变引起的黏性力相对于剪切黏性力可忽略不计，斯托克斯做如下假设：膨胀黏性系数 $\mu' = 0$. 应用斯托克斯假设后有 $\boldsymbol{P} = -p\mathbf{I} + \boldsymbol{P}'$，$p = -\dfrac{1}{3}P_{ii}$，即平均法应力. (56-6) 式中的偏应力张量 $\boldsymbol{P}'$ 的分量显然为

$$P'_{ij} = 2\mu \left(S_{ij} - \frac{1}{3}S_{kk}\delta_{ij}\right)$$

从而应力张量的广义牛顿公式可表示为

$$P_{ij} = -p\delta_{ij} + 2\mu \left(S_{ij} - \frac{1}{3}S_{kk}\delta_{ij}\right) \tag{56-7}$$

应用拉格朗日方程推导流体的运动方程，需要考虑流体的黏性耗散. 由于黏性力仅与偏应力张量有关，则黏性力为 $\boldsymbol{F} = \int_\tau (\mathrm{div}\boldsymbol{P}')\mathrm{d}\tau$，其中，积分核为

$$\begin{aligned}
\mathrm{div}\boldsymbol{P}' &= 2\mu\frac{\partial}{\partial x_j}\left(S_{ij} - \frac{1}{3}S_{kk}\delta_{ij}\right) = \mu\left(\frac{\partial^2 \dot{u}_i}{\partial x_j \partial x_j} + \frac{\partial^2 \dot{u}_j}{\partial x_i \partial x_j} - \frac{2}{3}\frac{\partial S_{kk}}{\partial x_i}\right)\\
&= \mu\left(\frac{\partial^2 \dot{u}_i}{\partial x_j \partial x_j} + \frac{1}{3}\frac{\partial S_{kk}}{\partial x_i}\right) = \mu\left[\frac{\partial^2 v_i}{\partial x_j \partial x_j} + \frac{1}{3}\frac{\partial}{\partial x_i}\left(\frac{\partial v_j}{\partial x_j}\right)\right]
\end{aligned}$$

则黏性力可表示为

$$F_i = \int_\tau \mathrm{d}\tau \left\{\mu\left[\frac{\partial^2 v_i}{\partial x_j \partial x_j} + \frac{1}{3}\frac{\partial}{\partial x_i}\left(\frac{\partial v_j}{\partial x_j}\right)\right]\right\} \tag{56-8}$$

将拉格朗日量 (56-3) 式和黏性力的 (56-8) 式代入第一类拉格朗日方程：$\dfrac{\mathrm{d}}{\mathrm{d}t}\left(\dfrac{\partial L}{\partial \dot{u}_i}\right) - \dfrac{\partial L}{\partial u_i} = F_i$，由于该方程左端的两项分别为

$$\begin{aligned}
\frac{\mathrm{d}}{\mathrm{d}t}\left(\frac{\partial L}{\partial \dot{u}_i}\right) &= \frac{\mathrm{d}}{\mathrm{d}t}\frac{\partial}{\partial \dot{u}_i}\int_\tau \mathrm{d}\tau\left[\frac{1}{2}\rho\dot{u}^2 - \int p\,\mathrm{d}(\mathrm{div}\boldsymbol{u}) - \int \mathbf{grad}\,p\cdot \mathrm{d}\boldsymbol{u} + \int \boldsymbol{f}\cdot \mathrm{d}\boldsymbol{u}\right]\\
&= \int_\tau \mathrm{d}\tau\frac{\mathrm{d}}{\mathrm{d}t}(\rho\dot{u}_i) = \int_\tau \mathrm{d}\tau\frac{\mathrm{d}}{\mathrm{d}t}(\rho v_i)
\end{aligned}$$

$$\begin{aligned}
\frac{\partial L}{\partial u_i} &= \frac{\partial}{\partial u_i}\int_\tau \mathrm{d}\tau\left[\frac{1}{2}\rho\dot{u}^2 - \int p\,\mathrm{d}(\mathrm{div}\boldsymbol{u}) - \int \mathbf{grad}\,p\cdot \mathrm{d}\boldsymbol{u} + \int \boldsymbol{f}\cdot \mathrm{d}\boldsymbol{u}\right]\\
&= \int_\tau \mathrm{d}\tau\left[-(\mathbf{grad}\,p)_i + \rho f_i\right]
\end{aligned}$$

由考虑黏性力的拉格朗日方程整理可得

$$\int_\tau \mathrm{d}\tau\frac{\mathrm{d}}{\mathrm{d}t}(\rho v_i) - \int_\tau \mathrm{d}\tau\left[-(\mathbf{grad}\,p)_i + \rho f_i\right] = \int_\tau \mathrm{d}\tau\left\{\mu\left[\frac{\partial^2 v_i}{\partial x_j \partial x_j} + \frac{1}{3}\frac{\partial}{\partial x_i}\left(\frac{\partial v_j}{\partial x_j}\right)\right]\right\}$$

即
$$\int_\tau \mathrm{d}\tau \left\{ \frac{\mathrm{d}}{\mathrm{d}t}(\rho v_i) + (\mathbf{grad}\, p)_i - \rho f_i - \mu \left[\frac{\partial^2 v_i}{\partial x_j \partial x_j} + \frac{1}{3}\frac{\partial}{\partial x_i}\left(\frac{\partial v_j}{\partial x_j}\right) \right] \right\} = 0$$

并由 $\mathrm{d}\tau$ 的任意性可由上式得到微分形式的纳维-斯托克斯方程:

$$\frac{\mathrm{d}}{\mathrm{d}t}(\rho v_i) + (\mathbf{grad}\, p)_i - \rho f_i - \mu \left[\frac{\partial^2 v_i}{\partial x_j \partial x_j} + \frac{1}{3}\frac{\partial}{\partial x_i}\left(\frac{\partial v_j}{\partial x_j}\right) \right] = 0$$

亦即

$$\frac{\mathrm{d}}{\mathrm{d}t}(\rho v_i) = -\frac{\partial p}{\partial x_i} + \mu \left[\frac{\partial^2 v_i}{\partial x_j \partial x_j} + \frac{1}{3}\frac{\partial}{\partial x_i}\left(\frac{\partial v_j}{\partial x_j}\right) \right] + \rho f_i$$

为了使读者更为方便地理解纳维-斯托克斯方程, 分别给出标注各项物理意义的矢量和分量形式:

$$\frac{\mathrm{d}(\rho \boldsymbol{v})}{\mathrm{d}t} = \underbrace{-\boldsymbol{\nabla} p}_{\substack{\text{单位质量流}\\\text{体压强梯度}}} + \underbrace{\mu \nabla^2 \boldsymbol{v}}_{\text{剪切黏性项}} + \underbrace{\frac{1}{3}\mu \boldsymbol{\nabla}(\boldsymbol{\nabla}\cdot\boldsymbol{v})}_{\substack{\text{膨胀或压缩的黏性项}\\\text{不可压缩时此项为零}}} + \underbrace{\boldsymbol{f}}_{\substack{\text{单位质量}\\\text{体积力}}}$$

$$\frac{\mathrm{d}}{\mathrm{d}t}(\rho v_i) = -\frac{\partial p}{\partial x_i} + \mu \frac{\partial^2 v_i}{\partial x_j \partial x_j} + \frac{1}{3}\mu\frac{\partial}{\partial x_i}\left(\frac{\partial v_j}{\partial x_j}\right) + \rho f_i \tag{56-9}$$

注意到方程含有速度 $\boldsymbol{v}$ (三个分量), 压强 p, 密度 ρ 等五个变量, 需结合连续性方程和压强与密度的本构方程求解.

纳维-斯托克斯方程存在如下三种简化情况:

(1) 不可压缩流体 $\rho = \mathrm{const}.$, 且 $\boldsymbol{\nabla}\cdot\boldsymbol{v} = 0$

$$\rho \frac{\mathrm{d}\boldsymbol{v}}{\mathrm{d}t} = \rho\left(\frac{\partial \boldsymbol{v}}{\partial t} + \boldsymbol{v}\cdot\boldsymbol{\nabla}\otimes\boldsymbol{v}\right) = \underbrace{-\boldsymbol{\nabla} p}_{\substack{\text{单位质量流}\\\text{体压强梯度}}} + \underbrace{\mu \nabla^2 \boldsymbol{v}}_{\text{剪切黏性项}} + \underbrace{\boldsymbol{f}}_{\substack{\text{单位质量}\\\text{体积力}}}$$

$$\rho \frac{\mathrm{d}v_i}{\mathrm{d}t} = \rho\left(\frac{\partial v_i}{\partial t} + v_j \frac{\partial v_i}{\partial x_j}\right) = -\frac{\partial p}{\partial x_i} + \mu \frac{\partial^2 v_i}{\partial x_j \partial x_j} + \rho f_i \tag{56-10}$$

(2) 欧拉方程: 不可压缩理想流体 $\rho = \mathrm{const}.$, $\boldsymbol{\nabla}\cdot\boldsymbol{v} = 0$ 且无黏性 $\mu = 0$

$$\rho \frac{\mathrm{d}\boldsymbol{v}}{\mathrm{d}t} = \rho\left(\frac{\partial \boldsymbol{v}}{\partial t} + \boldsymbol{v}\cdot\boldsymbol{\nabla}\otimes\boldsymbol{v}\right) = \underbrace{-\boldsymbol{\nabla} p}_{\substack{\text{单位质量流}\\\text{体压强梯度}}} + \underbrace{\boldsymbol{f}}_{\substack{\text{单位质量}\\\text{体积力}}}$$

$$\rho \frac{\mathrm{d}v_i}{\mathrm{d}t} = \rho\left(\frac{\partial v_i}{\partial t} + v_j \frac{\partial v_i}{\partial x_j}\right) = -\frac{\partial p}{\partial x_i} + \rho f_i \tag{56-11}$$

1966 年, 阿诺尔德 (Vladimir Igorevich Arnold, 1937—2010) 在所发表的题为 *Sur lagéométrie différentielle des groupes de Lie de dimension infinie et ses applications à l'hydrodynamique des fluides parfaits* 的论文中, 对于旋转刚体的欧拉方程

和流体动力学的欧拉方程给出了一种通用的几何解释,这是一个非常优美而深刻的刻画. 阿诺尔德把欧拉方程看作是保体积微分同胚组成的无穷维李群上的测地线方程, 不但清晰地揭示了流体运动内在不稳定性的几何根源, 而且为流体流动及其湍流相关的许多问题提供了数学解答.

(3) 阿基米德原理 (Archimedes' principle):对于静止的流体, $v = 0$

$$0 = \underbrace{-\nabla p}_{\text{单位质量流体压强梯度}} + \underbrace{f}_{\text{单位质量体积力}}$$

$$0 = -\frac{\partial p}{\partial x_i} + \rho f_i \tag{56-12}$$

对不可压缩流体积分可得

$$p - p_0 = \rho \boldsymbol{f} \cdot (\boldsymbol{x} - \boldsymbol{x}_0) \tag{56-13}$$

阿基米德 (如图 56.2 所示) 在大约公元前 250 年 (c. 250 BC) 的著作《论浮体》(*On Floating Bodies*)[11.18] 就已经阐明:"完全或部分地浸入液体中的任何物体,都受到与物体所排开的液体重量相等的力的支撑 (Any object, wholly or partially immersed in a fluid, is buoyed up by a force equal to the weight of the fluid displaced by the object)." 这就是阿基米德浮力原理, 简称阿基米德原理 (Archimedes' principle).

图 56.2 菲尔兹奖章正面的阿基米德头像, 并用拉丁文镌刻着:"超越人类极限, 做宇宙主人" 的格言

对于部分浸入液体中的物体, 阿基米德原理没有考虑液体的表面张力 (毛细力) 的影响[11.19]. 毛细力需要考虑固–气–液三相接触线 (triple contact line), 也就是说, 一旦物体完全地浸入到液体中时, 也就不存在毛细力的问题了.

阿基米德浮力原理, 奠定了流体静力学的基础. 传说希伦王召见阿基米德, 让他鉴定纯金王冠是否掺假. 他冥思苦想多日, 在跨进澡盆洗澡时, 从看见水面上升得到启示, 作出了关于浮体问题的重大发现, 并通过王冠排出的水量解决了国王的疑问. 在其著名的《论浮体》[11.18] 一书中, 他按照各种固体的形状和比重的变化来确定其浮于水中的位置, 并且详细阐述和总结了后来闻名于世的阿基米德原理: 放在液体中的物体受到向上的浮力, 其大小等于物体所排开的液体重量. 从此使人们对物体的沉浮有了科学的认识.

56.2 纳维–斯托克斯方程的无量纲化及无量纲数

在流体力学的实际计算中, 往往要用到无量纲化后的纳维–斯托克斯方程, 这样做的好处是使相关的理论分析有更高的概括性和更广泛的适用性. 方程的无量纲化 (nondimensionalization) 后自然会获得一些无量纲参数, 如雷诺数、马赫数等,这些无量纲量既反映物理本质，又是单纯的数字，不受尺寸、单位制、工程性质、实验装置类型的牵制和束缚. 因此，方程的无量纲化意义重大.

对于不可压缩流体，设体力为重力，则方程 (56-10) 可改写为如下形式：

$$\rho\frac{\mathrm{d}\boldsymbol{v}}{\mathrm{d}t} = \rho\left[\frac{\partial\boldsymbol{v}}{\partial t} + (\boldsymbol{v}\cdot\boldsymbol{\nabla})\boldsymbol{v}\right] = -\boldsymbol{\nabla}p + \mu\nabla^2\boldsymbol{v} + \rho\boldsymbol{g} \tag{56-14}$$

要将方程 (56-14) 式无量纲化，首先要找到相关参量所对应的特征参量 (characteristic parameter)，也就是标度参量 (scaling parameter). (56-14) 式相应的标度参数如表 56.1 所示.

表 56.1 对纳维−斯托克斯方程无量纲化的标度参量一览表

标度参量	物理意义	初始量纲
L，以对算子 $\boldsymbol{\nabla}$ 无量纲化	特征长度	L，长度的基本量纲
V，以对 $\boldsymbol{v}$ 无量纲化	特征速度	LT^{-1}
f，以对 t 无量纲化	特征频率	T^{-1}
$p_0 - p_\infty$，以对 p 无量纲化	参考压强差	$\mathrm{ML}^{-1}\mathrm{T}^{-2}$
g，以对 $\boldsymbol{g}$ 无量纲化	重力加速度	LT^{-2}

这样我们便可以定义如下带 * 号的无量纲变量 (nondimensional variables)：

$$\begin{cases} \boldsymbol{x}^* = \dfrac{\boldsymbol{x}}{L}, \quad t^* = ft, \quad \boldsymbol{v}^* = \dfrac{\boldsymbol{v}}{V} \\ \boldsymbol{\nabla}^* = L\boldsymbol{\nabla}, \quad p^* = \dfrac{p - p_\infty}{p_0 - p_\infty}, \quad \boldsymbol{g}^* = \dfrac{\boldsymbol{g}}{g} \end{cases} \tag{56-15}$$

从而有量纲的变量可等价地表示为

$$\begin{cases} \boldsymbol{x} = L\boldsymbol{x}^*, \quad t = \dfrac{t^*}{f}, \quad \boldsymbol{v} = V\boldsymbol{v}^* \\ \boldsymbol{\nabla} = \dfrac{\boldsymbol{\nabla}^*}{L}, \quad p = p_\infty + (p_0 - p_\infty)p^*, \quad \boldsymbol{g} = g\boldsymbol{g}^* \end{cases} \tag{56-16}$$

将 (56-16) 式代入 (56-14) 式，整理得到

$$\rho V f\frac{\partial\boldsymbol{v}^*}{\partial t^*} + \frac{\rho V^2}{L}(\boldsymbol{v}^*\cdot\boldsymbol{\nabla}^*)\boldsymbol{v}^* = -\frac{p_0 - p_\infty}{L}\boldsymbol{\nabla}^*p^* + \frac{\mu V}{L^2}\nabla^{*2}\boldsymbol{v}^* + \rho g\boldsymbol{g}^* \tag{56-17}$$

式中，每一项的量纲均为 $\mathrm{ML}^{-2}\mathrm{T}^{-2}$，故为了使 (56-17) 式无量纲化，则每一项应乘以 $\dfrac{L}{\rho V^2}$，原因是 $\left[\dfrac{L}{\rho V^2}\right] = \mathrm{M}^{-1}\mathrm{L}^2\mathrm{T}^2$，从而 (56-17) 式将变为

$$\underbrace{\frac{fL}{V}}_{St}\frac{\partial\boldsymbol{v}^*}{\partial t^*} + (\boldsymbol{v}^*\cdot\boldsymbol{\nabla}^*)\boldsymbol{v}^* = -\underbrace{\frac{p_0 - p_\infty}{\rho V^2}}_{Eu}\boldsymbol{\nabla}^*p^* + \underbrace{\frac{\mu}{\rho VL}}_{Re^{-1}}\nabla^{*2}\boldsymbol{v}^* + \underbrace{\frac{gL}{V^2}}_{Fr^{-2}}\boldsymbol{g}^* \tag{56-18}$$

第 11 章 流体动力学

纳维-斯托克斯方程最终的无量纲形式为

$$St\frac{\partial \boldsymbol{v}^*}{\partial t^*} + (\boldsymbol{v}^* \cdot \boldsymbol{\nabla}^*)\boldsymbol{v}^* = -Eu\boldsymbol{\nabla}^* p^* + \frac{1}{Re}\nabla^{*2}\boldsymbol{v}^* + \frac{1}{Fr^2}\boldsymbol{g}^* \tag{56-19}$$

上两式中，雷诺数 Re 及其物理意义为

$$Re = \frac{\text{惯性力}}{\text{黏性力}} = \frac{\rho V^2}{\mu \dfrac{V}{L}} = \frac{\rho V L}{\mu} \tag{56-20}$$

弗劳德数 (Froude number, Fr) 为

$$Fr = \sqrt{\frac{\text{惯性力}}{\text{重力}}} = \sqrt{\frac{\rho V^2}{\rho g L}} = \frac{V}{\sqrt{gL}} \tag{56-21}$$

斯特劳哈尔数 (Strouhal number, St) 为

$$St = \frac{\text{频率}}{\text{特征频率}} = \frac{fL}{V} \tag{56-22}$$

欧拉数 (Euler number, Eu) 为

$$Eu = \frac{\text{总压强}}{\text{惯性力}} = \frac{p_0 - p_\infty}{\rho V^2} \tag{56-23}$$

56.3 用快速匹配法获得纳维-斯托克斯方程的无量纲数

请先预习 §61.2 的内容.

56.3.1 雷诺数的获得

在不可压缩的纳维-斯托克斯方程 (56-14) 式中，$\rho\dfrac{d\boldsymbol{v}}{dt}$ 表示惯性项，$\mu\nabla^2\boldsymbol{v}$ 表示黏性耗散项，其粗略的数量级分别为

$$\left|\rho\frac{d\boldsymbol{v}}{dt}\right| \sim \rho\frac{V}{\tau}, \quad |\mu\nabla^2\boldsymbol{v}| \sim \mu\frac{V}{L^2} \tag{56-24}$$

式中，V 为特征速度，τ 为特征时间，L 为特征尺度. 惯性力和黏性力之比具有如下数量级：

$$\begin{aligned}\frac{|\rho\boldsymbol{a}|}{|\mu\nabla^2\boldsymbol{v}|} &\sim \frac{\rho V}{\tau}\frac{L^2}{\mu V} = \frac{\rho L}{\mu}\frac{L}{\tau} \\ &= \frac{\rho V L}{\mu} = Re\end{aligned} \tag{56-25}$$

在上式的推演中，利用了 $V = L/\tau$. 上式和 (56-20) 式的结果完全相同.

56.3.2 斯特劳哈尔数的获得

在 (56-14) 式中，比较局部加速度 $\dfrac{\partial \boldsymbol{v}}{\partial t}$ 和对流加速度 $(\boldsymbol{v} \cdot \boldsymbol{\nabla}) \boldsymbol{v}$ 的相对数量级，便可获得斯特劳哈尔数. 这两项加速度的数量级分别为

$$\left|\frac{\partial \boldsymbol{v}}{\partial t}\right| \sim \frac{V}{\tau}, \quad |(\boldsymbol{v} \cdot \boldsymbol{\nabla}) \boldsymbol{v}| \sim \frac{V^2}{L} \tag{56-26}$$

则局部加速度和对流加速度的量级之比为

$$\frac{|\partial \boldsymbol{v}/\partial t|}{|(\boldsymbol{v} \cdot \boldsymbol{\nabla}) \boldsymbol{v}|} \sim \frac{L}{\tau V} = \frac{fL}{V} = St \tag{56-27}$$

在上式的推演中，利用了特征时间和特征频率之间的关系式：$f \sim 1/\tau$. (56-27) 式和 (56-22) 式完全相同，该种方法的好处是，可以从另外一个侧面加深对斯特劳哈尔数的理解.

56.3.3 弗劳德数的获得

考察 (56-14) 式中的惯性项 $\rho \dfrac{\mathrm{d}\boldsymbol{v}}{\mathrm{d}t}$ 和重力项 $\rho \boldsymbol{g}$ 的相对重要性：

$$\sqrt{\frac{|\rho \boldsymbol{a}|}{|\rho \boldsymbol{g}|}} \sim \sqrt{\frac{V}{\tau g}} = \sqrt{\frac{V^2}{gL}} = \frac{V}{\sqrt{gL}} = Fr \tag{56-28}$$

在上式的推演过程中，利用了简单关系式：$\tau = L/V$.

56.3.4 欧拉数的获得

考察 (56-14) 式中压强的梯度项 $\boldsymbol{\nabla}p$ 和惯性项 $\rho \dfrac{\mathrm{d}\boldsymbol{v}}{\mathrm{d}t}$ 的相对重要性：

$$\frac{|\boldsymbol{\nabla}p|}{|\rho \boldsymbol{a}|} \sim \frac{(p-p_\infty)/L}{\rho V/\tau} = \frac{p-p_\infty}{\rho V^2} = Eu \tag{56-29}$$

在上式的推演中利用了 $V = L/\tau$.

56.4 相似律

设计大型机械 (诸如航空母舰、航天器、飞机等)、结构 (三峡大坝、港珠澳大桥) 时，通常需要先以模型实验来收集数据，校核理论和数值模拟的结果，以避免因设计失误而造成不可挽回的重大损失，这里的损失不单单是经济上的，在政治上所造成的恶劣影响往往更大.

然而，模型实验所获得的数据与所设计的原型之间有何关系？模型实验所获得的数据如何转换成实际工程中有用的数据？这完全要看模型和原型间是否满足相似律 (law of similarity).

流体动力学的相似律分为三种:

几何相似律: 模型和原型间的几何形状以一定的比例缩小或放大.

运动相似律: 一般而言, 若模型与原型之间的弗劳德数 (Fr)、马赫数 (Ma)、斯特劳哈尔数 (St) 都相等的话, 则运动相似律成立. 此时, 在所对应的时间下, 两者之间所对应点的速度方向都相同且速度大小都呈一定的比例.

动力相似律: 若模型与原型之间的雷诺数 (Re) 和韦伯数 (Weber number, We) 都相等的话, 动力相似律成立. 此时, 在所对应的时间下, 模型和原型之间所有对应点上的各种力方向都相同且大小都呈一定的比例.

这里, 韦伯数为惯性力和表面张力之比:

$$We = \frac{\rho V^2 L}{\gamma} \qquad (56\text{-}30)$$

式中, γ 为表面张力.

相似律还分为如下两种:

完全相似: 模型和原型间完全满足几何相似、运动相似和动力相似. 这就意味着所涉及的无量纲数都相等. 然而由于实际条件的限制, 完全相似通常不可能实现.

不完全相似: 根据实际情况, 设法使最重要的无量纲量保持相等, 而牺牲掉其他次要参数.

例 56.1　《自然》期刊 2018 年 10 月 18 日发表的一篇文章[11.20], 应用风洞揭示了蒲公英种子的飞行机制. 分析其实验是否满足相似律的要求.

解: 《自然》这篇文章的作者通过大量实验, 研究了蒲公英种子飞行背后的物理机制. 蒲公英是菊科多年生草本植物. 蒲公英为头状花序, 种子上有白色冠毛结成的绒球, 花开后随风飘到新的地方孕育新生命. 蒲公英的种子是用一种之前被科学家认为在现实世界中根本行不通的方式飞行的.

如图 56.3 和图 56.4 (a) 所示, 蒲公英的种子长有细长的花丝 (filament), 它们像自行车轮子上的辐条一样从一根中心柄上放射出来, 这一特征似乎是它们飞行的关键. 以前的研究已经发现蒲公英种子的细毛的根数具有一种一致性, 总是在 90—110 根之间.

蒲公英利用冠毛 (小绒毛) 来帮助种子飞行扩散. 冠毛会延缓种子的降落, 使种子飞行的距离超过水平风吹送的距离, 冠毛或许还可以影响种子降落的方向.

如图 56.4 所示, 英国爱丁堡大学的中山真美等构建了一个垂直风洞, 对自由飞行和固定的蒲公英种子的绕流做可视化处理. 通过长曝光摄影和高速成像, 研究

图 56.3　蒲公英利用冠毛 (小绒毛) 来帮助种子飞行扩散

人员发现了一个稳定的气泡——涡环，它与种子本体分离，但稳定地保持在冠毛下部固定距离的位置.

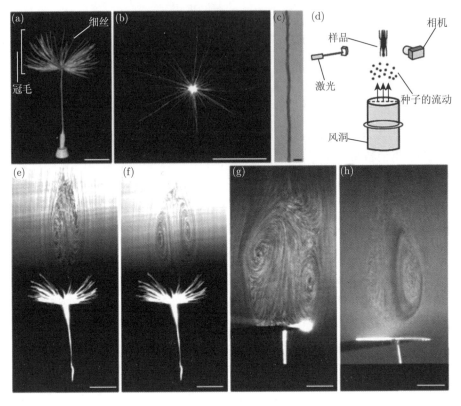

图 56.4 蒲公英种子和所产生的涡环

(a) 蒲公英种子的 μCT 扫描；(b) 冠毛自上而下的图像；(c) 一根细丝的光学显微镜图像；(d) 垂直风洞；(e) 种子终速的下游定常涡；(f) 60% 的终速时的分离涡环结构；(g) 实心圆盘的涡脱落；(h) 多孔圆盘的分离涡环. 图 (c) 中的比例尺为 5 μm, 其余为 50 mm

论文作者发现，如图 56.5 (b) 所示，种子辐条之间的空隙似乎是这些分离涡流保持稳定性的关键. 通过辐条运动的空气和在种子周围运动的空气之间的压力差产生了涡流环流. 不仅如此，蒲公英冠毛的孔隙度似乎受到精确调控以稳定涡环，并且与实心盘相比，冠毛产生的单位面积阻力是其四倍以上. 当该研究团队尝试设计小型硅制圆盘模仿这些辐条时，他们制作了一系列开口的模型——从固体圆盘到 92% 为空气的圆盘，就像蒲公英种子上的结构. 当研究人员在风洞中测试这些种子模型时，他们发现，只有最接近蒲公英种子的圆盘才能维持分离的涡流. 研究人员指出，如果圆盘上的开口数量比蒲公英种子少 10%，则涡流就会不稳定.

如图 56.5 所示，作者还在实验上测定了阻力系数和雷诺数 Re 的关系.

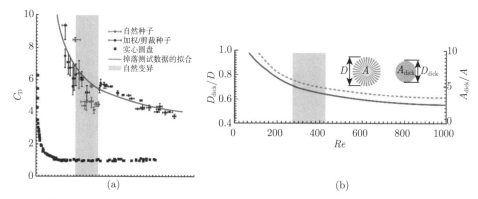

图 56.5 (a) 阻力系数 C_D 和雷诺数 Re 的关系；(b) 多孔和实心圆盘的影响

由于该问题结构十分简单，问题也很明确，因此，相似律比较容易实现。该文中所用的基本方程就是在 §58 给出的斯托克斯方程 (58-2) 式。

例 56.2　用张量的方法重新求解例 3.1.

解：速度矢量的梯度运算得到一个二阶张量：

$$\nabla \otimes \boldsymbol{v} = \left(\frac{\partial}{\partial x_j}\boldsymbol{e}_j\right) \otimes (v_i \boldsymbol{e}_i) = \frac{\partial v_i}{\partial x_j}\boldsymbol{e}_j \otimes \boldsymbol{e}_i \tag{56-31}$$

则有

$$\boldsymbol{v} \cdot \nabla \otimes \boldsymbol{v} = (v_k \boldsymbol{e}_k) \cdot \left(\frac{\partial v_i}{\partial x_j}\boldsymbol{e}_j \otimes \boldsymbol{e}_i\right) = v_k \frac{\partial v_i}{\partial x_j}\delta_{kj}\boldsymbol{e}_i$$
$$= v_j \frac{\partial v_i}{\partial x_j}\boldsymbol{e}_i \tag{56-32}$$

上式可用矩阵形式表示为 (3-8) 式。

例 56.3　不可压缩纳维-斯托克斯方程解的唯一性问题。

答：如图 56.1 图框中不可压缩的纳维-斯托克斯方程组，对于任意点在控制体积 (CV) 内的点 $\boldsymbol{x} \in \mathrm{CV}$ 初始条件为 $\boldsymbol{v}(\boldsymbol{x},0) = \boldsymbol{v}_0(\boldsymbol{x})$；对于控制表面 (CS)，边界条件为 $\boldsymbol{v} = \hat{\boldsymbol{v}}$。这里的速度场需要满足 C^2 类函数。对于不可压缩纳维-斯托克斯方程组，有如下唯一性定理 (uniqueness theorem)：设 $(\boldsymbol{v}_1, p_1)$ 和 $(\boldsymbol{v}_2, p_2)$ 是同一个黏性流问题的解，则满足

$$\begin{cases} \boldsymbol{v}_1 = \boldsymbol{v}_2 \\ p_1 = p_2 + \mathcal{P} \end{cases} \tag{56-33}$$

式中，$\mathcal{P}$ 为满足下式的对空间而言的常数：

$$\nabla \mathcal{P} = \boldsymbol{0} \tag{56-34}$$

§57. 马赫数、马赫锥、马赫角

著者在《力学讲义》[1.11] 的 §19.2 中，业已对力学大师和科学哲学家恩斯特 · 马赫 (Ernst Mach, 1838—1916) 在经典力学上的杰出贡献以及对爱因斯坦的极大启发，做了十分深入的介绍，请读者参阅. 为了连贯地介绍流体动力学中的无量纲数和相似律，本节主要讨论 §56.4 中业已引入的几何相似律中的马赫数.

1881 年，马赫在巴黎召开的国际展览会上，听到了一位比利时的炮师的报告，该报告讨论了从炮口喷出的压缩空气的破坏作用，并且认为炮弹携带的压缩空气超前于炮弹又可以引起像爆炸一样的机械破坏效应. 该问题引起了马赫的兴趣，1885 年他发表了一篇文章公布了 1884 年拍摄的四种不同波 (飞弹波、声波、火花波、抛体激波) 的照片，在这篇文章中，他引进了流速与声速之比的无量纲参量，这就是后人称为的马赫数 (Mach number)：

$$Ma = \frac{V}{c} \tag{57-1}$$

式中，V 为运动物体的速度，c 为当地声速.

1887 年，马赫在《通过空气投影的照相》(*Photographische Fixierung der durch Projektile in der Luft eingeleiten Vorgange*)[11.21] 的论文中，给出了通过纹影照相的到的超音速流动的相片和研究结果 (如图 57.1 所示)，这是最早对超音速流动的研究.

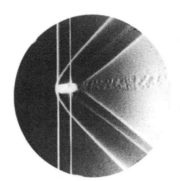

图 57.1　马赫锥 (左)；亚音速和超声速按照马赫数的分类 (右)

当扰动源运动速度高于波速时，扰动来不及传到扰动源的前面去，波面的包络面呈圆锥状，称为马赫锥 (Mach cone)，该方面开创性的工作是由马赫和合作者于 1877 年发表的 [11.22]. 体现为马赫锥的马赫波是一个位置固定的扰动源所发出的一系列扰动以超过波速的速度传播的波阵面. 以位置固定的扰动源为参考系, 当介质运动速度超过声速, 也就是马赫数大于 1 时, 扰动源发出的一个个扰动随介质以 V 的速度向下游移去, 同时扰动本身又以音速 c 向四面八方传播, 而扰动所能播及的区域必限于一个圆锥区域内, 这圆锥是一系列扰动球面的包络面, 称为马赫锥, 圆锥的半顶角称为马赫角 (Mach angle). 图 57.2 分别是 $Ma < 1$、$Ma = 1$ 和 $Ma > 1$ 条件下的波面图.

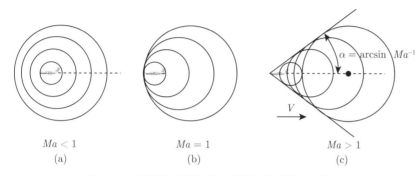

图 57.2　不同马赫数下的波面图, 马赫锥和马赫角

正如 Krehl 所指出的[11.23], "马赫锥" 和 "马赫角" 是力学大师普朗特 (Ludwig Prandtl, 1875—1953) 于 1913 年命名的[11.24]; 当然亦有文献[11.25]指出很可能是普朗特于 1907 年命名的, 但未给出文献出处. "马赫数 (Mach number, Ma)" 则是 Jakob Ackeret (1898—1981) 于 1929 年命名的[11.26].

§58. 斯托克斯阻力

1851 年, 斯托克斯推出计算一个小球在黏性流体中运动阻力的公式, 这是一个被广泛应用的公式, 称为斯托克斯阻力 (Stokes drag) 公式.

要推导小球的斯托克斯阻力, 必须了解斯托克斯流动 (Stokes flow).

58.1 斯托克斯流动

常见流体的控制方程中包括不可压缩的连续性方程 $\boldsymbol{v} \cdot \boldsymbol{\nabla} = 0$ 和纳维-斯托克斯方程:

$$\frac{\partial \boldsymbol{v}}{\partial t} + (\boldsymbol{v} \cdot \boldsymbol{\nabla}) \boldsymbol{v} = \frac{1}{\rho} \left(-\boldsymbol{\nabla} p + \mu \nabla^2 \boldsymbol{v} + \boldsymbol{f} \right) \tag{58-1}$$

当惯性力与黏性力相比居于次要位置时, 雷诺数 $Re \to 0$, 且不考虑体积力 $\boldsymbol{f}$ 的影响, (58-1) 式简化为

$$\begin{cases} \boldsymbol{v} \cdot \boldsymbol{\nabla} = 0 \\ \boldsymbol{\nabla} p = \mu \nabla^2 \boldsymbol{v} \end{cases} \tag{58-2}$$

方程组 (58-2) 称为斯托克斯方程. 满足斯托克斯方程的流动称之为斯托克斯流动.

对动量方程左右取旋度, 消去压强项, 得到关于涡量的方程 $\boldsymbol{\Omega} = \boldsymbol{\nabla} \times \boldsymbol{v}$:

$$\nabla^2 \boldsymbol{\Omega} = \boldsymbol{0} \tag{58-3}$$

满足拉普拉斯方程.

由于小球在黏性流体当中的流动为轴对称流动, 选择基于小球的球坐标系 (R, θ, ϕ):

$$u_R = \frac{1}{R^2 \sin \theta} \frac{\partial \psi}{\partial \theta}, \quad u_\theta = -\frac{1}{R \sin \theta} \frac{\partial \psi}{\partial R} \tag{58-4}$$

此时有

$$\nabla^2 \boldsymbol{\Omega} = \boldsymbol{0}, \quad \boldsymbol{\Omega} = -\frac{\boldsymbol{e}_\theta}{R} \mathrm{D}^2 \psi \tag{58-5}$$

其中

$$\mathrm{D}^2 = \frac{\partial^2}{\partial R^2} + \frac{\sin \theta}{R^2} \frac{\partial}{\partial \theta} \left(\frac{1}{\sin \theta} \frac{\partial}{\partial \theta} \right) \tag{58-6}$$

依据拉普拉斯方程得到关于流函数的双调和方程:

$$\mathrm{D}^2 \mathrm{D}^2 \psi = 0 \tag{58-7}$$

再依据流函数得到的速度场带回动量方程, 得到压强的关系为

$$\begin{aligned} \frac{\partial p}{\partial R} &= \frac{\mu}{R^2 \sin \theta} \frac{\partial}{\partial \theta} \left(\mathrm{D}^2 \psi \right) \\ \frac{\partial p}{\partial \theta} &= -\frac{\mu}{\sin \theta} \frac{\partial}{\partial R} \left(\mathrm{D}^2 \psi \right) \end{aligned} \tag{58-8}$$

58.2 斯托克斯阻力公式

取球坐标 (R, θ, ϕ), 坐标原点位于球心, 球半径为 a, 球的平移速度为 V_0, 指向 $\theta = 0$ 的方向, 此时流函数和边界条件为

$$\begin{aligned} &\mathrm{D}^2 \mathrm{D}^2 \psi = 0 \\ R = a: \quad &\frac{1}{a^2 \sin \theta} \frac{\partial \psi}{\partial \theta} = V_0 \cos \theta, \quad \frac{-1}{a \sin \theta} \frac{\partial \psi}{\partial R} = -V_0 \sin \theta \\ R \to \infty: \quad &\frac{1}{R^2 \sin \theta} \frac{\partial \psi}{\partial \theta} = \frac{-1}{R \sin \theta} \frac{\partial \psi}{\partial R} = 0 \end{aligned} \tag{58-9}$$

采用分离变量法求解问题: 设 $\psi(R, \theta) = F(R) G(\theta)$, 则 $R = a$ 的两边界条件给出

第 11 章 流体动力学

$$F(a)\,G'(\theta) = V_0 a^2 \sin\theta\cos\theta$$

$$F'(a)\,G(\theta) = V_0 a \sin^2\theta \tag{58-10}$$

可以看出解 $G(\theta)$ 必须正比于 $\sin^2\theta$, 即相当于设

$$\psi(R,\theta) = F(R)\sin^2\theta \tag{58-11}$$

则问题化为求解下列欧拉方程:

$$F'''' - \frac{4}{R^2}F'' + \frac{8}{R^3}F' - \frac{8}{R^4}F = 0 \tag{58-12}$$

它的一般解具有下列形式

$$F(R) = AR^4 + BR^2 + CR + \frac{D}{R} \tag{58-13}$$

式中待定常数 A, B, C, D 由下列边界条件确定:

$$F(a) = \frac{1}{2}V_0 a^2, \quad F'(a) = V_0 a, \quad \lim_{R\to\infty}\frac{1}{R^2}F = \lim_{R\to\infty}\frac{1}{R}F' = 0 \tag{58-14}$$

最后得到

$$F(R) = \frac{3}{4}V_0 aR - \frac{1}{4}V_0\frac{a^3}{R}$$

因而

$$\psi(R,\theta) = \frac{1}{4}V_0\sin^2\theta\left(3aR - \frac{a^3}{R}\right)$$

速度分布为

$$u_R = \frac{1}{2}V_0\cos\theta\left(\frac{3a}{R} - \frac{a^3}{R^3}\right), \quad u_\theta = -\frac{1}{4}V_0\sin\theta\left(\frac{3a}{R} + \frac{a^3}{R^3}\right) \tag{58-15}$$

根据之前斯托克斯流动的压强结果, 有

$$\frac{\partial p}{\partial R} = -3\frac{\mu V_0 a}{R^3}\cos\theta, \quad \frac{\partial p}{\partial \theta} = -\frac{3}{2}\frac{\mu V_0 a}{R^2}\sin\theta \tag{58-16}$$

再加上 $R\to\infty$ 时, $p = p_\infty$ 的条件可以解出

$$p = p_\infty + \frac{3}{2}\frac{\mu V_0 a}{R^2}\cos\theta \tag{58-17}$$

在球面 $R = a$ 上的压强分布为

$$p|_{r=a} = p_\infty + \frac{3}{2}\frac{\mu V_0}{a}\cos\theta \tag{58-18}$$

显然球面的流体压强不再像势流中那样具有前后对称性, 即存在压差阻力.

与势流不同的是，这时不仅有压差阻力，还有摩擦阻力，总阻力是流体表面正应力 τ_{RR} 和切应力 $t_{R\theta}$ 的合力

$$\tau_{RR}|_{R=a} = 2\mu \frac{\partial u_R}{\partial R}\bigg|_{R=a} = 0$$

$$\tau_{R\theta}|_{R=a} = \mu\left(\frac{1}{R}\frac{\partial u_R}{\partial \theta} + \frac{\partial u_\theta}{\partial R} - \frac{u_\theta}{R}\right)\bigg|_{R=a} = \frac{3}{2}\frac{\mu V_0}{a}\sin\theta \tag{58-19}$$

球所受的流体合力是

$$F_x = 2\pi \int_0^\pi [(-p+\tau_{RR})\cos\theta - \tau_{R\theta}\sin\theta]_{R=a} a^2 \sin\theta \mathrm{d}\theta$$
$$= \underbrace{-2\pi\mu V_0 a}_{\text{压阻占 } 1/3} \underbrace{-4\pi\mu V_0 a}_{\text{摩阻占 } 2/3} = -6\pi\mu V_0 a \tag{58-20}$$

因此得到球所受的总阻力为

$$F = -6\pi\mu V_0 a \tag{58-21}$$

其中压差阻力占 1/3，摩擦阻力占 2/3. 该公式就是著名的斯托克斯阻力公式.

1854 年，斯托克斯成为英国皇家学会的秘书. 这个新职位的工作分去了他许多精力，瑞利勋爵 (Lord Rayleigh, 1842—1919) 曾在斯托克斯的讣告中写道："论文汇编的读者会觉察到从那时以后作品发表的速度有一显著的下降. 这反映出科学家必须让他们从事科学研究，而不应该让他们担任过于繁重的行政职务." (In 1854, Stokes became secretary of the Royal Society. The work in this new position absorbed much of his energy and Lord Rayleigh, in Stokes' obituary notice, notes: "The reader of the Collected Papers can hardly fail to notice a marked falling off in the speed of production after this time. The reflection suggests itself that scientific men should be kept to scientific work, and should not be tempted to assume heavy administrative duties, at any rate until such time as they have delivered their more important messages to the world.")

有关斯托克斯的实验和讲课，瑞利勋爵评价道："他的很多发现是在他家餐厅后面一条狭窄的过道里做出来的 …… 他随时会将新鲜事物引入他的讲课中." ("Many of his discoveries were made in a narrow passage behind the pantry of his house … who was able to introduce into his lectures matter fresh from the anvil.")

斯托克斯从 1849 年起在剑桥大学担任卢卡斯讲座教授，1851 年当选为英国皇家学会会员 (FRS)，1854 年任皇家学会秘书，1885—1890 任皇家学会主席. 斯托克斯是继牛顿之后担任过这三个职务的第二人.

麦克斯韦、开尔文和斯托克斯都是"教科书式"的大科学家，在本书中也有很多篇幅谈及他们的杰出贡献。约瑟夫·拉莫尔 (Joseph Larmor, 1857—1942) 在其题为《乔治·加布里埃尔·斯托克斯爵士》[11.27] 的书中对如上三位大科学家做了如下令人深思的对比："麦克斯韦对于一切事物都富有幻想，斯托克斯却拘谨过度。麦克斯韦最喜欢构创想象的和实在的模型以及想象组成物质基础的分子活动；斯托克斯所发表的研究大部分是属于精细和表观的一类。以块体物质的性质和对称性为规准，在他的研究中很难引入分子的概念。开尔文大致介于他们两位之间。按照开尔文活动的主要特征，也是他本人所坚持的，他可以被正确地被说成是斯托克斯的一个门徒 (pupil of Stokes)。虽然他还缺乏勇气，不能大胆地驰骋于未知世界，但在重视实用的才能以外，他也有积极想象的能力。无疑地，他的想象能力在受到法拉第影响之后，主要是受麦克斯韦的启发。在麦克斯韦的著作中，想象的结果是极为成功的。"

铁摩辛柯在其著名的《材料力学史》[11.28] 中也转述了上述评价。

例 58.1 应用斯托克斯力开展 DNA 单分子操纵。

斯托克斯拖曳是通过水的流动将力施加于连有小球上用两块玻璃片密封形成样品腔，并在上面的玻璃片上开两个小孔连上细导管，以输入生化样品或更换缓冲液。通常将生物分子 (本例为 DNA) 的一端固定在被修饰过的玻璃基底上，而 (将 DNA 的) 另一端固定在小球上，然后在输液管接上微流泵以控制水流流速。力的大小用斯托克斯公式 (58-21) 式给出 [11.29]。

值得注意的是，斯托克斯阻力公式 (58-21) 式是球体浸没在无限大的液体中推导出的，由于本例中玻璃小球和基底的距离较近，此时的斯托克斯阻力公式需要进一步修正为

$$F = -6\pi\mu V_0 a \left(1 + \frac{9a}{16d}\right) \tag{58-22}$$

式中，d 为小球的球心和基底的距离。在该类实验中，一般针对 DNA 所施加的拉力在数十个 pN (10^{-12} N) 量级，雷诺数在 10^{-3} 的量级。

例 58.2 主宰斯托克斯方程 $\nabla p = \mu \nabla^2 \boldsymbol{v}$ 的无量纲数是什么？

答：类似于 §56.3 中的步骤，有 $|\nabla p| \sim (p - p_\infty)/L$，$|\mu \nabla^2 \boldsymbol{v}| \sim \mu V/L^2$，因而主宰斯托克斯方程的无量纲数为

$$\frac{p - p_\infty}{L} \frac{L^2}{\mu V} = \frac{p - p_\infty}{\mu V/L} \sim \frac{\text{压强差}}{\text{黏性应力}} \tag{58-23}$$

由 (56-20) 式中雷诺数的表达式和 (56-29) 式中欧拉数的表达式，可知 (58-23) 式中

的无量纲数可以表示为雷诺数和欧拉数的乘积:

$$Re \times Eu = \frac{\rho V^2}{\mu V/L} \times \frac{p - p_\infty}{\rho V^2} = \frac{p - p_\infty}{\mu V/L} \tag{58-24}$$

本书将该无量纲数命名为 "压黏数 (pressure-viscous number)".

例 58.3 斯托克斯阻力公式 (58-21) 在油滴实验 (oil-drop experiment) 中的应用.

答: 密立根 (Robert Millikan, 1868—1953) 在 1909—1917 年间所做的测量微小油滴上带的电荷的工作, 即所谓油滴实验, 堪称物理实验的典范之一. 密立根的博士生哈维·福莱柴尔 (Harvey Fletcher, 1884—1981, 1911 年获得芝加哥大学博士学位) 和李耀邦 (John Yiubong Lee, 1884—1940, 1914 年在芝加哥大学获得博士学位) 从一开始就参与了有关实验工作. 密立根因此而获得了 1923 年度诺贝尔物理学奖. 油滴实验取得所取得的结果包括: (1) 证明了电荷的不连续性, 所有电荷都是基本电荷 e 的整数倍; (2) 测量并得到了基本电荷即为电子电荷, 其值为 $e = 1.592 \times 10^{-19}$ 库仑.

值得特别注意的是, 密立根于 1910 年独立所发表的第一篇论文 (Millikan R A. The isolation of an ion, a precision measurement of its charge, and the correction of stokes's law. Science, 1910, 32: 436-448) 和一年后所发表的同名论文 (Millikan R A. The isolation of an ion, a precision measurement of its charge, and the correction of Stokes's law. Physical Review (Series I), 1911, 32: 349-397) 对经典的斯托克斯定律进行了修正.

在此实验中, 向下运动的油滴共受四个力的作用 (以向下为正): 一是向上的空气阻力

$$F_{\text{Stokes}} = -6\pi\mu v_1 a \tag{58-25}$$

式中, v_1 为油滴下落的终端速度 (terminal velocity for falling), a 为油滴的半径.

油滴所受到的向下的重力为

$$F_{\text{weight}} = \frac{4}{3}\pi a^3 \rho_{\text{oil}} g \tag{58-26}$$

油滴所受到的向上的浮力为

$$F_{\text{buoyancy}} = -\frac{4}{3}\pi a^3 \rho_{\text{air}} g \tag{58-27}$$

如图 58.1 所示, 在电场开启前, 油滴的运动达到其终端速度, 也就是油滴匀速

下降时, 上述三个力达到平衡, 此时可确定油滴的半径为

$$a = \sqrt{\frac{9\mu v_1}{2g(\rho_{\text{oil}} - \rho_{\text{air}})}} \tag{58-28}$$

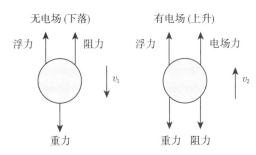

图 58.1　开启电场前和后液滴的受力分析

开启电场后, 油滴将受到一个向上的电场力作用:

$$F_{\text{electric}} = -qE = -q\frac{V}{d} \tag{58-29}$$

式中, V 为平板间的电位差, d 为平板间的距离. 如图 58.1 所示, 稍微再将电压 V 向上调升, 让油滴上升并得到一个新的终端速度 v_2, 再由液滴受到的电场力、重力、浮力和新的斯托克斯阻力 $F'_{\text{Stokes}} = -6\pi\mu v_2 a$ 的平衡, 得到所测的油滴电荷为

$$q = \frac{9\pi d\mu(v_1 + v_2)}{V}\sqrt{\frac{2\mu v_1}{g(\rho_{\text{oil}} - \rho_{\text{air}})}} \tag{58-30}$$

密立根在其于 1910 年发表的《科学》一文中, 认为当油滴的尺度 a 和气体分子的平均自由程 (mean free path) 可比时, 经典的斯托克斯公式 (58-21) 将不再适用, 密立根给出了斯托克斯阻力公式的修正形式为

$$F = -6\pi\mu V_0 a\left(1 + A\frac{l}{a}\right) \tag{58-31}$$

式中, l 为气体的平均自由程, A 为由实验确定的常数. 事实上, 由 (2-8) 式中努森数的定义, 修正后的斯托克斯阻力公式可通过努森数表示为

$$F = -6\pi\mu V_0 a(1 + A \cdot Kn) \tag{58-32}$$

例 58.4　奥森流 (Oseen flow) 和对斯托克斯阻力公式 (58-21) 的修正.

答: 对于低雷诺数的定常、不可压缩的流动, 在 (56-10) 式中, 由于 $\frac{\partial v_i}{\partial t} = 0$, 再

不考虑体积力时，(56-10) 式可退化为

$$\rho v_j \frac{\partial v_i}{\partial x_j} = -\frac{\partial p}{\partial x_i} + \mu \frac{\partial^2 v_i}{\partial x_j \partial x_j} \tag{58-33}$$

奥森 (Carl Wilhelm Oseen, 1879—1944) 于 1910 年对上述非线性方程做了进一步的简化，当物体以定常速度 V_0 流动时，(58-33) 式可进一步简化为如下奥森方程：

$$\rho V_0 \frac{\partial v_i}{\partial x_j} = -\frac{\partial p}{\partial x_i} + \mu \frac{\partial^2 v_i}{\partial x_j \partial x_j} \tag{58-34}$$

此时，斯托克斯阻力公式 (58-21) 式可通过雷诺数 Re 修正为

$$F = -6\pi\mu V_0 a \left(1 + \frac{3}{8} Re\right) \tag{58-35}$$

1957 年，Proudman 和 Pearson 给出了奥森流针对斯托克斯阻力公式的二阶修正：

$$F = -6\pi\mu V_0 a \left[1 + \frac{3}{8} Re + \frac{9}{40} (Re)^2 \log (Re) + \cdots \right] \tag{58-36}$$

例 58.5　小雷诺数下，球形颗粒在溶液中的扩散系数．

答：该扩散系数可表示为

$$D = \frac{k_B T}{6\pi\mu a} \tag{58-37}$$

该式被称为"斯托克斯-爱因斯坦关系式"．

§59. 从层流到湍流的转捩

59.1　雷诺 1883 年的经典论文

1868 年，时年 26 岁的雷诺 (Osborne Reynolds, 1842—1912) 在获得了欧文学院的教授职位后，和来自英国利兹的一位医生的女儿夏洛特 (Charlotte) 结婚[11.30]，十分不幸的是，第二年在他们的儿子诞生的 12 天后，雷诺的新婚妻子夏洛特便不幸去世了．然而更为不幸的是，1879 年 9 月 27 日，他们 10 岁的儿子也去世了！这个事件标志着雷诺在欧文学院第一个阶段的结束 (That event may be said to mark the end of the first phase of Reynolds career at Owens College).

1881 年的 12 月，时年 39 岁的雷诺和 22 岁的 Annie Charlotte Wilkinson 再婚 (In December 1881 Reynolds had re-married; his wife was Annie Charlotte Wilkinson, just 22 years old).

1883 年，雷诺在所发表的《平行渠道阻力的实验研究情况》(*An experimental investigation of the circumstances which determine whether the motion of water shall be direct or sinuous, and of the law of resistance in parallel channels*)[11.31] 这一流体力学经典论文中, 研究了层流 (laminar flow) 向湍流 (turbulent flow, turbulence) 的转捩 (transition), 如图 59.1 所示, 标志湍流研究的开端.

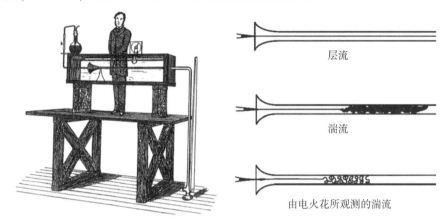

图 59.1　雷诺进行流体力学实验的情景再现 (左); 层流向湍流的转捩 (右)

雷诺在流体力学上最重要的贡献是发现了黏性流体流动的相似律 (similarity law). 他引入了 (56-20) 式的无量纲数. 正如在 §55.2 中所业已阐明的, 这个常数被德国物理学家索末菲于 1908 年提议称为 "雷诺数 Re".

雷诺数是流体力学中最为重要的无量纲数, 读者务必对该无量纲参数有十分透彻的理解. (56-20) 式中的所谓 "惯性力 (inertial force)" 事实上是具有应力量纲的动压 ρV^2, $\dfrac{V}{L}$ 为速度梯度的数量级, $\mu\dfrac{V}{L}$ 则为牛顿流体的剪应力, 也就是 (56-20) 式中的单位面积的黏性力 (viscous force). (56-20) 式还常常表示为

$$Re = \frac{VL}{\mu/\rho} = \frac{VL}{\nu} \tag{59-1}$$

式中, $\nu = \mu/\rho$ 称为运动黏度 (kinematic viscosity), 它和扩散系数量纲相同

$$[\nu] = \left[\frac{\mu}{\rho}\right] = \frac{\mathrm{ML}^{-1}\mathrm{T}^{-1}}{\mathrm{ML}^{-3}} = \mathrm{L}^2\mathrm{T}^{-1} \tag{59-2}$$

实验研究表明, 当 $Re < 2300$ 时, 流动为层流; 当 $2300 < Re < 4000$ 时, 流动为转捩流 (transient flow); 当 $Re > 4000$ 流动为湍流, 如图 59.2 所示.

事实上, 早在 1851 年, 斯托克斯已经注意到这个无量纲数的重要性. 1883 年雷诺通过管道中的流动实验发现, 在适当的雷诺数之下, 管道中的流动总是从层流转变为湍流. 说明这个无量纲数的重要性.

图 59.2　烟雾中的湍流

59.2　卡门涡街——科学与艺术结合的典范

卡门涡街 (von Kármán vortex street) 是一种自然界所广泛存在的现象. 如在一个平静流动的河流中有一个圆柱，在特定条件下圆柱下游的两侧，如图 59.3 左图所示，会产生两道非对称排列的旋涡，这两排旋涡的旋转方向相反，在空间上交错排列，就像街道两边的街灯一样，因而得名. 图 59.3 右图是大气流动经过一个障碍物时，所产生的大尺度的卡门涡街.

图 59.3　圆柱尾流中的卡门涡街 (左) 和大气流动中的卡门涡街 (右)

1911 年，冯·卡门在哥廷根大学当助教，普朗特教授当时的研究兴趣，主要集中在边界层问题上. 普朗特交给博士生哈依门兹 (Karl Hiemenz) 的任务，是设计一个水槽，使能观察到圆柱体后面的流动分裂，用实验来核对按边界层理论计算出来的分裂点. 为此，必须先知道在稳定水流中圆柱体周围的压力强度如何分布. 哈依门兹做好了水槽，但出乎意外的是在进行实验时，发现在水槽中的水流不断地发生激烈的摆动.

哈依门兹向普朗特教授报告这一情况后，普朗特告诉他："显然，你的圆柱体不够圆 (Obviously your cylinder is not circular)." 可是，当哈依门兹将圆柱体作了非常精细的加工后，水流还是在继续摆动. 普朗特又说："水槽可能不对称 (possible the channel was not symmetrical)."哈依门兹于是又开始细心地调整水槽，但仍不能解决问题.

冯·卡门当时所做的课题与哈依门兹的工作并没有关系，而他每天早上进实验室时总要跑过去问："哈依门兹先生，现在流动稳定了没有？"哈依门兹非常懊丧地回答："始终在摆动 (It always oscillates)."

这时冯·卡门想，如果水流始终在摆动，这个现象一定会有内在的客观原因. 在一个周末，冯·卡门用粗略的运算方法，试计算了一下涡系的稳定性. 他假定只有一个涡旋可以自由活动，其他所有的涡旋都固定不动. 然后让这一涡旋稍微移动一下位置，看看计算出来会有什么样的结果. 冯·卡门得到的结论是：如果是对称的排列，那么这个涡旋就一定离开它原来的位置越来越远；而对于反对称的排列，虽然也得到同样的结果，但当行列的间距和相邻涡旋的间距有一定比值时，这涡旋却停留在它原来位置的附近，并且围绕原来的位置作微小的环形路线运动.

星期一上班时，冯·卡门向普朗特教授报告了他的计算结果，并问普朗特对这一现象的看法如何？普朗特说："这里面有些道理，写下来罢，我把你的论文提交到学院去 (You have something, write it up and I will present your paper in the academy)."

通过涡街的这一重大发现，冯·卡门于 1912 年由一名助教，在数学大师克莱因的推荐下，担任了德国亚琛工业大学的正教授和航空研究所的主任，逐步成为国际力学大师.

冯·卡门后来回忆时，对此事写道："这就是我关于这一问题的第一篇论文[11.32]. 之后，我觉得，我的假定有点太武断. 于是又重新研究一个所有涡旋都能移动的涡系. 这样需要稍微复杂一些的数学计算. 经过几周后，计算完毕，我写出了第二篇论文[11.33]. 有人问我：'你为什么在三个星期内发表两篇论文 (publish two papers in three weeks) 呢？一定有一篇是错的罢.' 其实并没有错，我只是先得出个粗略的近似 (crude approximation)，然后再把它细致化 (afterward refined it)，基本上结果是一样的；只是得到的临界比的数值并不完全相同."

冯·卡门是针对哈依门兹的水槽实验，进行涡旋排列的研究的. 后来人们由于冯·卡门对其机理详细而又成功的研究，将它冠上了卡门的姓氏，称为卡门涡街.

冯·卡门自己后来在他所著的《空气动力学》(*Aerodynamics*)[11.34] 书中的第 68 页写道："我并不宣称，这些涡旋是我发现的 (I do not claim to have discovered these vortices). 早在我生下来之前，大家已知道有这样的涡旋. 我最早看到的是意大利博洛尼亚一个教堂 (in a church in Bologna) 中的一张图画. 图上画着圣·克里

斯朵夫 (St. Christopher) 抱着幼年的耶稣涉水过河. 画家在克里斯朵夫的赤脚后面, 画上了交错的涡旋."

在冯·卡门的一本题为《疾风及九霄之外: 西奥多·冯·卡门, 航空先驱和太空探路者》(*The Wind and Beyond: Theodore von Kármán, Pioneer in Aviation and Pathfinder in Space*)[5.4] 的自传体中, 也有针对该发现的类似描述.

遗憾的是, 冯·卡门在书中并没有给出令他产生发现 "卡门涡街" 灵感的意大利博洛尼亚教堂的这幅图画. 幸运的是, 四位学者于 2000 年在英国的《自然》期刊发专文 [11.35], 考证了此画. 如图 59.4 所示, 它是来自于意大利博洛尼亚圣·多米尼克教堂博物馆的一幅 14 世纪的壁画 (a fresco picture in the fourteenth century at the museum at the Church of St Dominic in Bologna, Italy), 该壁画的名称为 "*Madonna con bambino tra I Santi Dommenico, Pietro Martire e Critoforo*", 画家未知 (painted by an unknown artist).

"克里斯朵夫" 意为 "背负基督者". 传说一日一名儿童请求克里斯朵夫帮助他渡河, 克里斯朵夫却赫然发现那个小孩子竟然重到他几乎背不动. 在艰难的过河后他向小孩子说 "你重到让我有如世界被背在我肩上 (I do not think the whole world could have been as heavy on my shoulders as you were)", 而小孩子则回应 "你背负的不只是全世界, 还包含它的缔造者. 我就是基督, 你的君王 (You had on your shoulders not only the whole world but Him who made it. I am Christ your king)". 说完, 小孩便消失了.

英国格拉斯哥大学艺术史系的学者罗伯特·吉布斯针对上述的《自然》文章提出了不同的意见. 吉布斯认为冯·卡门所看到的博洛尼亚的圣·克里斯朵夫脚下交错的涡旋的图画并非来自于圣·多米尼克教堂, 而很可能来自于圣斯特凡诺修道院外回廊一些屋子里自从 1390 年代开始就长期展览的 Jacopo di Paolo 的画作 (The representation of St Christopher that Kármán is most likely to have seen is the triptych by Jacopo di Paolo from the 1390s that has long been shown in various rooms of the outer cloisters of San Stefano)[11.36].

冯·卡门是钱学森的博士导师 (普朗特–冯·卡门–钱学森师徒三代的合影如图 59.5 所示), 钱学森当然清楚卡门涡街在发现过程中, 博洛尼亚的圣·克里斯朵夫脚下交错的涡旋的艺术品所起的巨大的启发作用.

钱学森本人对他所提出的 "钱学森之问" 的回答是, 根据历史经验, 也根据他本人的体会, 就是我们的大学教育要实现科学与艺术的结合. 可以毫不夸张地说, 冯·卡门有关 "卡门涡街" 的科学发现就是科学与艺术结合的典范! 令人叹为观止! 应该说, 冯·卡门在自己所说的 "三个星期周写了两篇论文" 的时间段里, 应该处于韦伊所谈到的科学发现的 "亢奋状态"(见题表 1.13)!

图 59.4　抱着幼年的耶稣涉水的圣·克里斯朵夫的赤脚后面出现交错的涡旋给了冯·卡门发现的灵感

图 59.5　左起: 休·拉蒂默·德莱顿 (Hugh L. Dryden)、普朗特、冯·卡门、钱学森 (1945 年)

冯·卡门还说，在他之前，有一位英国科学家马洛克 (Henry Reginald Arnulpt Mallock, 1851—1933) 也已于 1907 年观察到平板后面交错的涡旋[11.37]，如图 59.6 所示.

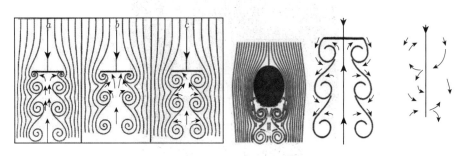

图 59.6　马洛克的平板后旋涡示意图

法国学者亨利·贝尔纳 (Henri Bénard, 1874—1939) 也于 1908 年作过关于这一问题的大量研究[11.38]. 只不过贝尔纳主要考察了黏性液体和胶悬溶液中的涡旋，并且其考察的角度是实验物理学的观点多于空气动力学的观点. 在 1926 年在苏黎世和 1930 年斯德哥尔摩召开的国际应用力学大会上，贝尔纳曾声明对该现象更早的观测的优先权 (claimed priority for earlier observation of the phenomenon). 冯·卡门风趣地说："我同意在柏林和伦敦被称为'卡门涡街'，而在巴黎则应称为'亨利·贝尔纳涡街 (Avenue de Henri Bénard)'." 在这些俏皮话 (wisecrack) 之后，冯·卡门和亨利·贝尔纳成了很好的朋友 (quite good friends).

亨利·贝尔纳是法国著名学者，见思考题 11.8.

冯·卡门认为他在 1911—1912 年，对这一问题研究的贡献主要是二个方面：一是发现涡街只有当涡旋是反对称排列，且仅当行列的距离对同行列内相邻两涡旋的间隔有一定的比值时才稳定；二是将涡系所携带的动量与阻力联系了起来.

判断一个力学研究工作是否重要可能有多重标准，但是如果一个研究工作不但激发了大量后续的理论、实验和数值模拟的研究，而且还在工程实践中得到了广泛的应用的话，无疑这就是一项开创性的彪炳史册的工作. 从这个标准来判断的话，卡门涡街无疑是一个足以彪炳于力学史册的工作.

20 世纪 40 年代，如图 59.7 所示的美国塔科玛峡谷桥 (Tacoma Narrow Bridge) 风毁事故的惨痛教训，使人们认识到卡门涡街对建筑安全上的重要作用.

1940 年，美国华盛顿州的塔科玛峡谷上花费 640 万美元，建造了一座主跨度 853.4 米的悬索桥. 建成四个月后，于同年 11 月 7 日碰到了一场风速为 19 米/秒的风. 虽风不算大，但桥却发生了剧烈的扭曲振动，且振幅越来越大 (接近 9 米)，直到桥面倾斜到 45 度左右，使吊杆逐根拉断导致桥面钢梁折断而塌毁，坠落到峡谷之中．当时正好有一支好莱坞电影队在以该桥为外景拍摄影片，记录了桥梁从开始振动到最后毁坏的全过程，它后来成为美国联邦公路局调查事故原因的珍贵资料．人们在调查这一事故收集历史资料时，惊异地发现：从 1818 年到 19 世纪末，由风引起的桥梁振动已至少毁坏了 11 座悬索桥．

图 59.7　塔科玛峡谷桥于 1940 年 11 月 7 日的坍塌事故

第二次世界大战结束后，人们对塔科玛桥的风毁事故的原因进行了研究．一开始，就有二种不同的意见在进行争论．一部分航空工程师认为塔科玛桥的振动类似于机翼的颤振；而以冯·卡门为代表的流体力学家认为，塔科玛桥的主梁有着钝头的 H 型断面，和流线型的机翼不同，存在着明显的涡旋脱落，应该用涡激共振机理来解释．冯·卡门 1954 年在《空气动力学》[11.34] 一书中写道：塔科玛峡谷桥的毁坏，是由周期性旋涡的共振 (resonance due to periodic vortices) 引起的．设计的人想建造一个较便宜的结构，采用了平钣来代替桁架作为边墙．不幸，这些平钣引起了涡旋的脱落 (shedding vortices)，使桥身开始扭转振动．这一大桥的破坏现象，是振动与涡旋发放发生共振而引起的.

塔科玛峡谷桥的风毁事故，是一定流速的流体流经边墙时，产生了卡门涡街；卡门涡街后涡的交替脱落，会在物体上产生垂直于流动方向的交变侧向力，迫使桥梁产生振动，当发放频率与桥梁结构的固有频率相耦合时，就会发生共振，造成破坏．

在冯·卡门 75 岁生日那天，德国西柏林工业大学的的维勒 (Rudolf Wille, 1911—1973) 教授赠送了他一张有 75 个涡的照片作为礼物，上面书写着：永无止境 (Prof. Wille presented to Kármán on his 75th birthday with 75 shed vortices, with the engaging caption: *Ad Infinitum*).

古代人发现，涡旋是对人有危险的最重要的信号，这看来是不奇怪的．《圣经》上讲："当心陌生女人的眼睛，她的眼睛就像个旋涡."(It is no wonder then that the ancients found in vortices the ultimate symbol of danger to men. "Beware of the eyes of strange woman," says a translation from the Bible. "Her eyes are like a vortex.")

59.3 费曼等对湍流的论述

费曼和其助教在其 20 世纪 60 年代出版的《费曼物理学讲义》中所说的对湍流研究的评价，直到今天可能仍然适用："最后，有一个物理问题在很多领域都很常见，那是一个很古老的仍未解决的问题 …… 这就是对循环流体或湍流流体的分析 …… 这个问题最简单的形式是用一根很长的管子，以很高的速度把水推进管子使其流动. 我们问：推动一定量的水流经管子的话，需要多大的压强？没有人能从第一性原理和水的性质来分析这个问题. 如果水流的十分的缓慢，或者我们用的是像蜂蜜一样的十分黏稠的液体的话，则我们可以很好地解决这个问题. 你可以从教科书中找到问题的解. 我们真正不能解决的是处理流经管道的实际的润湿水. 这是我们总有一天应该解决的核心问题，但我们还没有解决 (Finally, there is a physical problem that is common to many fields, that is very old, and that has not been solved … It is the analysis of circulating or turbulent fluids … The simplest form of the problem is to take a pipe that is very long and push water through it at high speed. We ask: to push a given amount of water through that pipe, how much pressure is needed? No one can analyze it from first principles and the properties of water. If the water flows very slowly, or if we use a thick goo like honey, then we can do it nicely. You will find that in your textbook. What we really cannot do is deal with actual, wet water running through a pipe. That is the central problem which we ought to solve some day, and we have not)."

59.4 理查森的串级

1922 年，英国气象学家、数值天气预报的奠基人刘易斯·弗莱·理查森 (Lewis Fry Richardson, 1881—1953) 在其专著《利用数值过程进行天气预报》(*Weather Prediction by Numerical Process*)[11.39] 中，提出湍流的能量串级模型，简称为"理查森的串级 (Richardson's cascade)". 该模型表明，大尺度涡通过湍动剪切从基本 (时均或平均) 流动中获取能量，然后再通过黏性耗散和色散 (失稳) 过程，这些大涡串级分裂成不同尺度的小涡，并在涡体的分裂破碎过程中将能量逐级传给小尺度涡，直至达到黏性耗散为止，如图 59.8 所示.

理查森用一首著名的关于串级的诗来描述湍流中能量的传递 (the famous poem of cascade to describe the energy transfer in turbulent flow)：

Big whirls have little whirls (大涡中有小涡)

that feed on their velocity (孕育于速度),

And little whirls have lesser whirls (小涡中有更小的涡)

and so on to viscosity (终止于黏滞).

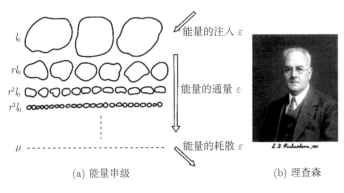

图 59.8　理查森的串级

标度不变性 (scale invariance) 是其中的一个重要概念. 所谓标度不变性是指, 在放大或缩小变换中所观察结构的不变的性质. 标度不变性具有一个重要的性质, 那就是没有特征尺度的 "自相似性 (self-similarity)". 如图 59.9 所示的分形结构便具有这种自相似的结构. 标度不变性对于理解理查森的能量串级现象十分重要.

图 59.9　我们身边的分形结构

如图 59.10 所示, 作为艺术品的俄罗斯套娃具有一种统计意义上的标度不变性.

59.5　柯尔莫哥洛夫的 K41 理论的 2/3 标度律 [11.40]

1941 年, 苏联著名数学家柯尔莫哥洛夫 (Andrey Kolmogorov, 1903—1987, 如图 59.11 所示) 运用量纲分析, 得到了不可压缩湍流的能谱的标度律, 简称 K41. 事实上, 柯尔莫哥洛夫在 1941 年曾连续发表了三篇相关文章 [11.40–11.42]. K41 表

图 59.10　俄罗斯套娃 (Russian dolls)

明, 不可压缩的湍流在一定的速度范围内, 速度场是自相似的 (self-similar). 该自相似性来源于由从大尺度 (大旋转) 涡到小尺度 (小旋转) 涡的无耗散能量的串级 (dissipationless cascade of energy) 产生的. 在柯尔莫哥洛夫理论中, 两点的平均速度差仅是两点距离 r 和单位质量的能量耗散率 ε 的函数, ε 的量纲为 $[\varepsilon] = \left[\dfrac{能量}{时间 \times 质量}\right] = L^2 T^{-3}$.

图 59.11　柯尔莫哥洛夫在授课 (左); 柯尔莫哥洛夫穿着游泳裤在滑雪 (中); 奥布霍夫 (右)

K41 理论认为, 无论一个湍流系统如何复杂, 其涡旋结构都有着相似性, 即涡的动能总是由外力作用施加给流场, 并注入最大尺度 (假设为 L) 的涡结构. 然后大尺度涡结构逐次瓦解并产生小型涡旋, 同时也将动能由大尺度逐级传向小尺度结构, 并依此类推. 但此过程并不会无限进行下去, 当涡结构尺度足够小 (假设为 η) 时, 流体黏性将占据主导地位, 动能转化为内能在该尺度上耗散掉, 而不会继续传向更小尺度的涡结构. 这个过程, 被称能级串过程.

柯尔莫哥洛夫以其深刻的物理直觉引入了一个大胆的假设, 将整个问题抽丝剥茧, 直至其本质完全暴露, 再辅以量纲分析便得到了描述湍流场涡结构动能的最有效的一个公式:

$$D_2(r) = \left\langle [\boldsymbol{v}(\boldsymbol{x}+\boldsymbol{r},t) - \boldsymbol{v}(\boldsymbol{x},t)]^2 \right\rangle = C\varepsilon^{2/3} r^{2/3} \tag{59-3}$$

式中, $r = |\boldsymbol{r}|$ 为湍流场中两点之间距离, ε 为由大尺度向小尺度的动能传递率, 亦等同于小尺度动能耗散率, C 为无量纲的柯尔莫哥洛夫常数. 而 $D_2(r)$ 为二阶结构函数, 其定义湍流场中相距为 r 两点间速度差 $\boldsymbol{v}(\boldsymbol{x}+\boldsymbol{r},t) - \boldsymbol{v}(\boldsymbol{x},t)$ 平方的 (空间或时间) 平均值. 此结构函数乃柯尔莫哥洛夫独创之统计量, 它度量了尺度小于等于 r 的所有涡结构之动能.

公式 (59-3) 式为 K41 理论的最核心部分, 被称为 "柯尔莫哥洛夫 2/3 标度律". 正如许多伟大的物理公式一样, 柯尔莫哥洛夫的 2/3 律形式优美, 结构简单. 而在这令人为之倾倒的美感背后,却隐藏着非凡的物理含义. 原来,湍流场中不同尺度 r 上的涡结构动能并非任意分布,而是要服从 $r^{2/3}$ 的幂函数形式. 因此,涡动能从大尺度到小尺度以幂函数形式衰减. 而由于柯尔莫哥洛夫引入的假设,2/3 律只在一定尺度范围 $\eta \ll r \ll L$ 内成立. 在此范围内,大尺度 L 的外力作用和小尺度 η 的黏性耗散均无影响,每个涡结构仅是一个动能传递的节点,其所见只是动能从大尺度传来,通过自己向小尺度传出. K41 理论问世之后成为了各学派争相进行实验对比的主题. 这些实验尺度大小有别,流动形态各异,而万变中的不变却是所测得的湍流 2/3 标度律,柯尔莫哥洛夫公式 (59-3) 中 2/3 这个数字已被无数实验所证实. 其中最惊人的当属卡尔·吉布森 (Carl H. Gibson) 1991 年所测得的星系湍流,其结构函数在近十个数量级的尺度范围内与湍流标度律相符 [11.43].

在 1942 年苏联科学院的一次学术研讨会上,柯尔莫哥洛夫进行了一次 K41 理论的专题报告会. 在到场的听众中,便有理论物理大师列夫·朗道 (Lev Landau, 1908—1968, 1962 年诺贝尔物理学奖获得者),柯尔莫哥洛夫和朗道这两位绝世高手之间惺惺相惜,朗道很快便理解了 K41 理论的精髓,尽管朗道平生为人狂傲,素来言辞尖锐,也不由得称赞 K41 理论之精妙. 朗道评论道:"柯尔莫哥洛夫是第一个正确理解了湍流小尺度结构的人 (Kolmogorov was the first to provide a correct understanding of the local structure of turbulent flow)."

我们最为关心的是,柯尔莫哥洛夫在 K41 理论方面重大创新的源头在哪里? 在科学发展史上,不乏有这样一些理论,它们的光芒绚丽夺目,犹如一盏指路明灯,驱散了当世的黑暗和阴霾,点亮了后世的发展道路. 然而,当你回首来时路,却发现理论的创立过程似乎平凡至极,K41 理论也正是如此. 柯尔莫哥洛夫以一个物理假设为起点,以能量级串模型为基础,将问题层层肢解,直击湍流的要害. 整个过程一气呵成,既无纷扰繁复的数学推导,亦无艰深晦涩的数学证明. K41 理论已经成为量纲分析之典范,是呀! 貌似平淡无奇的量纲分析背后竟然蕴涵着如此深厚的理论与实验之功,又如何不令人惊叹?!

柯尔莫哥洛夫在他 1985 年出版的专著《数学与力学》(*Mathematics and Mechanics*) 中是这样回忆自己初涉湍流领域时的想法的:"我很快便意识到想要建立一套严格的纯数学理论是不可能的. 因此,根据实验数据的分析来建立一些假设是湍流研究的必经之路 (it soon became clear to me that there was no chance to develop a closed purely mathematical theory. For lack of such a theory it was necessary to use some hypotheses based on the results of treatment of experimental data)."

在其专著《数学与力学》中，柯尔莫哥洛夫还回忆道："我在 30 年代后期兴趣转入了液体与气体的湍流研究. 从研究伊始我便意识到此领域的主要数学工具是有多个变量的随机函数（随机场）理论，而这在当时也刚刚被建立起来 (I took an interest in the study of turbulent flows of liquids and gases in the late thirties. It was clear to me from the very beginning that the main mathematical instrument in this study must be the theory of random functions of several variables (random fields) which had only then originated)." 随机场，这正是概率理论的用武之地，而此时的柯尔莫哥洛夫，在概率论上的造诣已独步天下，他在 1933 年出版的概率论巨著《概率论的基本概念》，首次将概率论建立在严格的公理基础上，就已经解决了希尔伯特第六问题中的概率部分，其著作一经问世便得到世界公认，为近代概率论的发展打下了坚实的基础. 由此看来，K41 理论的创立绝不是偶然的!

59.6 奥布霍夫的 −5/3 标度律，柯尔莫哥洛夫–奥布霍夫标度

柯尔莫哥洛夫的学生亚历山大·奥布霍夫 (Alexander Mikhailovich Obukhov, 1918—1989，如图 59.11 右图所示) 也于 1941 同年发表了关于湍流能量谱的文章 [11.44,11.45]. 基于柯尔莫哥洛夫前一年对概率论中随机过程能量谱的阐述，奥布科夫从纳维–斯托克斯方程出发建立了湍流能量谱的半经验式方程. 此方程描述了这样的物理过程：湍流能量在某一尺度上随时间的变化与不同尺度涡结构相互作用以及小尺度黏性耗散有关. 通过求解此方程，奥布霍夫得到了与柯尔莫哥洛夫 2/3 标度律所等价的湍流能量谱表述：

$$E(k) = C\varepsilon^{2/3}k^{-5/3} \tag{59-4}$$

式中，k 为波数，亦即尺度的倒数. 能量谱 (spectral energy density) 满足

$$\frac{1}{2}\langle \boldsymbol{v}(\boldsymbol{x},t)\cdot\boldsymbol{v}(\boldsymbol{x},t)\rangle = \int_0^\infty E(k)\,\mathrm{d}k \tag{59-5}$$

因此，$E(k)$ 可以解释为湍流的平均动能密度作为波数 ($0 \leqslant k < \infty$) 的函数，$E(k)$ 的量纲为 $[E(k)] = \mathrm{L}^3\mathrm{T}^{-2}$. 奥布霍夫和柯尔莫哥洛夫的两篇文章均发表于 1941 年. 由于柯尔莫哥洛夫的方法更为简洁明晰且适用度更为广泛，而且奥布霍夫的工作是基于柯尔莫哥洛夫的理论，世人对柯尔莫哥洛夫更为青睐，因而在一定程度上忽视了奥布霍夫的贡献. 值得注意的是，柯尔莫哥洛夫的学生 —— 格里戈里·巴伦布拉特 (Grigory Isaakovich Barenblatt, 1927—2018) —— 则将该标度律称为"柯尔莫哥洛夫–奥布霍夫标度 (Kolmogorov–Obukhov scaling)"[11.46].

阿基瓦·亚格洛姆 (Akiva Moiseevich Yaglom, 1921—2007) 在其有关柯尔莫哥洛夫湍流研究学派的评述文章中 [11.47]，详细地回顾了柯尔莫哥洛夫和奥布霍夫

第 11 章 流体动力学

1941 年彼此独立地确定了 $r^{2/3}$ 的 (59-3) 式和 $k^{-5/3}$ 的 (59-4) 式的历程. 亚格洛姆文中特别强调[11.47]: 柯尔莫哥洛夫对奥布霍夫的工作表现出高度尊敬 (demonstrated his high respect), 称 K41 理论是他与奥布霍夫所共同所创 (joint creation).

为了纪念奥布霍夫的杰出学术贡献, 俄罗斯科学院有以其姓名命名的研究所: 奥布霍夫大气物理研究所 (A. M. Obukhov Institute of Atmospheric Physics, Russian Academy of Sciences).

综上所述, 如果 1905 年被称为爱因斯坦的奇迹年的话, 1941 年则毫无疑问地可被称为柯尔莫哥洛夫湍流学派的奇迹年. 事实上, 1941 年对苏联来说是极为困难的年份, 请见思考题 11.13 和 11.14.

在文献中, $E(k) \propto k^{-5/3}$ 被广泛地称为柯尔莫哥洛夫频谱 (Kolmogorov spectrum), 如图 59.12 所示.

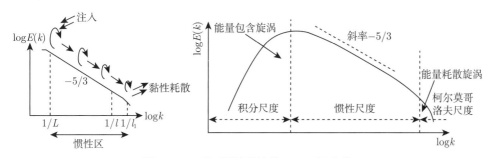

图 59.12 不可压缩湍流的 $-5/3$ 标度律

充分发展的湍流介质仅有两个量就可表征: (1) 量纲为 "能量/时间/质量" 的单位质量的能量耗散率, $[\varepsilon] = \mathrm{L}^2 \mathrm{T}^{-3}$; (2) 运动黏度, $[\nu] = \mathrm{L}^2 \mathrm{T}^{-1}$. 故由量纲分析的快速匹配法, 可迅速得到如下特征参量:

(1) 柯尔莫哥洛夫特征长度 (Kolmogorov length scale) 为
$$l_K = \left(\frac{\nu^3}{\varepsilon}\right)^{1/4} \tag{59-6}$$

(2) 柯尔莫哥洛夫特征时间 (Kolmogorov time scale) 为
$$\tau_K = \sqrt{\frac{\nu}{\varepsilon}} \tag{59-7}$$

(3) 柯尔莫哥洛夫特征速度 (Kolmogorov velocity scale) 为
$$v_K = (\nu \varepsilon)^{1/4} \tag{59-8}$$

以地球的大气层为例, 能量耗散率 $\varepsilon \sim 10 \text{ cm}^2 \cdot \text{s}^{-3}$, 运动黏度 $\nu \sim 0.1 \text{ cm}^2 \cdot \text{s}^{-1}$. 从而, 在地球的大气层中最小涡的柯尔莫哥洛夫特征长度的数量级为 $l_K \sim 0.1 \text{ cm}$,

这当然比空气分子的平均自由程 $\lambda \sim 10^{-4}$ cm 要大约 3 个数量级. 最小涡的旋转速度 $v_K \sim 1$ cm/s, 涡消失的衰减时间 (dying away with a decay time) 为 $\tau_K \sim 0.1$ s.

§60. 杨–拉普拉斯方程

60.1 谁最先提出了表面张力的概念？

约翰·塞格纳 (Johann Andreas von Segner, 1704—1777, 如图 60.1 所示), 匈牙利科学家. 约翰·塞格纳于 1725 年进入耶拿大学. 1729 年获得医学证书后回到普雷斯堡, 在那里及德布勒森开始了从医生涯, 其学术贡献主要在物理和数学两个方面.

图 60.1　匈牙利物理学家和数学家约翰·塞格纳

1751 年, 约翰·塞格纳通过和拉伸薄膜相类比, 第一次建议了液体的表面张力概念[11.48]. 在文献 [11.48] 的第 303 页, 塞格纳认为, 流体分子通过一种吸引力 (an attractive force, 原文: *vim attractricem*) 而聚集在一起, 他还认为, 这种吸引力的是如此的短程, 以至于没有人能感知它 (··· *ut nullo adhuc sensu percipi poterit*). 塞格纳尽管在表面张力的作用的数学描述方面没有能够成功, 但他为表面张力理论的后续发展奠定了基础.

表面张力被定义为新产生单位面积所需要的功, 表面张力是一个强度量 (intensive quantity, 见附录 B), 面积是广延量 (extensive quantity, 见附录 B), 表面张力和面积为功的共轭对 (work conjugate pair, 见附录 B). 功是标量, 而面积是矢量, 则表面张力是标量被矢量除, 按照在第 2 章开篇所提出的原则, 则表面张力为一矢量, 因此, 它有诸如液–气界面张力、固–液界面张力和固–气界面张力等分量.

杨–拉普拉斯方程主要反映的是液面曲率所产生的附加压强, 是由被称为 "最后一位知道一切的人" 的托马斯·杨 (Thomas Young, 1773—1829, 如 60.2 左图所示)[11.49] 和皮埃尔–西蒙·拉普拉斯侯爵 (Pierre-Simon marquis de Laplace, 1749—

1827, 如图 60.2 右图所示)[11.50] 于 1805 年独立提出的. 下面给出三种阐述方法.

60.2 应用能量法推导杨–拉普拉斯方程

对于如图 60.3 所示的任意曲率的界面元, 其面积为 $\mathrm{d}S = \mathrm{d}x\mathrm{d}y$, 如果沿界面的法线方向将界面元移动 $\mathrm{d}z$, 则长度元将分别变为

$$\mathrm{d}x' = \mathrm{d}x\left(1 + \frac{\mathrm{d}z}{R_1}\right), \quad \mathrm{d}y' = \mathrm{d}y\left(1 + \frac{\mathrm{d}z}{R_2}\right) \tag{60-1}$$

图 60.2 通才托马斯·杨 (左)、拉普拉斯 (右)

式中, R_1 和 R_2 分别为 x 和 y 方向的曲率半径. 移动 $\mathrm{d}z$ 后界面元的面积变为 $\mathrm{d}S' = \mathrm{d}x'\mathrm{d}y'$, 忽略高阶小量得到

$$\mathrm{d}S' = \mathrm{d}S\left(1 + \frac{\mathrm{d}z}{R_1} + \frac{\mathrm{d}z}{R_2}\right) \tag{60-2}$$

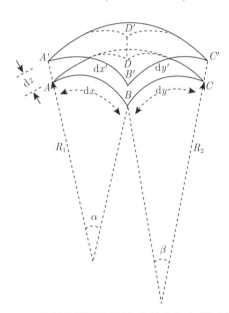

图 60.3 应用能量法推导杨–拉普拉斯方程示意图

界面元移动 $\mathrm{d}z$ 引起的表面能改变为 $\mathrm{d}E = \gamma(\mathrm{d}S' - \mathrm{d}S)$, 这里 γ 为表面张力. 此过程需外力克服弯曲表面上附加压力做功, 该力大小为界面元内外部压强差 Δp, 界面元沿法线方向移动 $\mathrm{d}z$ 所做的功为 $\Delta p \mathrm{d}S \mathrm{d}z$. 外力功与表面能改变相等, 联立可得杨–拉普拉斯方程:

$$\Delta p = \gamma\left(\frac{1}{R_1} + \frac{1}{R_2}\right) \tag{60-3}$$

特别地, 当液滴是半径为 $R_1 = R_2 = R$ 的球体时, (60-3) 式中的压强差退化为

$$\Delta p = \frac{2\gamma_{lv}}{R} \tag{60-4}$$

60.3 应用力平衡法推导杨–拉普拉斯方程

如图 60.4 所示，考虑表面上的一点 P，从 P 点向外拉出一条曲线，长度为 ρ.

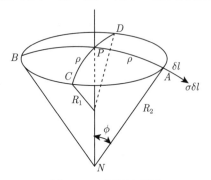

图 60.4 表面示意图

杨–拉普拉斯方程公式可以通过当 ρ 趋于 0 时来计算得到. 通过 P 点，可以获取两个主曲率截面 AB 和 CD，它们在 P 点处的曲率半径分别为 R_1 和 R_2. 在 A 点处取一微元，其沿切向方向的作用力 $\gamma\delta l$ 在竖直方向上的作用力分量可以表示为

$$\gamma\delta l\sin\phi \approx \gamma\delta l\phi = \gamma\delta l\frac{\rho}{R_2} \tag{60-5}$$

上面假定 φ 足够小，当考虑 A, B, C, D 四点出均有微元 δl 作用时，可以得到以下作用力的表达式：

$$2\rho\gamma\delta l\left(\frac{1}{R_1}+\frac{1}{R_2}\right) \tag{60-6}$$

上式为四个点处的作用力之和，故对上式取四分之一周长进行积分即可，积分后可得

$$\pi\rho^2\gamma\left(\frac{1}{R_1}+\frac{1}{R_2}\right) \tag{60-7}$$

作用在表面上的由内外压差引起的作用力可以表示为

$$\Delta p\pi\rho^2 \tag{60-8}$$

将以上两式相等，即可得到杨–拉普拉斯方程.

60.4 应用矢量法推导杨–拉普拉斯方程

假设液–气界面可以表示成 x 和 y 的关系式，$z = S = u(x, y)$，p 是 S 表面上的一点，则 $p \in u(x, y)$ 点处的平均曲率为

$$H = \frac{1}{2}(k_1 + k_2) = \frac{1}{2}\left(\frac{1}{R_1}+\frac{1}{R_2}\right) \tag{60-9}$$

液–气界面平均曲率 H 与界面单位法向量 $\boldsymbol{N}_s$ 有如下关系：

$$2H = \boldsymbol{\nabla} \cdot \boldsymbol{N}_s \tag{60-10}$$

在表面张力作用下，弯曲液面两边存在压强差，成为附加压强差，会产生拉普拉斯压强 (Laplace pressure)，附加压强差的方向总是指向曲率中心，以容器底面作为参考面，其与表面张力有如下关系：

$$\Delta p + \rho g u = \gamma \boldsymbol{\nabla} \cdot \boldsymbol{N}_s = 2H\gamma = \gamma \left(\frac{1}{R_1} + \frac{1}{R_2} \right) \tag{60-11}$$

上式即考虑重力作用时的杨–拉普拉斯方程.

对于容器中液体在壁面的爬升问题 (图 60.5)，为了以后章节叙述的方便，这里给出杨润湿方程和杨–拉普拉斯方程的矢量形式. 两倍的平均曲率 (mean curvature) 为

$$2H = \boldsymbol{\nabla} \cdot \boldsymbol{N}_S = \boldsymbol{\nabla} \cdot \frac{\boldsymbol{\nabla} u}{\sqrt{1 + (\boldsymbol{\nabla} u)^2}} \tag{60-12}$$

用两个主曲率半径 R_1 和 R_2 来表示的话，还可表示为 $2H = \left(R_1^{-1} + R_2^{-1} \right)$. 则杨润湿方程 (Young's wetting equation) 为

$$\boldsymbol{N}_S \cdot \boldsymbol{N}_\Gamma = \cos\theta_0 \tag{60-13}$$

杨–拉普拉斯方程为

$$\frac{\Delta p}{\gamma} = \boldsymbol{\nabla} \cdot \boldsymbol{N}_S = 2H = \boldsymbol{\nabla} \cdot \frac{\boldsymbol{\nabla} u}{\sqrt{1 + (\boldsymbol{\nabla} u)^2}} = \mathrm{div}\boldsymbol{T}u \tag{60-14}$$

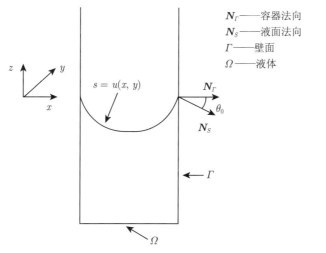

图 60.5 表面润湿示意图

例 60.1 滴水穿石的一种解释.

解：宋·罗大经《鹤林玉露·一钱斩吏》："乖崖援笔判曰：'一日一钱；千日千钱；绳锯木断；水滴石穿.'" 这便是"滴水穿石"的出处，比喻虽然力量比较小，但只要目标专一、持之以恒、坚持不懈，就一定能把艰难的事情办成.

如图 60.6 右图所示，高速摄影发现，在液滴和固体表面的撞击过程中，有一些十分微小的气泡产生.

 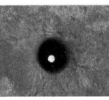 

图 60.6　生活中的滴水穿石 (左、中) 和液滴中的气泡 (右)

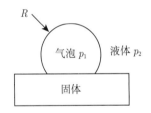

图 60.7　滴水穿石的微观机理解释

在如图 60.7 中，当气泡的曲率半径十分微小时，$R \to 0$，由 (60-4) 式可知，此时气泡施加给石头的附加压强趋于无穷大，这样当滴水的时间足够长的话，石头便会"一口一口地"被咬下，滴水就可以将石头打穿.

§61. 润滑近似和液滴铺展的动力学方程

61.1　润滑近似下膜厚方程的推导

为了数学上处理的方便，黏性液滴铺展过程的研究一般均采用润滑近似 (lubrication approximation)，该近似通常假定[11.51]：

(1) 黏性流体流动处于层流状态；
(2) 润滑膜的厚度远小于它的长度和宽度；
(3) 惯性力和重力均忽略不计，液滴铺展由毛细力主导；
(4) 液滴铺展是不可压缩牛顿流体等温流动；
(5) 在所述几何条件下可以忽略表面法线方向上的任何运动.

下面给出润滑近似下液滴铺展动力学方程推导方法.

应用上述润滑近似，液滴铺展过程可用两个量来描述：膜的高度 h；液体速度 $\boldsymbol{v}$ 水平分量 v^x 在垂直方向 z 上的平均量 v，也就是 $\int_0^h v^x \mathrm{d}z = hv$.

润滑近似下，液滴铺展的连续性方程为

$$\frac{\mathrm{D}}{\mathrm{D}t}\int h(x)\mathrm{d}x = 0 \tag{61-1}$$

应用雷诺输运定理 (30-5) 式，(61-1) 式变为

$$\int \left[\frac{\partial h}{\partial t} + \boldsymbol{\nabla} \cdot (h\boldsymbol{v})\right] \mathrm{d}x = 0 \tag{61-2}$$

可得液滴铺展薄膜微分形式的连续性方程：

$$\frac{\partial h}{\partial t} + \boldsymbol{\nabla} \cdot (h\boldsymbol{v}) = 0 \tag{61-3}$$

需要注意的是，由于上述方程中考虑的只是二维情况速度的平均量 v，故不再区分散度和梯度的区别，但若考虑的是三维问题且不是平均量时，则需要区分.

纳维–斯托克斯方程方程为

$$\rho \left[\frac{\partial \boldsymbol{v}}{\partial t} + (\boldsymbol{v} \cdot \mathbf{grad}) \boldsymbol{v}\right] = -\mathbf{grad}p + \mu \nabla^2 \boldsymbol{v} + \left(\zeta + \frac{1}{3}\mu\right) \mathbf{grad}\mathrm{div}\boldsymbol{v} \tag{61-4}$$

在润滑近似下，可忽略上面方程左边项的惯性力，当考虑液滴铺展过程中的不可压缩性 $\mathrm{div}\boldsymbol{v} = 0$，则上述纳维–斯托克斯方程简化为

$$\mathbf{grad}p = \mu\nabla^2 \boldsymbol{v} \tag{61-5}$$

针对于本问题，在润滑近似下，不失一般性可近似可到下式：

$$\boldsymbol{\nabla}p = \mu \frac{\partial^2 \boldsymbol{v}^x}{\partial z^2} \tag{61-6}$$

此时方程的位移变量为膜厚 h，当 $\boldsymbol{\nabla}h$ 为小量时，$(\boldsymbol{\nabla}h)^2$ 可被忽略，此时杨–拉普拉斯压强可写为

$$p = -\gamma \nabla^2 h \tag{61-7}$$

上式取 "–" 号的原因是该压强指向液体内部，为所建坐标系 z 方向的负向.

在润滑近似条件下，可认为 $\boldsymbol{\nabla}p$ 沿膜厚的变化略去不计，此时方程 (61-6) 的解为

$$\boldsymbol{v}^x = \frac{\boldsymbol{\nabla}p}{\mu}\left(\frac{1}{2}z^2 + C_1 z + C_2\right) \tag{61-8}$$

应用无滑移边界条件 $(v^x|_{z=0} = 0)$ 和自由面上的无剪切边界条件 $\left(\left.\frac{\partial v^x}{\partial z}\right|_{z=h} = 0\right)$，式中的两个积分常数确定为 $C_1 = -h$，$C_2 = 0$，则水平方向的速度分量为

$$\boldsymbol{v}^x = \frac{\boldsymbol{\nabla}p}{\mu}\left(\frac{1}{2}z^2 - hz\right) \tag{61-9}$$

应用 $\int_0^h v^x \mathrm{d}z = hv$，积分上式可得液体铺展速度的水平分量 v^x 在垂直方向 z 上

的平均量 v 为

$$\boldsymbol{v} = -\frac{\boldsymbol{\nabla} p}{3\mu} h^2 \tag{61-10}$$

将 (61-7) 式代入 (61-10) 式，可得

$$\boldsymbol{v} = \frac{\gamma h^2}{3\mu} \boldsymbol{\nabla} \left(\nabla^2 h \right) \tag{61-11}$$

将上式代回连续性方程 (61-3)，则得到在润滑近似下的液滴铺展过程中膜厚 h 的方程：

$$\frac{\partial h}{\partial t} + \frac{\gamma}{3\mu} \boldsymbol{\nabla} \cdot \left[h^3 \boldsymbol{\nabla} \left(\nabla^2 h \right) \right] = 0 \tag{61-12}$$

当只考虑二维问题时，(61-12) 式退化为

$$\frac{\partial h}{\partial t} + \frac{\gamma}{3\mu} \frac{\partial}{\partial x} \left(h^3 \frac{\partial^3 h}{\partial x^3} \right) = 0 \tag{61-13}$$

61.2 薄膜铺展的标度律

以下三个步骤是结合微分方程进行数量级估计 (order-of-magnitude estimate) 的关键：

(1) 首先要鉴别所进行数量级估计问题本身的物理、力学等过程；

(2) 对方程中的微分项进行合理简化. 在找准所研究问题的特征尺度、特征时间后，便可对微分项进行如下量纲上的简化：

$$\left[\frac{\mathrm{d} y}{\mathrm{d} x} \right] = \frac{[\mathrm{d} y]}{[\mathrm{d} x]} = \left[\frac{y}{x} \right] \tag{61-14}$$

式中 $[\cdot]$ 表示某物理量的量纲，对于二阶导数以及 n-阶导数还有

$$\left[\frac{\mathrm{d}^2 y}{\mathrm{d} x^2} \right] = \left[\frac{y}{x^2} \right], \quad \left[\frac{\mathrm{d}^n y}{\mathrm{d} x^n} \right] = \left[\frac{y}{x^n} \right] \tag{61-15}$$

如果 S 为一标量，$\boldsymbol{R}$ 为一矢量，而 L 为其特征尺度的话，则有

$$\begin{cases} [\mathbf{grad} S] = [\boldsymbol{\nabla} S] = \left[\frac{\partial S}{\partial x} \hat{\boldsymbol{i}} + \frac{\partial S}{\partial y} \hat{\boldsymbol{j}} + \frac{\partial S}{\partial z} \hat{\boldsymbol{k}} \right] = \left[\frac{S}{L} \right] \\ [\mathbf{curl} \boldsymbol{R}] = [\boldsymbol{\nabla} \times \boldsymbol{R}] = \left[\left(\hat{\boldsymbol{i}} \frac{\partial}{\partial x} + \hat{\boldsymbol{j}} \frac{\partial}{\partial y} + \hat{\boldsymbol{k}} \frac{\partial}{\partial z} \right) \times \left(R_x \hat{\boldsymbol{i}} + R_y \hat{\boldsymbol{j}} + R_z \hat{\boldsymbol{k}} \right) \right] = \left[\frac{|\boldsymbol{R}|}{L} \right] \\ [\mathrm{div} \boldsymbol{R}] = [\boldsymbol{\nabla} \cdot \boldsymbol{R}] = \left[\frac{\partial R_x}{\partial x} + \frac{\partial R_y}{\partial y} + \frac{\partial R_z}{\partial z} \right] = \left[\frac{|\boldsymbol{R}|}{L} \right] \\ [\mathrm{div} \mathbf{grad} S] = [\nabla^2 S] = \left[\frac{\partial^2 S}{\partial x^2} + \frac{\partial^2 S}{\partial y^2} + \frac{\partial^2 S}{\partial z^2} \right] = \left[\frac{S}{L^2} \right] \end{cases} \tag{61-16}$$

式中 $\left(\hat{\boldsymbol{i}}, \hat{\boldsymbol{j}}, \hat{\boldsymbol{k}} \right)$ 为直角坐标的三个单位矢量. 值得注意的是，对时间的导数或偏导数同样处理.

第 11 章　流体动力学

(3) 平衡方程等号两端的主项 (leading terms).

下面将通过例子说明如何应用上述三个步骤.

例 61.1　三维情况下液滴铺展的标度律

解：(61-12) 式是一个十分复杂的方程, 很难获得解析解. 但掌握一定规律后, 对其进行数量级估计, 从而进一步获得其标度律是可能的.

当液滴铺展的特征尺度为 L 的话, 则方程 (61-12) 式的两个主项的数量级分别为

$$\left[\frac{\partial h}{\partial t}\right] \sim \frac{h}{t}, \quad \left[\frac{\gamma}{3\mu}\boldsymbol{\nabla}\cdot\left[h^3\boldsymbol{\nabla}\left(\nabla^2 h\right)\right]\right] \sim \frac{\gamma}{\mu}\frac{1}{L}h^3\frac{1}{L}\frac{h}{L^2} = \frac{\gamma}{\mu}\frac{h^4}{L^4} \tag{61-17}$$

平衡两个主项：

$$\frac{h}{t} \sim \frac{\gamma}{\mu}\frac{h^4}{L^4} \quad \Rightarrow \quad \frac{L^4}{h^3} \sim \frac{\gamma}{\mu}t \tag{61-18}$$

对于三维情形, 密度一定, 体积守恒, 进行粗略数量级估计 (rough order-of-magnitude estimate) 时, 可将铺展的液滴视为一个半径为特征尺度 L, 高度为 h 的圆饼. 其体积守恒的条件为 $\pi L^2 h \sim \text{const}$, 将 $h \sim L^{-2}$ 的标度关系代入 (61-18) 式中第二式, 则有液滴在三维铺展时, 铺展半径随时间的标度律为

$$L \sim t^{1/10} \tag{61-19}$$

例 61.2　二维情况下液滴铺展的标度律

解：当液滴铺展的特征尺度为 L 的话, 则方程 (61-13) 的两个主项的数量级分别为

$$\left[\frac{\partial h}{\partial t}\right] \sim \frac{h}{t}, \quad \left[\frac{\gamma}{3\mu}\frac{\partial}{\partial x}\left(h^3\frac{\partial^3 h}{\partial x^3}\right)\right] \sim \frac{\gamma}{\mu}\frac{1}{L}h^3\frac{h}{L^3} = \frac{\gamma}{\mu}\frac{h^4}{L^4} \tag{61-20}$$

当平衡两个主项时, 仍得到 (61-18) 式中第二式. 对于二维情形, 密度一定, 体积守恒, 亦即 $Lh \sim \text{const}$, 将 $h \sim L^{-1}$ 的标度关系代入 (61-18) 式中第二式, 则有液滴在三维铺展时, 铺展半径随时间的标度律为

$$L \sim t^{1/7} \tag{61-21}$$

§62. 流体动力学中的不稳定性理论

62.1 里克特迈耶–梅什科夫 (RM) 不稳定性

两种不同密度的流体界面经瞬间加速而产生的不稳定性称为 "里克特迈耶–梅什科夫不稳定性 (Richtmyer–Meshkov instability, RMI, RM 不稳定性)". 通常这种

加速是由激波穿过所引起的. 不稳定性发展初期轻微扰动随时间线性增长, 之后会出现与瑞利－泰勒不稳定性中类似的由轻流体形成的 "气泡 (bubble)" 状结构与重流体形成的 "尖钉 (spike)" 状结构. 最终达到混沌状态, 两种流体充分混合, 如图 62.1 所示.

1960 年, 美国物理学家罗伯特·里克特迈耶 (Robert Davis Richtmyer, 1910—2003) 最早研究了这一现象 [11.52]. 1969 年, 苏联物理学家尤金·叶夫格拉福维奇·梅什科夫 (E. E. Meshkov) 在实验中证实 [11.53] 该现象, 因而得名.

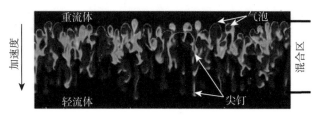

图 62.1 重和轻两种流体混合时所产生的气泡和尖钉结构

这种不稳定性出现于惯性约束聚变 (inertial confinement fusion, ICF)、超音速燃烧冲压发动机以及超新星爆发等物理过程中.

图 62.2 所示的是超新星爆发中两个物质壳碰撞 (collision between two shells of matter) 所引发的 RM 不稳定性 [11.54]. 大质量恒星在死亡之前一些年的时间里因为 "脉冲对不稳定性" 而先后喷发出两个物质壳 (two subsequent pulsational pair-instability supernova eruptions), 如果后面的壳的速度大于前面的, 就会追赶碰撞, 产生激波. 此图显示了两个物质壳碰撞过程模拟图的 "右上角". 物质壳 (包括了红色结点表示的碰撞碎片) 的半径是太阳-地球距离的 500 倍 (约 750 亿千米). 不同的颜色表示不同的密度, 最高密度是 10^{-11} g/cm^3 (红色表示), 最低密度是 10^{-16} g/cm^3 (暗蓝色表示).

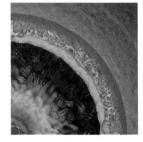

图 62.2 物质壳碰撞所产生的 RM 不稳定性

62.2 瑞利-泰勒 (RT) 不稳定性

瑞利-泰勒不稳定性 (Rayleigh–Taylor instability, RTI), 得名于瑞利 (Lord Rayleigh, 1842—1919) 于 1883 年 [11.55] 和杰弗里·泰勒 (Geoffrey Ingram Taylor, 1886—1975) 于 1950 年 [11.56] 的贡献, 简称 RT 不稳定性. 在任何时间都会发生在密集的重流体被轻的流体加速时. 这是发生在云与激波系统的事件, 或者当密度较高的流体浮在密度较低的液体, 像是漂浮在水上而密度较高的油.

无黏度的理想流体在平衡时, 所有的平面都是完全平行的, 但是由位能引起的轻微扰动, 像是较重的物质因为 (有效的) 重力作用而下沉, 并且轻的物质被替换而上升. 当不稳定发展时, 向下运动造成的不规则 (涟漪) 很快的就会被放大成为一系列的 "RT 指进"; 而向上升起的移动, 轻的物质会形成球状帽盖气泡.

这种过程在地质的形成上有许多的例子, 从盐丘到温度反转, 在天体物理和电动力学上也有."RT 指进" 在蟹状星云中特别明显, 在 1000 年前爆炸的超新星将物质喷发和扫掠过蟹状星云, 在爆炸中产生的脉冲风星云供给了蟹状星云的能量.

需要注意的是, 不要将喷射液体的 "瑞利不稳定性"(或普拉托–瑞利不稳定性) 与瑞利–泰勒不稳定性相混淆. 前者的不稳定性, 有时称为水龙软管 (firehose), 是由表面张力造成的, 表面张力作用于喷射的水柱上, 当水柱断裂成为一连串的水珠时, 会使水珠成为同样体积中表面积最小的.

62.3 开尔文–亥姆霍兹 (KH) 不稳定性

开尔文–亥姆霍兹不稳定性 (Kelvin–Helmholtz instability, KHI), 名称来自开尔文 (Lord Kelvin, 1824—1907, 如图 62.3 左图) 于 1871 年 [11.57] 和亥姆霍兹 (Hermann von Helmholtz, 1821—1894, 如图 62.3 右图) 于 1868 年 [11.58] 的贡献.

图 62.3 开尔文 (左) 和亥姆霍兹 (右)

KH 不稳定性是在有剪力速度的连续流体内部或有速度差的两个不同流体的界面之间发生的不稳定现象. 一个例子是风吹过水面时, 在水面上表面的波的不稳定. 而这种不稳定状况更常见于波状云 (billow clouds, 如图 62.4)、海洋、土星的云带、木星的大红斑 (Jupiter's great red spot, 如图 62.5 所示)[11.59]、太阳的日冕中.

本理论可预测不同密度的流体在不同的运动速度下的不稳定状态发生, 并且层流变成湍流的界限. 亥姆霍兹研究两种不同密度流体的动力学, 并发现小规模的扰动, 例如波发生时在不同流体间边界的反应.

在一些波长短到一定程度的状态下, 如果忽略表面张力, 以不同速度平行运动的两种不同密度流体的界面下, 在所有速度时都会不稳定. 然而, 表面张力可抵消短波长的不稳定状态, 而理论预测直到达到速度阈值以前都是稳定的. 包含表面张力的理论可大致预测在风吹过水面时产生波的界限.

在重力作用下，连续变化的密度和速度分布 (较轻的层在上方，所以流体是瑞利－泰勒稳定) 使开尔文–亥姆霍兹不稳定性的动力学是以泰勒－戈德斯坦方程描述. 而不稳定性开端可由理查逊数 (Richardson number, Ri) 得知. 通常情况下 $Ri < 0.25$ 就会不稳定. 这些效应常在云层中出现. 对于不稳定性的研究也可应用在等离子体物理学中，例如惯性约束聚变和等离子体–铍的界面.

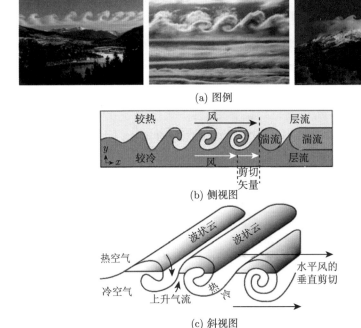

图 62.4 波状云中的 KH 不稳定性图例和起因示意图

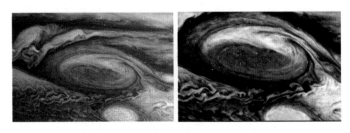

图 62.5 木星大红斑中的 KH 不稳定性

62.4 普拉托–瑞利 (PR) 不稳定性

普拉托–瑞利不稳定性 (Plateau-Rayleigh instability, PRI) 是描述水柱流动过程中的扰动现象，也就是水流最后断成不连续的液滴的情形.

第 11 章 流体动力学

约瑟夫·普拉托 (Joseph Plateau, 1801—1883, 如图 62.6 左图) 首先于 1873 年对该问题进行了测量 [11.60], 他注意到当水柱的长度超过水柱的直径约 3.13 倍时, 水柱将失稳成离散的液滴. 瑞利 (Lord Rayleigh, 1842—1919, 1909 年诺贝尔物理学奖获得者, 如图 62.6 右图) 随后于 1878 年对该问题建立了力学模型 [11.61].

图 62.6 普拉托 (左) 和瑞利 (右)

62.4.1 重力作用下圆柱水流的形状和速度分布

对 62.7 右图中的 A 和 B 两点应用伯努利方程, 有

$$\frac{1}{2}\rho V_0^2 + \rho g z + p_A = \frac{1}{2}\rho V^2 + p_B \tag{62-1}$$

式中, p_A 和 p_B 为和局部曲率相关的拉普拉斯压强. 水柱局部的曲率可以表示为

$$\nabla \cdot \boldsymbol{n} = \frac{1}{R_1} + \frac{1}{R_2} \approx \frac{1}{r} \tag{62-2}$$

式中, R_1 和 R_2 为两个主曲率, 由于圆形水柱的对称性, 故有了上式中的近似关系. 这样便可近似得到 A 和 B 两点处的拉普拉斯压强分别为

$$p_A \approx p_0 + \frac{\gamma}{a}, \quad p_B \approx p_0 + \frac{\gamma}{r} \tag{62-3}$$

将 (62-3) 式代入 (62-1) 式中, 整理得到

$$\frac{1}{2}\rho V^2 = \frac{1}{2}\rho V_0^2 + \rho g z + \frac{\gamma}{a}\left(1 - \frac{a}{r}\right) \tag{62-4}$$

为了对 (62-4) 式进行无量纲化, 上式两端每项均除以 $\frac{1}{2}\rho V_0^2$, 则有

$$\frac{V(z)}{V_0} = \sqrt{1 + \frac{2}{(Fr)^2}\frac{z}{a} + \frac{2}{We}\left(1 - \frac{a}{r}\right)} \tag{62-5}$$

式中, 两个无量纲数:

$$Fr = \frac{V_0}{\sqrt{ga}}, \quad We = \frac{\rho V_0^2 a}{\gamma} \tag{62-6}$$

分别为弗劳德数和韦伯数, 可参阅 (56-21) 和 (56-30) 两式, 这里的特征尺度 a 取的是水柱在 A 点的半径.

针对圆形水柱流动而言, 此时的质量守恒方程就是流量守恒 (flux conservation):

$$Q = \pi a^2 V_0 = \pi r^2 V(z) \quad \Rightarrow \quad \frac{V(z)}{V_0} = \left(\frac{a}{r(z)}\right)^2 \tag{62-7}$$

将 (62-7) 式的后半段代入 (62-5) 式中, 则得到水柱半径随着下落高度 z 的关系式为

$$\frac{r(z)}{a} = \left(\frac{V(z)}{V_0}\right)^{-1/2} = \left[1 + \frac{2}{(Fr)^2}\frac{z}{a} + \frac{2}{We}\left(1 - \frac{a}{r}\right)\right]^{-1/4} \tag{62-8}$$

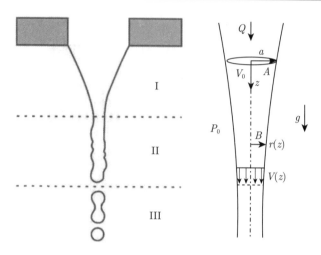

图 62.7 普拉托–瑞利不稳定性模型示意图

特别地,当韦伯数 $We \to \infty$,也就是惯性力和表面张力相比占主导地位时,水柱的半径分布和速度分布分别满足如下关系式:

$$\frac{r(z)}{a} = \left(1 + \frac{2gz}{V_0^2}\right)^{-1/4}, \quad \frac{V(z)}{V_0} = \left(1 + \frac{2gz}{V_0^2}\right)^{1/2} \tag{62-9}$$

62.4.2 重力作用下圆柱水流的稳定性分析

取如图 62.8 所示的一段具有代表性的无黏圆柱形水流,其半径为 R_0,表面张力为 γ,密度为 ρ,重力的影响可以忽略. 在定态 (stationary state) 时,圆形水柱内的压强满足杨–拉普拉斯方程:

$$p_0 = \gamma \boldsymbol{\nabla} \cdot \boldsymbol{n} = \frac{\gamma}{R_0} \tag{62-10}$$

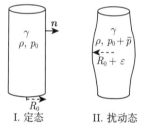

图 62.8 稳定性分析

如图 62.8 中右图的扰动态 (perturbated state) 所示,设圆柱水流的半径做如下扰动:

$$\tilde{R} = R_0 + \epsilon\, e^{\omega t + ikz} \tag{62-11}$$

式中,ϵ 为半径的扰动值,为一个小量: $\epsilon \ll R_0$,z 轴沿水流方向,k 为沿 z 轴的波数,相应的扰动波长为 $2\pi/k$,ω 则为不稳定性的增长速率. 用 $\tilde{V}_r$ 表示扰动速度的径向分量,$\tilde{V}_z$ 为扰动速速的沿 z 轴的分量,$\tilde{p}$ 为扰动压强. 此时,不考虑重力和流体黏性的影响时,理想流体的欧拉方程 (56-11) 式给出

$$\begin{aligned} r: & \quad \rho\left(\frac{\partial \tilde{V}_r}{\partial t} + \tilde{V}_r \frac{\partial \tilde{V}_r}{\partial r} + \tilde{V}_z \frac{\partial \tilde{V}_r}{\partial z}\right) = -\frac{\partial \tilde{p}}{\partial r} \\ z: & \quad \rho\left(\frac{\partial \tilde{V}_z}{\partial t} + \tilde{V}_r \frac{\partial \tilde{V}_z}{\partial r} + \tilde{V}_z \frac{\partial \tilde{V}_z}{\partial z}\right) = -\frac{\partial \tilde{p}}{\partial z} \end{aligned} \tag{62-12}$$

第 11 章 流体动力学

将 (62-12) 式线性化, 也就是去掉对流加速度项, 得到

$$\frac{\partial \tilde{V}_r}{\partial t} = -\frac{1}{\rho}\frac{\partial \tilde{p}}{\partial r}, \quad \frac{\partial \tilde{V}_z}{\partial t} = -\frac{1}{\rho}\frac{\partial \tilde{p}}{\partial z} \tag{62-13}$$

在柱坐标系下, 连续性方程已经由 (31-23) 式给出. 线性化后的连续性方程为

$$\frac{\partial \tilde{V}_r}{\partial r} + \frac{\tilde{V}_r}{r} + \frac{\partial \tilde{V}_z}{\partial z} = 0 \tag{62-14}$$

针对该问题, 可以设定径向 (r)、纵向 (z) 速度的扰动和压强的扰动解的形式和水柱半径的扰动形式相同,

$$\tilde{V}_r = \mathcal{R}(r)\,\mathrm{e}^{\omega t+\mathrm{i}kz} \tag{62-15}$$

$$\tilde{V}_z = \mathcal{Z}(r)\,\mathrm{e}^{\omega t+\mathrm{i}kz} \tag{62-16}$$

$$\tilde{p} = \mathcal{P}(r)\,\mathrm{e}^{\omega t+\mathrm{i}kz} \tag{62-17}$$

这种类型的解一般被称为 "设定某种特定形式的解 (ansatz solution)" 或简称为 "设定解", 亦称 "尝试解".

将 (62-15)—(62-17) 三式代入 (62-13) 和 (62-14) 三式中, 得到

$$\frac{\partial \mathcal{P}}{\partial r} = -\rho\omega\mathcal{R} \tag{62-18}$$

$$\mathrm{i}k\mathcal{P} = \rho\omega\mathcal{Z} \tag{62-19}$$

$$\frac{\partial \mathcal{R}}{\partial r} + \frac{\mathcal{R}}{r} + \mathrm{i}k\mathcal{Z} = 0 \tag{62-20}$$

对 (62-18) 和 (62-19) 两式对 r 进行微分, 对上述三式适当整理后可消去 $\mathcal{P}$ 和 $\mathcal{Z}$, 从而可以得到有关 $\mathcal{R}$ 的微分方程:

$$r^2\frac{\partial^2 \mathcal{R}}{\partial r^2} + r\frac{\partial \mathcal{R}}{\partial r} - (1+k^2r^2)\mathcal{R} = 0 \tag{62-21}$$

上式为一阶修正的贝塞尔方程 (modified Bessel equation of order 1), 其解为

$$\mathcal{R}(r) = C\mathrm{I}_1(kr) \tag{62-22}$$

式中, C 为待定常数, $\mathrm{I}_1(kr)$ 为一阶贝塞尔函数. 利用径向扰动速度的基本关系式:

$$\frac{\partial \tilde{R}}{\partial t} = \tilde{V}_r \quad \Rightarrow \quad \in\omega\mathrm{e}^{\omega t+\mathrm{i}kz} = \mathcal{R}(R_0)\mathrm{e}^{\omega t+\mathrm{i}kz} \quad \Rightarrow \quad \in\omega = C\mathrm{I}_1(kR_0) \tag{62-23}$$

便可获得 (62-22) 式中的积分常数：

$$C = \frac{\epsilon\,\omega}{\mathrm{I}_1\left(kR_0\right)} \tag{62-24}$$

问题的边界条件是扰动的压强需要满足如下杨-拉普拉斯方程：

$$p_0 + \tilde{p} = \gamma\left(\frac{1}{R_1} + \frac{1}{R_2}\right) \tag{62-25}$$

式中的主曲率之一满足 (62-11) 式：

$$R_1 = \tilde{R} = R_0 + \epsilon\,\mathrm{e}^{\omega t + \mathrm{i}kz} \tag{62-26}$$

由于 $\epsilon \ll R_0$，则第一主曲率满足如下近似关系式：

$$\frac{1}{R_1} = \frac{1}{R_0 + \epsilon\,\mathrm{e}^{\omega t + \mathrm{i}kz}} \approx \frac{1}{R_0}\left(1 - \frac{\epsilon}{R_0}\mathrm{e}^{\omega t + \mathrm{i}kz}\right) \tag{62-27}$$

第二主曲率则满足

$$\frac{1}{R_2} = -\frac{\partial^2 \tilde{R}}{\partial z^2} = \epsilon\,k^2\mathrm{e}^{\omega t + \mathrm{i}kz} \tag{62-28}$$

对于定态情形，初始压强所满足的杨-拉普拉斯方程为

$$p_0 = \frac{\gamma}{R_0} \tag{62-29}$$

将 (62-27)—(62-29) 三式代入 (62-25) 式，则扰动压强满足

$$\frac{\gamma}{R_0} + \tilde{p} = \gamma\left(\frac{1}{R_0} - \frac{\epsilon}{R_0^2}\mathrm{e}^{\omega t + \mathrm{i}kz} + \epsilon\,k^2\mathrm{e}^{\omega t + \mathrm{i}kz}\right) \quad \Rightarrow \quad \tilde{p} = -\frac{\gamma\,\epsilon}{R_0^2}\left(1 - k^2 R_0^2\right)\mathrm{e}^{\omega t + \mathrm{i}kz} \tag{62-30}$$

通过 (62-18) 式和 (62-24) 式再考虑

$$\mathrm{I}_1(x) = \mathrm{I}'_0(x) \tag{62-31}$$

则有

$$\mathcal{P}(r) = -\frac{\epsilon\,\omega^2 \rho}{\mathrm{I}_1(kR_0)}\int_0^r \mathrm{I}_1(kx)\,\mathrm{d}x = -\frac{\epsilon\,\omega^2 \rho}{k\mathrm{I}_1(kR_0)}\mathrm{I}_0(kr) \tag{62-32}$$

上式是 $\mathrm{I}_0(kr)$ 为零阶贝塞尔函数. 将上式代入 (62-17) 式，则扰动压强可确定为

$$\tilde{p} = -\frac{\epsilon\,\omega^2 \rho}{k\mathrm{I}_1(kR_0)}\mathrm{I}_0(kr)\,\mathrm{e}^{\omega t + \mathrm{i}kz} \tag{62-33}$$

令 (62-33) 和 (62-30) 两式中的扰动压强相等，则得到色散关系 (dispersion relation)：

$$\omega^2 = \frac{\gamma}{\rho R_0^3}kR_0\frac{\mathrm{I}_1(kR_0)}{\mathrm{I}_0(kR_0)}\left(1 - k^2 R_0^2\right) \tag{62-34}$$

$\omega^2 > 0$ 要求 $kR_0 < 1$. 方程 (62-34) 的函数关系如图 62.9 所示. ω 的最大值,也就是使水柱断裂成水滴的临界值,在 kR_0 取下列值时达到

$$kR_0 \approx 0.697 \tag{62-35}$$

从图 62.9 可以看出,频率 ω 的最大值为

$$\omega_{\max} = 0.34\sqrt{\frac{\gamma}{\rho R_0^3}} \tag{62-36}$$

则可从数量级上估算水柱断裂成水滴的时间为

$$t_{\text{breakup}} \sim \frac{1}{\omega} \approx 2.94\sqrt{\frac{\rho R_0^3}{\gamma}} \tag{62-37}$$

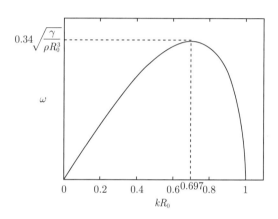

图 62.9 频率 ω 和 kR_0 之间的关系

有关普拉托-瑞利不稳定性的直观解释见思考题 11.16.

例 62.1 证明液滴表面张力 γ 的量纲和弹簧的劲度系数 k 相同,因此,弹簧劲度 k 的组合量纲的性质表面张力 γ 均满足.

证明:按照定义,弹簧的劲度系数是单位伸长所产生的力,所以弹簧劲度的量纲为

$$[k] = \text{MLT}^{-2}\text{L}^{-1} = \text{MT}^{-2}$$

而表面张力定义为新产生单位新的表面所需要的功,因此,有

$$[\gamma] = \text{ML}^2\text{T}^{-2}\text{L}^{-2} = \text{MT}^{-2}$$

可见弹簧的劲度系数和液滴的表面张力的量纲相同. 如弹簧振子的圆频率和周期分别为 $\omega \sim \sqrt{\frac{k}{m}}$ 和 $\tau \sim \sqrt{\frac{m}{k}}$,对于液滴而言其质量约为 $m \sim \rho R^3$,则 (62-36) 和 (62-37) 两式十分容易理解.

62.5 萨夫曼-泰勒 (ST) 不稳定性

62.5.1 泰勒和萨夫曼为何要研究黏性指进问题?

萨夫曼-泰勒不稳定性 (Saffman-Taylor instability, STI) 亦称为"萨夫曼-泰勒指进 (Saffman-Taylor fingering)"源于菲利普·萨夫曼 (Philip Geoffrey Saffman, 1931—2008, 如图 62.10 右图) 和其师爷泰勒 (Geoffrey Ingram Taylor, 1886—1975, 如图 62.10 左图) 于 1958 年所合作发表的论文 [11.62]. 泰勒的学生巴切勒 (George Keith Batchelor, 1920—2000, 如图 62.10 的中图) 是萨夫曼的导师.

泰勒在 1956 年访问美国 Humble 石油公司时, 了解到在石油开采中所面临的地下石油在岩层中黏性指进的真正挑战后, 从而产生了动机去研究这个 "赫尔-肖黏性指进问题 (Hele-Shaw problem of viscous fingering)". 赫尔-肖黏性指进的典型图案如图 62.11 右图所示.

图 62.10　泰勒 (左)、巴切勒 (中) 和萨夫曼 (右)

泰勒邀请萨夫曼参与该问题的数学建模和计算方面的研究, 但萨夫曼也亲自参与了实验研究. 萨夫曼的太太 Ruth 曾回忆起当年她丈夫和卡文迪许实验室研究助理 Walter Thompson 之间有关在该问题上的互动, 并且在剑桥的一条名为 "国王游行 (King's Parade)" 街道上的杂货店购买金糖浆 (golden syrup) 来搭建 "赫尔-肖盒 (Hele-Shaw cell)" —— 由两片玻璃和中间的一薄层黏性液体构成的三明治结构.

与泰勒的这次成功合作无疑给萨夫曼带来了极大荣誉. 萨夫曼以他所特有的谦逊态度, 在 1990 年代指出: "不应该叫 '萨夫曼-泰勒', 而应该叫 '泰勒-萨夫曼', 这是因为当时英国皇家学会只按作者姓氏的字母排序来发表文章" (It shouldn't be Saffman-Taylor. It should be Taylor-Saffman. That happened because at that time the Royal Society would only publish names alphabetically).

62.5.2 赫尔-肖黏性方程和指进问题

亨利·塞尔比·赫尔-肖 (Henry Selby Hele-Shaw, 1854—1941, 如图 62.11 左图)

家族的姓氏只是"Shaw（肖）"，他母亲娘家的姓氏是"Hele（赫尔）"，因受到母亲很大的影响，他在 20 岁出头时，将自己的姓氏改为复姓"Hele-Shaw（赫尔-肖）".

图 62.11　1922 年的赫尔-肖和典型的赫尔-肖黏性指进图案

1898 年，赫尔-肖研究了如图 62.12 所示的两片相距很近的玻璃板间的黏性流体的流动问题 [11.63]．下面我们来建立赫尔-肖方程．

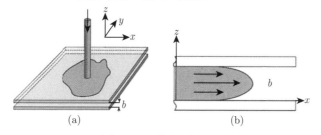

图 62.12　赫尔-肖流动

不考虑体力、不可压缩流体的纳维-斯托克斯方程如下：

$$\begin{cases} \rho\left[\dfrac{\partial \boldsymbol{v}}{\partial t} + (\boldsymbol{v}\cdot\boldsymbol{\nabla})\boldsymbol{v}\right] = -\boldsymbol{\nabla}p + \mu\nabla^2\boldsymbol{v} \\ \boldsymbol{\nabla}\cdot\boldsymbol{v} = 0 \end{cases} \tag{62-38}$$

由于两玻璃片之间距离狭窄，故做如下假设是合理的：

$$v_z = 0, \quad \dfrac{\partial \boldsymbol{v}}{\partial t} = \boldsymbol{0} \tag{62-39}$$

因此，(62-38) 式中第一式可进一步简写为

$$\rho\left[(\boldsymbol{v}\cdot\boldsymbol{\nabla})\boldsymbol{v}\right] = -\boldsymbol{\nabla}p + \mu\nabla^2\boldsymbol{v} \tag{62-40}$$

考虑到 $v_z = 0$，(62-40) 式的分量形式为

$$\begin{cases} \rho\left(v_x\dfrac{\partial v_x}{\partial x} + v_y\dfrac{\partial v_x}{\partial y}\right) = -\dfrac{\partial p}{\partial x} + \mu\nabla^2 v_x \\ \rho\left(v_x\dfrac{\partial v_y}{\partial x} + v_y\dfrac{\partial v_y}{\partial y}\right) = -\dfrac{\partial p}{\partial y} + \mu\nabla^2 v_y \\ 0 = -\dfrac{\partial p}{\partial z} \end{cases} \tag{62-41}$$

无滑移边界条件为

$$v_x\,|_{z=0,b} = v_y\,|_{z=0,b} = 0 \tag{62-42}$$

下面基于 (61-14)—(61-16) 三式对赫尔–肖流动做粗略的数量级估计. 由于流动在 x 和 y 方向是均等的, 故可设两个方向的速度的数量级为 $v_x \sim v_y \sim V$, 而流动在 x 和 y 方向的特征尺度为 L, 则有

$$v_x\frac{\partial v_x}{\partial x} \sim v_y\frac{\partial v_x}{\partial y} \sim v_x\frac{\partial v_y}{\partial x} \sim v_y\frac{\partial v_y}{\partial y} \sim \frac{V^2}{L} \tag{62-43}$$

下面对 (62-41) 式中的拉普拉斯算子 $\nabla^2 = \dfrac{\partial^2}{\partial x^2} + \dfrac{\partial^2}{\partial y^2} + \dfrac{\partial^2}{\partial z^2}$ 做数量级估计:

$$\frac{\partial^2}{\partial x^2} \sim \frac{\partial^2}{\partial y^2} \sim \frac{1}{L^2},\quad \frac{\partial^2}{\partial z^2} \sim \frac{1}{b^2} \tag{62-44}$$

由于赫尔–肖盒在几何上满足 $b \ll L$, 则有

$$\nabla^2 = \frac{\partial^2}{\partial x^2} + \frac{\partial^2}{\partial y^2} + \frac{\partial^2}{\partial z^2} \sim \frac{\partial^2}{\partial z^2} \tag{62-45}$$

下面再对 (62-41) 前两个式子中的等号的左端项进行数量级的估计, 我们来看看这两项和等号右端的第二项的相对大小:

$$\frac{\rho V^2/L}{\mu V/b^2} = \frac{\rho V b}{\mu}\frac{b}{L} = Re\,\frac{b}{L} \tag{62-46}$$

由于赫尔–肖流动为低雷诺数流动, 而且在 $b \ll L$ 的条件下, 通过如上的数量级估计可知, (62-41) 前两个式子中的等号的左端项可被忽略掉. 从而 (62-41) 式可退化为如下简单形式:

$$\begin{cases} \dfrac{\partial p}{\partial x} = \mu\dfrac{\partial^2 v_x}{\partial z^2} \\[4pt] \dfrac{\partial p}{\partial y} = \mu\dfrac{\partial^2 v_y}{\partial z^2} \\[4pt] \dfrac{\partial p}{\partial z} = 0 \end{cases} \tag{62-47}$$

式中的第三式表明 $p = p(x,y,t)$, (62-47) 前两个式子对 z 积分, 由边界条件 (62-42) 式可得到 x 和 y 方向的速度分布为

$$\begin{cases} v_x = \dfrac{1}{2\mu}\dfrac{\partial p}{\partial x}z(z-b) \\[4pt] v_y = \dfrac{1}{2\mu}\dfrac{\partial p}{\partial y}z(z-b) \end{cases} \tag{62-48}$$

式中的速度对液体膜厚 b 进行平均得到

$$\begin{cases} \bar{v}_x = \dfrac{1}{b}\int_0^b v_x \mathrm{d}z = -\dfrac{b^2}{12\mu}\dfrac{\partial p}{\partial x} \\ \bar{v}_y = \dfrac{1}{b}\int_0^b v_y \mathrm{d}z = -\dfrac{b^2}{12\mu}\dfrac{\partial p}{\partial y} \end{cases} \tag{62-49}$$

上式的矢量形式为

$$\bar{\boldsymbol{v}} = -\frac{b^2}{12\mu}\boldsymbol{\nabla} p \tag{62-50}$$

上式即为赫尔-肖方程 (Hele-Shaw equation). 再由液体的不可压缩条件 $\boldsymbol{\nabla}\cdot\boldsymbol{v}=0$, 得到赫尔-肖流动所满足的拉普拉斯方程:

$$\nabla^2 p = 0 \tag{62-51}$$

赫尔-肖流动亦被广泛地称为 "赫尔-肖自由边界问题 (Hele-Shaw free boundary problem)" 或 "拉普拉斯增长问题 (Laplacian growth problem)"[11.64].

亨利·达西 (Henry Darcy, 1803—1858) 于 19 世纪中叶所建立的达西定律 (Darcy's law) 表明, 在忽略体力时渗流通量 $\boldsymbol{q}$ 和压强梯度 $\boldsymbol{\nabla}p$ 之间满足

$$\boldsymbol{q} = -\frac{\kappa}{\mu}\boldsymbol{\nabla}p \tag{62-52}$$

而渗流速度 $\boldsymbol{v}$、渗流通量 $\boldsymbol{q}$ 和孔隙率 (porosity) φ 三者之间满足

$$\boldsymbol{v} = \frac{\boldsymbol{q}}{\varphi} \tag{62-53}$$

当孔隙率 $\varphi=1$ 时, 也就是赫尔-肖流动的情形, (62-50) 式事实上就是达西定律的一种具体形式:

$$\bar{\boldsymbol{v}} = -\frac{\kappa}{\mu}\boldsymbol{\nabla}p \tag{62-54}$$

式中, 达西系数为 $\kappa = b^2/12$.

62.5.3 萨夫曼-泰勒不稳定性

萨夫曼和泰勒通过保角变换, 在 $|\zeta|<1$ 情况下, 获得的解为

$$z = \frac{Vt}{\lambda} + \zeta + 2(1-\lambda) + \log\frac{1+\mathrm{e}^{-\zeta}}{2}, \quad 0<\lambda\leqslant 1 \tag{62-55}$$

模型分析表明参数的取值范围为 $0<\lambda\leqslant 1$, 从实验结果来看, 该参数取下列值理论和实验符合得很好:

$$\lambda = \frac{1}{2} \tag{62-56}$$

1958 年发表的那篇优美并极有影响力的论文 [11.62] 迄今已经被 google scholar 引用了 3000 多次, 但更重要的科学故事才刚刚开始. 在 1958 年的萨夫曼–泰勒解中尚有一个未解决的问题: 行进中的指进 (travelling finger) 的宽度和赫尔–肖盒的宽度之比, 在原文中被称作参数 λ, 是不确定的. 理论分析表明, λ 的取值范围在 0 和 1 之间, 而实验值表明该值通常取 $\lambda = 1/2$. 萨夫曼和泰勒虽然给出了 $\lambda = 1/2$ 的一些解释, 但该问题基本上没有得到完美的解决, 以至于后来引发了数学家、物理学家、力学家、材料学家和地质学家们的大量后续研究. 后来, 萨夫曼深情地回忆道, 当泰勒被问到他睡得好不好时, 泰勒回答说: 他通常睡得好, 除非是在思考赫尔–肖问题中理论和实验之间的差异时. 由此看来, 力学大师泰勒在睡觉时, 还在深入思考所研究的科学问题.

思考题和补充材料

11.1 约翰 · 伯努利对其兄雅克布 · 伯努利攻击时所用过的形容词 (an edifying list of adjectives that Johann uses to qualify his brother Jacob): Verbose, ambitious, greedy, secretive, misanthropist, envious, proud, and too imaginative $\cdots$

11.2 如题图 11.2 所示, 分析多孔介质流动和自行车骑行者的流线 (streamlines) 和阻力的方向.

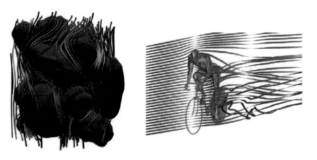

题图 11.2　多孔介质流动和自行车骑行者的流线

11.3 伯努利方程推导的第二种方法.

由理想流体微分形式运动方程推导伯努利方程

$$\rho \frac{\mathrm{d}\boldsymbol{v}}{\mathrm{d}t} = -\boldsymbol{\nabla} p + \rho \boldsymbol{f}$$

应用 (s2-25) 式, 于是有物质导数

$$\frac{\mathrm{d}\boldsymbol{v}}{\mathrm{d}t} = \frac{\partial \boldsymbol{v}}{\partial t} + \boldsymbol{v} \cdot (\boldsymbol{\nabla} \otimes \boldsymbol{v}) = \frac{\partial \boldsymbol{v}}{\partial t} + \mathbf{grad}\frac{v^2}{2} + \mathbf{rot}\boldsymbol{v} \times \boldsymbol{v}$$

很多教材和专著中用
$$\frac{\mathrm{D}\boldsymbol{v}}{\mathrm{D}t} = \frac{\partial \boldsymbol{v}}{\partial t} + \boldsymbol{v}\cdot(\boldsymbol{\nabla}\otimes\boldsymbol{v})$$
来表示物质导数. 运动方程可改写为
$$\frac{\partial \boldsymbol{v}}{\partial t} + \mathbf{grad}\frac{v^2}{2} + \mathbf{rot}\boldsymbol{v}\times\boldsymbol{v} = -\frac{1}{\rho}\mathbf{grad}p + \boldsymbol{f}$$
此即为理想流体兰姆–葛罗米柯 (Lamb-Gromeko) 形式的运动方程.

对于重力作用下的不可压缩流体, 则
$$\frac{1}{\rho}\mathbf{grad}p = \mathbf{grad}\frac{p}{\rho}$$
$$\boldsymbol{f} = -\mathbf{grad}(gz)$$
于是运动方程具有下列形式
$$\frac{\partial \boldsymbol{v}}{\partial t} + \mathbf{grad}\left(\frac{v^2}{2} + \frac{p}{\rho} + gz\right) + \mathbf{rot}\boldsymbol{v}\times\boldsymbol{v} = \boldsymbol{0}$$
只考虑定常运动, 有
$$\frac{\partial \boldsymbol{v}}{\partial t} = \boldsymbol{0}$$
于是运动方程变为
$$\mathbf{grad}\left(\frac{v^2}{2} + \frac{p}{\rho} + gz\right) + \mathbf{rot}\boldsymbol{v}\times\boldsymbol{v} = \boldsymbol{0}$$
上式两边点乘流线的切线单位向量 $\boldsymbol{s} = \dfrac{\boldsymbol{v}}{v}$ 得
$$\boldsymbol{s}\cdot\mathbf{grad}\left(\frac{v^2}{2} + \frac{p}{\rho} + gz\right) + \frac{\boldsymbol{v}}{v}\cdot(\mathbf{rot}\boldsymbol{v}\times\boldsymbol{v}) = 0$$
式中的第二项显然为零, 于是
$$\boldsymbol{s}\cdot\mathbf{grad}\left(\frac{v^2}{2} + \frac{p}{\rho} + gz\right) = 0$$
利用附录 D 中的 (D-8) 式, 得到沿流线的方向导数为
$$\frac{\partial}{\partial s}\left(\frac{v^2}{2} + \frac{p}{\rho} + gz\right) = 0$$
沿流线积分得
$$\frac{v^2}{2} + \frac{p}{\rho} + gz = C$$
其中 C 是积分常数, 沿同一条流线取同一常数值, 不同流线可以取不同的值. 此即伯努利方程. 其意义为, 重力作用下的不可压缩理想流体做定常运动, 总能量在流线上守恒. 左端各项分别代表单位质量内的动能, 势能和压力能.

11.4 对 §54 的补充: 从纳维–斯托克斯方程出发推导哈根–泊肃叶定律.

已知 N-S 方程的一般形式:
$$\frac{\partial \boldsymbol{v}}{\partial t} + (\boldsymbol{v} \cdot \boldsymbol{\nabla}) \boldsymbol{v} = \frac{1}{\rho} \left(-\boldsymbol{\nabla} p + \mu \nabla^2 \boldsymbol{v} + \boldsymbol{f} \right)$$

根据不可压缩流体假设 $\boldsymbol{\nabla} \cdot \boldsymbol{v} = 0$,

取柱坐标 (x, r, θ), 则沿着 x 方向的 N-S 方程可以简化为
$$\frac{1}{r}\frac{\mathrm{d}}{\mathrm{d}r}\left(r\frac{\mathrm{d}v}{\mathrm{d}r}\right) = -\frac{\Delta p}{\mu L} \Rightarrow v = -\frac{\Delta p}{4\mu L}r^2 + c_1 \ln r + c_2$$

结合边界条件: $v|_{r=R} = 0$ 和 $r \to 0$ 时速度的非奇异性, 得到
$$v = \frac{\Delta p}{4\mu L}\left(R^2 - r^2\right)$$

则
$$Q = \int_0^R \frac{\Delta p}{4\mu L}\left(R^2 - r^2\right) \cdot 2\pi r \mathrm{d}r = \frac{\Delta p}{8\mu L}\pi R^4$$
$$\Delta p = \frac{8\mu L Q}{\pi R^4}$$

11.5 大作业. 结合题图 11.5, 通过文献检索, 给出圆柱尾流从层流到湍流的 Re 数范围.

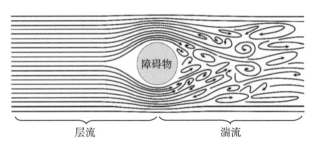

题图 11.5　圆柱尾流的从层流向湍流的转捩

11.6 前言中所提及的 1906 年诺贝尔物理学奖获得者汤姆逊对雷诺的回忆的英文原文为 [11.30]: "The Professor I had most to do with was Osborne Reynolds. He never did anything or expressed himself like anybody else. The result was that it was very difficult to take notes at his lectures so that we had to trust mainly to Rankine's text books. Occasionally in the higher classes he would forget all about having to lecture and after ten minutes or so we sent the janitor to tell him that the class was waiting. He would come rushing into the room pulling on his gown as he came through the door, take a volume of Rankine from the table, open it apparently at random, see some formula or other and say that it was wrong. He then went up to the blackboard to prove this. He wrote on the board with his back to us talking to himself, and every now and then rubbed it all out and said that it was wrong. He would then start afresh on a new line and so on. Generally, towards the end of the lecture he would finish one which he did not rub out and say that this proved Rankine was right after all."

11.7 上题中所提到的兰金 (William John Macquorn Rankine, 1820—1872, 题图 11.7 所示) 的教科书如下: (1) *Manual of Applied Mechanics* (1858); (2) *Manual of the Steam Engine and Other Prime Movers* (1859); (3) *Manual of Civil Engineering* (1861); (4) *Shipbuilding, Theoretical and Practical* (1866); (5) *Manual of Machinery and Millwork* (1869). 兰金的这些教材自 1850s 和 1860s 年代出版后, 一直被沿用了几十年.

其中, 兰金在《应用力学手册》(*A Manual of Applied Mechanics*) 中总结了固体强度方面的实验及理论结果, 提出以拉伸应力作为判据的强度理论, 在材料力学中被称为第一强度理论.

题图 11.7 兰金

11.8 亨利·贝纳尔 (Henri Bénard, 1880—1939, 如题图 11.8 左图所示) 1900 年在题为《加热薄层液体持续对流的漩涡胞》[11.65] 的文章中, 首次报道了由实验发现的底部加热薄层流体对流产生的漩涡胞, 被称为贝纳尔胞, 如题图 11.8 中和右图所示. 该种对流现象被称为 "瑞利–贝纳尔对流 (Rayleigh-Bénard convection)". 直到 2017 年, 仍有法国学者在期刊上发表论文[11.66], 系统地介绍贝纳尔在 "贝纳尔–卡门涡街" 中的贡献.

题图 11.8 贝纳尔胞和瑞利–贝纳尔对流

11.9 自然奇观中的贝纳尔胞 —— 石柱群. 分析题图 11.9 中石柱群的形成机理: 在火山喷发时, 由于瑞利–贝纳尔对流所形成的贝纳尔胞体. 石柱群的形成是一种典型的涌现或演生现象 (emergent phenomenon).

题图 11.9　爱尔兰的巨人堤道 (Giant's Causeway)(左) 和福建南碇岛石柱群 (右)

11.10 乔治·凯利 (George Cayley, 1773—1858, 如题图 11.10 左图所示), 航空之父 (the father of aviation). 1809—1810 年在《论空中航行》(*On Aerial Navigation*) 的三篇系列性论文中 [11.67], 开辟了从空气动力学来探讨飞行的道路. 凯利提出的科学论断包括: (1) 为作用在重于空气的飞行器上的四种力 —— 升力、重力、推力和阻力下定义; (2) 确定升力的机理是与推力机理分开的.

 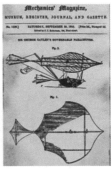

题图 11.10　乔治·凯利 (左) 和他的滑翔机 (右)

题图 11.11　冯·卡门纪念邮票

1853 年, 凯利又造了一架滑翔机 (glider, 如题图 11.10 右图所示), 并装上了灵巧的刹车杠杆, 进行历史上第一次有人乘坐的重于空气的航空器升空自由飞行. 这次他把家中的马车夫放在驾驶室里 (had his coachman aboard), 究竟飞了多远, 没有明确的记录. 有趣的是, 马车夫从飞机上下来后, 竟辞职不干了. 他说: "乔治爵士, 我想请你注意, 我是你雇来赶车的, 不是来飞行的 (After the short flight Cayley's coachman stated that he had been hired to drive a coach not to fly a glider)".

11.11 冯·卡门的祖国匈牙利于 1992 年发行了冯·卡门的纪念邮票, 如题图 11.11 所示,邮票的背景就是卡门涡街. 问题: 进一步比较该邮票上的涡街图案和图 59.4 中的图案的类似性.

11.12 在 §59.2 中曾多次提到普朗特的博士生卡尔·哈依门兹, 他于 1911 年获得哥廷根大学的博士学位, 博士论文题目:《*Die Grenzschichten an einem in den gleichfoermigen Fluessigkeitsstrom eingetauchten geraden Kreiszylinder*》.

11.13 柯尔莫哥洛夫建立不可压缩湍流标度律 K41 的 1941 年, 苏联发生了一场悲剧, 在整个世界战争史上都是令人震惊的. 1941 年 6 月 22 日纳粹德国发起 "闪击战" 后, 在五个月内不仅打到距莫斯科市区仅 10 公里处, 还成建制地歼灭了苏军 19 个集团军和 250 个师, 使苏联真到了生死存亡的边缘. 在 1941 年入伍, 20 世纪 80 年代担任总参谋长的谢·费·阿赫罗梅耶夫元帅曾有一句概括性的总结 —— "苏联军事思想的核心, 就是不使 1941 年的悲剧重演." 由于 1941 年苏联发生了这么重要的事件, 以至于有权威文献说, 柯尔莫哥洛夫的 K41 刚开始被学术界忘掉了. 但由于其极端重要性, 该结果 "回归" 学术界是迟早的事情.

11.14 柯尔莫哥洛夫曾于 1943 年说: "20 年内没有人会知道我们国家究竟发生了什么 (In twenty years no one will know what actually happened in our country)." 请思考下述问题: 为什么苏联这段十分困难时期 (战火摧残、饥寒交迫等等), 却成为数学和物理学家大师们的沃土?

11.15 当代数学物理界有杰出成就的威腾 (Edward Witten, 1951—), 在本科时念历史, 1990 年获得了数学的最高奖 —— 菲尔兹奖. 柯尔莫哥洛夫在莫斯科大学读本科开始时, 学习和研究历史, 后来改为数学.

11.16 如题图 11.16 所示, 从直观上对普拉托–瑞利不稳定性进行解释: 扰动较小时, 即波峰处压强较大, 波谷处压强较小, 可以恢复平衡 (想象此处压力较大, 把波峰压平); 当扰动较大时, 波谷处压强较大, 会掐断水柱, 形成水滴.

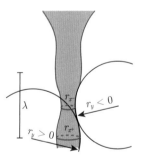

题图 11.16 普拉托–瑞利不稳定性

11.17 在思考题 10.8 中已经给出了 "简单物质" 的定义. 结合前言图 1, 大作业: (1) 阐明线弹性体和牛顿流体均是简单物质的特例; (2) 进一步探索和介质对称性相关的群 (迷向群、正交群、三斜群、特殊线性群亦即么模群) 在理性力学中的应用.

11.18 大作业: 为什么布满小坑 (dimple) 的高尔夫球比光滑球飞的更远? 光滑的高尔夫球一杆子能飞数十米, 而布满小坑的高尔夫球一竿子则可飞行 200 多米. 如题图 11.18 所示, 布满小坑的高尔夫球能够减阻的主要原因是, 小坑的存在加速了高尔夫球表面边界层由层流向湍流的转捩, 延缓了边界层和高尔球的分离点, 从而降低了高尔夫球飞行前后的压差阻力.

问题一: 如题图 11.18(c) 所示, 为什么在边界层内湍流可增加动量混合 (momentum mixing)?

问题二: 湍流和层流相比增加了动量混合, 亦即湍流将更多的高速气体的微团带到了近壁面, 这必然导致避免附近气体的速度梯度的增大, 也就是说黏性摩擦阻力部分增大了,解释:为何此时摩擦阻力和压差阻力相比是小量?

问题三:对于二维分离来说,分离点的判断是近壁面速度梯度由正转为负,从该判据出发,解释湍流边界层对分离点的延迟的机理.

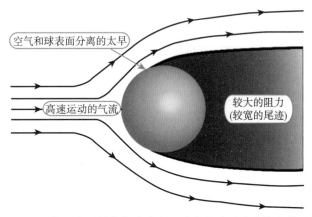

题图 11.18(a)　光滑表面的高尔夫球由于分离点过早造成较大的压差阻力

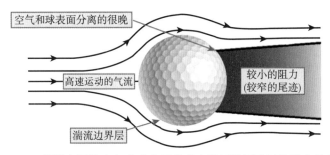

题图 11.18(b)　粗糙表面的高尔夫球由于分离点的延迟导致压差阻力的大幅下降

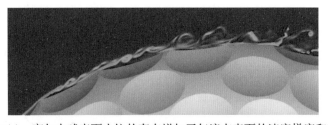

题图 11.18(c)　高尔夫球表面小坑的存在增加了气流在表面的速度梯度和动量交换

11.19　大作业:康达效应 (Coandǎ effect) 及其应用. 如题图 11.19 所示,所谓康达效应,也称之为边界层吸附效应. 当流体经过具有一定弯度的凸表面的时候, 有向凸表面吸附的趋向,该效应是以其发现者)——罗马尼亚学者亨利·康达 (Henri Coandǎ, 1886—1972)——命名的. 图 (a) 表明,空气的自由射流将空气分子从其直接环境中带走,在射流周围形成轴对称的低压区. 由于该低压区的轴对称性质,使得各个方向上的力是平衡的,射流稳定在一条直线上;图 (b) 表明,如果放置一固体表面在平行于射流的周围环境时,固体表面与射流之间的空气的夹带会导致射流在该区域的压力的降低且该侧的空气压力无法像射流"开口"的一侧的低压区那样得到迅速的平衡;图 (c) 表明,射流上下表面之间的压差使得射流向附近的表面偏移,然后附着在固体表面上;图 (d) 进一步表明,对于弯曲表面而言,射流更容易黏附. 这是由于表面方向的每一个增量变化都会导致射流向初始表面弯曲效果一样的弯曲;图 (e) 说明,在射流流过表面的地方增加一个挡板将会增强射流偏向表面的行为,这是由于在挡板于射流之间产生一个低压的涡,促进了射流向表面的倾角.

大作业一:伯努利原理在康达效应中的应用.

大作业二:康达效应在飞机航行中的应用,如图 (f) 所示.

大作业三:康达效应在飞碟设计中的应用.

大作业四:康达效应的流体力学模型及其解答.

大作业五:康达效应在微纳米尺度的应用.

大作业六:马格努斯效应 (Magnus effect) 和康达效应是否有关系?

大作业七:康达效应的局限性.

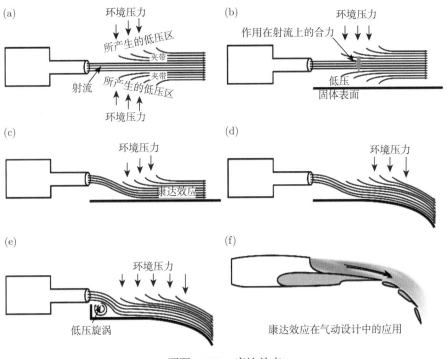

题图 11.19 康达效应

11.20 大作业:给出纳维-斯托克斯方程和对流-扩散方程的类比性和相互对应关系.

11.21 大作业:在 §55.2 中业已指出:普朗特的边界层 (boundary layer) 理论不仅解决了

达朗贝尔佯谬，而且还给出了计算物体在流体中运动时阻力的近似方法. 所谓边界层，是指高雷诺数绕流中紧贴物面的黏性力不可忽略的流动薄层，又称流动边界层、附面层，这个概念是由普朗特于 1904 年首先提出的. 如题图 11.21 所示，由无滑移边界条件 (non-slip boundary condition)，边界层内从物面当地速度为零开始，沿法线方向至速度与当地自由流速度 u_0 相等 (严格地说是等于 0.990 或 $0.995u_0$) 的位置之间的距离，记为 δ. 层流边界层的厚度为

$$\delta = 4.91\sqrt{\frac{\nu x}{u_0}} = 4.91\frac{x}{\sqrt{Re_x}} \qquad (\text{s}11\text{-}1)$$

式中，ν 为运动黏度，x 为和边界层开始的距离. 问题：(1) 由 (59-2) 式可知，运动黏度和扩散系数的量纲相同，(s11-1) 式还可表示为 $\delta = 4.91\sqrt{\nu\tau}$，这里 $\tau = x/u_0$ 为特征时间，是否可从热扩散的角度来理解边界层的厚度？(2) 进一步对湍流边界层进行深入调研.

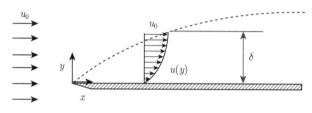

题图 11.21 边界层示意图

11.22 大作业：在思考题 1.2 中所述的七个千禧年大奖难题中，纳维–斯托克斯方程解的存在性与光滑性是唯一的一个和连续介质力学相关的问题. 为了有助于对该大奖难题的进一步理解，请系统调研该领域所已经获得解决的部分结果：

(1) 二维空间下的纳维–斯托克斯方程解的问题已在 1960 年代得证：存在光滑及全局定义解的解.

(2) 在初速 $v(x,t)$ 相当小时此问题也已得证：存在光滑及全局定义解的解.

(3) 若给定一初速 $v_0(x)$，且存在一有限、依 $v_0(x)$ 而变动的时间 T，使得在 $\mathbb{R}^3 \times (0,T)$ 的范围内，纳维–斯托克斯方程有平滑的解，还无法确定在时间超过 T 后，是否仍存在平滑的解.

(4) 法国数学家让·勒雷 (Jean Leray, 1906—1998) 在 1934 年 28 岁时证明了所谓纳维–斯托克斯问题弱解的存在，此解在平均值上满足纳维–斯托克斯问题，但无法在每一点上满足.

11.23 大作业：仍然是千禧年大奖难题和纳维–斯托克斯方程有关的问题. 许多纳维–斯托克斯方程解的基本性质都尚未被证明. 例如数学家就尚未证明在三维坐标，特定的初始条件下，纳维–斯托克斯方程是否有符合光滑性的解；也尚未证明若这样的解存

第 11 章　流体动力学

在时,其动能有其上下界,这就是 "纳维-斯托克斯存在性与光滑性" 问题. 问题: 对该难题进一步进行深入调研,看你是否有能力和毅力解决该难题.

11.24 大作业: 纳维-斯托克斯方程和麦克斯韦方程组之间的类比关系 [11.68–11.70], 物理量之间的详细类比见题表 11.24.

题表 11.24　流体动力学和电磁学物理量之间的类比关系

流体动力学量 (hydrodynamic quantities)	电磁量 (electromagnetic quantities)
比焓 (specific enthalpy) p/ρ	标量势 (scalar potential) V
速度矢量 $\boldsymbol{v}$	矢量势 (vector potential) $\boldsymbol{A}$
涡量 (vorticity) $\boldsymbol{\Omega}$	磁感应强度 (magnetic induction) $\boldsymbol{B}$
兰姆矢量 (Lamb vector) $\boldsymbol{l}$	电场强度 (electric field) $\boldsymbol{E}$
水动力电荷 (hydrodynamic charge) q_H	电荷 (electric charge) q_E

11.25 早在 1963 年, 物理学家就假定存在一种电子流动形成的量子流体 (quantum fluid): 这种量子流体来源于导电材料中的电子彼此之间的强烈相互作用, 电子可以在比人类头发宽度短一百倍的尺度上像水一样流动.

2019 年 4 月 12 日, 美国《科学》期刊连刊三篇文章 [11.71–11.73], 报道了石墨烯中发现量子流体的最新成果, 这是魔角石墨烯之后, 石墨烯领域迎来的又一重大突破! 曼彻斯特大学的石墨烯诺奖得主 A. K. Geim、D. A. Bandurin 团队 [11.73] 以及加州大学伯克利分校 Feng Wang 团队在《科学》发表文章 [11.69], 分别独立报道了在石墨烯中实验观测到二维电子流体的现象, 实验揭示了在水中无法观察到的量子流体流动, 可能会产生新的量子材料和电子学. 同时, 斯坦福大学 Andrew Lucas 教授还专门发表一篇展望文章 [11.71], 对石墨烯量子流体及最新发现进行系统阐述.

通过施加局部电压, 石墨烯可以通过少数外部电子充电, 类似于对电容器的一侧进行充电, 添加的电子很容易从一个原子移动到下一个原子. 由于量子效应, 这些电子在低温下会快速地相互移动, 当温度慢慢升高, 电子开始分散开来, 两个电子以两个水分子截然不同的方式发生相互反弹. 在每次碰撞中, 由于能量, 动量和电荷都是守恒的, 所以产生的电子流体的流动方式与水的流动方式大致相同.

一个关键的问题在于, 电子必须在一系列离子之间流动. 如果电子散射离子杂质或离子晶格振动, 就失去动量, 不能再像传统液体那样流动. 在普通金属中, 电子-晶格散射非常强, 因此完全无法实现电子流动. 只有在石墨烯这样的量子材料中, 移动的电子才能够足够快地彼此散射, 从而形成量子流体, 以类似流体的方式穿过石墨烯.

电子流体动力学 (electron hydrodynamics) 的基本方程包括不可压缩条件 (s11-2)、纳维-斯托克斯方程 (s11-3) [11.74,11.75]:

$$\nabla \cdot \boldsymbol{v}(\boldsymbol{r}) = 0 \tag{s11-2}$$

$$\underbrace{\frac{e}{m}\nabla\phi(\boldsymbol{r})}_{\text{体力项}}+\underbrace{\nu\nabla^2\boldsymbol{v}(\boldsymbol{r})}_{\text{黏性项}}=\underbrace{\frac{\boldsymbol{v}(\boldsymbol{r})}{\tau}}_{\text{平均加速度}} \tag{s11-3}$$

式中，$\boldsymbol{v}(\boldsymbol{r})$ 是线性化稳态流体单元速度 (linearized steady-state fluid-element velocity)，τ 是电子流体的弛豫时间，$\nu=\mu/(\bar{n}m)$ 表示流体的运动黏度，$\bar{n}$ 是电荷密度，m 是电荷的有效质量，$\phi(\boldsymbol{r})$ 为电荷电势.

11.26 英国《自然》期刊于 2019 年 12 月 5 日刊登了以色列魏兹曼科学研究所等多国科学家合作的题为《流体动力学电子泊肃叶流动的可视化》[11.76] 的文章，该文首次观测到电子像水一样流动的奇特行为. 以色列魏兹曼科学研究所团队研制出一种由碳纳米管晶体管制成的纳米级探测器，该探测器可以前所未有的灵敏度对流动电子成像，这种技术比其他方法至少灵敏 1000 倍，使对以前只能间接研究的现象进行了直接研究. 而该文的主要合作者英国曼彻斯特大学的 Geim 团队则研制出了引导电子流动的石墨烯通道，类似于引导水流的管道. 研究人员用碳纳米管晶体管探测器对其进行观察和成像，他们观察到了电子泊肃叶流动的主要特征，那就是石墨烯内电子在通道中心流动得更快，在壁上流动得更慢.

11.27 打喷嚏的流体动力学 (fluid dynamics of sneezing). 关于喷嚏后飞沫的播散，麻省理工学院 (MIT) 的课题组通过高速相机做了相关流体动力学的深入研究 [11.77]. 研究表明，打喷嚏后，大量的唾液和黏液 (绿色) 从口中喷出，但速度下降相对较快. 而一个携带更小液滴的湍流状态的飞沫 (红色)，将携带小液滴漂移长达数米的距离，说明戴口罩对于预防传染的必要性.

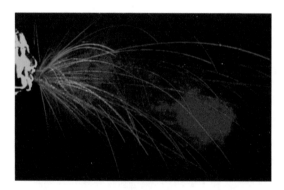

题图 11.27　高速视频捕捉到的喷嚏

参 考 文 献

[11.1] Harvey W. Exercitatio anatomica de motu cordis et sanguinis in animalibus (on the movement of the heart and blood in animals). Francofurti edition, 1628.

[11.2] Poiseuille J L M. Recherches sur la force du coeur aortique, 1828.

[11.3] Riva-Rocci S. Un nuovo sfigmomanometro. Gazzetta Medica di Torino, 1896, 47: 981–996.

[11.4] Korotkov N S. To the question of methods of study on blood (in Russian). Izv Imp Voen Med Acad, 1905, 11: 365–367.

[11.5] Guillen M. Five Equations that Changed the World: the Power and Poetry of Mathematics. Hachette Books, 2012.

[11.6] Bernoulli D. Hydrodynamica: sive de viribus et motibus fluidorum commentarii. Johannis Reinholdi Dulseckeri, 1738.

[11.7] Venturi J B. Récherches Experimentales sur le Principe de la Communication Laterale du Mouvement dans les Fluides appliqué a l'Explication de Differens Phenomènes Hydrauliques (Experimental investigations into the principle of the lateral communication of the movement in fluids applied to the explanation of different hydraulic phenomena). Paris: Houel et Ducros and Théophile Barrois, 1797.

[11.8] Poiseuille J L M. Experimental research on the movement of liquids in tubes of very small diameters. Mémoires presentés par divers savants a l'Académie Royale des Sciences de l'Institut de France, IX, 1846: 433-544.

[11.9] Hagen G. Über die Bewegung des Wassers in engen cylindrischen Röhren. Annalen der Physik und Chemie, 1839, 122: 423–442.

[11.10] Ostwald W. Über die geschwindigkeits-funktion der viskosität disperser systeme. I. Kolloid-Zeitschrift, 1925, 36: 99-117.

[11.11] Prandtl L, Tietjens O G. Applied Hydro- and Aeromechanics. New York: McGraw-Hill, 1934.

[11.12] Stokes G G. On the theories of the internal friction of fluids in motion, and of the equilibrium and motion of elastic solids. Transactions of the Cambridge Philosophical Society, 1845, 8: 287-341.

[11.13] Sutera S P, Skalak R. The history of Poiseuille's law. Annual Review of Fluid Mechanics, 1993, 25: 1-20.

[11.14] Prandtl L. Über Flüßigkeitsbewegung bei sehrkleiner Reibung. Verhandl III, Intern. Math. Kongr. Heidelberg, Auch: Gesammelte Abhandlungen, 1904, 2: 484-491.

[11.15] Sommerfeld A. The stability or instability of the steady motions of a perfect liquid and of a viscous liquid. Ein Beitrag zur hydrodynamischen Erklärung der turbulenten Flüssigkeitbewegungen, 1908, 27: 116-124.

[11.16] Navier C L M H. Memoire sur les lois du mouvement des fluides. Mem. Acad. Sci.

Inst. France, 1822, 6: 389-440.

[11.17] Stokes G G. On the theories of the internal friction of fluids in motion and of the equilibrium and motion of elastic solids. Transactions of the Cambridge Philosophical Society, 1845, 8: 287-319.

[11.18] Archimedes. On Floating Bodies.

[11.19] Falkovich G, Weinberg A, Denissenko P, Lukaschuk S. Floater clustering in a standing wave. Nature, 2005, 435: 1045-1046.

[11.20] Cummins C, Seale M, Macente A, Certini D, Mastropaolo E, Viola I M, Nakayama N. A separated vortex ring underlies the flight of the dandelion. Nature, 2018, 562: 414-418.

[11.21] Mach E, Salcher L. Photographische Fixierung der durch Projektile in der Luft eingeleiten Vorgange. Sitzungsber. Akad. Wiss. Wien, 1887, 95: 764-780.

[11.22] Mach E, Sommer J. Über die fortpflanzungsgeschwindigkeit von explosionsschallwellen. Sitzungsberichte Akademie der Wissenschaften in Wien, 1877, 75: 101-130.

[11.23] Krehl P O K. History of Shock Waves, Explosions and Impact: A Chronological and Biographical Reference. Berlin: Springer, 2008.

[11.24] Prandtl L. Gasbewegung. Handwörterbuch der Naturwissenschaften, 1913, 4: 544-560.

[11.25] Von Mises R, Geiringer H, Ludford G S S. Mathematical theory of compressible fluid flow. New York: Academic Press, 1958.

[11.26] Ackeret J. Der Luftwiderstand bei sehr grossen Geschwindigkeiten. Schweizerische Bauzeitung, 1929, 94: 179-183.

[11.27] Larmor J. Sir George Gabriel Stokes: Memoirs and Scientific Correspondence. Cambridge: Cambridge University Press, 1907.

[11.28] Timoshenko S P. History of Strength of Materials. New York: McGraw-Hill Book Co., 1953.

[11.29] Smith S B, Finzi L, Bustamante C. Direct mechanical measurements of the elasticity of single DNA molecules by using magnetic beads. Science, 1992, 258: 1122-1126.

[11.30] Launder B E. Horace Lamb & Osborne Reynolds: Remarkable Mancunians ⋯ and their Interactions. Journal of Physics: Conference Series, 2014, 530: 012001.

[11.31] Reynolds O. An experimental investigation of the circumstances which determine whether the motion of water shall be direct or sinuous, and of the law of resistance in parallel channels. Philosophical Transactions of the Royal Society of London, 1883, 174: 935-982.

[11.32] von Kármán T. Über den Mechanismus des Widerstandes, den ein bewegter Körper in einer Flüssigkeit erfährt. Nachrichten von der Gesellschaft der Wissenschaften zu Göttingen. Mathematisch-physikalische Klasse, 1911, 13: 509-517.

[11.33] von Kármán T. Über den Mechanismus des Widerstandes, den ein bewegter Körper in einer Flüssigkeit erfährt. Nachrichten von der Gesellschaft der Wissenschaften zu Göttingen. Mathematisch-physikalische Klasse, 1912, 14: 547-556.

[11.34] von Kármán T. Aerodynamics. Ithaca: Cornell University Press, 1954.

[11.35] Mizota T, Zdravkovich M, Graw K-U, Leder A. St Christopher and the vortex: A Kármán vortex in the wake of St Christopher's heels. Nature, 2000, 404: 226.

[11.36] Gibbs R. The search continues for Kármán's St Christopher. Nature, 2000, 406: 122.

[11.37] Mallock A. On the resistance of air. Proceedings of the Royal Society of London A, 1907, 79: 262–273.

[11.38] Bénard H. Formation de centres de gyration a l'arriere d'un obstacle en movement. Comptes rendus de l'Academie des sciences, 1908, 147: 839-842.

[11.39] Richardson L F. Weather Prediction by Numerical Process. Cambridge: Cambridge University Press, 1922.

[11.40] Kolmogorov A N. Local structure of turbulence in an incompressible liquid for ever large Reynolds numbers. Doklady Akademii Nauk SSSR (Proceedings of the USSR Academy of Sciences), 1941, 30: 299-303.

[11.41] Kolmogorov A N. Dissipation of energy in locally isotropic turbulence. Doklady Akademii Nauk SSSR (Proceedings of the USSR Academy of Sciences), 1941, 32(1): 16-18.

[11.42] Kolmogorov A N. Equations of turbulent motion in an incompressible fluid. Doklady Akademii Nauk SSSR (Proceedings of the USSR Academy of Sciences), 1941, 30: 299-303.

[11.43] Gibson C H. Kolmogorov similarity hypotheses for scalar fields: sampling intermittent turbulent mixing in the ocean and galaxy. Proceedings of the Royal Society of London A, 1991, 434: 149-164.

[11.44] Obukhov A. On the distribution of energy in the spectrum of turbulent flow, Doklady Akademii Nauk SSSR, 32 (1), 22-24, 1941.

[11.45] Obukhov A. Spectral energy distribution in a turbulent flow, Izvestiya Akademii Nauk SSSR, Seriya Geografii i Geofiziki, 5 (4-5), 453-466, 1941.

[11.46] Barenblatt G I, Chorin A J. New perspectives in turbulence: Scaling laws, asymptotics, and intermittency. SIAM review, 1998, 40: 265-291.

[11.47] Yaglom A M. A N Kolmogorov as a fluid mechanician and founder of a school in turbulence research. Annual Review of Fluid Mechanics, 1994, 26: 1-23.

[11.48] von Segner J A. De figuris superficierum fluidarum (On the shapes of liquid surfaces). Commentarii Societatis Regiae Scientiarum Gottingensis (Memoirs of the Royal Scientific Society at Göttingen), 1751, 1: 301-372.

[11.49] Young T. An essay on the cohesion of fluids. Philosophical Transactions of the Royal Society of London, 1805, 95: 65-87.

[11.50] Laplace P S. Traité de Mécanique Céleste, Volume 4, Paris: Courcier, 1805. Supplément au dixième livre du Traité de Mécanique Céleste, pages 1-79.

[11.51] 赵亚溥. 表面与界面物理力学. 北京: 科学出版社, 2012.

[11.52] Richtmyer R D. Taylor instability in shock acceleration of compressible fluids. Communications on Pure and Applied Mathematics, 1960, 13: 297-319.

[11.53] Meshkov E E. Instability of the interface of two gases accelerated by a shock wave. Soviet Fluid Dynamics, 1969, 4: 101-104.

[11.54] Heger A. Going supernova. Nature, 2013, 494: 46-47.

[11.55] Lord Rayleigh. Investigation of the character of the equilibrium of an incompressible heavy fluid of variable density. Proceedings of the London Mathematical Society. 1883, 14: 170-177.

[11.56] Taylor G I. The instability of liquid surfaces when accelerated in a direction perpendicular to their planes. Proceedings of the Royal Society of London A, 1950, 201: 192-196.

[11.57] Lord Kelvin. Hydrokinetic solutions and observations. Philosophical Magazine, 1871, 42: 362-377.

[11.58] von Helmholtz H. Über discontinuierliche Flüssigkeits-Bewegungen [On the discontinuous movements of fluids]. Monatsberichte der Königlichen Preussische Akademie der Wissenschaften zu Berlin. 1868, 23: 215-228.

[11.59] O'Donoghue J, Moore L, Stallard T S, Melin H. Heating of Jupiter's upper atmosphere above the Great Red Spot. Nature, 2016, 536: 190-192.

[11.60] Plateau J. Experimental and theoretical statics of liquids subject to molecular forces only. Paris: Gauthier-Villars, 1873.

[11.61] Rayleigh L. On the instability of jets. Proceedings of the London Mathematical Society, 1878, 1: 4-13.

[11.62] Saffman P G, Taylor G I. The penetration of a fluid into a porous medium or Hele-Shaw cell containing a more viscous liquid. Proceedings of the Royal Society of London A, 1958, 245: 312-329.

[11.63] Hele-Shaw H S. Investigations of the nature of surface resistance of water and of stream line motion under certain experimental conditions. Transactions of the Institution of Naval Architects, 1898, 40: 21-46.

[11.64] Bensimon D, Kadanoff L P, Liang S, Shraiman B I, Tang C. Viscous flows in two dimensions. Reviews of Modern Physics, 1986, 58: 977-999.

[11.65] Bénard H. Les tourbillons cellulaires dans une nappe liquide. Rev. Gen. Sci. Pures Appl., 1900, 11: 1261-1271.

[11.66] Wesfreid J E. Henri Bénard: Thermal convection and vortex shedding. Comptes Rendus Mecanique, 2017, 345: 446-466.

[11.67] Cayley G. On aerial navigation: Part 1, 2 & 3. Nicholson's Journal of Natural Philosophy, 1809-1810.

[11.68] Marmanis H. Analogy between the Navier-Stokes equations and Maxwell's equations: application to turbulence. Physics of Fluids, 1998, 10: 1428-1437.

[11.69] Sridhar S. Turbulent transport of a tracer: an electromagnetic formulation. Physical Review E, 1998, 58: 522-525.

[11.70] Rousseaux G, Seifer S, Steinberg V, Wiebel A. On the Lamb vector and the hydrodynamic charge. Experiments in Fluids, 2007, 42: 291-299.

[11.71] Lucas A. An exotic quantum fluid in graphene. Science, 2019, 364: 125.

[11.72] Gallagher P, Yang C S, Lyu T, Tian F, Kou R, Zhang H, Watanabe K, Taniguchi T, Wang F. Quantum-critical conductivity of the Dirac fluid in graphene. Science, 2019, 364: 158-162.

[11.73] Berdyugin A I, Xu S G, Pellegrino F M D, Kumar R K, Principi A, Torre I, Shalom M B, Taniguchi T, Watanabe K, Grigorieva I V, Polini M, Geim A K, Bandurin D A. Measuring Hall viscosity of graphene's electron fluid. Science, 2019, 364: 162-165.

[11.74] Torre I, Tomadin A, Geim A K, Polini M. Nonlocal transport and the hydrodynamic shear viscosity in graphene. Physical Review B, 2015, 92: 165433.

[11.75] Pellegrino F M D, Torre I, Geim A K, Polini M. Electron hydrodynamics dilemma: Whirlpools or no whirlpools. Physical Review B, 2016, 94: 155414.

[11.76] Sulpizio J A, Ella L, Rozen A, Birkbeck J, Perello D J, Dutta D, Ben-Shalom M, Taniguchi T, Watanabe K, Holder T, Queiroz R, Principi A, Stern A, Scaffidi T, Geim A K, Ilani S. Visualizing Poiseuille flow of hydrodynamic electrons. Nature, 2019, 576: 75-79.

[11.77] Bourouiba L, Dehandschoewercker E, Bush J W M. Violent expiratory events: on coughing and sneezing. Journal of Fluid Mechanics, 2014, 745: 537-563.

第 12 章 连续介质力学新发展
—— 思维动力学、金融动力学、社会动力学、管理动力学

§63. 脑科学中的首张连续介质力学张量图

63.1 左脑的批判性思维与右脑的创造性思维的对比

图 63.1 为斯佩里左右脑分工理论 1981 年获得诺贝尔生理学或医学奖所发行的邮票

美国生理学家罗杰·斯佩里 (Roger Wolcott Sperry, 1913—1994) 通过著名的割裂脑实验, 证实了大脑不对称性的 "左右脑分工理论", 因此荣获 1981 年诺贝尔生理学或医学奖. 图 63.1 为该奖所发行的邮票. 该图清晰地表明, 人的左脑是抽象脑和学术脑, 而人的右脑是艺术脑和创造脑.

如图 63.2 所示, 人的左脑负责批判性思维 (critical thinking), 而人的右脑则负责创造性思维 (creative thinking), 其详细对比如表 63.1 所示.

图 63.2 左右脑分别对应于批判性和创造性思维

爱因斯坦的大脑被公认为是世界上最著名的大脑. 爱因斯坦的大脑在未经其家人允许的情况下, 1955 年对爱因斯坦进行尸检的病理学家托马斯·斯托尔茨·哈维 (Thomas Stoltz Harvey, 1912—2007, 如 63.3 左图所示) 等切成二百多片保存在福尔马林中 (如 63.3 右图所示). 小木盒里面装着 46 片载玻片, 上面均放着一小片爱因斯坦的大脑切片.

研究表明 [12.1], 爱因斯坦大脑中胶质细胞与神经元的比例高于常人, 大脑左

顶叶的胶质细胞是正常数量的两倍, 右前额叶皮层厚度高于正常值.

表 63.1 批判性和创造性思维的对比

批判性思维 (Critical Thinking)	创造性思维 (Creative Thinking)
左脑 (left brain)	右脑 (right brain)
分析性的 (analytical)	创造性的 (generative)
汇聚性的 (convergent)	发散性的 (divergent)
垂直性的 (vertical)	横向性的 (lateral)
概率型的 (probability)	可能性的 (possibility)
判断型的 (judgment)	不做判断的 (suspended judgment)
聚焦的 (focused)	弥散的 (diffuse)
客观的 (objective)	主观的 (subjective)
确定性答案 (answer)	一种可能性的答案 (an answer)
语言性的 (verbal)	视觉性的 (visual)
线性的 (linear)	关联性的 (associative)
理智的 (reasoning)	丰富、新颖型的 (richness, novelty)
批判性肯定 (yes but)	延拓性肯定 (yes and)

图 63.3 对爱因斯坦大脑进行切片的哈维 (享年 94 岁)(左) 和装有爱因斯坦大脑切片的小木盒 (右)

63.2 大脑发育过程的首张连续介质力学张量图

2000 年的世纪之交, 英国《自然》期刊的一篇文章 [12.2] 在国际上首次用连续介质力学的张量图 (continuum mechanical tensor maps) 来揭示人大脑发育过程的生长模式, 研究大脑组织结构的动态变化, 对于研究脑发育、检测病变有重要意义. 这是一个具有开创性的工作. 对于力学工作者如何将连续介质力学特别是实验力学如何应用于脑科学的研究具有重要的启发意义. 该文迄今已被 google scholar 引用超过千次.

该文中的大脑生长分布图, 与前期研究结果相比, 具有空间分辨率高的特点. 这些结果基于连续介质力学的张量计算: 根据不同时间大脑磁共振成像 (magnetic resonance imaging, MRI) 扫描结果对比, 通过图像识别方法, 获得质点位移场, 进而构建脑组织三维变形场, 即变形张量图, 生成生长分布图.

可通过不同年龄脑组织对比, 获得局部生长速率分布图. 图 63.4 通过一个青少年四年前后大脑图像对比, 得到相应生长速率. 研究发现: (1) 3—6 岁, 大脑生长速率峰值在脑两半球前部, 关于保持精神警觉和建立新行为的区域; (2) 6—11 岁, 大脑生长速率峰值出现在胼胝体, 即关于空间辨识和语言能力的区域; (3) 11 岁后, 大脑生长速率峰值明显衰减; (4) 两周时间内, 大脑生长速率几乎为零 (以 12 岁女孩为例).

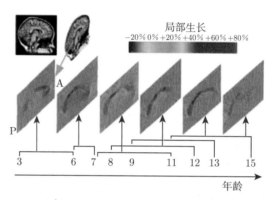

图 63.4　通过使用连续介质力学张量图检测发育中大脑的生长模式

图 63.5 和图 63.4 相类似, 通过不同年龄大脑图像的对比, 得到了大脑各个部位相应的生长速率. 对比不同时期大脑组织生长速率分布图, 以面尾区为例, 发现: (1) 3—6 岁, 大脑生长速率峰值在脑两半球前部, 关于保持精神警觉和建立新行为的区域; (2) 6—11 岁, 大脑生长速率峰值出现在胼胝体, 即关于空间辨识和语言能力的区域; (3) 11 岁后, 大脑生长速率峰值明显衰减. 以 12 岁女孩为例, 两周时间内, 大脑生长速率则几乎为零.

图 63.6 为脑组织整体生长速率分布构建图. 通过对比两个儿童四年间隔各自的脑组织生长速率, 发现 7—13 岁儿童脑组织生长主要发生在处理听觉讯息的中枢 —— 颞叶区.

图 63.7 给出了脑组织整体位移与体积变化情况. 研究表明, 脑组织的位移与体积变化信息, 共同决定脑组织的生长发育. 以 7—13 岁儿童的脑组织体积变化与位移为例. 脑组织主要体积膨胀区为脑室 (图 (a) 和图 (b)), 主要体积减少区为尾

状核区，相应位移为图 (c) 和图 (d). 尾状核负责大脑学习与记忆系统，其变形与脑整体变形相比，如图 (e) 和图 (f) 所示.

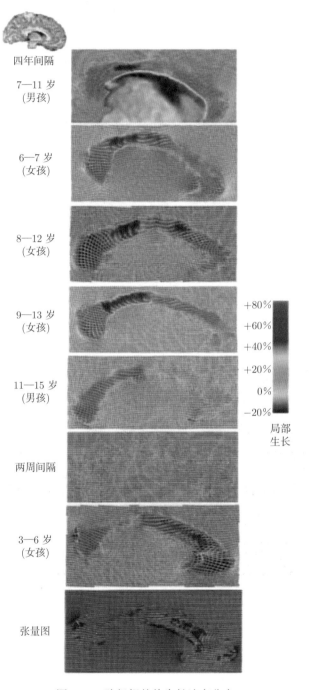

图 63.5 脑组织整体生长速率分布

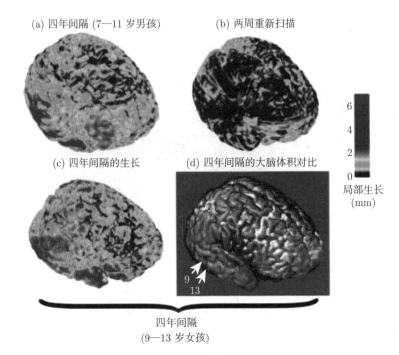

图 63.6 构建脑组织整体生长发育分布图

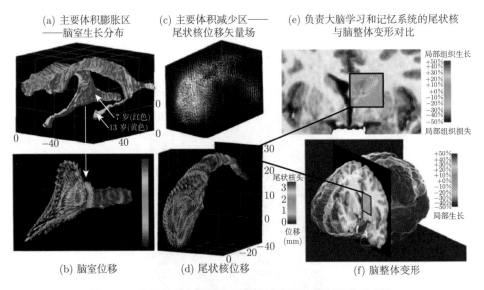

图 63.7 脑组织生长过程，不同区域有不同体积及位移变化

有关连续介质力学在大脑扩散张量成像中的系统介绍可参阅《近代连续介质力学》[1.6] 中第 32 章的内容.

63.3 思维动力学

脑科学和类脑智能技术研究早已成为各国必争战略前沿. 2013 年 4 月 2 日, 美国总统奥巴马宣布启动脑科学计划 (BRAIN Initiative), 随后欧盟几乎同时提出了脑科学研究计划, 日本紧随其后于 2014 年发起 Brain/MINDS 计划.

中国随后计划中的 "脑计划" 是一个 "一体两翼" 结构. 其中, "一体" 是指主体是基础研究, 理解人类大脑的认知功能是怎么来的. 就像我们看到计算机, 要分析它的功能就必须知道计算机的结构, 对于大脑的功能我们必须要知道大脑的网络结构, 这就叫做 "全脑介观神经联接图谱"; 而 "两翼", 一是指如何诊断和治疗重要的脑疾病, 人类大脑大约有 1000 亿个神经元, 它们如何连接、连接错误导致精神错乱或是出现严重的神经性疾病, 目前人类并没有弄清楚其中的奥秘. 随着全球人口老龄化时代的到来, 阿尔茨海默综合症、帕金森综合症等神经衰退性疾病的盛行, 人类迫切地希望知道: 大脑是如何工作的; 二是指发展人工智能与脑科学结合的脑机智能技术, 全力推动人工智能与脑科学的融合发展.

1992 年 3 月, 钱学森就提出 "要创建思维动力学". 应该客观地讲, 迄今思维动力学的范式仍未形成.

事实上, 不同领域的学者在不同的层次上在开展相关研究. 量子神经动力学认为, 意识是有序的量子过程. 意识的物质基础是相干波的干涉与全息.《力学讲义》[1,11] 在 §60 中已经介绍过欧亚鸲眼睛中的量子纠缠态以及量子罗盘. 俗话说: "眼睛是心灵的窗户", 人的意识和思维中是否存在量子纠缠态是一个十分有趣的问题.

脑电图又称为脑波图 (electroencephalography, EEG, 如图 63.8 所示) 是透过医学仪器脑电图描记仪, 将人体脑部自身产生的微弱生物电于头皮处收集, 并放大记录而得到的曲线图. EEG 所测量的是众多锥体细胞兴奋时的突触后电位的同步总和, EEG 测量来自大脑中神经元的离子电流产生的电压波动. EEG 早已用于辅助诊断脑部相关疾病. 脑波的种类见表 63.2 和图 63.9.

根据美国睡眠医学学会 (American Academy of Sleep Medicine, AASM) 在 2007 年修改的标准, 睡眠的四阶段特征如下:

阶段 I, 脑电从清醒状态的 alpha (α) 波转为 theta (θ) 波, 这个阶段可以描述为昏昏欲睡. 阶段 I 是睡眠的起始阶段, 在这个阶段, 肌肉开始放松, 人开始失去意识.

阶段 II 过程中, 脑电频率约为 12—16 Hz, 在这个阶段, 肌电比阶段 I 降低, 意识完全失去. 这一阶段约占睡眠的 45%—55% 时间.

阶段 III, delta (δ) 波，也称 δ 节律，0.5—4 Hz，构成阶段 III 50% 以下的脑电组成. 这一阶段是部分的深度睡眠，并向阶段 IV 过渡. 梦魇，梦游，尿床和梦话会在此阶段出现.

阶段 IV, delta (δ) 波占据 50% 以上的时间，是深度睡眠.

大量的研究表明，正常人的 EEG 具有明显的混沌特征.

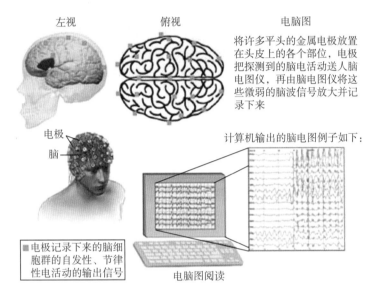

图 63.8　EEG 的采集

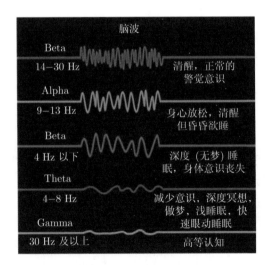

图 63.9　脑波种类示意图

表 63.2　脑波种类的详细说明

脑波种类		频率	人体特征
delta (δ)		0.1—3 Hz	深度睡眠且没有做梦时
theta (θ)		4—7 Hz	成人情绪受到压力时, 尤其是失望或挫折
alpha (α)		8—12 Hz	放松、平静、闭眼但清醒时
beta (β)	低范围	12.5—16 Hz	放松但精神集中
	中范围	16.5—20 Hz	思考、处理接收到外界讯息 (听到或想到)
	高范围	20.5—28 Hz	激动、焦虑
gamma (γ)		25—100 Hz (通常在 40 Hz)	提高意识、幸福感、减轻压力、冥想
lambda (λ)		诱发电位	眼睛受光刺激时 100 ms 后诱发 (又称作 P100)
P300		诱发电位	看到或听到脑中想像的东西时 300 ms 后诱发

思维过程具有自组织 (self-organization) 和涌现 (emergence) 的特征. 逻辑思维是大脑内神经元自组织活动的吸引子序列, 但这吸引子序列又是不规则的, 沿着错综复杂的曲线发展的, 它有复杂的时空花样. 这种吸引子序列又构成一个大的吸引子; 形象思维是大脑内神经元自组织活动的奇怪吸引子序列, 这种奇怪吸引子序列有更复杂的演化曲线和时空花样. 奇怪吸引子又称为混沌吸引子, 它具有分形的特征.

大量研究同时表明, 意识流也具有混沌的特征.

§64. 连续介质力学在人工智能中的应用
—— 张量流和张量网络

64.1　麦卡锡是如何受冯·诺依曼启发创始人工智能这一学科的?

在约翰·麦卡锡 (John McCarthy, 1927—2011, 1971 年图灵奖获得者) 在加州理工学院 (Caltech) 获得数学本科学位的那一年, 他参加了 1948 年 9 月 20—25 日在加州理工学院召开的 "Hixon 行为中的大脑机制研讨会" (1948 Hixon Symposium on Cerebral Mechanisms in Behavior). 研讨会上的演讲者代表了数学、心理学和新兴计算科学的交叉领域.

冯·诺依曼 (John von Neumann, 1903—1957) 作为 20 世纪最重要的数学家之一、"计算机之父" 和 "博弈论之父" 在这次研讨会上做了题为《自动机的一般逻辑理论》(The General and Logical Theory of Automata) 的报告. 冯·诺依曼的报告有如下几个主题: (1) 初步考虑 (Preliminary Considerations); (2) 计算机某些相关特性的讨论 (Discussion of Certain Relevant Traits of Computing Machines); (3) 计算机与生

物活体的比较 (Comparisons between Computing Machines and Living Organism); (4) 自动机的未来逻辑理论 (The Future Logical Theory of Automata); (5) 数字化的原则 (Principles of Digitalization); (6) 正式的神经网络 (Formal Neural Networks); (7) 复杂性和自我复制的概念 (The Concept of Complication and Self-Reproduction).

冯·诺依曼博学的、极其精彩的演说震惊了听众和其他演讲者人. 麦卡锡被迷住了. 从那时起, 他对开发能像人一样思考的机器的想法很感兴趣, 他受冯·诺依曼演讲的指引, 走上了他余生将要走的学术道路 (Von Neumann's erudite tour de force stunned audience members as well as fellow presenters. McCarthy was captivated. Now intrigued with the idea of developing machines that could think as people do, he was set upon the path which he would follow for the rest of his life). 现在, 麦卡锡被誉为 "人工智能之父", 可见, 冯·诺依曼就是人工智能学科当之无愧的 "启蒙者" 和 "引路人".

在获得数学学士学位后, 麦卡锡又继续在加州理工学院读了一年的研究生, 后来在普林斯顿大学完成了博士学位的剩余部分, 他认为普林斯顿是更适合研究数学的大学. 在访问普林斯顿高等研究院期间, 麦卡锡与热情的冯·诺依曼进行了一次讨论, 他分享了自己关于有限自动机交互作用的想法 —— 这一想法受到冯·诺依曼在 1948 年 Hixon 研讨会上演讲的启发. 尽管冯·诺依曼鼓励麦卡锡把这些想法写进论文, 但麦卡锡从未这样做过. 然而, 这些想法在接下来的十年里继续发展, 并以一种经过修改的形式出现在麦卡锡早期的人工智能论文中, 甚至 Lisp 编程语言本身. 麦卡锡 1951 年获得普林斯顿大学数学博士学位.

完成博士论文后, 受信息论的创始人克劳德·香农 (Claude Shannon, 1916—2001) 邀请, 麦卡锡和他的朋友马文·明斯基 (Marvin Minsky, 1927—2016, 1969 年图灵奖获得者) 在新泽西州的贝尔实验室工作一个夏天. 麦卡锡和香农合作编纂了一卷名为《自动机研究》(Automata Studies) 的论文集, 但麦卡锡最终还是觉得有些失望, 因为几乎没有论文涉及他的主要兴趣: 机器智能. 几年后, 他有机会提出一个夏季研究项目来解决这个问题, 他和 IBM 的信息研究主管向香农和明斯基提出了这个项目. 他们同意了, 一年后在新罕布什尔州达特茅斯校园举办了第一次人工智能研讨会.

1955 年, 麦卡锡担任了达特茅斯学院数学系的助理教授, 该年他向洛克菲勒基金会申请举办 "达特茅斯研讨会" (Dartmouth Workshop) 的会议资助费, 他在申请书中第一次使用了 "Artificial Intelligence (AI)" 这个词! 麦卡锡在 2006 年回忆道:"当我要写申请书时,我想到了这个名字,这个申请是为了得到洛克菲勒基金会对这次研讨会的资助. 说实话,取这个名称的原因是,我考虑的是参与者而不是资助者 (I came up with the name when I had to write the proposal to get research support for the conference from the Rockefeller Foundation. And to tell you the truth, the reason for the name is, I was thinking about the participants rather than the funder)." 这份题为《达特茅斯夏季人工智能研究项目建议书》(*A Proposal for the Dartmouth Summer Research Project on Artificial Intelligence*) 的建议人分别是 (如图 64.1 所示):麦卡锡、哈佛大学的明斯基,IBM 公司的罗切斯特 (Nathaniel Rochester, 1919—2001, IBM 701 总设计师) 和信息论的创始人香农.

图 64.1 人工智能的四位创始人

洛克菲勒基金会决定拨付 7000 美元资助 1956 年达特茅斯研讨会.

1956 年 6 月 18 日—8 月 17 日,"达特茅斯夏季人工智能研究计划" (Dartmouth Summer Research Project on Artificial Intelligence) (简称 "达特茅斯研讨会") 在达特茅斯学院举行. 该次达特茅斯研讨会标志着人工智能的诞生.

AI 的发展历程可大致划分为以下六个阶段:

一是起步发展期:1956 年—20 世纪 60 年代初期. AI 概念提出后,相继取得了一批令人瞩目的研究成果,如机器定理证明、跳棋程序等,掀起了 AI 发展的第一个高潮;

二是反思发展期:20 世纪 60 年代—70 年代初期. AI 发展初期的突破性进展大大提升了人们对 AI 的期望,人们开始尝试更具挑战性的任务,并提出了一些不切实际的研发目标. 然而,接二连三的失败和预期目标的落空,例如,无法用机器证明两个连续函数之和还是连续函数、机器翻译闹出笑话等,使 AI 的发展走入低谷.

三是应用发展期:20 世纪 70 年代初期—80 年代中期. 20 世纪 70 年代出现的

专家系统模拟人类专家的知识和经验解决特定领域的问题，实现了 AI 从理论研究走向实际应用、从一般推理策略探讨转向运用专门知识的重大突破. 专家系统在医疗、化学、地质等领域取得成功，推动 AI 走入应用发展的新高潮.

四是低迷发展期：20 世纪 80 年代中期—90 年代中期. 随着 AI 应用规模的不断扩大，专家系统存在的应用领域狭窄、缺乏常识性知识、知识获取困难、推理方法单一、缺乏分布式功能、难以与现有数据库兼容等问题逐渐暴露出来.

五是稳步发展期：20 世纪 90 年代中期—2010 年. 由于网络技术特别是互联网技术的发展，加速了 AI 的创新研究，促使 AI 技术进一步走向实用化. 1997 年 IBM 深蓝超级计算机战胜了国际象棋世界冠军卡斯帕罗夫，2008 年 IBM 提出"智慧地球"的概念. 以上都是这一时期的标志性事件.

六是蓬勃发展期：2011 年至今. 随着大数据、云计算、互联网、物联网等信息技术的发展，泛在感知数据和图形处理器等计算平台推动以深度神经网络为代表的 AI 技术飞速发展，大幅跨越了科学与应用之间的"技术鸿沟"，诸如图像分类、语音识别、知识问答、人机对弈、无人驾驶等人工智能技术实现了从"不能用、不好用"到"可以用"的技术突破，迎来爆发式增长的新高潮.

AI 可划分为弱人工智能和强人工智能两种.

弱人工智能 (weak AI) 亦称为应用人工智能 (applied AI) 或窄人工智能 (narrow AI, artificial narrow intelligence, ANI). ANI 只处理特定的问题和任务. 弱人工智能不需要具有人类完整的认知能力，甚至是完全不具有人类所拥有的感官认知能力，只要设计得看起来像有智慧就可以了. 面向特定任务 (如下围棋) 的弱人工智能系统由于任务单一、需求明确、应用边界清晰、领域知识丰富、建模相对简单，形成了 AI 领域的单点突破，在局部智能水平的单项测试中可以超越人类智能. AI 的近期进展主要集中在弱人工智能领域. 例如，阿尔法狗 (AlphaGo) 在围棋比赛中战胜人类冠军，人工智能程序在大规模图像识别和人脸识别中达到了超越人类的水平，人工智能系统诊断皮肤癌达到专业医生水平，等等.

强人工智能 (strong AI) 或通用人工智能 (artificial general intelligence, AGI) 是具备与人类同等智慧、或超越人类 AI，能表现正常人类所具有的所有智能行为. AGI 是 AI 研究的主要目标之一.

AGI 尚处于起步阶段，离现实还很远 (AGI is nowhere close to being a reality). 人的大脑是一个强大的通用智能系统，能举一反三、融会贯通，可处理视觉、听觉、判断、推理、学习、思考、规划、设计等各类问题，可谓"一脑万用". 真正意义上完备的 AI 系统应该是一个通用的智能系统. 当前的人工智能系统在信息感知、机器学习等"浅层智能"方面进步显著，但是在概念抽象和推理决策等"深层智能"方面的能力还很薄弱. 总体上看，目前的人工智能系统可谓有智能没智慧、有智商没情商、会计算不会"算计"、有专才而无通才. 因此，人工智能依旧存在明显的局限性，依然还有很多"不能"，与人类智慧还相差甚远.

2016 年 10 月 13 日，美国白宫科技政策办公室 (The Office of Science and Technology Policy, OSTP) 下属国家科学技术委员会 (The National Science and Technology Council, NSTC) 发布了《为人工智能的未来做好准备》(*Preparing for the Future of Artificial Intelligence*) 和《国家人工智能研究与发展战略计划》(*National Artificial Intelligence Research and Development Strategic Plan*) 两份报告. 前者探讨了 AI 的发展现状、应用领域以及潜在的公共政策问题；后者则提出了美国优先发展的 AI 七大战略方向及两方面建议，并提出在美国的 AI 中长期发展策略中要着重研究强人工智能.

AlphaGo 系统开发团队创始人戴密斯·哈萨比斯 (Demis Hassabis, 1976—) 提出朝着"创造解决世界上一切问题的通用人工智能"这一目标前进 (toward the creation of artificial general intelligence (AGI), it will solve all sorts of problem).

2019 年 3 月 27 日，国际计算机学会 (ACM) 宣布，有"深度学习三巨头"之称的约书亚·本吉奥 (Yoshua Bengio, 1964—)、杰弗里·辛顿 (Geoffrey Hinton, 1947—) 和杨乐昆 (Yann LeCun, 1960—) 共同获得了 2018 年的图灵奖，以表彰这三位深度学习之父给人工智能带来的重大突破，这些突破使深度神经网络成为计算的关键组成部分.

早在 1970 年，Wong 和 Bugliarello 就发表了题为《连续介质力学中的人工智能》[12.3] 的文章，作者在本文中介绍了一种将方程拆分计算的程序 CONFORM，程序的执行过程是将微分、积分以及张量点乘、叉乘等基本初等运算都以程序块的形式进行编程，在具体计算的时候根据需要按次序调用，这样就可以避免每次计算时都单独编程. 严格地来说，该文的思想并不属于如图 64.2 所示的人工智能的范畴，这是一种典型的面向对象编程：建立基本运算程序库，然后在需要的时候从库中调用程序，以此来提高程序的复用性，减少程序员的编程量. 这与 AI 最大的不同之处在于它不具有"学习"的能力，程序不能根据变化灵活的应对其领域内的问题，只能在程序库的框架内运行，程序库的丰富与否直接决定了整个程序的能力强弱. 该文虽然未在学术界产生重要的影响，但是其编程思想还是非常先进的，其前瞻性是值得肯定的 [12.4].

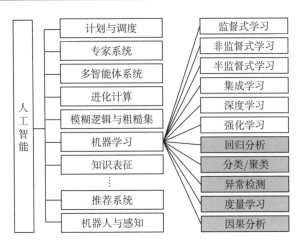

图 64.2 人工智能和机器学习的分类

64.2 机器学习在材料设计中的应用

图 64.3 刊登有从失败中学习的《自然》封面论文

2016 年 5 月 5 日，英国《自然》期刊发表了一篇"机器学习"算法改变材料发现方式的题为 Machine-learning-assisted materials discovery using failed experiments [12.5] 的封面文章 (如图 64.3 所示)，并提出"从失败中学习"。哈佛福德学院和普渡大学的研究人员利用机器学习算法，用失败或不成功的实验数据预测了新材料的合成，并且在实验中机器学习模型预测的准确率超过了经验丰富的化学家，这意味着机器学习将改变传统材料发现方式，发明新材料的可能性也大幅提高。

《自然》同期的新闻专稿栏目以《材料密码》(the material code) [12.6] 为题讨论了这一问题。一部分科研人员认为人工智能将给材料科学带来革命性的改变。该文章认为，通过计算机建模和机器学习技术，可以很快地根据人们所需要的性能预测出相应候选材料。因此，科学家们将不再需要"瞎猫撞死耗子"般地制作新材料，而是按照计算机计算结果的指导，制作出相应候选材料并测试，从而加快了新材料的研发的速度和效率。不过，人工智能变革材料科学研究方法尚存在不少问题：一是受制于材料数据、材料性能控制因素、计算能力，目前仅对少部分材料奏效。二是要计算机预测的材料，并不一定能够在实验室成功合成乃至规模量产，这个过程可能会很长。

64.3 深度学习

如图 64.4 左图所示，2016 年《自然》期刊封面文章 [12.7] 介绍在汉城围棋人机大战中击败李世石的 AlphaGo Lee 深度学习算法。该算法是基于监督式学习的卷积神经网络算法，算法训练分为三个部分：(1) 策略网络监督式学习阶段 (supervised learning of policy networks),在算法与人类棋手对弈的过程中,使算法能够尽可能的模拟人类棋手的行为,此阶段可以理解为是算法对围棋基本规则和技巧的学习;(2) 策略网络强化学习阶段 (reinforcement learning of policy networks),在监督式学习的基础上,算法开始自主对弈,并对其落子的策略网络进行优化,在这个阶段,算法不再以对人类棋手行为的模仿为目标,它的落子以棋局的输赢为目标,开始学习自己的落子策略;(3) 价值网络强化学习阶段 (reinforcement learning of value networks),在价值网络学习过程中,算法会对落子后棋局整体局势进行自我判断,并在下一步落子的模拟中预测获胜的概率,在所有的模拟落子情况中选择出最优解.

图 64.4 2016 年 1 月 28 日《自然》期刊封面 (左) 和 2018 年 12 月 7 日的《科学》期刊封面 (右)

2017 年,《自然》期刊上发表的一篇封面长文 [12.8] 震惊了整个世界. 该文报道, 刚刚开始时 AlphaGo Zero 的棋艺糟透了, 后来它逐渐成为一名缺乏经验的业余棋手, 最终进阶为围棋高手, 能够走出极具战略性的棋步. 这些进步仅花费了几天时间. 最初 10 小时内它就发现了一个定式. 随后不久它又领悟了一些棋法. 三天后, AlphaGo Zero 发现了人类专家正在研究的全新棋步.

仅三天时间, 具有自学能力的 AlphaGo Zero 自行掌握了围棋的下法, 还发明了更好的棋步. 这期间, 除了被告知围棋的基本规则, 它未获得人类的帮助. 经过三天的训练, 该系统能够击败 AlphaGo Lee, 后者是 2016 年击败了韩国选手李世石的 DeepMind 软件, 胜率是 100 比 0. 经过大约 40 天的训练, AlphaGo Zero 击败了 AlphaGo Master.

AlphaZero 下象棋的水平高于国际象棋大师们, 风格在国际象棋历史上闻所未闻. 对它来说, 区区几个小时的自我对弈后, 就达到了人类需要 1500 年才能达到的技能水平. AlphaZero 只被告知游戏的基本规则. 它的自我学习过程不涉及任何人类或是人为产生的数据.

2017 年 1 月, 美国卡耐基梅隆的科学家们制造人工智能软件, 在一对一的德州扑克对战中, 战胜了人类选手, 人类选手在整个过程中, 几乎没有胜算. 而 2016 年年底, 升级的 AlphaGo 也战胜了中国的顶尖围棋手. 对人脑神经网络的模拟, 已成为脑科学研究的一个出口. 不少脑科学研究都与类脑研究关系密切.

2018 年 12 月 7 日, 谷歌旗下的人工智能实验室 DeepMind 研究团队在《科学》期刊上发表封面论文 [12.9] (如图 64.4 右图), 公布了通用算法 AlphaZero 和测试数据. 《科学》期刊评价称, 通过单一算法就能够解决多个复杂问题, 是创建通用的机器学习系统、解决实际问题的重要一步.

2017 年 10 月 18 日, DeepMind 团队公布了最强版阿尔法狗, 代号 AlphaGo Zero. 彼时 DeepMind 表示, 棋类 AI 的算法主要基于复杂的枚举, 同时需要人工进行评估, 人们在过去几十年内已经将这种方法做到极致了. 而 AlphaGo Zero 在围棋中的超人表现, 则是通过与自己下棋练习出来的. 现在 DeepMind 研究团队将这种方法推广到 AlphaZero 的算法中, AlphaZero 最长花了 13 天 "自学成才", 随后与世界冠军级的棋类 AI 对决: (1) 在国际象棋中, AlphaZero 在 4 个小时后首次击败了第九季 TCEC 世界冠军 Stockfish; (2) 在日本将棋中, AlphaZero 在 2 小时后击败了将棋联盟赛世界冠军 Elmo; (3) 在围棋上, AlphaZero 经过 30 个小时的鏖战, 击败了李世石版 AlphaGo.

AlphaZero: 一个算法通吃三大棋类.

以上可以说明两点: 一是具有智力的机器人融入人类社会是必然的, 我们必须对此做好充分的准备, 但是人类不必惊恐; 二是脑本质上是一个庞大且复杂的信息处理系统, 据估计, 人脑中每秒完成的动态链接高达千万次量级, 可储存的信息量相当于美国国会图书馆藏书总量所包含信息的 50 倍. 不仅如此, 脑科学已在若干组构层次上, 揭示了大脑信息处理与传统计算机迥然不同的特点: 平行信息处理、神经元间信息的交互性传递、信息处理的高度可塑性等. 因此, 借鉴脑的运行原理来推进人工智能的研究, 有着广阔的前景.

自从 2016 年 AlphaGo 完胜李世石, 深度学习火了, 36 岁的李世石也主要是因为围棋 AI 不可战胜而于 2019 年 11 月宣布退役. 但似乎很少有人说得清深度学习的原理, 只是把它当作一个黑箱来使. 有人说, 深度学习就是一个非线性分类器. 有人说, 深度学习是对人脑的模拟.

深度学习是指多层神经网络上运用各种机器学习算法解决图像，文本等各种问题的算法集合. 深度学习从大类上可以归入神经网络, 不过在具体实现上有许多变化. 深度学习的核心是特征学习, 旨在通过分层网络获取分层次的特征信息, 从而解决以往需要人工设计特征的重要难题. 深度学习是一个框架, 包含多个重要算法:

- 卷积神经网络 (convolutional neural networks, CNN)
- 自动编码器 (auto-encoder)
- 稀疏编码 (sparse coding)
- 限制性玻尔兹曼机 (restricted boltzmann machine, RBM)
- 深信度网络 (deep belief networks, DBN)
- 多层反馈循环神经网络神经网络 (recurrent neural network, RNN)

对于不同问题 (图像、语音、文本), 需要选用不同网络模型才能达到更好效果.

此外, 最近几年强化学习 (reinforcement learning) 与深度学习的结合也创造了许多了不起的成果, AlphaGo 就是其中之一.

64.4 张量流

张量流 (TensorFlow)[12.10] 是谷歌基于 DistBelief 进行研发的第二代人工智能学习系统, 其命名来源于本身的运行原理. Tensor (张量) 意味着 N 维数组, Flow (流) 意味着基于数据流图的计算, TensorFlow 为张量从流图的一端流动到另一端计算过程. TensorFlow 是将复杂的数据结构传输至人工智能神经网中进行分析和处理过程的系统.

TensorFlow 可被用于语音识别或图像识别等多项机器学习和深度学习领域, 对 2011 年开发的深度学习基础架构 DistBelief 进行了各方面的改进, 它可在小到一部智能手机、大到数千台数据中心服务器的各种设备上运行. TensorFlow 将完全开源, 任何人都可以用.

64.5 维度诅咒与张量网络

在研究量子力学多体问题和人工智能神经元连接问题时往往会产生高阶张量 (在量子力学中维度可达到 $2^{10^{23}}$), 张量传统的爱因斯坦指标记法在解决这种问题时书写和计算都极不方便, 这个问题被理查德 · 贝尔曼 (Richard Ernest Bellman, 1920—1984) 在 1957 年出版的《动态规划》(*Dynamic Programming*) 一书的前言中称作张量的维度诅咒 (the curse of dimensionality). 沃尔特 · 科恩 (Walter Kohn, 1923—2016, 1998 年诺贝尔化学奖获得者) 在其诺贝尔奖获奖致辞中, 提出了和维数诅咒内容大致相同的指数墙 (exponential wall) 的概念，都是说在研究量子多体问题时，描述系统的总参数指数增长为天文数字无法求解的难题.

张量网络 (TensorNetwork) 就是为了解决维度诅咒或指数墙难题应运而生的.

20 世纪 50 年代，学术界出现了一股研究浪潮，试图量化并提供各种科学领域的几何模型，包括生物学和物理学. 生物学的几何化目的是将生物学的概念和原理简化为类似于物理学几十年前所做的几何概念那样. 当代物理学的几何化，最著名的就是广义相对论，爱因斯坦场方程的几何化，允许使用黎曼空间流形，将物体的运动轨迹建模为最佳路径，也就是测地线曲线，方程为 (10-6) 式或 (16-33) 式.

生物学和物理学几何化发展并行，含盖了人口、疾病暴发和进化等许多领域。同时，神经科学的几何化取得了一些进展. 为了更严格地研究它们，对脑功能进行量化变得越来越必要.

1979 年，Pellionisz 和 Llinas 由一个外在的张量网络来描述和建模内在的多维中枢神经系统，构建了大脑模型，张量网络方法由此提出 [12,11]. 张量网络方法的核心思想就是对复杂问题几何化，形象地表现各张量间的联系进而进行简化，必要时可以通过分解的方式将问题再次分解成便于求解的形式进行特征提取. 文中假设：(1) 神经元网络活动是矢量的；(2) 网络本身是张量组织的，大脑功能可以被量化并简单地描述为张量网络.

张量网络方法的另一个重要的应用领域就是在解决量子力学多体问题，量子力学问题中，随着量子比特的增加系统的密度矩阵阶次呈指数增长，这个问题方程的书写甚至求解带来极大不便，张量网络方法可以以几何的方式直观简介的表达多体问题中量子比特的纠缠等问题，并对简化求解提供便利.

利用张量网络方法发展出了许多优秀的算法，其中矩阵乘积态算法 (matrix product states, MPS) 应用最为广泛，该算法可以将高阶张量网络分解为许多低阶张量的组合，这样可以将总计算量由指数型增加降为线性增加，极大地提高了计算效率. MPS 算法源于人们对密度矩阵重整化群 (density matrix renormalization group, DMRG) 的原理探究，将高维张量分解为张量网络内的多个小的张量组合，这样张量的总参量增加时，总计算量线性增加而呈非指数增加. 例如对于 N 体问题中，单体有 d 阶，描述 N 体问题的张量需要 d^N 个参数 MPS 分解后，需要 Ndm^2 个参数，这里的 m 是内腿的数量，可见参数与 N 呈线性关系，而非指数关系.

张量网络方法与基于概率密度分布的玻尔兹曼机 (Boltzmann machine) 工作模式完全一致，该方法在人工智能领域的特征提取问题 (如图像识别、语音识别等) 中性能优异.

§65. 连续介质力学在社会动力学、金融动力学和管理学动力学中的应用

65.1 社会心理学中的场论和生活空间

复杂环境中人的行为，不规则且难以预测，社会心理学"场论"的创始人——库尔特·勒温 (Kurt Lewin, 1890—1947)，最早开展了有关群体动力学 (group dynamics) 的研究.

勒温在他的已被引用了 7000 多次的论文[12.12]中提出了矢量、动力场、拓扑心理学和生活空间 (life space) 等许多新概念，形成了他独创的心理学理论. 团体动力理论和场论是他对心理学理论的杰出贡献. 他把人的行为类比于气体分子运动，认为社会场的各个社会作用力是行为变化的操控者. 最著名之处还是他的力场分析法以及力场分析图. 正如个人在其生活空间里形成心理场一样，团体与其环境形成社会场. 团体的特点是有成员的动力相互依存性. 一个人的地位取决于他的区域，而他的区域又同别的区域 (团体成员) 相联系. 团体受制于内聚力和瓦解力. 当成员间阻隔交流的障碍太大时，便产生瓦解力. 团体构成一力场，个体之间或吸引或排斥，取决于团体内的引拒值.

勒温的场论最基本的概念是生活空间 (life space, 简称 Lsp)，也就是在特定时间所有个人心理现实 (psychological reality) 整体，包括个人 (person, 简称 P, 个人因素包括遗传、能力、性格、动机、情绪、健康状况等) 与环境 (environment, 简称 E, 环境因素则包括社会的与自然的一切条件) 两大部分，而且都被不同渗透型的边界 (boundaries) 划分为可能相互影响的区域 (regions)，如个人有需求、目标、希望、抱负等；环境则有家庭、友谊、职业、规范、禁忌等. 这些区域间只有相对位置或次序关系，没有距离或大小之别，拓扑心理学 (topological psychology) 就因此而得名.

勒温还将化合价 (valence) 的概念引入社会心理学的场论中，化合价是一种元素的一个原子与其他元素的原子化合即构成化合物时表现出来的性质. 勒温在场论中说每个生活区域都有潜在的引力或化合价，当个人有了需求或者意图时，相关区域或系统就会形成一种内在的张力 (tension)，并产生一股变化的力矢量 (force vector)，当合力不为零时便发生所谓的区域移位 (locomotion)，因而形成一种个人生活空间的路径.

生活空间是个人与环境相互作用的结果,而个人的行为又取决于特定的生活空间,所以个人的行为是个人与环境的函数. 这种个别差异与整体情境兼顾的解释行为观点,对后来心理学的发展影响巨大.

综上,库尔特·勒温认为,人作为一个场 (field),其行为 (B) 取决于个体的生活空间 ($L_{\rm sp}$). 通过勒温的基本公式可表示为

$$B = f(P, E) = f(L_{\rm sp}) \tag{65-1}$$

此外,勒温利用拓扑学和向量分析来解释心理现象. 他认为,拓扑学有助于了解个体在某个特定空间可能或不可能发生的事件;向量分析可以表明个体在某种情境里可能作出的各种行为有哪些将会成为现实.

在勒温的系统中,"场"这个术语的内涵与一般的解释略有不同. "场"不仅仅指知觉到的环境,而且还包括认知意义. 它既包括物质环境中的某些事件 (即被知觉到的物质环境),也包括个人的信念、感情和目的等. 简言之,勒温探讨的是认知场和知觉场. 他之所以借用"场论"的概念,是因为他认为它是一种分析关系的起因和建立科学体系的方法[12.13].

勒温的场论旨在预测个体的动机行为. 他认为,答案就在"生活空间"这个概念中. 如前所述,生活空间包括个体以及他的心理环境,生活空间是"决定个体在某一时间里的行为的全部事件的总和"[12.14]. 对生活空间的理解,关键在于如何理解"心理环境" (psychological environment) 这个概念.

图 65.1 给出了生活空间的示意图,心理环境十分重要. 人要实现其目标 (goal),总是存在着障碍 (barrier),两个矢量:驱动力 (driving force) 和抑制力 (restraining force),分别朝向和背向所需要实现的目标. 在生活空间的外部有一个用虚线表示外壳 (foreign hull),表示生活空间的渗透型和可扩展性.

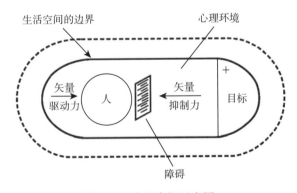

图 65.1 生活空间示意图

图 65.2 给出了更为精细的生活空间的示意图 [12.15]，和图 65.1 不同之处是对人的边界 (boundary of person) 又进行了划分，包括人的能力 (abilities) 和三个需要 (needs)，这三个需求分别用狄拉克符号表示，它们分别是：对历史求和 (sum over histories)、对场求和 (sum over fields) 以及对几何或拓扑求和 (sum over geometries/topologies)。

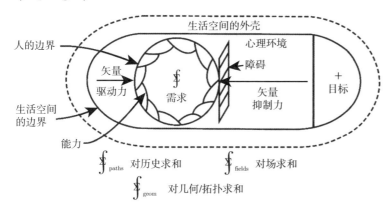

图 65.2 更为精细的生活空间示意图

65.2 连续介质力学在社会动力学中的应用

65.2.1 孔德所创立的社会物理学

1856 年，法国社会哲学家奥古斯特·孔德 (Auguste Comte, 1798—1857) 创立了 "社会物理学"(social physics or sociophysics)[12.16]。孔德将社会物理学定义为："社会物理学是一门研究社会现象的科学，社会现象与天文、物理、化学和生理现象一样，也服从自然和不变的法则，找出控制社会现象的法则就是其研究的一个极其重要的目标 (Social physics is that science which occupies itself with social phenomena, considered in the same light as astronomical, physical, chemical, and physiological phenomena, that is to say as being subject to natural and invariable laws, the discovery of which is the special object of its researches)."

正如法国学者雷蒙·阿隆 (Raymond Aron, 1905—1983) 在其所出版的《社会学主要思潮》[12.17] 一书中所评述的："静力学和动力学是奥古斯特·孔德的社会学的两大部分……社会静力学揭示人类社会的基本秩序；社会动力学则叙述这一基本秩序到实证主义这一最终阶段之前，所经过的曲折历程。"

简而言之，孔德是用牛顿力学中的静力学来分析社会结构，用动力学来分析社会演化.

英国哲学家和社会学家斯宾塞 (Herbert Spencer, 1820—1903) 进一步发展了孔德的思想, 形成了具体的"社会有机论"(social organism), 并于 1851 年出版了《社会静力学》(*Social Statics*) 一书. 所谓"静力学", 其目的就是要获得一个"完善社会的平衡状态".

斯宾塞在全书的开始论述了"大多数人的最大幸福"(the greatest happiness of the greatest number) 这样一个当时社会所热衷、追求的概念. 作者认为, 幸福就是指机体的所有功能都得到满足. 斯宾塞指出, 最大幸福只能间接的去寻求. 在对"最大幸福"的论述中, 斯宾塞还引出了有深远影响的"第一原理"(first principle): "每个人都有权要求运用它各种机能的最充分的自由, 前提是与所有其他人的同样自由不发生矛盾 (Every man has freedom to do all that he wills, provided he infringes not the equal freedom of any other man)." 因此, 斯宾塞提出了要实现最大幸福的四个条件: (1) 公正 (justice): 不减少他人活动范围, 保证他人的最大幸福; (2) 消极的善行 (negative beneficence): 不减少其他人的幸福; (3) 积极的善行 (positive beneficence): 为他人幸福的创造条件; (4) 最大限度追求自己的幸福.

斯宾塞的上述被称为同等自由法则的"第一原理"的提出具有深刻意义, 他是斯宾塞社会有机论的重要表现. 作为社会唯实论的支持者, 斯宾塞指出社会是一个由互相联系的社会成员所组成的. 各个成员在权利上互相制约, 每个人的权利或者自由的实现都是在一定的社会当中的所以它必然要牵涉到其他人的权利或自由, 因此斯宾塞认为"在不侵犯其他人权利的前提下实现自己最大的权利"是社会的"第一原理". 斯宾塞的理论对 19 世纪到 20 世纪的社会学领域影响是很大的, 他对社区、家庭、宗教、分工等问题的探讨, 一直到今天还是社会学研究的重要内容.

斯宾塞在其《社会静力学》中的最后一句话, 深刻地总结了个人与整体的关系: "在一切人都自由以前, 没有任何人能完全地自由; 在一切人都有道德之以前, 没有任何人能完全地有道德; 在一切人都幸福以前, 没有任何人能完全地幸福 (No one can be perfectly free till all are free; no one can be perfectly moral till all are moral; no one can be perfectly happy till all are happy)."

在当今社会, 社会物理学则必须充分挖掘大数据分析的作用.

65.2.2 社会作用力模型和 HMFV 模型

已有研究认为, 当行人感知所处的环境, 并根据个人的目标处理、评估感知信息时, 会最终做出行动的反应. 在模拟街道行人的集体行为以及建筑物内的行人如何寻找出口的行人动力学方面，主要有两大类人群行为模型：一类是离散空间模型或者元胞自动机模型，将行人看成是一个固定的或者自适应网格中具有相应位置的节点，而且这些行人所处位置的坐标通过离散的时间间隔来进行更新；另一类是连续空间模型，它又分成以下几种类型：第一种是将人群运动与液体或者气体的运动相类比而建立的流体动力学；第二种是在模拟过程中准许行人根据某种成本函数来选择最佳路径；第三种是前两种模型的综合，同时考虑流体动力学方法和成本函数. 下面主要以社会力相互作用的连续空间模型为例，介绍第一种模型的主要思想和结论. 在模拟行人的行为时引入社会和物理作用力，并把每一个行人看成是遵循牛顿力学定律的粒子，德国物理学家海耳宾 (Dirk Helbing) 在行人特定动力模型基础上提出了社会作用力模型 (the social force model, SFM)[12.18]. 在 HMFV (Helbing-Molnár-Farkas-Vicsek) 模型 [12.19,12.20] 中，每个行人受到的力以及对别人的作用力有两种，社会作用力和物理作用力. 社会作用力没有物理源头，是某一空间内行人为了避免和其他人或者墙壁等障碍物发生碰撞的意识反映，也包括行人在给定速度下向某个特定方向 (如出口运动方向) 的意识反映. 而物理作用力则是对人群之间的碰撞、摩擦等物理作用的直接描述，当人群密度高到足以迫使行人发生碰撞时，物理作用力的推力和阻力就体现出来. 具体的社会作用力模型表达如下：

$$m_i \frac{d\bm{v}_i}{dt} = \bm{F}_{\text{social}} + \bm{F}_{\text{push}} + \bm{F}_{\text{friction}} \tag{65-2}$$

式中，右边的第一项：

$$\bm{F}_{\text{social}} = A \exp\left(\frac{R_{ij} - d_{ij}}{B}\right) \bm{n}_{ij} \tag{65-3}$$

表示社会作用力. 其中，A 和 B 均为常数，R_{ij} 表示行人 i 和行人 j 的半径总和，是模型中给定的参数，d_{ij} 表示行人 i 和行人 j 之间的中心距离；$\bm{n}_{ij}$ 表示指向行人 i 的速度变化方向的单位矢量.

(65-2) 式右端的第二项：

$$\bm{F}_{\text{push}} = \kappa H(R_{ij} - d_{ij}) \bm{n}_{ij} \tag{65-4}$$

表示两个行人之间当距离足够近时所发生的推动作用力. 式中，κ 为常数，$H(R_{ij} - d_{ij})$ 为阶跃函数，当 $R_{ij} - d_{ij} > 0$ 时，$H(R_{ij} - d_{ij}) = R_{ij} - d_{ij}$，而当 $R_{ij} - d_{ij} \leqslant 0$ 时，$H(R_{ij} - d_{ij}) = 0$.

(65-2) 式右端的第三项：

$$\bm{F}_{\text{friction}} = \kappa |\bm{F}_{\text{push}}| \bm{n}_{ij} \tag{65-5}$$

表示两个行人之间当距离足够近时所发生的摩擦作用力. (65-2) 式右端的力的三项之矢量和即为行人选择路径时的总体影响因素, 其中, 阶跃函数的定义保证了当行人身体不发生接触时, 将不产生物理作用力.

海耳宾等在其《自然》期刊一文 [12.19] 中, 模拟出恐慌逃生的几种典型现象. 当所有人在冲撞时, 会在出口处形成弧形堵塞, 如图 65.3 所示. 当其崩溃时, 出现雪崩状撤离人群, 人们急于逃离, 导致"快者慢出"效应. 通过恐慌程度指数模拟群体行为和个体行为的结合, 模型认为, 当个体行为和群体行为结合会得到最佳的生存机会, 最好的策略是个体寻找出口的行为和跟随他人行为的折中, 如图 65.4 所示.

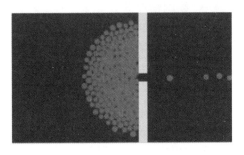

图 65.3　恐慌人群逃生形成的弧形堵塞

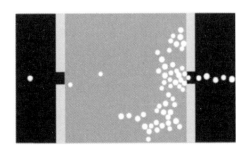

图 65.4　当恐慌系数太高时, 导致出口的低效利用

2009 年, 计算社会科学 (computational social science)[12.21] 创立.

2019 年初, 美国《科学》期刊刊文 [12.22], 研究人员通过观察数千名芝加哥马拉松运动员的集体性移动 (如图 65.5 所示) 发现, 大型人群的移动与液体流动类似, 可进行数学预测.

通过将流体模型应用于人群动态, 大型人群的集体行为可单用流体动力学理论原理来描述. 马拉松人群流动的连续性方程为

$$\frac{\partial \rho}{\partial t} + \nabla \cdot (\rho \boldsymbol{v}) = 0 \tag{65-6}$$

人群中的摩擦力矢量为

$$\boldsymbol{F} = -\Gamma_\parallel (v - v_0) \hat{\boldsymbol{x}} \tag{65-7}$$

式中，$\Gamma_\parallel$ 为人流纵向摩擦系数，$v - v_0$ 为人流速度和平均值的差，$\hat{\boldsymbol{x}}$ 表示人流方向的单位矢量. 描述人流的纳维–斯托克斯方程为

$$\frac{\partial v}{\partial t} + \rho_0 \frac{\mathrm{d} v_0}{\mathrm{d} \rho}\bigg|_{\rho_0} \frac{\partial v}{\partial x} = \frac{\rho_0 \beta}{\Gamma_\parallel} \frac{\partial^2 v}{\partial x^2} \tag{65-8}$$

式中，β 为人流纵向的可压缩性，ρ_0 为人流静止排队时的密度.

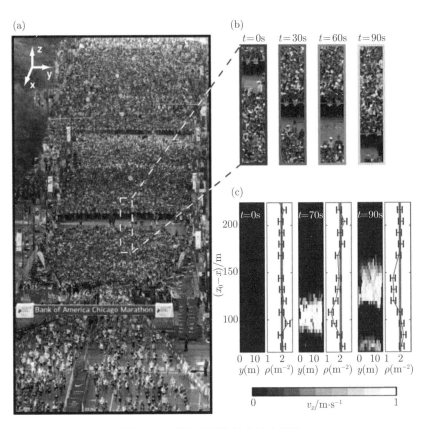

图 65.5　芝加哥马拉松中的人群波

(a) 刚开跑的情形；(b) 不同时刻的运动员在工作人员疏导下所形成的人链；
(c) 三个不同时刻的速度和密度场

研究表明，人群流体动力建模的预测能力或能为人群管理提供量化指导，这在发生事故或暴力等恐慌情况下尤其重要.

了解动物的集体行动多基于群组中个体间复杂的相互作用，这里的每个个体都有一组支配行为的"规则"和动机. 然而，对人而言，这一基于代理的方法受到

其描述人群活动能力的限制. Bain 和 Bartolo 介绍了一种不同的方法,他们忽略个体代理因子,而是将群体本身作为一种实体,旨在建立一种不含行为假设的大规模人群活动的流体动力学理论.

Bain 和 Bartolo 对马拉松选手观察后发现,选手们先是朝着芝加哥马拉松赛起始环形线慢慢移动,接着便以小群人停下并开始赛跑. 作者在每一个运动中都发现了人群密度和速度在起跑线前缘倾泻而下的波动. 更重要的是,这些波动会以恒定的速度在整个人群中传播,如图 65.5 所示. 这些动态变化在其他研究人员评估的比赛中也可预测性地建模.

65.3 金融动力学

美国波士顿大学的统计物理学家哈利·尤金·斯坦利 (Harry Eugene Stanley, 1941— ,2004 年当选美国科学院院士) 于 1994 年创造了 "经济物理学" 这个术语 (Stanley coined the term 'econophysics' in 1994),来作为处理经济现象的物理学领域 (the field of physics dealing with economic phenomena).

65.3.1 经典力学和经典物理在经济学中的应用

物理学和金融学哪个学科更难研究?华尔街第一代宽克 (quant) 中最著名的伊曼纽尔·德曼 (Emanuel Derman, 1946—) 比较道:"在物理中,你是在同上帝下棋,上帝并不经常改变他的规则,当你已经 '将' 注他时,他就会认输;而在金融中,你是在同上帝创造出的人类在下棋,这些人基于他们转瞬即逝的看法给资产估价. 他们不知道什么时候已经输掉了,只知道持续不断地努力下棋 (In physics you're playing against God, and He doesn't change his laws very often. When you've checkmated Him, He'll concede. In finance, you're playing against God's creatures, agents who value assets based on their ephemeral opinions. They don't know when they've lost, so they keep trying)." 因此,金融学和物理学相比更难. 这里的 "宽克" 指一群靠数学模型分析金融市场的物理学家和应用数学家.

亚里士多德曾告诉过我们,西方科学之父泰勒斯 (Thales) 靠自己的科学知识投机致富;伽利略离开帕多瓦大学为科西莫二世·德·美第奇 (Cosimo II de Medici, 1590—1621) 效力,写了《论骰子的发现》(On the Discoveries of Dice),是最早的宽克了.

1600 年前后,由于缺钱,物理学之父伽利略·伽利雷 (Galileo Galilei, 1564—1642) 靠给美第奇家族当家庭教师度日. 在科西莫二世·德·美第奇的资助下,伽利略发现了木星的几颗卫星. 他当即把这些星星命名为 "美第奇星" (Medicean Stars),

以向供养他的这个家族致敬.

意大利佛罗伦萨的美第奇家族是整个欧洲最显赫的商人贵族,家族里还出过三位教皇和两位法国王后,长期把持佛罗伦萨的市政. 包括比萨斜塔在内的浩大的教堂建筑群就是美第奇家族兴建的,也是他们的家族教堂之一;可以毫不夸张地说整个文艺复兴是靠美第奇家族资助的,该家族资助过包括文艺复兴后三杰在内的一大批人:达·芬奇 (Leonardo da Vinci, 1452—1519),米开朗基罗 (Michelangelo, 1475—1564),拉斐尔 (Raphael, 1483—1520). 美第奇家族也是有史以来欧洲最富裕的家族,达·芬奇的军事设计就是为讨好美第奇家族做的. 除了米开朗基罗,基本上所有文艺复兴时期出名的艺术家,科学家都是主动投靠美第奇家族的,因为他们给的薪水无人能比. 美第奇家族给伽利略的薪水是大学教授的几倍甚至几十倍. 因此,将伽利略称为 "第一个宽克" (the first quant) 是恰如其分的.

被朗道誉为 0 级的伟大科学家牛顿在 1720 年炒股的惨痛经历可给与德曼上述论述以最好的诠释. 牛顿于 1696 年担任英国皇家铸币厂的厂长,年薪 500 英镑 (Newton became Warden of the Royal Mint in April 1696 at £500 a year). 1700 年,牛顿继而担任了皇家铸币厂的总监 (Master of the Mint),不但收入更高,而且还是个闲职,他可以继续保留其在剑桥大学三一学院的教职.

如图 65.6 所示,英国南海公司的股票在 1720 年初疯涨近 10 倍. 牛顿当年

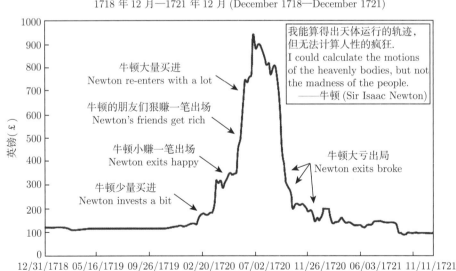

图 65.6　牛顿于 1720 年炒股破产的经历

4月入市5月清仓, 净赚了7000英磅. 后来股价持续疯涨, 牛顿的朋友们很赚了一大笔. 抵御不住市场诱惑的牛顿, 7月再次大举入市. 然而1周后, 股价在达到顶点后开始下挫, 随后一泄千里, 牛顿未及脱身, 共损失2万英磅.

1720年的2万英镑是什么概念? 在牛顿的建议下, 英国于1717年首先实行金本位制度, 黄金价值正式与英镑面值挂钩, 每盎司(纯度0.9)黄金定为3英镑17先令10便士. 在不考虑通货膨胀的前提下, 2019年5月13日每盎司黄金1238美元, 按1美元兑换6.85元人民币粗略估算, 300年前的2万英镑, 现在价值约4518.54万元人民币.

布阔伊(Georg Graf von Buquoy, 1781—1851, 如图65.7所示)是德国和捷克力学家, 他最大的力学成就是建立了变质量系统的力学方程:

$$\mathrm{d}(m\boldsymbol{v}) = m\mathrm{d}\boldsymbol{v} + \boldsymbol{v}\mathrm{d}m = \boldsymbol{F}\mathrm{d}t \tag{65-9}$$

经典力学中的"布阔伊问题"(Buquoy's problem)[12.23]是指: 一维均匀链垂直提拉与下落的变质量问题, 见《力学讲义》[1.11]第98页例22.2中链的下落和第120页链的提升.

布阔伊于1815年将经典牛顿力学的原理应用于经济学, 成为经济物理学中的里程碑式的工作.

图 65.7 布阔伊最早将经典力学应用于经济学研究

维弗雷多·帕雷托(Vilfredo Pareto, 1848—1923, 如图65.8所示)于1870获意大利都灵综合理工学院博士, 其博士论文为《固体平衡的基本原理》(*The Fundamental Principles of Equilibrium in Solid Bodies*), 这是典型的固体力学方向的博士论文, 为其后来从事经济学的研究奠定了坚实的理工科基础. 拥有固体力学博士学位的帕雷托转向了经济学研究, 他像物理学描述行星运动一样来运用牛顿力学来描述社会运动. 事实上, 帕雷托本人喜欢将开普勒定律与尚未发现的新经济定律之间进行比较.

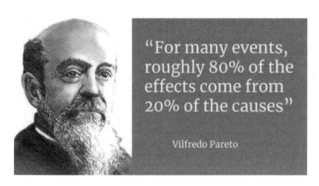

图 65.8 帕雷托和他的 80/20 法则

帕雷托最优 (Pareto optimality) 是经济学中的重要概念，并且在博弈论、工程学和社会科学中有着广泛的应用. 经济学理论认为，如果市场是完备的和充分竞争的，市场交换的结果一定是帕雷托最优的，并且会同时满足以下三个条件：

(1) 交换最优：即使再交易，个人也不能从中得到更大的利益. 此时对任意两个消费者，任意两种商品的边际替代率是相同的，且两个消费者的效用同时得到最大化.

(2) 生产最优：这个经济体必须在自己的生产可能性边界上. 此时对任意两个生产不同产品的生产者，需要投入的两种生产要素的边际技术替代率是相同的，且两个生产者的产量同时得到最大化.

(3) 产品混合最优：经济体产出产品的组合必须反映消费者的偏好. 此时任意两种商品之间的边际替代率必须与任何生产者在这两种商品之间的边际产品转换率相同.

帕雷托因对意大利 20%的人口拥有 80%的财产的观察而著名，并由此上升为 "帕雷托原理" (Pareto principle) 或 "80/20 法则 (rule)"，如图 65.8 所示. 80/20 法则最初只限定于经济学领域，后来这一法则也被推广到社会生活的各个领域，且深为人们所认同. 帕雷托法则是指在任何大系统中，约 80%的结果是由该系统中约 20%的变量产生的. 例如，在一个大的科研单位，约 80%的科研成果在统计意义上是由 20%的课题组完成的；在一个大的课题组，通常 80%的成果是由 20%的课题组成员取得的；一个大实验室的 80%的课题经费是由约 20%的实验室成员争取到的；在企业中，通常 80%的利润来自于 20%的项目或重要客户；心理学家认为，20%的人身上集中了 80%的智慧等；具体到时间管理领域是指大约 20%的重要项目能带来整个工作成果的 80%，并且在很多情况下，工作的头 20%时间会带来所有效益的 80%. 上述结论可再做如下引申，在任何特定群体中，重要的因子通常只占少数，而不重要的因子则占多数，因此只要能控制具有重要性的少数因子即能控制全局.

65.3.2 统计力学在经济学中的应用

罗萨里奥·曼泰格纳 (Rosario Nunzio Mantegna, 1960—) 于 1991 年通过米兰股票交易中的数据进行统计分析，得出了列维随机行走 (Lévy random walk) 会使股票交易超扩散 (superdiffusion) 的结论[12.24]. 曼泰格纳的这篇论文现在被认为是经济物理学新兴领域真正意义上的第一篇论文 (what is now considered to be the first paper that can truly be attributed to the newborn field of econophysics)[12.25].

曼泰格纳和斯坦利继而于 1995 年在《自然》期刊发文 [12.26]，他们关于股票市场高频数据的统计分析结果结果显示指数价格收益率的概率分布是尾部截断的列维分布.

65.3.3 湍流在经济学中的应用 —— 湍流经济学

如表 65.1 所示，Ghashghaie 等于 1996 年在《自然》期刊撰文 [12.27]，指出了充分发展的三维湍流 (fully developed three-dimensional turbulence) 和外汇交易 (foreign exchange, FX) 之间的类比性. 从而将湍流中能量串级的自相似和标度律的思想引入到外汇交易的分析中.

表 65.1 充分发展的湍流与外汇交易市场之间的类比性

水动力学的湍流 (hydrodynamic turbulence)	外汇交易 (foreign exchange, FX)
能量 (energy)	信息 (information)
空间距离 (spatial distance)	时间延迟 (time delay)
间歇性 (intermittency)	低和高波动的集群 (clusters of low and high volatility)
空间层次中的能量串级 (energy cascade in space hierachy)	时间层次中的信息串级 (information cascade in time hierachy)
$\langle (\Delta \boldsymbol{v})^n \rangle \sim (\Delta r)^{\xi_n}$	$\langle (\Delta x)^n \rangle \sim (\Delta t)^{\xi_n}$

1996 年，曼泰格纳和斯坦利针对 Ghashghaie 等的文章展开了进一步的讨论，他们对 S&P500 指数的波动与处于完全湍流状态的流液体速度波动进行了对比，他们的结果表明两者都表现出间歇性以及短暂的非高斯行为 [12.28]. 曼泰格纳和斯坦利的结果表明这两种过程的扩散指数却有着本质上的差异. S&P500 指数时间序列的扩散指数为 0.53，而处于完全湍流状态液体速度的时间序列的扩散指数为 0.33，其非常接近于理论值 1/3. 相对于湍流的情况，股票市场的扩散指数非常接近于当指数价格变化时间序列不具有自相关性时的扩散指数期望值 0.50，所以金融市场的指数价格波动类似于物质粒子的随机行走.

1999 年，曼泰格纳和斯坦利出版了《经济物理学导论》[12.29] 的专著，到 2019 年底已被 google scholar 引用了 4600 多次.

65.3.4 量子力学在经济学中的应用 —— 量子金融和量子经济学

随着爱因斯坦在 1905 年和 1915 年分别创立狭义和广义相对论，20 世纪 20 年代量子力学的创立，物理学进入了崭新的时代. 社会和经济学界不禁要问：如果旧的牛顿力学在解释像社会和经济这样的复杂系统时被发现是不充分的，那么新的物理学为科学地理解复杂性是否能开辟新的途径？

第 12 章 连续介质力学新发展 —— 思维动力学、金融动力学、社会动力学、管理动力学

意大利传奇物理学家埃托雷·马约拉纳 (Ettore Majorana, 1906—1938, 见题表 1.5) 在 1938 年失踪前的最后一篇题为《统计定律在物理和社会科学中的价值》[12.30] 中，提出将量子力学应用于社会科学的研究领域.

量子金融 (quantum finance)[12.31,12.32] 和量子经济 (quantum economics) 近 20 年来得到了很大关注. 一个典型的范式是：量子力学 → 路径积分 → 统计力学 (配分函数) → 金融，等等，其中，路径积分中的相因子差某个量纲意义上的哈密顿量.

总体来说，量子金融的两个大方向，一来用来解释传统基于方程的模型不符合实际测量结果，像量子干涉效应，假定随机过程和一个不在经典测度里的量子过程产生干涉；另一个就是利用哈密顿量 (Hamiltonian) 计算衍生品定价.

量子金融中的不确定性关系 (uncertainty principle)[12.33,12.34]、量子芝诺效应 (quantum Zeno effect)[12.35]、玻姆力学 (Bohmian mechanics) 和金融导航波 (financial pilot wave) 理论[12.36]、经济量子纠缠等问题得到了研究者的关注.

例 65.1 海森堡不确定性原理在零点能确定上的应用.

解：量子谐振器的哈密顿量为

$$H = \frac{p^2}{2m} + \frac{1}{2}m\omega^2 x^2 \tag{65-10}$$

在应用海森堡不确定性原理 (Heisenberg uncertainty principle) 确定零点能 (zero-point energy) 时，动量和位置这一对儿 (pair of) 作用量共轭变量的平均值满足如下关系：

$$\langle p \rangle = \langle x \rangle = 0 \tag{65-11}$$

从而，位置与动量的不确定度 Δx, Δp 可以用标准差 $\Delta x = \sqrt{\langle (x - \langle x \rangle)^2 \rangle}$ 量化，对标准差进行平方，得到

$$\begin{aligned}(\Delta x)^2 &= \langle (x - \langle x \rangle)^2 \rangle = \langle x^2 \rangle - \langle 2x\langle x \rangle \rangle + \langle \langle x \rangle^2 \rangle \\ &= \langle x^2 \rangle - 2\langle x \rangle \langle x \rangle + \langle x \rangle^2 \\ &= \langle x^2 \rangle - \langle x \rangle^2 \end{aligned} \tag{65-12}$$

应用上式，则知位置与动力的不确定度满足如下关系：

$$(\Delta x)^2 = \langle x^2 \rangle - \langle x \rangle^2 = \langle x^2 \rangle, \quad (\Delta p)^2 = \langle p^2 \rangle - \langle p \rangle^2 = \langle p^2 \rangle \tag{65-13}$$

对 (65-10) 式取平均，应用 (65-13) 式，有

$$\langle H \rangle = \frac{\langle p^2 \rangle}{2m} + \frac{1}{2}m\omega^2 \langle x^2 \rangle = \frac{\langle \Delta p \rangle^2}{2m} + \frac{1}{2}m\omega^2 \langle \Delta x \rangle^2 \tag{65-14}$$

海森堡不确定性原理的下限为

$$\Delta p \cdot \Delta x = \frac{\hbar}{2} \tag{65-15}$$

将上式中的动量的不确定度 $\Delta p = \dfrac{\hbar}{2\Delta x}$ 代入 (65-14) 式，得到用位移不确定度 Δx 表示的 $\langle H \rangle$：

$$\langle H \rangle = \frac{\hbar^2}{8m\langle \Delta x \rangle^2} + \frac{1}{2}m\omega^2 \langle \Delta x \rangle^2 \tag{65-16}$$

$\langle H \rangle$ 取最小值的条件是

$$\frac{\mathrm{d}\langle H \rangle}{\mathrm{d}\langle \Delta x \rangle} = -\frac{\hbar^2}{4m\langle \Delta x \rangle^3} + m\omega^2 \langle \Delta x \rangle = 0 \tag{65-17}$$

从上式解得

$$\langle \Delta x \rangle = \sqrt{\frac{\hbar}{2m\omega}} \tag{65-18}$$

将 (65-18) 式代回 (65-16) 式，则得到零点能的表达式：

$$\langle H \rangle_{\min} = \frac{\hbar^2}{8m\dfrac{\hbar}{2m\omega}} + \frac{1}{2}m\omega^2 \frac{\hbar}{2m\omega} = \frac{\hbar\omega}{4} + \frac{\hbar\omega}{4} = \frac{\hbar\omega}{2} \tag{65-19}$$

例 65.2 和量子力学中的海森堡不确定性原理相类比的量子金融中的不确定性关系.

答：量子力学中的海森堡不确定性原理的两种等价形式为

$$\Delta x \cdot \Delta p \geqslant \frac{\hbar}{2}, \quad \Delta E \cdot \Delta t \geqslant \frac{\hbar}{2} \tag{65-20}$$

作为类比，有学者将金融、股票市场中的不确定性表示为 [12.33]

$$\Delta\wp \cdot \Delta T \geqslant \frac{\hbar}{2} \tag{65-21}$$

式中，$\wp$ 为股票价格，$\Delta\wp$ 则为股票价格的不确定度；T 为股票价格的变化率：

$$T = m_0 \frac{\mathrm{d}\wp}{\mathrm{d}t} \tag{65-22}$$

式中，m_0 为股票市场的 "质量" (mass). 本书作者提出的疑问是，在量子力学的海森堡不确定性原理中，位置 x 和动量 p、能量 E 和时间 t 互为作用量的共轭变量. 在所谓的量子金融的不确定性关系 (65-21) 式中，如何保证股票价格 $\wp$ 和其变化率 T 为作用量的共轭变量？

65.4 弹性、刚性、柔性、韧性等概念在社会及管理动力学上的推广和延深

65.4.1 弹性和回弹概念的来源

霍林 (Crawford Stanley Holling, 1930—2019) 发表于 1973 年关于生态系统弹性或恢复力 (resilience) 的开创性论文 [12.37] 对生态学和其他自然和社会科学产生了重大影响. 如图 65.9 和图 65.10 所示, 弹性是一个系统在经历变化时吸收干扰和重组的能力, 以保持本质上相同的功能、结构、特性和反馈. 如图 65.9 所示, 弹性有三个关键方面: (1) 范围 (latitude): 一个系统在失去恢复能力之前所能改变的最大量; (2) 抵抗力 (resistance): 改变系统的难易程度, 也就是说系统对改变有多 "抗拒"; (3) 危险度 (precariousness): 系统当前状态与某个 "阈值" 的接近程度.

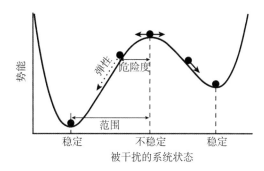

图 65.9 经典弹性或回弹力的能量地貌图

图 65.10 弹性 (resilience) 的确切含义

65.4.2 国家制度和政权的弹性

著者这一代人迄今所经历的最大的历史事件可能就是发生于 1989—1991 年的东欧剧变, 苏联和东欧等大部分社会主义政体发生崩溃, 为什么中国同样面临着深刻的危机, 却能及时变革并展现出巨大的 "弹性" (resilence)? 这无疑是一个巨大的政治、历史和社会科学的命题.

2013 年，季米特洛夫在剑桥大学出版社编辑出版了一部题为《为什么共产主义没有崩溃 —— 理解亚欧威权政体的弹性》的文集 [12.38]. 本书著者未必赞同该书中的某些观点，但十分赞同作者们从弹性的观点来阐述这一巨大的社会和历史的命题.

65.4.3 弹性思维、弹性社会、生态弹性、经济弹性等领域

弹性体发生变形并产生恢复力的这一概念逐渐被引入到生态学、心理学、社会-生态系统、社会-经济系统等研究领域中，其内涵也得以丰富和延伸. 但本质上都具有以下特性：

(1) 弹性是指系统和个人应对重大逆境或风险的能力 (Resilience is the capability of systems and individuals to cope with significant adversity or risk). 在承受或吸收一定外界干扰的情况下能够维持基本结构和功能特征不变；

(2) 能够建立自组织、自适应、自我恢复的能力.

在各类新闻报道中，我们耳熟能详的相关术语有：弹性思维 (resilience thinking)、弹性社会、弹性社区、弹性生态、心理弹性、经济弹性，等等.

近年来，弹性还被逐渐应用于城市规划领域，表征城市对社会经济发展的不确定性和复杂性的适应能力，并出现了 "弹性城市" "城市生态弹性" "城市工程弹性" "城市经济社会弹性" 等新理念. 用一句话来讲，就是要建设有弹性和抵御能力的社会.

65.4.4 刚性与柔性

刚性是指事物在外力作用下不易发生形变、难以通融和改变的性质.

刚性规划是指规划在战略思想、指标结构、编制程序、管制规则等方面所具有的权威性、固定性和指令性. 城市规划中的刚性体现在秉持以人为本、绿色低碳、历史传承、协调发展的规划理念，对道路、绿地、河道等城市 "五线" 实施空间开发管控，严格划定土地用途分区及其用途管制规则，逐级落实城市建设指令性、约束性指标，强化控规编制实施的执行力和规范化等方面.

65.4.5 韧性概念

韧性是物体受外力作用时产生变形但不折断的性质，兼具刚性力与弹性力. 韧性与弹性的概念有一定相似性，但实质却存在差异. 弹性强调物体在弹性极限内可产生变化并恢复原状的性质，基本可以认为不具有刚性，且当外力超过弹性极限时物体会发生断裂. 而韧性则是刚性和弹性的耦合，既能够保证自身在结构或功能方

面的稳定性,又能够以适当的手段吸收和缓冲外界变化产生一定形变,且当受到较大或突然性的外力作用时,韧性物体比弹性物体呈现出更好的稳定性和长期适应能力.

65.5 黏性概念在社会及管理动力学上的推广和延深

黏性为经济学中的一个广泛使用的概念,用来表示某一种变量变化较为缓慢. 较为典型的为 "黏性价格"(sticky prices),在微观经济学中指在动态的竞争市场模型下,由于市场压力等因素,其产品价格和工资仍然保持在一定的水平上. 其包括菜单成本、货币幻觉、不完全信息和公平性考虑等原因. 在总体经济学中,黏性价格可以用于解释为什么市场在短期内难以达到平衡.

价格黏性是指商品的价格不容易发生变动,而弹性就是指价格非常灵活. 一般认为,在市场经济条件下,价格随着供求变化而不断的发生变动,具有相当的弹性,从而可以促进资源实现优化配置;而在计划经济条件下,价格有官定,不随市场因素的变动而变动,因此具有黏性. 这只是简单意义上的阐述. 更深入一点地说,这是模型建立的两个相互对立的假设条件. 凯恩斯就认为在未实现充分就业的情况下,价格具有黏性,当实现了充分就业之后,随着货币供给的变化,价格就是弹性的.

黏性工资理论是指短期中名义工资的调整慢于劳动供求关系的变化.

价格黏性是新凯恩斯理论的基石. 货币非中性一般被认为是由于价格黏性导致的. 如果价格完全弹性,新凯恩斯理论预测货币政策就没有效果.

凯恩斯写道:"由于劳动者的流动性不够完善,从而工资不能精确地反映不同行业的真正的有利之处,所以任何个人和集体如果允许他们的货币工资做出相对于其他人的货币工资的削减,那么,削减就会使他们的实际工资相对地下降. 这已构成充分的理由来使他们抵抗货币工资的削减." 在凯恩斯看来,相对工资的下降会降低这部分工人的相对经济地位,会损害他们的利益,因此,工人会抵制这种相对工资的下降,从而使货币工资在向下方向上失去弹性.

为什么楼价在十多年的所谓泡沫预期中如此之 "黏"(黏性价格的特征),且在全球和中国经济下行周期中价格如此之 "刚"(刚性价格的特征) 呢? 楼价的价格之 "黏",这是地产和其上盖物产复合之后,形成的超长周期耐用消费品的价格特征,消费者是用自己的生命周期收入来形成有效需求的,这和商业类购买大宗产品是按商业周期 (短期收入) 来决策购买需求是不同的. 超长收入周期形成的刚性需求购买,是黏性的根源.

65.6 金融动力学中的艾略特波浪理论

艾略特波浪理论 (the Elliott wave principle) 是证券技术分析的主要理论之一,由拉尔夫·纳尔逊·艾略特 (Ralph Nelson Elliott, 1871—1948) 提出. 一些交易员用以分析金融市场周期,预测市场趋势. 艾略特作为专业会计师,于 1930 年提出分析方法,认为市场价格的走势具有特定的形态. 1938 年,艾略特出版专著《波浪理论》(The Wave Principle),正式提出此理论,并在 1946 年,他最后的主要著作《自然法则:宇宙的秘密》(Nature's Laws: The Secret of the Universe) 中进行了最为详细的阐述. 艾略特写道 "人受制于规律过程,关于人的活动的计算能被投射到未来的某个阶段,这一点我确信无疑".

因月球引力而造成的潮汐起伏可算最为自然之物,其中必然蕴含了很多丰富的自然和社会组织的道理. 艾略特在书中指出,这是他对证券行情走势 "波浪理论" 分析方法的思考源头. "艾略特波浪" (Elliott waves) 早已成为该领域的专有名词. 波浪理论对股市、外汇市场、货币、债券、黄金、石油等价格走势十分有效.

在艾略特看来,市场呈现出重复循环的过程,他指出,这种变化过程是投资者受外界因素影响导致其情绪变动所致,或在当时受到市场的主导心理影响. 艾略特解释称,价格的上行波动和下行波动是由市场集体心理状态引起,而这种心理始终在不断反复的形态中反映出来. 艾略特将价格的上行和下行波动称作 "波浪 (wave)".

如图 65.11 所示,艾略特波浪所具有的一种重要特性就是其具有分形结构. 和图中的海贝壳 (seashell)、云和闪电 (clouds and lightening)、雪花 (snowflake)、花椰菜 (broccoli)、分形山等分形结构类似,艾略特波浪也可能进一步细分为与其相似的更小层级艾略特波浪,因而具有一种自相似 (self-similar) 的结构.

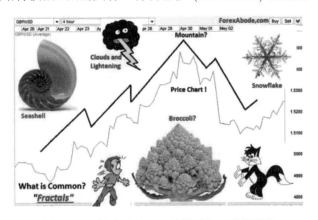

图 65.11 艾略特的 5-3 浪模式与分形的类比

艾略特认为，市场呈 5-3 浪形态波动，被称为 "5-3 浪模式"(the 5-3 wave pattern). 最初的 5 浪称为推动浪. 最后的 3 浪称为调整浪. 在该形态中, 第 1、3、5 浪是上升浪，意味着它们和总体上行趋势保持一致，而 2、4 浪为回调浪, 如图 65.12 所示.

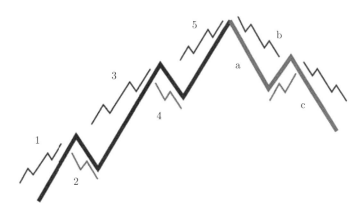

图 65.12 艾略特的 5-3 浪模式

下面对艾略特的 5-3 浪模式进行仔细分析:

第 1 浪. 股价最初呈现出上行趋势. 这通常是突然受到相对小部分投资者的买盘推动的结果 (因为不同的原因，真实的或人为臆测的), 这小部分投资者人为, 股价较为便宜，因此是时候买进. 这导致股价开始走高.

第 2 浪. 进入本阶段，相当一部分之前选择买入的投资者认为股价有些高估, 进而选择获利回吐. 这导致股价的回落. 不过, 股价并未跌破之前低点, 因为投资者在股价经过回调后, 认为是时候再次买入.

第 3 浪. 浪 3 通常是 5 浪中最长也是最强劲的一波浪. 该股已经受到市场上绝大多数投资者的关注. 更多的投资者已经发现该股的投资价值, 并打算买入. 这也促使该股不断走高. 该波段浪通常超过浪 1 尾期所创出的高度.

第 4 浪. 投资者在这一阶段再次选择获利回吐, 因为他们再次认为股价被高估. 但该波段浪的回调较为疲弱, 因为市场上相对多数的投资者仍然看好该股票前景, 并打算逢低买入.

第 5 浪. 在本阶段，市场上的绝大多数投资者都将投资的触角伸向该股，股价也是受到市场狂热的追捧. 你通常会看到, 该公司 CEO 通常作为年度人物出现在某知名财经杂志封面上. 交易者和投资者们开始疯狂追涨该股票, 当有人和他们持不同意见时, 他们会展开强烈的反驳. 而此时, 股票价值也是高估最为严重的阶段. 市场上的反对人士开始做空该股, 于是, abc 调整浪出现了.

abc 回调. 汇价在经历一轮推动浪走势后, 随即出现 3 浪回调. 这次我们用字母而非数字来标示回调浪.

艾略特波浪的分形属性是指: 每个波浪都是由更小级别的艾略特波浪组成, 如图 62.6 中的细线所示, 1、3 和 5 浪是由一组小级别的艾略特推动浪形成, 而 2 和 4 浪则是由小级别的一组修正浪形成, 这些就是分形.

艾略特波浪理论按波浪的级数可分为超级大周期 (多个世纪)、超级周期 (约 40—70 年)、常规周期 (一年至数年)、主要周期 (数月到数年)、中级周期 (数周至数月)、次级周期 (数周)、小型周期 (数天)、微型周期 (数小时)、迷你周期 (数分钟).

65.7 金融数学

金融数学 (Financial Mathematics) 是 20 世纪 80 年代末、90 年代初兴起的数学与金融学的交叉学科. 金融数学主要运用现代数学理论和方法 (如随机分析、随机最优控制、组合分析、非线性分析、多元统计分析、数学规划、现代计算方法等) 对金融 (除银行功能之外, 还包括投资、债券、基金、股票、期货、期权等金融工具和市场) 的理论和实践进行数量的分析研究. 其核心问题是不确定条件下的最优投资策略的选择理论和资产的定价理论. 套利, 最优和均衡是其中三个主要概念. 近二十几年来, 金融数学不仅对金融工具的创新和对金融市场的有效运作产生直接的影响, 而且对公司的投资决策和对研究开发项目的评估 (如实物期权) 以及在金融机构的风险管理中得到广泛应用.

金融系统由于非线性、不确定性而成为复杂系统, 为金融数学提出了较高的要求, 尤其金融市场的特性: 波动性、突发事件、市场不完全、信息不对称等成为金融数学当前面临的重要课题. 金融市场上的波动现象一般可归结为随机的问题, 像几何布朗运动, 然后进行随机分析. 但是金融市场多数情况下并不满足稳定的假设, 时常出现异常的波动.

金融数学的历史可以追溯到 1900 年法国数学家巴谢利耶 (Louis Bachelier, 1870—1946) 的博士论文《投机的理论》(*Théorie de la spéculation*), 这宣告了金融数学的诞生. 在文中他首次用布朗运动来描述股票价格的变化, 他认为在资本市场中有买有卖, 买者看涨、卖者看跌, 其价格的波动是布朗运动其统计分布是正态分布. 然而, 巴谢利耶的工作没有引起金融学界的重视达半个世纪.

巴谢利耶的博士导师是数学力学、物理学、哲学大师庞加莱. 巴谢利耶是于 1900 年 3 月 29 日在巴黎大学进行的博士论文答辩. 巴切利耶的博士论文首次引入了布朗运动的数学模型, 并将其用于股票期权估值, 这是历史上第一篇将高等数学

第 12 章　连续介质力学新发展 —— 思维动力学、金融动力学、社会动力学、管理动力学

用于金融研究的论文. 因此, 巴切利耶被认为是数学金融学的鼻祖和随机过程研究的先驱 (forefather of mathematical finance and a pioneer in the study of stochastic processes).

但由于巴谢利耶的博士论文试图把数学应用到数学家不熟悉的领域, 虽然他的导师庞加莱给与了正面的评价 (positive feedback), 但当时巴谢利耶的博士论文并未受到数学家们的欢迎, 在当时的法国社会, 巴谢利耶的博士论文工作还不足以使他找到一个直接的教学职位.

20 世纪 50 年代初, 萨缪尔森 (见本章思考题 12.3) 通过统计学家萨维奇重新发现了巴谢利耶的工作, 这标志了现代金融学的开始. 现代金融学随后经历了两次主要的革命, 第一次是在 1952 年. 那年, 25 岁的马尔柯维茨 (Harry Markowitz, 1927—) 发表了他的博士论文, 提出了资产组合选择的均值方差理论. 它的意义是将原来人们期望寻找 "最好" 股票的想法引导到对风险和收益的量化和平衡的理解上来. 给定风险水平极大化期望收益, 或者给定收益水平极小化风险, 这就是上述均值方差理论的主要思想. 稍后, 夏普 (William Forsyth Sharpe, 1934—) 和林特纳 (John Lintner, 1916—1983) 进一步拓展了马尔柯维茨的工作, 提出了资本资产定价模型 (capital asset pricing model, CAPM). 美国经济学家弗兰科·莫迪利安尼 (Franco Modigliani, 1918—2003, 1985 年诺贝尔经济学获得者) 和默顿访桌 (Merton Miller, 1923—2000) 于 1958 年发表的《资本成本、公司财务和投资管理》一书中, 提出了最初莫迪利安尼-米勒 (Modigliani-Miller, MM) 公司财务理论, 引发了第一次 "华尔街革命", 是金融数学的开端. 马尔柯维茨、米勒、夏普也因他们在金融数学中的开创性贡献而获得 1990 年诺贝尔经济学奖.

1973 年, 费雪·布莱克 (Fischer Black, 1938—1995) 和迈伦·斯克尔斯 (Myron Scholes, 1941—) 用数学方法给出了 "布莱克-斯克尔斯期权定价公式"(Black-Scholes option pricing formula), 以及稍后, 罗伯特·默顿 (Robert Cox Merton, 1944—) 对该公式的发展和深化, 亦被广泛地称为 "布莱克-斯克尔斯-默顿期权定价公式" (Black-Scholes-Merton option pricing formula) 期权定价公式给金融交易者和银行家在衍生金融资产的交易中带来了便利, 推动了期权交易的发展, 期权交易很快成为世界金融市场的主要内容, 成为第二次 "华尔街革命". 默顿和斯克尔斯获得了 1997 年度诺贝尔经济学奖.

两次 "华尔街革命" 避开了一般经济均衡的理论框架, 形成了一门新兴的交叉学科, 即金融数学. 马尔科维茨-夏普理论和布莱克-斯克尔斯公式一起构成了蓬勃发展的新学科 —— 金融数学的主要内容, 同时也是研究新型衍生证券设计的新学

科——金融工程的理论基础，从而使这两次革命的先驱者分别在 1990 年和 1997 年获得了诺贝尔经济学奖. 美国经济学家罗伯特·恩格尔 (Robert Fry Engle III, 1942—) 和英国经济学家克莱夫·格兰杰 (Clive Granger, 1934—2009) 对时间序列理论在经济和金融的研究中取得重要成果，也于 2003 年获得诺贝尔经济学奖，可以认为这是金融数学的研究第三次获得诺贝尔经济学奖. 金融数学这门新兴的交叉学科已经成为国际金融界的一枝奇葩.

布莱克-斯克尔斯-默顿公式为一抛物线型的偏微分方程：

$$\frac{\partial V}{\partial t} + \frac{1}{2}\sigma^2 S^2 \frac{\partial^2 V}{\partial S^2} + rS\frac{\partial V}{\partial S} - rV = 0 \tag{65-23}$$

式中，$V(S,t)$ 为期权的价格 (the price of the option as a function of the underlying asset, S at time, t)；t 为以年为单位的时间 (a time in years)；σ 为股票回报率的标准差 (the standard deviation of the stock's returns)；$S(t)$ 为所交易金融资产现价 (the price of the underlying asset at time t)；r 为年化无风险利率 (the annualized risk-free interest rate).

思考题和补充材料

12.1 历史机遇与重大的科学发现之间的关系是学术界十分关系的问题，狄拉克 (Paul Adrien Maurice Dirac, 1902—1984, 1933 年诺贝尔物理学奖获得者，时年 31 岁) 在描述 20 世纪 20 年代量子力学大发展时曾精辟地指出："打个恰当的比方，这就是一局棋，一局非常有趣的棋. 每当一个人解决了一个小问题，就可以写一篇关于它的论文. 在那些日子里，任何二流的物理学家都很容易做出一流的工作. 此后就没有这么辉煌的时代了. 现在，一流的物理学家很难做出二流的工作 (It was a good description to say that it was a game, a very interesting game one could play. Whenever one solved one of the little problems, one could write a paper about it. It was very easy in those days for any second-rate physicist to do first-rate work. There has not been such a glorious time since then. It is very difficult now for a first-rate physicist to do second-rate work)." 据本书著者所考，狄拉克是在 20 世纪 70 年代中期在新西兰举行的一场学术报告会上讲这番话的. 狄拉克在理论物理界以十分奇特的性格著称. 1926 年，爱因斯坦在一封信中曾对狄拉克评价道："在天才与疯狂之间令人眼花缭乱的道路上，这种平衡是可怕的 (This balancing on the dizzying path between genius and madness is awful)."

12.2 继续关于历史机遇和重大科学发现关系的话题. 朗道经常唉声叹气地说自己出生晚了好几年，因而无法参与量子力学的创始性工作. 1929 年，在柏林召开的一个理论

第 12 章　连续介质力学新发展 —— 思维动力学、金融动力学、社会动力学、管理动力学

物理学研讨会上，他风趣地说道："所有的好女孩都被抢去结婚了，所有的好问题都解决了．所有剩下的我都不喜欢"(Landau often lamented that he had been born several years too late to participate in the pioneering work in quantum mechanics, and at a colloquium on theoretical physics in Berlin in 1929, remarked, "All the nice girls have been snapped up and married, and all the nice problems have been solved. I don't really like any of those that are left.")

12.3 保罗·萨缪尔森 (Paul Samuelson, 1915—2009, 1970 年获得第二届诺贝尔经济学奖，是第一位得到诺贝尔经济学奖的美国学者) 论获得诺贝尔奖的五个条件. 1970 年 12 月 10 日，萨缪尔森在斯德哥尔摩的诺奖致辞中，谈到了"轻松"(it is quite easy) 获得诺贝尔奖的五个条件：(1) 你必须做的第一件事是有伟大的老师 (The first thing you must do is to have great teachers)；(2) 其次，你一定也有很多优秀的同事、合作者和同学 (Second, you must also have been blessed with great colleagues, collaborators, and fellow students)；(3) 第三，你必须有伟大的学生 (Thirdly, you must have great students)；(4) 第四个必要条件，也是一个重要的学术观点，你必须阅读大师的作品 (A fourth necessary condition, and an important one from a scholarly point of view, you must read the works of the great masters)；(5) 现在已经列举了学术成功的五个必要条件中的四个. 为了避免耽误你们跳舞，让我赶快说出最后一个必要条件，它是完成一个完整解的充分条件的必要条件. 最后一个因素当然是运气 (Four out of the five necessary conditions for scholarly success have now been enumerated. Lest I delay your dancing, let me hasten to name the final necessary condition that serves to complete the sufficient conditions for a complete solution. The final element is, of course, luck).

12.4 芬兰经济学者 Matti Estola 于 2017 年出版了题为《牛顿微观经济学》(Newtonian Microeconomics) 的专著 [12.39]. 书中借鉴牛顿力学，专门列出一章介绍了所建立的经济学中的量纲体系. 在自然科学中，物理量分为基本物理量 (见附录 B 中的表 B.1) 和导出物理量 (见附录 B 中的表 B.3). Matti Estola 提出，经济学有四个基本量纲：(1) 商品量 (the amount of goods) [R]，商品量由于为实物 (real)，故用英文字母 R 表示；(2) 货币价值 (the monetary values of various things) [M]；(3) 时间 (time) [T]；(4) 满意度 (satisfaction, 亦称为"效用 (utility)") [S]. 经济学的很多事件都可以基于这四个基本量纲来度量，其他变量的量纲都可以用这些基本量纲来导出.

12.5 谁首次将熵的概念引入到经济学中？尼可拉斯·乔治斯库–罗根 (Nicholas Georgescu-Roegen, 1906—1994) 是罗马尼亚裔美国数学家，统计学家和经济学家. 他最主要的贡献是在 1971 年出版的巨著 (magnum opus)《熵定律和经济过程》(The Entropy

Law and the Economic Process)[12.40] 将熵的概念首次引入经济学. 作为经济学的鼻祖和范式创始人 (a progenitor and a paradigm founder in economics), 乔治斯库–罗根的工作对于将生态经济学 (ecological economics) 确立为经济学的一个独立的分支学科具有开创性意义.

12.6 脑科学和量子力学之间的关系. 量子意识 (quantum consciousness) 或量子心智 (quantum mind) 认为经典力学无法解释意识这种现象, 而量子力学中的现象、如: 量子纠缠、量子叠加可能在大脑的功能方面起着重要作用, 因而可以成为解释意识现象的理论基础. 现在学界专家提出的量子意识的研究包括下列方面: (1) 量子物理和生物学的关系; (2) 量子物理和神经系统的关系; (3) 精神失常是否可以用量子物理上变化解释?

12.7 物理学家彭罗斯 (Roger Penrose, 1931—   , 2020 年诺贝尔物理学奖获得者) 和麻醉学家哈默洛夫 (Stuart Hameroff, 1952—   ) 共同提出来的谐客观化归 (orchestrated objective reduction, Orch-OR) 假说. 而彭罗斯的思想来自于哥德尔的 "不完备性定理", 在 1989 年出版的《皇帝的新脑》(*The Emperor's New Mind*) 的有关人工智能的科普读物中, 彭罗斯指出: 一个形式系统无法证明它自身的不一致性, 而哥德尔的无法证明的结论却可以由人类的数学家证明. 他利用这种差异说明人类的数学家本身并不是形式系统, 因此 [人的大脑] 不可能等同于一个可计算的算法.

这里的中心思想就是: 人的大脑并不等于一个精致、超高速运算的计算机, 而这个观点正好和目前流行的人工智能学说和部分语言学理论相反, 后者将人的大脑看做是运行某种算法的计算设备 (computational device).

12.8 杨卫[12.41,12.42] 采用连续介质力学的方法论, 在学科空间和历史维度组成的时空构架下定量地阐述研究生教育学. 文中提出研究生教育的两项本质特征是与学科建设和知识创新的关系; 提出中国研究生教育的四项基本假设, 即: 学科个体假设、资源贫乏假设、产品继承假设、控制递减假设. 文中针对资源、产品、结构三类变量, 初步建构了由五类场方程控制的研究生教育动力学框架.

12.9 杨卫等[12.43] 提出了 "X-力学" 的概念. 交叉力学是力学学科具有包容力的新生长点, 并赋予力学固有领地新的高度和广度. 力学横跨理工, 是研究万物之间交叉作用的科学, 力学是交叉性最强的学科. 在理科方面, 力学与数、理、化、天、地、生均有交集; 在工科方面, 力学在机械、材料、航空航天、土木、电机、数据与计算、生物医学等领域均有运用.

12.10 《流体力学年鉴》于 2019 年发表了题为《数据时代的湍流模型》的综述文章[12.44]. 机器学习方法出现以前, 学者主要通过数学建模来描述湍流模型, 而后利用实验和模拟得到的数据对模型进行简单修正. 机器学习方法的出现使得科研人员能够直接使用数据对模型进行修正甚至重构.

模型对物理问题描述的误差主要分四种: Level 1 (L1): 使用粗粒化方法对物理场数据采集导致数据细节失真; Level 2 (L2): 对物理问题理解不深刻导致独立变量选择不全面; Level 3 (L3): 数学模型对数据拟合引起的误差; Level 4 (L4): 模型系数选择引入的误差.

传统研究中, 修正雷诺平均模型是从三个方面入手: 模型误差修正 (L2、L3)、参数误差 (L4) 以及模型失效区.

模型误差修正思路主要大致有两种, 其一是通过逼近物理问题理论的边界来预测可能的行为区间; 另一种是通过加入可实现的物理约束扰动来探索雷诺平均模型的边界区间. 这两种都是对雷诺平均模型本身的误差进行的研究 (L3), 另外还有学者在模型中加入了修正项进行局部平均时导致细节失真的项来修正偏差点 (L2).

在理想状态下, 模型中的参数应该使用最优的, 但是复杂问题的最优参数往往无法获得, 因此往往采用对简单问题求解得到的参数应用于复杂问题, 这样就造成了参数误差的出现.

模型失效区指在利用模型解决问题时, 某些区域会出现与理论不符合乃至相背离的情况 (如得到负的黏度系数), 目前有通过在模型中加入标记函数和使用机器学习两种方法对此问题进行了探索.

最开始数据驱动模型的研究是利用确定形状流道和流动状态的数据来确定模型参数 (L4) 或者对模型函数进行重构 (L3), 但是这种方式得到的模型只在特定条件下可用, 此后研究者利用使用机器学习的方法对流场空间的每一个点进行分析来修正模型, 得到了更精确的结果.

理论上, 机器学习的方法在数据和待研究问题之间构建了一个桥梁, 学者们可以将机器学习的方法视为一个 "黑箱" 对现有的模型进行修正 (L3、L4), 也可以利用算法对大量的数据进行学习直接得到待研究问题的显式模型 (L2), 目前使用机器学习的方法也得到了非常令人欢欣鼓舞的进展, 但是仍有许多问题需要解决, 主要有如下五个方面: (1) 数据集和待研究问题之间必须具有关联性; (2) 数据与模型必须匹配. 由于不同模型采用了不同假设, 倘若训练所用的数据和待修正模型不匹配就无法得到很好的结果; (3) 研究目标的设定. 一般来说, 研究者需要对待研究问题涉及的特征量进行设置, 使其满足物理约束和数学约束, 但是应该设置多少特征量, 对于不同的问题, 最优设置是什么这些问题都需要解决; (4) 所得模型的可信度. 训练和验证模型的数据都具有一定的局限性, 这就导致所得到的模型在一部分问题中适用, 但在其他问题中不适用; (5) 依赖机器学习和物理约束等条件可以重构出新的模型, 也可能得到错误的结果, 如何保持数据和模型之间平衡是一个待解决的问题.

参 考 文 献

[12.1] Diamond M C, Scheibel A B, Murphy Jr G M, Harvey T. On the brain of a scientist: Albert Einstein. Experimental Neurology, 1985, 88: 198-204.

[12.2] Thompson P M, Giedd J N, Woods R P, MacDonald D, Evans A C, Toga A W. Growth patterns in the developing brain detected by using continuum mechanical tensor maps. Nature, 2000, 404: 190-193.

[12.3] Wong A K C, Bugliarello G. Artificial intelligence in continuum mechanics. Journal of the Engineering Mechanics Division, 1970, 96: 1239-1265.

[12.4] Rao D K. Discussion of "Artificial Intelligence in Continuum Mechanics by Andrew Ka Ching Wong and George Bugliarello". Journal of the Engineering Mechanics Division, 1971, 97: 1583.

[12.5] Raccuglia P, Elbert K C, Adler P D F, Falk C, Wenny M B, Mollo A, Zeller M, Friedler S A, Schrier J, Norquist A J. Machine-learning-assisted materials discovery using failed experiments. Nature, 2016, 533: 73-76.

[12.6] Nosengo N. The material code. Nature, 2016, 533: 22-25.

[12.7] Silver D, Huang A, Maddison C J, Guez A, Sifre L, van den Driessche G, Schrittwieser J, Antonoglou I, Panneershelvam V, Lanctot M, Dieleman S, Grewe D, Nham J, Kalchbrenner N, Sutskever I, Lillicrap T, Leach M, Kavukcuoglu K, Graepel T, Hassabis D. Mastering the game of Go with deep neural networks and tree search. Nature, 2016, 529: 484-489.

[12.8] Silver D, Schrittwieser J, Simonyan K, Antonoglou I, Huang A, Guez A, Hubert T, Baker L, Lai M, Bolton A, Chen Y, Lillicrap T, Hui F, Sifre L, van den Driessche G, Graepel T, Hassabis D. Mastering the game of Go without human knowledge. Nature, 2017, 550: 354-359.

[12.9] Silver D, Hubert T, Schrittwieser J, Antonoglou I, Lai M, Guez A, Lanctot M, Sifre L, Kumaran D, Graepel T, Lillicrap T, Simonyan K, Hassabis D. A general reinforcement learning algorithm that masters chess, shogi, and Go through self-play. Science, 2018, 362: 1140-1144.

[12.10] Abadi M, Barham P, Chen J, Chen Z, Davis A, Dean J, Devin M, Ghemawat S, Irving G, Isard M, Kudlur M, Levenberg J, Monga R, Moore S, Murray D G, Steiner B, Tucker P, Vasudevan V, Warden P, Wicke M, Yu Y, Zheng X. Tensorflow: A system for large-scale machine learning//Proceedings of the 12th USENIX Symposium on Operating Systems Design and Implementation (OSDI'16). Savannah, GA, USA, November 2–4, 2016, 16: 265-283.

[12.11] Pellionisz A, Llinás R. Brain modeling by tensor network theory and computer simulation. The cerebellum: Distributed processor for predictive coordination. Neuroscience, 1979, 4: 323-348.

[12.12] Lewin K. Action research and minority problems. Journal of Social Issues, 1946, 2: 34-46.

[12.13] Sahakian W S. History and systems of social psychology. 1982. (中译本：萨哈金. 社会心理学的历史与体系 (周晓虹等译). 贵阳: 贵州人民出版社, 1991.)

[12.14] Lewin K. Principles of Topological Psychology. New York: McGraw-Hill, 1936.

[12.15] Ivancevic V, Aidman E. Life-space foam: A medium for motivational and cognitive dynamics. Physica A, 2007, 382: 616-630.

[12.16] Comte A. A General View of Positivism. London: Trübner and Co., 1865.

[12.17] Aron R. Les Étapes de la pensée sociologique, Paris: Gallimard, 1967 (英文版: Main Currents in Sociological Thought, London: Weidenfeld & Nicolson, 1965. 中译本: 雷蒙·阿隆. 社会学主要思潮 (葛智强、王沪宁、胡秉诚译). 上海: 上海译文出版社, 2005.)

[12.18] Helbing D, Molnár P. Social force model for pedestrian dynamics. Physical Review E, 1995, 51: 4282-4287.

[12.19] Helbing D, Farkas I, Vicsek T. Simulating dynamical features of escape panic. Nature, 2000, 407: 487-490.

[12.20] Lakoba T I, Kaup D J, Finkelstein N M. Modifications of the Helbing-Molnár-Farkas-Vicsek social force model for pedestrian evolution. Simulation, 2005, 81: 339-352.

[12.21] Lazer D, Pentland A, Adamic L, Aral S, Barabási A L, Brewer D, Christakis N, Contractor N, Fowler J, Gutmann M, Jebara T, King G, Macy M, Roy D, van Alstyne M. Computational social science. Science, 2009, 323: 721-723.

[12.22] Bain N, Bartolo D. Dynamic response and hydrodynamics of polarized crowds. Science, 2019, 363: 46-49.

[12.23] Šíma V, Podolský J. Buquoy's problem. European Journal of Physics, 2005, 26: 1037-1045.

[12.24] Mantegna R N. Lévy walks and enhanced diffusion in Milan stock exchange. Physica A, 1991, 179: 232-242.

[12.25] Slanina F. Essentials of Econophysics Modelling. Oxford: Oxford University Press, 2013.

[12.26] Mantegna R N, Stanley H E. Scaling behaviour in the dynamics of an economic index. Nature, 1995, 376: 46-49.

[12.27] Ghashghaie S, Breymann W, Peinke J, Talkner P, Dodge Y. Turbulent cascades in

[12.28] Mantegna R N, Stanley H E. Turbulence and financial markets. Nature, 1996, 383: 587-588.

[12.29] Mantegna R N, Stanley H E. Introduction to Econophysics: Correlations and Complexity in Finance. Cambridge: Cambridge University Press, 1999.

[12.30] Majorana E, Mantegna R N. The value of statistical laws in physics and social sciences//Ettore Majorana Scientific Papers. Springer, Berlin, Heidelberg, 2006: 237-260; Majorana E. Il valore delle leggi statistiche nella fisica e nelle scienze sociali. Scientia, 1942, 36: 58-66.

[12.31] Schaden M. Quantum finance. Physica A, 2002, 316: 511-538.

[12.32] Baaquie B E. Quantum Finance: Path Integrals and Hamiltonians for Options and Interest Rates. Cambridge: Cambridge University Press, 2007.

[12.33] Zhang C, Huang L. A quantum model for the stock market. Physica A, 2010, 389: 5769-5775.

[12.34] Pedram P. The minimal length uncertainty and the quantum model for the stock market. Physica A, 2012, 391: 2100-2105.

[12.35] Piotrowski E W, Słdkowski J. Quantum market games. Physica A, 2002, 312: 208-216.

[12.36] Choustova O A. Quantum Bohmian model for financial market. Physica A, 2007, 374: 304-314.

[12.37] Holling C S. Resilience and stability of ecological systems. Annual Review of Ecology and Systematics, 1973, 4: 1-23.

[12.38] Dimitrov M K. Why communism did not collapse: Understanding authoritarian regime resilience in Asia and Europe. Cambridge: Cambridge University Press, 2013.

[12.39] Estola M. Newtonian Microeconomics. New York: Palgrave Macmillan, 2017.

[12.40] Georgescu-Roegen N. The Entropy Law and the Economic Process. Cambridge, Massachusetts: Harvard University Press, 1971.

[12.41] 杨卫. 研究生教育动力学——理论框架初探. 学位与研究生教育, 2006 (2): 1-9.

[12.42] 杨卫. 研究生教育动力学——定性讨论与案例构想. 学位与研究生教育, 2006 (4): 8-14.

[12.43] Yang W, Wang H T, Li T F, Qu S X. X-Mechanics — An endless frontier. Science China Physics, Mechanics & Astronomy, 2019, 62: 14601.

[12.44] Duraisamy K, Iaccarino G, Xiao H. Turbulence modeling in the age of data. Annual Review of Fluid Mechanics, 2019, 51: 357-377.

附录 A 和本书内容相关的科学大事年表

年代	事件	本书章节
250 BC	阿基米德 *On Floating Bodies*，阿基米德原理	§56.1
1600 年	威廉·吉尔伯特 (William Gilbert) 出版第一部磁学专著 *De Magnete*	
1609 年	托马斯·哈里奥特 (Thomas Harriot) 用望远镜观测月地球，首次创作完成了月球表面的地形图	
1614 年	约翰·纳皮尔 (John Napier) 提出 "对数" (logarithms)	
1619 年	开普勒提出彗星尾巴总是指向并远离太阳的原因是受到了来自阳光的辐射压	
1628 年	哈维《心血运动论》	§50
1637 年	笛卡儿坐标系创立 费马大定理提出	§15 思考题 1.3
1638 年	伽利略《关于两门新科学的对话》	§48
1644 年	笛卡儿《哲学原理》讨论碰撞问题，引入动量的概念	
1646 年	帕斯卡水桶实验	§41.1
1670 年	马略特开展 "牛顿摇篮" 的实验	§55.1
1678 年	胡克定律	题表 3.1
1686 年	莱布尼兹引入活力 (vis viva, mv^2) 的概念 马略特《论水和其他流体的运动》，得出流体阻力与流速成正比	§51.1 §55.1
1687 年	牛顿《自然哲学的数学原理》	§1.1, §53.2, §55.1
1694 年	雅克布·伯努利提出梁的平截面假定，伯努利梁模型	§4.3
1696 年	罗必塔法则	§28.1
1717 年	以牛顿的建议为基础，英国实行金本位制度	§65.3
1733 年	黑尔斯测量马的血压	§50
1738 年	丹尼尔·伯努利《水动力学》	§51.2.4
1743 年	达朗贝尔《论动力学》	
1747 年	达朗贝尔建立弦的振动方程，现代偏微分方程经典文献	§4.1
1749 年	达朗贝尔伴谬	§55.2
1750 年	欧拉 "微元体"	§4.2
1751 年	塞格纳提出表面张力概念	§60.1
1765 年	欧拉提出 "转动惯量" (moment of inertia)	§9.1
1782 年	黎卡提利用弯曲实验确定材料的弹性模量	§40
1788 年	拉格朗日《分析力学》	§4.1
1797 年	文丘里效应；文丘里第一次指出：达·芬奇不仅仅是位艺术家，还是一位科学家	§51.2
1807 年	托马斯·杨引入能量的概念	
1815 年	布阔伊将经典力学原理引入到经济学	§65.3
1821 年	纳维提出弹性体的一般运动方程	§40

续表

年代	事件	本书章节
1822 年	柯西应力原理与基本定理，9 月 30 日在巴黎科学院宣布	§46.4
	纳维应用分子理论推导出流体的动力学方程	§56.1
1826 年	纳维出版国际上第一部《材料力学》	§48
1827 年	高斯绝妙定理，内蕴微分几何	§21.1
1829 年	泊松比	§40
1831 年	科里奥利引入力做功的概念，在活力 (mv^2) 前面引入 1/2 即动能	
1839 年	格林应变	§27.1
	哈根的圆管流动实验	§54
1845 年	焦耳定律，焦耳提出"内能"的概念	
	斯托克斯完善"纳维–斯托克斯方程"	§56
1847 年	李斯亭引入拓扑学	§22.1
1851 年	斯托克斯阻力公式	§58
	斯宾塞《社会静力学》	§65.2
1852 年	拉梅引入曲线坐标	例 15.11
1853 年	凯利进行历史上第一次有人乘坐的重于空气的航空器升空自由飞行	思考题 11.10
1854 年	黎曼几何	§21.2
1855 年	圣维南原理	§48
1856 年	孔德创立社会物理学	§65.2
1859 年	拉梅关于曲纹坐标系的描述方法，拉梅系数	§15
1864 年	麦克斯韦建立功的互等定理	思考题 9.2
1865 年	麦克斯韦方程组创立	例 8.4
1866 年	克罗内克 δ 符号	§8.1
1868 年	黎曼于 1854 年的论文得以发表	§21.2
1869 年	克里斯托费尔符号 (Christoffel symbols)	§16
1870 年	克劳修斯建立位力定理	
1872 年	贝蒂定理 (Betti's theorem)	§47.1
	玻尔兹曼动理学方程	§E.1
1873 年	普拉托水流失稳离散成液滴的实验	§62.4
1874 年	两种构形面积微元变换的南森公式	§29.1
1878 年	瑞利对水流失稳离散成液滴建立模型，普拉托–瑞利不稳定性	§62.4
1883 年	雷诺发表《平行渠道阻力的实验研究情况》，研究层流向湍流的转捩	§59.1
1885 年	马赫数	§57
	亥维赛将麦克斯韦方程组的 20 个方程简化到 4 个	思考题 2.10
1896 年	意大利内科和儿科医生里瓦罗基袖带水银血压计	§50
1898 年	沃耳德玛·福格特 (Woldemar Voigt) 引入现代张量的标记法	
	赫尔–肖流动	§62.5.2

续表

年代	事件	本书章节
1900 年	里奇和他的学生列维–奇维塔《绝对微分法及其应用》 贝尔纳胞,贝尔纳对流 巴谢利耶的博士论文《投机的理论》宣告了金融数学的诞生	§8.2 思考题 11.8 §65.7
1901 年	庞加莱提出分析力学的对偶原理 (duality principle)	
1903 年	莱特兄弟第一架飞机试飞成功 雷诺输运定理	题表 3.1 §30
1904 年	普朗特建立边界层理论	§55.2
1905 年	柯洛克夫加上了听诊器的现代血压测量方法 庞加莱预言引力波以光速传播	§50 §15
1907 年	马洛克 (Henry Reginald Arnulpt Mallock) 观察到流体流经平板后所产生的交错涡旋	§59.2
1908 年	索末菲建议使用无量纲数 "雷诺数 R" 贝尔纳涡街	§55.2 §59.2
1909 年	科恩不等式	例 12.7
1910 年	奥森流和奥森方程	例 58.4
1911 年	阿尔曼西应变 努森数 Kn 卡门涡街	§27.2 §2.2 §59.2
1915 年	索末菲引入 "精细结构常数"	思考题 3.8
1916 年	爱因斯坦求和约定 史瓦西度规	§8.1 §15
1918 年	诺特定理	§20.3
1919 年	爱丁顿利用超长日全食对爱因斯坦广义相对论的光线引力偏折效应进行成功检测	§10
1922 年	理查森能量串级	§59.4
1923 年	爱丁顿《相对论的数学理论》,张量方程的量纲一致性原理	§10
1928 年	亨奇提出对数应变	§27.1
1929 年	Jakob Ackeret 命名 "马赫数" Ma	§57
1934 年	比奥应变 泰勒引入毛细数 勒雷证明纳维–斯托克斯方程弱解的存在性	§28.2 附录 C 思考题 11.22
1939 年	狄拉克符号	§5
1940 年	美国塔科马峡谷桥倒塌,成为典型风载作用下结构失效的经典案例	§59.2
1941 年	柯尔莫哥洛夫关于不可压缩湍流的 2/3 标度律 (K41) 奥布霍夫关于不可压缩湍流的 $-5/3$ 标度律	§59.5 §59.6
1943 年	博戈柳博夫级联	§2.3
1946 年	勒温创立社会心理学中的场论和生活空间	§65.1
1948 年	里夫林建立新胡克体构关系	§42
1950 年	奥尔德罗伊德本构公理	§39.1

续表

年代	事件	本书章节
1954 年	Yang-Mills 规范场	§10
1956 年	达特茅斯研讨会召开，人工智能 (AI) 诞生	§64.1
1958 年	诺尔本构三公理	§39.1
	萨夫曼–泰勒不稳定性	§62.5
1961 年	赛斯应变度量	§28.2
1965 年	摩尔定律	题表 1.10
1968 年	希尔应变度量	§28.1
	功共轭	§29.3
1970 年	保罗·萨缪尔森获得诺贝尔经济学奖并提出获得诺贝尔奖的五个条件	思考题 12.3
1971 年	尼可拉斯·乔治斯库–罗根首次将熵的概念引入经济学	思考题 12.5
1972 年	索利斯和科斯特里兹提出拓扑相变	§22.2
	贝尔定律	题表 1.10
1973 年	布莱克–斯克尔斯期权定价公式	§65.7
1994 年	斯坦利创造"经济物理学"术语	§65.3
1995 年	安德鲁·怀尔斯发表有关费马大定理的证明	思考题 1.3
1996 年	湍流理论被应用于经济学领域	§65.3
2000 年	人脑发育生长的张量图诞生	§63
2002 年	维拉尼证明科恩不等式与玻尔兹曼统计理论中熵增的关系	例 12.7
2006 年	深度学习诞生	§64.2
2013 年	美国脑科学计划 (BRAIN Initiative) 启动	§63.3
2016 年	索利斯·霍尔丹和科斯特里兹因拓扑相变获得	§22.2
	诺贝尔物理学奖 AlphaGo 诞生	§64.2
	TensorFlow	§64.3
2017 年	AlphaZero 诞生	§64.2
2018 年	Arthur Ashkin 因基于辐射压的概念发明光学镊子获得诺贝尔物理学奖	

附录 B 连续介质力学中的相关物理量

物理量可以从多个不同的角度进行分类. 一方面, 可分为基本量 (base quantity) 和导出量 (derived quantity), 基本量如图 B.1 所示, 其相互依赖关系如图 B.2 所示; 另一方面, 还可分为广延量 (extensive quantity) 和强度量 (intensive quantity). 再一方面, 可分为标量、矢量 (向量) 和张量. 表 B.1 给出了基本物理量一览表, 熟悉这些量对进行量纲分析, 特别是进行快速量纲匹配十分重要.

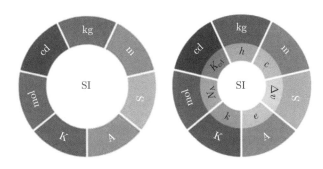

图 B.1 基本物理量, 右图为 2018 年 11 月新修订的国际单位制

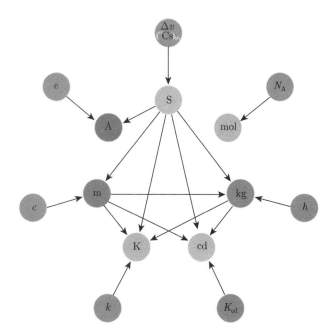

图 B.2 基本物理量的相互依赖关系

2018 年 11 月 16 日，第 26 届国际计量大会 (CGPM) 经各个成员国表决，通过了关于"修订国际单位制 (SI)"的决议. 根据决议，国际单位制基本单位中的 4 个 —— 千克、安培、开尔文、摩尔分别改由普朗克常数、基本电荷常数、玻尔兹曼常数、阿佛加德罗常数定义. 决议将于 2019 年国际计量日，即 5 月 20 日期正式生效.

表 B.1 基本物理量一览表

基本量	符号	描述	国际基本单位	量纲	评论
长度 (length)	l	物体的 1 维尺度	米 (meter, m)	L	广延量、标量
质量 (mass)	m	物体抵抗加速度的度量	千克 (kilogram, kg)	M	广延量、标量
时间 (time)	t	事件的持续时间	秒 (second, s)	T	标量
电流 (electric current)	I	每单位时间的电荷流量	安培 (ampere, A)	I	
温度 (temperature)	T	系统每个自由度的动能	开尔文 (kelvin, K)	Θ	强度量、标量
物质的量 (amount of substance)	n	与 0.012 千克的 ^{12}C 中原子的数量相等粒子的数量	摩尔 (mole, mol)	N	广延量、标量
照度 (luminous intensity)	L	单位固体角发射光的波长加权功率	坎德拉 (candela, cd)	J	标量

乘积具有功或能量的量纲的一对由强度量和广延量所组成的参量称为共轭参量对，如表 B.2 所示. 这里的乘积是针对标量而言的；对于向量而言，是内积；对于张量则是双缩并.

作用量 (action) 的共轭变量对为：能量和时间；动量和位置；角动量和旋转角度.

表 B.2 功共轭参量对一览表

公式	强度量	广延量
$F = U - TS$, 亥姆霍兹自由能等于内能减去束缚能	温度 T	熵 S
$H = U + pV$, 焓等于内能加上压强和体积的乘积	压强 p	体积 V
$dE = TdS - pdV + \mu dN$	化学势 μ	粒子数 N
$dE = TdS - pdV + \mu dN + \gamma dA$	表面张力 γ	表面积 A

常用的导出物理学由表 B.3 给出.

附录 B 连续介质力学中的相关物理量

表 B.3　导出物理量一览表 (以物理量的英文排序)

导出量 derived quantity	符号 symbol	描述 description	国际单位 SI unit	量纲 dimension	评论 comments
加速度 (acceleration)	$\boldsymbol{a}$	速度的时间变化率	米/秒^2 (ms^{-2})	LT^{-2}	矢量
角加速度 (angular acceleration)	$\boldsymbol{\alpha}$	角速度的时间变化率	弧度/秒^2 (rad s^{-2})	T^{-2}	
角动量 (angular momentum)	$\boldsymbol{L}$	位置矢量和动量的叉积	千克·米^2/秒^2 (kg m^2 s^{-1})	$M L^2 T^{-1}$	赝矢量
角速度 (angular speed or velocity)	$\boldsymbol{\omega}$	角位移的时间变化率	弧度/秒 (rad s^{-1})	T^{-1}	赝矢量
面积 (area)	A	表面的范围	米^2 (m^2)	L^2	矢量
电容 (capacitance)	C	单位电势的存储电荷	法拉 (F, s^4·A^2·m^{-2}·kg^{-1})	$M^{-1} L^{-2} T^4 I^2$	标量
化学势 (chemical potential)	μ	每单位物质变化的能量	焦耳/摩尔 (J mol^{-1})	$M L^2 T^{-2} N^{-1}$	强度量
爆裂声 (crackle)	$\boldsymbol{c}$	位置的五阶时间导数	米/秒^5 (ms^{-5})	$L T^{-5}$	矢量
电流密度 (current density)	$\boldsymbol{J}$	单位面积上的电流	安培/米^2 (A/m^2)	$A L^{-2}$	矢量
密度 (density)	ρ	单位体积的质量	千克/米^3 (kg/m^3)	ML^{-3}	强度量
动力黏度 (dynamic viscosity)	μ	不可压缩牛顿流体切应力和应变率之比	Pa·s	$M L^{-1} T^{-1}$	
电荷 (electric charge)	Q	单位电场强度的力	库仑 (C, As)	TI	广延量
电荷密度 (electric charge density)	ρ	单位体积电荷	库仑/米^3 (C/m^3)	$L^{-3}TI$	强度量
电导 (electric conductance)	G	电流通过材料容易程度的度量	西门子 (S, A^{2}s^{3}kg^{-1}m^{-2})	$M^{-1}L^{-2}T^3I^2$	标量
电导率 (electric conductivity)	σ	材料传导电流能力的度量	西门子/米 (Sm^{-1})	$M^{-1}L^{-3}T^3I^2$	标量
电位移 (electric displacement)	$\boldsymbol{D}$	电位移的强度	库仑/米^2 (C/m^2)	$L^{-2}TI$	矢量
电场强度 (electric field strength)	$\boldsymbol{E}$	电场的强度	伏/米 (V/m)	$MLT^{-3} I^{-1}$	矢量

续表

导出量 derived quantity	符号 symbol	描述 description	国际单位 SI unit	量纲 dimension	评论 comments
电势（电位） (electric potential)	V	单位电荷所具有的电势能	伏特 (V, kg m² A⁻¹ s⁻³)	$M L^2 T^{-3} I^{-1}$	广延量
电阻 (electric resistance)	R	单位电流的电势	欧姆 (Ω, kg m² A⁻² s⁻³)	$M L^2 T^{-3} I^{-2}$	广延量
电阻率 (electric resistivity)	ρ	电阻与横截面积的乘积与长度的比值，材料本身的性质	欧姆·米 ($\Omega\cdot$m, kg m³ A⁻² s⁻³)	$M L^3 T^{-3} I^{-2}$	强度量
能量 (energy)	E	物体做功的能力	焦耳 (J, kg m² s⁻²)	$M L^2 T^{-2}$	广延量
能量密度 (energy density)		单位体积的能力	焦耳/米³ (J/m³)	$M L^{-1} T^{-2}$	强度量
熵 (entropy)	S	体系混乱程度的度量	焦耳/开尔文 (J K⁻¹)	$M L^2 T^{-2} \Theta^{-1}$	广延量
力 (force)	$\boldsymbol{F}$	单位时间所传递的动量	牛顿 (N, kg m s⁻²)	$M L T^{-2}$	矢量
频率 (frequency)	f	单位时间发生的次数	赫兹 (Hz, s⁻¹)	T^{-1}	标量
热 (heat)	Q	热能	焦耳 (J)	$M L^2 T^{-2}$	广延量
热容 (heat capacity)	C_p	单位温度变化的能量	焦耳/开尔文 (J/K)	$M L^2 T^{-2} \Theta^{-1}$	广延量
热流密度 (heat flux density)		单位时间单位面积的热流	瓦/米² (J/m²)	$M T^{-3}$	
冲量 (impulse)	$\boldsymbol{J}$	所传递的动量	牛顿·秒 (N·s, kg m s⁻¹)	$M L T^{-1}$	矢量
加加速度 (jerk) 急动度	$\boldsymbol{j}$	加速度为时间的导数；位置对时间的三次导数	米/秒³ (ms⁻³)	$L T^{-3}$	矢量
加加加速度 (jounce or snap)	$\boldsymbol{s}$	位置对时间的四次导数	米/秒⁴ (ms⁻⁴)	$L T^{-4}$	矢量
运动黏度 (kinematic viscosity)	ν	动力黏度和密度之比	m²s⁻¹	$L^2 T^{-1}$	
质量扩散率 (mass diffusivity)	D	分子通量和分子浓度梯度之间的比例	米²/秒 (m²s⁻¹)	$L^2 T^{-1}$	
转动惯量 (moment of inertia)	I	相当于角加速度的惯性	千克·米² (kg·m²)	$M L^2$	张量、标量
动量 (momentum)	$\boldsymbol{p}$	物体质量和速度的乘积	牛顿·秒 (N·s)	$M L T^{-1}$	矢量
介电常数 (permittivity)	ε	外部电场影响材料极化的度量	法拉/米 (Fm⁻¹)	$M^{-1} L^{-3} T^4 I^2$	强度量

附录 B 连续介质力学中的相关物理量

续表

导出量 derived quantity	符号 symbol	描述 description	国际单位 SI unit	量纲 dimension	评论 comments
砰然声 (pop)	$\boldsymbol{p}$	位置的六阶时间导数	米/秒^5 (ms^{-6})	$L\,T^{-6}$	矢量
功率 (power)	P	单位时间所做的功	瓦特 (W)	$M\,L^2\,T^{-3}$	广延量
压强 (pressure)	p	单位面积的力	帕斯卡 (kg m^{-1} s^{-2})	$M\,L^{-1}\,T^{-2}$	强度量
应变 (strain)	ε	单位长度的伸长	无量纲		二阶张量
应力 (stress)	$\boldsymbol{\sigma}$	单位面积的力	帕斯卡	$M\,L^{-1}\,T^{-2}$	二阶张量
表面张力 (surface tension)	γ	产生单位新表面 所需的能量	焦耳/米^2 (Jm^{-2})	$M\,T^{-2}$	强度量
热导率 (thermal conductivity)	κ	物质导热能力的度量	瓦/(米·开尔文) (W m^{-1} K^{-1})	$M\,L\,T^{-3}\Theta^{-1}$	强度量
热扩散率 (thermal diffusivity)	α	$\alpha=\kappa/(\rho c_p)$	米^2/秒 (m^2/s)	$L^2\,T^{-1}$	
波数 (wavenumber)	k	单位距离内波的个数	1/米 (m^{-1})	L^{-1}	标量
波矢量 (wavevector)	$\boldsymbol{k}$		1/米 (m^{-1}) 有方向	L^{-1}	矢量
功 (work)	W	力矢量和距离矢量的点积	焦耳 (joule, J)	$M\,L^2\,T^{-2}$	标量
杨氏模量 (Young's modulus)	E	一维时，各向同性线弹性 材料的应力和应变之比	帕斯卡 (pascal, Pa) 单位面积上的力	$M\,L^{-1}\,T^{-2}$	标量

附录 C 连续介质力学中的无量纲数

无量纲数名称	提出者	表达式和符号意义	说明
阿基米德数 Archimedes number	以阿基米德 (Archimedes, c.287—c.212 BC) 命名	$Ar = \dfrac{重力}{黏性力} = \dfrac{gL^3 \rho_l (\rho - \rho_l)}{\mu^2}$	可用来判别因密度差异造成的流体运动
阿伦尼乌斯数 Arrhenius number	以阿伦尼乌斯 (Svante Arrhenius, 1859—1927) 命名	$\alpha = \dfrac{激活能}{热能} = \dfrac{E_0}{RT}$	用于传质，特别用于反应速率的计算
宾厄姆数 Bingham number	以宾厄姆 (Eugene Cook Bingham, 1878—1945) 命名	$Bm = \dfrac{屈服应力}{黏性应力} = \dfrac{\tau_y L}{\mu V}$	用于流变力学
毕奥数 Biot number	Jean-Baptiste Biot 1774—1862	$Bi = \dfrac{传热阻力}{对流阻力} = \dfrac{hL}{\kappa}$	用于传热学 h 为传热系数
邦德数 Bond number	以邦德 (Wilfrid Noel Bond, 1897—1937) 命名	$Bo = \dfrac{重力}{表面张力} = \dfrac{\rho g L^2}{\gamma}$	用于液滴动力学
毛细数 Capillary number	泰勒 (G I Taylor, 1886—1975) 于 1934 年引入	$Ca = \dfrac{黏性力}{表面张力} = \dfrac{\mu V}{\gamma}$	用于液滴动力学
柯西数 Cauchy number	以柯西 (Augustin-Louis Cauchy, 1789—1857) 命名	$Ca = \dfrac{惯性力}{弹性模量} = \dfrac{\rho V^2}{E}$	弹性模量表征的是固体材料的可压缩性，用于冲击动力学
空化数 Cavitation number	由 Thoma 和 Leroux 于 1923—1925 年间提出	$\sigma_C = \dfrac{压强差}{惯性力} = \dfrac{2(p - p_v)}{\rho V^2}$	用于固-液界面动力学
损伤数 Damage number	William Johnson (1922—2010) 于 1972 年建议	$Da = \dfrac{惯性力}{屈服应力} = \dfrac{\rho V^2}{\sigma_y}$	用于冲击动力学
达姆科勒数 Damköhler number	Gerhard Damköhler 1908—1944	$Da = \dfrac{流动特征时间}{化学反应特征时间} = \dfrac{t_f}{t_r}$	应用于化学动力学及反应工程学
达西数 Darcy number	以达西 (Henry Darcy, 1803—1858) 命名	$Da = \dfrac{渗透率}{特征长度^2} = \dfrac{\kappa}{L^2}$	用于多孔介质
德博拉数 Deborah number	以圣经中女预言家德博拉 (Deborah) 命名	$De = \dfrac{弛豫时间}{观测时间} = \dfrac{\tau}{T}$	由莱纳 (Markus Reiner, 1886—1976) 引入用于介质划分

附录 C 连续介质力学中的无量纲数

续表

无量纲数名称	提出者	表达式和符号意义	说明
埃克曼数 Ekman number	Vagn Walfrid Ekman (1874—1954)	$Ek = \dfrac{\text{黏滞力}}{\text{科氏力}} = \dfrac{\nu}{\Omega L^2}$	描述黏滞力和科氏力 平衡的边界层(埃克曼层), Ω 为角速度
厄缶数 Eötvös number	以厄缶 (Loránd Eötvös, 1848—1919) 命名	和邦德数相同, 在欧洲常称为厄缶数, 在世界其他地方常称邦德数	
欧拉数 Euler number	以欧拉 (Leonhard Euler, 1707—1783) 命名	$Eu = \dfrac{\text{压强差}}{\text{惯性力}} = \dfrac{\Delta p}{\rho V^2}$	见 §56.3 和 §56.4
傅里叶数 Fourier number	以傅里叶 (Joseph Fourier, 1768—1830) 命名	$Fo = \dfrac{\text{时间}}{\text{热传导特征时间}} = \dfrac{\alpha t}{L^2}$ $Fo = \dfrac{\text{时间}}{\text{扩散特征时间}} = \dfrac{Dt}{L^2}$	均可理解为无量纲时间
弗劳德数 Froude number	William Froude 1810—1879	$Fr = \sqrt{\dfrac{\text{惯性力}}{\text{重力}}} = \dfrac{V}{\sqrt{gL}}$	
伽利略数 Galilei number	以伽利略 (Galileo Galilei, 1564—1642) 命名	$Ga = \dfrac{\text{重力}}{\text{黏性力}} = \dfrac{gL^3}{\nu^2}$	用于黏性流和化工
格拉晓夫数 Grashof number	以格拉晓夫 (Franz Grashof, 1826—1893) 命名	$Gr = \dfrac{\text{浮力}}{\text{黏性力}} = \dfrac{gL^3 \alpha \Delta T}{\nu^2}$	α 为体积热膨胀系数 ΔT 为温度差
哈特曼数 Hartmann number	Julius Hartmann 1881—1951	$Ha = \sqrt{\dfrac{\text{磁场力}}{\text{黏性力}}} = BL\sqrt{\dfrac{\sigma}{\mu}}$	用于磁流体力学 B 为磁场; σ 为导电率
努森数 Knudsen number	Martin Knudsen 1871—1949	$Kn = \dfrac{\text{气体平均自由程}}{\text{物体特征尺度}} = \dfrac{\lambda}{L}$	1911 年丹麦 物理学家努森提出
拉普拉斯数 Laplace number	以拉普拉斯 (Pierre-Simon Laplace, 1749—1827) 命名	$La = \dfrac{\text{表面张力}}{\text{动量传递力}} = \dfrac{\gamma \rho L}{\mu^2}$	$La = (Oh)^{-2}$
马赫数 Mach number	Ernst Mach 1838—1916	$Ma = \dfrac{\text{运动物体速度}}{\text{当地声速}} = \dfrac{V}{c}$	1885 年马赫提出, Jacob Ackeret 于 1929 年建议使用 Mach number 的称谓
努塞尔特数 Nusselt number	Wilhelm Nusselt 1882—1957	$Nu = \dfrac{\text{对流传热}}{\text{热传导}} = \dfrac{hL}{\kappa}$	用于边界层 h 为传热系数
奥内佐格数 Ohnesorge number	Wolfgang von Ohnesorge 1901—1976	$Oh = \dfrac{\text{黏性毛细时间}}{\text{瑞利时间}} = \dfrac{\mu}{\sqrt{\rho L \gamma}}$	于 1936 年在其博士 论文中提出. $t_{vis} = \mu L/\gamma$, $t_R = \sqrt{\rho L^3/\gamma}$

续表

无量纲数名称	提出者	表达式和符号意义	说明
佩克莱数 Péclet number	以佩克莱 (Jean Claude Eugène Péclet, 1793—1857) 命名	$Pe = \dfrac{\text{对流输运率}}{\text{热扩散输运率}} = \dfrac{LV}{\alpha}$	$Pe = Re \cdot Pr$
普朗特数 Prandtl number	以普朗特 (Ludwig Prandtl, 1875—1953) 命名	$Pr = \dfrac{\text{运动黏性系数}}{\text{热扩散率}} = \dfrac{\nu}{\alpha}$	用于传热传质
雷诺数 Reynolds number	Osborne Reynolds 1842—1912	$Re = \dfrac{\text{惯性力}}{\text{黏性力}} = \dfrac{\rho VL}{\mu} = \dfrac{VL}{\nu}$	1883 年雷诺提出,1908 年索末菲建议使用 Reynolds number 的称谓
理查森数 Richardson number	以理查森 (Lewis Fry Richardson, 1881—1953) 命名	$Ri = \dfrac{\text{浮力}}{\text{惯性力}} = \dfrac{\Delta \rho g h}{\rho V^2}$	用于大气和海洋流动
响应数 Response number	由著者于 1998 年建议	$Rn = \dfrac{\rho V^2}{\sigma_y} \left(\dfrac{L}{H}\right)^2$	在国际上被称为赵无量纲数 (Zhao dimensionless number) 用于冲击动力学
罗什科数 Roshko number	以罗什科 (Anatol Roshko, 1923—2017) 命名	$Ro = St \cdot Re = \dfrac{fL^2}{\nu}$	用于震荡流
施密特数 Schmidt number	以施密特 (Ernst Heinrich Wilhelm Schmidt, 1892—1975) 命名	$Sc = \dfrac{\text{动量扩散率}}{\text{质量扩散率}} = \dfrac{\nu}{D}$	用于传热传质
斯特劳哈尔数 Strouhal number	Vincenc Strouhal 1850—1922 捷克物理学家	$St = \dfrac{\text{频率}}{\text{特征频率}} = \dfrac{fL}{V}$	Strouhal first investigated the steady humming or singing of telegraph wires in 1878.
韦伯数 Weber number	Moritz Weber 1871—1951 德国力学家	$We = \dfrac{\text{惯性力}}{\text{表面张力}} = \dfrac{\rho V^2 L}{\gamma}$	见 §56.4
魏森贝格数 Weissenberg number	以魏森贝格 (Karl Weissenberg, 1893—1976) 命名	$Wi = \text{剪切率} \times \text{弛豫时间} = \dot{\gamma}\tau$	用于流变学
沃姆斯莱数 Womersley number	以沃姆斯莱 (John Ronald Womersley, 1907—1958) 命名	$Wo = \sqrt{\dfrac{\text{瞬态力}}{\text{黏性力}}} = L\sqrt{\dfrac{\omega}{\nu}}$	用于生物脉动流 ω 为圆频率 为罗什科数的开方

注: 表中的主要符号

ρ—密度, γ—表面张力, μ—动力黏度系数, ν—运动黏度系数, f—频率, g—重力加速度, L—特征尺度, V—速度.

附录 D 弗雷歇导数和加托导数

本附录是 §18、§43 和 §44 等内容的数学基础.

D.1 可微、可偏导、连续的关系

D.1.1 全微分、偏微分、连续性

首先讨论全微分. 函数 $z = f(x,y)$ 在点 $X_0 = (x_0, y_0)$ 的全增量为

$$\begin{aligned}\Delta z &= f(x+\Delta x, y+\Delta y) - f(x,y) \\ &= a\Delta x + b\Delta y + o\left(\sqrt{\Delta x^2 + \Delta y^2}\right)\end{aligned} \tag{D-1}$$

则称函数 $z = f(x,y)$ 在点 $X_0 = (x_0, y_0)$ 处可微，且全微分定义为

$$\mathrm{d}z \triangleq a\Delta x + b\Delta y \tag{D-2}$$

将 Δx, Δy 分别记为 $\mathrm{d}x$, $\mathrm{d}y$, 可将全微分表示为 $\mathrm{d}z = a\mathrm{d}x + b\mathrm{d}y$, 其极限形式为

$$\lim_{\Delta x, \Delta y \to 0} \frac{\Delta z - (a\Delta x + b\Delta y)}{\sqrt{\Delta x^2 + \Delta y^2}} = 0 \tag{D-3}$$

然后讨论偏微分. 函数 $z = f(x,y)$ 在 $X_0 = (x_0, y_0)$ 处可微，按照全微分的定义 $\mathrm{d}z = a\mathrm{d}x + b\mathrm{d}y$, 可定义函数关于 x 的偏微分为 $\mathrm{d}_x z \triangleq \frac{\partial z}{\partial x}\Delta x = \frac{\partial z}{\partial x}\mathrm{d}x$, 定义函数关于 y 的偏微分为 $\mathrm{d}_y z \triangleq \frac{\partial z}{\partial y}\Delta y = \frac{\partial z}{\partial y}\mathrm{d}y$, 函数的两个偏导数为 $a = \frac{\partial z}{\partial x}$, $b = \frac{\partial z}{\partial y}$ 均存在. 函数 $z = f(x,y)$ 的全微分可以通过偏导数表示为

$$\mathrm{d}z = \frac{\partial z}{\partial x}\mathrm{d}x + \frac{\partial z}{\partial y}\mathrm{d}y \tag{D-4}$$

函数的连续性可由极限 $\lim\limits_{\Delta x, \Delta y \to 0} \Delta z = 0$ 表示.

D.1.2 方向导数

三维欧几里得空间中存在函数 $u = f(x,y,z)$, 若点 X 沿射线 l 的方向趋于点 $X_0 = (x_0, y_0, z_0)$ 的极限 $\lim\limits_{X \to X_0} \dfrac{f(X) - f(X_0)}{\|X - X_0\|}$ 存在，则在 X_0 处沿 l 方向的方向导数存在，且称 $\dfrac{\partial u}{\partial l} = \lim\limits_{X \to X_0} \dfrac{f(X) - f(X_0)}{\|X - X_0\|}$ 为函数 $u = f(x,y,z)$ 在 X_0 处沿 l 方向的方向导数，记为

$$\frac{\partial u}{\partial l} = f'_l(X_0) = \lim_{X \to X_0} \frac{f(X) - f(X_0)}{\|X - X_0\|} \tag{D-5}$$

对于方向导数的计算，进一步假设 l 方向为

$$l = (\cos\alpha, \cos\beta, \cos\gamma) \tag{D-6}$$

则射线方程表示为

$$x = x_0 + t\cos\alpha, \quad y = y_0 + t\cos\beta, \quad z = z_0 + t\cos\gamma \tag{D-7}$$

即 $X = X_0 + tl$，其中 $l = (\cos\alpha, \cos\beta, \cos\gamma)$，$t = \|X - X_0\|$.

根据函数 $u = f(x, y, z)$ 的全增量，方向导数可表示为

$$\left.\frac{\partial u}{\partial l}\right|_{X=X_0}$$
$$= \lim_{X \to X_0} \frac{f(X) - f(X_0)}{\|X - X_0\|}$$
$$= \lim_{X \to X_0} \frac{\frac{\partial u}{\partial x}\Delta x + \frac{\partial u}{\partial y}\Delta y + \frac{\partial u}{\partial z}\Delta z + o(\|X - X_0\|)}{\|X - X_0\|}$$
$$= \lim_{X \to X_0} \left[\frac{\partial u}{\partial x}\frac{\Delta x}{\|X - X_0\|} + \frac{\partial u}{\partial y}\frac{\Delta y}{\|X - X_0\|} + \frac{\partial u}{\partial z}\frac{\Delta z}{\|X - X_0\|} + \frac{o(\|X - X_0\|)}{\|X - X_0\|}\right]$$
$$= \frac{\partial u}{\partial x}\cos\alpha + \frac{\partial u}{\partial y}\cos\beta + \frac{\partial u}{\partial z}\cos\gamma$$

即方向导数与偏导数的关系为

$$\frac{\partial u}{\partial l} = \frac{\partial u}{\partial x}\cos\alpha + \frac{\partial u}{\partial y}\cos\beta + \frac{\partial u}{\partial z}\cos\gamma \tag{D-8}$$

例 D.1 函数 $z = \sqrt{x^2 + y^2}$ 在坐标原点，两个偏导数都不存在，但是在该点沿任何方向的方向导数都存在：

$$\frac{\partial z}{\partial l} = \lim_{X \to X_0} \frac{\sqrt{x^2 + y^2} - 0}{\sqrt{x^2 + y^2}} = 1 \tag{D-9}$$

对 D.1 节总结如下：

(1) 函数 $f(X)$ 在 X_0 处可微，则必在点 X_0 处连续，且可偏导；

(2) 反之，若偏导数 $\frac{\partial z}{\partial x}, \frac{\partial z}{\partial y}$ 在点 X_0 处连续，则函数 $f(X)$ 在 X_0 处可微；

(3) 方向导数存在，偏导数不一定存在；

(4) 可微是方向导数存在的充分条件，而不是必要条件.

D.2 弗雷歇导数、加托导数

将数学分析中的全微分和方向导数概念推广到巴拿赫空间 (Banach space), 即为弗雷歇导数 (Fréchet derivative) 和加托导数 (Gâteaux derivative). 巴拿赫空间就是完备的赋范向量空间, 见表 1.3 的介绍.

弗雷歇导数以法国数学家莫里斯·弗雷歇 (Maurice Fréchet, 1878—1973) 命名; 而加托导数则以法国英年早逝的数学家勒内·加托 (René Eugène Gateaux, 1889—1914, 见题表 1.5 的介绍) 命名.

首先对于 $x(t):[a,b] \to \mathbb{R}$ 是一个函数, 其中 $t \in [a,b] \subset \mathbb{R}$ 为实数值, $x \in \mathbb{R}$ 为实数, 对 $t_0 \in [a,b]$, 若存在 $x'_0 \in \mathbb{R}$, 使得

$$\lim_{\Delta t \to 0} \left\| \frac{x(t_0 + \Delta t) - x(t_0)}{\Delta t} - x'_0 \right\| = 0 \tag{D-10}$$

则称 $x(t)$ 在 $t = t_0$ 可微, x'_0 叫做 $x(t)$ 在 t_0 处的导数.

同理将之推广到巴拿赫空间, 设 $g(x): K \subset \mathcal{B}_1 \to \mathcal{B}_2$, 其中 $\mathcal{B}_1$, $\mathcal{B}_2$ 为巴拿赫空间, K 是 $\mathcal{B}_1$ 中某个开集, 若存在 $F \in (\mathcal{B}_1 \to \mathcal{B}_2)$, 使得在 x_0 附近

$$g(x_0 + h) - g(x_0) = Fh + o(\|h\|) \tag{D-11}$$

则称 $g(x)$ 在 x_0 处弗雷歇可微, Fh 记作 $\mathrm{D}[g(x_0)h] = Fh = g'(x_0)h$ 称为 $g(x)$ 在 x_0 处的弗雷歇微分, $g'(x_0)$ 为 x_0 处弗雷歇导数. 即

$$\lim_{\|h\| \to 0} \left\| \frac{g(t_0 + h) - g(t_0)}{h} - A'(x_0) \right\| = 0 \tag{D-12}$$

若由 m 个 n 元函数 $f_i(x_1, \cdots, x_n)\,(i=1,2,\cdots,m)$, $\mathbb{R}^n$ 和 $\mathbb{R}^m$ 都是赋范向量空间. 如果存在一个线性算子 $A: \mathbb{R}^n \to \mathbb{R}^m$ 使得

$$\lim \frac{\|f(x+h) - f(x) - Ah\|}{\|h\|} = 0 \tag{D-13}$$

称函数 $f: \mathbb{R}^n \to \mathbb{R}^m$ 为弗雷歇可微. 其中点 $x = (x_1, \cdots, x_n)$, 增量为 $h = (h_1, \cdots, h_n)$. $\mathrm{D}f(x) = A$ 为映射 f 在 x 的弗雷歇导数.

设 $\mathbb{R}^n$ 和 $\mathbb{R}^m$ 都取自然基, $f: \mathbb{R}^n \to \mathbb{R}^m$ 在点 $x \in \mathbb{R}^n$ 为弗雷歇可微. 线性算子 $A: \mathbb{R}^n \to \mathbb{R}^m$ 可用 $m \times n$ 矩阵表示为

$$D[f(x)h] = Ah = \begin{bmatrix} \dfrac{\partial f_1(x)}{\partial x_1} & \dfrac{\partial f_1(x)}{\partial x_2} & \cdots & \dfrac{\partial f_1(x)}{\partial x_n} \\ \dfrac{\partial f_2(x)}{\partial x_1} & \dfrac{\partial f_2(x)}{\partial x_2} & \cdots & \dfrac{\partial f_2(x)}{\partial x_n} \\ \vdots & \vdots & \cdots & \vdots \\ \dfrac{\partial f_m(x)}{\partial x_1} & \dfrac{\partial f_m(x)}{\partial x_2} & \cdots & \dfrac{\partial f_m(x)}{\partial x_n} \end{bmatrix} \begin{bmatrix} h_1 \\ \vdots \\ h_n \end{bmatrix} \tag{D-14}$$

其中, A 为雅可比矩阵.

将方向导数的概念推广至巴拿赫空间, 设 $g(x): K \subset \mathcal{B}_1 \to \mathcal{B}_2$, 若对任何 $h \in \mathcal{B}_1$, 使得在 x_0 附近极限:

$$d[g(x_0)h] = \lim_{\alpha \to 0} \frac{1}{\alpha}[g(x_0 + \alpha h) - g(x_0)] \tag{D-15}$$

都存在且 $d[g(x_0)h] \in \mathcal{B}_2$, 则称 $d[g(x_0)h]$ 为 $g(x)$ 在 x_0 处的加托微分. 若 $d[g(x_0)h] = bh = g'(x_0)h$, 则称 $g(x)$ 在 x_0 处具有有界线性的加托导数.

在欧氏空间中, 常涉及标量、矢量与张量的函数导数, 下面分别予以讨论.

D.2.1 标量的标量、矢量或张量函数的导数

对于 $t \in \mathbb{R}$ 为实数值, 函数值 $\boldsymbol{g}$ 可以为欧几里得空间 $\mathcal{E}$ 中的标量, 矢量或者张量, 即映射 $\boldsymbol{g}(\cdot): \mathbb{R} \to \mathcal{E}$, 若 $\boldsymbol{g}(t)$ 对于 t 的导数存在, 则有

$$\boldsymbol{g}(t+\alpha) = \boldsymbol{g}(t) + \alpha \dot{\boldsymbol{g}}(t) + o(\alpha) \tag{D-16}$$

导数 (弗雷歇导数) 定义为

$$\dot{\boldsymbol{g}}(t) = \frac{d}{dt}\boldsymbol{g}(t) = \lim_{\alpha \to 0} \frac{1}{\alpha}[\boldsymbol{g}(t+\alpha) - \boldsymbol{g}(t)] \tag{D-17}$$

D.2.2 矢量的标量、矢量或张量函数的导数

若映射 $\boldsymbol{g}(\cdot): \mathcal{V} \to \mathcal{E}$ 在 $\boldsymbol{x} \in \mathcal{V}$ 可微, 则对于 $\boldsymbol{u} \in \mathcal{V} \to \boldsymbol{0}$:

$$\boldsymbol{g}(\boldsymbol{x}+\boldsymbol{u}) = \boldsymbol{g}(\boldsymbol{x}) + D[\boldsymbol{g}(\boldsymbol{x})\boldsymbol{u}] + o(\|\boldsymbol{u}\|) \tag{D-18}$$

即

$$\lim_{\|\boldsymbol{u}\| \to 0} \left\| \frac{\boldsymbol{g}(\boldsymbol{x}+\boldsymbol{u}) - \boldsymbol{g}(\boldsymbol{x}) - D[\boldsymbol{g}(\boldsymbol{x})\boldsymbol{u}]}{\|\boldsymbol{u}\|} \right\| = \lim_{\|\boldsymbol{u}\| \to 0} \left\| \frac{o(\|\boldsymbol{u}\|)}{\|\boldsymbol{u}\|} \right\| = 0 \tag{D-19}$$

其中, 弗雷歇微分 $D[\boldsymbol{g}(\boldsymbol{x})\boldsymbol{u}] = \dfrac{d\boldsymbol{g}(\boldsymbol{x})}{d\boldsymbol{x}} \cdot \boldsymbol{u} = \boldsymbol{g}'(\boldsymbol{x}) \cdot \boldsymbol{u}$, $\dfrac{d\boldsymbol{g}(\boldsymbol{x})}{d\boldsymbol{x}} = \boldsymbol{g}'(\boldsymbol{x})$ 为 $\boldsymbol{g}(\boldsymbol{x})$ 的

弗雷歇导数. 若该导数存在, 其大小等价于加托方向导数:

$$\mathrm{D}\left[g\left(x\right)u\right] = \mathrm{d}\left[g\left(x\right)u\right] = \frac{\mathrm{d}}{\mathrm{d}\alpha}g\left(x+\alpha u\right)|_{\alpha=0} = \lim_{\substack{\alpha \to 0 \\ \alpha \in \mathbb{R}}} \frac{1}{\alpha}\left[g\left(x+\alpha u\right) - g\left(x\right)\right] \quad \text{(D-20)}$$

例 D.2 若存在如下矢量的标量函数

$$\phi\left(v\right) = v \cdot v \quad \text{(D-21)}$$

其中, $\phi\left(\cdot\right) : \mathcal{V} \to \mathbb{R}$, 则

$$\phi\left(v+u\right) = \left(v+u\right) \cdot \left(v+u\right) = v \cdot v + 2v \cdot u + u \cdot u$$
$$= \phi\left(v\right) + 2v \cdot u + o\left(\|u\|\right) \quad \text{(D-22)}$$

则 $\mathrm{D}\left[\varphi\left(v\right)u\right] = \varphi\left(v+u\right) - \varphi\left(v\right) = 2v \cdot u$, 即弗雷歇导数为 $\dfrac{\mathrm{d}\varphi\left(v\right)}{\mathrm{d}v} = 2v$. 由于矢量 v 有三个分量, 因此, $\dfrac{\mathrm{d}\varphi\left(v\right)}{\mathrm{d}v} = 2v$ 表示为偏导数形式 $\dfrac{\partial\varphi\left(v\right)}{\partial v} = 2v$ 更为合理些.

D.2.3 矢量的标量、矢量或张量函数的导数

若映射 $S\left(\cdot\right) : \mathrm{Lin} \to \mathcal{E}$ 在 $A \in \mathrm{Lin}$ 可微, 则对于 $U \in \mathrm{Lin} \to \mathbf{0}$:

$$S\left(A+U\right) = S\left(A\right) + \mathrm{D}\left[S\left(A\right)U\right] + o\left(\|U\|\right) \quad \text{(D-23)}$$

其中, 弗雷歇微分 $\mathrm{D}\left[S\left(A\right)U\right] = \dfrac{\mathrm{d}S\left(A\right)}{\mathrm{d}A} : U$, 此处 $\dfrac{\mathrm{d}S\left(A\right)}{\mathrm{d}A}$ 为弗雷歇导数. 此处, "Lin" 表示的是 "所有张量的集合 (the set of all tensors)".

例 D.3 若存在如下张量函数:

$$S\left(A\right) = A \cdot A \quad \text{(D-24)}$$

其中, $S\left(\cdot\right) : \mathrm{Lin} \to \mathrm{Lin}$, 则

$$S\left(A+U\right) = \left(A+U\right) \cdot \left(A+U\right) = A \cdot A + A \cdot U + U \cdot A + U \cdot U$$
$$= S\left(A\right) + A \cdot U + U \cdot A + o\left(U\right) \quad \text{(D-25)}$$

则弗雷歇微分为

$$\mathrm{D}\left[S\left(A\right)U\right] = S\left(A+U\right) - S\left(A\right) = A \cdot U + U \cdot A \quad \text{(D-26)}$$

例 D.4 张量迹的运算:

$$\varphi\left(A\right) = \mathrm{tr}A = \mathbf{I} : A \quad \text{(D-27)}$$

其中，$\varphi(\cdot): \text{Lin} \to \mathbb{R}$，则其弗雷歇微分表示为

$$\begin{aligned} \text{D}\left[\varphi(\boldsymbol{A})\boldsymbol{U}\right] &= \frac{\text{d}\varphi(\boldsymbol{A})}{\text{d}\boldsymbol{A}}:\boldsymbol{U} \\ &= \varphi(\boldsymbol{A}+\boldsymbol{U}) - \varphi(\boldsymbol{A}) \\ &= \mathbf{I}:(\boldsymbol{A}+\boldsymbol{U}) - \mathbf{I}:\boldsymbol{A} \\ &= \mathbf{I}:\boldsymbol{U} \end{aligned} \tag{D-28}$$

其加托方向导数为

$$\begin{aligned} \text{d}\left[\varphi(\boldsymbol{A})\boldsymbol{U}\right] &= \lim_{\substack{\alpha \to 0 \\ \alpha \in \mathbb{R}}} \frac{1}{\alpha}\left[\phi(\boldsymbol{A}+\alpha\boldsymbol{U}) - \phi(\boldsymbol{A})\right] = \lim_{\alpha \to 0} \frac{\mathbf{I}:(\boldsymbol{A}+\alpha\boldsymbol{U}) - \mathbf{I}:\boldsymbol{A}}{\alpha} \\ &= \mathbf{I}:\boldsymbol{U} \\ &= \text{D}\left[\boldsymbol{S}(\boldsymbol{A})\boldsymbol{U}\right] \end{aligned} \tag{D-29}$$

即其弗雷歇导数与加托导数相等，都表示为

$$\varphi'(\boldsymbol{A}) = \frac{\text{d}\varphi(\boldsymbol{A})}{\text{d}\boldsymbol{A}} = \mathbf{I} \tag{D-30}$$

正如在例 17.2 中业已详细讨论过的，由于张量 $\boldsymbol{A}$ 有多个分量，(D-30) 式表示为偏导数形式 $\dfrac{\partial \varphi(\boldsymbol{A})}{\partial \boldsymbol{A}} = \mathbf{I}$ 更为合理些.

定理：若 $g(x)$ 在 x_0 处弗雷歇可微，则在 x_0 处必具有加托导数，且 $\text{D}\left[g(x_0)h\right] = \text{d}\left[g(x_0)h\right]$，即弗雷歇导数与加托导数相等，都表示为 $g'(x_0)$.

若 $g(x)$ 在 x_0 处某邻域内具有加托导数且在 x_0 处连续，则 $g(x)$ 在 x_0 处弗雷歇可微.

注：$g(x)$ 在 x_0 处某邻域内具有加托导数，一般无法推出 $g(x)$ 在 x_0 处弗雷歇可微.

对 D.2 节总结如下：

(1) 弗雷歇可微，一定存在加托微分，且两个导数大小相等；

(2) 具有加托导数无法推出弗雷歇可微；

(3) 邻域内加托微分 + 连续性 = 弗雷歇可微；

(4) 在理性连续介质力学中，欧几里得空间下物理量场 (标量，矢量，张量) 对时间以及物理量的导数，皆为弗雷歇导数，其大小可通过加托导数求出.

附录 E 玻尔兹曼动理学方程、BBGKY 级联、利用玻尔兹曼方程对连续介质力学守恒律的证明

E.1 玻尔兹曼动理学方程 (1872)

我们在本科一年级的《力学》课上讲过，牛顿第二定律、拉格朗日方程、哈密顿方程、刘维尔方程、薛定谔方程等均具有时间反演不变性，也称为"卡西米尔微观可逆性原理"，见《力学讲义》[1.11] 第 225 页. 时间反演相当于速度方向的反转，即运动方向的反转，而不是时间倒流.

而本附录所推导的玻尔兹曼动理方程 (Boltzmann's kinetic equation) 破坏了时间反演不变性.

热力学第二定律告诉我们，不可逆性是宏观系统的基本性质，非平衡统计力学 (non-equilibrium statistical mechanics) 的主要目的与基础工作，就是要从可逆性的微观运动规律导出不可逆的宏观运动规律，把一个微观保守系统的运动规律变为宏观耗散系统的规律.

非平衡统计力学的发展始于 1872 年，也就是始于我们本附录所推导的玻尔兹曼方程，它是描述稀薄气体非平衡现象的重要方程. 玻尔兹曼时年 28 岁.

定义一个概率密度函数 (probability density function) $f(\boldsymbol{q},\boldsymbol{p},t)$，这里 $\boldsymbol{q}$ 为广义坐标，$\boldsymbol{p}$ 为广义动量，我们考虑的是由广义坐标和广义动量所组成的相空间，为简单起见，我们考虑的是一个六维的相空间，亦即，广义坐标和广义动量各为三维. 因此，概率密度函数对三维动量积分的结果得到的是单位体积的数密度：

$$n = \int f(\boldsymbol{q},\boldsymbol{p},t)\,\mathrm{d}^3\boldsymbol{p} \tag{E-1}$$

式中，$\mathrm{d}^3\boldsymbol{p} = \mathrm{d}p_1\mathrm{d}p_2\mathrm{d}p_3$. 单位体积的数密度再对三维广义坐标积分则得到总粒子数：

$$N = \int n\,\mathrm{d}^3\boldsymbol{q} = \int \mathrm{d}^3\boldsymbol{q}\int f(\boldsymbol{q},\boldsymbol{p},t)\,\mathrm{d}^3\boldsymbol{p} \tag{E-2}$$

在从时刻 t 变到时刻 $t+\mathrm{d}t$ 时，概率密度函数将从 $f(\boldsymbol{q},\boldsymbol{p},t)$ 变为 $f\left(\boldsymbol{q}+\dfrac{\boldsymbol{p}}{m}\mathrm{d}t,\boldsymbol{p}+\boldsymbol{F}\mathrm{d}t,t+\mathrm{d}t\right)$，因此有

$$f\left(\boldsymbol{q}+\frac{\boldsymbol{p}}{m}\mathrm{d}t,\boldsymbol{p}+\boldsymbol{F}\mathrm{d}t,t+\mathrm{d}t\right) - f(\boldsymbol{q},\boldsymbol{p},t) = \mathrm{d}f \tag{E-3}$$

对 (E-3) 式左端的第一项进行一阶泰勒展开, 有

$$\frac{\partial f}{\partial t} + \frac{\boldsymbol{p}}{m} \cdot \frac{\partial f}{\partial \boldsymbol{q}} + \boldsymbol{F} \cdot \frac{\partial f}{\partial \boldsymbol{p}} = \frac{\mathrm{d}f}{\mathrm{d}t} = \left(\frac{\partial f}{\partial t}\right)_{\text{force}} + \left(\frac{\partial f}{\partial t}\right)_{\text{diffusion}} + \left(\frac{\partial f}{\partial t}\right)_{\text{collision}} \tag{E-4}$$

式中, $\dfrac{\boldsymbol{p}}{m}$ 为广义速度, $\dfrac{\partial f}{\partial \boldsymbol{q}}$ 为概率密度函数对广义坐标 $\boldsymbol{q}$ 的梯度, 可表示为 $\dfrac{\partial f}{\partial \boldsymbol{q}} = \boldsymbol{\nabla}_{\boldsymbol{q}} f$; 同理, $\dfrac{\partial f}{\partial \boldsymbol{p}} = \boldsymbol{\nabla}_{\boldsymbol{p}} f$. $\left(\dfrac{\partial f}{\partial t}\right)_{\text{force}}$ 为外力的贡献项, $\left(\dfrac{\partial f}{\partial t}\right)_{\text{diffusion}}$ 为扩散的贡献项, $\left(\dfrac{\partial f}{\partial t}\right)_{\text{collision}}$ 则为粒子间碰撞的贡献项. (E-4) 式可进一步表示为

$$\frac{\partial f}{\partial t} + \frac{\boldsymbol{p}}{m} \cdot \boldsymbol{\nabla}_{\boldsymbol{q}} f + \boldsymbol{F} \cdot \boldsymbol{\nabla}_{\boldsymbol{p}} f = \frac{\mathrm{d}f}{\mathrm{d}t} \tag{E-5}$$

(E-4) 式或 (E-5) 式即为玻尔兹曼动理学方程, 简称玻尔兹曼方程. 若不用相空间的概念, 而只用欧氏空间的概念, (E-4) 式或 (E-5) 式还经常性地表示为

$$\frac{\partial f}{\partial t} + \boldsymbol{v} \cdot \frac{\partial f}{\partial \boldsymbol{r}} + \frac{\boldsymbol{F}}{m} \cdot \frac{\partial f}{\partial \boldsymbol{v}} = \frac{\mathrm{d}f}{\mathrm{d}t} \quad \text{或} \quad \frac{\partial f}{\partial t} + \boldsymbol{v} \cdot \boldsymbol{\nabla}_{\boldsymbol{r}} f + \frac{\boldsymbol{F}}{m} \cdot \boldsymbol{\nabla}_{\boldsymbol{v}} f = \frac{\mathrm{d}f}{\mathrm{d}t} \tag{E-6}$$

式中, $\boldsymbol{v}$ 为速度矢量, $\boldsymbol{r}$ 为位矢.

E.2 BBGKY 级联

在统计力学中, BBGKY 级联指的是 "Bogoliubov–Born–Green–Kirkwood–Yvon hierarchy" (见图 E.1 所示的五位学者), 有时亦简称博戈柳博夫级联 (Bogoliubov hierarchy). BBGKY 级联是一组描述大量相互作用粒子的系统动力学的方程. BBGKY 层次结构中 s-粒子概率密度函数 (分布函数) 的方程包括 (s + 1)-粒子的分布函数, 从而形成耦合的方程链.

图 E.1 图 E.1 从左至右: 博戈柳博夫 (Nikolay Nikolayevich Bogoliubov, 1909—1992); 玻恩 (Max Born, 1882—1970); 玻恩在爱丁堡大学的博士生格林 (Herbert Sydney Green, 1920—1999); 柯克伍德 (John Gamble Kirkwood, 1907—1959); 伊万 (Jacques Yvon, 1903—1979)

一个没有涨落的由 N 个粒子组成的系统，其概率密度函数 $f_N(\boldsymbol{q}_1,\cdots,\boldsymbol{q}_N,\boldsymbol{p}_1,\cdots,\boldsymbol{p}_N,t)$ 满足

$$\frac{\partial f_N}{\partial t}+\sum_{i=1}^{N}\frac{\boldsymbol{p}_i}{m}\cdot\frac{\partial f_N}{\partial \boldsymbol{q}_i}+\sum_{i=1}^{N}\boldsymbol{F}_i\cdot\frac{\partial f_N}{\partial \boldsymbol{p}_i}=0 \tag{E-7}$$

作用在第 i 个质点上的力为

$$\boldsymbol{F}_i=-\sum_{j=1\neq i}^{N}\frac{\partial \Phi_{ij}}{\partial \boldsymbol{q}_i}-\frac{\partial \Phi_i^{\text{ext}}}{\partial \boldsymbol{q}_i} \tag{E-8}$$

其中，Φ_{ij} 为粒子间的对势 (pair potential)，Φ_i^{ext} 为外场势 (external field potential).

下面讨论 s-粒子和 $(s+1)$-粒子的级联. s-粒子的归一化条件为

$$\int_{\text{位置}}\int_{\text{动量}} f_s(\boldsymbol{q}_1,\cdots,\boldsymbol{q}_s;\boldsymbol{p}_1,\cdots,\boldsymbol{p}_s;t)\mathrm{d}^s\boldsymbol{q}\mathrm{d}^s\boldsymbol{p}=1 \tag{E-9}$$

相应地，N-粒子的归一化条件为

$$\int_{\text{位置}}\int_{\text{动量}} f_N(\boldsymbol{q}_1,\cdots,\boldsymbol{q}_N;\boldsymbol{p}_1,\cdots,\boldsymbol{p}_N;t)\mathrm{d}^N\boldsymbol{q}\mathrm{d}^N\boldsymbol{p}=1 \tag{E-10}$$

由 (E-9) 和 (E-10) 两式相等有

$$\int f_s(\boldsymbol{q}_1,\cdots,\boldsymbol{q}_s;\boldsymbol{p}_1,\cdots,\boldsymbol{p}_s;t)\mathrm{d}\boldsymbol{q}_1\cdots\mathrm{d}\boldsymbol{q}_s\mathrm{d}\boldsymbol{p}_1\cdots\mathrm{d}\boldsymbol{p}_s$$
$$=\int f_N(\boldsymbol{q}_1,\cdots,\boldsymbol{q}_N;\boldsymbol{p}_1,\cdots,\boldsymbol{p}_N;t)\mathrm{d}\boldsymbol{q}_1\cdots\mathrm{d}\boldsymbol{q}_N\mathrm{d}\boldsymbol{p}_1\cdots\mathrm{d}\boldsymbol{p}_N$$
$$=\int f_N\mathrm{d}\boldsymbol{q}_1\cdots\mathrm{d}\boldsymbol{q}_s\mathrm{d}\boldsymbol{q}_{s+1}\cdots\mathrm{d}\boldsymbol{q}_N\mathrm{d}\boldsymbol{p}_1\cdots\mathrm{d}\boldsymbol{p}_s\mathrm{d}\boldsymbol{p}_{s+1}\cdots\mathrm{d}\boldsymbol{p}_N$$
$$=\int f_N\mathrm{d}\boldsymbol{q}_{s+1}\cdots\mathrm{d}\boldsymbol{q}_N\mathrm{d}\boldsymbol{p}_{s+1}\cdots\mathrm{d}\boldsymbol{p}_N(\mathrm{d}\boldsymbol{q}_1\cdots\mathrm{d}\boldsymbol{q}_s\mathrm{d}\boldsymbol{p}_1\cdots\mathrm{d}\boldsymbol{p}_s) \tag{E-11}$$

由上式可得到 s-粒子和 N-粒子概率密度函数的级联关系：

$$f_s=\int f_N\mathrm{d}\boldsymbol{q}_{s+1}\cdots\mathrm{d}\boldsymbol{q}_N\mathrm{d}\boldsymbol{p}_{s+1}\cdots\mathrm{d}\boldsymbol{p}_N \tag{E-12}$$

式中，令 $N=s+1$，则得到 s-粒子和 $(s+1)$-粒子间的递推关系：

$$f_s(q_1,\cdots,q_s;p_1,\cdots,p_s;t)=\int f_{s+1}(q_1,\cdots,q_s;p_1,\cdots,p_{s+1};t)\mathrm{d}q_{s+1}\mathrm{d}p_{s+1} \tag{E-13}$$

在 N-粒子系统中，对 s-粒子中粒子作用的有 s-粒子中粒子间相互作用的对势，外场势，以及 $(N-s)$-粒子对粒子的对势，则力矢量为

$$F_i=-\sum_{j=1\neq i}^{s}\frac{\partial \Phi_{ij}}{\partial q_i}-\frac{\partial \Phi_i^{ext}}{\partial q_i}-(N-s)\frac{\partial \Phi_{i\,s+1}}{\partial q_i} \tag{E-14}$$

将 (E-14) 式代回 (E-7) 式，得到

$$\frac{\partial f_s}{\partial t} + \sum_{i=1}^{s} \frac{p_i}{m} \frac{\partial f_s}{\partial q_i} - \sum_{i=1}^{s} \left(\sum_{j=1\neq i}^{s} \frac{\partial \Phi_{ij}}{\partial q_i} + \frac{\partial \Phi_i^{\text{ext}}}{\partial q_i} \right) \frac{\partial f_s}{\partial p_i} = (N-s) \sum_{i=1}^{s} \frac{\partial \Phi_{i\,s+1}}{\partial q_i} \frac{\partial f_s}{\partial p_i} \tag{E-15}$$

将 (E-13) 式，亦即 s 与 $s+1$ 间概率密度函数的递推关系代入 (E-15) 式，整理便得到 BBGKY 级联方程：

$$\frac{\partial f_s}{\partial t} + \sum_{i=1}^{s} \frac{\boldsymbol{p}_i}{m} \cdot \frac{\partial f_s}{\partial \boldsymbol{q}_i} - \sum_{i=1}^{s} \left(\sum_{j=1\neq i}^{s} \frac{\partial \Phi_{ij}}{\partial \boldsymbol{q}_i} + \frac{\partial \Phi_i^{\text{ext}}}{\partial \boldsymbol{q}_i} \right) \cdot \frac{\partial f_s}{\partial \boldsymbol{p}_i}$$

$$= (N-s) \sum_{i=1}^{s} \int \frac{\partial \Phi_{i\,s+1}}{\partial \boldsymbol{q}_i} \frac{\partial f_{s+1}}{\partial \boldsymbol{p}_i} \mathrm{d}q_{s+1} \mathrm{d}p_{s+1} \tag{E-16}$$

BBGKY 方程 (E-16) 的连锁性在于 f_1 中含有 f_2，f_2 中含有 f_3，余此类推.

E.3 应用玻尔兹曼方程对连续介质力学守恒律的证明

对于保守系统，(E-1) 式已经给出了有数密度 $n = \int f \mathrm{d}^3 \boldsymbol{p}$，当数密度足够大时，对任意 A 都有其平均值：

$$\langle A \rangle = \frac{1}{n} \int A f \mathrm{d}^3 \boldsymbol{p} \tag{E-17}$$

则有

$$\int A f \mathrm{d}^3 \boldsymbol{p} = n \langle A \rangle \tag{E-18}$$

假设 A 是广义动量 p_i 的函数，力 F_i 是位置的函数，进一步可以得到

$$\begin{aligned}
\int A \frac{\partial f}{\partial t} \mathrm{d}^3 \boldsymbol{p} &= \int \frac{\partial (Af)}{\partial t} \mathrm{d}^3 \boldsymbol{p} - \int f \frac{\partial A}{\partial t} \mathrm{d}^3 \boldsymbol{p} \\
&= \frac{\partial}{\partial t} \int A f \mathrm{d}^3 \boldsymbol{p} - 0 \\
&= \frac{\partial}{\partial t} (n \langle A \rangle)
\end{aligned} \tag{E-19}$$

$$\begin{aligned}
\int \frac{p_j A}{m} \frac{\partial f}{\partial x_j} \mathrm{d}^3 \boldsymbol{p} &= \frac{1}{m} \left[\int \frac{\partial}{\partial x_j} (p_j A f) \mathrm{d}^3 \boldsymbol{p} - \int f \frac{\partial (p_j A)}{\partial x_j} \mathrm{d}^3 \boldsymbol{p} \right] \\
&= \frac{1}{m} \frac{\partial}{\partial x_j} \int (p_j A f) \mathrm{d}^3 \boldsymbol{p} - 0 \\
&= \frac{1}{m} \frac{\partial}{\partial x_j} (n \langle p_j A \rangle)
\end{aligned} \tag{E-20}$$

附录 E　玻尔兹曼动理学方程、BBGKY 级联、利用玻尔兹曼方程对连续介质力学···

$$\int AF_j \frac{\partial f}{\partial p_j} \mathrm{d}^3 \boldsymbol{p}$$
$$= \int \frac{\partial}{\partial p_j}(AF_j f)\mathrm{d}^3\boldsymbol{p} - \int \frac{\partial (AF_j)}{\partial p_j} f \mathrm{d}^3\boldsymbol{p}$$
$$= \int \frac{\partial}{\partial p_j}(AF_j f)\mathrm{d}^3\boldsymbol{p} - \int F_j \frac{\partial A}{\partial p_j} f \mathrm{d}^3\boldsymbol{p}$$
$$= \int \left[\frac{\partial (AF_1 f)}{\partial p_1} + \frac{\partial (AF_2 f)}{\partial p_2} + \frac{\partial (AF_3 f)}{\partial p_3}\right]\mathrm{d}p_1 \mathrm{d}p_2 \mathrm{d}p_3 - F_j \int \frac{\partial A}{\partial p_j} f \mathrm{d}^3\boldsymbol{p}$$
$$= \frac{1}{2}\int\iint \left[\frac{\partial (AF_1 f)}{\partial p_1} + \frac{\partial (AF_2 f)}{\partial p_2}\right]\mathrm{d}p_1 \mathrm{d}p_2 \cdot \mathrm{d}p_3$$
$$+ \frac{1}{2}\int\iint \left[\frac{\partial (AF_2 f)}{\partial p_2} + \frac{\partial (AF_3 f)}{\partial p_3}\right]\mathrm{d}p_2 \mathrm{d}p_3 \cdot \mathrm{d}p_1$$
$$+ \frac{1}{2}\int\iint \left[\frac{\partial (AF_1 f)}{\partial p_1} + \frac{\partial (AF_3 f)}{\partial p_3}\right]\mathrm{d}p_1 \mathrm{d}p_3 \cdot \mathrm{d}p_2 - F_j \int \frac{\partial A}{\partial p_j} f \mathrm{d}^3\boldsymbol{p} \quad \text{(E-21)}$$

根据格林公式 $\iint\left(\frac{\partial Q}{\partial x} - \frac{\partial P}{\partial y}\right)\mathrm{d}x\mathrm{d}y = \oint P\mathrm{d}x + Q\mathrm{d}y$，同时对保守系统与积分路线无关有

$$\int AF_j \frac{\partial f}{\partial p_j}\mathrm{d}^3\boldsymbol{p}$$
$$= \frac{1}{2}\int\iint\left[\frac{\partial (AF_1 f)}{\partial p_1} + \frac{\partial (AF_2 f)}{\partial p_2}\right]\mathrm{d}p_1 \mathrm{d}p_2 \cdot \mathrm{d}p_3$$
$$+ \frac{1}{2}\int\iint\left[\frac{\partial (AF_2 f)}{\partial p_2} + \frac{\partial (AF_3 f)}{\partial p_3}\right]\mathrm{d}p_2 \mathrm{d}p_3 \cdot \mathrm{d}p_1$$
$$+ \frac{1}{2}\int\iint\left[\frac{\partial (AF_3 f)}{\partial p_3} + \frac{\partial (AF_1 f)}{\partial p_1}\right]\mathrm{d}p_1 \mathrm{d}p_3 \cdot \mathrm{d}p_2 - F_j\int\frac{\partial A}{\partial p_j}f\mathrm{d}^3\boldsymbol{p}$$
$$= \frac{1}{2}\int\oint(-AF_2 f\mathrm{d}p_1 + AF_1 f\mathrm{d}p_2)\cdot \mathrm{d}p_3 + \frac{1}{2}\int\oint(-AF_3 f\mathrm{d}p_2 + AF_2 f\mathrm{d}p_3)\cdot \mathrm{d}p_1$$
$$+ \frac{1}{2}\int\oint(-AF_1 f\mathrm{d}p_3 + AF_3 f\mathrm{d}p_1)\cdot \mathrm{d}p_2 - nF_j\left\langle\frac{\partial A}{\partial p_j}\right\rangle$$
$$= -\frac{1}{2}\iint AF_2 f\mathrm{d}p_1\mathrm{d}p_3 + \frac{1}{2}\iint AF_1 f\mathrm{d}p_2\mathrm{d}p_3 - \frac{1}{2}\iint AF_3 f\mathrm{d}p_2\mathrm{d}p_1$$
$$+ \frac{1}{2}\iint AF_2 f\mathrm{d}p_3\mathrm{d}p_1 - \frac{1}{2}\iint AF_1 f\mathrm{d}p_3\mathrm{d}p_2 + \frac{1}{2}\iint AF_3 f\mathrm{d}p_1\mathrm{d}p_2 - nF_j\left\langle\frac{\partial A}{\partial p_j}\right\rangle$$
$$= -nF_j\left\langle\frac{\partial A}{\partial p_j}\right\rangle \quad \text{(E-22)}$$

又 A 在碰撞中守恒有

$$\int A\left(\frac{\partial f}{\partial t}\right)_{\text{coll}}\mathrm{d}^3\boldsymbol{p} = 0 \quad \text{(E-23)}$$

则式 $\dfrac{\partial f}{\partial t} + \dfrac{\boldsymbol{p}}{m} \cdot \boldsymbol{\nabla} f + \boldsymbol{F} \cdot \dfrac{\partial f}{\partial \boldsymbol{p}} = \left(\dfrac{\partial f}{\partial t}\right)_{\text{coll}}$ 两边同时乘上函数 A 并积分后得到

$$\int A \left(\frac{\partial f}{\partial t} + \frac{\boldsymbol{p}}{m} \cdot \boldsymbol{\nabla} f + \boldsymbol{F} \cdot \frac{\partial f}{\partial \boldsymbol{p}}\right) \mathrm{d}^3 \boldsymbol{p} = \int A \left(\frac{\partial f}{\partial t}\right)_{\text{coll}} \mathrm{d}^3 \boldsymbol{p} \tag{E-24}$$

对上式进行平均，并将 (E-19) 式、(E-20) 式、(E-22) 式和 (E-23) 式代入进行平均后的 (E-24) 式，得到如下重要方程：

$$\frac{\partial}{\partial t}\left(n\left\langle A\right\rangle\right) + \frac{1}{m}\frac{\partial}{\partial x_j}\left(n\left\langle p_j A\right\rangle\right) - nF_j\left\langle\frac{\partial A}{\partial p_j}\right\rangle = 0 \tag{E-25}$$

E.3.1 质量守恒

在 (E-25) 式中令 $A = m$，有

$$\frac{\partial}{\partial t}\left(n\left\langle m\right\rangle\right) + \frac{1}{m}\frac{\partial}{\partial x_j}\left(n\left\langle mp_j\right\rangle\right) - nF_j\left\langle\frac{\partial m}{\partial p_j}\right\rangle = 0 \tag{E-26}$$

注意到密度可以通过数密度表示为 $\rho = nm$. 考虑到质量 m 和动量 p_j 的无关性，则 $\dfrac{\partial m}{\partial p_j} = 0$. (E-26) 式可进一步化简为

$$\frac{\partial \rho}{\partial t} + \frac{\partial}{\partial x_j}\left(\rho\left\langle\frac{p_j}{m}\right\rangle\right) - 0 = \frac{\partial \rho}{\partial t} + \frac{\partial}{\partial x_j}\left(\rho\left\langle v_j\right\rangle\right) = 0 \tag{E-27}$$

采用记号：$V_j = \langle v_j \rangle$，(E-27) 式即为连续性方程：

$$\frac{\partial \rho}{\partial t} + \frac{\partial}{\partial x_j}(\rho V_j) = 0 \quad \text{或} \quad \frac{\partial \rho}{\partial t} + \boldsymbol{\nabla} \cdot (\rho \boldsymbol{V}) = 0 \tag{E-28}$$

E.3.2 动量守恒

在 (E-25) 式中令 $A = p_i$，则有

$$\begin{aligned}
&\frac{\partial}{\partial t}\left(n\left\langle mv_i\right\rangle\right) + \frac{1}{m}\frac{\partial}{\partial x_j}\left(n\left\langle mv_i p_j\right\rangle\right) - nF_j\left\langle\frac{\partial p_i}{\partial p_j}\right\rangle \\
&= \frac{\partial}{\partial t}(\rho V_i) + \frac{\partial}{\partial x_j}\left(\rho\left\langle v_i\frac{p_j}{m}\right\rangle\right) - nF_j\left\langle\frac{\partial p_i}{\partial p_j}\right\rangle \\
&= \frac{\partial}{\partial t}(\rho V_i) + \rho\frac{\partial}{\partial x_j}\left\langle v_i v_j\right\rangle - nF_i \\
&= \frac{\partial}{\partial t}(\rho V_i) + \rho\frac{\partial}{\partial x_j}\left\langle(v_j - V_j)(v_i - V_i) + V_i V_j\right\rangle - nF_i \\
&= \frac{\partial}{\partial t}(\rho V_i) + \frac{\partial}{\partial x_j}\left[\rho V_i V_j + \rho\left\langle(v_j - V_j)(v_i - V_i)\right\rangle\right] - nF_i
\end{aligned} \tag{E-29}$$

其中，因为 $V_j = \langle v_j \rangle$ 和 $V_i = \langle v_i \rangle$，有

$$\langle v_i v_j\rangle = \langle v_i v_j - v_i V_j - v_j V_i + V_i V_j + V_i V_j\rangle$$

$$= [V_i V_j + \langle (v_j - V_j)(v_i - V_i) \rangle] \tag{E-30}$$

令

$$P_{ij} = \rho \langle (v_j - V_j)(v_i - V_i) \rangle \tag{E-31}$$

为压力张量则有

$$\begin{aligned}
&\frac{\partial}{\partial t}(n \langle m v_i \rangle) + \frac{1}{m}\frac{\partial}{\partial x_j}(n \langle m v_i p_j \rangle) - n F_j \\
&= \frac{\partial}{\partial t}(\rho V_i) + \frac{\partial}{\partial x_j}[\rho V_i V_j + \rho \langle (v_j - V_j)(v_i - V_i) \rangle] - n F_i \\
&= \frac{\partial}{\partial t}(\rho V_i) + \frac{\partial}{\partial x_j}(\rho V_i V_j + P_{ij}) - n F_i \\
&= 0
\end{aligned} \tag{E-32}$$

得到动量守恒方程如下：

$$\frac{\partial}{\partial t}(\rho V_i) + \frac{\partial}{\partial x_j}(\rho V_i V_j + P_{ij}) - n F_i = 0 \tag{E-33}$$

式中，左端最后一项 $n\boldsymbol{F}$ 为单位体积的体力.

E.3.3 能量守恒

在 (E-25) 式中令 $A = \dfrac{p_i p_i}{2m}$，此时 A 显然为哈密顿量，则有

$$\begin{aligned}
&\frac{\partial}{\partial t}\left(n \left\langle \frac{p_i p_i}{2m} \right\rangle\right) + \frac{1}{m}\frac{\partial}{\partial x_j}\left(n \left\langle \frac{p_i p_i}{2m} p_j \right\rangle\right) - n F_j \left\langle \frac{\partial \frac{p_i p_i}{2m}}{\partial p_j} \right\rangle \\
&= \frac{\partial}{\partial t}\left(\frac{1}{2}\rho \langle v_i v_i \rangle\right) + \frac{\partial}{\partial x_j}\left(\frac{1}{2}\rho \langle v_i v_i v_j \rangle\right) - n F_i \left\langle \frac{p_i}{m} \right\rangle \\
&= \frac{1}{2}\frac{\partial}{\partial t}(\rho [V_i V_i + \langle (v_i - V_i)(v_i - V_i) \rangle]) + \frac{1}{2}\rho \frac{\partial}{\partial x_j}\langle v_i v_i v_j \rangle - n F_i V_i
\end{aligned} \tag{E-34}$$

式中，由于

$$\begin{aligned}
&\langle v_i v_i v_j \rangle \\
&= \langle (v_i - V_i)(v_i - V_i)(v_j - V_j) + 2 v_i v_j V_i - v_j V_i^2 + v_i^2 V_j - 2 v_i V_j V_i + V_j V_i^2 \rangle \\
&= \langle (v_i - V_i)(v_i - V_i)(v_j - V_j) \rangle + \langle 2 v_i v_j V_i - v_j V_i^2 + v_i^2 V_j - 2 v_i V_j V_i + V_j V_i^2 \rangle \\
&= \langle (v_i - V_i)(v_i - V_i)(v_j - V_j) \rangle + 2 V_i \langle v_i v_j \rangle + \langle V_j V_i^2 - v_j V_i^2 \rangle + \langle v_i^2 V_j - 2 v_i V_j V_i \rangle \\
&= \langle (v_i - V_i)(v_i - V_i)(v_j - V_j) \rangle + 2 V_i [V_i V_j + \langle (v_j - V_j)(v_i - V_i) \rangle] + V_i V_i V_j \\
&\quad + \langle (v_i - V_i)(v_i - V_i) \rangle V_j - 2 V_i V_j V_i
\end{aligned}$$

$$= \langle (v_i - V_i)(v_i - V_i)(v_j - V_j) \rangle + 2V_i \langle (v_j - V_j)(v_i - V_i) \rangle + V_i V_i V_j$$
$$+ \langle (v_i - V_i)(v_i - V_i) \rangle V_j \tag{E-35}$$

其中，
$$\left\langle (v_i - V_i)^2 (v_j - V_j) \right\rangle = \left\langle v_i^2 v_j - 2v_i v_j V_i + v_j V_i^2 - v_i^2 V_j + 2v_i V_j V_i - V_j V_i^2 \right\rangle \tag{E-36}$$

令
$$u = \frac{1}{2}\rho \langle (v_i - V_i)(v_i - V_i) \rangle, \quad J_{qi} = \frac{1}{2}\rho \langle (v_i - V_i)(v_k - V_k)(v_k - V_k) \rangle \tag{E-37}$$

分别为动能密度 (kinetic thermal energy density) 和热通量 (thermal flux). 则有

$$\frac{\partial}{\partial t}\left(n\left\langle \frac{p_i p_i}{2m} \right\rangle\right) + \frac{1}{m}\frac{\partial}{\partial x_j}\left(n\left\langle \frac{p_i p_j}{2m} p_j \right\rangle\right) - nF_j \left\langle \frac{\partial \left(\frac{p_i p_i}{2m}\right)}{\partial p_j} \right\rangle$$
$$= \frac{1}{2}\frac{\partial}{\partial t}\left(\rho \left[V_i V_i + \langle (v_i - V_i)(v_i - V_i) \rangle\right]\right) + \frac{1}{2}\rho \frac{\partial}{\partial x_j} \langle v_i v_i v_j \rangle - nF_i V_i$$
$$= \frac{\partial}{\partial t}\left(\frac{1}{2}\rho V_i V_i + u\right) + \frac{1}{2}\rho \frac{\partial}{\partial x_j}[\langle (v_i - V_i)(v_i - V_i)(v_j - V_j) \rangle$$
$$+ 2V_i \langle (v_j - V_j)(v_i - V_i) \rangle + V_i V_i V_j + \langle (v_i - V_i)(v_i - V_i) \rangle V_j] - nF_i V_i$$
$$= \frac{\partial}{\partial t}\left(\frac{1}{2}\rho V_i V_i + u\right) + \frac{\partial}{\partial x_j}\left[J_{qj} + V_i P_{ij} + \frac{1}{2}\rho V_i V_i V_j + uV_j\right] - nF_i V_i$$
$$= 0 \tag{E-38}$$

由上式则得到能量守恒方程：
$$\frac{\partial}{\partial t}\left(\frac{1}{2}\rho V_i V_i + u\right) + \frac{\partial}{\partial x_j}\left[J_{qj} + V_i P_{ij} + \frac{1}{2}\rho V_i V_i V_j + uV_j\right] - nF_i V_i = 0 \tag{E-39}$$

E.4 结 束 语

本书名为 "教程" 理应适当地探讨些在教学上的先进理念.

著者在《近代连续介质力学》[1.6] 第 5 页, 曾讨论过玻恩 (Max Born, 1882—1970, 1954 年诺贝尔物理学奖得主) 所创立的哥廷根量子力学学派. 既然本附录的内容和玻恩有密切的关系, 这里的讨论就围绕着玻恩的教育理念展开. 当年玻恩在格廷根聚集了一批特别年轻的学生, 被物理学界戏称为 "玻恩幼儿园"（Born's Kindergarten）, 那些 "孩子" 中受到过玻恩影响而获得诺贝尔奖的物理学家就有 13 人之多, 玻恩因而是 20 世纪著名物理学家中培养出获得诺贝尔奖的徒弟最多的物理学教育大师之一. 这也充分说明, 研究型大学的教育之美就在于师生在学术前沿的共同探索.

玻恩在研究特别是在教育上巨大成就的取得当然也离不开"天时、地利、人和".

据著者的粗浅认识, 当年玻恩的"天时"和"地利"主要有三点: (1) 本书在思考题 12.1 中业已讨论过的, 在量子力学的创始年代, 正像狄拉克所指出的那样"在那些日子里, 任何二流的物理学家都很容易做出一流的工作"; (2) 当时的德国仍然是世界的科技中心; (3) 在 20 世纪初, 哥廷根大学已成为无可争辩的世界数学中心和圣地. 当时全世界学数学的学生中, 最响亮的口号就是"打起你的背包, 到哥廷根去!"

当然"人和"更是不可或缺的软实力. 这和玻恩的"知人善任"是密不可分的! 玻恩不仅能在工作中认识到这些学生和助手的才能, 同时还能"让那些巨星的光芒能盖过他; 对于那些不甚有天赋的, 他也能耐心地分配给他们可观却又可行的工作"(Born not only recognized talent to work with him, but he "let his superstars stretch past him; to those less gifted, he patiently handed out respectable but doable assignments").

如图 E.1 所示, 在 BBGKY 级联中的 G (Herbert Sydney Green, 格林) 是玻恩在爱丁堡大学于 1947 年进行博士论文答辩的学生. 玻恩于 1964 年 10 月, 在为格林于 1965 年出版的为澳大利亚阿德莱德大学三年级本科生所撰写的教材 *Matrix Mechanics* (《矩阵力学》)[E.2] 所写的前言中指出: "我印象中现在存在这样一种倾向, 即 (人们在教学中) 忽视历史根由, 而将理论建立在后来才发现的基础之上. 这种方法毫无疑问能够迅速接近现代问题, 也很适合培养能够应用这些知识的专家. 但是, 我怀疑这是否是培养做原创性研究的学生的好的教学方法, 因为这种方法不能展示先驱者在无序事实的丛林中, 以及隐晦、含糊的理论尝试中, 是如何发现他的道路的."(I have the impression that there is a tendency to neglect the historic roots and to build the theory on foundations which have actually been discovered later. This methods undoubtedly arrives quickly at modern problems and is well suited for producing experts able to apply what they have learned. But I doubt whether it is a good training for doing original research, because it fails to show how the pioneer finds his way in the jungle of unordered facts and obscure theoretical attempts.)

在玻恩上述"大师的洞见"中, 读者不难发现玻恩教学的一个鲜明特点是注意厘清知识的来龙去脉, 通过再现物理学家做出发现的历史, 给学生更多学习思考解决问题的方法的机会. 事实上, 我们教书育人的根本宗旨是, 通过再现那些各个学科的先驱们如何发现和解决问题的过程, 来帮助学生深刻领会研究方法, 达到培养学生研究能力的目的, 并将科学发现的豪迈的激情注入到学生的心灵之中!

著者相信, 只有当我国涌现出一批像希尔伯特、柯尔莫哥洛夫、玻尔、朗道、费曼、玻恩、索末菲等科学大师兼教育大师时, 我国的科技界才会真正出现令世人瞩目的人才和成果辈出的局面, 才能真正实现从科技大国向科技强国的转变, 世界的科技中心才有可能向中国转移.

参 考 文 献

[E.1] Pitaevskii L P, Lifshitz E M. Physical Kinetics. Vol. 10. Butterworth-Heinemann, 2012. (中文版: 栗弗席兹, 皮塔耶夫斯基. 物理动理学. 徐锡申, 徐春华, 吴京民译. 北京: 高等教育出版社, 2008)

[E.2] Green H S. Matrix Mechanics. Groningen: Noordhoff, 1965.

索　引

A

AGI　446, 447

AI　303, 443–448

Almansi 应变　213, 214, 217

AlphaGo　446–450

AlphaGo Lee　448, 449

AlphaGo Master　449

AlphaGo Zero　449, 450

AlphaZero　450

ansatz solution　413

阿贝尔群　153, 163, 164

阿贝尔交换群　153

阿尔曼西应变　213, 214, 217

阿尔曼西应变率　224

阿尔法狗　446–450

阿基米德原理　286, 371

埃克曼数　491

埃克曼层　491

艾略特波浪　470–472

爱丁顿张量　91

爱因斯坦场方程　82, 452

爱因斯坦求和约定　84, 451

奥尔-索末菲方程　366

奥森方程　386

奥森流　385, 386

B

BBGKY　500, 507

Betti 定理　311–313

BKT 相变　175

巴拿赫空间　13, 15, 495, 496

保持方向映射　199, 227

保角变换　419

保角映射　351, 353

保守力　82, 123, 272

保守系统　499, 502, 503

贝蒂定理　311–313

贝尔定律　7, 50

贝纳尔胞　423, 424

贝塞尔方程　413

贝塞尔函数　413, 414

本构关系

　　超弹性　224, 288–300

　　牛顿流体　287, 357–360, 368, 387, 404

　　线弹性　278–285, 325–331

本征分解　104

本征向量　104–108, 112

本征值　104–107, 112, 134

毕奥数　491

毕奥应变　217

边界层　351, 352, 366, 388, 425–428

边界层吸附效应, 426

变分　305-311

变换

　　勒让德　40, 41, 49, 254

　　洛伦兹　114, 260, 274

　　平移　154, 260

　　旋转　106, 154, 163, 260

变换群
    伽利略　260
    正交　161
变形梯度　134, 198–235, 238, 242, 243, 246, 247, 255, 261–263, 292–294, 297–299, 305–309, 322, 331, 335–337
变形协调方程　335, 336
表面张力　348, 371, 375, 400–407, 411–415
标度
    动力学　25
    动理学　25
    流体力学　25
标度不变性　395
标度律　前言 iii, 57, 395–399, 406, 407, 425, 464
标量　13, 15, 28, 62, 64, 65, 76, 87, 92, 95, 98, 99, 104, 109, 111, 238, 242, 258-263, 299, 400, 406, 429, 485–489, 496–498
标量函数　80, 129–137, 214, 215, 240, 292, 293, 306, 310, 331, 351, 497
标准差　465, 474
别列津斯基–科斯特利茨–索利斯相变　47, 175
玻尔兹曼常数　289–291, 486
玻尔兹曼动理学方程　499, 500
玻尔兹曼机　452
博戈柳博夫级联　25, 500
伯努利
    方程　93, 342–349, 411, 420, 421
    机械手指　349, 350
    夹持　349
    梁　37, 38
    吸盘　349

波动方程　33, 73, 110, 120, 320–322, 332, 333
波函数　18, 160
波粒二象性　14, 147
波矢　19, 489
波矢空间　13, 19, 147
波数　111, 332, 398, 412, 489
波状云　409, 410
泊松比　279, 283, 284
泊松方程　82, 109, 267–269
泊松括号　73, 74, 193
泊松引力势　82
玻姆力学　465
不变性
    标度　395
    反射　269, 270
    伽利略　265, 270
    空间平移　161, 243, 244
    空间旋转　246
    扩展伽利略　270–272
    时间平移　161, 248
不变量　134, 135, 168, 270, 276, 295–297, 299
不等式
    科恩　102–104
    克劳修斯–迪昂　252–254, 276
    克劳修斯–普朗克　253, 254
    勒让德–阿达玛　331-333
不可压缩条件　265, 267, 335, 348, 351, 353, 358, 360, 367, 370–373, 377, 379, 385, 395, 396, 399, 404, 405, 417, 419, 421, 422, 425, 429, 487
不确定度　465, 466
不确定性　465, 466, 468, 472

索　引

不完备性定理　476
不稳定性
　　开尔文-亥姆霍兹　409, 410
　　里克特迈耶-梅什科夫　407, 408
　　普拉托-瑞利　409–415, 425
　　瑞利-泰勒　408, 409
　　萨夫曼-泰勒　416–420
布尔巴基学派　11
布莱克-斯克尔斯-默顿公式　473, 474
布朗运动　472
布辛涅斯克问题　323

C

CAPM　473
Cof　134, 135, 150, 163, 218, 220, 295, 331
猜想
　　贝赫和斯维讷通-戴尔　39
　　霍奇　39
　　哥德巴赫　10
　　开普勒　10
　　黎曼　10, 39, 48, 49
　　庞加莱　10, 39, 170, 217
材料标架无差异性原理　258
材料客观性原理　258
参考构形　63, 199–227, 231–235, 242–247, 251, 254, 262, 263, 299, 305–311, 322, 335–337
残余应力　232
测地线　114, 119, 127–129, 452
测地线方程　93, 127, 128, 150, 151, 371, 452
层流　348, 349, 358, 361, 366, 386, 387, 404, 409, 422, 425, 428
场论　453–455
　　非线性　1, 258
　　规范　1, 93, 120, 171, 195

经典　74, 78
　　量子力学　74
超扩散　463
超弹性　224, 288–300
超新星爆发　408
乘法分解　231, 299
尺寸效应　24, 316
弛豫时间　430, 492
重整化群　452
初始应力　317, 337, 338
串级　394–396, 464
创造性思维　前言 iii, 436, 437
纯剪切　108, 284, 285, 322, 336
丛投影　17
粗糙表面　426
粗略数量级估计　前言 ii, 21, 373, 389, 407, 418, 462
存在性　2, 12, 39, 364, 428

D

DNA　22, 47, 158, 172, 383
DMRG　452
达朗贝尔佯谬　364, 365, 427
达朗贝尔算子　73, 120
达西定律　419
代数拓扑　169, 171
单射　199, 226, 227
单位张量
　　二阶　16, 88–90, 95, 138, 148, 162
　　四阶　62, 92, 138–141, 281, 282
当前构形　199–205, 210, 211, 216, 218–221, 224–227, 231, 233, 238, 242, 243, 245, 250, 251, 253–255, 262, 263, 299, 305–311, 335, 336
德博拉数　284, 490

德布罗意关系　19, 147

德西特空间　92

德西特宇宙　92

等容波　76, 111

等势线　352

狄拉克符号　63, 64, 69, 90, 120, 232, 233, 455

 右矢　63, 64, 69

 左矢　63, 64

第二宇宙速度　21

第一原理　456

点积　13, 16, 17, 28, 78, 80, 85, 86, 88, 89, 94, 104, 107, 109, 125, 130, 132, 133, 135, 148, 161, 162, 165, 199–203, 223, 224, 244, 245, 260, 262, 269, 297, 307, 309–311, 489

电磁张量　76, 190, 191

电荷密度　240, 256, 430, 487

电荷守恒方程　240, 256

电流密度　240, 256, 487

电子流体动力学　429

叠加原理　前言 ii, 280, 317, 319, 320

定常流动　225

定常运动　348, 421

定理

 贝蒂功的互等　313, 314

 动量矩　321

 动能　250, 251

 高斯–奥斯特罗格拉德斯基　187

 高斯绝妙　168

 蝴蝶　9

 开尔文–斯托克斯　187

 凯莱–哈密顿　296

 雷诺输运　238, 240, 249, 251, 252, 255, 405

麦克斯韦–贝蒂功的互等　313, 314

诺特　前言 ii, 147, 159, 161, 243, 246, 248, 282

散度　81, 104, 187, 247, 255, 310

斯托克斯　186, 187

托里拆利　364

旋度　187

定律

 贝尔　7, 50

 伯努利　343

 达西　419

 动量守恒　245, 365

 高斯引力　81, 82

 广义胡克　279–284, 303

 哈根–泊肃叶　361–364, 422

 胡克　142, 279–284, 303

 活力守恒　343, 344, 346

 机械能守恒　255, 343

 迈特卡夫　51

 摩尔　7, 50

 能量守恒　前言 ii, 161, 248–251, 348, 505, 506

 万有引力　81, 82

 质量守恒　37, 240

动力学

 管理　前言 i, 467

 金融　前言 i, 460, 470

 社会　前言 i, 455

 思维　前言 i, 441

动量　19, 30, 74, 159, 244, 359, 392, 429, 465, 466, 486–488, 504

动量混合　425

动量矩　76, 89, 321

动量矩定理　321

索引

动量空间 13, 18, 19, 49, 50, 157
动量扩散 269, 359, 492
动量算子 74, 75, 159–161
动量通量 359
度规
 闵可夫斯基 118
 史瓦西 117, 118
度量空间 12–15, 170
度量 (度规) 张量 112–121, 124, 125, 128, 150, 179, 194, 233, 235
短程线 119, 128
对称建立 159
对称破缺 159
对称群 276, 323
对称性
 大 282, 312, 313, 338
 小 282, 312, 313, 338
对流-扩散方程 427
对流导数 27–29
对偶空间 15, 87, 181
对势 501
对数应变 206
多重线性 87, 192
多重线性映射 86, 87
多孔介质 420, 490
多体问题 451, 452

E

EEG 441, 442
Epsilon 符号 70, 91
俄罗斯套娃 395
厄米共轭 69
二次型 56, 113

F

Finger 张量 203
Föppl-von Kármán 应变 210, 211

法则
 莱布尼兹 192
 链式 8, 27, 124, 136, 137, 150, 225, 297–299
 罗必塔 216
 右手 8, 64
反对称性 76, 100, 183, 259, 263
范式 1, 279, 315, 317, 441, 465, 476
范数 13, 19
方程
 爱因斯坦场 82
 奥森 386
 波动 33, 73, 110, 120, 320–322, 332, 333
 泊松 82, 109, 267–269
 测地线 93, 127, 128, 150, 151, 371, 452
 场 82, 120, 198, 238
 电荷守恒 240, 256
 对流-扩散 427
 格罗斯-皮塔耶夫斯基 18
 赫尔-肖 417, 419
 柯西第一运动 310
 拉普拉斯 109, 351, 352, 355, 356, 380, 419
 连续性 240, 242, 243, 245, 251, 255, 256, 260, 261, 268, 269, 348, 351, 370, 379, 404–406, 413, 458, 504
 纳维-斯托克斯 3, 28, 39, 42, 80, 265, 361, 364, 366, 367, 370–373, 377, 379, 398, 405, 417, 422, 427, 428, 429, 459
 欧拉 274, 367, 370, 371, 381, 412
 欧拉-拉格朗日方程 30, 31, 34, 35, 40, 49, 50, 124, 127, 244, 248, 369, 499

熵平衡　251, 252
双调和　380
斯托克斯　377, 380, 383
平衡　110, 245, 247, 264, 265, 324, 325, 331, 332
薛定谔　18, 50, 159, 160, 499
杨–拉普拉斯　400–403
应变协调　335
运动　245, 246, 250, 270, 310, 311
主　255
方向导数　180, 194, 421, 493–498
仿射变换　164, 165
仿射量　164, 165
仿射群　164, 165
仿射空间　13, 165, 270, 273
菲尔兹奖　39, 47, 103, 195, 425
非阿贝尔规范场　93
非阿贝尔群　153
非保守力　272
非牛顿流体　277, 357, 358
非欧几何　2, 113, 169
非平衡统计力学　499
非线性方程　18, 386
芬格变形张量　202, 295
芬格应变　217
芬格张量　203, 262
分离变量法　380
分配律　13, 85, 183
分析力学　1, 25, 30, 31, 33, 40, 167, 178
风洞　375, 376
弗劳德数　372–375, 411, 491
弗雷歇导数　200, 495–498
弗普尔–冯·卡门应变　210, 211
浮力　371, 384, 385, 491, 492

复变函数　351
复合映射　17, 178, 180, 192
复速度势　352
赋范向量空间　13, 15, 495

G

$GL^+(3)$　322
概率密度函数　499–502
刚度张量　前言 ii, 282, 287, 313, 337, 338
刚化原理　323
纲领
　　埃尔朗根　6
　　朗兰兹　6, 38
刚体　30, 34, 64, 65, 86, 89, 99, 100, 157, 259, 276, 344, 370
高斯–奥斯特罗格拉德斯基定理　187
高斯绝妙定理　168
高斯曲率　168
高斯引力定律　81, 82
各向同性　24, 99, 110, 246, 279, 283, 286, 288, 317, 330, 331, 333, 358, 489
格林公式　503
格林应变　43, 204, 207, 209–213, 217, 222, 232, 293, 294, 303, 309, 335, 336
格林应变率　223, 224, 250
公理
　　本构　275, 277
　　等存在　276
　　记忆　277
　　客观性　276
　　邻域　277
　　皮亚诺　6
　　欧氏几何　6
　　物质不变性　276
　　相容性　277

索引

因果性 276
功的互等定理 313, 314
工程应变 205, 206, 236, 284
共轭 95, 198, 221, 222, 254, 293, 298, 351, 400, 465, 466, 486
共轭空间 15, 87
共轭调和函数 351
共形 16
共旋率 264
共振 393, 438
构形
　　参考 63, 199–227, 231–235, 242–247, 251, 254, 262, 263, 299, 305–311, 322, 335–337
　　当前 199–205, 210, 211, 216, 218–221, 224–227, 231, 233, 238, 242, 243, 245, 250, 251, 253–255, 262, 263, 299, 305–311, 335, 336
　　初始 199, 231, 243, 250, 308, 309, 335–337
　　生长 231
　　无应力 231
　　中间 231, 232, 299
股票 461, 463, 464, 466, 471–474
固有时 127
观察者 168, 258, 261–263, 274
管理动力学　前言 i, 467
惯性矩 318
惯性力 373, 375, 379, 387, 412, 490–492
光滑函数 17, 180, 192
光滑流形 18
光滑性 17, 18, 39, 364, 428
光速 2, 20, 21, 44, 73, 118, 120, 148–150, 274
光子 21

广延量 249, 400, 485–489
广义胡克定律 279–284, 303
广义量纲 93, 94
广义量纲原理 91, 93
广义速度 127, 500
广义相对论 1, 21, 50, 71, 77, 92, 112, 119, 128, 171, 452, 464
广义坐标 14, 15, 30, 49, 127, 499, 500
规范场 1, 93, 120, 171, 195
　　非阿贝尔 93
　　杨–米尔斯 39, 93, 171
归一化条件 501

H

HMFV 模型 457
哈根–泊肃叶定律 361–364, 422
哈密顿方程 30, 499
海森堡不确定性原理 465, 466
豪斯多夫空间 14, 170
盒子算子 73, 120
赫尔–肖方程 417, 419
赫尔–肖盒 416, 418, 420
赫尔–肖流动 417–419
赫尔–肖自由边界问题 419
黑洞 21, 118
黑森矩阵 130
黑森算子 109
恒等映射 16, 17, 98
横波 76, 111
横波波速 33, 112, 322, 333
横观各向同性 328, 329
亨奇应变 206, 217
虹吸管 365
蝴蝶定理 9
胡克定律 142, 279–284, 303

互等定理 313, 314
华尔街革命 473
化合价 453
环量 356
黄金分割 22
混沌吸引子 443
混合积 71, 72, 91, 97, 246
混合欧拉–拉格朗日型张量 200, 218, 220, 238, 292
混合拉格朗日–欧拉型张量 200, 201
活力 (vis viva) 343, 344
活力守恒定律 343, 344, 346
霍奇对偶 190, 191
霍奇星号 188–190

I

i (虚数) 50, 74, 159–161, 351–353
ICF 408

J

Jaumann-Zaremba 客观率 264
几何
    黎曼 1, 30, 44, 50, 70, 119, 169, 171
    欧几里得 3, 167
    辛 1, 30
吉尔德定律 50
机器学习 447–451, 476, 477
机械能守恒定律 255, 343
基本流 352
    均匀来流 (uniform flow) 354
    偶极子流 (doublet flow) 354
    涡流 (vortex flow) 356
    源流 (source flow) 355
基尔霍夫应力 220–222
迹 76, 90, 95, 96, 99, 120, 132–136, 141, 148, 225, 280, 282, 497
极惯性矩 318, 321

极坐标系 122, 123, 355
级联 25, 500–502, 507
计算社会科学 458
伽利略变换 260, 270, 273
伽利略变换群 260
伽利略不变性 265, 270
加速度
    角 260, 271, 272, 487, 488
    绝对 67
    科里奥利 260, 271
    离心 260, 271, 272
    牵连 68, 260
    重力 372, 492
加托导数 46, 495–498
假设
    各向同性 317
    均匀性 317
    连续性 317
    无初始应力 317, 337
    小变形 206, 317, 336
简单剪切 108, 322, 335
简单物质 337, 425
剪切
    纯 108, 284, 285, 322, 336
    简单 108, 322, 335
剪切流 224, 225
剪切率 225, 357, 358, 492
剪切致稠 357, 358
剪切致稀 358
交换律 13, 85
角动量 74–76, 86, 161, 246, 282, 486, 487
角动量算子 75, 161
角加速度 260, 271, 272, 487, 488
角加速度力 271, 272

索　引

结合律　13, 85, 153, 182
阶跃函数　457, 458
金本位　462
金融动力学　前言 i, 460, 470
金融导航波　465
金融数学　472–474
晶圆　349
精细结构常数　92, 148–150
经济量子纠缠　465
经济弹性　468
经济物理学　460, 462–464
静力等效　323
静水压　28, 99, 108, 244, 249, 281, 283, 286,
　　292, 293, 300, 360
纠缠态　314, 441
矩阵
　　伴随　163
　　酉　158
　　余子式　134, 150, 163, 331
　　正交　158, 236
矩阵乘积态算法　452
矩阵力学　159, 507
绝妙定理　168

K

k-form (k 次形式)　182, 189, 190
K41　前言 iii, 395–399, 425
KH 不稳定性　409, 410
Kelvin-Helmholtz 不稳定性　409, 410
KT 相变　175
Kirchhoff 应力　220–222
卡门涡街　前言 ii, 388–393, 423, 424
卡诺循环　43
开尔文–亥姆霍兹不稳定性　409, 410
开尔文–斯托克斯定理　187

开集　前言 iv, 12, 17, 170, 178, 179, 495
开球　178
凯莱–哈密顿定理　296
康达效应　426, 427
科里奥利加速度　260, 271
科里奥利力　271, 272
科斯特里兹–索利斯相变　175
可加性　14, 317
客观率
　　Jaumann-Zaremba　264
客观性
　　流体动力学　前言 i, 265–272
　　欧几里得　258–263
　　平衡方程　264, 265
柯尔莫哥洛夫
　　频谱　399
　　特征长度　399
　　特征时间　399
　　特征速度　399
　　湍流研究学派　398
柯尔莫哥洛夫–奥布霍夫标度　398
柯西–黎曼条件　351
柯西变形张量　202, 204, 214
柯西第一运动方程　310
柯西面力　262
柯西应变　100, 101, 193, 206–208, 210, 215,
　　309, 310, 335
柯西应力　62, 220, 245, 247, 256, 262, 264,
　　282, 292, 294, 310, 331, 337
克劳修斯–迪昂不等式　252–254, 276
克劳修斯–普朗克不等式　253, 254
克里斯托费尔符号　124, 128, 194
　　第二类　125, 126, 242
空间

巴拿赫　13, 15, 495, 496

波矢　13, 19, 147

德西特　92

动量　13, 18, 19, 49, 50, 157

度量　12–15, 170

对偶　15, 87, 181

仿射　13, 165, 270, 273

赋范向量　13, 15, 495

共轭　15, 87

豪斯多夫　14, 170

积　198

闵可夫斯基　45, 187, 274

内积　11, 13, 14, 198

欧几里得　2, 11, 13, 14, 19, 25, 30, 50,
　　63, 64, 69, 84, 85, 114, 116, 168,
　　170, 177–181, 187–189, 192–194,
　　198, 199, 258, 493, 496, 498, 500

切　179–181, 194, 200

生活　453–455

实　19

索伯列夫　12, 13, 17

拓扑　12, 14, 177

位形　14, 30, 49, 50, 119, 124, 127, 178

位置　13, 18, 19

希尔伯特　14, 63, 157

线性　11–14, 19, 104, 260

相　15, 30, 50, 178, 499, 500

向量　13–19, 62, 87, 104, 165, 181, 188

余切　180, 181, 194

坐标　19

空间平移不变性　161, 243, 244

空间旋转不变性　246

孔隙率　419

控制表面　35, 255, 377

控制体积　35, 255, 377

宽克 (quant)　460, 461

亏格　171, 173

扩散　303, 375, 427, 428, 464, 491, 500

扩散系数　386, 387, 428, 488, 489, 492

扩散张量成像　440

扩展伽利略不变性　270–272

L

Levi-Civita 符号　70, 82

Lin　497, 498

拉回　179, 192, 193, 224

拉格朗日乘子　293

拉格朗日方程　30, 31, 34, 35, 40, 49, 50, 124,
　　127, 244, 248, 369, 499

拉格朗日量 (函数)　1, 30, 31, 33–35, 49, 124,
　　154, 244, 246, 248, 367–369

拉格朗日描述　26, 27, 29, 35, 137, 204, 211,
　　212, 215, 217, 242, 243, 245, 251, 255,
　　311

拉格朗日型张量　200, 202, 203, 204, 214,
　　220, 293

拉梅常数　282

拉梅矢量　150

拉梅系数　122

拉普拉斯方程　109, 351, 352, 355, 356, 380,
　　419

拉普拉斯算子　18, 109, 418

拉普拉斯压强　403, 405

拉伸　168, 170, 173, 205, 288, 318, 423

莱布尼兹法则　192

兰道阶符号　130

兰姆矢量　80, 82, 429

勒让德变换　40, 49, 254

雷诺数　372, 373, 375–377, 386, 387, 418,
　　428, 492

索　引

雷诺输运定理　238, 240, 249, 251, 252, 255, 405
类比　2, 19, 29, 51, 119, 160, 161, 170, 181, 194, 195, 287, 303, 346, 347, 400, 427, 429, 453, 457, 464, 466, 470
黎曼度量张量　112–121, 124, 125, 128, 150, 179, 194, 233, 235
黎曼几何　1, 30, 44, 50, 70, 119, 169, 171
黎曼假设 (猜想)　10, 39, 44, 48, 49
黎曼空间　119, 120, 167
黎曼流形　17, 44, 50, 194, 452
黎曼曲面　47
离心加速度　260, 271, 272
离心力　271, 272
李导数　193
李群　152, 157
里克特迈耶–梅什科夫不稳定性　407, 408
理想宾厄姆塑性体　358
理想固体　357, 358
理想流体　274, 285, 287, 347, 348, 357, 358, 367, 370, 408, 412, 420, 421
理性力学　1, 275
理性热力学　277
力–化耦合　303
力学
　　材料　37, 205, 279, 280, 315–318, 423
　　分析　1, 25, 30, 31, 33, 40, 167, 178
　　固体　前言 i, 前言 ii, 29, 315, 462
　　理性　1, 275
　　连续介质　前言 ii, 前言 iii, 2–4, 12, 13, 15, 18, 23–26, 28, 32, 34–36, 38, 39, 62, 498, 499
　　流变　357, 358
　　流体　前言 i, 前言 ii, 23, 25, 29, 340

弹性　3, 12, 42, 102, 108, 109, 110, 205, 264, 278, 279, 317, 324, 332, 337, 338
统计　2, 103, 288, 463, 465, 499, 500
连续性方程　240, 242, 243, 245, 251, 255, 256, 260, 261, 268, 269, 348, 351, 370, 379, 404–406, 413, 458, 504
连续映射　170
联络　129, 171, 194, 195
链式法则　8, 27, 124, 136, 137, 150, 225, 297–299
两点张量　200, 220, 238, 245, 262, 263, 292, 293, 332
量纲分析
　　快速匹配法　前言 ii, 364 373 399
量纲一致性原理　93, 363
量子金融　464–466
量子经济学　464
量子纠缠　441, 465, 476
量子凝聚　174
量子心智　476
量子意识　476
量子芝诺效应　465
量子力学　1, 2, 14, 18, 19, 31, 63, 73–75, 149, 157, 160, 161, 175, 451, 452, 464–466, 474, 476, 506, 507
量子力学场论　74
量子流体　429
量子谐振器　465
列维–奇维塔符号　70, 82, 360
列维分布　464
列维随机行走　463
邻域　216, 253, 275, 277, 498
零点能　465, 466

零矢量　13, 65

零张量　131, 132

流函数　351–356, 380

流量　348, 353, 356, 361–365, 411

流体

　　不可压缩　367, 370, 379, 380, 405, 417, 419, 422, 429

　　非牛顿　357, 358

　　可压缩　前言 ii, 348

　　理想　274, 285, 287, 347, 348, 357, 358, 367, 370, 408, 412, 420, 421

　　牛顿　287, 358

流体力学　前言 i, 前言 ii, 23, 25, 29, 340

流线　194, 195, 348, 352-354, 356, 420, 421

流形　2, 3, 16–18, 25, 50, 119, 129, 158, 167, 170, 177–181, 192–196, 452

流形上的张量分析　前言 i, 167, 194

罗必塔法则　216

螺旋度　83

洛伦兹变换　114, 260, 274

洛伦兹群　157

M

maplet arrow　15

MFI　258

mmHg　341

Mooney-Rivlin 本构　299, 301

MPS　452

马尔科维茨–夏普理论　473

马格努斯效应　427

马赫波　378

马赫角　379

马赫数　375, 378, 379, 491

马赫锥　378, 379

马略特瓶　365

麦克斯韦–贝蒂功的互等定理　313, 314

麦克斯韦方程组　1, 44, 73, 78, 79, 154, 190, 192, 196, 256, 429

毛细力　371, 404

迈特卡夫定律　51

么模群　前言 vi, 425

迷向群　前言 vi, 425

密度　21, 25, 28, 33, 34, 81, 90, 240, 259, 317, 341, 348, 370, 407–412, 459, 487, 488, 490, 492

密度矩阵重整化群　452

闵可夫斯基度规　118

闵可夫斯基空间　45, 187, 274

莫尔圆　285

摩尔定律　7, 50

穆尼–里夫林本构　299–301

N

Nanson 公式　218

Navier-Stokes 方程　39, 265, 367

纳维–斯托克斯方程　3, 28, 39, 42, 80, 265, 361, 364, 366, 367, 370–373, 377, 379, 398, 405, 417, 422, 427, 428, 429, 459

南森公式　218

脑　436–443

脑波　441–443

内积　11, 13, 14, 16, 19, 27, 63, 69, 96, 120, 161, 486

内积空间　11, 13, 14, 198

内蕴微分几何　168

能量串级　394, 395, 464

能量耗散率　396, 399

能量守恒　前言 ii, 161, 248–251, 348, 505, 506

能量算子　159–161

逆变　86, 87, 92–94, 120–126, 181, 234, 235

黏性工资理论　469

黏性价格　469

黏性力　369, 373, 379, 387, 428, 490–492

黏性系数

    动力　287, 358, 487, 488, 492

    运动　387, 399, 428, 430, 488, 492

黏性张量　287

牛顿第二定律　28, 32, 33, 62, 63, 499

牛顿流体　287, 358

牛顿微观经济学　475

扭转　318, 320–322

扭转刚度　318

努森数　22, 23, 385, 491

诺贝尔奖　前言 i-iv, 32, 48, 57, 63, 112, 147, 149, 150, 173–175, 277, 361, 384, 397, 411, 422, 436, 451, 473, 474, 475, 506

诺特定理　前言 ii, 147, 159, 161, 243, 246, 248, 282

O

Ogden 模型　299

Orr-Sommerfeld 方程　366

欧几里得

    变换　259

    几何　3, 167

    客观性　259, 261–263

    空间　2, 11, 13, 14, 19, 25, 30, 50, 63, 64, 69, 84, 85, 114, 116, 168, 170, 177–181, 187–189, 192–194, 198, 199, 258, 493, 496, 498, 500

欧拉–伯努利梁　37, 38

欧拉–拉格朗日方程　30, 31, 34, 35, 40, 49, 50, 124, 127, 244, 248, 369, 499

欧拉方程　274, 367, 370, 371, 381, 412

欧拉描述　26, 27, 29, 35, 137, 211–213, 215, 217, 240, 244, 245, 247, 248, 251

欧拉数　372–374, 491

欧拉型张量　200, 202, 204, 215, 262, 263

P

PDE　12, 33

PK1　62, 220–222, 245–247, 262, 263, 292–294, 331, 332, 337

PK2　220–222, 246, 250, 262, 263, 292–294, 298, 303

Piola-Kirchhoff 应力　62, 220

帕雷托原理　463

帕雷托最优　463

帕斯卡定律　42, 286

帕斯卡水桶实验　286

帕斯卡桶裂实验　286

庞加莱猜想　10, 39, 170, 217

批判性思维　前言 iv, 436, 437

皮奥拉–基尔霍夫应力　62, 220

偏微分方程

    抛物型　474

    双曲型　33, 322, 332, 333

    椭圆型　332

平截面假定　38

平均曲率　402, 403

平均自由程　23, 25, 385, 400, 491

谱定理　104, 107, 108, 110, 112, 203, 333

谱分解　104, 105, 106

普拉托–瑞利不稳定性　409–415, 425

普朗克长度　20, 119

普朗克常数　18, 20, 74, 148–150, 157, 159, 486

普朗克时间　20

Q

齐次性　318

奇怪吸引子　443

期权定价　473

期权交易　473

千禧年大奖难题　39, 364, 428

牵连加速度　68, 260

牵连速度　66, 67

强度量　249, 287, 400, 485–489

强勒让德–阿达玛条件　332, 333

强椭圆性条件　332, 333

切丛　17, 18, 180, 192

切空间　179–181, 194, 200

切映射　192, 200

屈服应力　490

曲率

 高斯　168

 平均　402, 403

曲纹坐标系　113, 120, 121, 124–126, 242

曲线坐标系　121–124, 150

全微分　49, 135, 183, 194, 209, 219, 291, 493, 495

群

 阿贝尔　153, 163, 164

 变换　161, 260

 对称　276, 323

 仿射　164, 165

 非阿贝尔　153

 公理　153

 紧致　164

 李　153, 157–159, 164, 171, 195, 371

 么模　前言 vi, 425

 迷向　前言 vi, 425

 三斜　前言 vi, 425

 特殊线性　前言 vi, 425

 同伦　171

 旋转　158, 159, 164

 一般线性　322

 正交　161, 162

R

RM 不稳定性　407, 408

Richtmyer-Meshkov 不稳定性　407, 408

RT 不稳定性　408, 409

RTT　238, 249, 251, 255

Rayleigh-Taylor 不稳定性　408, 409

热流通量　251

热扩散　428, 489, 492

人工智能

 强　446, 447

 弱　446

柔度张量　前言 ii, 62, 92, 281

软物质　209, 283, 284, 303

瑞利–贝纳尔对流　423, 424

瑞利–泰勒不稳定性　408, 409

润滑近似　404–406

S

SL(3)　前言 vi, 425

SO(2)　158, 159, 163, 164

SO(3)　158, 159, 162

ST 不稳定性　416–420

STP　23

SVP　323

Sym(3)　322, 323

萨夫曼–泰勒不稳定性　416–420

赛斯应变度量　216, 217

赛斯–希尔应变度量　214

三重积　65, 72, 73, 81, 238

三相接触线　371

三斜群　前言 vi, 425

散度定理　81, 104, 187, 247, 255, 310

索　引

色散关系　414
熵　103, 173, 174, 251–254, 277, 288–291, 475, 476, 486, 488
熵平衡方程　251, 252
社会动力学　前言 i, 455
社会静力学　455, 456
社会弹性　468
社会物理学　455, 456
社会作用力模型　456, 457
伸长比　205, 214, 215, 299
深度学习　87, 447–451
神经元　436, 441, 443, 450, 451, 452
渗流通量　419
生成元　159, 260
生活空间　453–455
生态弹性　468
生长构形　231
生长张量　231, 232
声学张量　110, 112, 333
圣经　57, 167, 393, 490
圣维南–基尔霍夫材料　303
圣维南原理　323, 324, 338
时间平移不变性　161, 248
石墨烯　429, 430
石柱群　424
事件　13, 195, 258, 425
事件视界　118, 119
势函数　43, 350–356
势流　350–352, 381
史瓦西半径　21, 118
史瓦西度规
　　逆变　118
　　协变　118
收缩压　340, 341
守恒律　前言 ii, 7, 238–256, 502

舒张压　340, 342
数量级估计　前言 ii, 21, 406, 407, 418
双射　16
双调和方程　380
双线性映射　16
水力压裂　286
水动力学　286, 347, 464
水静力学　286
思维
　　创造性　前言 iii, 436, 437
　　批判性　前言 iv, 436, 437
思维动力学　前言 i, 441
四阶单位张量　62, 92, 138–141, 281, 282
四阶投影张量　141
斯特劳哈尔数　373, 374, 375, 492
斯特林近似　288
斯托克斯方程　377, 380, 383
斯托克斯流动　379–381
斯托克斯流体　287
斯托克斯阻力公式　379–386
　　修正的　383, 385, 386
速度　259, 265, 267, 269, 274, 287, 346, 348, 350–356
速度势　350–353
速度梯度　28, 221, 223, 225, 238, 263, 270, 274, 285, 287, 357–359, 368, 387, 425, 426
算子
　　达朗贝尔　73, 120
　　动量　74, 75, 159–161
　　盒子　73, 120
　　黑森　109
　　霍奇星号　188–190
　　角动量　75, 161

拉普拉斯　18, 109, 418
能量　159–161
梯度　80, 81, 109, 233–235, 242, 308
线性　18, 160, 495
能量　159–161
梯度　80, 81, 109, 233–235, 242, 308
线性　18, 160, 495
随机场
索伯列夫空间　12, 13, 17

T

TensorFlow　前言 i, 87, 451
TensorNetwork　前言 i, 452
塔科玛峡谷桥　392, 393
泰勒展开　前言 ii, 32, 100, 129, 130, 131, 199, 201, 215, 235, 321, 337, 500
弹性
　动力学　110, 112
　力学　3, 12, 42, 102, 108, 109, 110, 205, 264, 278, 279, 317, 324, 332, 337, 338
　社会　468
　社区　468
　思维　468
特殊线性群　前言 vi, 425
特征尺度　20, 23–25, 338, 373, 395, 406, 407, 411, 418, 491, 492
特征时间　25, 373, 374, 399, 406, 428, 490, 491
梯度算子　80, 81, 109, 233–235, 242, 308
体模量　281, 283
体应变　283
调和函数　351
　共轭　351
听诊器　341, 342

同构　16, 19, 92, 177
同胚　170–172, 177
同胚映射　170
投影张量
　二阶　88, 91
　四阶　141
图灵奖　47, 443, 444, 447
湍流　37, 57, 141, 303, 476
　标度律　397, 425
　充分发展　399, 464
湍流经济学　464
推前　179, 192, 193, 224
椭圆性条件　332
拓扑空间　12, 14, 177
拓扑心理学　453
拓扑学　2, 71, 169–174, 177, 454
拓扑相变　173–176

U

酉矩阵　158

V

vis viva　343, 344, 346

W

弯曲　102, 278, 318, 319
弯曲刚度　318
弯曲空间　119, 120, 169
完全正交变换群　161
万有引力常数　20, 81
万有引力定律　81, 82
外汇交易　464
外积　64, 89
外微分　167, 181–186, 188–192
微分几何　2, 3, 128, 129, 167, 168, 171
微分流形　前言 i, 2, 153, 167, 171, 176–180, 192, 195, 198

索　引

微分同胚　172, 176, 177, 198, 371

微观状态数　288–291

唯一性　2, 227, 317, 337, 338, 377

维度诅咒　451, 452

位矢　64–66, 120, 160, 258, 500

位形空间　14, 30, 49, 50, 119, 124, 127, 178

位置空间　13, 18, 19

位移梯度　102, 212, 231, 235, 306, 308, 332, 337

文丘里

　　管　349

　　效应　349

涡环　376

涡量　79, 80, 83, 350, 351, 360, 380

涡旋　175, 389–393, 396

无滑移边界　23, 91, 362, 405, 418, 428

无量纲数

　　Ar　490

　　Bi　490

　　Bm　490

　　Bo　490

　　Ca　490

　　Da　490

　　De　490

　　Ek　491

　　Eu　372–374, 491

　　Fo　491

　　Fr　372–375, 411, 491

　　Ga　491

　　Gr　491

　　Ha　491

　　Kn　22, 23, 385, 491

　　La　491

　　Ma　375, 378, 379, 491

　　Nu　491

　　Oh　491

　　Pe　492

　　Pr　492

　　Re　372, 373, 375–377, 386, 387, 418, 428, 492

　　Ri　492

　　Rn　492

　　Ro　492

　　Sc　492

　　St　372–375, 492

　　We　375, 411, 492

　　Wi　492

　　Wo　492

无旋波　76, 111

无旋流动　350, 353

无旋性　82

物质导数　193, 194, 420, 421

物质流形　196, 198, 227

物质梯度　309

X

X-力学　476

希尔伯特第六问题　2, 398

希尔伯特空间　14, 63, 157

希尔应变度量　214, 215

吸引子　443

狭义相对论　1, 76, 118, 119, 260

纤维丛　1, 195

线性变换　16, 161, 164, 165

线性方程　18, 320

线性空间　11–14, 19, 104, 260

线性算子　18, 160, 495

线性映射　15, 16, 18, 87, 196, 227, 317

相变

BKT 175

KT 175

拓扑 173-176

相互贯穿 227

相空间 15, 30, 50, 178, 499, 500

相似律

 动力相似 375

 几何相似 315, 375, 378

 运动相似 375

向量空间 13–19, 62, 87, 104, 165, 181, 188

楔积 182–184

协变 86, 90, 92–94, 118–125, 181, 233, 235

协调方程 101, 335

心理环境 454

新胡克本构模型 291

新胡克体 288

辛几何 1, 30

虚功 64, 310, 311

虚功原理 305–311, 315

虚拟力 271, 272

虚数 50, 74, 159–161, 351–353

虚位移 64, 65, 305–310

旋率 221, 224, 225, 259, 263

旋转群 158, 159, 164

学说 5, 340, 344, 476

血压 340–342, 346, 361

血压计 340–342, 346, 361

血液循环 340, 344, 346

薛定谔方程 18, 50, 159, 160, 499

Y

Young-Laplace 方程 400–403

哑标 68, 84, 86, 128, 200, 209, 337

雅可比行列式 44, 186, 259

演生 424

赝塑性体 358

赝张量 70

杨–拉普拉斯方程 400-403

杨–米尔斯场 39, 93, 171

杨润湿方程 403

杨氏模量 278, 283, 489

尧曼–扎伦巴客观率 264

一般线性群 322

引理 9, 182, 184

引力半径 21

引力势 82

应变

 阿尔曼西 213, 214, 217

 毕奥 217

 对数 206

 芬格 217

 格林 43, 204, 207, 209-213, 217, 222, 232, 293, 294, 303, 309, 335, 336

 亨奇 206, 217

 柯西 100, 101, 193, 206-208, 210, 215, 309, 310, 335

 皮奥拉 217

应变度量

 希尔 214, 215

 赛斯 216, 217

应变率 221, 224, 225, 263, 285, 287, 298, 357, 360, 487

应变能 221, 222, 227, 283-285, 292, 293, 297-299, 303, 326, 331

应变协调方程 335

应力 62

映射

 保持方向 199, 227

 单射 199, 226, 227

索 引

等距　17, 162, 259
恒等　16, 17, 98
连续　170
逆　170, 176, 201
双射　16
双线性　16
同胚　170
线性　15, 16, 18, 87, 196, 227, 317
一对一　121, 226, 258
正交　17

涌现　424, 443
酉矩阵　158
油滴实验　384, 385
右柯西-格林变形张量　202, 203, 212, 213, 224, 262, 263, 294, 297
右矢　63, 64, 69
余切空间　180, 181, 194
余子式矩阵　134, 150, 163, 331
宇宙半径　21
原理

阿基米德　286, 371
材料标架无差异性　258
材料客观性　258
叠加　前言 ii, 280, 317, 319, 320
刚化　323
广义量纲　91
海森堡不确定性　465, 466
量纲一致性　93, 363
圣维南　323, 324, 338
虚功　305-311, 315
虚位移　310
最小作用量　1, 8, 32, 34, 119

约化普朗克常数　18, 20, 74, 148, 159

Z

增广矩阵　165
张量

处理器　87
单位　16, 88-90, 92, 95, 138-141, 148, 162, 231, 281, 282
电磁　76, 190, 191
对称　97, 99, 104, 107, 221, 327, 360
反对称　97, 99, 221, 225, 360
混合欧拉-拉格朗日型　200, 218, 220, 238, 292
混合拉格朗日-欧拉型　200, 201
刚度　前言 ii, 282, 287, 313, 337, 338
拉格朗日型　200, 202, 203, 204, 214, 220, 293
两点　200, 220, 238, 245, 262, 263, 292, 293, 332
流　前言 i, 87, 443
欧拉型　200, 202, 204, 215, 262, 263
柔度　前言 ii, 62, 92, 281
声学　110, 112, 333
生长　231, 232
图　前言 i, 437, 438
网络　前言 i, 452
相容　232, 255
协调　231, 232
正常正交　162, 259, 331
正交　161, 162, 203, 258-263
转动　100, 102, 162, 203

真实流体　348
正交

变换群　161
矩阵　158, 236
群　161, 162
映射　17

张量　161, 162, 203, 258-263
正则方程　30
正则对易关系　74
整体坐标系　198
指进　409, 416, 417, 420
指数墙　451, 452
质量间隔　39
置换符号　70-72, 74, 91, 188, 189, 256, 270, 335-337
质量扩散　488, 492
质量流密度　240
质量守恒　37, 240
中间构形　231, 232, 299
重力　244, 348, 372-374, 384, 385, 403, 408, 410-412, 421, 424, 490, 491
重力加速度　372, 492
主方程　255
主应力　107, 284, 285
驻值　31, 35
转动惯量　86, 89, 90, 321, 488
转换　387, 422, 425
转置　17, 85, 97, 102, 130, 140, 199-204
自接触　227
自同构　16, 157
自同态　16
自由度　30, 178, 486
自由分子流　23
自由能　254, 289, 290, 312, 486
自相似　395, 396, 464, 470
资本资产定价模型　473
纵波　76, 110-112
纵波波速　112

阻力系数　376, 377
左柯西–格林变形张量　202, 204, 214, 224, 262, 295, 297, 298, 299
左矢　63, 64
坐标空间　19
坐标图册　178, 179
坐标系
　　笛卡儿　28, 65, 90, 109, 121, 122, 162, 198, 200, 242, 243, 352
　　极　122, 123, 355
　　球　115, 116, 234, 235, 243, 353, 380
　　曲纹　113, 120, 121, 124-126, 242
　　圆柱　115, 233, 234, 243, 352, 413
　　直角　70, 87, 91, 112, 115, 116, 120, 121, 124, 132, 162, 198, 212
最概然分布　288
最速下降线　343, 344
最小作用量原理　1, 8, 32, 34, 119
作用量　31, 34, 465, 466, 486

其他

0-form (0 次形式)　182, 188, 189
1-form (1 次形式)　182, 189
2-form (2 次形式)　182, 189, 190
3-form (3 次形式)　182, 189
4-form (4 次形式)　189
$1/\alpha \approx 137$　148
5-3 浪模式　470-472
2/3 标度律　395-397
$-5/3$ 标度律　前言 iii, 398, 399
80/20 法则　462, 463
M^3　198

人 像 索 引

A

阿贝尔 (Abel N H) 166

阿达玛 (Hadamard J) 332

阿基米德 (Archimedes) 371

爱迪生 (Edison T) 58

爱丁顿 (Eddington A) 93

埃林根 (Eringen A C) 273

爱因斯坦 (Einstein A) 52, 54, 58, 71, 77, 93, 117, 148

埃伦费斯特 (Ehrenfest P) 93

奥布霍夫 (Obukhov A M) 396

B

巴切勒 (Batchelor G K) 前言 v, 416

贝蒂 (Betti E) 313

贝多芬 (Beethoven L) 55

贝尔 (Bell G) 50

贝尔纳 (Bernard C) 54

贝纳尔 (Bénard H) 423

别列津斯基 (Berezinskii V L) 176

博戈柳博夫 (Bogoliubov N N) 500

伯努利 (Bernoulli D) 347

伯努利 (Bernoulli Jacob) 343

伯努利 (Bernoulli Johann) 343

玻恩 (Born M) 500

泊肃叶 (Poiseuille J L M) 361

布阔伊 (von Buquoy G G) 462

布拉格 (Bragg W L) 59

布辛涅斯克 (Boussinesq J) 334

C

陈省身 (Chern S S) 195

D

达·芬奇 (da Vinci L) 36, 53, 54

达朗贝尔 (d'Alembert) 367

戴森 (Dyson F J) 60

德莱顿 (Dryden H L) 391

德西特 (de Sitter W) 93

狄拉克 (Dirac P A M) 53

F

费马 (de Fermat P) 40

费曼 (Feynman R) 113

费米 (Fermi E) 149

冯·布劳恩 (von Braun W) 56

冯·卡门 (von Kármán T) 前言 ii, 56, 147, 391, 424

冯·诺依曼 (von Neumann J) 56

弗普尔 (Föppl A O) 210

G

高斯 (Gauss C F) 113

格林 (Green H S) 500

格廷 (Gurtin M E) 273

H

哈代 (Hardy G H) 52

哈根 (Hagen G) 361

哈密顿 (Hamilton R) 30, 51, 148

哈维 (Harvey T S) 437

哈维 (Harvey W) 340

海森堡 (Heisenberg W K) 367

亥姆霍兹 (Helmholtz H)　409
亥维赛 (Heaviside O)　78
赫尔-肖 (Hele-Shaw H S)　417
赫维兹 (Hurwitz A)　77
赫胥黎 (Huxley A)　58
黑尔斯 (Hales S)　341

J

伽罗瓦 (Galois É)　166
基尔霍夫 (Kirchhoff G R)　221
吉尔德 (Gilder G)　50

K

开尔文 (Lord Kelvin)　409
凯利 (Cayley G)　424
康托尔 (Cantor G)　68
柯恩 (Korn A)　102
柯尔莫哥洛夫 (Kolmogorov A)　396
柯克伍德 (Kirkwood J G)　500
柯洛特克夫 (Korotkov N S)　342
柯瓦列夫斯卡娅 (Kovalevskaya S)　52
柯西 (Cauchy A)　228
科斯特里兹 (Kosterlitz J M)　176
克里斯托费尔 (Christoffel E B)　124
克罗内克 (Kronecker L)　58, 68
肯尼迪 (Kennedy J F)　147

L

拉格朗日 (Lagrange J L)　26, 30, 41
拉梅 (Lamé G)　279
拉普拉斯 (Laplace P S)　41, 109, 401
兰道 (Landau E)　130
兰金 (Rankine W J M)　423
勒让德 (Legendre A M)　41
雷诺 (Reynolds O)　239, 387
黎曼 (Riemann B)　113
李 (Lie S)　153

里夫林 (Rivlin R S)　59, 302
里奇 (Ricci-Curbastro G)　71, 148
里瓦罗基 (Riva-Rocci S)　341
理查森 (Richardson L F)　395
列维 (Lévy M)　334
列维-奇维塔 (Levi-Civita T)　71, 148
洛伦兹 (Lorentz H A)　93
罗切斯特 (Rochester N)　445

M

马赫 (Mach E)　378
迈特卡夫 (Metcalfe R)　51
麦卡锡 (McCarthy J)　445
麦克斯韦 (Maxwell J C)　51
明斯基 (Minsky M)　445
摩尔 (Moore G)　50
穆尼 (Mooney M)　302

N

纳维 (Navier C L M H)　367
牛顿 (Newton I)　30, 53
努森 (Knudsen M)　23
诺尔 (Noll W)　258
诺特 (Noether E)　159

O

欧拉 (Euler L)　26, 172, 367

P

帕雷托 (Pareto V)　462
帕斯卡 (Pascal B)　229
泡利 (Pauli W)　59, 367
培根 (Bacon F)　51
佩雷尔曼 (Perelman G Y)　39
皮奥拉 (Piola G)　221
皮普金 (Pipkin A C)　273
普拉托 (Plateau J)　411
普朗克 (Planck M)　52

人像索引

普朗特 (Prandtl L)　前言 i, 366, 391

Q
钱学森 (Tsien H S)　前言 ii, 391
切比雪夫 (Chebyshev P)　54

R
瑞利 (Lord Rayleigh)　411

S
萨夫曼 (Saffman P G)　416
塞格纳 (von Segner J A)　400
赛斯 (Seth B R)　216
圣·克里斯朵夫 (St Christopher)　391
圣维南 (de Saint-Venant A J C B)　334
史瓦西 (Schwarzschild K)　117
斯梅尔 (Smale S)　195
斯托克斯 (Stokes G G)　367
索利斯 (Thouless D J)　176
索末菲 (Sommerfeld A)　149, 367

T
泰勒 (Taylor G I)　前言 i, 416
汤姆森 (Thompson D W)　52

汤姆逊 (Thomson G P)　前言 iv
汤姆逊 (Thomson J J)　前言 iv, 59
特鲁斯德尔 (Truesdell C A)　273
铁摩辛柯　317

W
外尔 (Weyl H)　56
韦伊 (Weil A)　55, 59
魏尔斯特拉斯 (Weierstrass K)　51
文丘里 (Venturi G B)　349
沃尔泰拉 (Volterra V)　71

X
希尔 (Hill R)　216
希尔伯特 (Hilbert D)　54
香农 (Shannon C)　445

Y
亚里士多德 (Aristotle)　58
杨 (Young T)　401
杨振宁 (Yang C N)　195
伊万 (Yvon J)　500